# The ARRL Antenna Book

**Editor**
Gerald L. Hall, K1TD

**Assistant Editors**
George Woodward, W1RN
Charles L. Hutchinson, K8CH

**Contributing Editors**
Doug DeMaw, W1FB
Edward P. Tilton, W1HDQ
Paul M. Wilson, W4HHK

**Published by**
**The American Radio Relay League, Inc.**
**Newington, CT 06111**

Cover Photo by Paul M. Wilson, W4HHK

This work is Publication No. 15 of the Radio Amateur's Library, published by the League. 

Printed in USA



Fourteenth Edition
3rd Printing

Library of Congress Catalog Card Number:
55-8966

$8.00 in USA
$8.50 elsewhere

ISBN: 0-87259-414-9

# Foreword

Generations of radio amateurs have known that the most important part of their station is the *antenna.* The finest equipment hooked to an inadequate or improper antenna system will produce great frustration for the operator, while the simplest gear, when used in conjunction with a well-engineered antenna installation, can provide great satisfaction.

Even today, when factory-built equipment adorns most ham shacks and "homebrew" stations are the exception, it's a rare radio amateur who hasn't built an antenna at one time or another. But this book is not written just for those who intend to build their own. Rather, it is aimed at anyone who wants to understand what happens to a radio signal from the time it leaves the transmitter to the time it is amplified and detected in a distant receiver. Armed with a knowledge of radio propagation and of the fundamentals of transmission lines and antennas, an amateur can choose the antenna system that best meets his or her operating objectives. A big, elaborate and expensive system may not be necessary; depending upon what one wants to accomplish, it may be possible to fashion some bits of wire into an antenna that will perform as well as a commercial product costing hundreds of dollars!

The purpose of this book is to assemble theoretical and practical material on wave propagation, transmission lines and antennas into the form most useful for radio amateurs. The book has three principal divisions. Chapters 1 through 7 deal with the principles of antennas and transmission lines, wave propagation and its relationship to antenna design, and the performance characteristics of directive antenna systems. Beginning with Chapter 8, there is a series of chapters in which complete data are given on specific designs for the various amateur frequency bands, including those suitable for space communications — satellites, moonbounce, etc. — and for direction-finding. The remaining chapters deal with highly important related subjects such as measurements, test equipment and determining the best orientation for your antenna. Tables of latitude and longitude coordinates for 474 locations around the world are included in Chapter 16 for those who wish to use computer programs to generate their own, personalized charts of beam headings and distances. Finally, there is an appendix with a glossary of terms and useful tabulated information, and an extensive index.

In eight years on the market, some 215,000 copies of the previous edition of *The ARRL Antenna Book* were sold. This new edition, the first to be published in the new, larger format, promises to be even more popular. We hope you will share with us your suggestions for making the next edition even better.

David Sumner, K1ZZ
*General Manager*
*Newington, CT*

# Contents

# Metric Equivalents, Gain Reference

Throughout this book distances and dimensions are usually expressed in English units — the mile, the foot, and the inch. Conversions to metric units may be made by using the following formulas:

km = mi. × 1.609
m = ft (') × 0.3048
mm = in. (") × 25.4

An inch is 1/12 of a foot. Tables in the appendix provide information for accurately converting inches and fractions to decimal feet, and vice versa, without the need for a calculator.

Also throughout this book, gain is customarily referenced to a dipole antenna in decibels (dB). In some sections, gain is referenced to an isotropic radiator in decibels, designated as dBi. A dipole has a gain of 2.14 dBi.

# Chapter 1

# Wave Propagation

Because radio communication is carried on by means of electromagnetic waves traveling through the earth's atmosphere, it is important to understand the nature of these waves and their behavior in the propagation medium. Most antennas will radiate the power applied to them efficiently, but if they are not erected in a manner that allows the energy they radiate to reach desired destinations, the time and money that they represent will not have been well spent. No antenna can do all things well under all circumstances. Whether you design and build your own antennas, or buy them and have them put up by a professional, you'll need propagation know-how for best results.

## THE NATURE OF RADIO WAVES

You probably have some familiarity with the concept of electric and magnetic fields. A radio wave is a combination of both, with the energy divided equally between them. If the wave could originate at a point source in free space, it would spread out in an ever-growing sphere, with the source at the center. No antenna can be designed to do this, but the theoretical *isotropic antenna* is useful in explaining and measuring the performance of practical antennas we *can* build. It is, in fact, the basis for any discussion or evaluation of antenna performance.

Our theoretical spheres of radiated energy would expand very rapidly — at the same speed as the propagation of light — approximately 186,000 miles or 300,000,000 meters per second. These values are close enough for practical purposes, and will be used throughout this book.

The path of a ray from the source to any point on the spherical surface is considered to be a straight line — a radius of the sphere. An observer on the surface of the sphere would think of it as flat, just as the earth seems flat to us. A radio wave far enough from its source to appear flat is called a *plane wave*. We will be discussing primarily plane waves from here on.

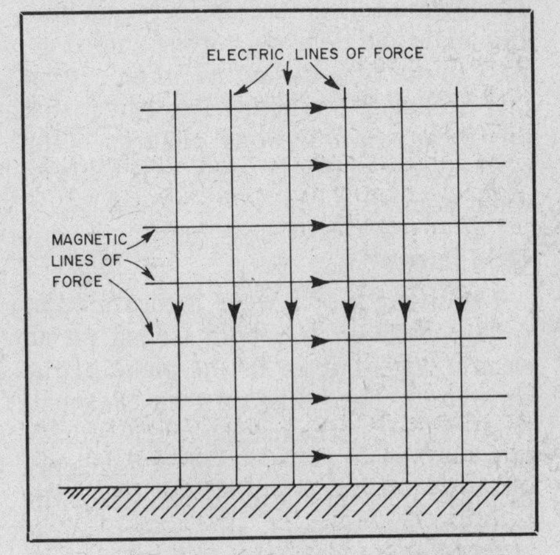


Fig. 1 — Representation of the magnetic and electric fields of a vertically polarized plane wave traveling along the ground. The arrows indicate instantaneous directions of the fields for a wave traveling perpendicularly out of the page toward the reader. Reversal of the direction of one set of lines reverses the direction of travel. There is no change in direction when both sets are reversed. Such a dual reversal occurs, in fact, once each half cycle.

It helps to understand the radiation of electromagnetic energy if we visualize a plane wave as being made up of electric and magnetic forces as shown in Fig. 1. The nature of wave propagation is such that the electric and magnetic lines of force are always perpendicular. The plane containing the sets of crossed lines represents the wave front. The direction of travel is always perpendicular to the wave front; "forward" or "backward" is determined by the relative directions of the electric and magnetic forces.

The speed of travel of a wave through anything but a vacuum is always less than 300,000,000 meters per second. How much less depends on the medium. If it is air the reduction in propagation speed can be ignored in most discussions of propagation at frequencies below 30 MHz. In the vhf range and higher, temperature and moisture content of the medium have increasing effects on the communications range, as will be discussed later. In solid insulating materials the speed is considerably less. In distilled water (a good insulator) the speed is 1/9 that in free space. In good conductors the speed is so low that the opposing fields set up by the wave front occupy practically the same space as the wave itself, and thus cancel it out. This is the reason for "skin effect" in conductors at high frequencies, making metal enclosures good shields for electrical circuits working at radio frequencies.

### Phase and Wavelength

Because the velocity of wave propagation is so great we tend to ignore it. Only 1/7 of a second is needed for a radio wave to travel around the world — but in working with antennas the *time* factor is extremely important. The wave concept developed because an alternating current flowing in a wire (antenna) sets up moving electric and magnetic fields. We can hardly discuss antenna theory or performance at all without involving travel time, consciously or otherwise.

Waves used in radio communication may have frequencies from about 10,000 to several billion hertz (Hz). Suppose the frequency is 30,000,000 Hz, more commonly called 30 megahertz (MHz). One cycle, or period, is completed in 1/30,000,000 second. The wave is traveling at 300,000,000 meters per second, so it will move only 10 meters during the time that the current is going through one complete period of alternation. The electromagnetic field 10 meters away from the antenna is caused by the current that was flowing one period earlier in time. The field 20 meters away is caused by the current that was flowing two periods earlier, and so on.

If each period of the current is simply a repetition of the one before it, the currents at corresponding instants in each period will be identical, and the fields caused by those currents will also be identical. As the fields move outward from the antenna they become more thinly spread over larger and larger surfaces; their amplitudes decrease with distance from the antenna, but they do not lose their identity with respect to the instant of the period at which they were generated. They are, and they remain, *in phase*. In the example above, at intervals of 10 meters, measured outward from the antenna, the phase of the waves at any given instant is identical.

From this information we can define both "wave front" and "wavelength." The wave front is a surface in every part of which the wave is in the same phase. The wavelength is the distance between two wave fronts having the same phase at any given instant. This distance must be measured perpendicular to the wave fronts — along the line that represents the direction of travel. Expressed in a formula, the length of a wave is

$$\lambda = \frac{v}{f}$$

where
$\lambda$ = wavelength
v = velocity of wave
f = frequency of current causing the wave

The wavelength will be in the same length units as the velocity when the frequency is expressed in the same *time* units as the velocity. For waves traveling in free space (and near enough for waves traveling through air) the wavelength is

$$\lambda \text{ (meters)} = \frac{300}{f \text{ (MHz)}}$$

There will be few pages in this book where phase, wavelength and frequency do not come into the discussion. It is essential, therefore, to have a clear understanding of their meaning before we discuss the design, installation, adjustment or use of antennas, matching systems or transmission lines in detail. In essence, "phase" means "time." When something goes through periodic variations, as an alternating current does, corresponding instants in succeeding periods are in phase.

The points A, B and C in Fig. 2 are all in phase. They are corresponding instants in the current flow, at one-wavelength intervals. This is a conventional view of a sine-wave alternating current, with time progressing to the right. It also represents a "snapshot" of the intensity of the traveling fields, if distance is substituted for time in the horizontal axis. The distance between A and B or B and C is one wavelength. The field-intensity distribution follows the sine curve, in both amplitude and polarity, corresponding exactly to the time variations in the current that produced the fields. Remember that this is an *instantaneous* picture — the wave moves outward much as a wave in water does.

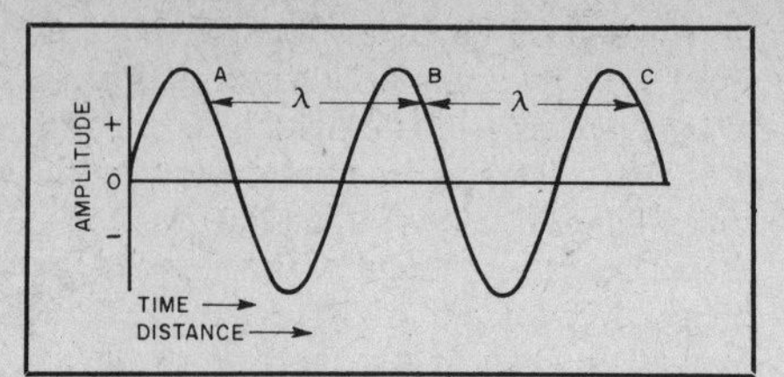


Fig. 2 — The instantaneous amplitude of both fields (electric and magnetic) varies sinusoidally with time as shown in this graph. Since the fields travel at constant velocity, the graph also represents the instantaneous distribution of field intensity along the wave path. The distance between two points of equal phase, such as A-B and B-C, is the length of the wave.

### Field Intensity

The strength of a wave is measured as voltage between two points lying on an electric line of force in the plane of the wave front. The voltage in a wave is usually quite low, so the measurement is made in microvolts per meter. The voltage goes through time variations like those of the current that caused the wave. It is measured like any other ac voltage — in terms of the effective value or, sometimes, the peak value.

It is fortunate that in amateur work it is not necessary to measure actual field strength, as the equipment required is elaborate. We need to know only if an adjustment has been beneficial, so only relative measurements are required. These can be made quite easily, with equipment that is often homebuilt.

### Polarization

A wave like that in Fig. 1 is said to be polarized in the direction of the electric lines of force. The polarization here is vertical, because the electric lines are perpendicular to the earth. It is one of the laws of electromagnetic action that electric lines touching the surface of a good conductor must do so perpendicularly. Most ground is a rather good conductor at frequencies below about 10 MHz, so waves of these frequencies, over good ground, are mainly vertically polarized. Over partially conducting ground there may be a forward tilt to the wave front, the tilt in the electric lines of force increasing as the energy loss in the ground becomes greater.

Waves traveling in contact with the surface of the earth are of little use in amateur communication because, as the frequency is raised, the distance over which they will travel without excessive energy loss becomes smaller and smaller. The surface wave is most useful at low frequencies and through the standard broadcast band. At high frequencies the wave reaching the receiving antenna has had little contact with the ground, and its polarization is not necessarily vertical.

If the electric lines of force are horizontal, the wave is said to be horizontally polarized. Horizontally and vertically polarized waves may be classified generally under linear polarization. Linear polarization can be anything between horizontal and vertical. In free space, "horizontal" and "vertical" have no meaning, the reference of the seemingly horizontal surface of the earth having been lost.

In many cases the polarization of waves is not fixed, but rotates continually, somewhat at random. When this occurs the wave is said to be elliptically polarized. A gradual shift in polarization in a medium is known as Faraday rotation. For space communications, circular polarization is commonly used to overcome the effects of Faraday rotation. A circularly polarized wave rotates its polarization through 360° as it travels a distance of one wavelength in the propagation medium. The direction of rotation as viewed from the transmitting antenna defines the direction of circularity, right-hand (clockwise) or left-hand (counterclockwise). Linear and circular polarization may be considered as special cases of elliptical polarization.

### Attenuation

In free space the field intensity of the wave varies inversely with the distance from the source. If the field strength at one mile from the source is 100 millivolts per meter, at two miles it will be 50 millivolts per meter, at 100 miles it will be 1 millivolt per meter, and so on. The relationship between field intensity or field strength and power density is similar to that for voltage and power in ordinary circuits. They are related by the impedance of free space, which has been determined to be 377 ohms. The power density therefore varies with the square root of the field intensity, or inversely with the *square* of the distance.

The decrease in strength is caused by the spreading of the wave energy over ever larger spheres, as the distance from the source increases. It will be important to remember this spreading loss when antenna performance is discussed later. Gain can come only from narrowing the radiation pattern of an antenna, to concentrate the radiated energy in the desired direction. There is no "antenna magic" by which the total energy radiated can be increased.

In practice, attenuation of the wave

energy may be much greater than the "inverse-distance" law would indicate. The wave does not travel in a vacuum, and the receiving antenna seldom is situated so that there is a clear line of sight. The earth is spherical and the waves do not penetrate its surface appreciably, so communication beyond visual distances must be by some means that will bend the waves around the curvature of the earth. These means involve additional energy losses that increase the path attenuation with distance, over that for theoretical spreading loss in a vacuum.

### Bending of Radio Waves

Radio waves and light waves are both propagated as electromagnetic energy; their major difference is in wavelength, though radio-reflecting surfaces are usually much smaller in terms of wavelength than those for light. In a material of a given electrical conductivity, long waves penetrate farther than short ones, and so require a thicker mass for good reflection. Thin metal is a good reflector of radio waves of even quite long wavelength. With poorer conductors, such as the earth's crust, long waves may penetrate quite a few feet below the surface.

*Reflection* occurs at any boundary between materials of differing dielectric constant. Familiar examples with light are reflections from water surfaces and window panes. Both water and glass are transparent for light, but their dielectric constants are very different from that of air. Light waves, being very short, seem to "bounce off" both surfaces. Radio waves, being much larger, are practically unaffected by glass, but their behavior upon encountering water may vary, depending on the purity of that medium. Distilled water is a good insulator; salt water is a relatively good conductor.

Depending on their length (or frequency), radio waves may be reflected by buildings, trees, vehicles, the ground, water, ionized layers in the outer atmosphere, or at boundaries between air masses having different temperatures and moisture content. Most of these factors can affect antenna performance. Ionospheric and atmospheric conditions are important in practically all communication beyond purely local ranges.

*Refraction* is the bending of a ray as it passes from one medium to another at an angle. The appearance of bending of a straight stick, where it is made to enter water at an angle, is an example of light refraction known to us all. The degree of bending of radio waves at boundaries between air masses increases with the radio frequency. There is some slight atmospheric effect in our hf bands. It becomes noticeable at 28 MHz, more so at 50 MHz, and it is much more of a factor in the higher vhf range and in uhf and microwave propagation.

*Diffraction* of light over a solid wall prevents total darkness on the far side from the light source. This is caused largely by scattering of the light by impurities in the air. (Remember photos showing the sharp black shadows, when men walked in the airless environment of the moon?) The dielectric constant of the surface of the obstruction, and the condition of the air above it, may effect what happens to our radio waves when they encounter terrestrial obstructions — but the radio "shadow area" is never totally "dark."

The three terms, reflection, refraction and diffraction, were in use long before the radio age began. Radio propagation is nearly always a mix of these phenomena, and it may not be easy to identify or separate them in practical communications experience. We will tend to rely on the words *bending* and *scattering* in our discussions, with appropriate modifiers as needed. The important thing is to remember that any alteration of the path taken by energy as it is radiated by our antennas is almost certain to have some effect on our on-the-air results — which is why we have a *propagation* chapter in an *antenna* book.

## THE GROUND WAVE

Radio waves, as we have already seen, are affected in many ways by the media through which they travel. This has led to some confusion of terms in literature concerning wave propagation. Waves travel close to the ground in several ways, some of which involve relatively little contact with the ground itself. The term *ground wave* has had several meanings in antenna literature, but more or less by common consent it has come to be applied to any wave that stays close to the earth, reaching the receiving point without leaving the earth's lower atmosphere. It could be traveling in actual contact with the ground, as in Fig. 1. It could travel directly between the transmitting and receiving antennas, when they are high enough so that they can "see" each other. In this text we include waves that are made to follow the earth's curvature by bending in the earth's lower atmosphere, or troposphere, usually no more than a few miles above the ground. Often called *tropospheric bending,* this mode is a major factor in amateur communication above 50 MHz.

### The Surface Wave

The surface wave travels in contact with the earth's surface. It can provide coverage up to about 160 kilometers in the standard broadcast band in daytime, but attenuation is quite high. As can be seen from Fig. 3, the attenuation increases with frequency. The surface wave is of little value in amateur communication, except possibly at 1.8 MHz. Vertical antennas must be used, which tends to limit amateur surface-wave communication except where large vertical systems can be erected.

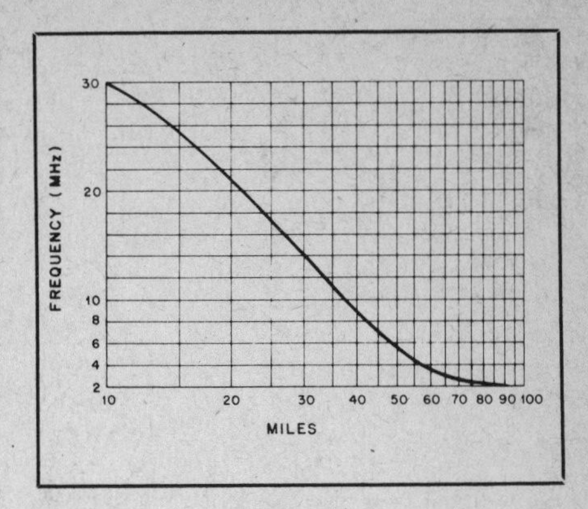


Fig. 3 — Typical hf ground-wave range as a function of frequency.

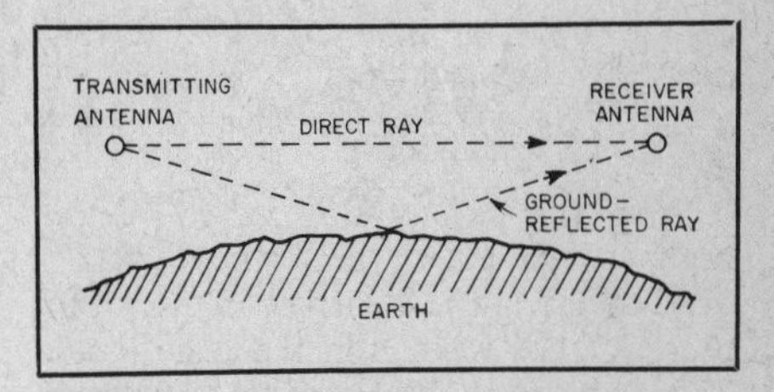


Fig. 4 — The ray traveling directly from the transmitting antenna to the receiving antenna combines with a ray reflected from the ground to form the space wave.

### The Space Wave

Propagation between two antennas situated within line-of-sight of each other is shown in Fig. 4. Energy traveling directly between the antennas is attenuated to about the same degree as in free space. Unless the antennas are very high or quite close together, an appreciable portion of the energy is reflected from the ground. This combines with direct radiation to produce the actual signal received.

In most communication between stations on the ground (not air-to-ground or air-to-air), the angle at which the wave strikes the ground will be rather small. For a horizontally polarized signal, such a reflection reverses the phase of the wave. If the distances traveled by both parts of the wave were the same the two would arrive out of phase, and would therefore cancel each other.

The ground-reflected ray has to travel a little farther, so the phase difference between the two depends on the lengths of the paths, measured in wavelengths. From this it can be seen that the wavelength in use is important in determining the useful signal strength in this type of communication. If the difference in path length is 3 meters, the phase difference with

360-meter waves would be only 3 degrees. This is a negligible difference from the 180-degree shift caused by the reflection, so the change in effective signal strength over the path would be very small. But with 6-meter radio waves the phase shift with the same difference in path length would be 180 degrees — and the two rays would add. Thus, the space wave is a negligible factor at low frequencies, but it is increasingly useful as the frequency is raised. It is a dominant factor in local amateur communication at 50 MHz and higher.

Interaction between the direct and reflected waves is the principle cause of "mobile flutter" observed in local vhf communication between fixed and mobile stations. The flutter effect decreases once the stations are separated enough so that the reflected ray becomes inconsequential. The reflected energy can also confuse the results of field-strength measurements during tests on antennas for vhf use.

As with most propagation explanations, the space-wave picture presented here is simplified, and practical considerations dictate modifications. There is energy loss when the wave is reflected from the ground, and the phase of the ground-reflected wave is not shifted exactly 180 degrees, so the waves never cancel completely. At uhf ground-reflection losses can be greatly reduced or eliminated by using highly directive antennas. By confining the antenna pattern to something approaching a flashlight beam, nearly all the energy is in the direct wave. The resulting energy loss is low enough so that microwave relays, for example, can operate with moderate power levels over hundreds and even thousands of miles. Thus we see that, while the space wave is inconsequential below about 20 MHz, it is a prime asset in the vhf realm and higher.

## VHF Propagation Beyond Line-of-Sight

From Fig. 4 it appears that use of the space wave depends on direct line-of-sight between the antennas of the communicating stations. This is not literally true, though that belief was common in the early days of amateur communication on frequencies above 30 MHz. When equipment was built that operated efficiently and antenna techniques improved, it soon became clear that vhf waves were bent or scattered in several ways, permitting reliable communication somewhat beyond visual distances between the two stations. This was found true with low power and simple antennas. The average communications range can be approximated by assuming that the waves travel in straight lines, but the earth's radius is increased by one-third. The distance to the "radio horizon" is then given as

$$D\ (mi) = 1.415 \sqrt{H\ (ft)}$$

or

$$D\ (km) = 4.124 \sqrt{H\ (m)}$$

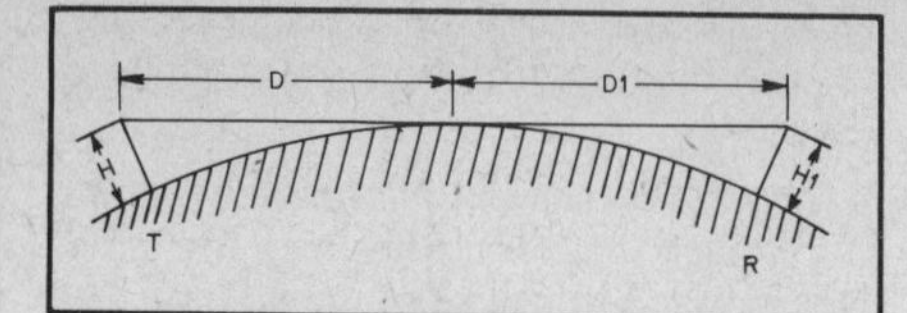


Fig. 5 — The distance, D, to the horizon from an antenna of height H is given by formulas as in the text. The maximum line-of-sight distance between two elevated antennas is equal to the sum of their distances to the horizon, as indicated here.

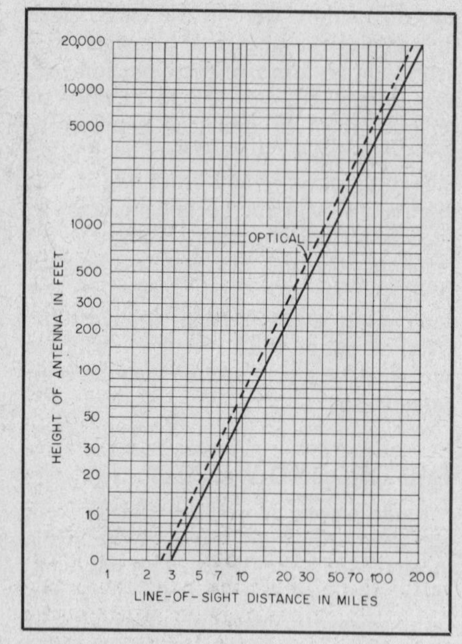


Fig. 6 — Distance to the horizon from an antenna of given height. The solid curve includes the effect of atmospheric refraction. The optical line-of-sight distance is given by the broken curve.

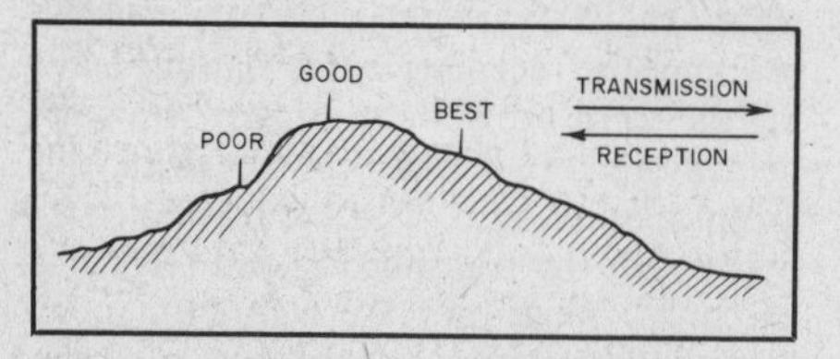


Fig. 7 — Propagation conditions are generally best when the antenna is located slightly below the top of a hill on the side that faces the distant station. Communication is poor when there is a sharp rise immediately in front of the antenna in the direction of communication.

where H is the height of the transmitting antenna, as shown in Fig. 5. The formula assumes that the earth is smooth out to the horizon, so any obstructions along the path must be taken into consideration. For an elevated receiving antenna the communications distance is equal to D + D1, that is, the sum of the distances to the horizon of both antennas. Radio horizon distances are given in graphical form in Fig. 6. Two stations on a flat plain, one with its antenna 60 feet above ground and the other 40 feet, could be up to about 20 miles apart for strong-signal line-of-sight communication (11 + 9 mi). The terrain is almost never completely flat, and variations along the way may add to or subtract from the distance for reliable communication. Energy is absorbed, reflected or scattered in many ways, in nearly all communications situations. The formula or the chart will be a good guide for estimating the potential radius of coverage for a vhf fm repeater, assuming the users are mobile or portable with simple, omnidirectional antennas. Coverage with optimum home-station equipment, high-gain directional arrays, and ssb or cw is quite a different manner. A much more detailed method for estimating coverage on frequencies above 50 MHz is given later in this chapter.

For maximum use of the ordinary space wave it is important to have the antenna as high as possible above nearby buildings, trees, wires and surrounding terrain. A hill that rises above the rest of the countryside is a good location for an amateur station of any kind, and particularly so for extensive coverage on the frequencies above 50 MHz. The highest point on such an eminence is not necessarily the best location for the antenna. In the example shown in Fig. 7, the hilltop would be a good site in all directions. But if maximum performance to the right was the objective, a point just below the crest might do better. This would involve a trade-off with reduced coverage in the opposite direction. Conversely, an antenna situated on the left side, lower down the hill, might do well to the left, but almost certainly would be inferior in performance to the right.

Selection of a home site for its radio potential is a complex business, at best. A vhf enthusiast dreams of the highest hill. The DX-minded may be more attracted by a dry spot near a salt marsh. A wide salt-water horizon almost *smells* of DX. In shopping for ham radio real estate, a mobile or portable rig for the frequencies you're most interested in can provide useful clues.

## Antenna Height and Polarization

If effective communication over long distances were the only consideration, we would be concerned mainly with radiation of energy at the lowest possible angle above the horizon. However, being engaged in a residential avocation often imposes practical restrictions on our antenna projects. As an example, our 80- and 160-meter bands are used primarily for short-distance communication because they serve that purpose with antennas that are not difficult or expensive to put up. Out to a few hundred miles, simple wire antennas for these bands do quite well, even though their radiation is mostly at high angles above the horizon. Vertical systems might be better for long-distance use, but they require extensive ground systems for optimum performance.

Horizontal antennas that radiate well at low angles are more easily erected for 7 MHz and higher frequencies, so horizontal wires and arrays are almost standard practice work on 40 through 10 meters. Vertical antennas are used in this frequency range, such as a single omnidirectional antenna of multiband design. An antenna of this type may be a good solution to the space problem for a city dweller on a small lot, or even for the resident of an apartment building.

High-gain antennas are almost always used at 50 MHz and higher frequencies, and most of them are horizontal. The principal exception is mobile communication with fm, through repeaters, discussed earlier. The height question is answered easily for vhf enthusiasts — the higher the better.

The theoretical and practical effects of height above ground are treated in detail in Chapter 2. Note that it is the height in *wavelengths* that is important — a good reason to think in the metric system, rather than in feet and inches.

In working locally on any amateur frequency band, best results will be obtained with the same polarization at both stations, except for rather rare polarization shift caused by terrain obstructions or reflections from buildings. Where such shift is observed, mostly above 100 MHz or so, horizontal polarization tends to work better than vertical. This condition is found primarily on short paths, so it is not too important. Polarization shift may occur on long paths where tropospheric bending is a factor, but here the effect tends to be random. Long-distance communication by way of the ionosphere produces random polarization effects routinely, so polarization matching is of little or no importance. This is fortunate for the hf mobile enthusiast, who will find that even his short, inductively loaded whips work very well at all distances other than local.

Because it responds to all plane polarizations equally, circular polarization may pay off on circuits where the arriving polarization is random, but it exacts a 3-dB penalty when used with a single-plane polarization of any kind. Circular systems find greatest use in work with orbiting satellites. It should be remembered that "horizontal" and "vertical" are meaningless terms in space, where the plane-earth reference is lost.

## Polarization Factors Above 50 MHz

In most vhf communication over short distances, the polarization of the space wave tends to remain constant. Polarization discrimination is quite high, usually in excess of 20 dB, so the same polarization should be used at both ends of the circuit. Horizontal, vertical and circular polarization all have certain advantages, above 50 MHz, so there has never been complete standardization on any one of them.

Horizontal systems are popular, in part because they tend to reject man-made noise, much of which is vertically polarized. There is some evidence that vertical polarization shifts to horizontal in hilly terrain, more readily than the reverse. With large arrays, horizontal systems may be easier to erect, and they tend to give higher signal strengths over irregular terrain, if any difference is observed.

Practically all work with vhf mobiles is now handled with vertical systems. For use in a vhf repeater system, the vertical antenna can be designed to have gain without losing the desired omnidirectional quality. In the mobile station a small vertical whip has obvious esthetic advantages. Often a telescoping whip used for broadcast reception can be pressed into service for the 2-meter fm rig. A car-top mount is preferable, but the broadcast whip is a practical compromise. Tests with at least one experimental repeater have shown that horizontal polarization can give a slightly larger service area, but mechanical advantages of vertical systems have made them the almost unanimous choice in vhf fm communication. Except for the repeater field, horizontal is the standard vhf system almost everywhere.

In communication over the earth-moon-earth (EME) route the polarization picture is blurred, as might be expected with such a diverse medium. If the moon were a flat target we could expect a 180- degree phase shift from the moon reflection process. But it is not flat, and the moon's libration and wave travel both ways through the earth's entire atmosphere and magnetic field provides other variables that confuse the phase and polarization issue. Building a huge array that will track the moon, and give gains in excess of 20 dB, is enough of a task so that most EME enthusiasts tend to take their chances with phase and polarization problems. Where rotation of the element plane has been tried it has helped to stabilize signal levels, but it is not widely employed.

## Tropospheric Propagation of VHF Waves

The effects of changes in the dielectric constant of the propagation medium were discussed earlier. Varied weather patterns over most of the earth's surface can give rise to boundaries between air masses of very different temperature and humidity characteristics. These boundaries can be anything from local anomalies to air-circulation patterns of continental proportions.

Under stable weather conditions, large air masses can retain their individual characteristics for hours or even days at a time. See Fig. 8. Stratified warm dry air over cool moist air, flowing slowly across the Great Lakes region to the Atlantic Seaboard, can provide the medium for east-west communication on 144 MHz and higher amateur frequencies over as much as 1200 miles, but more commonly over a third to half that distance.

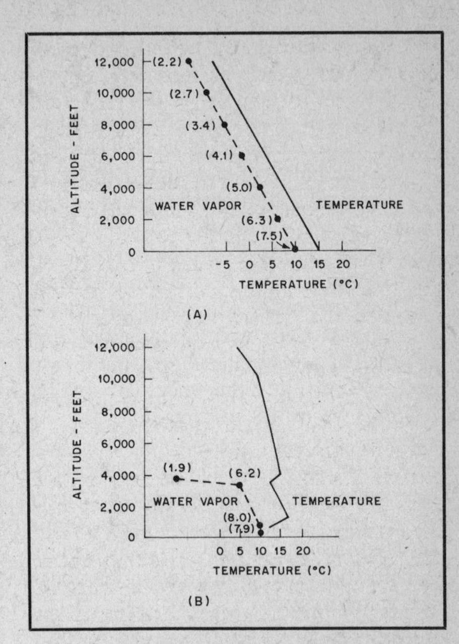


Fig. 8 — Upper air conditions that produce extended range communication on the vhf bands. At the top is shown the U.S. Standard Atmosphere temperature curve. The humidity curve (dotted) is what would result if the relative humidity were 70 percent from the ground level to 12,000 feet elevation. There is only slight refraction under this standard condition. At the bottom is shown a sounding that is typical of marked refraction of vhf waves. Figures in parentheses are the "mixing ratio" — grams of water vapor per kilogram of dry air. Note the sharp break in both curves at about 4000 feet.

A similar *inversion* along the Atlantic Seaboard as a result of a tropical storm air-circulation pattern may bring vhf and uhf openings extending from the Maritime Provinces of Canada to the Carolinas. Propagation across the Gulf of Mexico, sometimes with very high signal levels, enlivens the vhf scene in coastal areas from Florida to Texas. The California Coast, from below the Bay Area to Mexico, is blessed with a similar propagation aid during the warmer months. Tropical storms moving west, across the Pacific below the Hawaiian Islands, may provide a transpacific long-distance vhf medium. This was first exploited by amateurs on 144, 220 and 432 MHz, in 1957. It has been used fairly often in the summer months since, though not yearly.

The examples of long-haul work cited above may occur infrequently, but lesser extensions of the minimum operating range are available almost daily. Under minimum conditions there may be little more than increased signal strength over paths that are workable at any time.

There is a diurnal effect in temperate climates. At sunrise the air aloft is warmed more rapidly than that near the earth's surface, and as the sun goes lower late in the day the upper air is kept warm, while the ground cools. In fair, calm weather the sunrise and sunset temperature inversions can improve signal strength over paths beyond line-of-sight as much as 20 dB over levels prevailing during the hours of high sun. The diurnal inversion may also extend the operating range for a given strength by some 20 to 50 percent. If you would be happy with a new antenna, try it first around sunrise!

There are other short-range effects of local atmospheric and topographical conditions. Known as *subsidence,* the flow of cool air down into the bottom of a valley, leaving warm air aloft, is a familiar summer-evening pleasure. The daily inshore-offshore wind shift along a seacoast in summer sets up daily inversions that make coastal areas highly favored as vhf sites. Ask any jealous 2-meter operator who lives more than a few miles inland!

Tropospheric effects can show up at any time, in any season. Late spring and early fall are the most favored periods, though a winter warming trend can produce strong and stable inversions that work vhf magic almost equal to that of the more familiar spring and fall events.

Regions where the climate is influenced by large bodies of water enjoy the greatest degree of tropospheric bending. Hot, dry desert areas see little of it, at least in the forms described above.

### Tropospheric Ducting

Tropospheric propagation of vhf and uhf waves can influence signal levels at all distances from purely local to something beyond 4000 km (2500 mi). The outer limits are not well known. At the risk of over-simplification we will divide the modes into two classes — extended local and long distance. This concept must be modified depending on the frequency under consideration, but in the vhf range the extended-local effect gives way to a form of propagation much like that of microwaves in a waveguide, called *ducting.* The transition distance is ordinarily somewhere around 200 miles. The basic difference lies in whether the atmospheric condition producing the bending is localized or continental in scope. Remember, we're concerned here with frequencies in the vhf range, and perhaps up to 500 MHz. At 10 GHz, for example, the scale is much smaller.

In vhf propagation beyond a few hundred miles, more than one weather front is probably involved, but the wave is propagated between the inversion layers and ground, in the main. On long paths over the ocean (two notable examples are California to Hawaii and Ascension Island to Brazil), propagation is likely to be between two atmospheric layers. On such circuits the communicating station antennas must be in the duct, or capable of propagating strongly into it. Here again, we see that the positions and radiation angles of the antennas are important. As with microwaves in a waveguide, the low-frequency limit for the duct is quite critical. In long-distance ducting it is also very variable. Airborne equipment has shown that duct capability exists well down into the hf region in the stable atmosphere west of Ascension Island. Some contacts between Hawaii and Southern California on 50 MHz are believed to have been by way of tropospheric ducts. Probably all work over these paths on 144 MHz and higher bands is because of duct propagation.

Amateurs have played a major part in the discovery and eventual explanation of tropospheric propagation. In recent years they have shown that, contrary to beliefs widely held in earlier times, long-distance communication using tropospheric modes is possible to some degree on all amateur frequencies from 50 to at least 10,000 MHz.

## SKY-WAVE PROPAGATION

From earliest times we've had trouble with radio propagation language. We were never sure how long or short a "short wave" or a "long wave" was. Changing to *frequency* as a measurement term in place of *wavelength* was a useful innovation, but with technology outstripping terminology, adding modifiers such as *low, medium, high, very high, ultrahigh* and *superhigh* — all relative terms — didn't clarify much. A more recent change, dropping the descriptive *cycle per second* (or merely cycle) for the honorary *hertz* as a unit of frequency measurement hardly aided the communication process.

There is similar confusion in the accepted names for propagation modes and media. The term "ground wave" is commonly applied to propagation that is confined to the earth's lower atmosphere, but that definition allows use of the name for a mode that has been shown to work up to at least 4000 km, on occasion. Now we are about to use "skywave" to describe a mode that covers the same distances, but by a different medium, and at different frequencies, with different degrees of reliability.

### THE IONOSPHERE

There will be inevitable "grey areas" in our discussion of the earth's atmosphere and the changes wrought in it by the sun and by the associated changes in the earth's magnetic field. This is not a story that can be told in neat equations, or numbers carried out to a satisfying number of decimal places. But it must be told, and understood — with its well-known limitations — if we are to put up good antennas and make them serve us well.

Thus far we've been concerned with what might be called our above-ground living space — that portion of the total atmosphere wherein we can survive without artificial breathing aids, or up to about 6 km (4 miles). The boundary area is a broad one, but life (and radio propagation) undergo basic changes beyond this life and weather zone. Somewhat farther out, but still technically within the earth's atmosphere, the role of the sun in the wave propagation picture is a dominant one.

This is the *ionosphere* — a region where the air pressure is so low that free electrons and ions can move about for some time without getting close enough to recombine into neutral atoms. A radio wave entering this rarified atmosphere, a region of many free electrons, is affected in the same way as in entering a medium of different dielectric constant — its direction of travel is altered.

Ultraviolet radiation from the sun is the primary cause of ionization in the outer atmosphere. The degree of ionization does not increase uniformly with distance from the earth's surface. Instead there are relatively dense regions (layers) of ionization, each quite thick and more or less parallel to the earth's surface, at fairly well-defined intervals outward from about 40 to 300 km (25 to 200 miles).

Ionization is not constant within each layer, but tapers off gradually on either side of the maximum at the center of the layer. The total ionizing energy from the sun reaching a given point, at a given time, is never constant, so the height and intensity of the ionization in the various regions will also vary. Thus, the practical effect on long-distance communication is an almost continuous variation in signal level, related to the time of day, the season of the year, the distance between the earth and the sun, and both short-term and long-term variations in solar activity. It would seem from all this that only the very wise or the very foolish would attempt to predict radio propagation conditions, but it is now possible to do so with a fair chance of success — if one has the time and will for the job.

#### Layer Characteristics

The lowest known ionized region, called the D layer, lies between 60 and 92 km (37 to 57 miles) above the earth. In this relatively low and dense part of the atmosphere, atoms broken up into ions by sunlight recombine quickly, so the ionization level is directly related to sunlight. It begins at sunrise, peaks at local noon and disappears at sundown. When electrons in this dense medium are set in motion by a passing wave, collisions between particles are so frequent that a major portion of their energy may be used up as heat.

The probability of collisions depends on the distance an electron travels under the influence of the wave — in other words, on the wavelength. Thus, our 160- and 80-meter bands suffer the highest daytime absorption loss, particularly for waves that enter the medium at the lowest

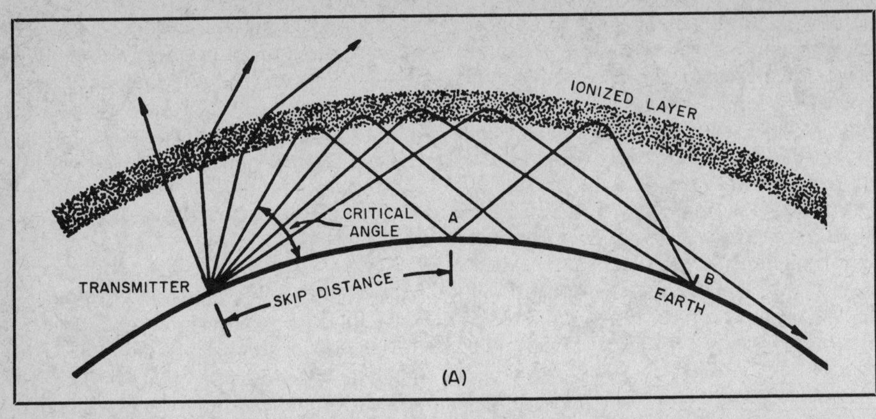


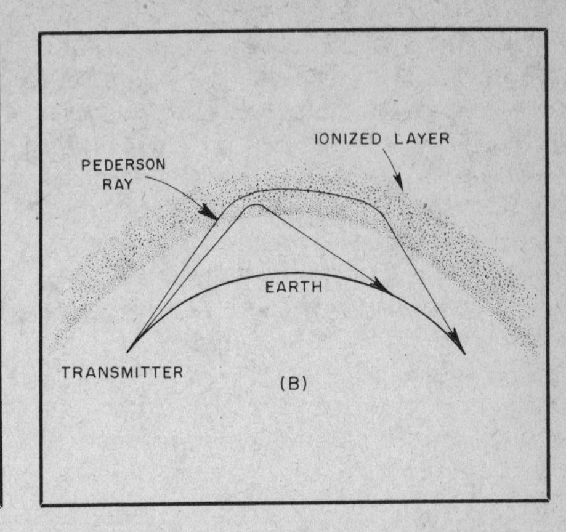


Fig. 9 — Behavior of waves encountering the ionosphere. As shown at A, rays entering the ionized region at angles above the critical angle are not bent enough to be returned to earth, and are lost to space. Waves entering at angles below the critical angle reach the earth at increasingly greater distances as the angle approaches the horizontal. The maximum distance that may normally be covered in a single hop is 4000 km. Greater distances may be covered with multiple hops. Studies have shown that under some conditions, rays entering the layer at intermediate angles will propagate farther than those entering at lower angles, as shown at B. Evidence indicates that communications distances of many thousand kilometers are covered by such "ducting," perhaps in combination with hopping.

angles. But at times of high solar activity (peak years of the solar cycle) even waves entering the D layer vertically suffer almost total energy absorption around midday, making these bands almost useless for communication over appreciable distances during the hours of high sun. They "go dead" quickly in the morning, but come alive again the same way in late afternoon. The diurnal D region effect is less at 7 MHz (though still quite marked), slight at 14 MHz and inconsequential on higher amateur frequencies.

The D layer is ineffective in bending hf waves back to earth, so its role in long-distance communication by amateurs is largely a negative one. It is the principal reason why our frequencies up through the 7-MHz band are useful mainly for short-distance communication, in the high-sun hours.

The lowest portion of the ionosphere that is useful for long-distance communication by amateurs is the E region, about 100 to 115 km (62 to 71 miles) above the earth. In the E layer, at intermediate atmospheric density, ionization varies with the sun angle above the horizon, but solar ultraviolet radiation is not the sole ionizing agent. Solar X-rays and meteors entering this portion of the earth's atmosphere also play a part. Ionization increases rapidly after sunrise, reaches maximum around noon local time, and drops off quickly after sundown. The minimum is after midnight, local time. As with the D region, the E layer absorbs wave energy in the lower frequency amateur bands when the sun angle is high, around mid day. The other varied effects of E region ionization will be discussed later.

Most of our long-distance communications capability stems from the tenuous outer reaches of the earth's atmosphere known as the F region. At these heights ions and electrons recombine more slowly, so the observable effects of the sun angle develop more slowly. Also, the region holds its ability to reflect wave energy back to earth well into the night. The maximum usable frequency (muf) for F-layer propagation on east-west paths thus peaks just after noon at the midpoint, and the minimum occurs after midnight.

Using the F region effectively is by no means that simple, however. The layer height may be from 210 to 420 km (130 to 260 miles), depending on the season of the year, the latitudes of the communicating stations, the time of day and, most capricious of all, what the sun has been doing in the last few minutes and in perhaps the last three days before the attempt is made. The muf between Eastern USA and Europe, for example, has been anything from 7 to 70 MHz, depending on the conditions mentioned above, plus the point in the long-term solar activity cycle at which the check is made. Nevertheless, the muf for a given circuit can be estimated with fair accuracy after one "gets the feel" for it.

Easy-to-use prediction charts appear monthly in *QST*. Propagation information tailored to amateur needs is transmitted in all information bulletin periods by the ARRL Headquarters station, W1AW. Finally, solar and geomagnetic field data, transmitted hourly and updated eight times daily, is given in brief bulletins carried by the National Bureau of Standards station, WWV. More on these services later.

During the day the F region may split into two layers. The lower and weaker F1 layer, about 160 km (100 miles) up, has only a minor role, acting more like the E than the F2 region. At night the F1 layer disappears and the F2 layer height drops somewhat.

### Bending in the Ionosphere

The degree of bending of a wave path in an ionized layer depends on the density of the ionization and the length of the wave (or its frequency). The bending at any given frequency or wavelength will increase with increased ionization density. With a given ionization density, bending increases with wavelength (decreases with frequency). Two extremes are thus possible. If the intensity of the ionization is sufficient and the frequency low enough, even a wave entering the layer perpendicularly will be reflected back to earth. Conversely, if the frequency is high enough or the ionization decreases to a low enough density, a condition is reached where the wave angle is not affected enough by the ionosphere to cause a useful portion of the wave energy to return to the earth. This basic principle has been used for many years to "sound" the ionosphere for determining its communication potential at various wave angles and frequencies.

A simplified example, showing only one layer, is given in Fig. 9A. The effects of additional layers are shown in Fig. 10. The simple case of Fig. 9 illustrates several important facts about antenna design for long-distance communication. At the left we see three waves that will do us no good — they all take off at angles high enough so that they pass through the layer and are lost in space. Note that, as the angle of radiation decreases (i.e. the wave is closer to the horizon), the amount of bending needed for skywave communication also decreases. The fourth wave from the left takes off at what is called the *critical angle* — the highest that will return the wave to earth at a given density of ionization in the layer for the frequency under consideration.

We can communicate with point A but, at this frequency, not much closer to our transmitter site. Nor under this set of conditions as to layer height, layer density and wave angle can we communicate much farther than point A. But suppose

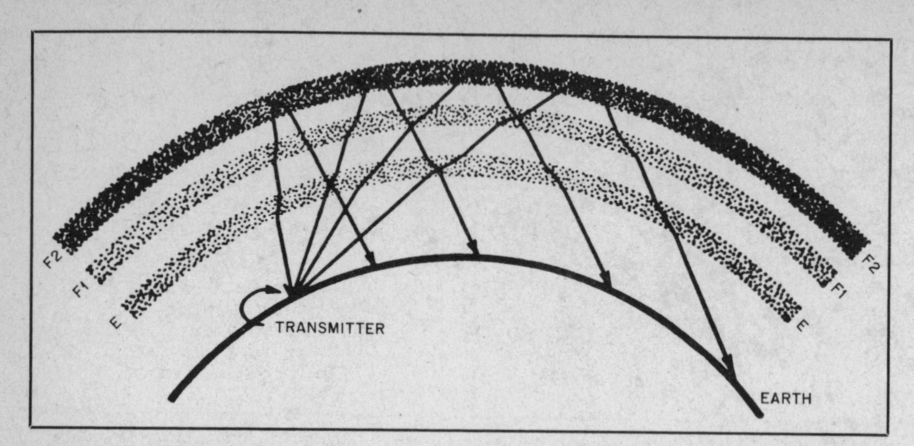


Fig. 10 — Typical daytime propagation of high frequencies (14 to 28 MHz). The waves are partially bent in going through the E and F1 layers, but not enough to be returned to earth. The actual reflection is from the F2 layer.

we install an antenna that radiates at a lower angle, as with the fifth wave from the left. This will bring our signal down to earth appreciably farther away than the higher (critical) angle did. Perhaps we can accomplish even more if we can achieve a very low radiation angle. Our sixth wave, with its radiation angle lower still, comes back to earth much farther away. Note also that its arrival point, B, is in the same area that the fourth wave will reach on its "second hop," *if* it manages to make a second hop by reflection from the ground at point A. Assuming that both waves do reach point B, the low-angle wave will almost certainly contain more energy because it suffers only one reflection loss, compared with three for the higher angle route. More than two hops are possible, however. In each case the distance at which a ray reaches the earth in a single hop depends on the angle at which it left the transmitting antenna. This radiation angle will come into play throughout this book.

**Skip Distance**

When the critical angle is less than 90 degrees there will always be a region around the transmitting site where the ionospherically propagated signal cannot be heard, or is heard weakly. This area lies between the outer limit of the ground-wave range and the inner edge of energy return from the ionosphere. It is called the *skip zone*, and the distance between the originating site and the beginning of the ionospheric return is called the *skip distance*. The signal may often be heard to some extent within the skip zone, through various forms of scattering, but it will ordinarily be marginal. When the skip distance is short, both ground-wave and sky-wave signals may be received at distances not far from the transmitter. In such instances the sky wave frequently is stronger than the ground wave, even as close as a few miles from the transmitter. The ionosphere is an efficient communications medium under average or favorable conditions. Comparatively, the ground wave is not.

**Single and Multihop Propagation**

The information in Fig. 9A is greatly simplified in the interest of explanation and example. In actual communication the picture is complicated by many factors. One is that the transmitted energy spreads over a considerable area after it leaves the antenna. Even with an antenna array having the sharpest practical beam pattern, there is what might be described as a cone of radiation centered on the wave lines shown in the drawing. The "reflection" in the ionosphere is also varied, and is the cause of considerable spreading and scattering.

Modification of the wave path is even more complex when more than one "hop" is involved. A signal may take two or more hops in its travel between distant locations of the earth's surface. However, present propagation theory holds that for communications distances of many thousands of kilometers, signals do not hop in relatively short increments along the entire path. Instead, the wave is thought to propagate inside the ionosphere throughout some portion of the path length, tending to be ducted in the ionized layer. This theory is supported by the results of propagation studies that have shown that a medium-angle ray sometimes reaches the earth at a greater distance from the transmitter than a low-angle ray, as shown in Fig. 9B. This higher angle ray, named the Pederson ray, is believed to penetrate the layer farther than lower angle rays. In the less densely ionized upper edge of the layer, the amount of refraction nearly equals the curvature of the layer itself as it encircles the earth. This non-hopping theory is further supported by studies of propagation times for signals that travel completely around the world. The time required is significantly less than would be necessary to hop between the earth and the ionosphere 10 or more times while circling the earth.

Propagation between two points thousands of kilometers apart may consist of a combination of ducting and hopping. Whatever the exact mechanics of long-distance wave propagation may be, the signal must first enter the ionosphere at some point. The amateur wanting to work great distances should have the lowest possible wave angle, for years of amateur experience have shown this to be a decided advantage under all usual conditions.

Despite all the complex factors involved, most long-distance propagation can be seen to follow certain rules. Thus, much commercial and military point-to-point communication over long distances employs antennas designed to make maximum use of known radiation angles and layer heights, even on paths where multihop propagation is assumed.

In amateur work we try for the lowest practical radiation angle, hoping to keep reflection losses to a minimum. The geometry of propagation via the F2 layer limits our maximum distance along the earth's surface to about 4000 km (2500 miles) for a single hop. With higher radiation angles this same distance may require two or more hops (with higher reflection loss), so the fewer hops the better, in most cases. If you have a near neighbor who consistently outperforms you on the longer paths, a radiation angle difference in his favor may be the reason.

**Virtual Height and Critical Frequency**

Ionospheric sounding devices have been in service at enough points over the world's surface so that a continuous record of ionospheric propagation conditions going back many years is available for current use, or for study. The sounding principle is similar to that of radar, making use of travel time to measure distance. The sounding is made at vertical incidence, to measure the useful heights of the ionospheric layers. This can be done at any one frequency, but the sounding usually is done over a frequency range wider than the expected return-frequency spread, so that information related to the maximum usable frequency (muf) is also obtained.

The distance so measured, called the *virtual height*, is that from which a pure reflection would have the same effect as the rather diffused refraction that actually happens. The method is illustrated in Fig. 11. Some time is consumed in the refraction process, so the virtual height is slightly more than the actual.

The sounding procedure involves pulses of energy at progressively higher frequencies, or transmitters with the output frequency swept at many kilohertz per second. As the frequency rises, the returns show an area where the virtual height seems to increase rapidly, and then cease. The highest return, at what is known as the *critical frequency*, can be used to determine the maximum usable frequency for long-distance communication by way

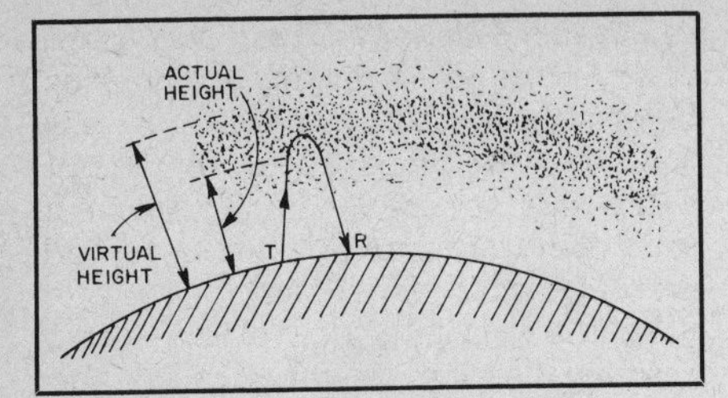


Fig. 11 — The "virtual" height of the refracting layer is measured by sending a wave vertically to the layer and measuring the time it takes for it to come back to the receiver. The actual height is somewhat less because of the time required for the wave to "turn around" in the ionized region.

**Table 1**

**Time and Frequency Stations Useful for Propagation Monitoring**

| *Call* | *Frequency (kHz)* | *Location* |
|---|---|---|
| WWV | 2500, 5000, 10,000 15,000 20,000 | Ft. Collins, Colorado |
| WWVH | Same as WWV, but no 20,000 | Kekaha, Kauai, Hawaii |
| CHU | 3330, 7335, 14,670 | Ottawa, Ontario, Canada |
| RID | 5004, 10,004, 15,004 | Irkutsk, USSR* |
| RWM | 4996, 9996, 14,996 | Novosibirsk, USSR* |
| ZUO | 2500, 5000 | Pretoria, South Africa |
| VNG | 7500 | Lyndhurst, Australia |
| BPV | 5000, 10,000, 15,000 | Shanghai, China |
| JJY | 2500, 5000, 10,000 15,000 | Tokyo, Japan |
| LOL | 5000, 10,000, 15,000 | Buenos Aires, Argentina |

*The call, taken from an international table, may not be that used during actual transmission. Locations and frequencies appear to be as given.

of the layer, at that time. As shown in Fig. 9A, the bending needed decreases as the radiation angle decreases. At the lowest practical angle the range for a single hop reaches the 4000-km limit.

## Maximum Usable and Optimum Working Frequencies

The critical frequency is the maximum usable frequency for local skywave communications. It is also useful in the selection of optimum working frequencies and the determination of the maximum usable frequency for distant points at a given time. The abbreviations *muf* and *owf* will be used hereafter. *Owf* is often designated *fot*, for the French equivalent, *frequence optimum de travail.*

The critical frequency ranges between about 1 and 4 MHz for the E layer, and 2 and 13 MHz for the F2 layer. The lowest figures are for nighttime conditions in the lowest years of the solar cycle. The highest are for the daytime hours in the years of high solar activity. These are average figures. Critical frequencies have reached something around 20 MHz briefly during exceptionally high solar activity.

The muf for a 4000-km distance is about 3.5 times the critical frequency existing at the path midpoint. For 1-hop signals, if a uniform ionosphere is assumed, the muf decreases with shorter distances along the path. This is true because the higher frequency waves are not bent sufficiently to reach the earth at closer ranges, and so a lower frequency (where more bending occurs) must be used. Precisely speaking, a maximum usable frequency or muf is defined for communication between two specific points on the earth's surface, for the conditions existing at the time. At the same time and with the same conditions, the muf from either of these two points to a third point may be quite different. The muf cannot therefore be expressed broadly as a single frequency, even for any given location at a particular time. The ionosphere is never uniform, and in fact at a given time and for a fixed distance, the muf changes significantly with changes in compass direction for almost any point on the earth. Under usual conditions, the muf will always be highest in the direction toward the sun — to the east in the morning, to the south at noon (from northern latitudes), and to the west in the afternoon and evening.

Especially where the limited power levels of the Amateur Radio Service are concerned, it is important to work fairly near the muf, where signals suffer the least loss. The mufs can be estimated with sufficient accuracy by using the prediction charts that appear monthly in *QST*. They can also be *observed*, with the use of a continuous-coverage communications receiver. Frequencies up to the mufs are in round-the-clock use today. When you "run out of signals" while tuning up from your favorite ham band, you have a pretty good clue as to which band is going to work well, right then. Of course it helps to know the direction to the transmitters whose signals you are hearing. Shortwave broadcasters know what frequencies to use, and you can hear them anywhere, if conditions are good. Time-and-frequency stations are excellent indicators, since they operate around the clock. See Table 1. WWV is also a reliable source of propagation data, *hourly*, as discussed in more detail later in this chapter.

The value of working near the muf is two-fold. Under average conditions, the absorption loss decreases with higher frequency. Perhaps more important, the hop distance is considerably greater as the muf is approached. A transcontinental contact is much more likely to be made on a single hop on 28 MHz than on 14, so the higher frequency will give the stronger signal most of the time. The strong-signal reputation of the 10-meter band is founded on this fact.

There is also a lower limit to the range of frequencies that provide useful communication between two given points by way of the ionosphere. Lowest usable frequency is abbreviated *luf*. If it were possible to start near the muf and work gradually lower in frequency, the signal would decrease in strength and eventually disappears into the ever-present "background noise." This happens because the absorption increases at lower frequencies. The frequency nearest the point where reception became unusable would be the luf. It is not likely that you would want to work at the luf, though reception could be improved if the station could increase power by a considerable amount. By contrast, if there is a wide range between the luf and the muf, operation near the muf can be quite satisfactory, even with a transmitter power of 1 or 2 watts.

Frequently, the "window" between the luf and the muf for two fixed points is very narrow, and there may be no amateur frequencies available inside the window. On occasion the luf may be ***higher*** than the muf between two points. This means that, for the highest possible frequency that will propagate through the ionosphere for that path, the absorption is so great as to make even that frequency unusable. Under these conditions it is not possible to establish amateur skywave communications between those two points, no matter what frequency is used. (It would normally be possible, however, to communicate between either point and other points on *some* frequency under the existing conditions.) Conditions when amateur skywave communications are impossible occur commonly for long distances where the total path is in darkness, and for very great distances in the daytime during periods of low solar activity.

## Transmission Distance and Layer Height

It was shown in connection with Fig. 9A that the distance at which a ray returns to earth depends on the angle at which it left the earth (the wave angle). Though it is not shown specifically in that drawing, distance also depends on the layer height at the time.

There is a large difference in the distance covered in a single hop, depending on whether the E or the F2 layer is used. The maximum distance via the E layer is about 2000 km (1250 miles) — or about half the maximum distance via the F2 layer. Practical communicating distances for single-hop E or F layer work at

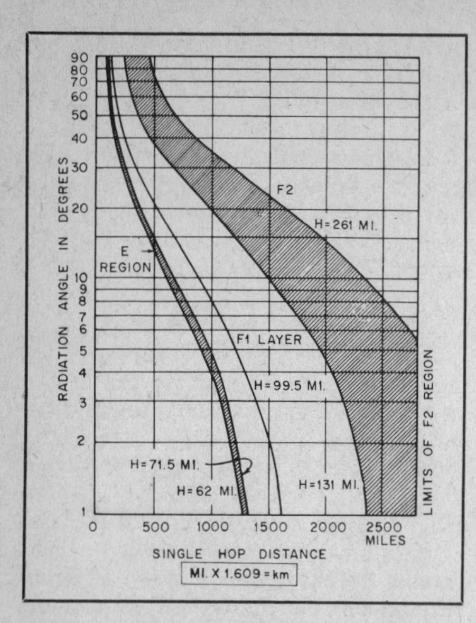


Fig. 12 — Distance plotted against wave angle (one-hop transmission) for the nominal range of virtual heights for the E and F2 layers, and for the F1 layer.

various wave angles are shown in graph form in Fig. 12.

For various reasons, actual communications experience will not fit these patterns exactly on all occasions. The particular amateur band in use will complicate the picture, in part because the E layer will be a factor much more often on our lower frequencies. Application of the chart to multihop paths is difficult, in part because both E and F layers may be involved. Also, the lowest wave angles are more practical at the higher frequencies.

Data on angle-of-arrival of signals in four amateur bands for a typical transatlantic communications circuit are shown in Table 2. Note that only rarely did the wave angle approach the really low angles (3 degrees) needed for the maximum F2-layer distance. From these figures it is apparent that much work was via two or more hops. Even where the same antennas and frequency are used over a test period, the angle of arrival will change over a wide range, with accompanying large variations in signal level.

**Table 2**

**Measured Vertical Angles of Arrival of Signals from England at Receiving Location in New Jersey**

| *Freq. MHz* | *Angle below which signals arrived 99% of the time* | *Angle above which signals arrived 50% of the time* | *Angle above which signals arrived 99% of the time* |
|---|---|---|---|
| 7 | 35° | 22° | 10° |
| 14 | 17° | 11° | 6° |
| 21 | 12° | 7° | 4° |
| 28 | 9° | 5° | 3° |

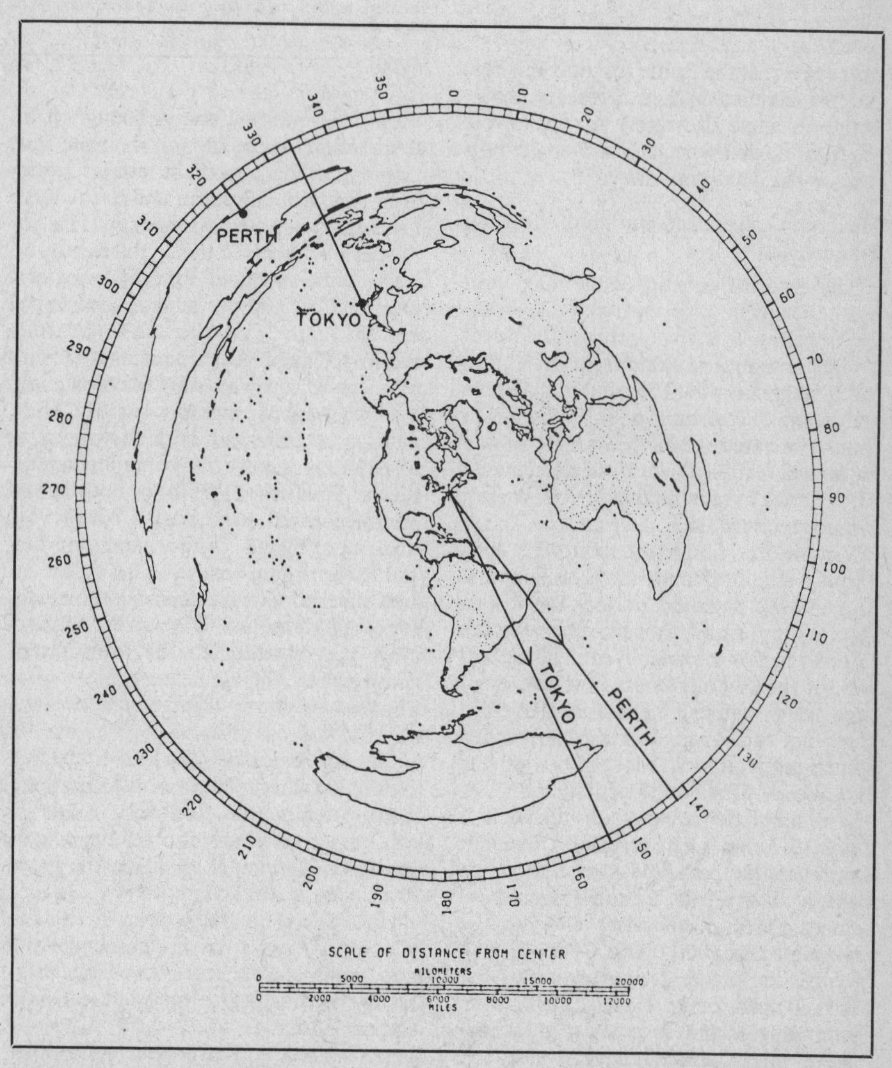


Fig. 13 — N5KR's computer-generated azimuthal-equidistant projection centered on Newington, Connecticut. (See chapter 16 for ordering information.) Information showing long paths to Perth and Tokyo have been added. Notice that the paths in both cases lie almost entirely over water, rather than over land masses.

### One-Way Propagation

On occasion a signal may be started on the way back toward the earth by reflection from the F layer, only to come down into the *top* of the E region and be reflected back up again. This set of conditions is one explanation for the often-reported phenomenon called *one-way skip*. The reverse path may not necessarily have the same multilayer character, and the effect is more often a difference in the signal strengths, rather than a complete lack of signal in one direction. It is important to remember this possibility, when a long-path test with a new antenna system yields apparently conflicting evidence. Even many tests, on paths of different lengths and headings, may provide data that are difficult to understand. Communication by way of the ionosphere is not always a source of consistent answers to antenna questions.

### Short or Long Path?

Propagation between any two points on the earth's surface is usually by the shortest direct route — the great-circle path found by stretching a string tightly between the two points on a globe. If a rubber band going completely around the globe in a straight line is substituted for the string, it will show another great-circle path that may serve for communication over the desired circuit, when conditions are favorable along the longer route. Especially if there is knowledge of this potential at the other end, long-path communication may work very well. Cooperation is almost essential, because both the antenna aiming and the timing of the attempts must be right for any worthwhile result.

Sunlight is a required element in long-distance communication via the F layer. This fact tends to define long-path timing and antenna aiming. Both are essentially the reverse of the "normal" for a given circuit. We know also that salt-water paths work better than overland ones. This can be significant in long-path work.

We can better understand several aspects of long-path propagation if we become accustomed to thinking of the earth as a ball. This is easy if we use a globe frequently. A flat map of the world, of the azimuthal-equidistant projection type, is a useful substitute. The ARRL World Map is one, centered on Kansas City. A somewhat similar world map prepared by N5KR and centered on Newington, Connecticut, is shown in Fig. 13. These help to clarify paths involving those areas of the world.

### Long-Path Examples

There are numerous long-path routes well known to "DX"-minded amateurs, those who continually seek foreign countries. Two long paths that work frequently and well on 28 MHz from northeastern USA are New England to Perth, Western Australia; and New England to Tokyo. Though they represent different beam headings and distances, they share some favorable conditions. By long path, Perth is close to half way around the world; Tokyo is about three quarters of the way. On 28 MHz, both areas come through in the early daylight hours, Eastern Time, but not necessarily on the same days. Both paths are at their best around the equinoxes. (The sunlight is more uniformly distributed over transequatorial paths at these times.) Probably the factor that most favors both is the nature of the first part of the trip at this end. To work Perth via long path, northeastern USA antennas are aimed southeast, out over salt water for thousands of miles — the best low-loss start a signal could have. It is salt water essentially all the way, and the distance, about 13,000 miles, is not too much greater than the "short" path.

The long path to Japan is more toward the south, but still with no major land mass at the early reflection points. It is much longer, however, than that to Western Australia. Japanese signals are more limited in number on the long path than on the short, and signals average somewhat weaker, probably because of the greater distance.

On the short path an amateur in the Perth area is looking at the worst conditions — away from the ocean, and out across a huge land mass unlikely to provide low-loss ground reflections. That this is a poor path is shown by the rarity of VK6s in northeastern USA at night, while much of the rest of the VK areas are workable with fair ease, most of the year, on 28 MHz. But on mornings when their long path is working at all, VK6s are almost as strong as the other Australians are on the short path, at night.

The short paths to both Japan and Western Australia, from most of the eastern half of North America, are hardly favorable. The first hop comes down in various western areas likely to be desert or mountains, or both, and not favored as reflection points.

A practical factor in determining our results is the Amateur Radio population within our own region, and along the line nearer to our objective. This is especially true in work beyond the single-hop distance, where the competing station one or more hops closer to the objective may have a decisive advantage, even when we are hearing the foreign stations well. In both examples described above, the station in the Northeast works under an interference handicap on the short path much of the time, but he may have the best shot of all, the long way around. In fact, it may well be that his area will be the only one being heard at the other end.

A word of warning: Don't count on the long-path signals always coming in on the same beam heading. There can be noticeable differences in the line of propagation via the ionosphere on even relatively short distances. There can be more on long path, especially on circuits close to halfway around the world. Remember, for a point exactly halfway around, all directions of the compass represent great circle paths.

### Fading

Taking into account all the variable factors in long-distance communication, it is not surprising that signals vary in strength during almost every contact beyond the local range. In vhf communication we can encounter some fading, at any distances greater than just to the visible horizon. These are mainly the result of changes in the temperature and moisture content of the air in the first few thousand feet above the ground.

On paths covered by ionospheric modes, the causes of fading are very complex — constantly changing layer height and density, random polarization shift, portions of the signal arriving out of phase, and so on. The energy arriving at the receiving antenna has components which have been acted upon differently by the ionosphere. Often the fading is very different for small changes in frequency. With a signal of a wideband nature, such as high-quality fm, or even double-sideband am, the sidebands may have different fading rates from each other, or from the carrier. This causes severe distortion, as a result of what is termed *selective fading*. The effects are greatly reduced when single-sideband suppressed-carrier (ssb) is used. Wideband reception can be improved as to degree or fading (but not as to the distortion induced by selective fading) by using two or more receivers on separate antennas, preferably with different polarizations, and combining the receiver outputs in what is known as a *diversity receiving system.*

## THE ROLE OF THE SUN

Everything that happens in radio propagation, as with all life on earth, is the result of radiation from the sun. The variable nature of radio propagation reflects the ever-changing intensity of ultraviolet and X-ray radiation, the ionizing agents in solar energy. For probably some 4 billion years, solar dynamics have been turning hydrogen into helium, releasing an unimaginable blast of energy into space in the process.

Why the intensity of this release changes, and what the sun is going to do at any time in the future, are known only in a general way. (See Fig. 14.) Still, tremendous progress has been made within our lifetimes in understanding the nature of the sun and predicting its future course.

### Sunspots and the Solar Flux

The most readily observed characteristic of the sun, other than its blinding brilliance, is its tendency to have grayish-black blemishes, seemingly at random times and at random places, on its fiery surface. There are written records of naked-eye sightings of sunspots in the Orient back to more than 2000 years ago. As far as is known, the first indication that sunspots are actually part of the sun was the result of observations by Galileo in the early 1600s, not long after he developed one of the first practical telescopes.

Galileo also developed the projection method for observing the sun safely, but probably not before he had suffered severe eye damage by trying to look at the sun directly. (He was afflicted with blindness in his last years.) His drawings of sunspots, indicating their variable nature and position, are the first such record known to have been made. His reward for this brilliant work was immediate condemnation by church authorities of the time, which probably set back progress in learning more about the sun for generations. But one way or another, observation of the sun continued, mostly in secret.

Systematic study of the sun began around 200 years ago, so the fairly reliable record of sunspot numbers goes back about that far. This record shows clearly that the sun is always in a state of change. It never looks exactly the same from one day to the next. The most obvious daily change is the movement of visible activity centers (sunspots or groups thereof)

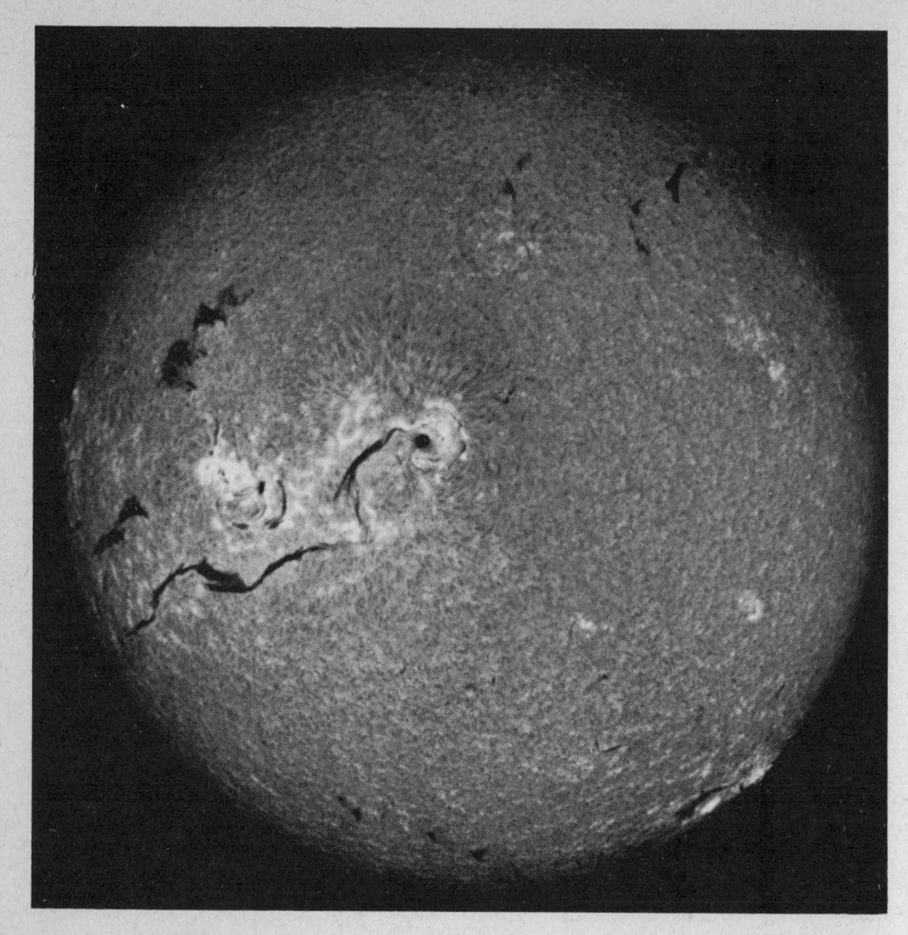

Fig. 14 — Much more than sunspots can be seen when the sun is viewed through selective optical filters. This photo was taken through a hydrogen-alpha filter that passes a narrow light segment at 6562 angstroms. The bright patches are active areas around and often between sunspots. Dark irregular lines are filaments of activity having no central core. Faint magnetic field lines are visible around a large sunspot group near the disc center. (*photo courtesy of Sacramento Peak Observatory, Sunspot, New Mexico*)

across the solar disc, from east to west, at a constant rate. This movement was soon found to be the result of the rotation of the sun, at a rate of approximately four weeks for a complete round. The average is about 27.5 days.

The daily and seasonal variations in the ionized layers that result from changes in the amount of ultraviolet light received from the sun have already been mentioned. A so-called 11-year sunspot cycle affects propagation conditions because there is a direct correlation between sunspot activity and ionization. The 11-year figure for the time between successive peaks of sunspot activity is only an average; any given cycle may vary a few years either way. The peak in 1968 (solar cycle 20) was of average stature, having a maximum smoothed sunspot number not much in excess of 110. By contrast, the peak which occurred in 1957-1958 (solar cycle 19) had a smoothed number of sunspots exceeding 200 and was to date the highest ever recorded. At this writing the peak of solar cycle 21 has yet to be determined. Unusually high levels of solar activity have persisted well into 1981, and on occasions the F2 muf has risen well into the vhf portion of the spectrum.

Daily sunspot counts are recorded, and monthly and yearly *averages* determined. The smoothed sunspot number (also called the 12-month running average) for a given month is the mean for the preceding 6 months and the succeeding 6 months about the month in question. The sunspot number is not the actual number of spots observed, but rather a weighted figure that takes into account such factors as the number of *groups* of sunspots, the number of individual spots counted, and the equipment used to make the measurements. The result is known as the Wolf number, R (after the man who derived the system for standardizing the sunspot count), and has been in use since the mid 18th century. Because international records are kept in Zurich, Switzerland, Wolf numbers are also known as Zurich smoothed sunspot numbers.

Individual sunspots may vary in size and appearance, or even disappear totally, within a single day. In general, larger active areas persist through several rotations of the sun. Some have been identified over periods up to about two years.

For more than 50 years it has been well known that radio propagation phenomena vary with the number and size of sunspots — but not in any very precise way.

A major step forward was made with the development of various methods for observing narrow portions of the sun's spectrum. Narrowband light filters that can be used with any good telescope perform a visual function very similar to the aural function of a sharp filter added to a communications receiver. This enables the observer to see the actual area of the sun doing the radiating of the ionizing energy, in addition to the sunspots, which are more a byproduct than a cause. Studies of the ionosphere with instrumented probes, and later with satellites, manned and unmanned, have added greatly to our knowledge of the effects of the sun on radio communication. From these advances has come the opportunity for an ambitious amateur to understand and predict propagation variations with a worthwhile degree of accuracy.

### The Solar Flux — A Better Tool

One of the most useful developments in the study of the sun's effects on radio propagation came as an offshoot from the World War II radar program in Canada, largely as a result of the work of a former Canadian amateur, Arthur E. Covington, ex-VE5CC, under the auspices of the National Research Council of Canada. It had been known for many years that noise from the sun could be heard over a very wide frequency range. Noise amplitude at 2800 MHz was found to be closely related to the ionization density of the F layer, and thus to variations in the muf for long-distance radio communication.

As soon as an accurate measuring system could be developed, continuous monitoring of what is now known as the solar flux was begun at an observatory near Ottawa, Ontario. The solar flux has been recorded at many other frequencies, from 240 to over 15,000 MHz (and probably higher) in the ensuing years. Our interest lies mainly with the 2800-MHz data, for which there is a continuous daily record since early 1947.

For some years the Canadian 2800-MHz figure (taken near Ottawa at 1700 UT) has been sent automatically to Boulder, Colorado, for transmission by the National Bureau of Standards station WWV. It is part of a brief propagation bulletin carried at 18 minutes past each hour. The solar flux number is updated daily on their 1800 bulletin, so the information is only a bit more than an hour old at 1818 UT.

### Using WWV Propagation Data

The hourly bulletin gives the solar flux, geomagnetic A-Index, Boulder K-Index, and a brief statement of solar and geomagnetic activity in the past and coming 24-hour periods, in that order. The solar flux and A-Index are changed with the 1800 bulletin, the rest every three hours — 0000, 0300, 0600 UT and so on.

Radio amateurs are probably the largest audience for this service. Their use of the data runs from an occasional check on "the numbers" to round-the-clock recording and detailed charting of the data, usually by months, or by solar-rotation periods. The latter method helps to identify recurrent activity on the sun. Detailed suggestions for using WWV data are contained in the wave propagation chapter of recent editions of *The Radio Amateur's Handbook*, and in *QST* articles referenced at the end of this chapter.

The solar flux tracks well with smoothed sunspot numbers, as shown in Fig. 15. The great usefulness of the solar flux is its immediacy, and its direct bearing on our field of interest. Sunspot numbers are valuable mainly as a long-term link with the past, but they are vague, at best, and old when they become available. Weekly sunspot information in W1AW propagation bulletins is the freshest information available. Bulletin details appear below.

The WWV A-Index is a daily figure for the state of activity of the earth's magnetic field. It is updated with the 1800 UT bulletin. The K-Index (new every three hours) reflects Boulder readings on the geomagnetic field in the hours just preceding the bulletin data changes. It is the nearest thing to current data on radio propagation available. With new data every three hours, K-Index *trend* is important. Rising is bad news; falling is good, especially related to propagation on paths involving latitudes above 30° north.

The A-Index tells you mainly how yesterday was, but it is very revealing when charted regularly, because geomagnetic disturbances nearly always recur at four-week intervals. The statement of past and future solar and geomagnetic conditions is couched in such vague terms that it may be of slight practical value.

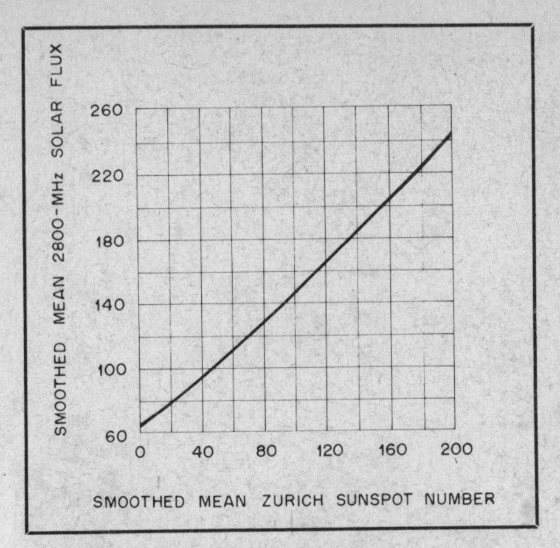


Fig. 15 — Relationship between smoothed mean Zurich sunspot number and smoothed mean 2800-MHz solar flux.

### Propagation Estimations Made Easily

Any propagation estimation service involving mailing, or even weekly radio bulletins, takes time in preparation and dissemination. Amateurs who want the most current forecast can do the job themselves, with some investment in time and practice. Daily observation of the sun is a great help, especially if detailed records are kept. Drawings (not necessarily fancy) showing the location of sunspot areas and their approximate size are useful. Charting of at least the WWV solar flux and A-Index is almost a must. It's nice to copy several bulletins per day, but even one is useful. The 1800 UT bulletin is freshest, but others will do, preferably any fairly soon after the 1800. A simple chart such as that of Fig. 16 will show solar- and geomagnetic-activity trends clearly. Supplemented by a visual record of the sun, it will constitute a good base for forecasting future solar and propagation trends, a few days to two or three months in advance. Try it, but be aware that you may find it so interesting that it cuts into your on-the-air activity!

### Using *QST* Propagation Charts

For some years *QST* has carried easy-to-use charts that can be a great help in planning your operating time and band usage for maximum return. These charts are based on the kinds of data that were formerly employed manually in connection with maps and relevant information obtained from the U.S. Government Printing Office. It was an interesting method, but it entailed some detailed and laborious work. (See references.)

Preparation of the new *QST* charts involves computer methods applied to the data formerly used in the manual-prediction method. The result is a series of 30 sets of 3 curves each. These represent an estimation of the average, best and worst conditions that can be expected during the last half of one month and the first half of the next — over about any path an American amateur might be interested in. Of course they can also be used by a resident of any of the areas at the other ends.

Propagation bulletins are transmitted from W1AW several times a day. The bulletins contain information on solar and geomagnetic activity. This information helps to make interpretation of the monthly charts in *QST* more accurate for a given week in the four-week forecast period of the charts. The bulletins are normally prepared each Friday, but they can be revised at any time if their content does not match current conditions. The propagation information is part of the regular W1AW bulletin service, on all suitable modes. See April and October *QST*, yearly, for W1AW summer and winter schedules. Comments and suggestions are welcomed.

## MINOR PROPAGATION MODES

In propagation literature there is a tendency to treat the various propagation modes as if they were separate and distinct phenomena. This they may be at times, but often there is a shifting from one to another, or a mixture of two or more kinds of propagation affecting communication at one time. In the upper part of the usual frequency range for F-layer works, for example, there may be enough tropospheric bending at one end (or both ends) to have an appreciable effect on the usable path length. There is the frequent combination of E and F in long-distance work, already discussed. And, in the case of the E layer, there are various causes of ionization that have very different effects on communication. Finally, there are weak-signal variations of both tropospheric and ionospheric modes, lumped under the term "scatter." We look at these phenomena separately here, but in practice we may have to deal with them in combination, more often than not.

### Sporadic-E ($E_s$)

First note that this is E-subscript-s, a usefully descriptive term, wrongly written "Es" so often that it is sometimes called "ease" — which is certainly *not* descriptive. Sporadic E is ionization at E-layer height, but of quite different origin and communications potential from the E layer that affects mainly our lower amateur frequencies.

The formative mechanism for sporadic E is believed to be wind shear. This explains ambient ionization being redistributed and compressed into a ledge of high density, without the need for production of extra ionization. Neutral winds of high velocity, flowing in opposite directions at slightly different altitudes, produce shears. In the presence of the magnetic field the ions are collected at a particular altitude, forming a thin, overdense layer. Data from rockets entering $E_s$ regions confirm the electron density, wind velocities and height parameters.

The ionization is formed in clouds of high density, lasting only a few hours at a time and distributed randomly. They vary in density and move rapidly from southeast to northwest, in the middle latitudes in the northern hemisphere. Though $E_s$ can develop at any time, it is most prevalent in the northern hemisphere between May and August, with a minor season about half as long beginning in December (the summer and winter solstices). The seasons and distribution in the southern hemisphere are not so well known. Australia and New Zealand seem to have conditions much like those in the U.S., but with the length of the seasons reversed, of course. Much of what is

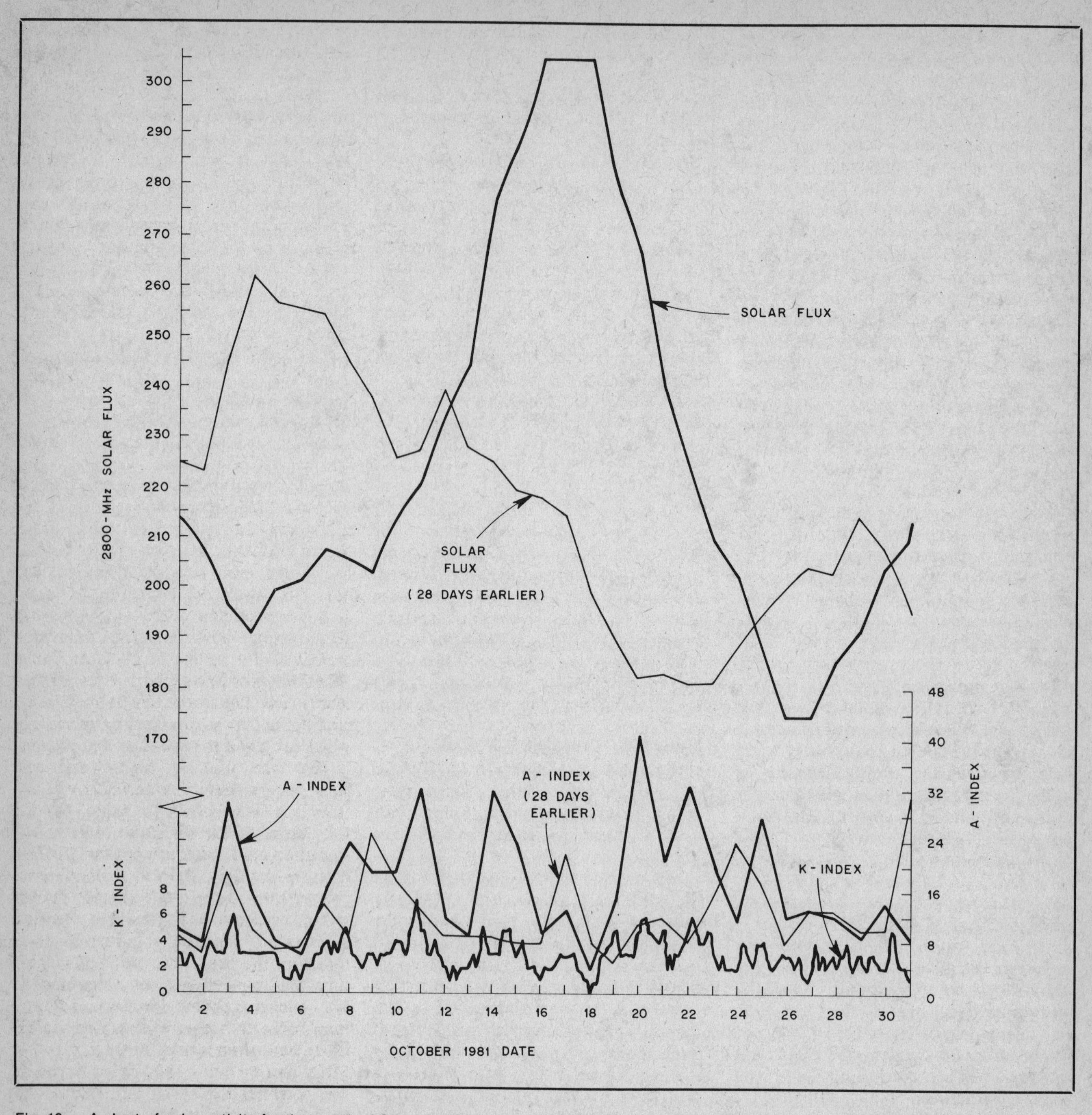


Fig. 16 — A chart of solar activity for the month of October 1981 (heavy lines). Data from WWV propagation bulletins are plotted for the A-Index, K-Index and 2800-MHz solar flux. To help in forecasting propagation conditions, data are also plotted for the previous period of solar rotation, 28 days earlier (light lines.) Trends from the previous period may be observed, but note that sometimes the correlation may be negative rather than positive. It has been observed that broad peaks in the solar flux level generally have a lifetime of approximately four months from buildup to decay, although exceptions may certainly occur.

known about $E_s$ came as the result of amateur pioneering in the vhf range. Historical and technical references at the end of this chapter are interesting reading.

Correlation of $E_s$ openings with observed natural phenomena, including sunspot activity, is not readily apparent, though there is a meteorological tie-in with high-altitude winds. There is also a form of $E_s$, mainly in the northern part of the north temperate zone, that is associated with auroral phenomena.

At the peak of the long $E_s$ season, most commonly in late June and early July, ionization becomes extremely dense and widespread. This extends the usable range from the more-common "single-hop" maximum of about 1400 miles to "double-hop" distances, mostly 1400 to 2500 miles. With 50-MHz techniques and interest improving in recent years, it has been shown that distances considerably beyond 2500 miles can be covered. There is also an $E_s$ "link-up" possibility with other modes, believed to be involved in some 50-MHz work between antipodal points, or even long-path communication beyond 12,500 miles.

The muf for $E_s$ is not known precisely. It was long thought to be around 100 MHz, but in the last 20 years or so there have been thousands of 144-MHz contacts during the summer $E_s$ season.

Presumably, the possibility exists at 220 MHz, also. The skip distance at 144 MHz does average much longer than at 50 MHz, and the openings are usually brief and extremely variable.

The terms "double" and "single" hop may not be accurate technically, since it is likely that cloud-to-cloud paths are involved. There may also be "no-hop" Es. At times the very high ionization density produces critical frequencies up to the 50-MHz region, with no skip distance at all. It is often said that the Es mode is a great equalizer; with the reflecting region practically overhead, even a simple dipole close to the ground may do as well over a few hundred miles as a large stacked antenna array designed for low-angle radiation. It's a great mode for low power and simple antennas on 28 and 50 MHz.

**The Scatter Modes**

The term "skip zone" should not be taken literally. Two stations communicating over a single ionospheric hop can be heard to some degree at almost any point along the way, unless they are running quite low power and using simple antennas. Some of the wave energy is scattered in all directions, including back to the starting point and farther. The wave energy of vhf stations is not gone after it reaches the radio horizon, described early in this chapter. It is scattered, but it can be heard to some degree for hundreds of miles. Everything on earth, and in the regions of space up to at least 100 miles, is a potential scattering agent.

*Tropospheric scatter* is always with us. Its effects are often hidden, masked by more effective propagation modes on the lower frequencies. But beginning in the vhf range, scatter from the lower atmosphere extends the reliable range quite markedly if we make use of it. Called "tropo scatter," this is what produces that nearly flat portion of the curves given in the following section on reliable vhf coverage. We are not out of business at somewhere between 50 and 100 miles, on the vhf and even uhf bands, especially if we don't mind weak signals and something less than 99 percent reliability. As long ago as the early 1950s, vhf enthusiasts found that vhf contests could be won with high power, big antennas and a good ear for signals deep in the noise. They still can.

*Ionospheric scatter* works much the same as the tropo version, except that the scattering medium is the E region of the ionosphere, with some help from the D and F. Ionospheric scatter is useful mainly *above* the muf, so its useful frequency range depends on geography, time of day, season and the state of the sun. With near maximum legal power, good antennas and quiet locations, ionospheric scatter can fill in the skip zone with marginally readable signals scattered from ionized trails of meteors, small areas of random ionization, cosmic dust, satellites and whatever may come into the antenna patterns at 50 to 150 miles or so above the earth. It's mostly an E-layer business, so it works at E-layer distances. Good antennas and keen ears help.

*Backscatter* is a sort of ionospheric radar. Because it is mainly scattering from the earth at the point where the strong ionospherically propagated signal comes down, it is, in fact, a part of long-distance radar techniques. It is also a great "filler-inner" of the skip zone, particularly in work near the muf where propagation is best. It was proved by amateurs using sounding techniques that you can tell to what part of the world a band is usable (single-hop F) by probing the backscatter with a directive antenna, even when the earth-contact point is open ocean. In fact, that's where the mode is at its best. (See references.)

Backscatter is very useful on 28 MHz, particularly when that band *seems* dead simply because nobody is active in the right places. The mode keeps the 10-meter band lively in the low years of the solar cycle, thanks to the never-say-die attitude of some long-time users. The mode is also an invaluable tool of 50-MHz DX aspirants, in the *high* years of the sunspot cycle, for the same reasons. On a high-muf morning hundreds of 6-meter beams may zero in on a hot spot somewhere in the Caribbean or South Atlantic, where there is no *land*, let alone other 6-meter stations — keeping in contact while they wait for the band to open to a place where there *is* somebody.

*Sidescatter* is similar to backscatter, except the ground scatter zone is merely somewhat off the direct line between the participants. A typical example, often observed on 28 MHz during the lowest years of the solar cycle, is communication on 28 MHz between Eastern USA (and adjacent areas of Canada) and much of the European continent. Often, this may start as "backscatter chatter" between stations on the west side of the Atlantic. Then suddenly the Europeans join the fun, perhaps for only a few minutes, but sometimes much longer. Duration of the game can be extended, at times, by careful reorientation of antennas at both ends, as with backscatter. The secret, of course, is to keep hitting the highest-muf area of the ionosphere and the most favorable ground-reflection points.

The favorable route is usually, but not always, south of the direct great-circle heading. There can be sidescatter from the auroral regions. This was first demonstrated on 50 MHz by amateur stations operating in the far north regions of the Arctic Ocean, in the late 1950s, after the peak of solar cycle 19 had passed. Sidescatter signals are likely to be stronger than backscatter signals using the same general area of ground scattering. A study of the *QST* propagation charts will indicate areas where backscatter and sidescatter are likely to occur, when a direct path may not work.

*Transequatorial scatter (TE)* was an amateur 50-MHz discovery in the years around the peak of solar cycle 18, 1946-1947. It was turned up almost simultaneously on three separate north-south paths, by amateurs of all continents, who tried to communicate on 50 MHz, even though the *predicted* muf was around 40 MHz for the favorable daylight hours. The first success came at night, when the muf was thought to be even lower.

A remarkable research program inaugurated by amateurs in Europe, Cyprus, Southern Rhodesia (now Zimbabwe) and South Africa eventually provided technically sound theories to explain the then-unknown mode. Nearly 25 years later, the work continues, having become almost a lifetime project for some of the principals. In recent years they have shown that the TE mode works on 144 MHz, and even to some degree at 432 MHz! It was in the 1940s that amateurs first demonstrated there was something very wrong with existing prediction techniques for certain north-south paths. What was not known then is condensed into a few lines here. Everyone should read the full story, principal references for which are given at the end of this chapter.

That the muf is higher and less seasonally variable on transequatorial circuits had been known for years, but the full extent of the difference was hardly suspected until amateur work on 50 MHz brought it to light. Briefly, the ionosphere over equatorial regions is higher, thicker and more dense than elsewhere. Because of its more constant exposure to solar radiation the equatorial belt has nighttime muf possibilities not demonstrated until amateurs probed for them on a long-term basis. It is now known that the TE mode can often work marginally at 144 MHz, and even at 432 MHz on occasion. The potential muf varies with solar activity, but not to the extent that conventional F-layer propagation does. It is a late-in-the-day mode, taking over about when normal F-layer propagation goes out.

The TE range is usually within about 4000 km (2500 miles) either side of the geomagnetic equator. The earth's magnetic axis is tilted with respect to the geographical axis so the TE belt appears as a curving band on conventional flat maps of the world. See Fig. 17. As a result, TE has a different latitude coverage in the Americas from that shown in the drawing. Our TE belt just reaches into southern USA. Puerto Rico, Mexico and even the northern parts of South America encounter the mode more often than do most U.S. areas. It is no accident that TE was discovered as a result of 50-MHz work in Mexico City and Buenos Aires.

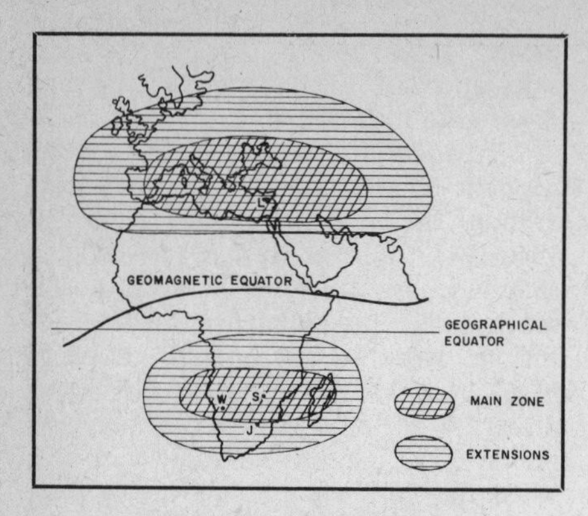


Fig. 17 — Main and occasional zones of transequatorial 50-MHz propagation as described by ZE2JV, show Limassol, Cyprus, and Salisbury, Zimbabwe, to be almost ideally positioned with respect to the curving geomagnetic equator. Windhoek, Namibia, is also in a favorable spot; Johannesburg somewhat less so.

Within its optimum regions of the world, the TE mode extends the usefulness of the 50-MHz band far beyond that of conventional F-layer propagation, since the practical TE muf runs around 1.5 times that of normal F2. Both its seasonal and diurnal characteristics also are extensions of what is considered normal for 50-MHz propagation. The existence of TE affects the whole character of band usage, in that part of the Americas south of about 20° N. latitude, especially in years of high solar activity.

**Auroral Propagation**

Sudden bursts of solar activity are accompanied by ejection of charged particles from the sun. These travel in various directions, and some enter the earth's atmosphere, usually 24 to 36 hours after the event. Some information on the major outbursts is obtainable from the WWV propagation bulletins discussed earlier. The timing of major flares is announced, usually in the next bulletin period, and the flare is given a rating. Some hours later, if the particles are intercepted by our geostationary satellites that are designed for that purpose, a *proton event* is announced. Around 12 to 24 hours later the solar particles enter the earth's atmosphere. Here they may react with the magnetic field to produce a visible or radio aurora, the former if their time of entry is after dark.

The visible aurora is, in effect, fluorescense at E-layer height — a curtain of ions capable of refracting radio waves in the frequency range above about 20 MHz. D-region absorption increases on lower frequencies during auroras. The exact frequency ranges depend on many factors: time, season, position with relation to the earth's auroral regions and the level of solar activity at the time, to name a few.

Auroral effect on vhf waves is another amateur discovery, this one dating back to the 1930s, coming coincidentally with improved transmitting and receiving techniques for what was then called "uhf." The returning signal is diffused in frequency by the diversity of the auroral curtain as a refracting (scattering) medium. The result is a modulation of a cw signal, from just a slight burbling sound to what is best described as a keyed roar. Before ssb took over in vhf work, voice was all but useless for auroral paths. A sideband signal suffers, too, but its narrower bandwidth helps to retain some degree of understandability. Distortion induced by a given set of auroral conditions increases with the frequency in use. In general, 50-MHz signals are much more intelligible than those on 144 MHz on the same path at the same time. On 144 MHz, cw is almost mandatory for effective auroral communication.

The number of auroras that can be expected per year varies with the *geomagnetic* latitude. Drawn with respect to the earth's magnetic poles instead of the geographical ones, these latitude lines in the U.S. tilt upward to the northwest. For example, Portland, Oregon, is 2 degrees farther north (*geographic* latitude) than Portland, Maine. But the Maine city's geomagnetic latitude line crosses the Canadian border before it gets as far west as its Oregon namesake. In terms of auroras intense enough to produce vhf propagation results, Portland, Maine, is likely to see about 10 times as many per year. Oregon's auroral prospects are more like those of southern New Jersey or central Pennyslvania.

The antenna requirements for auroral work are mixed. High gain helps, but the area of the aurora yielding the best returns sometimes varies quite rapidly, so very high directivity can be a disadvantage. So could a very low radiation angle, or a beam pattern very sharp in the vertical plane. Experience seems to indicate that few amateur antennas are sharp enough in either plane to present a real handicap. The beam heading for maximum signal can change, however, so a bit of scanning in azimuth may turn up some interesting results. A very large array such as is commonly used for moonbounce (with el-az control) should be worthwhile.

The incidence of auroras, their average intensity, and their geographical distribution as to visual sightings and vhf propagation effects, all vary to some extent with solar activity. There is some indication that the peak period for auroras lags the sunspot-cycle peak by a year or two. But like sporadic E, an auroral humdinger can come at any season. There is a marked diurnal swing in the number of auroras. Favored times are late afternoon and early evening, late evening through early morning, and early afternoon, in about that order. Major auroras often start in early afternoon and carry through to early morning the next day. Since auroras are related to solar flares and the periods of increased geomagnetic activity that follows them, we now have an aurora-alert system of sorts in the WWV propagation bulletins. Major flares don't always produce major geomagnetic disturbances and major auroras, but they are a good warning that an aurora is likely, usually 24 to 36 hours after the flare.

The WWV K-Index is a timely clue to aurora possibilities, a *Boulder* reading of geomagnetic activity that may not correlate closely with conditions in other areas. Values of 3, and rising, warn that conditions associated with auroras and degraded hf propagation are present in the Boulder area at the time of the bulletin's preparation, but the timing and severity may be different elsewhere.

The A-Index, a 24-hour figure, is more useful as a reference for the future. A curve drawn from the daily A figures (see Fig. 16) is very useful for anticipating periods of high geomagnetic activity on the next solar rotation, 26 to 28 days later. The valleys between the A-Index peaks foretell geomagnetic quiet (good hf propagation) four weeks ahead.

If the solar flux and the A Index are plotted one above the other it will become apparent that there is a subtle relationship between them that is a good short-term propagation indicator. A sharp rise in the solar flux (a jump of 10 points or more in a day) means new solar activity. The muf will rise coincidentally, but there will also be an increase in the charged-particle emission from the sun that causes auroras and degraded hf conditions. The solar flux rise is thus a two-day warning (average) that these mixed blessings are on the way. The WWV bulletin is not an infallible crystal ball, but it is the best we have.

## RELIABLE VHF COVERAGE

In the preceding pages we discussed means by which our bands above 50 MHz may be used intermittently for communication far beyond the visual horizon. In emphasizing distance we should not neglect a prime asset of the vhf bands: reliable communication over relatively short distances. The vhf region is far less subject to disruption of local communication than are frequencies below about 30 MHz. Since much amateur communication is essentially local in nature, our vhf assignments could carry a greater load then they presently do, and this would help solve interference problems on lower frequencies.

Possibly some amateur unwillingness to migrate to the vhf bands arises from misconceptions about the coverage obtainable. This reflects the age-old idea that vhf waves travel only in straight lines, except when the DX modes described above happen to be present. Let us survey the picture in the light of modern wave-propagation knowledge and see what the bands above 50 MHz are good for on a day-to-day basis, ignoring the anomalies that may result in extensions of normal coverage.

It is possible to predict with fair accuracy how far you should be able to work consistently on any vhf or uhf band, provided a few simple facts are known. The factors affecting operating range can be reduced to graph form, as described in *QST* by D.W. Bray, K2LMG. To estimate your station's capabilities, two basic numbers must be determined: station gain and path loss. Station gain is made up of eight factors: receiver sensitivity, transmitted power, receiving antenna gain, receiving antenna height gain, transmitting antenna gain, transmitting antenna height gain and required signal-to-noise ratio. This looks complicated but it really boils down to an easily made evaluation of receiver, transmitter, and antenna performance. The other number, path loss, is readily determined from the nomogram, Fig. 18. This gives path loss over smooth earth, for 99 percent reliability.

For 50 MHz, lay a straightedge from the distance between stations (left side) to the appropriate distance at the right side. For 1296 MHz, use the full scale, right center. For 144, 220 and 432, use the dot in the circle, square or triangle, respectively. Example: At 300 miles the path loss for 144 MHz is 214 dB.

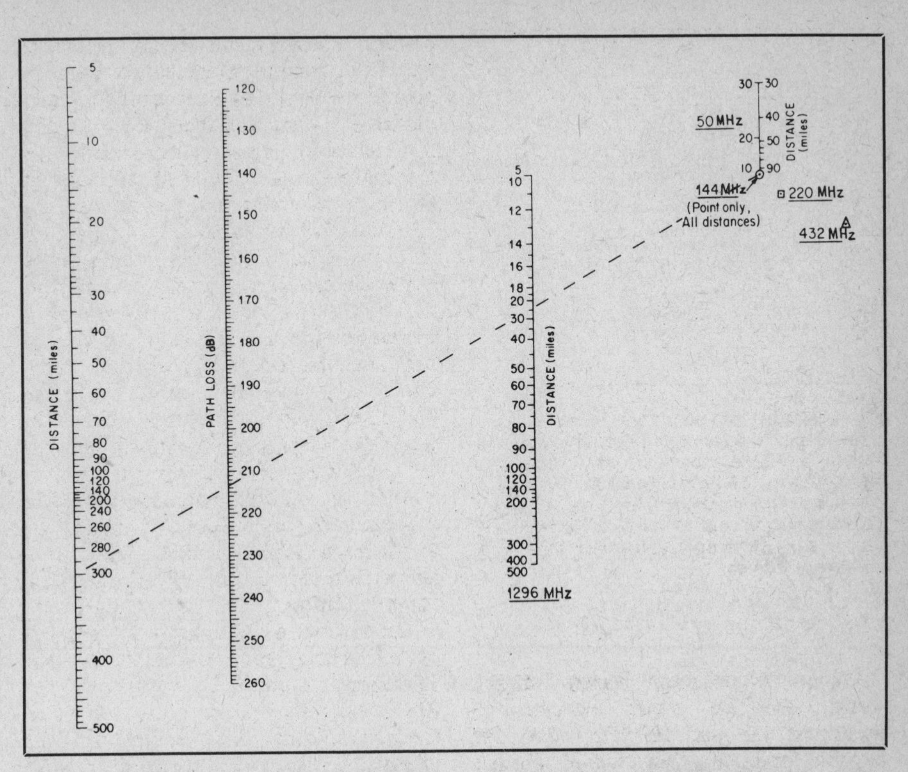


Fig. 18 — Nomogram for finding the capabilities of stations on amateur bands from 50 to 1300 MHz. Either the path loss for a given distance or vice versa may be found if one of the two factors is known.

### Station Gain

The largest of the eight factors involved in station gain is receiver sensitivity. This is obtainable from Fig. 19, if you know the approximate receiver noise figure and transmission-line loss. If you can't measure noise figure, assume 3 dB for 50 MHz, 5 for 144 or 220, 8 for 432 and 10 for 1296, if you know that your equipment is working moderately well. These noise figures are well on the conservative side for modern solid-state receivers, particularly at 432 MHz. Line loss can be taken from information in Chapter 3 for the line in use, if the antenna system is fed properly. Lay a straightedge between the appropriate points at either side of Fig. 19, to find effective receiver sensitivity in decibels below 1 watt (dbW). Use the narrowest bandwidth that is practical for the emission intended, with the receiver you will be using. For cw, an average value for effective work is about 500 Hz. Phone bandwidth can be taken from the receiver instruction manual.

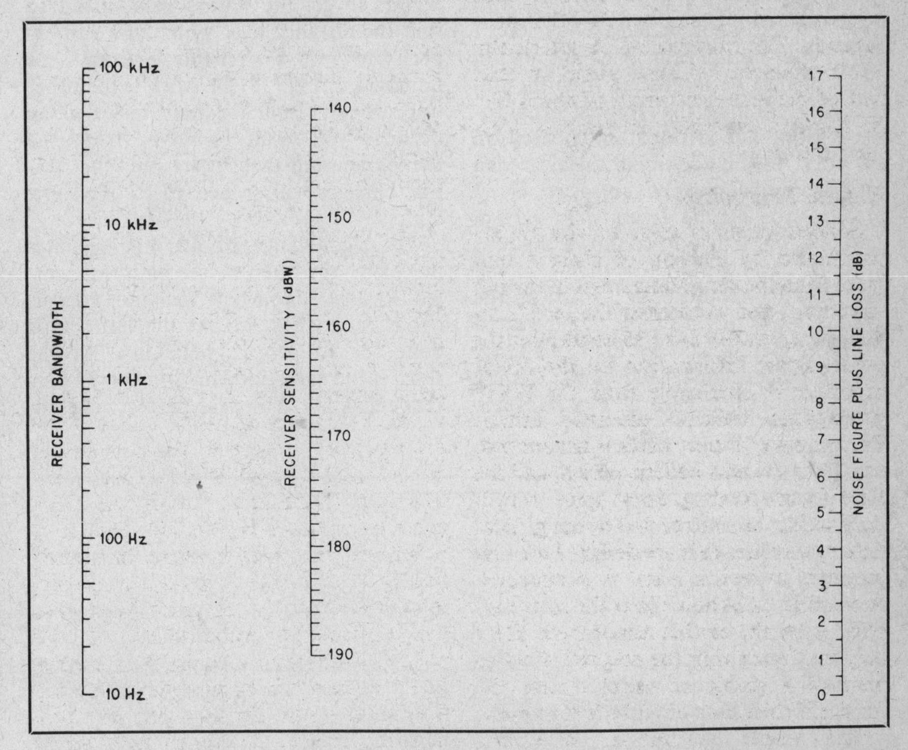


Fig. 19 — Nomogram for finding effective receiver sensitivity.

Antenna gain is next in importance. Gains of amateur antennas are often exaggerated. For well-designed Yagis they run close to 10 times the boom length in wavelengths. (Example: A 24-foot Yagi on 144 MHz is 3.6 wavelengths long; 3.6 × 10 = 36, or about 15-1/2 dB.) Add 3 dB for stacking, where used properly. Add 4 dB more for ground-reflection gain. This varies in amateur work, but averages out near this figure. We have one more plus factor: antenna height gain, obtainable from Fig. 20. Note that this is greatest for short distances. The left edge of the horizontal center scale is for 0 to 10 miles, the right edge for 100 to 500 miles. Height gain for 10 to 30 feet is assumed to be zero. It will be seen that for 50 feet the height gain is 4 dB at 10 miles, 3 dB at 50 miles, and 2 dB at 100 miles. At 80 feet the height gains are roughly 8, 6 and 4 dB for

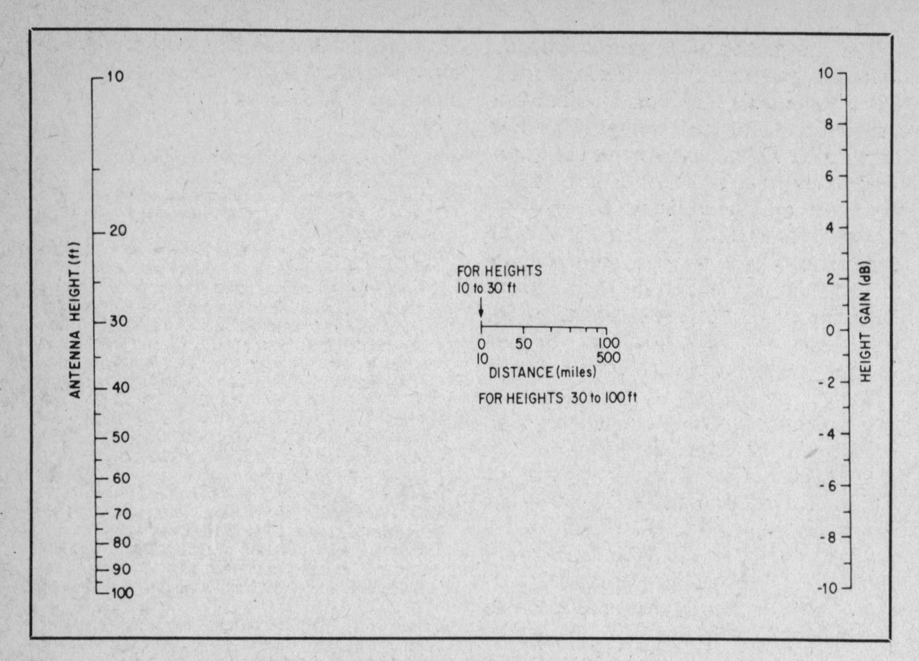


Fig. 20 — Nomogram for determining antenna-height gain.

these distances. Beyond 100 miles the height gain is nearly uniform for a given height, regardless of distance.

Transmitter power output must be stated in decibels above 1 watt. If you have 500 watts output, add 500/1, or 27 dB, to your station gain. The transmission line loss must be subtracted from the station gain. So must the required signal-to-noise ratio. The information is based on cw work, so the additional signal needed for other modes must be subtracted. Use 3 dB for ssb. Fading losses must be accounted for. It has been shown that for distances beyond 100 miles the signal will vary plus or minus about 7 dB from the average level, so 7 dB must be subtracted from the station gain for high reliability. For distances under 100 miles, fading diminishes almost linearly with distance. For 50 miles, use minus 3.5 dB for fading.

### What It All Means

After adding all the plus-and-minus factors to get the station gain, use it to find the distance over which you can expect to work reliably, from the nomogram, Fig. 18. Or work it the other way around: Find the path loss for the distance you want to cover from the nomogram and then figure out what station changes will be needed to overcome it.

The significance of all this becomes more obvious when we see path loss plotted against frequency for the various bands, as in Fig. 21. At the left this is done for 50 percent reliability. At the right is the same information for 99 percent reliability. For near perfect reliability, a path loss of 195 dB (easily countered at 50 or 144 MHz) is involved in 100-mile communication. But look at the 50 percent reliability curve: The same path loss takes us out to well over 250 miles. Few amateurs demand near-perfect reliability. By choosing our times, and accepting the necessity for some repeats or occasional loss of signal, we can maintain communication out to distances far beyond those usually covered by vhf stations.

Working out a few typical amateur vhf station setups with these curves will show why an understanding of these factors is important to any user of the vhf spectrum. Note that path loss rises very steeply in the first 100 miles or so. This is no news to vhf operators; locals are very strong, but stations 50 or 75 miles away are much weaker. What happens beyond 100 miles is not so well known to some of us.

From the curves of Fig. 21, we see that path loss levels off markedly at what is the approximate limit of working range for average vhf stations using wideband modulation modes. Work out the station gain for a 50-watt station with an average receiver and moderate-sized antenna, and you'll find that it will come out around 180 dB. This means about a 100-mile working radius in average terrain, for good but not perfect reliability. Another 10 dB may extend the range to as much as 250 miles. Just changing from a-m phone to ssb and cw made a major improvement in daily coverage on the vhf bands. A bigger antenna, a higher one if your present beam is not at least 50 feet up, an increase in power to 500 watts from 50, an improvement in receiver noise figure if it is presently poor — any of these things can make a big improvement in reliable coverage. Achieve all of them, and you will have very likely tripled your sphere of influence, thanks to that hump in the path-loss curves. This goes a long way toward explaining why using a 10-watt

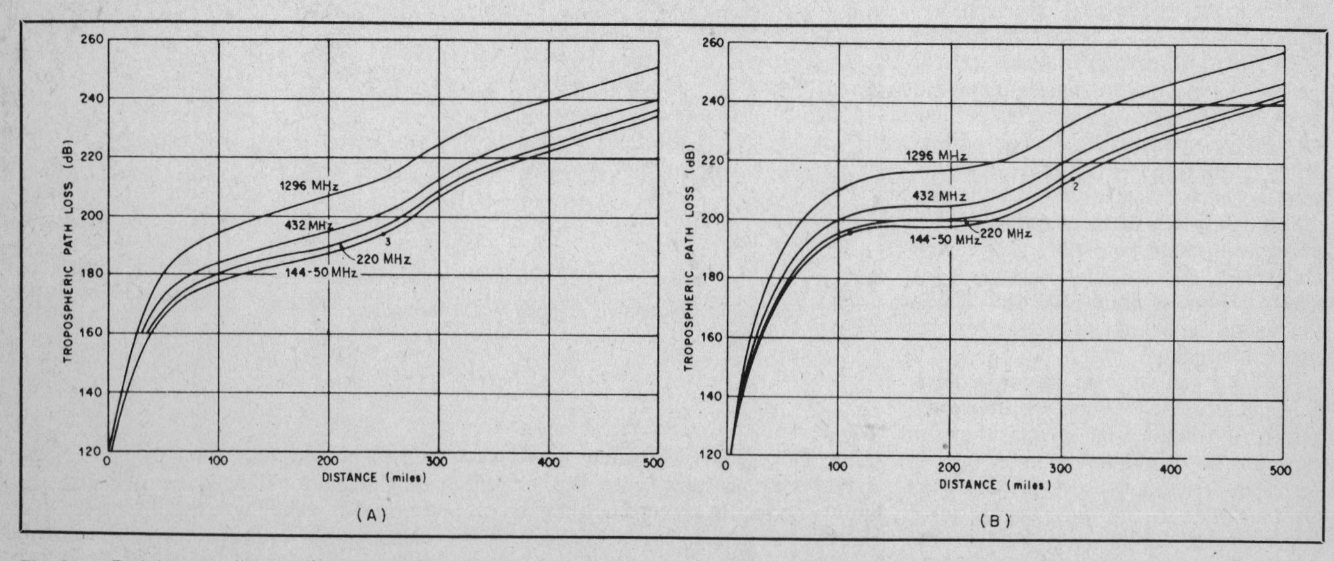


Fig. 21 — Path loss vs. distance for amateur frequencies above 50 MHz. At A are curves for 50 percent of the time; at B, for 99 percent. The curves at A are more representative of Amateur Radio requirements.

packaged station with a small antenna, fun though it may be, does not begin to show what the vhf bands are *really* good for.

**About Terrain**

The coverage figures derived from the above are for average terrain. What of stations in mountainous country? Though an open horizon is generally desirable for the vhf station site, mountain country should not be considered hopeless. Help for the valley dweller often lies in the optical phenomenon known as knife-edge refraction. A flashlight beam pointed at the edge of a partition does not cut off sharply at the partition edge, but is refracted around it, partially illuminating the shadow area. A similar effect is observed with vhf waves passing over ridges; there is a shadow effect, but not a complete blackout. If the signal is strong where it strikes the mountain range, it will be heard well in the bottom of a valley on the far side.

This is familiar to all users of vhf communications equipment who operate in hilly terrain. Where only one ridge lies in the way, signals on the far side may be almost as good as on the near. Under ideal conditions (a very high and sharp-edged obstruction near the midpoint of a path long enough so that signals would be weak over average terrain), knife-edge refraction may yield signals even stronger than would be possible with an open path.

The obstruction must project into the radiation patterns of the antennas used. Often mountains that look formidable to the viewer are not high enough to have an appreciable effect, one way or the other. Since the normal radiation from a vhf array is several degrees above the horizontal, mountains that are less than three degrees above the horizon, as seen from the antenna, are missed by the radiation from the array. Moving them out of the way would have substantially no effect on vhf signal strength in such cases.

Rolling terrain, where obstructions are not sharp enough to produce knife-edge refraction, still does not exhibit a complete shadow effect. There is no complete barrier to vhf propagation — only attenuation, which varies widely as the result of many factors. Thus, even valley locations are usable for vhf communication. Good antenna systems, preferably as high as possible, the best available equipment, and above all, the willingness and ability to work with weak signals may make possible outstanding vhf work, even in sites that show little promise by casual inspection.

## REFERENCES

No single book or paper is likely to cover the whole subject of radio propagation adequately. Below are recommended sources of propagation information, listed in approximately the order of their appearance in this chapter. ARRL publications contain detailed bibliographies, both historical and technical.


*The Radio Amateur's Handbook,* ARRL, 1976 and later editions preferred.

*The Radio Amateur's VHF Manual,* ARRL, 1965 through 1972. Out of print, but can be found in many libraries.

Davies, K., *Ionospheric Radio Propagation,* National Bureau of Standards. Also known as NBS Monograph 80. Out of print, but may be found in technical libraries. Excellent technical reference.

Hall, J., "HF Propagation Estimations for the Radio Amateur," *QST,* March 1972. Printed predictions mentioned are no longer available, but basic information applies to propagation charts now printed in *QST.*

Hull, R., "Air-Mass Conditions and the Bending of UHF Waves," *QST,* June 1935 and May 1937. Classic first work in the field of tropospheric propagation in the vhf region.

Collier, J., "Upper-Air Conditions for 2-Meter DX," *QST,* September 1955. More information on the conditions shown in Fig. 8 of this chapter.

"The World Above 50 MHz," *QST,* September 1957, August 1959 and September 1960. First vhf communication across the Pacific on 144, 220 and 432 MHz.

Pierce, J. A., "Interpreting 56-Mc. DX," *QST,* September 1938. Probably the first explanation of sporadic-E propagation in the vhf range.

Wilson, M. S., "Midlatitude Intense Sporadic-E Propagation," *QST,* December 1970 and March 1971.

Tilton, E., "The DXer's Crystal Ball," *QST,* June, August and September 1975.

Cracknell, R. G., Whiting, R. A., and others, on transequatorial propagation of vhf signals, *QST,* December 1959, August 1960 and April 1963.

Cracknell, R. G., and others, "The Euro-Asia to Africa VHF Transequatorial Circuit During Solar Cycle 21," *QST,* November and December 1981. A detailed report on recent TE work on 50, 144 and 432 MHz.

"Twenty-One Years of TE," *Radio Communication* (RSGB), June-July 1980.

Bray, D., "A Method of Determining VHF Station Capabilities," *QST,* November 1961.

# Chapter 2

# Antenna Fundamentals

An antenna is an electric circuit of a special kind. In the ordinary type of circuit the dimensions of coils, capacitors and connections usually are small compared with the wavelength that corresponds to the frequency in use. When this is the case most of the electromagnetic energy stays in the circuit itself and either is used up in performing useful work or is converted into heat. But when the dimensions of wiring or components become appreciable compared with the wavelength, some of the energy escapes by *radiation* in the form of electromagnetic waves. When the circuit is intentionally designed so that the major portion of the energy is radiated, we have an *antenna.*

Usually, the antenna is a straight section of conductor, or a combination of such conductors. Very frequently the conductor is a wire, although rods and tubing also are used. In this chapter we shall use the term "wire" to mean any type of conductor having a cross section that is small compared with its length.

The strength of the electromagnetic field radiated from a section of wire carrying radio-frequency current depends on the length of the wire and the amount of current flowing. It would also be true to say that the field strength depends on the voltage across the section of wire, but it is generally more convenient to measure current. The electromagnetic field consists of both magnetic and electric energy, with the total energy equally divided between the two. One cannot exist without the other in an electromagnetic wave, and the voltage in an antenna is just as much a measure of the field intensity as the current. Other things being equal, the field strength will be directly proportional to the current. It is therefore desirable to make the current as large as possible, considering the power available. In any circuit that contains both resistance and reactance, the largest current will flow (for a given amount of power) when the reactance is "tuned out" — in other words, when the circuit is made *resonant* at the operating frequency. So it is with the common type of antenna; the current flowing in the antenna will be largest, and the radiation therefore greatest, when the antenna is resonant.

In an ordinary circuit the inductance is usually concentrated in a coil, the capacitance in a capacitor, and the resistance is principally concentrated in resistors, although some may be distributed around the circuit wiring and coil conductors. Such circuits are said to have *lumped constants.* In an antenna, on the other hand, the inductance, capacitance and resistance are distributed along the wire. Such a circuit is said to have *distributed constants.* Circuits with distributed constants are so frequently straight-line conductors that they are customarily called linear circuits.

## RESONANCE IN LINEAR CIRCUITS

The shortest length of wire that will resonate to a given frequency is one just long enough to permit an electric charge to travel from one end to the other and then back again in the time of one rf cycle. If the speed at which the charge travels is equal to the velocity of light, 299,793,077 meters per second, the distance it will cover on one cycle or period will be equal to this velocity divided by the frequency in hertz, or approximately

$$\lambda = \frac{300{,}000{,}000}{f}$$

in which λ is the wavelength in meters. Since the charge traverses the wire *twice*, the length of wire needed to permit the charge to travel a distance λ in one cycle is λ/2, or one-half wavelength. Therefore the shortest *resonant* wire will be a half-wavelength long.

The reason for this length can be made clear by a simple example. Imagine a trough with barriers at each end. If an elastic ball is started along the trough from one end, it will strike the far barrier, bounce back, travel along to the near barrier, bounce again, and continue until the energy imparted to it originally is all dissipated. If, however, whenever it returns to the near barrier it is given a new push just as it starts away, its back-and-forth motion can be kept up indefinitely. The impulses, however, must be *timed* properly; in other words, the rate or frequency of the impulses must be adjusted to the length of travel and the rate of travel. Or, if the timing of the impulses and the speed of the ball are fixed, the length of the trough must be adjusted to "fit."

In the case of the antenna, the speed is essentially constant, so we have the alternatives of adjusting the frequency to a given length of wire, or adjusting the length of wire to a given operating frequency. The latter is usually the practical condition.

By changing the units in the equation just given and dividing by 2, the formula

$$\ell = \frac{492}{f_{MHz}}$$

is obtained. In this case $\ell$ is the length *in feet* of a *half* wavelength for a frequency f, given in megahertz, when the wave travels with the velocity of light. This formula is the basis upon which several significant lengths in antenna work are developed. It represents the length of a half wavelength in space, when no factors that modify the speed of propagation exist. To determine a half wavelength in meters, the relationship is

$$\ell = \frac{150}{f_{MHz}}$$

### Current and Voltage Distribution

If the wire in an antenna were infinitely long, the charge (voltage) and the current (an electric current is simply a charge in motion) would both slowly decrease in amplitude with distance from the source. The slow decrease would result from dissipation of energy in the form of radio waves and in heating the wire because of its resistance. When the wire is short the charge is reflected when it reaches the far end, however, just as the ball bounced back from the barrier. With radio-frequency excitation of a half-wave antenna, there is of course not just a single charge but a continuous supply of energy, varying in voltage according to a sine-wave cycle. We might consider this as a series of charges, each of slightly different amplitude than the preceding one. When a charge reaches the end of the antenna and is reflected, the direction of current flow reverses, since the charge is now traveling in the opposite direction. The next charge is just reaching the end of the antenna, however, so we have two currents of practically the same amplitude flowing in opposite directions. The resultant current at the end of the antenna therefore is zero. As we move farther back from the end of the antenna the magnitudes of the outgoing and returning currents are no longer the same because the charges causing them have been supplied to the antenna at different parts of the rf cycle. There is less cancellation, therefore, and a measurable current exists. The greatest difference — that is, the largest resultant current — will be found to exist a quarter wavelength away from the end of the antenna. As we move back still farther from this point the current will decrease until, a half wavelength away from the end of the antenna, it will reach zero again. Thus, in a half-wave antenna the current is zero at the ends and maximum at the center.

This resultant current distribution along a half-wave wire is shown in Fig. 1. The distance measured vertically from the antenna wire to the curve marked "current," at any point along the wire, represents the relative amplitude of the current as measured by an ammeter at that point. This is called a *standing wave* of current. The *instantaneous* value of current at any point varies sinusoidally at the applied frequency, but its amplitude is different at every point along the wire as shown by the curve. The standing-wave curve itself has the shape of a half sine wave, at least to a good approximation.

The voltage along the wire will behave differently; it is obviously greatest at the end since at this point we have two practically equal charges adding. As we move back along the wire, however, the outgoing and returning charges are not equal and their sum is smaller. At the quarter-wave point the returning charge is of equal magnitude but of opposite sign to the outgoing charge, since at this time the polarity of the voltage wave from the source has reversed (one-half cycle). The two voltages therefore cancel each other and the resultant voltage is zero. Beyond the quarter-wave point, away from the end of the wire, the voltage again increases, but this time with the opposite polarity.

It will be observed, therefore, that the voltage is maximum at every point where the current is minimum, and vice versa. The polarity of the current or voltage reverses every half wavelength along the wire, but the reversals do not occur at the same points for both current and voltage; the respective reversals occur, in fact, at points a quarter wave apart. A maximum point on a standing wave is called a loop (or antinode); a minimum point is called a node.

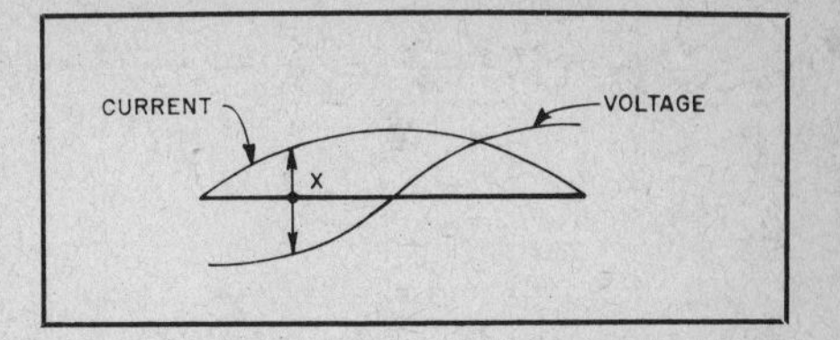


Fig. 1 — Current and voltage distribution on a half-wave wire. The wire is represented by the heavy line. In this conventional representation the distance at any point (X, for instance) from the wire to the curve gives the relative intensity of current or voltage at that point. The relative direction of current flow (or polarity of the voltage) is indicated by drawing the curve either above or below the line representing the antenna. The voltage curve here, for example, indicates that the instantaneous polarity in one half of the antenna is opposite to that in the other half.

## Harmonic Operation

If there is reflection from the end of a wire, the number of standing waves on the wire will be equal to the length of the wire divided by a half wavelength. Thus, if the wire is two half waves long there will be two standing waves; if three half waves long, three standing waves, and so on. These longer wires, each multiples of a half wave in length, will also be resonant, therefore, at the same frequency as the single half-wave wire. When an antenna is two or more half waves in length at the operating frequency it is said to be *harmonically resonant*, or to operate at a harmonic. The number of the harmonic is the number of standing waves on the wire. For example, a wire two half waves long is said to be operating on its *second* harmonic, one three half waves long on its third *harmonic,* and so on.

Harmonic operation of a wire is illustrated in Fig. 2. Such operation is often used in antenna work because it permits operating the same antenna on several harmonically related amateur bands. It is also an important principle in the operation of certain types of directive antennas.

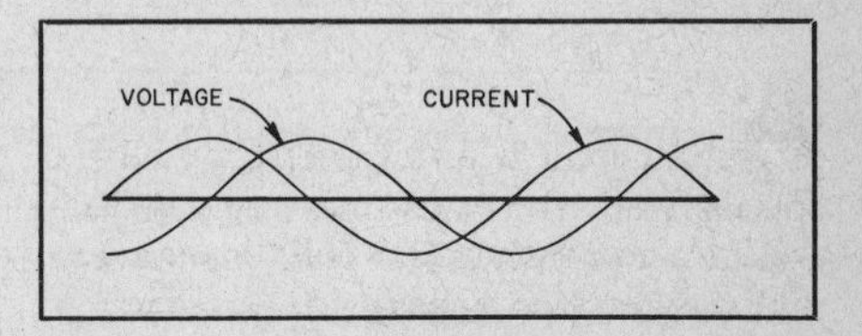


Fig. 2 — Harmonic operation of a long wire. The wire is long enough to contain several half waves. The current and voltage curves cross the heavy line representing the wire to indicate that there is reversal in the direction of the current, and a reversal in the polarity of the voltage, at intervals of a half wavelength. The reversals of current and voltage do not coincide, but occur at points a quarter wavelength apart.

## Electrical Length

The *electrical* length of a linear circuit such as an antenna wire is not necessarily the same as its *physical* length in wavelengths or fractions of a wavelength. Rather, the electrical length is measured by the *time* taken for the completion of a specified phenomenon.

For instance, we might imagine two linear circuits having such different characteristics that the speed at which a charge travels is not the same in both. Suppose we wish to make both circuits resonant at the same frequency, and for that purpose adjust the physical length of each until a charge started at one end travels to the far end, is reflected and completes its return journey to the near end in exactly the time of one rf cycle. Then it will be found that the *physical* length of the circuit with the lower velocity of propagation is shorter than the physical length of the other. The *electrical* lengths, however, are identical, each being a half wavelength.

In alternating-current circuits the instantaneous values of current or voltage are determined by the instant during the cycle at which the measurement is made (assuming, of course, that such a measurement could be made rapidly enough). If the current and voltage follow a sine curve — which is the usual case — the time, for any instantaneous value, can be specified in terms of an angle. The sine of the angle gives the instantaneous value when multiplied by the *peak* value of the current or voltage. A complete sine curve occupies the 360 degrees of a circle and represents one cycle of alternating current or voltage. Thus, a half cycle is equal to 180 degrees, a quarter cycle to 90 degrees, and so on.

It is often convenient to use this same form of representation for linear circuits. When the electrical length of a circuit is such that a charge *traveling in one direction* takes the time of one cycle or period to traverse it, the length of the circuit is said to be 360 degrees. This corresponds to one wavelength. On a wire a half wave in electrical length, the charge completes a one-way journey in one half cycle and its length is said to be 180 degrees. The angular method of measurement is quite useful for lengths that are not simple fractions or multiples of such fractions. To

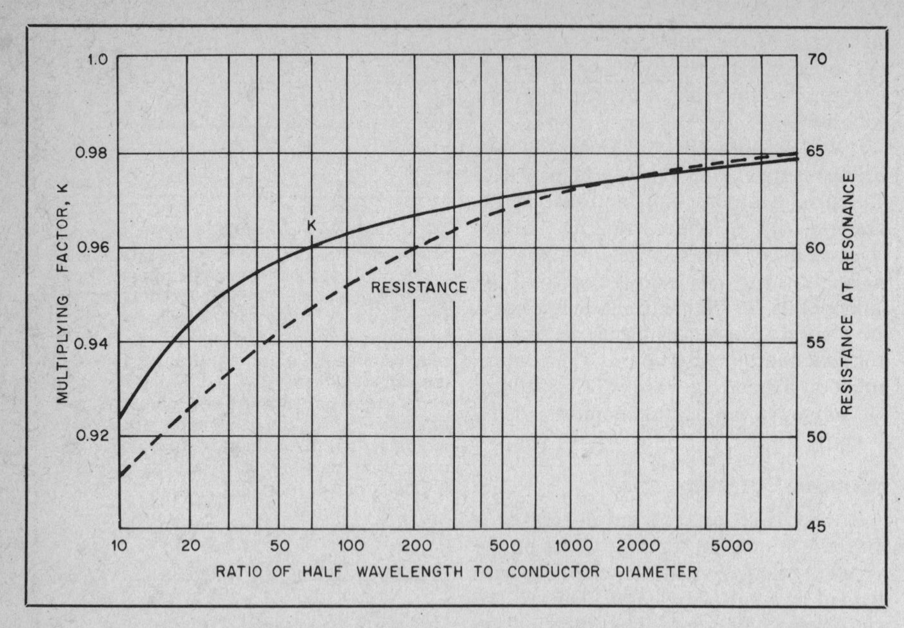


Fig. 3 — The solid curve shows the factor, K, by which the length of a half wave in free space should be multiplied to obtain the physical length of a resonant half-wave antenna vs. the length/diameter ratio. This curve does not take end effect into account. The broken curve shows how the radiation resistance of a half-wave antenna varies with the length/diameter ratio.

convert fractional wavelengths to angular lengths, multiply the fraction by 360°.

### Velocity of Propagation

The velocity at which electromagnetic waves travel through a medium depends upon the dielectric constant of the medium. At rf the dielectric constant of air is practically unity, so the waves travel at essentially the same velocity as light in a vacuum. This is also the velocity, very closely, of the charge traveling along a wire.

If the dielectric constant is greater than 1, the velocity of propagation is lowered. Thus, the introduction of insulating material having a dielectric constant greater than 1 will cause the wave to slow down. This effect is encountered in practice in connection with antennas and transmission lines. It causes the electrical length of the line or antenna to be greater than the actual physical length would indicate.

### Length of a "Half-Wave" Antenna

Even if the antenna could be supported by insulators that did not cause the electromagnetic fields traveling along the wire to slow down, the physical length of a practical antenna always is somewhat less than its electrical length. That is, a "half-wave" antenna is not one having the same length as a half wavelength in space. It is one having an *electrical* length equal to 180 degrees. Or, to put it another way, it is one with a length which has been adjusted to "tune out" any reactance, so it is a *resonant* antenna.

The antenna length required to resonate at a given frequency (independently of any dielectric effects) depends on the ratio of the length of the conductor to its diameter. The smaller this ratio (or the "thicker" the wire), the shorter the antenna must be for a given electrical length. This effect is shown in Fig. 3 as a factor (K) by which a free-space half wavelength must be multiplied to find the resonant length. K is a function of the ratio of the free-space half wavelength to conductor diameter, known as the *length/diameter ratio*. The curve is based on theoretical considerations and is useful as a guide to the probable antenna length for a given frequency. It applies only to conductors of uniform diameter (tapered elements such as those used in some types of beam antennas will generally be longer, for the same frequency) and does not include any effects introduced by the method of supporting the conductor.

A length/diameter ratio of 10,000 is roughly average for wire antennas (actually, it is approximately the ratio for a 7-MHz half-wave antenna made of no. 12 wire). In this region K changes rather slowly, and a half-wave antenna made of wire is about 2 percent shorter than a half wavelength in space.

The shortening effect is most pronounced when the length/diameter ratio is 100 or less. An antenna constructed of 1-inch diameter tubing for use on 144 MHz for example, would have a length/diameter ratio of about 40 and would be almost 5 percent shorter than a free-space half wavelength.

If the antenna is made of rod or tubing and is not supported near the ends by insulators, the following formula will give the required physical length of a half-wave antenna based on Fig. 3.

$$\text{Length}_{\text{feet}} = \frac{492 \times K}{f_{\text{MHz}}}$$

or

$$\text{length}_{\text{inches}} = \frac{5902 \times K}{f_{\text{MHz}}}$$

or

$$\text{length}_{\text{meters}} = \frac{150 \times K}{F_{\text{MHz}}}$$

where K is taken from Fig. 3 for the particular length/diameter ratio of the conductor used.

### End Effect

If the formulas of the preceding section are used to determine the length of a wire antenna, the antenna will resonate at a somewhat lower frequency than is desired. The reason is that an additional "loading" effect is caused by the insulators used at the ends of the wires to support the antenna. These insulators and the wire loops that tie the insulators to the antenna add a small amount of capacitance to the system. This capacitance helps to tune the antenna to a slightly lower frequency, in much the same way that additional capacitance in any tuned circuit will lower the resonant frequency. In an antenna it is called *end effect*. The current at the ends of the antenna does not quite reach zero because of the end effect, as there is some current flowing into the end capacitance.

End effect increases with frequency and varies slightly with different installations. However, at frequencies up to 30 MHz (the frequency range over which wire antennas are most likely to be used), experience shows that the length of a half-wave antenna is of the order of 5 percent less than the length of a half wave in space. As an average, then, the physical length of a resonant half-wave antenna may be taken to be

$$\ell_{\text{(feet)}} = \frac{492 \times 0.95}{f_{\text{MHz}}} \approx \frac{468}{f_{\text{MHz}}}$$

Similarly

$$\ell_{\text{meters}} \approx \frac{143}{f_{\text{MHz}}}$$

These formulas are reasonably accurate for finding the physical length of a half-wave antenna for a given frequency, but do not apply to antennas longer than a half wave in length. In the practical case, if the antenna length must be adjusted to exact frequency (not all antenna systems require it) the length should be "pruned" to resonance.

## ANTENNA IMPEDANCE

In the simplified description given earlier of voltage and current distribution along an antenna it was stated that the

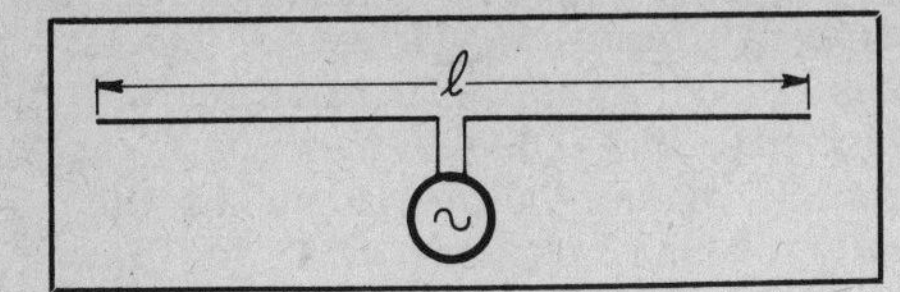


Fig. 4 — The center-fed antenna. It is assumed that the leads from the source of power to the antenna have zero length.

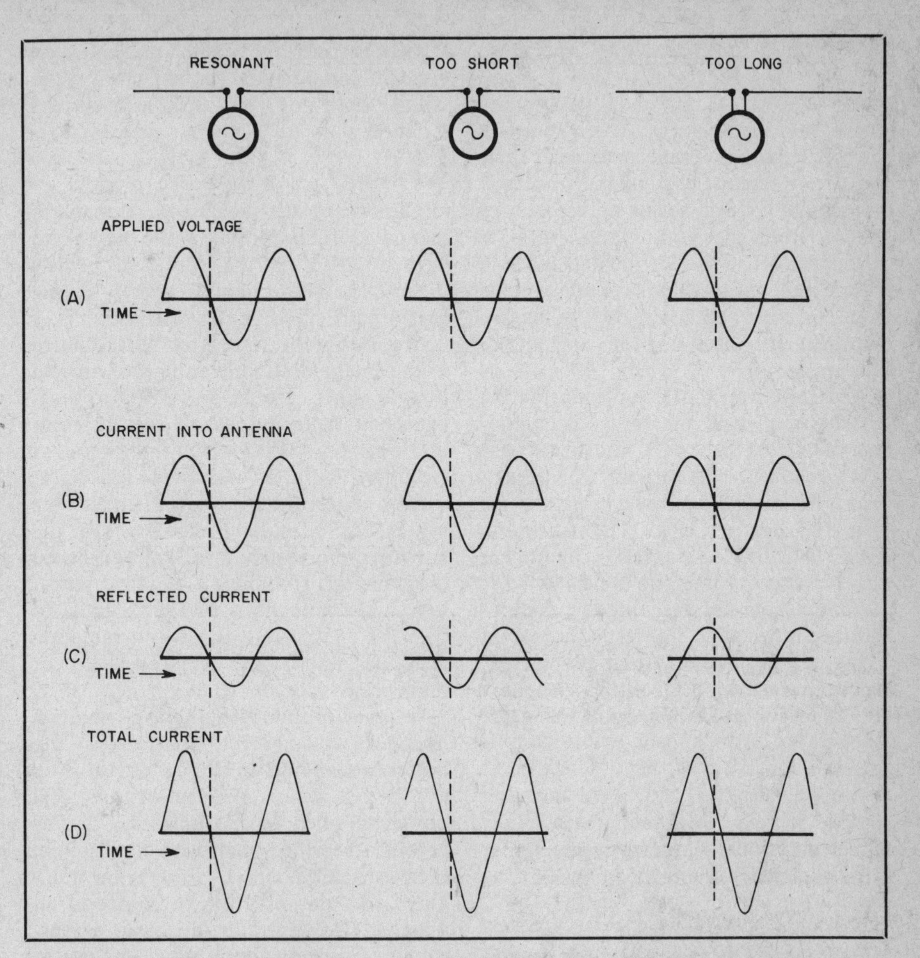


Fig. 5 — Current flow in resonant and off-resonant antennas. The initial current flow, B, caused by the source of power, is in phase with the applied voltage, A. This is the outgoing current discussed in the text. The reflected current, C, combines with the outgoing current to form the resultant current, D, at the input terminals of the antenna.

voltage was zero at the center of a half-wave antenna (or at any current loop along a longer antenna). It is more accurate to say that the voltage reaches a *minimum* rather than zero. Zero voltage with a finite value of current would imply that the circuit is entirely without resistance. It would also imply that no energy is radiated by the antenna, since a circuit without resistance would take no real power from the driving source.

Actually, an antenna, like any other circuit, consumes power. The current that flows in the antenna therefore must be supplied at a finite value of voltage. The *impedance* of the antenna is simply equal to the voltage applied to its terminals divided by the current flowing into those terminals. If the current and voltage are exactly in phase the impedance is purely resistive. This is the case when the antenna is resonant. If the antenna is not exactly resonant the current will be somewhat out of phase with the applied voltage and the antenna shows reactance along with resistance.

Most amateur transmitting antennas are operated at or close to resonance so that reactive effects are in general comparatively small. They are nevertheless present, and must be taken into account whenever an antenna is operated at other than the exact design frequency.

In the following discussion it is assumed that power is applied to the antenna by opening the conductor at the center and applying the driving voltage across the gap. This is shown in Fig. 4. While it is possible to supply power to the antenna by other methods, the selection of different driving points leads to different values of impedance; this can be appreciated after study of Fig. 1, which shows that the ratio of voltage to current (which by definition is the impedance) is different at every point along the antenna. To avoid confusion, it is desirable to use the conditions at the center of the antenna as a basis.

### The Antenna as a Circuit

If the frequency applied at the center of a half-wave antenna is varied above and below the resonant frequency, the antenna will exhibit much the same characteristics as a conventional series-resonant circuit. Exactly at resonance the current at the input terminals will be in phase with the applied voltage. If the frequency is on the low side of resonance the phase of the current will lead the voltage; that is, the reactance of the antenna is capacitive. When the frequency is on the high side of resonance the opposite occurs; the current lags the applied voltage and the antenna exhibits inductive reactance.

It is not hard to see why this is so. Consider the antennas shown in Fig. 5 — one resonant, one too short for the applied frequency, and one too long. In each case the applied voltage is shown at A, and the instantaneous current going *into* the antenna because of the applied voltage is shown at B. Note that this current is always in phase with the applied voltage, regardless of the antenna length. For the sake of simplicity only the current flowing in one leg of the antenna is considered; conditions in the other leg are similar.

In the case of the resonant antenna, the current travels out to the end and back to the driving point in one half cycle, since one leg of the antenna is 90 degrees long and the total path out and back is therefore 180 degrees. This would make the phase of the *reflected* component of current differ from that of the outgoing current by 180 degrees, since the latter current has gone through a half cycle in the meantime. However, it will be remembered that there is a phase shift of 180 degrees at the end of the antenna, because the direction of current reverses at the end. The *total* phase shift between the outgoing and reflected currents, therefore, is 360 degrees. In other words, the reflected component arrives at the driving point exactly in phase with the outgoing component. The reflected component, shown at C, adds to the outgoing component to form the resultant or total current at the driving point. The resultant current is shown at D, and in the case of the resonant antenna it is easily seen that the resultant is exactly in phase with the applied voltage. This being the case, the load seen by the source of power is a pure resistance.

Now consider the antenna that is too short to be resonant. The outgoing component of current is still in phase with the applied voltage, as shown at B. The reflected component, however, gets back to the driving point *too soon*, because it travels over a path less than 180 degrees, out and back. This means that the maximum value of the reflected component occurs at the driving point ahead of (in time) the maximum value of the outgoing component, since that particular charge took less than a half cycle to get back. Including the 180-degree reversal at the end of the antenna, the total phase shift is therefore less than 360 degrees. This is shown at C, and the resultant current is the combination of the outgoing and reflected components as given at D. It can be seen that the resultant current leads the applied voltage, so the antenna looks like a resistance in series with a capacitance. The shorter the antenna, the greater the phase shift between voltage and current; that is, the capacitive reactance increases as the antenna is shortened.

When the antenna is too long for the applied frequency, the reflected component of current arrives too late to be exactly in phase with the outgoing component because it must travel over a path more than 180 degrees long. The maximum value of the reflected component therefore occurs later (in time) than the maximum value of the outgoing component, as shown at C. The resultant current at the antenna input terminals therefore lags behind the applied voltage. The phase lag increases as the antenna is made longer. That is, an antenna that is too long shows inductive reactance along with resistance, and this reactance increases with an increase in antenna length over the length required for resonance.

If the antenna length is increased to 180 degrees on each leg, the go-and-return path length for the current becomes 360 degrees. This, plus the 180-degree reversal at the end, makes the total phase shift 540 degrees, which is the same as a 180-degree shift. In this case the reflected current arrives at the input terminals exactly *out* of phase with the outgoing component, so the resultant current is very small. The resultant is in phase with the applied voltage, so the antenna impedance is again purely resistive. The resistance under these conditions is very high, and the antenna has the characteristics of a parallel-resonant circuit. A *voltage* loop, instead of a *current* loop, appears at the input terminals when each leg of the antenna is 180 degrees long.

The amplitude of the reflected component is less than that of the component of current going into the antenna. This is the result of energy loss by radiation as the current travels along the wire. It is perhaps easier to understand if, instead of thinking of the electromagnetic fields as being brought into being by the current flow, we adopt the more fundamental viewpoint that *current flow along a conductor is caused by a moving electromagnetic field.* When some of the energy escapes from the system because the field travels out into space, it is not hard to understand why the current becomes less as it travels farther. There is simply less energy left to cause it. The difference between the outgoing and reflected current amplitudes accounts for the fact that the current does not go to zero at a voltage loop, and a similar difference between the applied and reflected voltage components explains why the voltage does not go to zero at a current loop.

**Resistance**

The energy supplied to an antenna is dissipated in the form of radio waves and in heat losses in the wire and nearby dielectrics. The radiated energy is the useful part, but so far as the antenna is concerned it represents a loss just as much as the energy used in heating the wire is a loss. In either case the dissipated power is equal to $I^2R$. In the case of heat losses, R is a real resistance, but in the case of radiation R is an assumed resistance, which, if present, would dissipate the power that is actually radiated from the antenna. This assumed resistance is called the radiation resistance. The total power loss in the antenna is therefore equal to $I^2(R_o + R)$, where $R_o$ is the radiation resistance and R the real resistance, or ohmic resistance.

In the ordinary half-wave antenna operated at amateur frequencies the power lost as heat in the conductor does not exceed a few percent of the total power supplied to the antenna. This is because the rf resistance of copper wire even as small as no. 14 is very low compared with the radiation resistance of an antenna that is resonably clear of surrounding objects and is not too close to the ground. Therefore it can be assumed that the ohmic loss in a resonably well-located antenna is negligible, and that all of the resistance shown by the antenna is radiation resistance. As a radiator of electromagnetic waves, such an antenna is a highly efficient device.

The value of radiation resistance, as measured at the center of a half-wave antenna, depends on a number of factors. One is the location of the antenna with respect to other objects, particularly the earth. Another is the length/diameter ratio of the conductor used. In "free space" — with the antenna remote from everything else — the radiation resistance of a resonant antenna made of an infinitely thin conductor is approximately 73 ohms. The concept of a free-space antenna forms a convenient basis for calculation because the modifying effect of the ground can be taken into account separately. If the antenna is at least several wavelengths away from ground and other objects, it can be considered to be in free space insofar as its own electrical properties are concerned. This condition can be met with antennas in the vhf and uhf range.

The way in which the free-space radiation resistance varies with the length/diameter ratio of a half-wave antenna is shown by the broken curve in Fig. 3. As the antenna is made thicker the radiation resistance decreases. For most wire antennas it is close to 65 ohms. It will usually lie between 55 and 60 ohms for antennas constructed of rod or tubing.

The actual value of the radiation resistance — at least so long as it is 50 ohms or more — has no appreciable effect on the radiation efficiency of the antenna. This is because the ohmic resistance is only of the order of 1 ohm with the conductors used for thick antennas. The ohmic resistance does not become important until the radiation resistance drops to very low values — say less than 10 ohms — as may be the case when several antennas are coupled together to form an array of elements.

The radiation resistance of a resonant antenna is the "load" for the transmitter

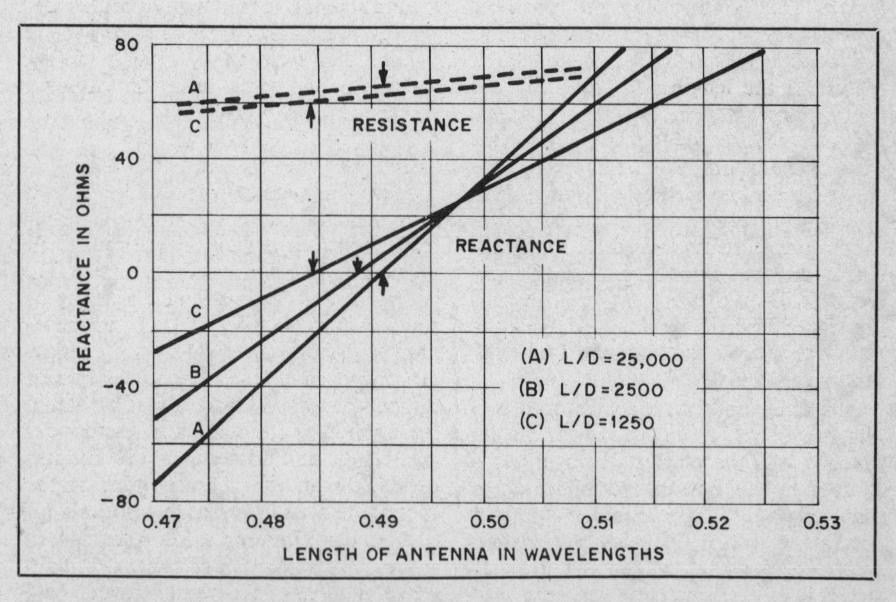


Fig. 6 — Resistance and reactance at the input terminals of a center-fed antenna as a function of its length near a half wavelength. As shown by curves A, B and C, the reactance is affected more by the length/diameter ratio of the conductor than is the radiation resistance.

or for the rf transmission line connecting the transmitter and antenna. Its value is important, therefore, in determining the way in which the antenna and transmitter or line are coupled together.

The resistance of an antenna varies with its length as well as with the ratio of its length to its diameter. When the antenna is approximately a half wave long, the resistance changes rather slowly with length. This is shown by the curves of Fig. 6, where the change in resistance as the length is varied a few percent on either side of resonance is shown by the broken curves. The resistance decreases somewhat when the antenna is slightly short, and increases when it is slightly long. These curves also illustrate the effect of changing the frequency applied to an antenna of fixed length, since increasing the frequency above resonance is the same thing as having an antenna that is too long, and vice versa.

The range covered by the curves in Fig. 6 is representative of the frequency range over which a fixed antenna is operated between the limits of an amateur band. At greater departures from the resonant length the resistance continues to decrease somewhat uniformly as the antenna is shortened, but tends to increase rapidly as the antenna is made longer. The resistance increases very rapidly when the length of a leg exceeds about 135 degrees, 3/8 wavelength, and reaches a maximum when the length of one side is 180 degrees. This is considered in more detail in later sections of this chapter.

#### Reactance

The rate at which the reactance of the antenna increases as the length is varied from resonance depends on the length/diameter ratio of the conductor. The thicker the conductor the smaller the reactance change for a given change in length. This is shown by the reactance curves in Fig. 6. Curves for three values of length/diameter ratio are shown; L represents the length of a half wave in space, approximately, and D is the diameter of the conductor in the same units as the length. The point where each curve crosses the zero axis (indicated by an arrow in each case) is the length at which an antenna of that particular length/diameter ratio is resonant. The effect of L/D on the resonant length also is illustrated by these curves; the smaller the ratio, the shorter the length at which the reactance is zero.

It will be observed that the reactance changes about twice as rapidly in the antenna with the smallest diameter (A) as it does in the antenna with the largest diameter (C). With still larger diameters the rate at which the reactance changes would be even smaller. As a practical matter, it is advantageous to keep the reactance change with a given change in length as small as possible. This means that when the antenna is operated over a small frequency band centered on its resonant frequency, the reactance is comparatively low and the impedance change with frequency is small. This simplifies the problem of supplying power to the antenna when it must be used at frequencies somewhat different from its resonant frequency.

At lengths considerably different from the resonant length the reactance changes more rapidly than in the curves shown in Fig. 6. As in the case of the resistance change, the change is most rapid when the length exceeds 135 degrees (3/8 wavelength) and approaches 180 degrees (1/2 wavelength) on a side. In this case the reactance is inductive and reaches a maximum at a length somewhat less than 180 degrees. Between this maximum point and 180 degrees of electrical length the reactance decreases very rapidly, becoming zero when the length is such as to be parallel resonant.

Very short antennas have a large capacitive reactance. It was pointed out in the preceding section that with antennas shorter than 90 degrees on a side the resistance decreases at a fairly uniform rate, but this is not true of the reactance. It increases rather rapidly when the length of a side is shortened below about 45 degrees.

The behavior of antennas with different length/diameter ratios corresponds to the behavior of ordinary resonant circuits having different Qs. When the Q of a circuit is low, the reactance is small and changes rather slowly as the applied frequency is varied on either side of the resonant frequency. If the Q is high, the converse is true. The response curve of the low-Q circuit is "broad"; that of the high-Q circuit "sharp." So it is with antennas; a thick antenna works well over a comparatively wide band of frequencies while a thin antenna is rather sharp in tuning. The Q of the thick antenna is low; the Q of the thin antenna is high, assuming essentially the same value of radiation resistance in both cases.

#### Coupled Antennas

A conventional tuned circuit far enough away from all other circuits so that no external coupling exists can be likened to an antenna in free space, in one sense. That is, its characteristics are unaffected by its surroundings. It will have a Q and resonant impedance determined by the inductance, capacitance and resistance of which it is composed, and those quantities alone. But as soon as it is coupled to another circuit its Q and impedance will change, depending on the characteristics of the other circuit and the degree of coupling.

A similar situation arises when two or more "elementary" antennas — half-wave antennas, frequently called *half-wave dipoles* — are coupled together. This coupling takes place merely by having the two antennas in proximity to each other. The sharpness of resonance and the radiation resistance of each "element" of the system are affected by the mutual interchange of energy between the coupled elements. The exact effect depends on the degree of coupling (that is, how close the antennas are to each other in terms of wavelength, and whether the wires are parallel or not) and the tuning condition (whether tuned to resonance or slightly off resonance) of each element. Analysis is extremely difficult and even then has to be based on some simplifying assumptions that may not be true in practice. Only a few relatively simple cases have been analyzed. Such data as are available for even moderately complicated systems of coupled antennas are confined to a few types and are based on experimental measurements. They are, therefore, subject to the inaccuracies that accompany any measurements in a field where measurement is difficult at best.

Antenna systems consisting of coupled elements will be taken up later in this book. At this point it is sufficient to appreciate that the free-space values that have been discussed in this chapter may be modified drastically when more than one antenna element is involved in the system. It has already been pointed out that the presence of the ground, as well as nearby conductors or dielectrics, also will modify the free-space values. The free-space characteristics of the elementary half-wave dipole are only the point of departure for a practical antenna system. In other words, they give the basis for understanding antenna principles but cannot be applied too literally in the practical case.

It is of interest to note that the comparison between an isolated tuned circuit and an antenna in free space is likewise not to be taken too literally. In one sense the comparison is wholly misleading. The tuned circuit is usually so small, physically, in comparison with the wavelength, that practically no energy escapes from it by radiation. An antenna, to be worthy of the name is always so large in comparison with the wavelength that practically *all* the energy supplied to it escapes by radiation. Thus, the antenna can be said to be very tightly coupled to space, while the tuned circuit is not coupled to anything. This very fundamental difference is one reason why antenna systems cannot be analyzed as readily, and with as satisfactory results in the shape of simple formulas, as ordinary electrical circuits.

## HARMONICALLY OPERATED ANTENNAS

An antenna operated at a harmonic of its fundamental frequency has considerably different properties than the half-wave dipole previously discussed. It must first be emphasized that harmonic

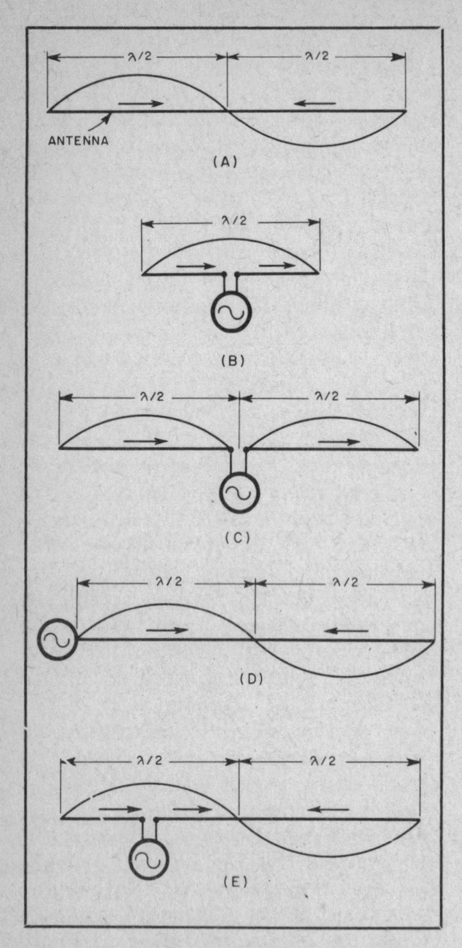


Fig. 7 — Moving the feed point makes a difference in current distribution along the antenna. With center feed, increasing the length of each side of the antenna keeps the current flowing in the same direction in the two halves, up to the point where each side is a half-wavelength long. For harmonic operation, the antenna must be fed in such a way that the current direction reverses in alternate half-wavelength sections. Suitable methods are shown at D and E.

operation implies that there is a reversal of the direction of current flow in alternate half-wave sections of the antenna, as shown in Fig. 2 and again at A in Fig. 7. In Fig. 7A, the curve shows the standing wave of current intensity along the wire; the curve is above the line to indicate current flow in one direction (assumed to be to the right, in the direction of the arrow) and below the line to indicate current flow in the opposite direction in the other half-wave section. (During the next radio-frequency half cycle the current flow in the left-hand half-wave section would be toward the left, and in the right-hand half-wave section to the right; this alternation in direction takes place in each succeeding half cycle. However, the direction of current flow in adjacent half-wave sections is at all times opposite.) The antenna in this drawing is one wavelength long and is operating on its second harmonic.

Now consider the half-wave antenna shown at Fig. 7B. It is opened in the center and fed by a source of rf power through leads that are assumed to have zero length. Since one terminal of the generator is positive at the same instant that the other terminal is negative, current flows *into* one side of the generator while it is flowing *out* at the other terminal. Consequently the current flows in the same direction in both sections of the half-wave antenna. It has the amplitude distribution shown by the curve over the antenna wire.

If we now increase the length of the wire on each side of the generator in Fig. 7B to one half wavelength, we have the situation shown in Fig. 7C. At the instant shown, current flows into the generator from the left-hand half-wave section, and out of the generator into the right-hand half-wave section. Thus the currents in the two sections are in the same direction, just as they were in Fig. 7B. The current distribution in this case obviously is not the same as in Fig. 7A. Although the overall lengths of the antennas shown at A and C are the same, the antenna at A is operating on a harmonic but the one in C is not.

For true harmonic operation it is necessary that the power be fed into the antenna at the proper point. Two methods that result in the proper current distribution are shown at D and E in Fig. 7. If the source of power is connected to the antenna at one end, as in D, the direction of current flow will be reversed in alternate half-wave sections. Or if the power is inserted at the center of a half-wave section, as in E, there will be a similar reversal of current in the next half-wave section. For harmonic operation, therefore, the antenna should be fed either at the end or at a current loop. If the feed point is at a current node the current distribution will not be that expected on a harmonic antenna.

**Length of a Harmonic Wire**

The physical length of a harmonic antenna is not exactly the same as its electrical length, for the same reasons discussed earlier in connection with the half-wave antenna. The physical length is somewhat shorter than the length of the same number of half waves in space because of the length/diameter ratio of the conductor and the end effects. Since the latter are appreciable only where insulators introduce additional capacitance at a high-voltage point along the wire, and since a harmonic antenna usually has such insulation only at the ends, the end-effect shortening affects only the half-wave sections at each end of the antenna. It has been found that the following formulas for the length of a harmonic antenna of the usual wire sizes work out very well in practice:

$$\text{length (feet)} = \frac{492\,(N - 0.05)}{f_{MHz}}$$

or

$$\text{length (meters)} = \frac{150\,(N - 0.05)}{f_{MHz}}$$

where N is the number of *half* waves on the antenna.

Because the number of half waves varies with the harmonic on which the antenna is operated, consideration of the formulas together with that for the half-wave antenna (the fundamental frequency) will show that the relationship between the antenna fundamental frequency and its harmonics is not exactly integral. That is, the "second-harmonic" frequency to which a given length of wire is resonant is not exactly twice its fundamental frequency; the "third-harmonic" resonance is not exactly three times its fundamental and so on. The actual resonant frequency on a harmonic is always a little higher than the exact multiple of the fundamental. A full-wave (second-harmonic) antenna, for example, must be a little longer than twice the length of a half-wave antenna.

Frequently it is desired to determine the electrical length of a harmonically operated wire antenna of fixed physical length for a given frequency. With a rearrangement of the terms of the above formulas, the following equation is useful for making these determinations:

$$\lambda = \frac{fL\text{ (feet)}}{984} + 0.025$$

$$= \frac{fL\text{ (meters)}}{300} + 0.025$$

where $\lambda$ is the length of the wire in *wavelengths* at the frequency, f, in megahertz, and L is the physical length of the wire.

**Impedance of Harmonic Antennas**

A harmonic antenna can be looked upon as a series of half-wave sections placed end to end (collinear) and supplied with power in such a way that the currents in alternate sections are out of phase. There is a certain amount of coupling between adjacent half-wave sections. Because of this coupling, as well as the effect of radiation from the additional sections, the impedance as measured at a current loop in a half-wave section is not the same as the impedance at the center of a half-wave antenna.

Just as in the case of a half-wave antenna, the impedance consists of two main components, radiation resistance and reactance. The ohmic or loss resistance is low enough to be ignored in the practical case. If the antenna is exactly resonant there will be no reactance at the input terminals and the impedance consists only of the radiation resistance. The value of the radiation resistance depends on the number of half waves on the wire and, as in the case of the half-wave antenna, is modified by the presence of nearby conductors and dielectrics, particularly the earth. As a point of departure, however, it is of interest to know the order of magnitude of the radiation resistance of a theoretical harmonic antenna consisting of an infinitely thin conductor in free space, with its length adjusted to exact harmonic resonance. The radiation resistance of such an antenna having a length of one wavelength is approximately 90 ohms, and as the antenna length is increased the resistance also increases. At 10 wavelengths it is approximately 160 ohms, for example. The way in which the radiation resistance of a theoretical harmonic wire varies with length is shown by curve A in Fig. 20 in a later section of this chapter. It is to be understood that the radiation resistance is always measured at a current loop.

When the antenna is operated at a frequency slightly off its exact resonant frequency, reactance as well as resistance will appear at its input terminals. In a general way, the reactance varies with applied frequency in much the same fashion as in the case of the half-wave antenna already described. However, the reactance varies *at a more rapid rate* as the applied frequency is varied; on a harmonic antenna a given percentage change in applied frequency causes a greater change in the phase of the reflected current as related to the outgoing current than is the case with a half-wave antenna. This is because, in traveling the greater length of wire in a harmonic antenna, the reflected current gains the same amount of time in *each* half-wave section, if the antenna is too short for resonance, and these gains add up as the current travels back to the driving point. When the antenna is too long, the reverse occurs and the reflected current progressively drops behind in phase as it travels back to the point at which the voltage is applied. This effect increases with the length of the antenna, and the change of phase can be quite rapid when the frequency applied to an antenna operated on a high-order harmonic is varied.

Another way of looking at it is this: Consider the antenna of Fig. 8A, driven at the end by a source of power having a frequency of f/2, where f is the fundamental or half-wave resonant frequency of the antenna. When the frequency f/2 is applied, there is one quarter wavelength on the wire, with the current distribution as shown. At this frequency the antenna is resonant, and it appears as a pure resistance of low value to the source of power because the current is large and the voltage is small at the feed point.

If the frequency is now increased slightly the antenna will be too long and the resultant current at the input terminals will lag behind the applied voltage (as explained by Fig. 5), and the antenna will have inductive reactance along with resistance. As we continue to raise the frequency, the value of reactance increases to a maximum and then decreases, reaching zero when the frequency is f, such that the wire is a half wavelength long, shown in Fig. 8B. On further increasing the frequency, the reactance becomes capacitive, increases to a maximum, and then decreases, reaching zero when the wire is an odd number of quarter wavelengths long. As the frequency is increased still further, the reactance again becomes inductive, reaches a maximum, and again goes to zero at frequency 2f. At this point there are two complete standing waves of current (two half waves or one wavelength) and the wire is exactly resonant on its second harmonic. This last condition is shown in Fig. 8C.

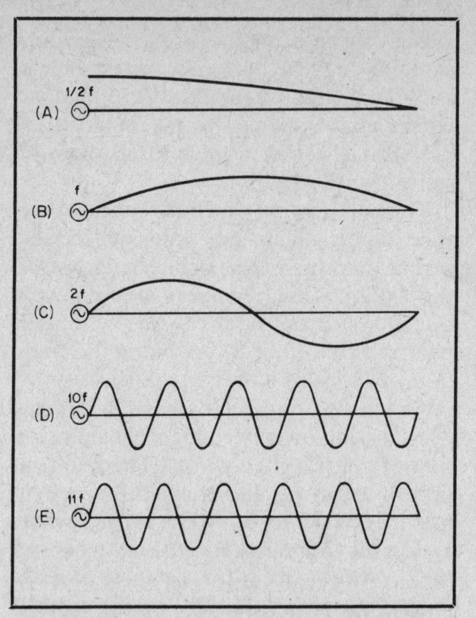

Fig. 8 — The percentage frequency change from one high-order harmonic to the next (for example, between the 10th and 11th harmonics shown at C and D) is much smaller than between the fundamental and second harmonic (A and B). This makes impedance variations more rapid as the wire becomes longer in terms of wavelength.

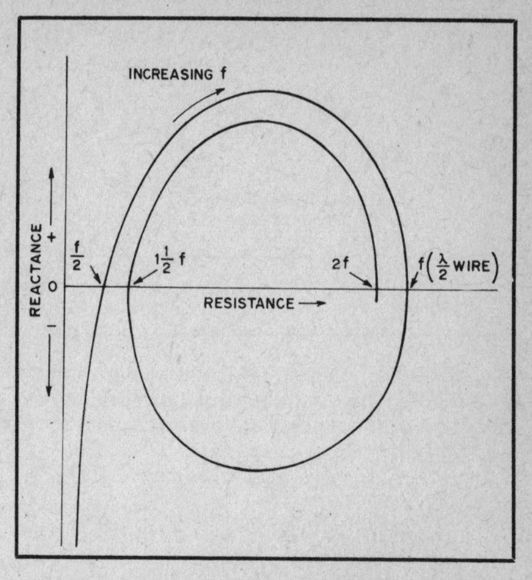

Fig. 9 — This drawing shows qualitatively the way in which the reactance and resistance of an end-fed antenna vary as the frequency is increased from one half the fundamental (f/2) to the second harmonic (2f). Relative resistance values are shown on the horizontal axis, and reactance values on the vertical axis; increasing frequency appears in a clockwise direction on the curve. Actual values of resistance and reactance and the frequencies at which the reactances are maximum will depend on the size of the conductor and the height of the antenna above ground.

In varying the frequency from f/2 to 2f, the resistance seen by the source of power also varies. This resistance increases as the frequency is raised above f/2 and reaches a maximum when the wire is a half wavelength long, decreases as the frequency is raised above f, reaching a minimum when the wire is an odd number of quarter wavelengths long, then rises again with increasing frequency until it reaches a new maximum when the frequency is 2f.

This behavior of reactance and resistance with frequency is shown in Fig. 9. A similar change in reactance and resistance occurs when the frequency is moved from any harmonic to the *next adjacent* one, as well as between the fundamental and second harmonic shown in the drawing. That is, the impedance goes through a cycle, starting with a high value of pure resistance at f then becoming capacitive and decreasing, passing through a low value of pure resistance, and then becoming inductive and increasing until it again reaches a high value of pure resistance at 2f. This cycle occurs as the frequency is continuously varied from any harmonic to the next higher one.

Look now at Fig. 8D. The frequency has been increased to 10f, 10 times its original value, so the antenna is operated on its 10th harmonic. Raising the frequency to 11f, the 11th harmonic, causes the impedance of the antenna to go through the complete cycle described above. But 11f is only 10% higher than 10f, so a 10% change in frequency has caused a complete impedance cycle. In contrast, changing from f to 2f is a 100% increase in frequency — for the same impedance cycle. The impedance changes 10 times as fast when the frequency is varied about the 10th harmonic as it does when the frequency is varied to the same percentage about the fundamental.

To offset this, the actual impedance change — that is, the ratio of the maximum to the minimum impedance through the impedance cycle — is not as great at the higher harmonics as it is near the fundamental. This is because the radiation resistance increases with the order of the harmonic, raising the minimum point on the resistance curve and also lowering the maximum point. If the curve of Fig. 9 were continued through higher order harmonics, it would thus spiral inward, toward a central point. This comes about because the reflected current returning to the input end of a long harmonic wire is not as great as the outgoing current since energy has been lost by radiation; this is not taken into account in the theoretical pictures of current distribution so far discussed.

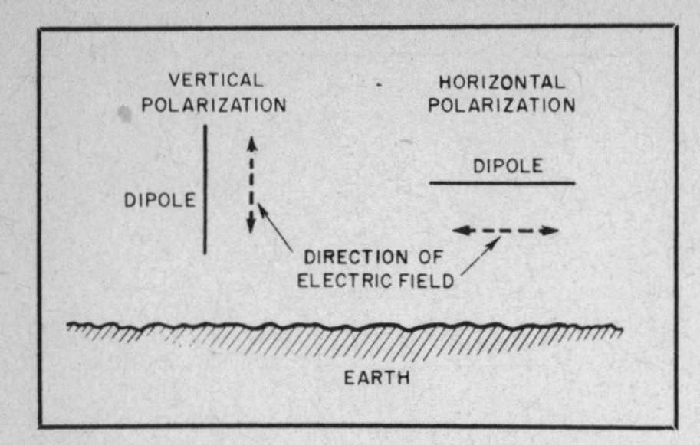


Fig. 10 — Vertical and horizontal polarization of a dipole. The direction of polarization is the direction of the electric field with respect to earth.

The overall result, then, is that the *magnitude* of the impedance variations becomes less as the wire is operated at increasingly higher harmonics. Nevertheless, the impedance reaches a maximum at each adjacent harmonic and a minimum halfway between, independently of the actual values of impedance.

## OTHER ANTENNA PROPERTIES

### Polarization

The polarization of a half-wave dipole is the same as the direction of its axis. That is, at distances far enough from the antenna for the waves to be considered as plane waves (see Chapter 1) the direction of the electric component of the field is the same as the direction of the antenna wire. Vertical and horizontal polarization, the two most commonly used for antennas, are indicated in Fig. 10.

Antennas composed of a number of half-wave elements with all arranged so that their axes lie in the same or parallel directions will have the same polarization as that of any one of the elements. A system composed of a group of horizontal dipoles, for example, will be horizontally polarized. If both horizontal and vertical elements are used and radiate in phase, the polarization will be the *resultant* of the contributions made by each set of elements to the total electromagnetic field at a distance. In such a case the resultant polarization will be *linear*, tilted between horizontal and vertical. If vertical and horizontal elements are fed out of phase, where the beginning of the rf period applied to the feed point of the vertical elements is not in time coincidence with that applied to the horizontal, the resultant polarization will be either *elliptical* or *circular*. With circular polarization, the wave front, as it appears in passing to a fixed observer, rotates every quarter period between vertical and horizontal, making a complete 360-degree rotation once every period. Field intensities are equal at all instantaneous polarizations. Circular polarization is frequently used for space communications.

Harmonic antennas also are polarized in the direction of the wire axis. In some combinations of harmonic wires such as the V and rhombic antennas described in a later chapter, however, the polarization becomes elliptical in most directions with respect to the antenna.

The polarization of a half-wave sloping-wire dipole, with one end appreciably higher than the other, is linear. However, the polarization changes with compass direction. Broadside to the wire, the polarization will be tilted, but in the direction of the wire it will be vertical.

As pointed out in Chapter 1, sky-wave transmission usually changes the polarization of the traveling waves. The polarization of receiving and transmitting antennas in the 3- to 30-MHz range, where almost all communications is by means of the sky wave (except for distances of a few miles), therefore need not be the same at both ends of a communication circuit. In this range the choice of polarization for the antenna is usually determined by the factors such as the height of available antenna supports, the polarization of man-made rf noise from nearby sources, probable energy losses in nearby houses, wiring, etc., and the liklihood of interfering with neighborhood broadcast or TV reception.

### Reciprocity in Receiving and Transmitting

The basic conditions existing when an antenna is used for radiating power are not the same as when it is used for receiving a distant signal. In the transmitting case the electromagnetic field originates with the antenna and the waves are not plane-polarized in the immediate vicinity. In the receiving case the antenna is always far enough away from the transmitter so that the waves that the antenna intercepts are plane-polarized. This causes the current distribution in a receiving antenna to be different than in a transmitting antenna except in a few special cases. These special cases, however, are those of most interest in amateur practice, since they occur when the antenna is resonant and is delivering power to a receiver.

For all practical purposes, then, the properties of a resonant antenna used for reception are the same as its properties in transmission. It has the same directive pattern in both cases, and so will deliver maximum signal to the receiver when the signal comes from the direction in which the antenna transmits best. The impedance of the antenna is the same, at the same point of measurement, in receiving as in transmitting.

In the receiving case, the antenna is to be considered as the *source* of power delivered to the receiver, rather than as the *load* for a source of power as in transmitting. Maximum output from the receiving antenna is secured when the load to which the antenna is connected is matched to the impedance of the antenna. Under these conditions half of the total power picked up by the antenna from the passing waves is delivered to the receiver and half is reradiated into space.

"Impedance matching" in the case of a receiving antenna does not have quite the same meaning as in the transmitting case. This is considered in later chapters.

The power gain (defined later in this chapter) in receiving is the same as the gain in transmitting, assuming that certain conditions are met. One such condition is that both the antenna under test and the comparison antenna (usually a half-wave antenna) work into load impedances matched to their own impedances so that maximum power is delivered in both cases. In addition, the comparison antenna should be oriented so that it gives maximum response to the signal used in the test; that it, it should have the same polarization as the incoming signal and should be placed so that its direction of maximum gain is toward the signal source.

In long-distance transmission and reception via the ionosphere the relationship between receiving and transmitting may not be exactly reciprocal. This is because the waves do not take exactly the same paths at all times and so may show considerable variation in alternate transmission and reception. Also, when more than one layer is involved in the wave travel it is sometimes possible for transmission to be good in one direction and reception to be poor in the other, over the same path. In addition, the polarization of the waves is shifted in the ionosphere, as pointed out in Chapter 1. The tendency is for the arriving wave to be elliptically polarized, regardless of the polarization of the transmitting antenna, and a vertically polarized antenna can be expected to show no more difference between transmission and reception than a horizontal antenna. On the average, an antenna that transmits well in a certain direction will give favorable reception from the same direction, despite ionosphere variations.

### Pickup Efficiency

Although the transmitting and receiving properties of an antenna are, in general, reciprocal, there is another fundamental difference between the two cases that is of very great practical importance. In the transmitting case all the power supplied to the antenna is radiated (assuming negligible ohmic resistance) regardless of the physical size of the antenna system. For example, a 300-MHz half-wave radiator, which is only about 500 *mm* (19 *inches*) long, radiates every bit as efficiently as a 3.5-MHz half-wave antenna, which is about 41 *meters* (134 *feet*) long. But in receiving, the 300-MHz antenna does not abstract anything like the amount of energy from passing waves that the 3.5-MHz antenna does.

This is because the section of wave front from which the antenna can draw energy extends only about a quarter wavelength from the conductor. At 3.5 MHz this represents an area roughly 1/2 wavelength or 41 meters in diameter, but at 300 MHz the diameter of the area is only about 1/2 meter. Since the energy is evenly distributed throughout the wave front regardless of the wavelength, the effective area that the receiving antenna can utilize varies directly with the *square* of the wavelength. A 3.5-MHz half-wave antenna therefore picks up something like 7000 times as much energy as a 300-MHz half-wave antenna, the field strength being the same in both cases.

The higher the frequency, consequently, the less energy a receiving antenna has to work with. This, it should be noted, does not affect the *gain* of the antenna. In making gain measurements, both the antenna under test and the comparison antenna are working at the same frequency. Both therefore are under the same handicap with respect to the amount of energy that can be intercepted. Thus the effective area of an antenna at a given wavelength is directly proportional to its gain. Although the pickup efficiency decreases rapidly with increasing frequency, the smaller dimensions of antenna systems in the vhf and uhf regions make it relatively easy to obtain high gain. This helps to overcome the loss of received signal energy.

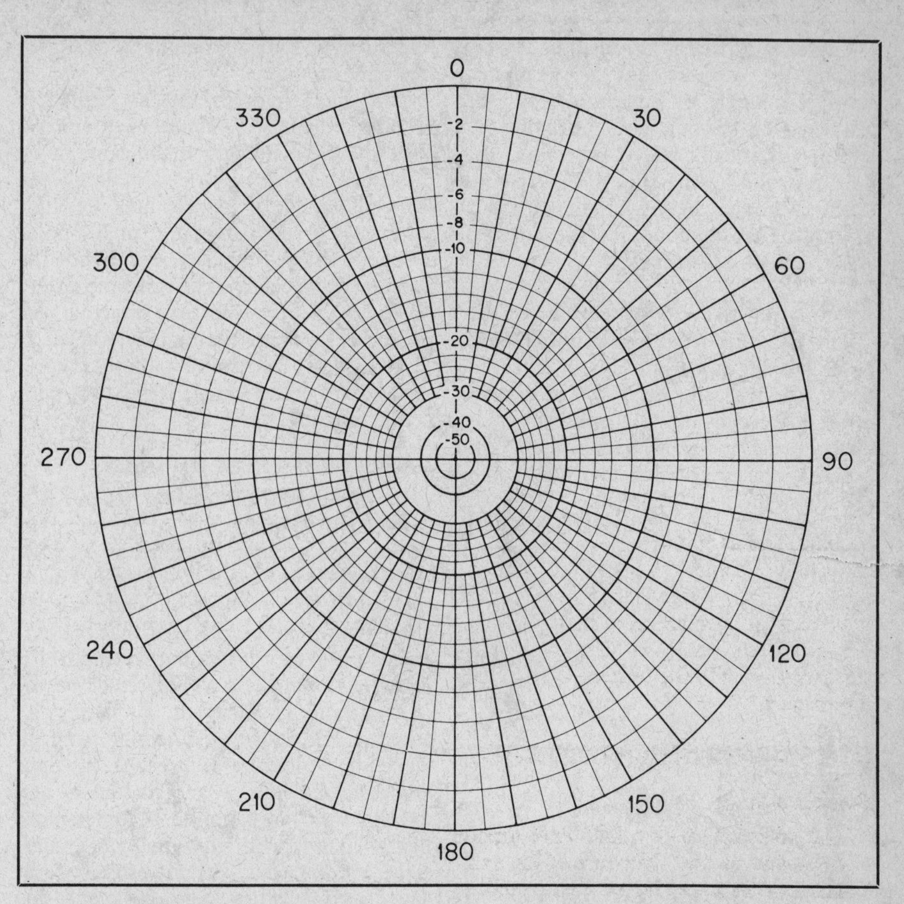


Fig. 11 — The coordinate system for plots of antenna patterns in this publication. The concentric circles are graduated in decibels, with the reference, 0 dB, at the outer edge. The outer azimuth scale is indicated in degrees. Worksheets for the plotting of antenna patterns on this grid are available from ARRL Hq., 100 for $3 at the time of this writing. A worksheet measures 8-1/2 × 11 inches, with a 6-inch-diameter 0-dB circle. The sheet also has blocks for entering related antenna information. See Fig. 48, Chapter 15. (Chart design by G. Hall, K1TD; drafting by Sue Fagan.)

### The Induction Field

Throughout this chapter the fields we have been discussing are those forming the traveling electromagnetic waves — the waves that go long distances from the antenna. These are the radiation fields. They are distinguished by the fact that their intensity is inversely proportional to the distance and that the electric and magnetic components, although perpendicular to each other in the wave front, are in phase in time. Several wavelengths from the antenna, these are the only fields that need to be considered.

Close to the antenna, however, the situation is much more complicated. In an ordinary electric circuit containing inductance or capacitance the magnetic field is a quarter cycle out of phase (in time) with the electric field. The intensity of these fields decreases in a complex way with distance from the source. These are the induction fields. The induction field exists about an antenna along with the radiation field, but dies away with much greater rapidity as the distance from the antenna is increased. At a distance equal to the wavelength divided by $2\pi$, or slightly less than 1/6 wavelength, the two types of field have equal intensity.

Although the induction field is of no importance insofar as effects at a distance are concerned, it *is* important when antenna elements are coupled together, particularly when the spacing between elements is small. Also, its existence must be kept in mind in making field-strength measurements about an antenna. Error may occur if the measuring equipment is set up too close to the antenna system.

## RADIATION PATTERNS AND DIRECTIVITY

A graph showing the actual or relative intensity at a fixed distance, as a function of the direction from the antenna system, is called a radiation pattern. At the outset it must be realized that such a pattern is a three-dimensional affair and therefore cannot be represented in a plane drawing. The "solid" radiation pattern of an antenna in free space would be found by measuring the field strength at every point on the surface of an imaginary sphere having the antenna at its center. The information so obtained is then used to construct a solid figure such that the distance from a fixed point (representing the antenna) to the surface, in any direction, is proportional to the field strength from the antenna in that direction.

For ease of pattern interpretation, antenna-pattern plots in this publication are made on a non-linear polar coordinate system graduated in decibels, Fig. 11. (Decibel units are discussed in detail in a later section of this chapter.) A log-periodic system is used for the coordinates, where the graduations vary periodically with the logarithm of the field intensity in voltage units. The constant of periodicity is 0.89. To explain, the scale distance covered by the outermost 2-dB increment, 0 to −2 dB, is a particular length, approximately 1/10 the radius of the chart. The scale distance for the next 2-dB increment is slightly less, 89.0 percent of the first, to be exact. The scale distance for the third 2-dB increment is 89 percent of the second, and so on, each 2-dB increment becoming progressively smaller toward chart center. The scale is constructed so the progression ends with −100 dB at exact chart center. In amateur practice, signals of 50 or 60 dB and more below the reference level will normally be so weak as to be insignificant, so the small size of the −50 dB innermost circle is considered quite acceptable for the intent of this publication. (This coordinate system may not be suitable for some laboratory work, however, where better definition at greater dynamic ranges may be desirable.)

The azimuth scale, indicated in degrees around the outside edge of the chart, shows the angle of departure from the reference or starting point, usually 0°. Unless specifically stated otherwise, the 0-dB reference (outer edge of the chart) is taken as the field strength in the direction of maximum radiation for the antenna system under consideration.

## THE ISOTROPIC RADIATOR

The radiation from a practical antenna never has the same intensity in all directions. The intensity may even be zero in some directions from the antenna; in others it may be greater than one would expect from an antenna that *did* radiate equally well in all directions. But even though no actual antenna radiates with equal intensity in all directions, it is nevertheless useful to assume that such an antenna exists. It can be used as a "measuring stick" for comparing the properties of actual antenna systems.

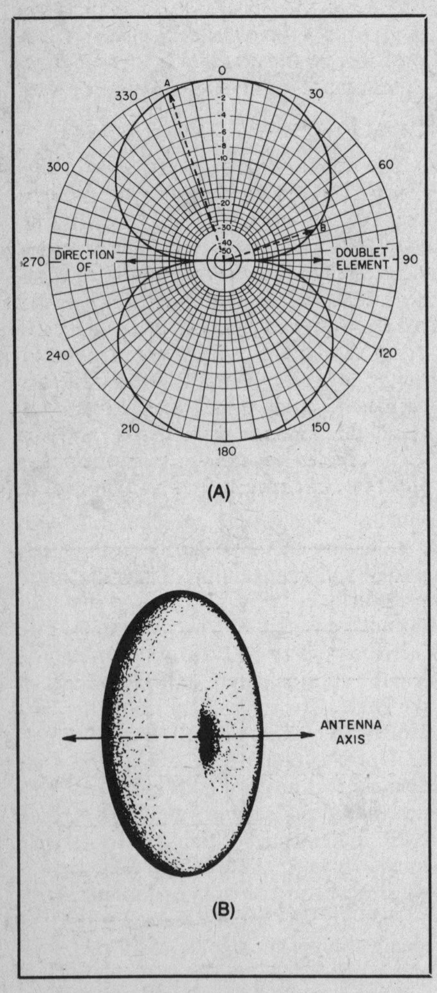


Fig. 12 — At A, directive diagram of an elementary doublet in the plane containing the wire axis. The length of each dash-line arrow represents the relative field strength in that direction, referenced to the direction of maximum radiation. At B, the solid pattern of the same antenna. These same diagrams apply to any antenna considerably less than a half wavelength long.

Such a hypothetical antenna is called an *isotropic radiator*.

The solid pattern of an isotropic radiator, therefore, would be a sphere, since the field strength is the same in all directions. In any plane containing the isotropic antenna (which may be considered as a point in space, or a "point source") the pattern is a circle with the antenna at its center. The isotropic antenna has the simplest possible directive pattern; that is, it has no directivity at all.

An infinite variety of pattern shapes, some quite complicated, is possible with actual antenna systems.

## RADIATION FROM DIPOLES

In the analysis of antenna systems it is convenient to make use of another fictitious type of antenna called an *elementary doublet* or *elementary dipole*. This is just a very short length of conductor, so short that it can be assumed that the current is the same throughout its length. (In an actual antenna, it will be remembered, the current is different all along its length.) The radiation intensity from an elementary doublet is greatest at right angles to the line of the conductor, and decreases as the direction becomes more nearly in line with the conductor until, right off the ends, the intensity is zero. The directive pattern in a single plane, one containing the conductor, is shown in Fig. 12A. If the pattern were drawn on a linear coordinate system showing field intensity, it would consist of two tangent circles. The solid pattern is the doughnut-shaped figure which results when the plane shown in the drawing is rotated about the conductor as an axis, Fig. 12B.

The radiation from an elementary doublet is not uniform in all directions because there is a definite direction to the current flow along the conductor. It will be recalled that a similar condition exists in the ordinary electric and magnetic fields set up when current flows along any conductor; the field strength near a coil, for example, is greatest at the ends and least on the outside of the coil near the middle of its length. There is nothing strange, therefore, in the idea that the field strength should depend on the direction in which it is measured from the radiator.

When the antenna has appreciable length, so that the current in every part is not the same at any given instant, the shape of the radiation pattern changes. In this case the pattern is the summation of the fields from *each* elementary doublet of which the antenna may be assumed to consist, strung together in chain fashion. If the antenna is short compared with a half wavelength there is very little change in the pattern, but at a half wavelength the pattern takes the shape shown in cross section in Fig. 13. The intensity decreases somewhat more rapidly, as the angle with the wire is made smaller when compared with the elementary doublet. In the case of the elementary doublet, the half-power (or −3 dB) points on each lobe occur 90° apart; for the half-wavelength dipole they are 78° apart. If the wire length is increased further, this tendency continues, with a somewhat wider null appearing in the pattern off the ends of the wire as the antenna approaches a full wavelength. (The antenna is assumed to be driven at the center, as in Fig. 7B and 7C.) The solid pattern from a half-wave wire is formed, just as in the case of the doublet, by rotating the plane diagram shown in Fig. 13 about the wire as an axis.

The single-plane diagrams just discussed are actually cross sections of the solid pattern, cut by planes in which the axis of the antenna lies. If the solid pattern is cut by any other plane the diagram will be different. For instance, imagine a plane passing through the center of the wire at right angles to the axis. The cross section of the pattern for either the elementary doublet or the half-wave antenna will simply be a circle in that case. This is shown in Fig. 14 where the dot at the center represents the antenna as viewed "end on," as if one were looking into the side of the doughnut of Fig. 12B. In other words, the antenna is perpendicular to the page. This means that in any direction in a plane at right angles to the wire, the field intensity is exactly the same at the same distance from the antenna. At right angles to the wire, then, an antenna a half wave or less in length is *nondirectional*. Also, at every point on such a circle the field is in the *same phase*.

## E- AND H-PLANE PATTERNS

The solid pattern of an antenna cannot adequately be shown with field-strength data on a flat sheet of paper. For this purpose, cross-sectional or *plane diagrams* are very useful. Two such diagrams, as in Figs. 13 and 14, one in the plane con-

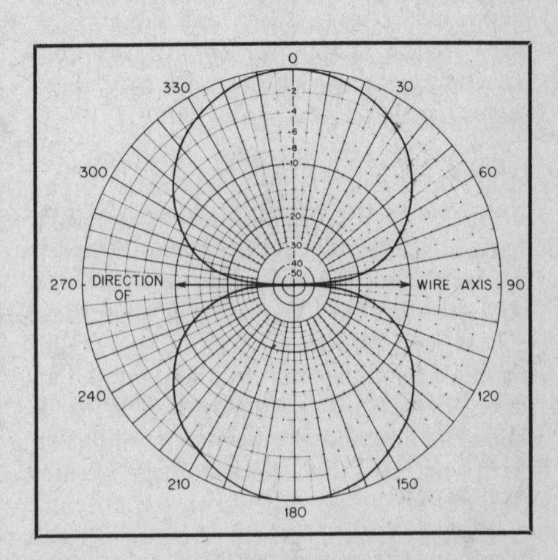


Fig. 13 — Plane directive diagram (E plane) of a half-wave antenna. The solid line shows the direction of the wire, although the antenna itself is considered to be merely a point at the center of the diagram. As explained in the text, a diagram such as this is simply a cross section of the solid figure that describes the relative radiation in all possible directions.

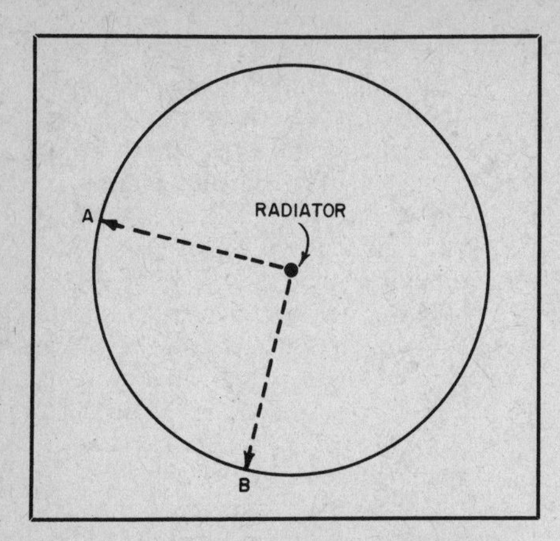


Fig. 14 — Directive diagram of a doublet or dipole in the plane perpendicular to the wire axis (H plane). The direction of the wire is into or out of the page.

taining the axis of the antenna and one in the plane perpendicular to the axis, can give a great deal of information. The former is called the E-plane pattern and the latter the H-plane pattern. These designations are used because they represent the planes in which the electric (symbol E), and the magnetic (symbol H) lines of force lie, respectively. The E lines are taken to represent the polarization of the antenna, consistent with the description of antenna polarization given earlier. The electromagnetic field pictured in Fig. 1 of Chapter 1, as an example, is the field that would be radiated from a vertically polarized antenna; that is, an antenna in which the conductor is mounted vertically over the earth.

After a little practice, and with the exercise of a little imagination, the complete solid pattern can be visualized with fair accuracy from inspection of just the two diagrams. Plane diagrams are plotted on polar-coordinate paper, as described earlier. The points on the pattern where the radiation is zero are called nulls, and the curved section from one null to the next on the plane diagram, or the corresponding section on the solid pattern, is called a lobe or ear.

## HARMONIC-ANTENNA PATTERNS

In view of the change in radiation patterns as the length of the antenna is increased from the elementary doublet to the half-wave dipole, it is to be expected that further pattern changes will occur as the antenna is made still longer. We find, as a matter of fact, that the patterns of harmonic antennas differ very considerably from the pattern of the half-wave dipole.

As explained earlier in this chapter, a harmonic antenna consists of a series of half-wave sections with the currents in adjacent sections always flowing in opposite directions. This type of current flow causes the pattern to be split up into a number of lobes. If there is an *even* number of half-waves in the harmonic antenna there is always a null in the plane at right angles to the wire; this is because the radiation from one half-wave section cancels the radiation from the next, in that particular direction. If there is an *odd* number of half waves in the antenna, the radiation from all but one of the sections cancels itself out in the plane perpendicular to the wire. The "leftover" section radiates like a half-wave dipole, and so a harmonic antenna with an odd number of half wavelengths does have some radiation at right angles to its axis.

The greater the number of half waves in a harmonic antenna, the larger the number of lobes into which the pattern splits. A feature of all such patterns is the fact that the "main" lobe — the one that gives the largest field strength at a given distance — always is the one that makes the smallest angle with the antenna wire. Furthermore, this angle becomes smaller as the length of the antenna is increased. Fig. 15 shows how the angle which the main lobe makes with the axis of the antenna varies with the antenna length in wavelengths. The angle shown by the solid curve is the maximum point of the lobe; that is, the direction in which the field strength is greatest. The broken curve shows the angle at which the first null (the one that occurs at the smallest angle with the wire) appears. There is also a null in the direction of the wire itself (0 degrees) and so the total width of the main lobe is the angle between the wire and the first null. It can be seen from Fig. 15 that the width of the lobe decreases as the wire becomes longer. At 1 wavelength, for example, it has a width of 90 degrees, but at 8 wavelengths the width is slightly less than 30 degrees.

A plane diagram of the radiation pattern of a 1-wavelength harmonic wire is shown in Fig. 16. This is a free-space diagram in the plane containing the wire axis (E plane), corresponding to the diagrams for the elementary doublet and half-wave dipole shown in Figs. 12 and 13. It is based on an infinitely thin antenna conductor with ideal current distribution, and in a practical antenna system will be modified by the presence of the earth and other effects that will be considered later.

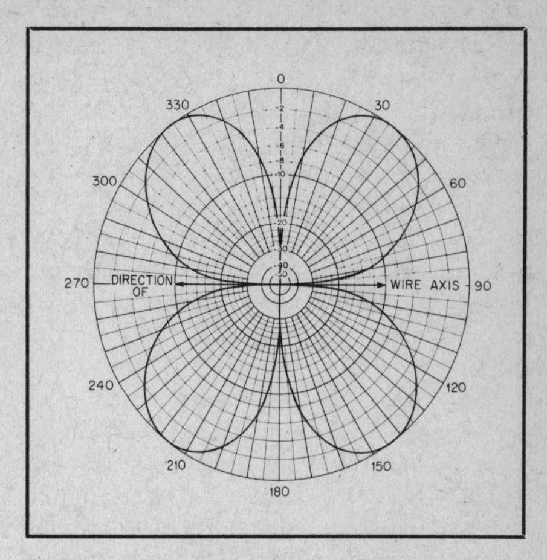


Fig. 16 — Free-space directive diagram of a 1-wavelength harmonic antenna in the plane containing the wire axis (E plane).

## HOW PATTERNS ARE FORMED

The radiation pattern of the half-wave dipole is found by summing up, at every point on the surface of a sphere with the antenna at its center, the field contributions of all the elementary dipoles that can be imagined to make up the full-size dipole. Antenna systems often are composed of a group of half-wave dipoles arranged in various ways, in which case each half-wave dipole is called an antenna element. An antenna having two or more such dipoles is called a multielement antenna. (A harmonic antenna can be

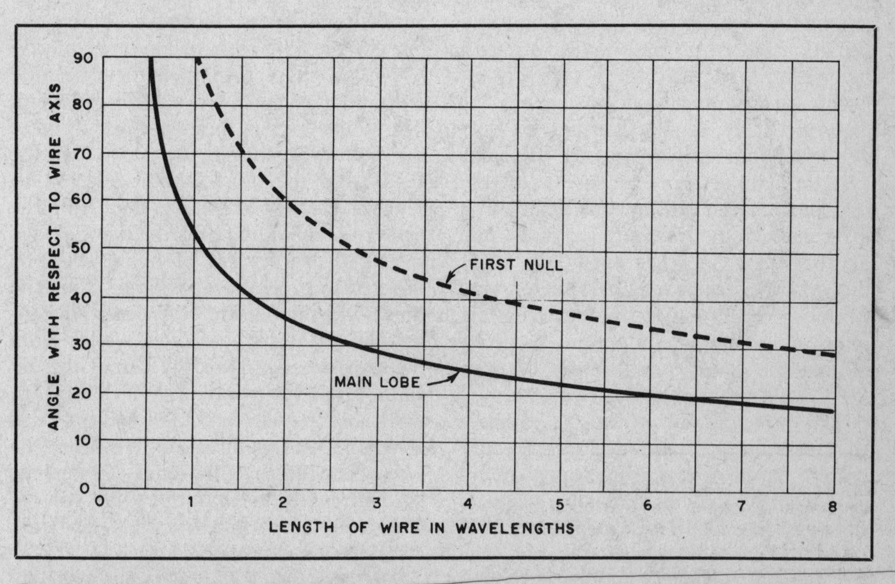


Fig. 15 — Angle at which the field intensity from the main lobe of a harmonic antenna is maximum, as a function of the wire length in wavelengths. The curve labeled "First Null" locates the angle at which the intensity of the main lobe decreases to zero. The null marking the other boundary of the main lobe always is at zero degrees with the wire axis.

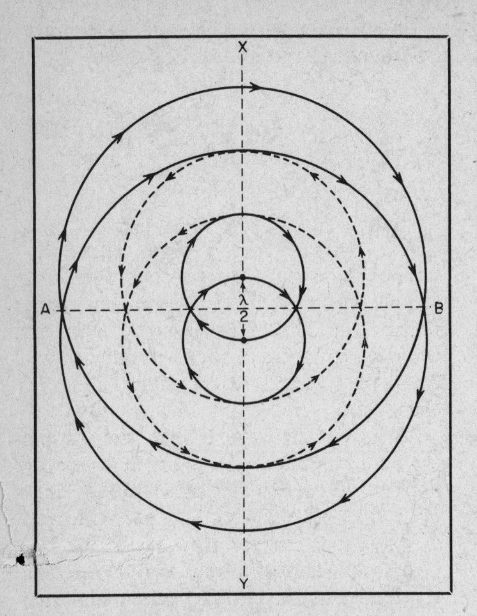


Fig. 17 — Interference between waves from two separate radiators causes the resultant directional effects to differ from those of either radiator alone. The two radiators shown here are separated one-half wavelength. The radiation fields of the two cancel along the line XY but, at distances which are large compared with the separation between the radiators, add together along line AB. The resultant field decreases uniformly as the line is swung through intermediate positions from AB to XY.

considered to be constructed of a number of such elements connected in series and fed power appropriately, as described earlier, but is not usually classed as a multielement antenna.)

In a multielement antenna system the overall radiation pattern is determined by the way in which the fields at a distant point from the separate antenna elements combine. With two antenna elements, for example, the field strength at a given point depends on the amplitudes and phase relationship of the fields from each antenna. A requirement in working out a radiation pattern is that the field strength be measured or calculated at a *distant* point — distant enough so that, if the elements carry equal currents, the field strength from each is exactly the same even though the size of the antenna system may be such that one antenna element is a little nearer the measuring point than another. On the other hand, this slight difference in distance, even though it may be only a small fraction of a wavelength in many miles, is very important in determining the *phase* relationships between the fields from the various elements.

The principle on which the radiated fields combine to produce the directive pattern, in the case of multielement antennas, is illustrated in the simple example shown in Fig. 17. In this case it is assumed that there are two antenna elements, each having a circular directive pattern. The two elements, therefore, could be half-wave dipoles oriented perpendicular to the page (which gives the plane pattern shown in Fig. 14). The separation between the two elements is assumed to be a half wavelength and the currents in them are assumed to be equal. Furthermore, the two currents are in phase; that is, they reach their maximum values in the same polarity at the same instant.

Under these conditions the fields from the two antennas will be in the same phase at any point that is equally distant from both antenna elements. At the instant of time selected for the drawing of Fig. 17 the solid circles having the upper antenna at their centers represent, let us say, the location of all points at which the field intensity is maximum and has the direction indicated by the arrowheads. The *distance* between each pair of concentric solid circles, measured along a radius, is equal to one wavelength because, as described earlier in this chapter, it is only at intervals of this distance that the fields are in phase. The broken circle locates the points at which the field intensity is the same as in the case of the solid circles, but is *oppositely* directed. It is, therefore, 180 degrees out of phase with the field denoted by the solid circles, and the distance between the solid and broken circles is therefore one-half wavelength.

Similarly, the solid circles centered on the lower antenna locate all points at which the field intensity from that antenna is maximum and has the same direction as the solid circles about the upper antenna. In other words, these circles represent points in the same phase as the solid circles around the upper antenna. The broken circle having the lower antenna at its center likewise locates the points of opposite phase.

Considering now the fields from both antennas it can be seen that along the line AB the fields from the two always are exactly in phase, because every point along AB is equally distant from both antenna elements. However, along the line XY the field from one antenna always is *out* of phase with the other, because every point along XY is a half wavelength nearer one element than the other. It takes one-half cycle longer, therefore, for the field from the more distant element to reach the same point as the field from the nearer antenna, and thus the one field arrives 180 degrees out of phase with the other. Since we have assumed that the points considered are sufficiently distant so that the amplitudes of the fields from the two antennas are the same, the *resultant* field at any point along XY is zero and the antenna combination shown will have a null in that direction. However, the two fields add together along AB and the field strength in that direction will be twice the amplitude of the field from either antenna alone.

The drawing of Fig. 17 is not quite accurate because it cannot be made large enough. Actually, the two fields along AB do not have exactly the same direction

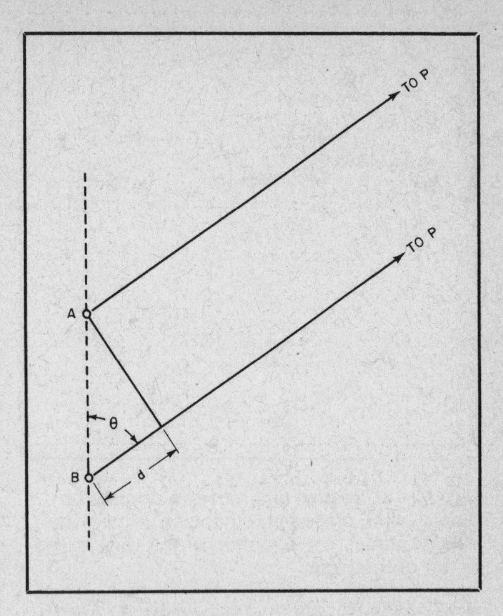


Fig. 18 — Graphical construction to determine the relative phase, at a distant point, of waves originating at two antennas, A and B. The phase is determined by the additional distance, d, that the wave from B has to travel to reach the distant point. This distance will vary with the angle that the direction to P makes with the axis of the antenna system, θ.

until the distance to the measuring point is large enough, compared with the dimensions of the antenna system, so that the waves become plane. In a drawing of limited size the waves are necessarily represented as circles — that is, as representations of a *spherical* wave. The reader, therefore, should imagine Fig. 17 as being so much enlarged that the circles crossing AB are substantially straight lines in the region under discussion.

### Pattern Construction

The drawing of Fig. 17 does not tell us much about what happens to the field strength at points that do not lie on either AB or XY although we could make the reasonable guess that the field strength at intermediate points probably would decrease as the point was moved along the arc of a circle farther away from AB and near to XY. To construct an actual pattern it is necessary to use a different method. It is simple in principle and can be done with a ruler, protractor and pencil, or by trigonometry.

In Fig. 18 the two antennas, A and B, are assumed to have circular radiation patterns, and to carry equal currents in the same phase. (In other words, the conditions are the same as in Fig. 17.) The relative field strength at a distant point P is to be determined. Here again the limitations of the printed page make it necessary to use the imagination, because we assume that P is far enough from A and B so that the lines AP and BP are, for all practical purposes, parallel. When this is so, the distance d, between B and a perpendicular dropped to BP from A, will be equal to the difference in length between the distance from A to P and the distance

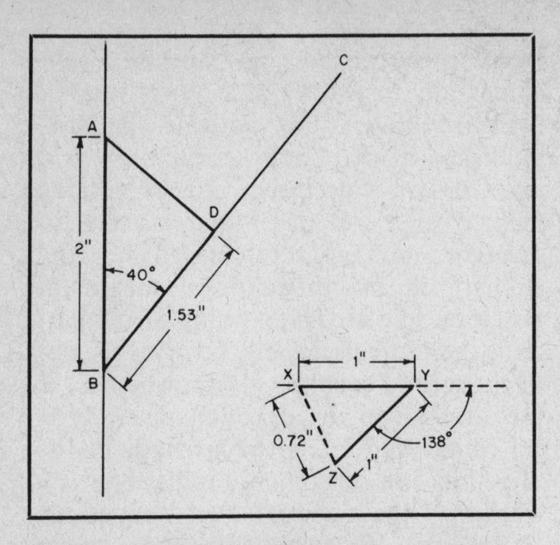


Fig. 19 — Graphical construction in the example discussed in the text (25.4 × inches = mm).

from B to P. The distance d thus measures the difference in the distance the waves from A and B have to cover to reach P; d is also, therefore, a measure of the difference in the time of arrival or *phase* of the waves at P.

Under the assumed conditions, the relative field strengths easily can be combined graphically. The phase angle in degrees between the two fields at P is equal to

$$\frac{d}{\lambda} \times 360$$

where $\lambda$ is the wavelength and d is found by constructing a figure similar to that shown in Fig. 18 for P in any desired direction. The angle $\theta$ is the angle between a line to P and the line drawn between the two antenna elements, and is used simply to identify the direction of P from the antenna system; $\lambda$ and d must be expressed in the same length units.

For example, let us assume that $\theta$ is 40 degrees. We then arbitrarily choose a scale such that 4 inches is equal to one wavelength — a scale large enough for reasonable accuracy but not too large to be unwieldy. Since the two antenna elements are assumed to be a half-wavelength apart, we start the drawing by placing two points 2 inches apart and connecting them by a line, as shown in Fig. 19. Then, using B as a center and employing the protractor, we lay off an angle of 40 degrees and draw the line BC. The next step is to drop a perpendicular from A to BC; this may be done with the 90-degree mark on the protractor but the corner of an ordinary sheet of paper will do just about as well. The distance BD is then measured, preferably with a ruler graduated in tenths of inches rather than the more usual eighths. By actual measurement distance BD is found to be 1.53 inches. The phase difference is therefore $d/\lambda \times 360 = 1.53/4 \times 360 = 138$ degrees.

The relative field strength in the direction given by $\theta$ (40 degrees in this example) is found by arbitrarily selecting a line length to represent the strength of the field from each antenna, and then combining them "vectorially." One inch is a convenient length to select. XY, Fig. 19, is such a line, representing the strength of the field from antenna element A. We then measure off an angle of 138 degrees from XY, using Y as a center, and draw YZ one inch long to represent the strength and phase of the field from antenna element B. The angle is measured off clockwise from XY because the field from B lags behind that from A. The distance from X to Z then represents the relative field strength resulting from the combination of the separate fields from the two antennas, and measurement shows it to be approximately 0.72 inch. In the direction $\theta$, therefore, the field strength is 72% as great as the field from either antenna alone. Using trigonometry, the determination may be made by using the equation,

$$\text{field strength} = 2\cos\left(\frac{S}{2}\cos\theta\right)$$

where S = spacing between elements in electrical degrees.

By selecting different values for $\theta$ and proceeding as above in each case, the complete pattern can be determined. When $\theta$ is 90 degrees, the phase difference is zero and YZ and XY are simply end-to-end along the same line. The maximum field strength is therefore twice that of either antenna alone. When $\theta$ is zero, YZ lies on top of XY (phase difference 180 degrees) and the distance XZ is therefore zero; in other words the radiation from B cancels that from A at such an angle.

The patterns of more complex antenna systems can readily be worked out by this method, although more labor is required if the number of elements is increased. But whether or not actual patterns are worked out, an understanding of the method will do much to make it plain why certain combinations of antenna elements result in specific directive patterns.

The illustration above is a very simple case, but it is only a short step to systems in which the antenna elements do not carry equal currents or currents in the same phase. A difference in current amplitude is easily handled by making the lengths of lines XY and YZ proportional to the current in the respective elements; if the current in B is one half that in A, for example, YZ would be drawn one half as long as XY. If B's current leads the current in A by 25 degrees, then after the angle determined by the distance d is found the line YZ is simply rotated 25 degrees in the *counterclockwise* direction before measuring the distance XZ. The rotation would be *clockwise* for any line representing a lagging current. The lead or lag of current always has to be referred to the current in *one* element of the system, but any desired element can be chosen as the reference.

For two elements fed out of phase but having equal currents, the relationship

$$\text{field strength} = 2\cos\left(\frac{\phi - S\cos\theta}{2}\right)$$

may be used, where $\phi$ is the phase difference between the two fed elements. Simple trigonometric equations are insufficient for determining array patterns when the currents in the elements are unequal or when there are more than two elements.

It should be noted that the simple methods described above for determining pattern shapes do not take mutual coupling between elements into account, i.e., the fact that current flowing in one element will induce a voltage and therefore a resultant current into the other, and vice versa. When mutual coupling is taken into account the shape of the pattern remains the same for a given condition of element spacing and phasing, but the magnitudes of the resultant vectors used in plotting points for various values of $\theta$ are altered by a fixed factor. The "fixed" factor will vary with changes in spacing and phasing of the elements. Therefore, a *direct* comparison of the *sizes* of different patterns obtained by these simple procedures cannot be used for determining, say, the gain of one antenna system over another, even though both patterns were derived by using the same scale. Mutual coupling is covered in more detail in a later chapter.

## DIRECTIVITY AND GAIN

It has been stated that all antennas, even the simplest types, exhibit directive effects in that the intensity of radiation is not the same in all directions from the antenna. This property of radiating more strongly in some directions than in others is called the *directivity* of the antenna. It can be expressed quantitatively by comparing the solid pattern of the antenna under consideration with the solid pattern of the isotropic antenna. The field strength (and thus power per unit area or "power density") will be the same everywhere on the surface of an imaginary sphere having a radius of many wavelengths and having an isotropic antenna at its center. At the surface of the same imaginary sphere around an actual antenna radiating the same total power, the directive pattern will result in greater power density at some points and less at others. The ratio of the maximum power density to the average power density taken over the entire sphere (the latter is the same as from the isotropic antenna under the specified conditions) is the numerical measure of the directivity of the antenna. That is,

$$D = \frac{P}{P_{av}}$$

where D is the directivity, P is the power density at its maximum point on the surface of the sphere, and $P_{av}$ is the average power density.

### Gain

The gain of an antenna is closely related to its directivity. Since directivity is based solely on the *shape* of the directive pattern, it does not take into account any power losses that may occur in an actual antenna system. To determine gain, these losses must be subtracted from the power supplied to the antenna. The loss is normally a constant percentage of the power input, so the antenna gain is

$$G = k \frac{P}{P_{av}}$$

where G is the gain expressed as a power ratio, k is the efficiency (power radiated divided by power input) of the antenna, and P and $P_{av}$ are as above. For many of the antenna systems used by amateurs the efficiency is quite high (the loss amounts only to a few percent of the total), and in such cases the gain is essentially equal to the directivity.

The more the directive diagram is compressed — or, in common terminology, the "sharper" the lobes — the greater the power gain of the antenna. This is a natural consequence of the fact that as power is taken away from a larger and larger portion of the sphere surrounding the radiator it is added to the smaller and smaller volume represented by the lobes. The power is therefore concentrated in some directions at the expense of others. In a general way, the smaller the volume of the solid radiation pattern, compared with the volume of a sphere having the same radius as the length of the largest lobe in the actual pattern, the greater the power gain.

As stated above, the gain of an antenna is related to its directivity, and directivity is related to the shape of the directive pattern. A commonly used index of directivity, and therefore the gain of an antenna, is a measure of the width of the major lobe (or lobes) of the plotted pattern. The width is expressed in degrees at the half-power or $-3$ dB points, and is often called the beamwidth. The reader should be aware, however, that this information provides only a general idea of relative gains, rather than an exact measure. This is because an absolute measure involves knowing the power density at every point on the surface of a sphere, while a single diagram shows the pattern shape in only one plane of that sphere. It should be customary to examine at least the E-plane and the H-plane patterns before making any comparisons between antennas.

Gain referred to an isotropic radiator is necessarily theoretical; that is, it must be calculated rather than measured because the isotropic radiator has no existence. In practice, measurements on the antenna being tested usually are compared with measurements made on a half-wave dipole. The latter should be at the same height and have the same polarization as the antenna under test, and the reference field — that from the half-wave dipole comparison antenna — should be measured in the most favored direction of the dipole. The data can be secured either by measuring the field strengths produced at the same distance from both antennas when the same power is supplied to each, or by measuring the power required in each antenna to produce the same field strength at the same distance.

A half-wave dipole has a theoretical gain of $10 \log (1 + 2/\pi) = 2.14$ dB over an isotropic radiator. Thus the gain of an actual antenna over a half-wave dipole can be referred to isotropic by adding 2.14 dB to the measured gain, or if the gain is expressed over an isotropic antenna it can be referred to a half-wave dipole by subtracting 2.14 dB.

It should be noted that the field strength (voltage) produced by an antenna at a given point is proportional to the square root of the power. That is, when the two are expressed as ratios (the usual case),

$$\frac{F1}{F2} = \sqrt{\frac{P1}{P2}}$$

### The Decibel

As a convenience the power gain of an antenna system is usually expressed in decibels. The decibel is an excellent practical unit for measuring power ratios because it is more closely related to the actual effect produced than the power ratio itself. One decibel represents a just-detectable change in signal strength, regardless of the actual value of the signal voltage. A 20-decibel (20-dB) increase in signal, for example, represents 20 observable "steps" in increased signal. The power ratio (100 to 1) corresponding to 20 dB would give an entirely exaggerated idea of the improvement in communication to be expected. The number of decibels corresponding to any power ratio is equal to 10 times the common logarithm of the power ratio, or

$$dB = 10 \log \frac{P1}{P2}$$

If the *voltage* ratio is given, the number of decibels is equal to 20 times the common logarithm of the ratio. That is,

$$dB = 20 \log \frac{E1}{E2}$$

When a voltage ratio is used both voltages must be measured in the same value of impedance. Unless this is done the decibel figure is meaningless, because it is fundamentally a measure of a *power* ratio.

Even though the antenna patterns published in this book appear on a grid marked in decibels, the patterns themselves are in terms of relative field strength (voltage), referenced to the strength in the direction of maximum radiation (0 dB). The plotting of the patterns therefore necessarily involves the use of the second equation above, where E1 is the strength in the direction under consideration and E2 is the strength in the direction of maximum radiation. By working the above two equations backwards, field-strength and power ratios may be determined from the published patterns.

Table 2 of the appendix shows the number of decibels corresponding to various power and voltage ratios. One advantage of the decibel is that successive power gains expressed in decibels may simply be added together. Thus a gain of 3 dB followed by a gain of 6 dB gives a total gain of 9 dB. In ordinary power ratios, the ratios would have to be multiplied together to find the total gain. Furthermore, a *reduction* in power is handled simply by subtracting the requisite number of decibels. Thus reducing the power to 1/2 is the same as *subtracting* 3 decibels. We might, for example, have a power gain of 4 in one part of a system and a reduction of 1/2 in another part, so that the total power gain is $4 \times 1/2 = 2$. In decibels, this would be $6 - 3 = 3$ dB. A power reduction or "loss" is simply indicated by putting a negative sign in front of the appropriate number of decibels.

### Power Gains of Harmonic Antennas

In splitting off into a series of lobes, the solid radiation pattern of a harmonic antenna is compressed into a smaller volume as compared with the single-lobed pattern of the half-wave dipole. This means that there is a concentration of power in certain directions with a harmonic antenna, particularly in the main lobe. The result is that a harmonic antenna will produce an increase in field strength, in its most favored direction, over a half-wave dipole in *its* most favored direction, when both antennas are supplied with the same amount of power.

The power gain from harmonic operation is small when the antenna is small in terms of wavelengths, but is quite appreciable when the antenna is fairly long. The theoretical power gain of harmonic antennas or "long wires" is shown by curve B in Fig. 20, using the half-wave dipole as a base. A 1-wavelength or "second harmonic" antenna has only a slight poer gain, but an antenna 9 wavelengths long will show a power gain of nearly 7 dB over the dipole. This gain is secured in one direction by reducing or eliminating the power radiated in other directions; thus the longer the wire the more directive the

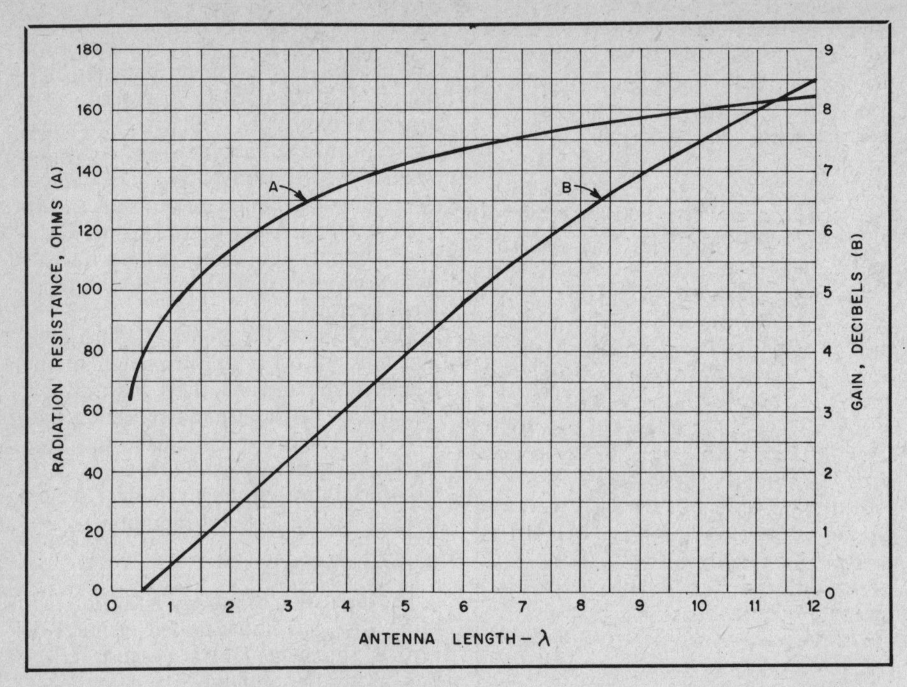


Fig. 20 — The variation in radiation resistance and power in the major lobe of long-wire antennas. Curve A shows the change in radiation resistance with antenna length, as measured at a current loop, while curve B shows the power gain in the lobes of maximum radiation for long-wire antennas as a ratio to the maximum of a half-wave antenna.

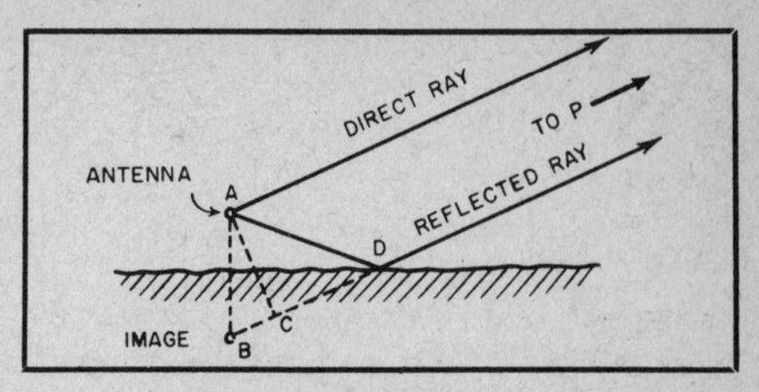


Fig. 21 — At any distant point, P, the field strength will be the resultant of two rays, one direct from the antenna, the other reflected from the ground. The reflected ray travels farther than the direct ray by the distance BC, where the reflected ray is considered to originate at the "image" antenna.

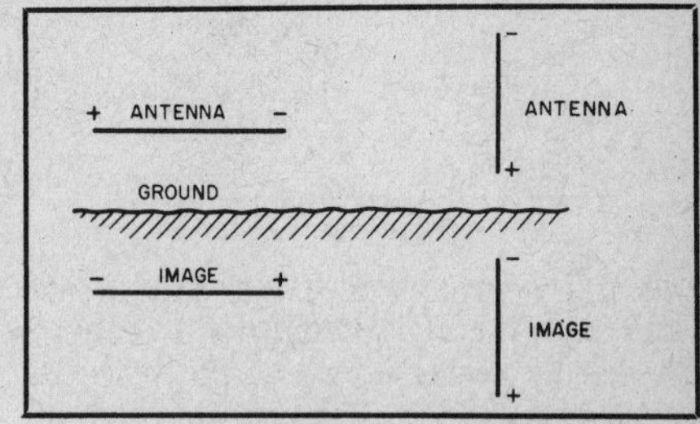


Fig. 22 — Horizontal and vertical half-wave antennas and their images.

antenna becomes.

Curve A in Fig. 20 shows how the radiation resistance, as measured at a current loop, varies with the length of a harmonic antenna.

## GROUND EFFECTS

The performance of an antenna, particularly with respect to its directive properties, is considerably modified by the presence of the earth underneath it. The earth acts as a huge reflector for those waves that are radiated from the antenna at angles lower than the horizon. These downcoming waves strike the surface and are reflected by a process very similar to that by which light waves are reflected from a mirror. As in the case of light waves, the angle of reflection is the same as the angle of incidence, so that a wave striking the surface at an angle of, for instance, 15 degrees, is reflected upward from the surface at the same angle.

The reflected waves combine with the direct waves (those radiated at angles above the horizontal) in various ways, depending on the orientation of the antenna with respect to earth, the height of the antenna, its length and the characteristics of the ground. At some vertical angles above the horizontal the direct and reflected waves may be exactly in phase — that is, the maximum field strengths of both waves are reached at the same time at the same spot, and the directions of the fields are the same. In such a case the resultant field strength is simply equal to the sum of the two. (This represents an increase of 6 dB in strength at these angles.) At other vertical angles the two waves may be completely out of phase — that is, the fields are maximum at the same instant and the directions are opposite, at the same spot. The resultant field strength in that case is the *difference* between the two. At still other angles the resultant field will have intermediate values. Thus the effect of the ground is to increase the intensity of radiation at some vertical angles and to decrease it at others.

The effect of reflection from the ground is shown graphically in Fig. 21. At a sufficiently large distance, two rays converging at the distant point can be considered to be parallel. However, the reflected ray travels a greater distance in reaching P than the direct ray does, and this difference in path length accounts for the effect described in the preceding paragraph. If the ground were a perfect conductor for electric currents, reflection would take place without a change in phase when the waves are vertically polarized. Under similar conditions there would be a complete reversal (180 degrees) of phase when a horizontally polarized wave is reflected. The actual earth is, of course, not a perfect conductor, but is usually assumed to be one for purposes of calculating the vertical pattern of an antenna.

As an example, when the path of the reflected ray is exactly a half wave longer than the path of the direct ray, the two waves will arrive out of phase if the polarization is vertical. This corresponds to the condition illustrated in Fig. 17 along the line XY. If the path of the reflected ray is just a wavelength longer than that of the direct ray, however, the two rays arrive in phase.

### Image Antennas

It is often convenient to use the concept of an image antenna to show the effect of reflection. As Fig. 21 shows, the reflected ray has the same path length (AD equals BD) that it would if it originated at a second antenna of the same characteristics as the real antenna, but situated below the ground just as far as the actual antenna is above it. Like an image in a mirror, this image antenna is "in reverse," as shown in Fig. 22.

If the real antenna is horizontal, and is instantaneously charged so that one end is positive and the other negative, then the image antenna, also horizontal, is oppositely poled; the end under the positively charged end of the real antenna is negative, and vice versa. Likewise, if the lower end of a half-wave vertical antenna is instantaneously positive, the end of the vertical image antenna nearest the surface is negative. Now if we look at the antenna and its image from a remote point on the surface of the ground, it will be obvious that the currents in the horizontal antenna and its image are flowing in opposite directions, or are 180 degrees out of phase, but the currents in the vertical antenna and its image are flowing in the *same* direction, or are *in* phase. The effect of the ground reflection, or the image antenna, is therefore different for horizontal and vertical half-wave antennas. The physical reason for this difference is the fact that vertically polarized waves are reflected from a perfectly con-

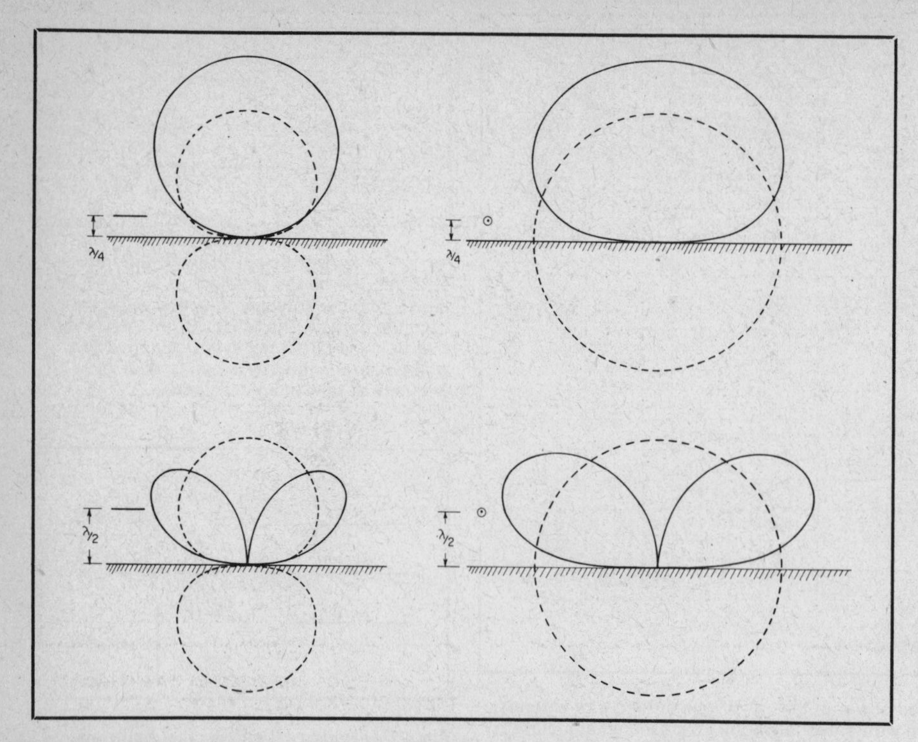


Fig. 23 — Effect of the ground on the radiation from a horizontal half-wave antenna, for heights of one-fourth and one-half wavelength. Dashed lines show what the pattern would be if there were no reflection from the ground.

ducting earth with no change in phase, but that horizontally polarized waves have their phase shifted by 180 degrees on reflection.

By using techniques similar to those discussed earlier for determining patterns of antennas containing more than one element, it is possible to derive the vertical pattern which will result from an antenna and its image beneath the surface of the earth. The resultant pattern is a modification of the free-space pattern of the antenna. Fig. 23 shows how such modification takes place for a horizontal half-wave antenna. The patterns at the left show the relative radiation when one views the antenna from the side; those at the right show the radiation pattern when one looks at the end of the antenna. Changing the height from one-fourth to one-half wavelength makes quite a difference in the upward radiation — that is, the radiation at high angles. (The *radiation angle* or *wave angle* is measured from the ground up.)

### Reflection Factor

The effect of reflection from the ground can be expressed as a pattern factor, given in decibels. For any given vertical angle, adding this factor algebraically to the value for that angle from a free-space pattern for the antenna gives the resultant radiation value at that same angle. The limiting conditions are those represented by the direct ray and reflected ray being exactly in phase and exactly out of phase when both, assuming there are no ground losses, have exactly equal amplitudes. Thus, the resultant field strength at a distant point may be either 6 dB greater than the field strength from the antenna alone (twice the field strength), or zero.

The way in which pattern factors vary with height for horizontal antennas is shown in Figs. 24 through 35. These factors are based on perfectly conducting ground, and apply to horizontal antennas of any length. It must be remembered that these graphs are *not* plots of vertical radiation patterns of antennas, but simply represent pattern factors resulting from ground reflection. (Only in one special case do these graphs represent radiation patterns. That case is for a single horizontal wire, such as a dipole antenna. In that case the graphs may be taken to represent the vertical radiation pattern when the wire is viewed from one end.)

Fig. 36 illustrates the use of these pattern factors. At Fig. 36A is a typical free-space H-plane pattern of a multielement array, such as a Yagi antenna. The forward direction of the array is to the right, and it has a 30 dB front-to-back ratio. Assume this antenna is placed one wavelength above the earth. To determine the vertical radiation pattern, factors from Fig. 31 are to be applied. (Note that 6 dB is to be added to the factors charted in Figs. 24 through 35.) The resultant vertical-plane pattern for the multielement array when placed one wavelength above a perfect conductor is shown in Fig. 36B. (Worksheets for pattern plotting are available from ARRL Hq. See Fig. 11.)

It should be understood that these pattern factors apply at vertical angles only. The ground, if of uniform characteristics, makes no distinction between geographical directions — that is, horizontal angles from the antenna — in reflecting waves. This is portrayed in Fig. 23, where the resultant radiation patterns are shown for both the E plane and H plane. For a given height, the same pattern factors have been applied to obtain the two solid-line patterns.

Fig. 37 shows the angles at which nulls and maxima occur as a function of the height of the antenna. This chart gives a rough idea of the ground-reflection pattern for heights intermediate to those shown in detail in Figs. 24 through 35. It also facilitates selecting the right height for any desired angle of radiation.

### Vertical Antennas

The discussion of ground reflection factors so far has not included mention of vertical antennas. This is because reflection from the ground for a vertically polarized wave occurs with no change in phase, while for a horizontally polarized wave there is a 180° phase shift. Therefore the effects of the earth on a vertical antenna are entirely different than for a horizontal antenna, as illustrated in Fig. 22. The effect of adding an image antenna beneath a vertical element placed close to the earth is to double its length. Current in the real antenna and in its image are in phase. (See Fig. 45.) Therefore, a quarter-wavelength vertical element above a perfect conductor will have the same vertical radiation pattern as the broadside pattern of a half-wave dipole in free space, Fig. 13. Of course there can be no radiation below the surface of the earth, so the resultant pattern will be that of the left half of Fig. 13, rotated 90° clockwise.

Similarly, a half-wave vertical element above a perfect conductor will exhibit the same vertical radiation pattern as one-half the broadside pattern of two half waves in phase in free space. Antenna arrays such as two half waves in phase are discussed in Chapter 6. If a vertical element is placed some distance above the earth's surface, the resultant pattern is the same as that of a collinear array with spacing between the elements, also discussed in Chapter 6.

## GROUND CHARACTERISTICS

As already explained, the charts of Figs. 24 through 35 are based on the assumption that the earth is a perfect conductor. The actual ground is far from being "perfect" as a conductor of electricity. Its behavior depends considerably on the transmitted frequency and the angle at which the ray strikes the earth. At low frequencies — through the standard broadcast band, for example — most types of ground do act very much like a good conductor at all angles. At these frequencies the waves can penetrate for quite a distance and thus find a large cross sec-

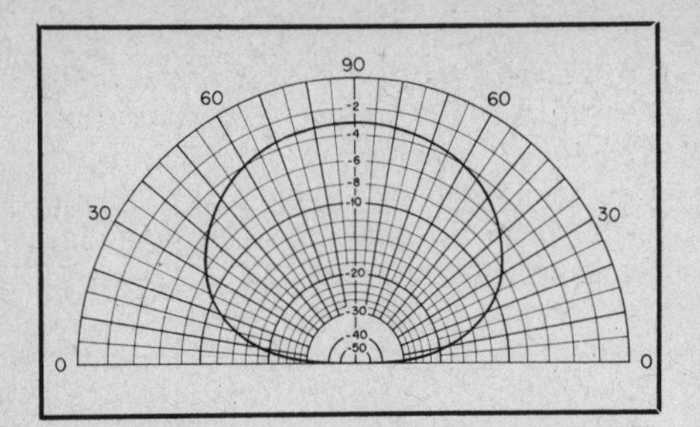


Fig. 24 — Horizontal antennas 1/8 wavelength high. Add 6 dB to values shown.

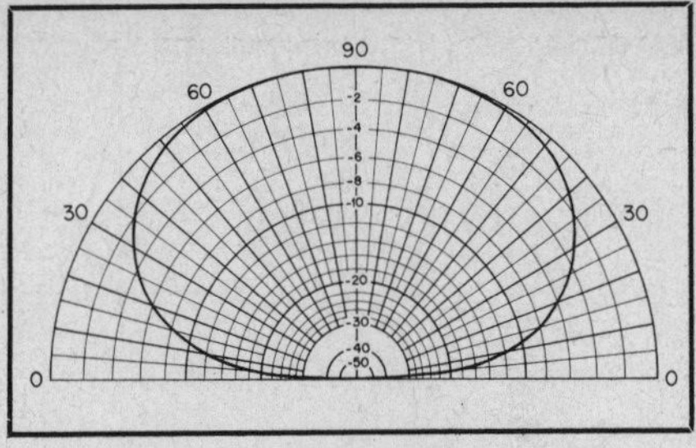


Fig. 25 — Horizontal antennas 1/4 wavelength high. Add 6 dB to values shown.

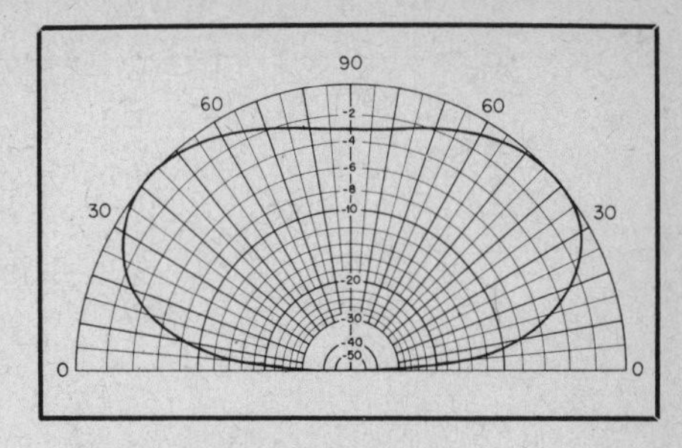


Fig. 26 — Horizontal antennas 3/8 wavelength high. Add 6 dB to values shown.

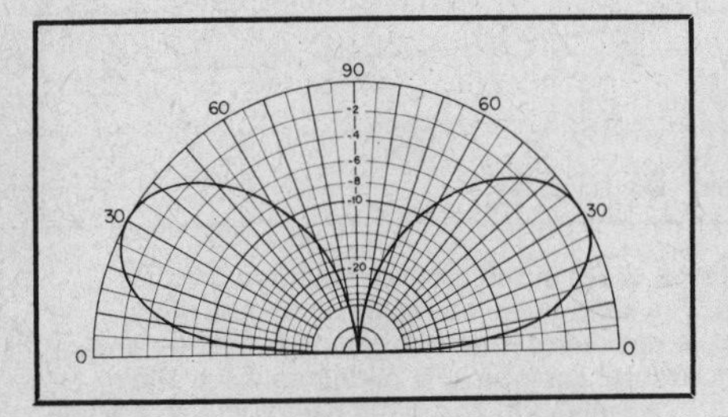


Fig. 27 — Horizontal antennas 1/2 wavelength high. Add 6 dB to values shown.

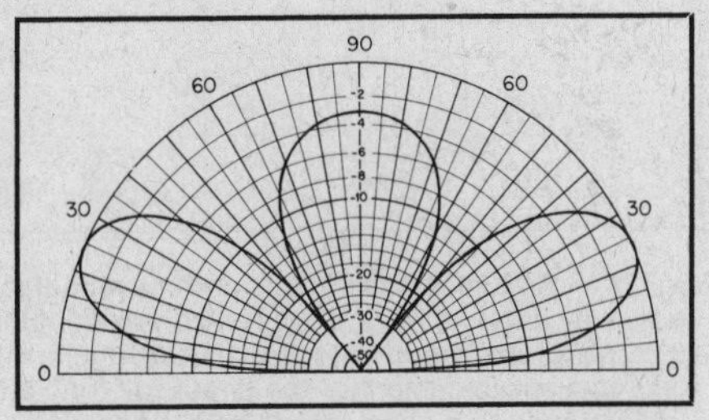


Fig. 28 — Horizontal antennas 5/8 wavelength high. Add 6 dB to values shown.

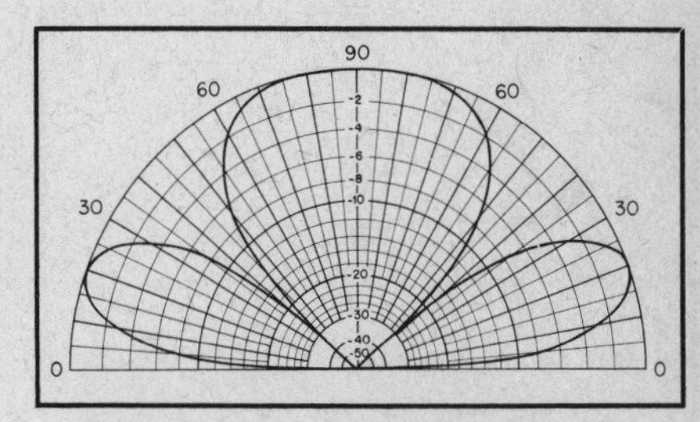


Fig. 29 — Horizontal antennas 3/4 wavelength high. Add 6 dB to values shown.

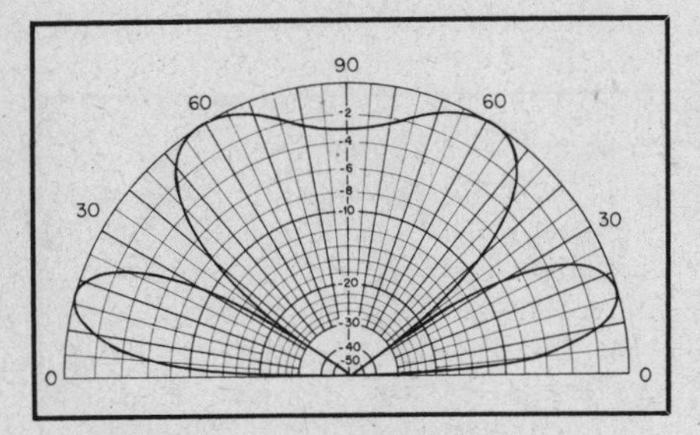


Fig. 30 — Horizontal antennas 7/8 wavelength high. Add 6 dB to values shown.

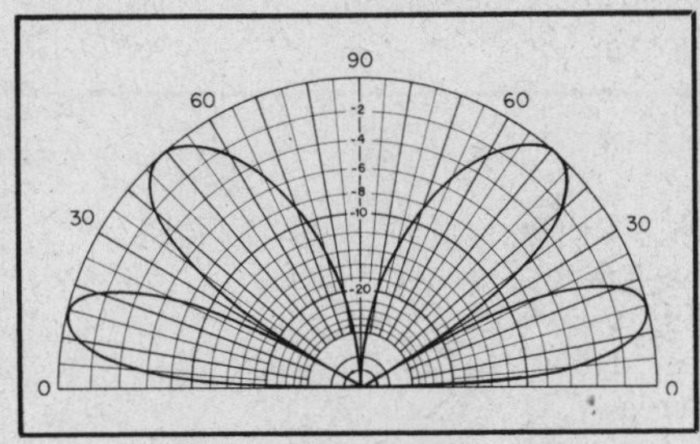


Fig. 31 — Horizontal antennas 1 wavelength high. Add 6 dB to values shown.

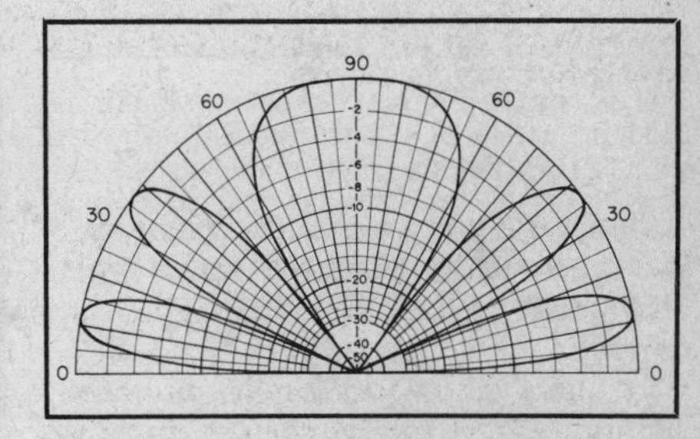


Fig. 32 — Horizontal antennas 1-1/4 wavelengths high. Add 6 dB to values shown.

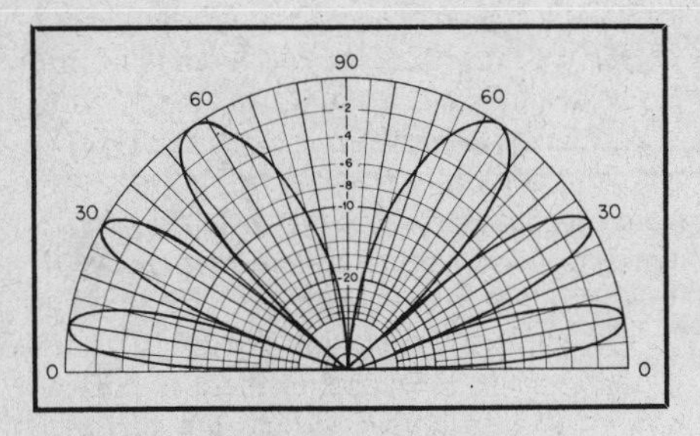


Fig. 33 — Horizontal antennas 1-1/2 wavelengths high. Add 6 dB to values shown.

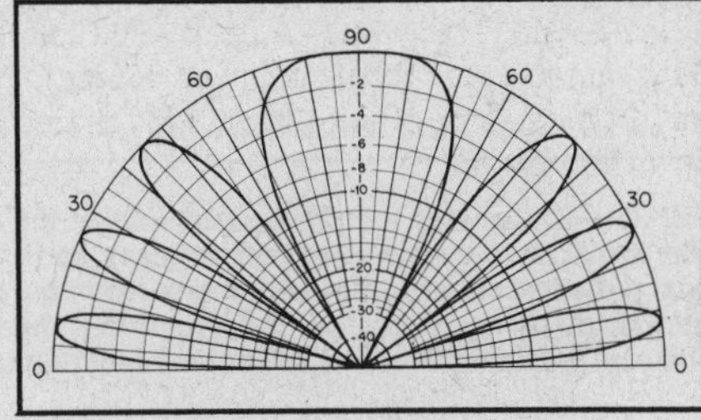


Fig. 34 — Horizontal antennas 1-3/4 wavelengths high. Add 6 dB to values shown.

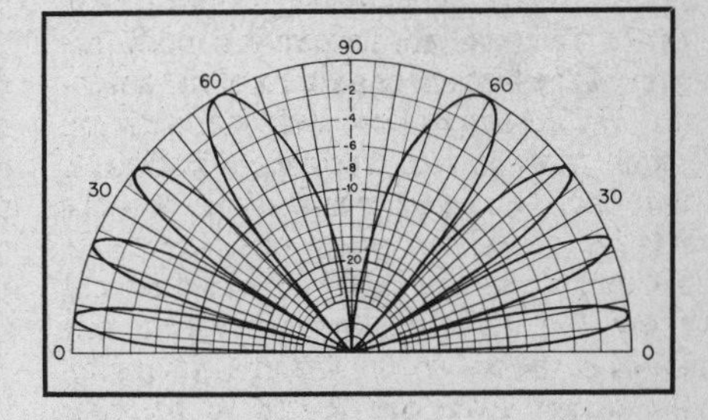


Fig. 35 — Horizontal antennas 2 wavelengths high. Add 6 dB to values shown.

Factors to which the free-space radiation pattern of a horizontal antenna should be added to include the effect of reflection from perfectly conducting ground. These factors affect only the vertical angle of radiation (wave angle) for horizontal antennas.

tion in which to cause current flow along their paths. The resistance of even a moderately good conductor will be low if its cross section is large enough. The ground acts as a fairly good conductor even at frequencies as high as the 3.5-MHz band, and so the pattern-factor charts give a rather good approximation of the effect of the ground at this frequency.

In the higher frequency region the penetration decreases and the ground may even take on the characteristics of a lossy dielectric, rather than a good conductor. The chief effect of this change is to absorb most of the energy which would otherwise be reflected at high angles, in the frequency region from about 7 to 25 MHz. In general, the reflection factor for horizontal antennas will be lower than given by the charts (a higher negative number) at angles greater than about 15°. At grazing angles, the charts provide reliable information, even at the higher frequencies.

For vertical antennas, the absorption is lowest above about 15°. Below 5°, reflection from vertically polarized waves is very small compared with that at higher angles. The "zero angle" reflection factor with a vertical half-wave antenna, for example, is theoretically 6 dB. In practice

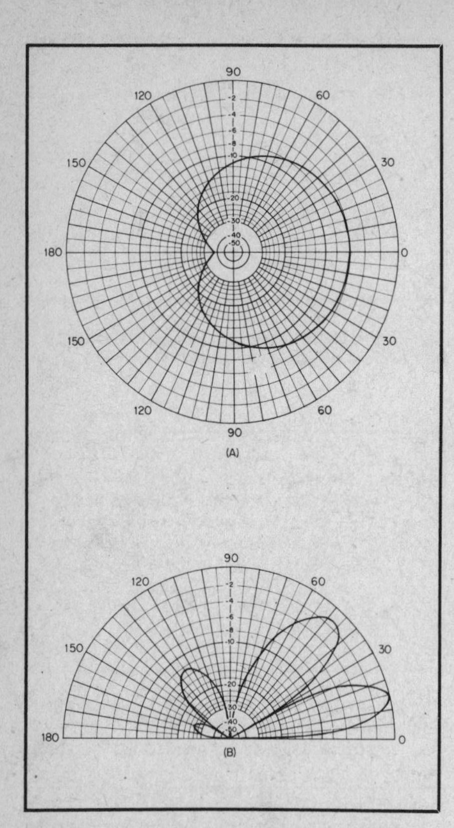
Fig. 36 — At A, a typical free-space H-plane radiation pattern of a horizontal multielement array, such as a Yagi antenna. The forward direction of the array is to the right. At B, the free-space pattern as modified by placing the antenna one wavelength above a perfect conductor (pattern factors from Fig. 31).

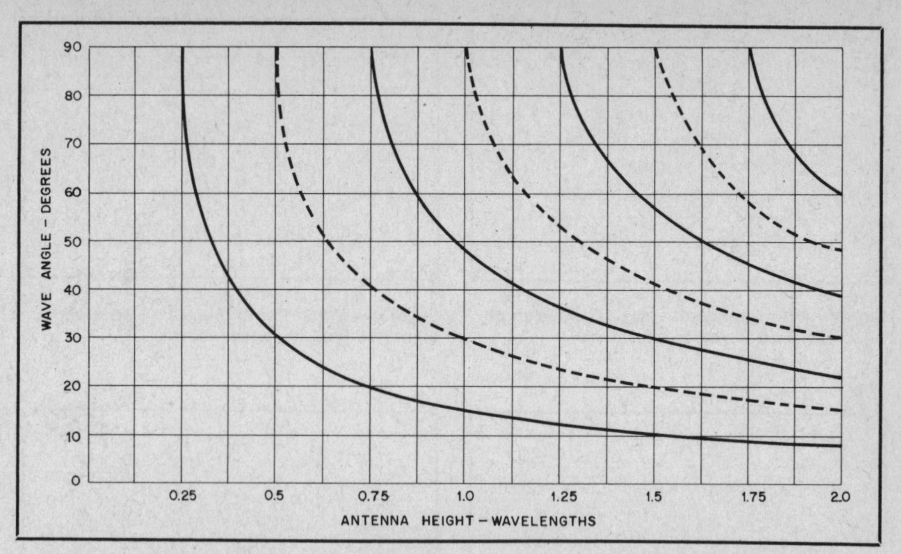

Fig. 37 — Angles at which nulls and maxima (factor = 6 dB) in the ground-reflection factor appear for antenna heights up to two wavelengths. The solid lines are maxima, dashed lines nulls, for all horizontal antennas. For example, if it is desired to have the ground reflection give maximum reinforcement of the direct ray from a horizontal antenna at a 20-degree wave angle (angle of radiation) the antenna height should be 0.75 wavelength. The same height will give a null at 42 degrees and a second maximum at 90 degrees., Values may also be determined from the trigonometric relationship $\theta$ = arc sin (A/4h), where $\theta$ is the wave angle and h is the antenna height expressed in wavelengths. For the first maxima, A has a value of 1; for the first null A has a value of 2, for the second maxima 3, for the second null 4, and so on.

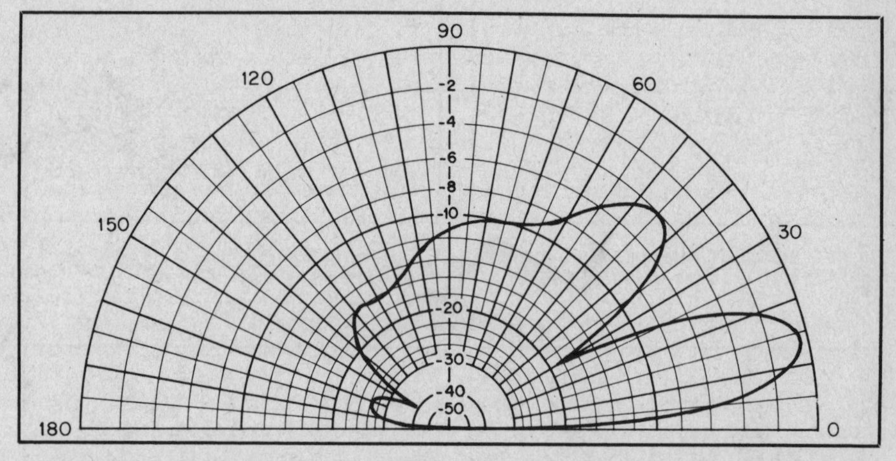

Fig. 38 — Calculated resultant pattern when the antenna producing the free-space pattern of Fig. 36A is placed one wavelength above lossy earth. At mid-hf, 10 to 20 MHz, a typical measured vertical radiation pattern will more nearly resemble this drawing than that of Fig. 36B. Calculations were made by assuming that reflection from the earth was near 100% at grazing angles, decreasing with the square of the cosine function of the angle to zero at 90°.

this factor is near zero, so the apparent advantage of the vertical antenna for very low-angle radiation is not realized in practice at the higher frequencies.

The "effective reflecting plane" of the ground — that is, the surface from which the reflection is considered to take place at the heights given in the charts — seldom coincides with the actual surface of the ground. Usually it will be found that this plane appears to be a few feet below the surface; in other words, the height of the antenna taken for purposes of estimating reflection is a few feet more than the actual height of the antenna. A great deal depends on the character of the ground, and in some cases the reflecting plane may be "buried" a surprising distance. Thus in some instances the charts will not give an accurate indication of the effect of reflection. On the average, however, they will give a reasonably satisfactory representation of reflection effects, with the qualifications with respect to high frequencies and low angles mentioned above.

In general, the effects of placing the antenna over real earth, rather than a perfect conductor, are to decrease the magnitude of the lobes of the pattern and to fill in the nulls. These effects are shown in Fig. 38.

In the vhf and uhf region (starting in the vicinity of the 28-MHz band) a different situation exists. At these frequencies little, if any, use is made of the part of the wave that travels in contact with the ground. The antennas, both transmitting and receiving, usually are rather high in terms of wavelength. The wave that is actually used — at least for line-of-sight communication — is in most cases several wavelengths above the surface of the ground. At such a height there is no consequential loss of energy; the direct ray travels from the transmitter to the receiver with only the normal attenuation caused by spreading, as explained in Chapter 1. The loss of energy in the reflected ray is beneficial rather than otherwise, as also explained in that chapter. The net result is that radiation at very low angles is quite practicable in this frequency region. Also, there is little practical difference between horizontal and vertical polarization.

## GROUND REFLECTION AND RADIATION RESISTANCE

Waves radiated from the antenna directly downward reflect vertically from the ground and, in passing the antenna on their upward journey induce a current into it. The magnitude and phase of this induced current depends upon the height of the antenna above the reflecting surface.

The total current in the antenna thus consists of two components. The amplitude of the first is determined by the power supplied by the transmitter and the free-space radiation resistance of the antenna. The second component is induced in the antenna by the wave reflected from the ground. The second component, while considerably smaller than the first at most useful antenna heights, is by no means inappreciable. At some heights the two components will be in phase, so the total current is larger than would be expected from the free-space radiation resistance. At other heights the two components are out of phase, and at such heights the total current is the difference between the two components.

Thus, merely changing the height of the antenna above ground will change the amount of current flow, assuming that the power input to the antenna is held constant. A higher current at the same value of power means that the effective resistance of the antenna is lower, and vice versa. In other words, the radiation resistance of the antenna is affected by the height of the antenna above ground. Fig. 39 shows the way in which the radiation resistance of a horizontal half-wave antenna varies with height, in terms of wavelengths, over perfectly conducting ground. Over actual ground the variations will be smaller, and tend to become negligible as the height approaches a half wavelength. The antenna on which this chart is based is assumed to have an infinitely thin conductor, and thus has a somewhat higher free-space value of radiation resistance (73 ohms) than an antenna constructed of wire or tubing. (See Fig. 6.)

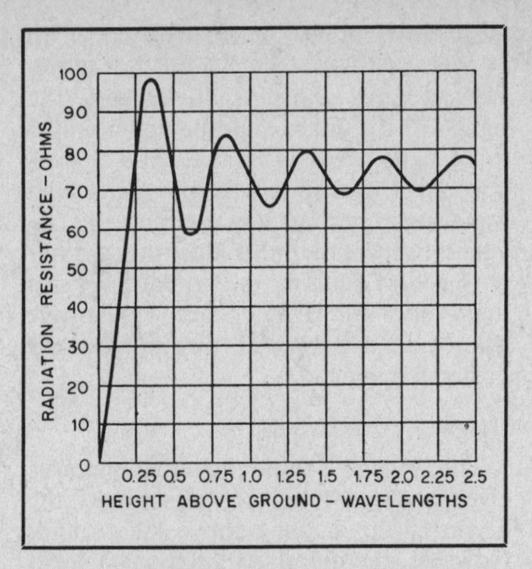


Fig. 39 — Variation in radiation resistance of a horizontal half-wave antenna with height above perfectly conducting ground.

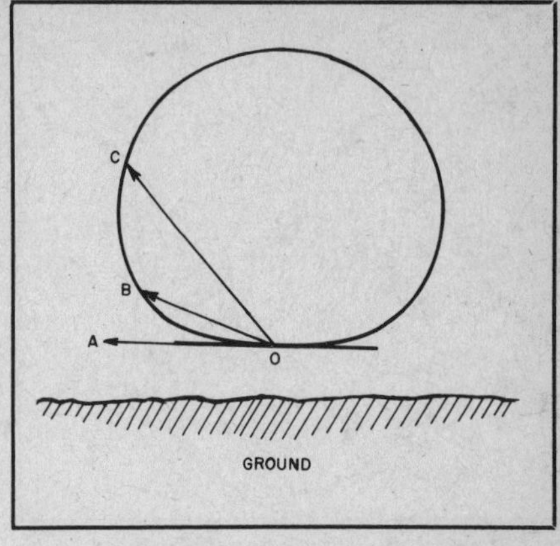


Fig. 40 — The effective directive pattern of the antenna depends upon the angle of radiation considered. As shown by the arrows, the field strength in a given compass direction will be quite different at different vertical angles.

### Ground Screens

The effect of a perfectly conducting ground can be simulated under the antenna by installing a metal screen or mesh (such as chicken wire or hardware cloth) near or on the surface of the ground. The screen should extend at least a half wavelength in every direction from the antenna. Such a screen will effectively establish the height of the antenna insofar as radiation resistance is concerned, since it substitutes for the actual earth underneath the antenna.

For vertical quarter-wave antennas the screen also reduces losses in the ground near the antenna, since if the screen conductors are solidly bonded to each other the resistance is much lower than the resistance of the ground itself. With other types of antennas — e.g., horizontal — at heights of a quarter wavelength or more, losses in the ground beneath the antenna are much less important.

Ground screens will affect only the very high-angle rays from horizontal antennas, and will not appreciably modify the effect of the earth itself at the lower radiation angles which ordinarily are used for long-distance communication. At low wave angles, the vertical radiation pattern is affected by ground conditions out to distances as great as 15 to 20 wavelengths, making it impractical to use ground screening for improvement.

## DIRECTIVE DIAGRAMS AND THE WAVE ANGLE

In the discussion of radiation patterns or directive diagrams of antennas, it was brought out that such patterns are always three-dimensional affairs, and that it is difficult to show, on a plane sheet of paper, more than a cross section of the solid pattern. The cross sections usually selected are the E plane, that plane containing the wire axis, and the H plane, that plane perpendicular to the wire axis.

If the antenna is horizontal, the E-plane pattern represents the horizontal or azimuthal radiation pattern of the antenna when the wave leaves the antenna (or arrives) at a zero angle of elevation above the earth. In the case of the vertical antenna, the horizontal radiation pattern at zero wave angle is given by the H-plane pattern.

However, with two exceptions — surface waves at low frequencies and space waves at vhf and higher — the wave angle used for communication is not zero. What we are interested in, then, is the directive pattern of the antenna at a wave angle which is of value in communications.

### Effective Directive Diagrams

The directive diagram for a wave angle of zero elevation (purely horizontal radiation) does not give an accurate indication of the directive properties of a horizontal antenna at wave angles above zero. For example, consider the half-wave dipole pattern in Fig. 13. It shows that there is no radiation directly in line with the antenna itself, and this is true at zero wave angle. However, if the antenna is horizontal and some wave angle other than zero is considered, *it is not true at all.*

The reason why will become clear on inspection of Fig. 40, which shows a horizontal half-wave antenna with a cross section of its free-space radiation pattern, cut by a plane that is vertical with respect to the earth and which contains the axis of the antenna conductor. In other words, the view is looking broadside at the antenna wire. (For the moment, reflections from the ground are neglected.) The lines OA, OB and OC all point in the same *geographical* direction (the direction in which the wire itself points), but make different angles with the antenna in the vertical plane. In other words, they correspond to different wave angles or angles of radiation, with all three rays aimed along the same line on the earth's surface. So far as compass directions are concerned, *all three waves are leaving the end of the antenna.*

The purely horizontal wave OA has zero amplitude, but at a somewhat higher angle corresponding to the line OB the field strength is appreciable. At a still higher angle corresponding to the line OC the field strength is still greater. In this particular pattern, the higher the wave angle the greater the field strength in the compass direction OA. It should be obvious that it is necessary, in plotting a directive diagram that purports to show the behavior of the antenna in different compass directions, to specify the angle of radiation for which the diagram applies. When the antenna is horizontal the shape of the diagram will be altered considerably as the wave angle is changed.

As described in Chapter 1, the wave angles that are useful depend on two things — the distance over which communication is to be carried on, and the height of the ionospheric layer that does the reflecting. Whether the E or F2 layer (or a combination of the two) will be used depends on the operating frequency, the time of day, season, and the sunspot cycle. The same half-wave antenna,

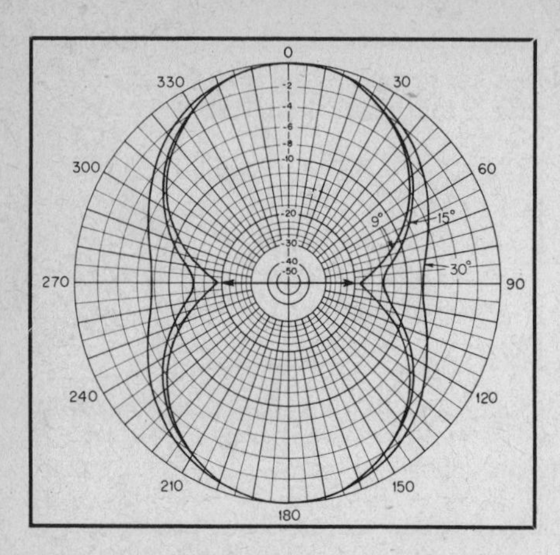


Fig. 41 — Directive patterns of a horizontal half-wave antenna at three radiation angles, 9, 15 and 30 degrees. The direction of the antenna itself is shown by the arrow. All three patterns are plotted to the same maximum, but the actual amplitudes at the various angles will depend upon the antenna height, as described in the text. The patterns shown here indicate only the shape of the directive diagram as the angle is varied.

operating on the same frequency, may be almost nondirectional for distances of a few hundred miles but will give substantially better results broadside than off its ends at distances of the order of 1000-1200 miles during the day when transmission is by the E layer. In the evening, when the F layer takes over, the directivity may be fairly well marked at long distances and not at all pronounced at 1000 miles or less. From this it might seem that it would be impossible to predict qualifications. However, it is possible to get a very good idea of the directivity by choosing a few angles that, on the average, are representative for different types of work. With patterns for such angles available it is fairly simple to interpolate for intermediate angles. Combined with some knowledge of the behavior of the ionosphere, a fairly good estimate of the directive characteristics of a particular antenna can be made for the particular time of day and distance of interest.

In the directive patterns given in Fig. 41, the wave angles considered are 9, 15 and 30 degrees. These represent, respectively, the median values of a range of angles that have been found to be effective for communication at 28, 21, 14 and 7 MHz. Because of the variable nature of ionospheric propagation, the patterns should not be considered to be more than general guides to the sort of directivity to be expected.

Since one S unit on the signal-strength scale is roughly 5 or 6 dB, it is easy to get an approximate idea of the operation of the antenna. For example, off the ends of a half-wave antenna the signal can be expected to be "down" between 2 and 3 S units compared with its strength at right angles or broadside to the antenna, at a wave angle of 15 degrees. This would be fairly representative of its performance on 14 MHz at distances of 500 miles or more. With a wave angle of 30 degrees, the signal off the ends would be down only 1 to 2 S units, while at an angle of 9 degrees it would be down 3 to 4 S units. Since high wave angles become less useful as the frequency is increased, this illustrates the importance of running the antenna wire in the proper direction if best results are wanted in a particular direction at the higher frequencies.

### Height Above Ground

The *shapes* of the directive patterns given in Fig. 41 are not affected by the height of the antenna above the ground. However, the *amplitude* relationships between the patterns of a given antenna for various wave angles are modified by the height. In the figure, the scale is such that the same field intensity is assumed in the direction of maximum radiation, regardless of the wave angle. To make best use of the patterns, the effect of the ground-reflection factor should also be included.

Assume a horizontal half-wave antenna is placed a half wavelength above perfectly conducting ground. The graph of the ground-reflection factors for this height is given in Fig. 27. For angles of 9, 15 and 30 degrees the values of the factor as read from the curve (with 6 dB added) are −0.5, 3.2 and 6.0 dB, respectively. These factors are applied to field strength. For convenience, take the 9° angle as reference. Then at a wave angle of 15° the field strength will be 3.7 dB greater than the field strength at 9°, *in any compass direction,* and at a wave angle of 30° will be 6.5 dB greater than the field strength at 9°, in any compass direction. To put it another way, at a wave angle of 30 degrees the antenna is about 1 to 1-1/2 S units better than it is at 9 degrees. There is about a half S-unit difference between 9 and 15 degrees, and between 15 and 30 degrees. If we wished, we could subtract 2.8 dB from every point on the 15-degree graph in Fig. 41, and 6.5 dB from every point on the 9-degree graph, and thus show graphically the comparison in amplitude as well as shape of the directive pattern at the three angles. This has been done in the graph of Fig. 42. However, it is generally unnecessary to take the trouble to draw separate graphs because it is so easy to add or subtract the requisite number of decibels as based on the appropriate ground-reflection factor.

It should be emphasized again that the patterns are based on idealized conditions not realized over actual ground. Nevertheless, they are useful in indicating about what *order* of effect to expect.

## RADIATION RESISTANCE AND GAIN

The field strength produced at a distant point by a given antenna system is directly proportional to the current flowing in the antenna. In turn, the amount of current that will flow, when a fixed amount of power is applied, will be inversely proportional to the square root of the radiation resistance. Lowering the radiation resistance will increase the field strength and raising the radiation resistance will decrease it. This is not to be interpreted broadly as meaning that a low value of radiation resistance is good and a high value is bad, regardless of circumstances. That is far from the actual case. What it means is that with an antenna of given dimensions, a change that reduces the radiation resistance *in the right way* will be accompanied by a change in the directive pattern that in turn will increase the field strength in some directions at the expense of reduced field strength in other directions. This principle is used in certain types of directive systems described in detail in Chapter 6.

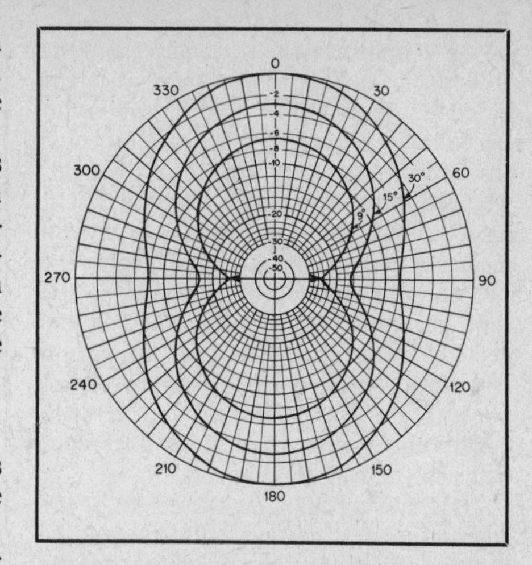


Fig. 42 — These diagrams compare the relative amplitudes of radiation at wave angles of 9, 15 and 30 degrees from a horizontal half-wave antenna when the height is 1/2 wavelength above a perfect conductor.

The shape of the directional pattern in the vertical plane is, as previously described, modified by the height of the antenna above ground. The effect of height on radiation resistance has been shown in Fig. 39 for the horizontal half-wave dipole. The plots of ground-reflection factors shown in Figs. 24 to 35, inclusive, show the actual shape of the H-plane pattern of such a half-wave dipole. That is, they show the variation in intensity with wave angle in the direction broadside to the antenna. In an approximate way, the radiation resistance is smaller as the area of the pattern is less, as may be seen by comparing the ground reflective patterns with the curve of Fig. 39.

Varying the height of a horizontal half-wave antenna while the power input is held constant will cause the current in the antenna to vary as its radiation resistance changes. Under the idealized conditions represented in Fig. 39 (an infinitely thin conductor over perfectly conducting ground) the field intensity at the optimum wave angle for each height will vary as shown in Fig. 43. In this figure the relative field intensity is expressed in decibels, using the field when the radiation resistance is 73 ohms as a reference (0 dB). From this cause alone, there is a gain of about 1 dB when the antenna height is 5/8 wavelength as compared with either 1/2 or 3/4 wavelength.

The gain or loss from the change in radiation resistance should be combined with the reflection factor for the particular wave angle and antenna height considered, in judging the overall effect of height on performance. For example, Fig. 44 shows the reflection factor, plotted in decibels, for a wave angle of 15 degrees (solid curve). This curve is based on data from Figs. 24 to 35, inclusive. Taken alone, it would indicate that a height of slightly less than 1 wavelength is optimum for this wave angle. However, when the values taken from the curve of Fig. 43 are added, the broken curve results. Because of the change in radiation resistance, there is a maximum near a height of 5/8 wavelength that is very nearly as good as the next maximum at a height of 1 to 1-1/4 wavelength. The change in radiation resistance also has the effect of steepening the curve at the lower heights and flattening it in the optimum region. Thus it would be expected that, for this wave angle, increasing the height of a half-wave dipole is very much worthwhile up to about 5/8 wavelength, but that further increases would not result in any material improvement. At 14 MHz, where a 15-degree wave angle is taken to be average, 5/8 wavelength is about 45 feet.

There is, of course, some difficulty in applying the information obtained in this fashion because of the uncertainty as to just where the ground plane is. One possibility, if the antenna can be raised and lowered conveniently, is to measure the current in it while changing its height, keeping the power input constant. Starting with low heights, the current should first go through a minimum (at a theoretical height of about 3/8 wavelength) and then increase to a maximum as the height is increased. The height at which this maximum is obtained is the optimum.

It should be kept in mind that no one wave angle does all the work. Designing for optimum results under average conditions does not mean that best results will be secured for all types of work and under all conditions. For long-distance work, for example, it is best to try for the lowest possible angle — 10 degrees or less is better for multihop propagation at 14 MHz, for example. However, an antenna that radiates well at such low angles may not be as good for work over shorter distances as one having a broader lobe in the vertical plane.

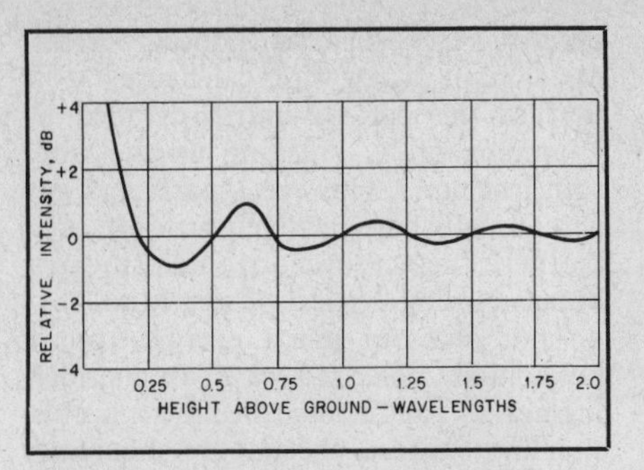


Fig. 43 — Gain or loss in decibels because of change in antenna current with radiation resistance, for fixed power input. Perfectly conducting ground is assumed.

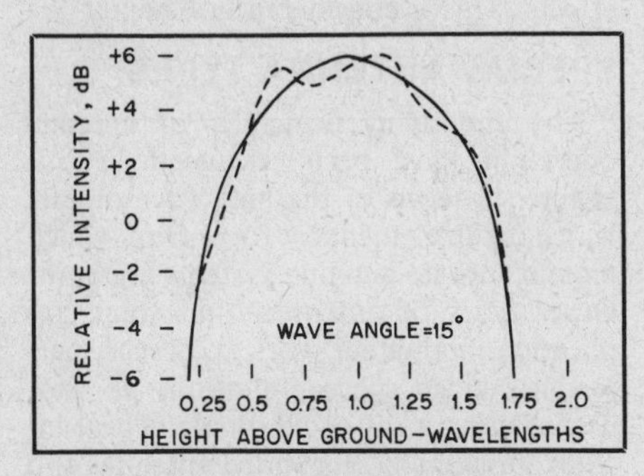


Fig. 44 — Solid curve, relative intensity vs. height at a wave angle of 15 degrees, because of reflection from perfect ground. Broken curve, height and effect of change in radiation resistance (Fig. 43) combined.

The effect of radiation resistance is somewhat more marked at the lower frequencies. To cover a distance of 350 miles at night (F-layer propagation) requires a wave angle of 60 degrees. As shown by Fig. 37, optimum antenna height for this wave angle is somewhat above 1/4 wavelength. However, it is in the region below 1/4 wavelength that the radiation resistance decreases most rapidly. At a height of 1/8 wavelength there is a gain of 3.5 dB over a height of 1/4 wavelength (Fig. 43) because of lowered radiation resistance. To offset this, the ground-reflection factor for a wave angle of 60 degrees is about 2 dB at 1/8 wavelength (Fig. 24) as compared with not quite 6 dB for 1/4 wavelength (Fig. 25); this is a loss of 4 dB. There is thus a difference of only 1-1/2 dB, which is not observable, between 1/8 and 1/4 wavelength. At 3.5 MHz, this is a considerable difference in actual height, since 1/8 $\lambda$ is about 35 feet and 1/4 $\lambda$ is about 70 feet. For *short*-distance work the cost of the supports required for the greater height would not be justified.

Information on the variation in radiation resistance with height for antenna types other than the half-wave dipole is not readily available. A harmonic antenna can be expected to show such variations, but in general an antenna system that tends to minimize the radiation directly toward the ground under the antenna can be expected to have a lesser order of variation in radiation resistance with height than is the case with the half-wave dipole.

## SOME PRACTICAL CONSIDERATIONS

It is stressed that the results from a practical antenna cannot be expected to be exactly according to the theoretical performance outlined in this chapter. The theory that leads to the impedances, radiation patterns, and power-gain figures discussed is necessarily based on idealized assumptions that cannot be exactly realized, although they may be approached, in practice.

The effect of imperfectly conducting earth has been mentioned several times. It will cause the actual radiation resistance of an antenna to differ somewhat from the theoretical figure at a given height. In addition, there is the effect of the length/diameter ratio of the conductor to be considered. Nevertheless, the theoretical figure will approximate the actual radiation resistance closely enough for most practical work. The value of radiation resistance is of principal importance in determining the proper method for feeding power to the antenna through a transmission line, and a variation of 10 or even 20 percent will not be serious. Adjustments can easily be made to compensate for the discrepancy between practice and theory.

So far as radiation patterns are concerned, the effect of imperfect earth is to decrease the amplitude of the reflected ray and to introduce some phase shift on reflection. The phase shift is generally small with horizontal polarization. Both effects combine to make the maximum reflection factor somewhat less than 6 dB, and to prevent complete cancellation of radiation in the nulls in the theoretical patterns, as shown in Fig. 38. There may also be a slight change in the wave angle at which maximum reinforcement occurs, as a result of the phase shift.

Finally, the effect of nearby conductors and dielectrics cannot, of necessity, be included in the theoretical patterns. Conductors such as power and telephone lines, house wiring, piping, etc., close to the antenna can cause considerable distortion of the pattern if currents of appreciable magnitude are induced in them. Under similar conditions they can also have a marked effect on the radiation resistance. Poor dielectrics such as green foliage near the antenna can introduce loss, and may make a noticeable difference between summer and winter performance.

The directional effects of an antenna will conform more closely to theory if the antenna is located in clear space, at least a half wavelength from anything that might affect its properties. In cities, it may be difficult to find such a space at low frequencies. The worst condition arises when nearby wires or piping happen to be

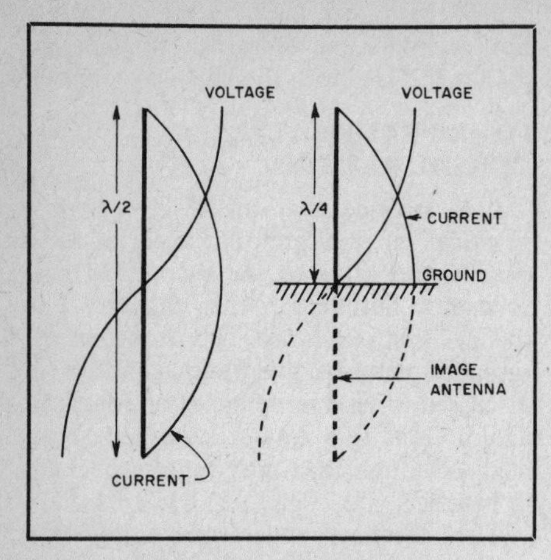


**Fig. 45 — The half-wave antenna and its grounded quarter-wave counterpart. The missing quarter wavelength can be considered to be supplied by the image in ground of good conductivity.**

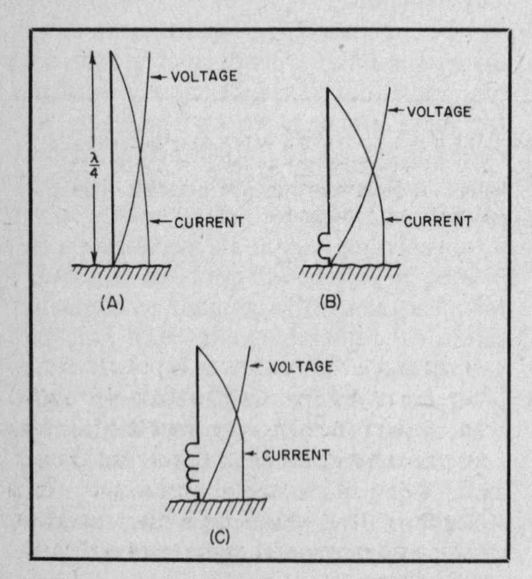


**Fig. 46 — Current and voltage distribution on a grounded quarter-wave antenna (A) and on successively shorter antennas loaded to resonate at the same frequency.**

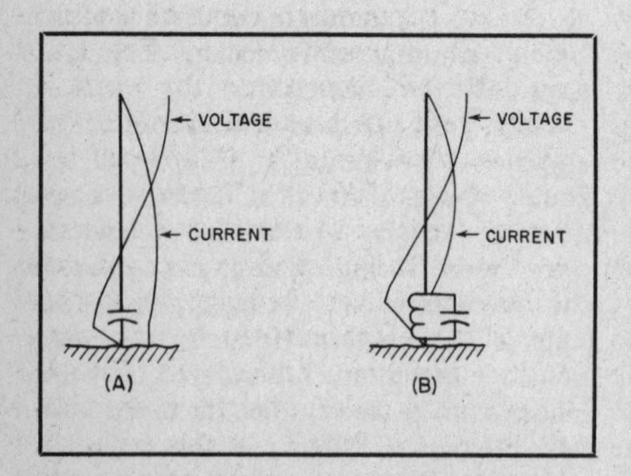


**Fig. 47 — Current and voltage distribution on grounded antennas longer than 1/4 wavelength. (A) between 1/4 and 3/8 wave, approximately; (B) half wave.**

resonant, or nearly so, at the operating frequency. Such resonances often can be destroyed by bonding pipes or BX coverings at trial points, checking with a diode-detector wavemeter or rf current probe to determine the measures necessary to reduce the induced current. Metal masts or guy wires can cause distortion of the pattern unless detuned by grounding or by breaking up the wires with insulators. However, masts and guy wires usually have relatively little effect on the performance of horizontal antennas because, being vertical or nearly so, they do not pick up much energy from a horizontally polarized wave. In considering nearby conductors, too, the transmission line that feeds the antenna should not be overlooked. Under some conditions that are rather typical with amateur antennas, currents will be induced in the line by the antenna, leading to some undesirable effects. This is considered in Chapter 5.

## SPECIAL ANTENNA TYPES

The underlying principles of antenna operation have been discussed in this chapter in terms of the half-wave dipole, which is the elementary form from which more elaborate antenna systems are built. Other types of antennas find some application in amateur work, however, particularly when space limitations do not permit using a full-sized dipole. These include, principally, grounded antennas and loops.

### THE GROUNDED ANTENNA

In cases where vertical polarization is required — for example, when a low wave angle is desired at frequencies below 4 MHz — the antenna must be vertical. At low frequencies the height of a vertical half-wave antenna would be beyond the constructional reach of most amateurs. A 3.5-MHz vertical half wave would be 133 feet high, for instance.

However, if the lower end of the antenna is grounded it need be only a quarter wave high to resonate at the same frequency as an ungrounded half-wave antenna. The operation can be understood when it is remembered that ground having high conductivity acts as an electrical mirror, and the missing half of the antenna is supplied by the mirror image. This is shown in Fig. 45.

The directional characteristic of a grounded quarter-wave antenna will be the same as that of a half-wave antenna in free space. Thus, a vertical grounded quarter-wave antenna will have a circular radiation pattern in the horizontal or azimuthal plane. In the vertical plane the radiation will decrease from maximum along the ground to zero directly overhead.

The grounded antenna may be much smaller than a quarter wavelength and still be made resonant by "loading" it with inductance at the base, as in Fig. 46 at B and C. By adjusting the inductance of the loading coil, even very short wires can be tuned to resonance.

The current along a grounded quarter-wave vertical wire varies practically sinusoidally, as is the case with a half-

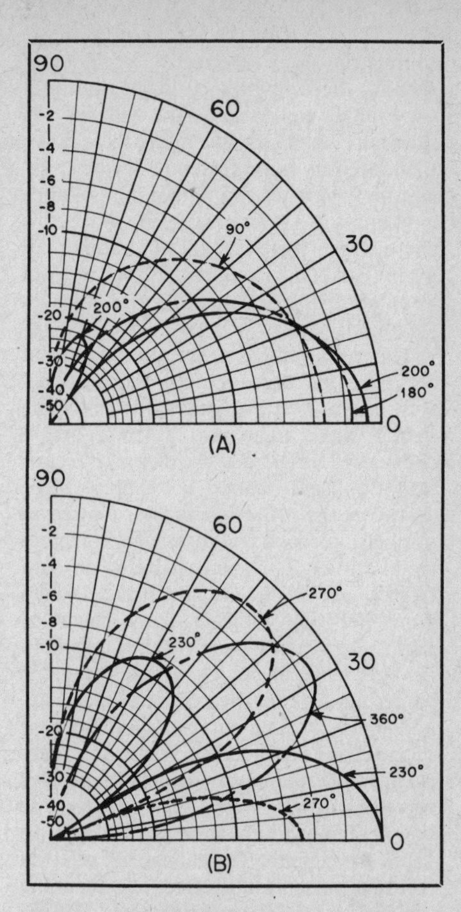


**Fig. 48 — Vertical-plane radiation patterns of vertical antennas for several values of antenna height, *ℓ*. The amplitude responses in both A and B are to the same scale, relative to the peak response of the 230° (0.64 λ) vertical plotted in B. A perfect conductor beneath the antennas and zero loss resistances are assumed.**

wave wire, and is highest at the ground connection. The rf voltage, however, is highest at the open end and minimum at the ground. The current and voltage distribution are shown in Fig. 46A. When the antenna is shorter than a quarter wave but is loaded to resonance, the current and voltage distribution are part sine waves along the antenna wire. If the loading coil is substantially free from distributed capacitance, the voltage across it will increase uniformly from minimum at the ground, as shown at B and C, while the current will be the same throughout.

Extremely short antennas are used, of necessity, in mobile work on the lower frequencies such as the 3.5-MHz band. These may be "base loaded" as shown at B and C in Fig. 46, but there is a small advantage to be realized by placing the loading coil at the center of the antenna. In neither case, however, is the current uniform throughout the coil, since the inductance required is so large that the coil tends to act as a linear circuit rather than as a "lumped" inductance.

If the antenna height is greater than a quarter wave but less than a half wave, the antenna shows inductive reactance at its

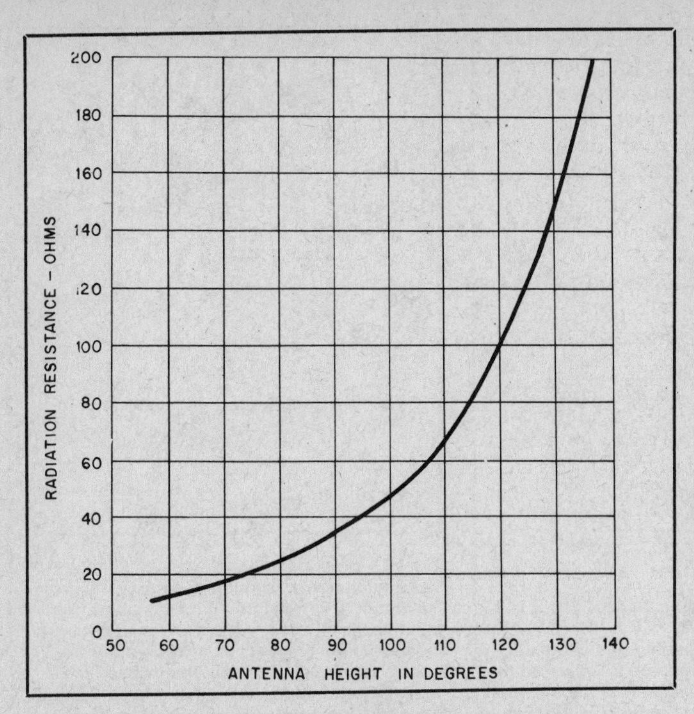


Fig. 49 — Radiation resistance vs. free-space antenna height in electrical degrees for a vertical antenna over perfectly conducting ground, or over a highly conducting ground plane. This curve also may be used for center-fed antennas (in free space) by multiplying the radiation resistance by two; the height in this case is half the actual antenna length.

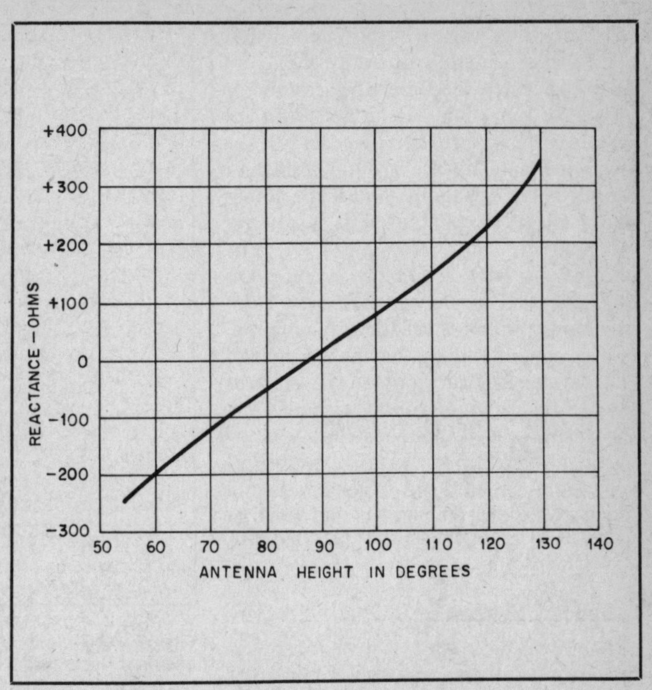


Fig. 51 — Approximate reactance of a vertical antenna over perfectly conducting ground and having a length/diameter ratio of about 1000. Actual values will vary considerably with length/diameter ratio. The remarks under Fig. 49 also apply to this curve.

terminals and can be tuned to resonance by means of a capacitance of the proper value. This is shown in Fig. 47A. As the length is increased progressively from 1/4 to 1/2 wavelength the current loop moves up the antenna, always being at a point 1/4 wavelength from the top. When the height is 1/2 wavelength the current distribution is as shown at B in Fig. 47. There is a voltage loop (current node) at the base, and power can be applied to the antenna through a parallel-tuned circuit, resonant at the same frequency as the antenna, as shown in the figure.

Up to a little more than 1/2 wavelength, increasing the height compresses the directive pattern in the vertical plane; this results in an increase in field strength for a given power input at the very low radiation angles. The theoretical improvement is about 1.7 dB for a half-wave antenna when compared with a quarter-wave antenna, as shown in Fig. 48A.

When the height of a vertical antenna is increased beyond a half wavelength, secondary lobes appear in the pattern, Fig. 48A and B. These become major lobes at relatively high angles when the length approaches 3/4 wavelength. At one wavelength the low-angle lobe disappears and a single lobe remains at approximately 35°.

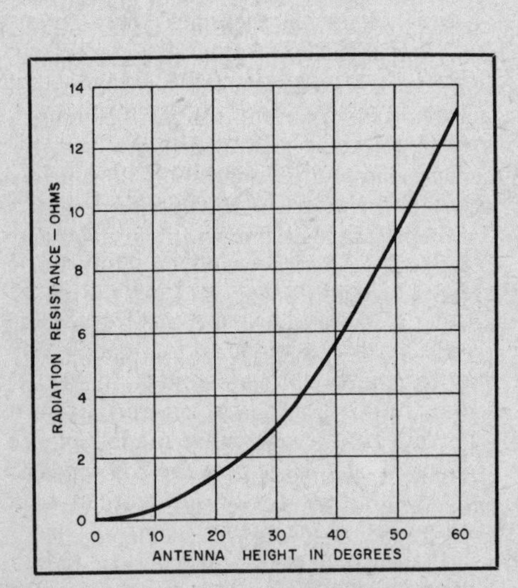


Fig. 50 — Same as Fig. 49, for heights below 60 degrees.

### Radiation Resistance

The radiation resistance of a grounded vertical antenna, as measured between the base of the antenna and ground, varies as shown in Fig. 49 as a function of the antenna height. The word "height" as used in this connection has the same meaning as "length" as applied to a horizontal antenna. This curve is for an antenna based on (but not directly connected to) ground of perfect conductivity. The height is given in electrical degrees, the 60 to 135 degree range shown corresponding to heights from 1/6 to 3/8 wavelength. The antenna is approximately self-resonant at a height of 90 degrees (1/4 wavelength). The actual resonant length will be somewhat less because of the length/diameter ratio mentioned earlier.

In the range of heights covered by Fig. 49 the radiation resistance is practically independent of the length/diameter ratio. At greater heights the length/diameter ratio is important in determining the actual value of radiation resistance. At a height of 1/2 wavelength the radiation resistance may be as high as several thousand ohms.

The variation in radiation resistance with heights below 60 degrees is shown in Fig. 50. The values in this range are very low.

A very approximate curve of reactance vs. height is given in Fig. 51. The actual reactance will depend on the length/diameter ratio, so this curve should be used only as a rough guide. It is based on a ratio of about 1000 to 1. Thicker antennas can be expected to show lower reactance at a given height, and thinner antennas should show more. At heights below and above the range covered by the curve, large reactance values will be encountered, except for heights in the vicinity of 1/2 wavelength. In this region the reactance decreases, becoming zero when the antenna is resonant.

### Efficiency

The *efficiency* of the antenna is the ratio of the radiation resistance to the total resistance of the system. The total resistance includes radiation resistance, resistance in conductors and dielectrics, including the resistance of loading coils if used, and the resistance of the grounding system, usually referred to as "ground resistance."

It was stated earlier in this chapter that a half-wave antenna operates at a very high efficiency because the conductor resistance is negligible compared with the radiation resistance. In the case of the grounded antenna the ground resistance usually is not negligible, and if the antenna is short (compared with a quarter wavelength) the resistance of the necessary loading coil may become appreciable. To attain an efficiency comparable with that of a half-wave antenna, in a grounded antenna having a height of 1/4 wavelength or less, great care must be used to reduce both ground resistance and the resistance of any required loading inductors. Without a fairly elaborate grounding system, the efficiency is not likely to exceed 50 percent and may be much less, particularly at heights below 1/4 wavelength.

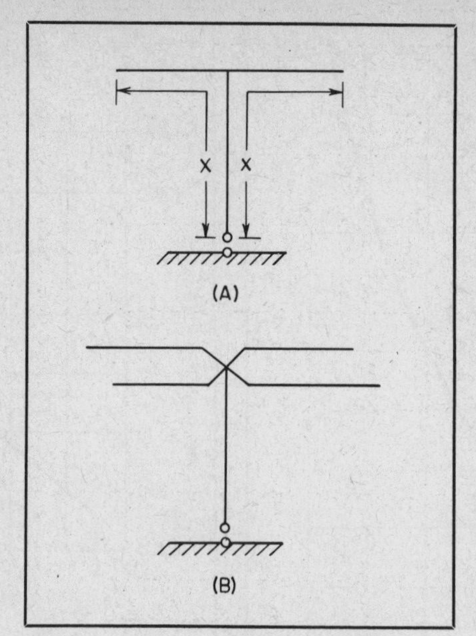


Fig. 52 — Simple top loading of a vertical antenna. The antenna terminals, indicated by the small circles, are the base of the antenna and ground, and should not be taken to include the length of any lead-ins or connecting wires.

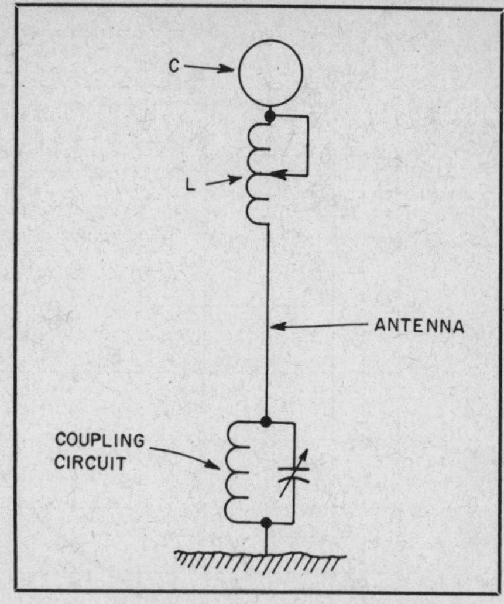


Fig. 53 — Top loading with lumped constants. The inductance, L, should be adjusted to give maximum field strength with constant power input to the antenna. A parallel-tuned circuit, independently resonant at the operating frequency, is required for coupling to the transmitter when the top loading is adjusted to bring a current node at the lower end of the antenna.

## Grounding Systems

The ideal grounding system for a vertical grounded antenna would consist of about 120 wires, each at least 1/2 wavelength long, extending radially from the base of the antenna and spaced equally around a circle. Such a system is the practical equivalent of perfectly conducting ground and has negligible resistance. The wires can either be laid directly on the surface of the ground or buried a few inches below.

Such a system would be practical in an amateur installation in very few cases. Unfortunately, the resistance increases rapidly when the number of radials is reduced, and if at all possible at least 15 radials should be used. Experimental measurements have shown that even with this number the resistance is such as to decrease the antenna efficiency to about 50 percent at a height of 1/4 wavelength.

It has also been found that as the number of radials is reduced the length required for optimum results with a particular number of radials also decreases; in other words, if only a small number of radials can be used there is no point in extending them out a half wavelength. With 15 radials, for example, a length of 1/8 wavelength is sufficient. With as few as two radials the length is almost unimportant, but the efficiency of a quarter-wave antenna with such a grounding system is only about 25 percent. It is considerably lower with shorter antennas.

In general, a large number of radials, even though some or all of them have to be short, is preferable to a few long radials. The conductor size is relatively unimportant, no. 12 to no. 28 copper wire being suitable.

The measurement of ground resistance at the operating frequency is difficult. The power loss in the ground depends on the current concentration near the base of the antenna, and this depends on the antenna height. Typical values for small radial systems (15 or less) have been measured to be from about 5 to 30 ohms, for antenna heights from 1/16 to 1/4 wavelength.

## Top Loading

Because of the difficulty of obtaining a really low-resistance ground system, it is always desirable to make a grounded vertical antenna as high as possible, since this increases the radiation resistance. (There is no point in going beyond a half wavelength, however, as the radiation resistance decreases with further increases in height.) At the low frequencies where a grounded antenna is generally used, the heights required for the realization of high radiation resistance usually are impractical for amateur work. The objective of design of vertical grounded antennas that are necessarily 1/4 wavelength or less high is to make the current loop come near the top of the antenna, and to keep the current as large as possible throughout the length of the vertical wire. This requires "top loading," which means replacing the missing height by some form of electrical circuit having the same characteristics as the missing part of the antenna, so far as energy traveling up to the end of the antenna is concerned.

One method of top loading is shown in Fig. 52. The vertical section of the antenna terminates in a "flat-top", which supplies a capacitance at the top into which current can flow. The simple single-conductor system shown at A is more readily visualized as a continuation of the antenna so that the dimension X is essentially the overall length of the antenna. If this dimension is a half wavelength, the resistance at the antenna terminals (indicated by the small circles, one being grounded) will be high. A disadvantage of this system is that the horizontal portion also radiates to some extent, although there is cancellation of radiation in the direction at right angles to the wire direction, since the currents in the two portions are flowing in opposite directions.

A multiwire system such as is shown in Fig. 52B will have more capacitance than the single-conductor arrangement, and thus will not need to be as long, for resonating at a given frequency, but requires extra supports for the additional wires. Ideally, an arrangement of this sort should be in the form of a cross, but parallel wires separated by several feet will give a considerable increase in capacitance over a single wire. With either system shown in Fig. 52, dimension X, the length from the base of the antenna along one conductor to the end, should be not more than one-half wavelength nor less than one-fourth wavelength.

Instead of a flat top, it is possible to use a simple vertical wire with concentrated capacitance and inductance at its top to simulate the effect of the missing length. The capacitance used is not the usual type of capacitor, which would be ineffective since the connection is one-sided, but consists of a metallic structure large enough to have the necessary self-capacitance, Fig. 53. Practically any sufficiently large metallic structure can be used for the purpose, but simple geometric forms such as the sphere, cylinder and disc are preferred because of the relative ease with which their capacitance can be calculated. The inductance may be the usual type of rf coil, with suitable protection from the weather.

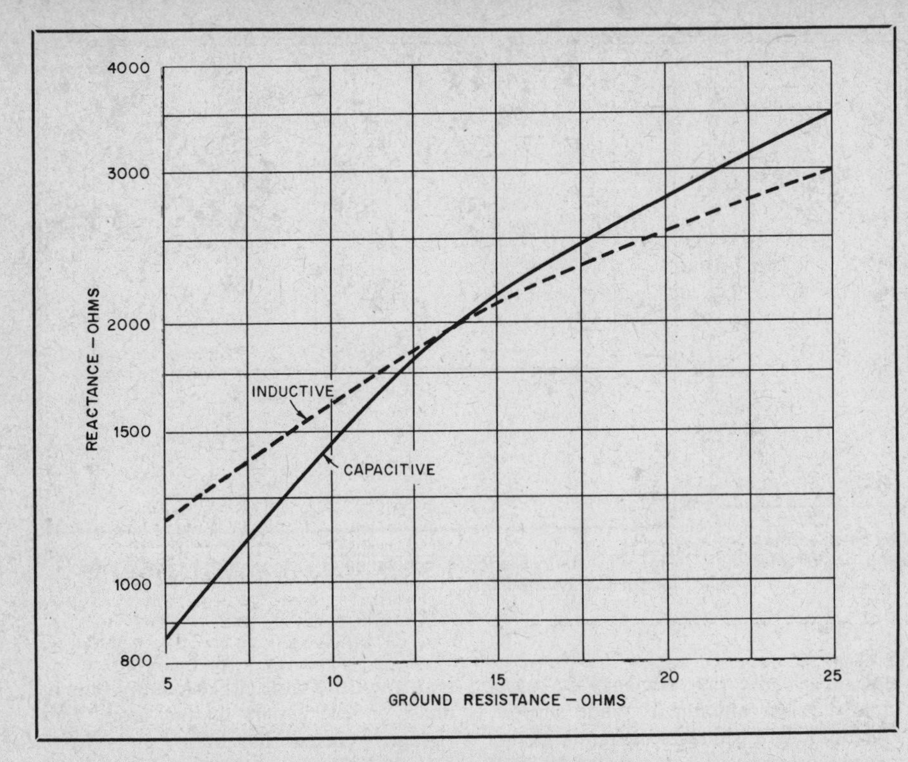


Fig. 54 — Inductive and capacitive reactance required for top loading a grounded antenna by the method shown in Fig. 53. The reactance values should be converted to inductance and capacitance, using the usual formulas, at the operating frequency.

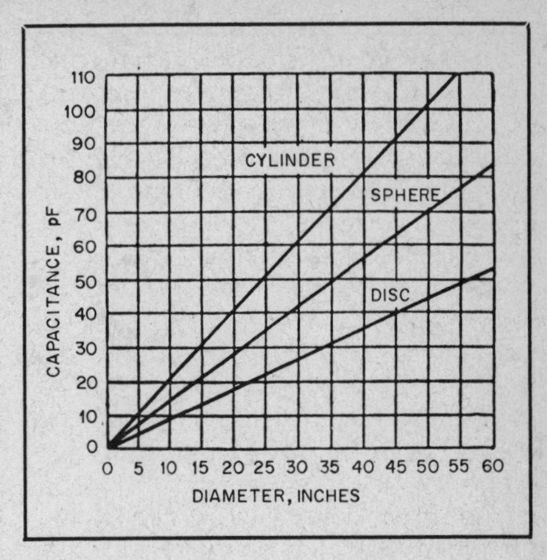


Fig. 55 — Capacitance of sphere, disc and cylinder as a function of their diameters. The cylinder length is assumed equal to its diameter.

The minimum value of capactive reactance required depends principally upon the ground resistance. Fig. 54 is a set of curves giving the reactances required under representative conditions. These curves are based on obtaining 75 percent of the maximum possible increase in field strength over an antenna of the same height without top loading, and apply with sufficient accuracy to all antenna heights. An inductance coil of reasonably low-loss construction is assumed. The general rule is to use as large a capacitance (low capacitive reactance) as the circumstances will permit, since an increase in capacitance will cause an improvement in the field strength. It is particularly important to do this when, as is usually the case, the ground resistance is not known and cannot be measured.

The capacitance of three geometric forms is shown by the curves of Fig. 55 as a function of their size. For the cylinder, the length is specified equal to the diameter. The sphere, disc and cylinder can be constructed from sheet metal, if such construction is feasible, but the capacitance will be practically the same in each case if a "skeleton" type of construction with screening or networks of wire or aluminum tubing is used.

## GROUND-PLANE ANTENNAS

Instead of being actually grounded, a 1/4-wave antenna can work against a simulated ground called a *ground plane*. Such a simulated ground can be formed from wires 1/4 wavelength long radiating from the base of the antenna, as shown in Fig. 56. It is obvious that with 1/4-wave radials the antenna and any one radial have a total length of 1/2 wavelength and therefore will be a resonant system. However, with only one radial the directive pattern would be that of a half-wave antenna bent into a right angle at the center; if one section is vertical and the other horizontal, this would result in equal components of horizontal and vertical polarization and a nonuniform pattern in the horizontal plane. This can be overcome by using a ground plane in the shape of a disc with a radius of 1/4 wavelength. The effect of the disc can be simulated, with simpler construction, by using at least four straight radials equally spaced around the circle, as indicated in the drawing.

The ground-plane antenna is widely used at vhf, for the purpose of establishing a "ground" for a vertical antenna mounted many wavelengths above actual ground. This prevents a metallic antenna support from carrying currents that tend to turn the system into the equivalent of a vertical long-wire antenna and thus raising the wave angle.

At frequencies of the order of 14 to 30 MHz, a ground plane of the type shown in Fig. 56 will permit using a quarter-wave vertical antenna (for nondirectional low-angle operation) at a height that will let the antenna be clear of its surroundings. Such short antennas mounted on the ground itself are frequently so surrounded by energy-absorbing structures and trees as to be rather ineffective. Since the quarter-wave radials are physically short at these frequencies, it is quite practical to mount the entire system on a roof top or pole. A ground plane at a height of a half wavelength or more closely approximates perfectly conducting earth, and the resistance curve of Fig. 49 applies with reasonable accuracy. The antenna itself may be any desirable height and does not have to be exactly a quarter wavelength. The radials, however, preferably should be quite close to a quarter wave in length.

A ground plane also can be of considerable benefit at still lower frequencies provided the radials and the base of the antenna can be a moderate height above the ground. In order to take over the function of the actual ground connection, the ground plane must be so disposed that the field of the antenna prefers to travel along the ground plane wires rather than letting it flow in lossy earth. This is particularly necessary where the current is greatest, i.e., close to the antenna. If the ground plane has to be near the earth, the number of wires should be increased, using as many as is practicable and spacing them as evenly as possible in a circle around the antenna. If the construction of a multiwire ground plane is impracticable, a better plan is to bury as many radials as possible in the ground as described earlier.

## SHORT ANTENNAS IN GENERAL

Although the discussions in this chapter have principally been in terms of self-resonant antennas, particularly those a half wavelength long, it would be a mistake to assume that there is anything sacred about resonance. The resonant length happens to be one that is convenient to analyze. At other lengths the directive properties will be different, the radiation resistance will be different, and the impedance looking into the terminals

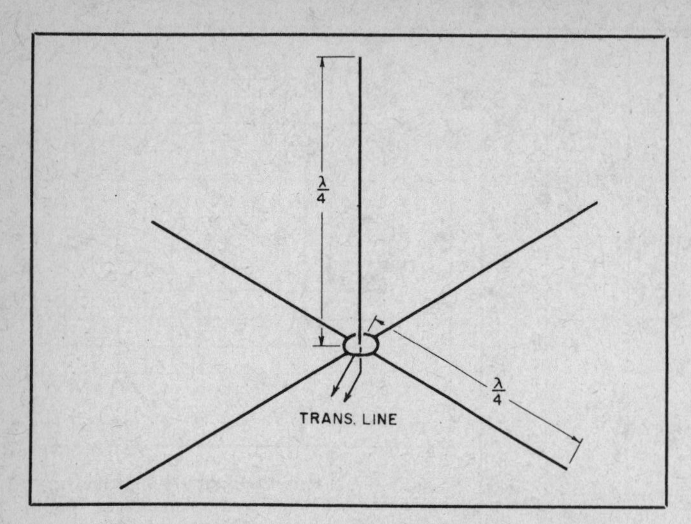


Fig. 56 — The ground-plane antenna. Power is applied between the base of the antenna and the center of the ground plane, as indicated in the drawing.

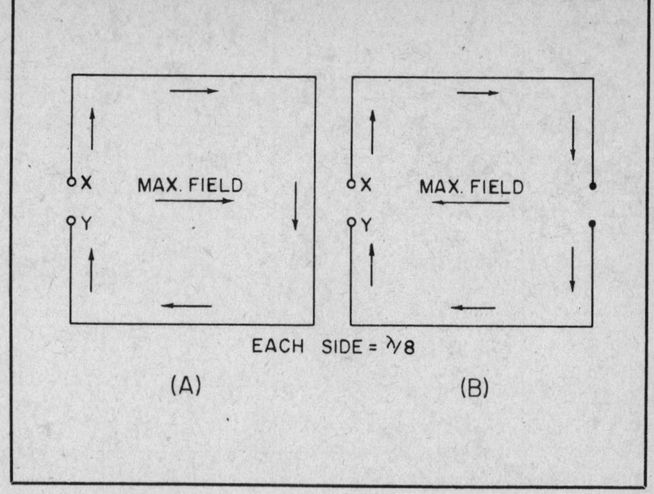


Fig. 57 — Half-wave loops, consisting of a single turn having a total length of 1/2 wavelength.

of the antenna will contain reactance as well as resistance. The reactance presents no problem, because it can easily be "tuned out" and the antenna *system* thereby resonated, even though the antenna itself is *not* resonant. The purpose of resonating either the antenna or the system as a whole is simply to facilitate feeding power to the antenna, and such resonating or tuning does not affect the antenna's radiating properties.

Physical conditions frequently make it necessary to use antennas shorter than a half wavelength. The directive pattern of a short antenna does not differ greatly from that of a half-wave antenna, and in the limit the pattern approaches that shown in Fig. 12. The difference in the field strength that this shift in pattern shape causes is negligible. The most important difference is the decrease in radiation resistance and its effect on the efficiency of the antenna. This has been discussed in the preceding section on grounded antennas. The curve of Figs. 49 and 50 can be used for any center-fed nongrounded antenna by using half the actual antenna length and multiplying the corresponding radiation resistance by two. For example, a center-fed dipole antenna having an actual length of 120 degrees (1/3 wavelength) has a half length of 60 degrees and a radiation resistance of about 13 ohms per side or 26 ohms for the whole antenna. The reactance, which will be capacitive and of the order of 400 ohms (Fig. 51, same technique as above), can be tuned out by a loading coil or coils. As described earlier, low-resistance coils must be used if the antenna efficiency is to be kept reasonably high. However, the ground resistance loss can be neglected in a horizontal center-fed antenna of this type if the height is a quarter wavelength or more.

A short antenna should not be made shorter than the physical circumstances require, because the efficiency decreases rapidly as the antenna is made shorter. For example, a center-fed antenna having an overall length of 1/4 wavelength (half length 45 degrees) will have a radiation resistance of $2 \times 7 = 14$ ohms, as shown by Fig. 50. Depending on the length/diameter ratio, the impedance probably will have a resistance of 3 to 6 ohms, so the probable efficiency will be 70 to 80 percent, or a loss of 1 to 1.5 dB. While this is not too bad, further shortening not only further decreases the radiation resistance but enters a length region where the reactance increases very rapidly, so that the coil resistance quickly becomes larger than the radiation resistance. Where the antenna must be short, a small length/diameter ratio (thick antenna) is definitely desirable as a means of keeping down the reactance and thus reducing the size of loading inductance required.

## LOOP ANTENNAS

A loop antenna is a closed-circuit antenna — that is, one in which a conductor is formed into one or more turns so that its two ends are close together. Loops can be divided into two general classes, those in which both the total conductor length and the maximum linear dimension of a turn are very small compared with the wavelength, and those in which both the conductor length and the loop dimensions begin to be comparable with the wavelength.

A "small" loop can be considered to be simply a rather large coil, and the current distribution in such a loop is the same as in a coil. That is, the current is in the same phase and has the same amplitude in every part of the loop. To meet this condition the total length of conductor in the loop must not exceed about 0.1 wavelength. Small loops are discussed further in Chapter 14.

A "large" loop is one in which the current is not the same either in amplitude or phase in every part of the loop. This change in current distribution gives rise to entirely different properties as compared with a small loop.

### Half-Wave Loops

The smallest size of "large" loop generally used is one having a conductor length of 1/2 wavelength. The conductor is usually formed into a square, as shown in Fig. 57, making each side 1/8 wavelength long. When fed at the center of one side the current flows in a closed loop as shown at A. The current distribution is approximately the same as on a half-wave wire, and so is maximum at the center of the side opposite the terminals X-Y, and minimum at the terminals themselves. This current distribution causes the field strength to be maximum in the plane of the loop and in the direction looking from the low-current side to the high-current side. If the side opposite the terminals is opened at the center as shown at B (strictly speaking, it is then no longer a loop because it is no longer a closed circuit) the direction of current flow remains unchanged but the maximum current flow occurs at the terminals. This reverses the direction of maximum radiation.

The radiation resistance at a current antinode (which is also the resistance at X-Y in Fig. 57B) is on the order of 50 ohms. The impedance at the terminals in A is a few thousand ohms. This can be reduced by using two identical loops side by side with a few inches spacing between them and applying power between terminal X on one loop and terminal Y on the other.

Unlike a half-wave dipole or a small loop, there is no direction in which the radiation from a loop of the type shown in Fig. 57 is zero. There is appreciable radiation in the direction perpendicular to the plane of the loop, as well as to the "rear"

— the opposite direction to the arrows shown. The front-to-back ratio is of the order of 4 to 6 dB. The small size and the shape of the directive pattern result in a loss of about 1 dB when the field strength in the optimum direction from such a loop is compared with the field from a half-wave dipole in its optimum direction.

The ratio of the forward radiation to the backward radiation can be increased and the field strength likewise increased at the same time to give a gain of about 1 dB over a dipole, by using inductive reactances to "load" the sides joining the front and back of the loop. This is shown in Fig. 58. The reactances, which should have a value of approximately 360 ohms, decrease the current in the sides in which they are inserted and increase it in the side having terminals. This increases the directivity and thus increases the efficiency of the loop as a radiator.

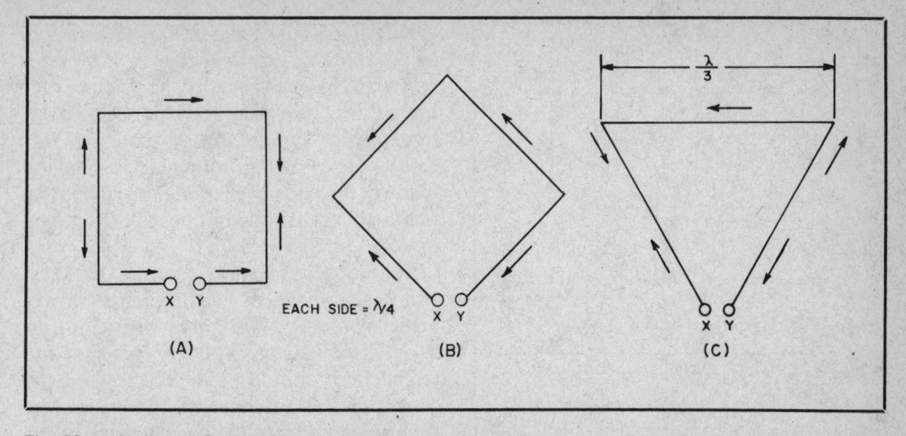


Fig. 59 — At A, and B, loops having sides 1/4 wavelength long, and at C having sides 1/3 wavelength long (total conductor length one wavelength). The polarization depends on the orientation of the loop and the point at which the terminals X-Y are located.

### One-Wavelength Loops

Loops in which the conductor length is one wavelength have different characteristics than half-wave loops. Three forms of one-wavelength loops are shown in Fig. 59. At A and B the sides of the squares are equal to 1/4 wavelength, the difference being in the point at which the terminals are inserted. At C the sides of the triangle are equal to 1/3 wavelength. The relative direction of current flow is as shown in the drawings. This direction reverses halfway around the perimeter of the loop, since such reversals always occur at the junction of each half-wave section of wire.

The directional characteristics of loops of this type are opposite in sense to those of a small loop. That is, the radiation is maximum perpendicular to the plane of

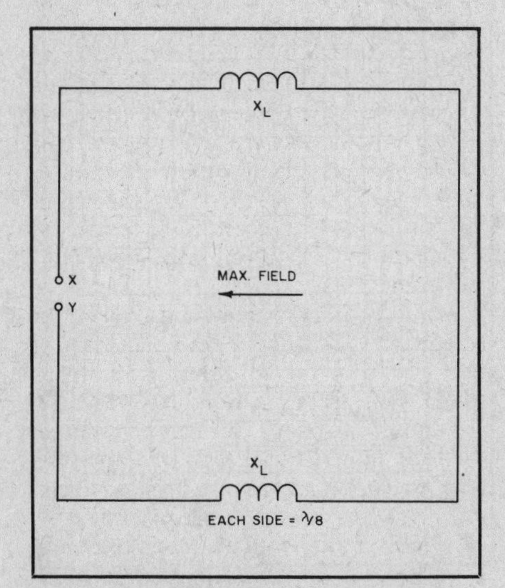


Fig. 58 — Inductive loading in the sides of a half-wave loop to increase the directivity and gain. Maximum radiation or response is in the plane of the loop in the direction shown by the arrow.

the loop and is minimum in any direction in the plane containing the loop. If the three loops shown in Fig. 59 are mounted in a vertical plane with the terminals at the bottom, the radiation is horizontally polarized. When the terminals are moved to the center of one vertical side in A, or to a side corner in B, the radiation is vertically polarized. If the terminals are moved to a side corner in C, the polarization will be diagonal, containing both vertical and horizontal components.

In contrast to straight-wire antennas, the electrical length of the circumference of a one-wavelength loop is *shorter* than the actual length. For loops made of wire and operating at frequencies below 30 MHz or so, where the ratio of conductor length to wire diameter is large, the loop will be close to resonance when

$$\text{length (ft)} = \frac{1005}{f_{MHz}}$$

or

$$\text{length (m)} = \frac{306.3}{f_{MHz}}$$

The radiation resistance of a resonant one-wavelength loop is approximately 100 ohms, when the ratio of conductor length to diameter is large. As the loop dimensions are comparable with those of a half-wave dipole, the radiation efficiency is high.

In the direction of maximum radiation (that is, broadside to the plane of the loop, regardless of the point at which it is fed) the one-wavelength loop will show a small gain over a half-wave dipole. Theoretically, this gain is about 2 dB, and measurements have confirmed that it is of this order.

The one-wavelength loop is more frequently used as an element of a directive antenna array (the quad and delta-loop antennas described in later chapters) than singly, although there is no reason why it cannot be used alone. In the quad and delta loop, it is nearly always driven so that the polarization is horizontal.

## FOLDED DIPOLES

In the diagram shown in Fig. 60, suppose for the moment that the upper conductor between points B and C is disconnected and removed. The system is then a simple center-fed dipole, and the direction of current flow along the antenna and line at a given instant is as shown by the arrows. Then if the upper conductor between B and C is restored, the current in it will flow away from B and toward C, in accordance with the rule for reversal of direction in alternate half-wave sections along a wire. However, the fact that the second wire is "folded" makes the currents in the two conductors of the antenna flow in the *same* direction. Although the antenna physically resembles a transmission line, it is not actually a line from the standpoint of antenna currents but is merely two conductors in parallel. The connections at the ends of the two are assumed to be of negligible length.

A half-wave dipole formed in this way will have the same directional properties and total radiation resistance as an ordinary dipole. However, the transmission line is connected to only *one* of the conductors. It is therefore to be expected that the antenna will "look" different, with respect to its input impedance, as viewed by the line.

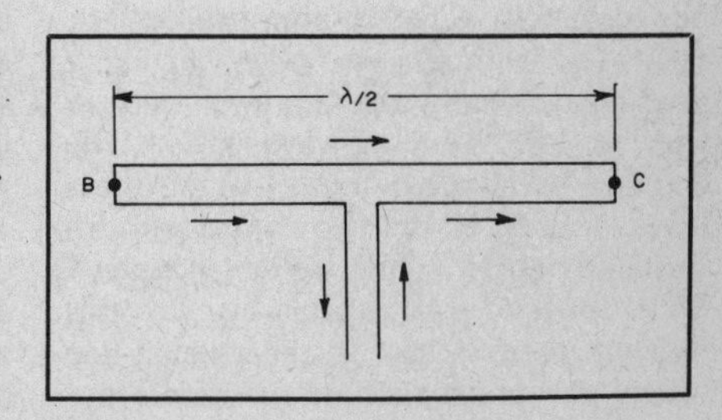


Fig. 60 — Direction of current flow in a folded dipole.

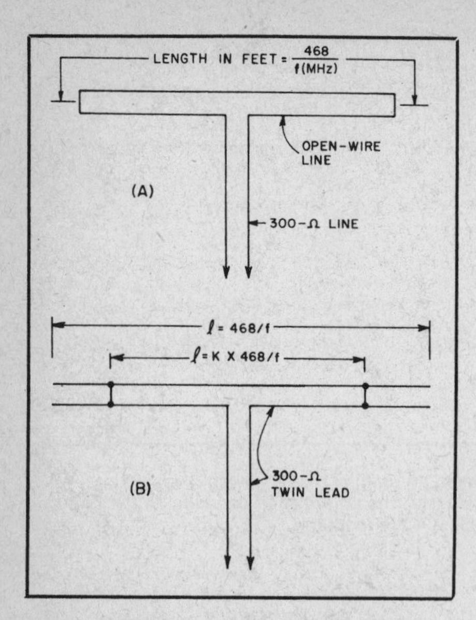


**Fig. 61 — Folded half-wave dipoles. If solid-dielectric transmission line is used for the radiating element, the shorting connections should be placed inward from the ends of the antenna, as shown at B. K = velocity factor of the line.**

The effect on the impedance at the antenna input terminals can be visualized quite readily. The center impedance of the dipole *as a whole* is the same as the impedance of a single-conductor dipole — that is, approximately 70 ohms. A given amount of power will therefore cause a definite value of current, I. In the ordinary half-wave dipole this current flows at the junction of the line and antenna. In the folded dipole the same total current also flows, but is equally divided between two conductors in parallel. The current in each conductor is therefore I/2. Consequently, the line "sees" a higher impedance because it is delivering the same power at only half the current. It is easy to show that the new value of impedance is equal to four times the impedance of a simple dipole. If more wires are added in parallel the current continues to divide between them and the terminal impedance is raised still more. This explanation is a simplified one based on the assumption that the conductors are close together and have the same diameter.

The two-wire system in Fig. 61A is an especially useful one because the input impedance is so close to 300 ohms that it can be fed directly with 300-ohm twin-lead or open line without any other matching arrangement, and the line will operate with a low standing-wave ratio. The antenna itself can be built like an open-wire line, that is, the two conductors can be held apart by regular feeder spreaders. TV "ladder" line is suitable. In this antenna there is also a transmission-line effect; the impedance of the dipole appears in parallel with the impedance of the shorted transmission-line sections. The value of 468 appearing in Fig. 61A results in an antenna length which is 95% of a half wave in free space. If the velocity factor of the antenna, looking at it as a transmission line, is approximately of this value, the shorting connections may be made at the ends of the antenna. (Velocity factors of transmission lines are discussed in detail in Chapter 3.)

Solid-dielectric lines, such as 300-ohm twin-lead, have a significantly lower velocity factor than open-wire line. In such cases the position of the shorting connections should be moved toward the center of the antenna, as shown in Fig. 61B, to avoid introducing reactance at the feed point when the antenna is cut to a resonant physical length.

The folded dipole has a somewhat "flatter" impedance-vs.-frequency characteristic than a simple dipole. That is, the reactance varies less rapidly, as the frequency is varied on either side of resonance, than with a single-wire antenna. The transmission-line effect mentioned above accounts for this phenomena. At a frequency away from resonance, the reactance of the dipole is of the opposite type from that of the shorted line, and there is some reactance cancellation at the feed point as a net result.

## Harmonic Operation

A folded dipole will not accept power at twice the fundamental frequency, or any even multiples of the fundamental. At such multiples the folded section simply acts like a continuation of the transmission line. No other current distribution is possible if the currents in the two conductors of the actual transmission lines are to flow in opposite directions.

On the third and other odd multiples of the fundamental the current distribution is correct for operation of the system as a folded antenna. Since the radiation resistance of a 3/2-wave antenna is not greatly different from that of a half-wave antenna, a folded dipole can be operated on its third harmonic.

## Multi- and Unequal-Conductor Folded Dipoles

Larger impedance ratios than 4 to 1 are frequently desirable when the folded dipole is used as the driven element in a directive array because the radiation resistance is frequently quite low. A wide choice of impedance step-up ratios is available by varying the relative size and spacing of the conductors, and by using more than two. Fig. 62 gives design information of this nature for two-conductor folded dipoles and Fig. 63 is a similar chart for three-conductor dipoles. Fig. 63 assumes that the three conductors are in the same plane and that the two that are not directly connected to the transmission line are equally spaced from the driven conductor.

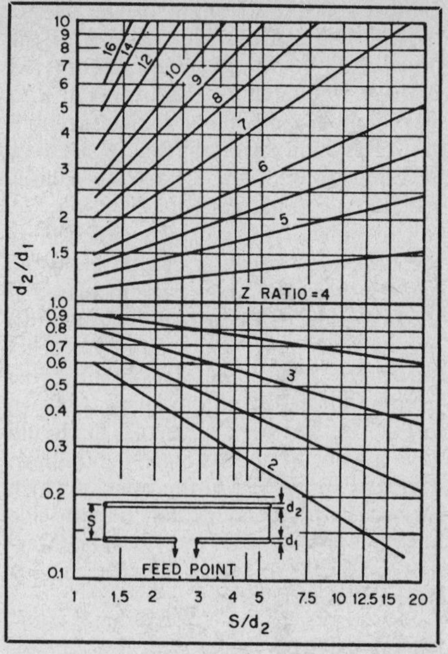

Fig. 62 — Impedance step-up ratio for the two-conductor folded dipole, as a function of conductor diameters and spacing. Dimensions $d_1$ $d_2$ and S are shown in the inset drawing. The step-up ratio, r, may also be determined from:

$$r = \left(1 + \frac{\log \frac{2S}{d_1}}{\log \frac{2S}{d_2}}\right)^2$$

In computing the length of a folded dipole using thick conductors — i.e., tubing such as is used in rotary beam antennas — it should be remembered that the resonant length may be appreciably less than that of a single-wire antenna cut for the same frequency. Besides the shortening required with thick conduc-

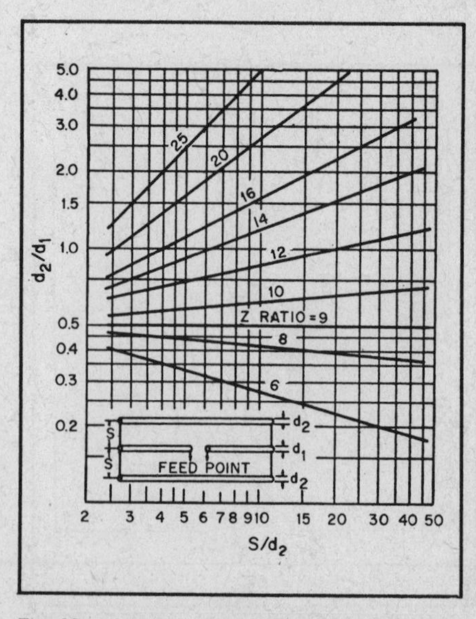


**Fig. 63 — Impedance step-up ratio for the three-conductor folded dipole. The conductors that are not directly driven must have the same diameter, but this diameter need not be the same as that of the driven conductor. Dimensions are indicated in the inset.**

tors, as discussed earlier in this chapter, the parallel conductors tend to act like the boundaries of a conducting sheet of the same width as the spacing between the conductors. The "effective diameter" of the folded dipole will lie somewhere between the actual conductor diameter and the maximum distance between conductors. The relatively large effective thickness of the antenna reduces the rate of change of reactance with frequency, so the tuning becomes relatively broad and the antenna length is not too critical for a given frequency.

Further information on the folded dipole, as pertains to feeding and matching, is contained in Chapter 4.

## OTHER TYPES OF ANTENNAS

The half-wave dipole and the few special types of antennas described in this chapter form the basis for practically all antenna systems in amateur use at frequencies from the vhf region down. Other fundamental types of radiators are applicable at microwaves, but they are not used at lower frequencies because the dimensions are such as to be wholly impractical when the wavelength is measured in meters rather than centimeters.

## BIBLIOGRAPHY

Source material and more extended discussion of topics covered in this chapter can be found in the references given below.


Brown, Lewis and Epstein, "Ground Systems as a Factor in Antenna Efficiency," *Proc. I.R.E.*, June 1937.

Carter, Hansell and Lindenblad, "Development of Directive Transmitting Antennas by R.C.A. Communications," *Proc. I.R.E.*, October 1931.

Dome, "Increased Radiating Efficiency for Short Antennas," *QST*, September 1934.

Grammer, "More on the Directivity of Horizontal Antennas," *QST*, March 1937.

King, Mimno and Wing, *Transmission Lines, Antennas and Wave Guides* (New York: McGraw-Hill Book Co.).

Landskov, H. K., "Pattern Factors for Elevated Horizontal Antennas Over Real Earth," *QST*, November 1975, p. 19.

Lindsay, "Quads and Yagis," *QST*, May 1968.

Reinartz, "Half-Wave Loop Antennas," *QST*, October 1937.

Terman, *Radio Engineering* (New York: McGraw-Hill Book Co.).

Williams, "Radiating Characteristics of Short-Wave Loop Aerials," *Proc. I.R.E.*, October 1940.

**Textbooks on Antennas and Transmission Lines**

Jasik, *Antenna Engineering Handbook* (New York: McGraw-Hill Book Co.).

Johnson, *Transmission Lines and Networks* (New York: McGraw-Hill Book Co.).

Jordan, *Electromagnetic Waves and Radiation Systems* (Englewood Cliffs, New Jersey: Prentice-Hall, Inc.).

King, *The Theory of Linear Antennas* (Cambridge, Mass.: Harvard University Press).

Laport, *Radio Antenna Engineering* (New York: McGraw-Hill Book Co.).

Kraus, *Antennas* (New York: McGraw-Hill Book Co.).

Schelkunoff, *Advanced Antenna Theory* (New York: John Wiley & Sons, Inc.).

Shelkunoff and Friis, *Antenna Theory and Practice* (New York: John Wiley & Sons, Inc.).

Skilling, *Electric Transmission Lines* (New York: McGraw-Hill Book Co.).

Slurzburg and Osterheld, *Essentials of Radio* (New York: McGraw-Hill Book Co.).

Southworth, *Principles and Applications of Wave-Guide Transmission* (New York: D. Van Nostrand Co., Inc.).

# Chapter 3

# Transmission Lines

The desirability of installing an antenna in a clear space, not too near buildings or power and telephone lines, cannot be stressed too strongly. On the other hand, the transmitter that generates the rf power for driving the antenna is usually, as a matter of necessity, located some distance from the antenna terminals. The connecting link between the two is the rf transmission line, feeder or feed line. Its sole purpose is to carry rf power from one place to another, and to do it as efficiently as possible. That is, the ratio of the power *transferred* by the line to the power *lost* in it should be as large as the circumstances will permit.

At radio frequencies every conductor that has appreciable length compared with the wavelength in use will *radiate* power. That is, every conductor becomes an antenna. Special care must be used, therefore, to minimize radiation from the conductors used in rf transmission lines. Without such care, the power radiated by the line may be much larger than that which is lost in the resistance of conductors and dielectrics (insulating materials). Power loss in resistance is inescapable, at least to a degree, but loss by radiation is largely avoidable.

**Preventing Radiation**

Radiation loss from transmission lines can be prevented by using two conductors so arranged and operated that the electromagnetic field from one is balanced everywhere by an equal and opposite field from the other. In such a case the resultant field is zero everywhere in space; in other words, there is no radiation.

For example, Fig. 1A shows two parallel conductors having currents I1 and I2 flowing in opposite directions. If the current I1 at point Y on the upper conductor has the same amplitude as the current I2 at the corresponding point X on the lower conductor, the fields set up by the two currents will be equal in magnitude. Because the two currents are flowing in opposite directions, the field from I1 at Y will be 180 degrees out of phase with the field from I2 at X. However, it takes a measurable interval of time for the field from X to travel to Y. If I1 and I2 are alternating currents, the phase of the field from I1 at Y will have changed in such a time interval, and so at the instant the field from X reaches Y the two fields at Y are not exactly 180 degrees out of phase. The two fields will be exactly 180 degrees out of phase at every point in space only when the two conductors occupy the same space — an obviously impossible condition if they are to remain separate conductors.

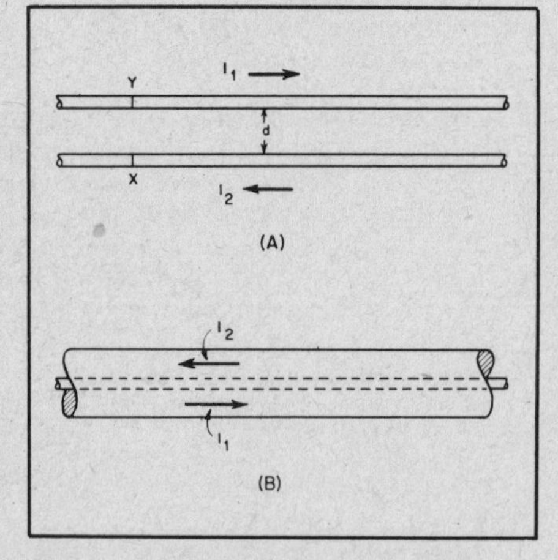


Fig. 1 — The two basic types of transmission lines.

The best that can be done is to make the two fields cancel each other as completely as possible. This can be accomplished by making the distance, d, between the two conductors small enough so that the time interval during which the field from X is moving to Y is a very small part of a cycle. When this is the case the phase difference between the two fields at any given point will be so close to 180 degrees that the cancellation is practically complete.

Practicable values of d, the separation between the two conductors, are determined by the physical limitations of line construction. A separation that meets the condition of being "very small" at one frequency may be quite large at another. For example, if d is 6 inches, the phase difference between the two fields at Y will be only a fraction of a degree if the frequency is 3500 kHz. This is because a distance of 6 inches is such a small fraction of a wavelength (one wavelength = 360 degrees) at 3500 kHz. But at 144 MHz the phase difference would be 26 degrees, and at 420 MHz it would be 73 degrees. In neither of these cases could the two fields be considered to "cancel" each other. The separation must be very small in comparison with the wavelength used; it should never exceed 1 percent of the wavelength, and smaller separations are desirable. Transmission lines consisting of two parallel conductors as in Fig. 1A are called parallel-conductor lines, or open-wire lines, or two-wire lines.

A second general type of line construction is shown in Fig. 1B. In this case one of the conductors is tube-shaped and encloses the other conductor. This is called a coaxial line ("coax") or concentric line. The current flowing on the inner conductor is balanced by an equal current flowing in the opposite direction on the inside surface of the outer conductor. Because of skin effect the current on the inner surface of the tube does not penetrate far enough to appear on the outer surface. In fact, the total electromagnetic field outside the coaxial line, as a result of currents flowing on the conductors inside, always is zero because the tube acts as a shield at radio frequencies. The separation between the inner conductor and the outer conductor is therefore unimportant from the standpoint of reducing radiation.

## CURRENT FLOW IN LONG LINES

In Fig. 2, imagine that the connection between the battery and the two wires is made instantaneously and then broken. During the time the wires are in contact with the battery terminals, electrons in wire no. 1 will be attracted to the positive battery terminal and an equal number of electrons in wire no. 2 will be repelled from the negative terminal. This happens only near the battery terminals at first, since electrical effects do not travel at infinite speed, so some time will elapse before the currents become evident at more extreme parts of the wires. By ordinary standards the elapsed time is very short, since the speed of travel along the wires may be almost 300,000,000 meters per second, so it becomes necessary to measure time in millionths of a second (microseconds) rather than in more familiar time units.

For example, suppose that the contact with the battery is so short that it can be measured in a very small fraction of a microsecond. Then the "pulse" of current that flows at the battery terminals during this time can be represented by the vertical line in Fig. 3. At the speed of light this pulse will travel 30 meters along the line in 0.1 microsecond; 30 meters more, making a total of 60 meters in 0.2 microsecond; a total of 90 meters in 0.3 microsecond, and so on for as far as the line reaches. The current does not exist all along the wires but is only present at the point that the pulse has reached in its travel; at this point it is present in both wires, with the electrons moving in one direction in one wire and in the other direction in the other wire. If the line is infinitely long and has no resistance or other cause of energy loss, the pulse will travel undiminished forever.

Extending the example of Fig. 3, it is not hard to see that if instead of one pulse we started a whole series of them on the line at equal time intervals, they would travel along the line with the same time and distance spacing between them, each pulse independent of the others. In fact, each pulse could have a different amplitude, if the battery voltage were varied to that end. Furthermore, the pulses could be so closely spaced that they touched each other, in which case we would have current present everywhere along the line simultaneously.

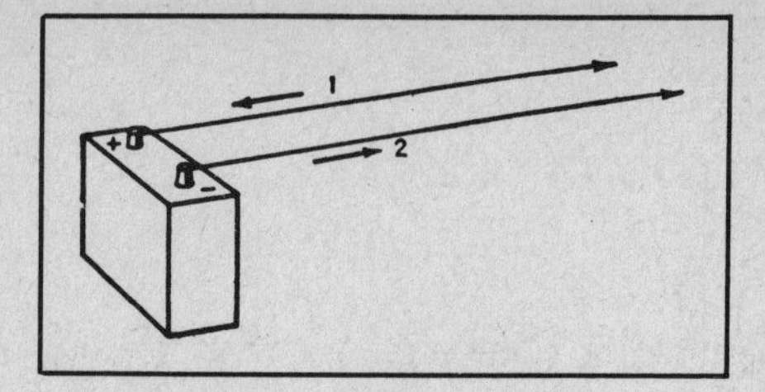


Fig. 2 — Illustrating current flow on a long transmission line.

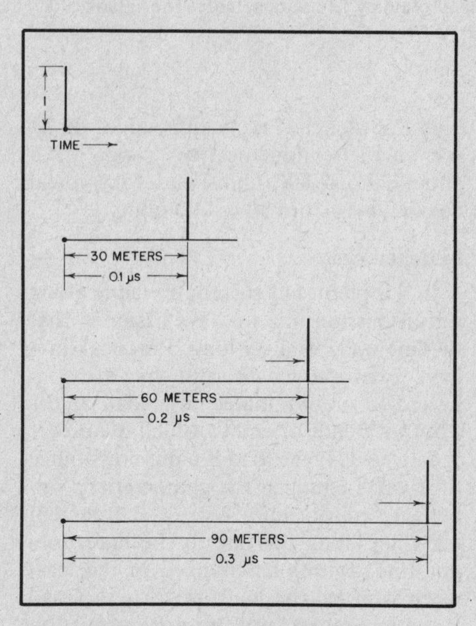


Fig. 3 — A current pulse traveling along a transmission line at the speed of light would reach the successive positions shown at intervals of 0.1 microsecond.

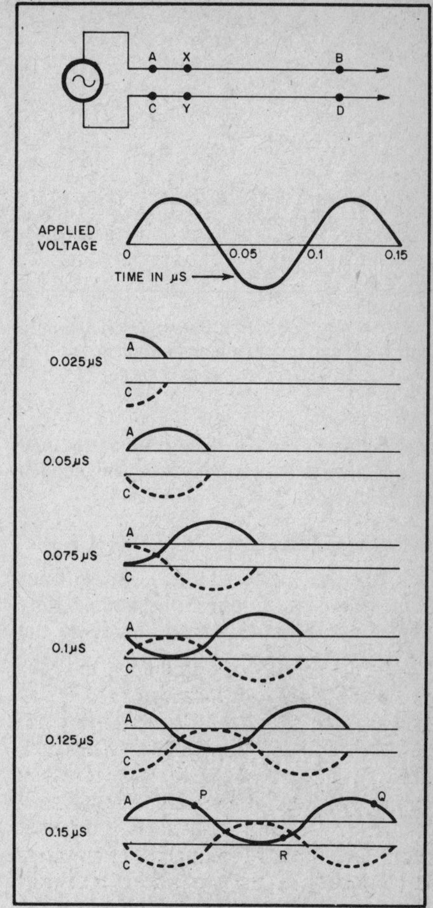

Fig. 4 — Instantaneous current along a transmission line at successive time intervals. The frequency is such that the time of one cycle is 0.1 microsecond.

### Wavelength

It follows from this that an alternating voltage applied to the line would give rise to the sort of current flow shown in Fig. 4. If the frequency of the ac voltage is 10,000,000 hertz (cycles per second) or 10 MHz, each cycle will occupy 0.1 microsecond, so a complete cycle of current will be present along each 30 meters of line. This is a distance of one wavelength. Any currents observed at B and D occur just one cycle later in time than the currents at A and C. To put it another way, the currents initiated at A and C do not appear at B and D, one wavelength away, until the applied voltage has had time to go through a complete cycle.

Since the applied voltage is always changing, the currents at A and C are changing in proportion. The current a short distance away from A and C — for instance, at X and Y — is not the same as the current at A and C because the current at X and Y was caused by a value of voltage that occurred slightly earlier in the cycle. This is true all along the line; at any instant the current anywhere along the line from A to B and C to D is different from the current at every other point in that same distance. The series of drawings shows how the instantaneous currents might be distributed if we could take snapshots of them at intervals of one-quarter cycle. The current travels out from the input end of the line in waves.

At any selected point on the line the current goes through its complete range of ac values in the time of one cycle just as it does at the input end. Therefore (if there are no losses) an ammeter inserted in either conductor would read exactly the same current at any point along the line, because the ammeter averages the current over a whole cycle. The phases of the currents at any two separated points would be different, but an ammeter cannot show phase.

### Velocity of Propagation

In the example above it was assumed that energy traveled along the line with the velocity of light. The actual velocity is very close to that of light in a line in which the insulation between conductors is solely air. The presence of dielectrics other than air reduces the velocity, since electromagnetic waves travel more slowly in dielectrics than they do in a vacuum. Because of this the wavelength as measured along the line will depend on the velocity factor that applies in the case of the particular type of line in use. The velocity factor V is related to the dielectric constant $\varepsilon$ by

$$V = \frac{1}{\sqrt{\varepsilon}}$$

See a later section in this chapter for ac-

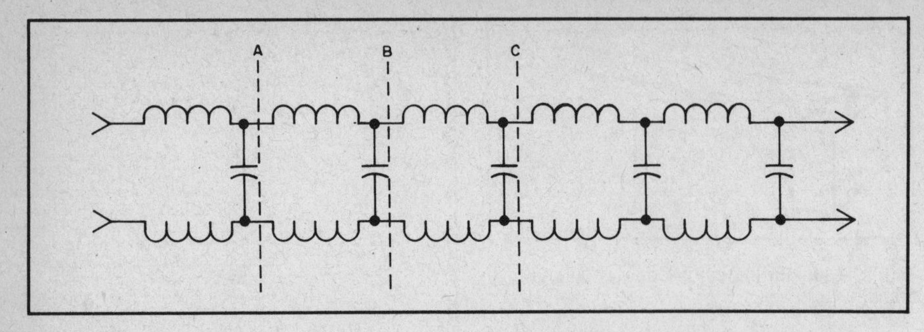


Fig. 5 — Equivalent of a transmission line in terms of ordinary circuit constants. The values of L and C depend on the line construction.

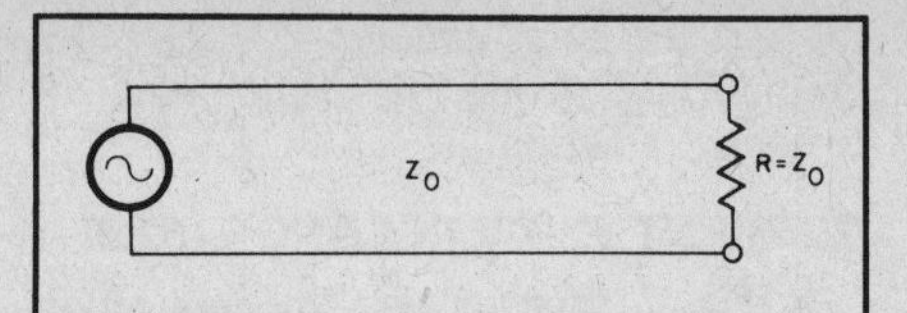


Fig. 6 — A transmission line terminated in a resistive load equal to the characteristic impedance of the line.

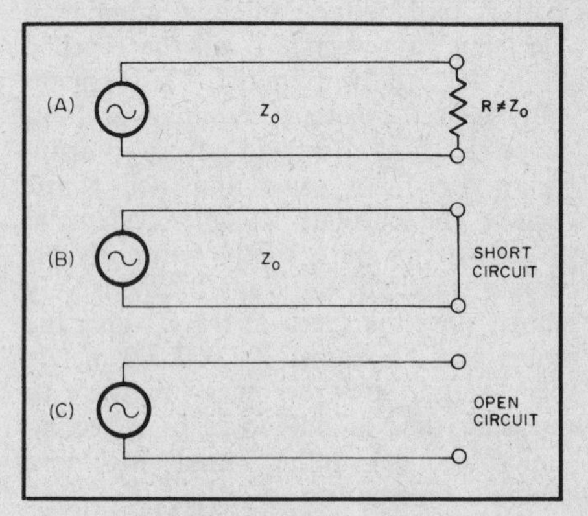


Fig. 7 — Mismatched lines. A — termination not equal to $Z_o$; B — short-circuited line; C — open-circuited line.

tual figures. The wavelength in a practical line is always shorter than the wavelength in free space.

## CHARACTERISTIC IMPEDANCE

If this is a "perfect" line — one without resistance — a question immediately comes up: What is the amplitude of the current in the pulse? Will a larger voltage result in a larger current, or is the current theoretically infinite for any applied voltage, as we would expect from applying Ohm's Law to a circuit without resistance? The answer is that the current does depend directly on the voltage, just as though resistance were present.

The reason for this is that the current flowing in the line is something like the charging current that flows when a battery is connected to a capacitor. That is, the line has capacitance. However, it also has inductance. Both of these are "distributed" properties. We may think of the line as being composed of a whole series of small inductors and capacitors connected as in Fig. 5, where each coil is the inductance of an extremely small section of wire and the capacitance is that existing between the same two sections. Each inductance limits the rate at which each immediately following capacitor can be charged, and the effect of the chain is to establish a definite relationship between current and voltage. Thus the line has an apparent "resistance," called its characteristic resistance — or, a more general term, its characteristic impedance or surge impedance. The conventional symbol for characteristic impedance is $Z_o$.

## TERMINATED LINES

The value of the characteristic impedance is equal to $\sqrt{L/C}$ in a perfect line — i.e., one in which the conductors have no resistance and there is no leakage between them — where L and C are the inductance and capacitance, respectively, per unit length of line. The inductance decreases with increasing conductor diameter, and the capacitance decreases with increasing spacing between the conductors. Hence a line with large conductors closely spaced will have relatively low characteristic impedance while one with thin conductors widely spaced will have high impedance. Practicable values of $Z_o$ for parallel-conductor lines range from about 200 to 800 ohms and for typical coaxial lines from 50 to 100 ohms.

### Matched Lines

In this picture of current traveling along a transmission line we have assumed that the line was infinitely long. Practical lines have a definite length, and they are connected to or terminated in a load at the "output" end, or end to which the power is delivered. If the load is a pure resistance of a value equal to the characteristic impedance of the line, Fig. 6, the current traveling along the line to the load does not find conditions changed in the least when it meets the load; in fact, the load just "looks like" still more transmission line of the same characteristic impedance.

The reason for this can perhaps be made a little clearer by considering it from another viewpoint. In flowing along a transmission line, the power is handed from one of the elementary sections in Fig. 5 to the next. When the line is infinitely long this power transfer always goes on in one direction — away from the source of power. From the standpoint of Section B, Fig. 5, for instance, the power it has handed over to section C has simply disappeared in C. So far as section B is concerned, it makes no difference whether C has absorbed the power itself or has in turn handed it along to more line. Consequently, if we substitute something for section C that has the same electrical characteristics, section B will not know the difference. A pure resistance equal to the characteristic impedance of C, which is also the characteristic impedance of the line, meets this condition. It absorbs all the power just as the infinitely long line absorbs all the power transferred by section B.

A line terminated in a purely resistive load equal to the characteristic impedance is said to be matched. In a matched transmission line, power travels outward along the line from the source until it reaches the load, where it is completely absorbed. Thus with either the infinitely long line or its matched counterpart the impedance presented to the source of power (the line-input impedance) is the same regardless of the line length. It is simply equal to the characteristic impedance of the line. The current in such a line is equal to the applied voltage divided by the characteristic impedance, and the power put into it is $E^2/Z_o$ or $I^2Z_o$, by Ohm's Law.

### Mismatched Lines

Now take the case where the terminating resistance, R, is *not* equal to $Z_o$, as in Fig. 7. The load R no longer "looks like" more line to the section of line immediately adjacent. Such a line is said to be *mismatched*. The more R differs from $Z_o$, the greater the mismatch. The power reaching R is not totally absorbed, as it was when R was equal to $Z_o$, because R requires a different voltage-to-current ratio than the one at which the power is traveling along the line. The result is that R absorbs only part of the power reaching it (the *incident* power); the remainder acts as though it had bounced off a wall and starts back along the line toward the source. This is *reflected power,* and the greater the mismatch the larger the percentage of the incident power that is reflected. In the extreme case where R is zero (a short circuit) or infinity (an open circuit) *all* of the power reaching the end of the line is reflected.

Whenever there is a mismatch, power is traveling in both directions along the line. The voltage-to-current ratio is the same for the reflected power as for the incident power, since this ratio is determined by the $Z_o$ of the line. The voltage and current travel along the line in both directions in the same sort of wave motion shown in Fig. 4. When the source of power is an ac generator, the outgoing or incident voltage and the returning or reflected voltage are simultaneously present all along the line, so the actual voltage at any point along the line is the sum of the two components, taking phase into account. The same is true of the current.

The effect of the incident and reflected components on the behavior of the line can be understood more readily by considering first the two limiting cases — the short-circuited line and the open-circuited line. If the line is short-circuited as in Fig. 7B, the voltage at the end must be zero. Thus the incident voltage must disappear suddenly at the short. It can do this only if the reflected voltage is opposite in phase and of the same amplitude. This is shown by the vectors in Fig. 8. The current, however, does not disappear in the short circuit; in fact, the incident current flows through the short and there is in addition the reflected component in phase with it and of the same amplitude. The reflected voltage and current must have the same amplitudes as the incident voltage and current because no power is used up in the short circuit; all the power starts back toward the source. Reversing the phase of *either* the current or voltage (but not both) will reverse the direction of power flow; in the short-circuited case the phase of the voltage is reversed on reflection but the phase of the current is not.

If the line is open-circuited (Fig. 7C) the current must be zero at the end of the line. In this case the reflected current is 180 degrees out of phase with the incident current and has the same amplitude. By reasoning similar to that used in the short-circuited case, the reflected voltage must be in phase with the incident voltage, and must have the same amplitude. Vectors for the open-circuited case are shown in Fig. 9.

Where there is a finite value of resistance at the end of the line, Fig. 7A, only part of the power reaching the end of the line is reflected. That is, the reflected voltage and current are smaller than the incident voltage and current. If R is less than $Z_o$ the reflected and incident voltage are 180 degrees out of phase, just as in the case of the short-circuited line, but the amplitudes are not equal because all of the voltage does not disappear at R. Similarly, if R is greater than $Z_o$ the reflected and incident currents are 180 degrees out of phase, as they were in the open-circuited line, but all of the current does not appear in R so the amplitudes of the two components are not equal. These two cases are shown in Fig. 10. Note that the resultant current and voltage are in phase in R, since R is a pure resistance.

### Reflection Coefficient

The ratio of the reflected voltage to the incident voltage is called the *reflection*

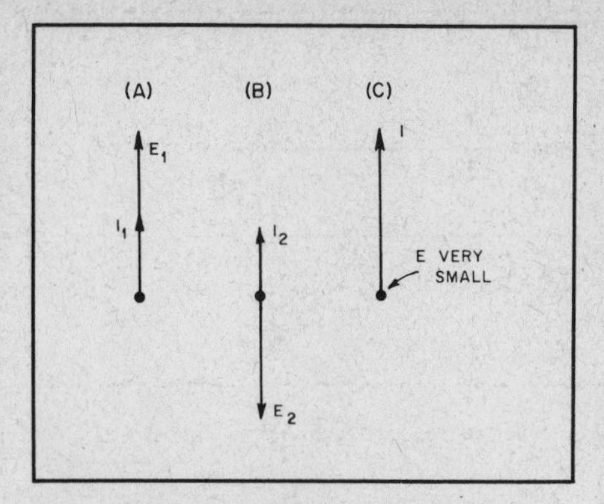


Fig. 8 — Voltage and current at the short circuit on a short-circuited line. These vectors show how the outgoing voltage and current (A) combine with the reflected voltage and current (B) to result in high current and very low voltage in the short-circuit (C).

*coefficient*. Thus

$$\rho = \frac{E_r}{E_f}$$

where $\rho$ is the reflection coefficient, $E_r$ is the reflected voltage, and $E_f$ is the incident or forward voltage. The reflection coefficient is determined by the relationship between the line $Z_o$ and the actual load at the terminated end of the line. For any given line and load it is a constant if the line has negligible loss in itself. The coefficient can never be larger than 1 (which indicates that all the incident power is reflected) nor smaller than zero (indicating that the line is perfectly matched by the load).

*If the load is purely resistive,* the reflection coefficient can be found from

$$\rho = \frac{R - Z_o}{R + Z_o}$$

where R is the resistance of the load terminating the line. In this expression $\rho$ is positive if R is larger than $Z_o$ and negative if R is smaller than $Z_o$. The change in signs accompanies the change in phase of the reflected voltage described above.

## STANDING WAVES

As might be expected, reflection cannot occur at the load without some effect on the voltages and currents all along the line. A detailed description tends to become complicated, and what happens is most simply shown by vector diagrams. Fig. 11 is an example in the case where R is less than $Z_o$. The voltage and current vectors at the load, R, are shown in the reference position; they correspond with the vectors in Fig. 10A. Going back along the line from R toward the power source, the incident vectors, E1 and I1, lead the vectors at the load according to their position along the line measured in electrical degrees. (The corresponding distances in fractions of a wavelength also are shown.) The vectors representing reflected voltage and current, E2 and I2, successively lag the same vectors at the load. This lag and

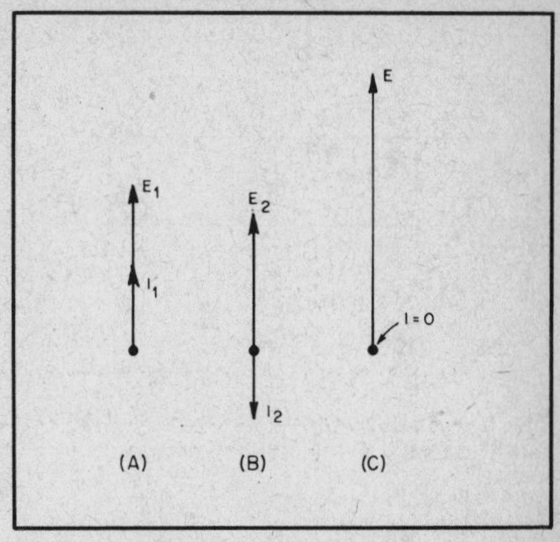


Fig. 9 — Voltage and current at the end of an open-circuited line. A — outgoing voltage and current; B — reflected voltage and current; C — resultant.

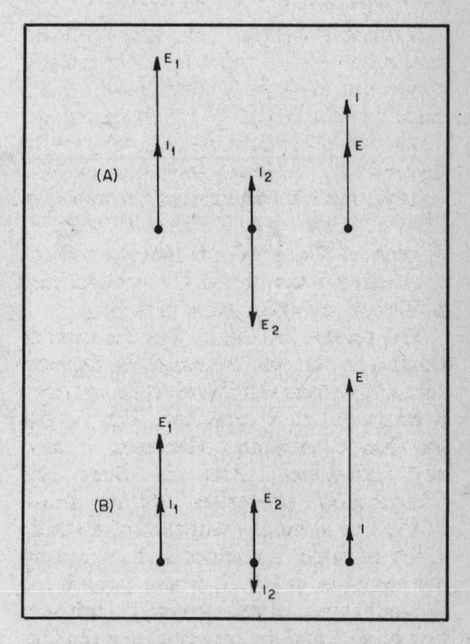


Fig. 10 — Incident and reflected components of voltage and current when the line is terminated in a pure resistance not equal to $Z_o$. In the case shown, the reflected components have half the amplitude of the incident components. A — R less than $Z_o$; B — R greater than $Z_o$.

lead is the natural consequence of the direction in which the incident and reflected components are traveling, together with the fact that it takes time for the power to travel along the line. The resultant voltage, E, and current, I, at each of these positions are shown dotted. Although the incident and reflected components maintain their respective amplitudes (the reflected component is shown at half the incident-component amplitude in this drawing) their phase relationships vary with position along the line. The phase shifts cause both the amplitude and phase of the *resultants* to vary with position on the line.

If the amplitude variations (disregarding phase) of the resultant voltage and current are plotted against position along the line, graphs like those of Fig. 12A will

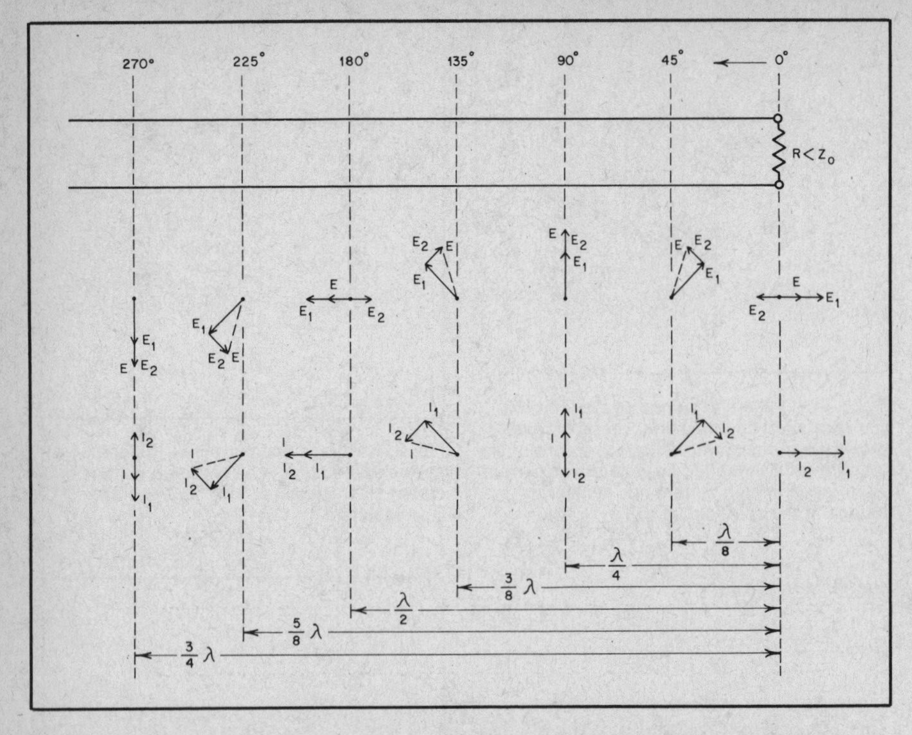


Fig. 11 — Incident and reflected components at various positions along the line, together with resultant voltages and currents at the same positions. The case shown is for R less than $Z_o$.

result. If we could go along the line with a voltmeter and ammeter plotting the current and voltage at each point, we should find that the collected data gives curves like these. In contrast, if the load matched the $Z_o$ of the line, similar measurements along the line would show that the voltage is the same everywhere (and similarly for the current). The mismatch between load and line is responsible for the variations in amplitude which, because of their wave-like appearance, are called standing waves.

From the earlier discussion it should be clear that when R is greater than $Z_o$ the voltage will be largest and the current smallest at the load. This is just the reverse of the case shown in Fig. 12A. In such case the curve labeled E would become the I (current) curve, while the current curve would become the voltage curve.

Some general conclusions can be drawn from inspection of the standing-wave curves: At a position 180 degrees (1/2 wavelength) from the load, the voltage and current have the same values they do at the load. At a position 90 degrees from the load the voltage and current are "inverted." That is, if the voltage is lowest and current highest at the load (R less than $Z_o$), then 90 degrees or 1/4 wavelength from the load the voltage reaches its highest value and the current reaches its lowest value. In the case where R is greater than $Z_o$, so that the voltage is highest and the current lowest at the load, the voltage has its lowest value and the current its highest value at a point 90 degrees from the load.

Note that the conditions existing at the 90-degree point also are duplicated at the 270-degree point (3/4 wavelength). If the graph were continued on toward the source of power it would be found that this duplication occurs at every point that is an odd multiple of 90 degrees (odd multiple of a quarter wavelength) from the load. Similarly, the voltage and current are the same at every point that is a multiple of 180 degrees (any multiple of one-half wavelength) as they are at the load.

### Standing-Wave Ratio

The ratio of the maximum voltage along the line to the minimum voltage — that is, the ratio of $E_{max}$ to $E_{min}$ in Fig. 12A is called the *voltage standing-wave ratio* (abbreviated VSWR) or simply the *standing-wave ratio* (SWR). The ratio of the maximum current to the minimum current ($I_{max}/I_{min}$) is the same as the VSWR, so either current or voltage can be measured to determine the standing-wave ratio. Fig. 13 contains a convenient nomograph from which SWR can be determined in accordance with forward and reflected readings on an rf wattmeter.

The standing-wave ratio is an index of many of the properties of a mismatched line. It can be measured with good accuracy with fairly simple equipment, and so is a convenient quantity to use in making calculations on line performance. *If the load contains no reactance,* the SWR is numerically equal to the ratio between the load resistance, R, and the characteristic impedance of the line; that is,

$$SWR = \frac{R}{Z_o}$$

when R is greater than $Z_o$, and

$$SWR = \frac{Z_o}{R}$$

when R is less than $Z_o$. The smaller quantity is always used in the denominator of the fraction so the SWR will be a number larger than 1.

This relationship shows that the greater the mismatch — that is, the greater the difference between $Z_o$ and R — the larger the SWR. In the case of open- and short-circuited lines the SWR becomes infinite. On such lines the voltage and current become zero at the minimum points ($E_{min}$ and $I_{min}$) since total reflection occurs at the end of the line and the incident and reflected components have equal amplitudes.

## INPUT IMPEDANCE

The relationship between the voltage and current at any point along the line (including the effects of both the incident and reflected components) becomes more clear when only the resultant voltage and current are shown, as in Fig. 12B. Note that the voltage and current are in phase not only at the load but also at the 90-degree point, the 180-degree point, and the 270-degree point. This is also true at every point that is a multiple of 90 degrees from the load.

Suppose the line were cut at one of these points and the generator or source of power were connected to that portion terminated in R. Then the generator would "see" a pure resistance, just as it would if it were connected directly to R. However, the value of the resistance it sees would depend on the line length. If the length is 90 degrees, or an odd multiple of 90 degrees, where the voltage is high the current low, the resistance seen by the generator would be greater than R. If the length is 180 degrees or a multiple of 180 degrees, the voltage and current relationships are the same as in R, and therefore the generator "sees" a resistance equal to the actual load resistance at these line lengths.

With a resistive load the current and voltage are exactly in phase only at points that are multiples of 90 degrees from the load. At all other points the current either leads or lags the voltage, and so the load seen by the generator when the line length is not an exact multiple of 90 degrees is not a pure resistance. The input impedance of the line — that is, the impedance seen by the generator connected to the line — in such a case has both resistive and reactive components. When the current lags behind the voltage the reactance is inductive; when it leads the voltage the reactance is capacitive. The upper drawing in Fig. 12B shows that when the line is terminated in a resistance smaller than $Z_0$ the reactance is inductive in the first 90 degrees of line moving from the load toward the generator, is capacitive in the second 90 degrees, inductive in the third 90 degrees, and so on every 90 degrees or quarter wavelength. The lower drawing illustrates the case where R is greater than $Z_0$. The voltage and current vectors are merely interchanged, since, as explained in connection with Fig. 11, in this case the vector for the reflected current is the one that is reversed in phase on reflection. The reactance becomes capacitive in the first 90 degrees, inductive in the second, and so on.

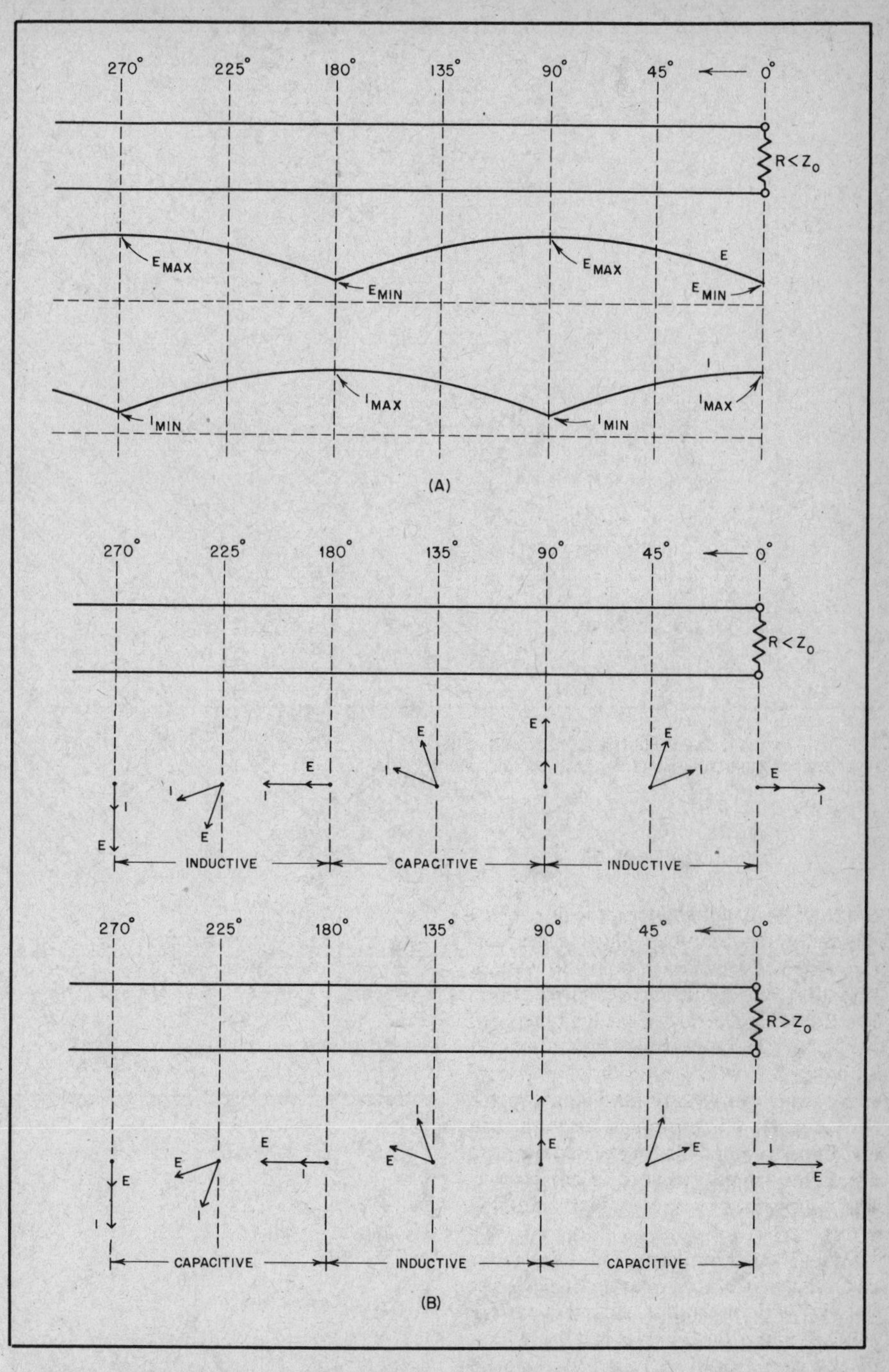


Fig. 12 — Standing waves of current and voltage along the line, for R less than $Z_0$ (A). Resultant voltages and currents along a mismatched line are shown at B. Above — R less than $Z_0$; below — R greater than $Z_0$.

### Factors Determining the Input Impedance

The magnitude and phase angle of the input impedance depend on the SWR, the line length, and the $Z_0$ of the line. If the SWR is small, the input impedance is principally resistive at all line lengths; if the SWR is high, the reactive component may be relatively large. The input impedance of the line can be represented by a series circuit of resistance and reactance, as shown in Fig. 14 where $R_s$ is the resistive component and $X_s$ is the reactive component. Frequently the "s" subscripts are omitted, and the series-equivalent impedance denoted as R + jX. The j is an operator function, used to indicate that the values for R and X cannot be added directly, but that vector addition must be used if the overall impedance is to be determined. (This is analogous to solving a right triangle for the length of its hypotenuse, where R and X represent the length of its two sides. The length of the hypotenuse represents Z, the overall impedance.) By convention, a plus sign is assigned to j when the reactance is inductive (R + jX), and a minus sign is used when the reactance is capacitive (R − jX).

### Equivalent Circuits for the Input Impedance

The series circuits shown in Fig. 14 are equivalent to the actual input impedance of the line because they have the same total impedance and the same phase angle. It is also possible to form a circuit with resistance and reactance in parallel that will have the same total impedance and phase angle as the line. This equivalence is shown in Fig. 15. The individual values in the parallel circuit are not the same as those in the series circuit (although the overall result is the same) but are related to the series-circuit values by the equations shown in the drawing.

Either of the two equivalent circuits may be used, depending on which happens to be more convenient for the particular purpose. These circuits are important from the standpoint of designing coupling networks so that the desired amount of power will be taken from the source.

## REACTIVE TERMINATIONS

So far the only type of load considered has been a pure resistance. In general, the load will be fairly close to being a pure resistance, since most transmission lines used by amateurs are connected to resonant antenna systems, which are principally resistive in nature. Consequently, the resistive load is an important practical case.

However, an antenna system is purely resistive only at one frequency, and when it is operated over a band of frequencies without readjustment — the usual condition — its impedance will contain a certain amount of reactance along with resistance. The effect of such a combination is to increase the standing-wave ratio — that is, as between two loads, one having only resistance of, say, 100 ohms as compared with a reactive load having the same total impedance, 100 ohms, the SWR will be higher with the reactive load than with the purely resistive load. Also as between

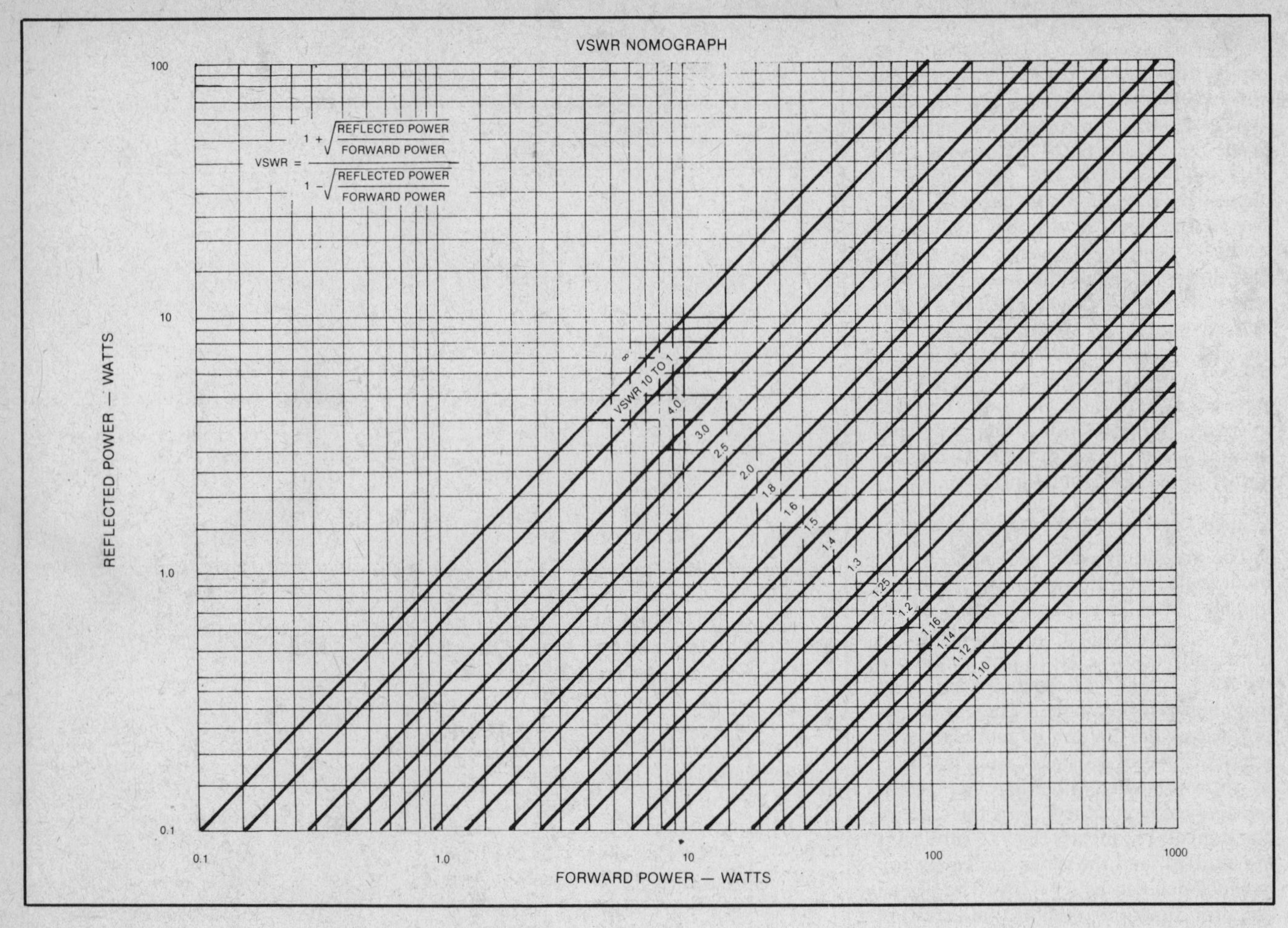


Fig. 13 — VSWR as a function of forward and reflected power.

two loads containing the same value of resistance but one being without reactance while the other has a reactive component in addition, the SWR will be higher with the one having the reactance.

The effect of reactance in the load is to shift the phase of the current with respect to the voltage both in the load itself and in the reflected components of voltage and current. This in turn causes a shift in the phase of the resultant current with respect to the resultant voltage. The net result is to shift the points along the line at which the various effects already described will occur. With a load having inductive reactance the point of maximum voltage and minimum current is shifted toward the load. The reverse occurs when the reactance in the load is capacitive.

## SMITH-CHART TRANSMISSION-LINE CALCULATIONS

It has already been stated that the input impedance, or the impedance seen when "looking into" a length of line, is dependent upon the SWR, the length of the line, and the $Z_o$ of the line. The SWR, in turn, is dependent upon the load which terminates the line. There are complex mathematical relationships which may be used to calculate the various values of impedances, voltages, currents, and SWR values which exist in the operation of a particular transmission line. However, it is much easier to determine such parameters graphically, with the aid of a very useful device, the Smith Chart. If the terminating impedance is known, it is a simple matter to determine the input impedance of the line for any length by means of the chart. Conversely, with a given line length and a known (or measured) input impedance, the load impedance may be determined by means of the chart — a convenient method of

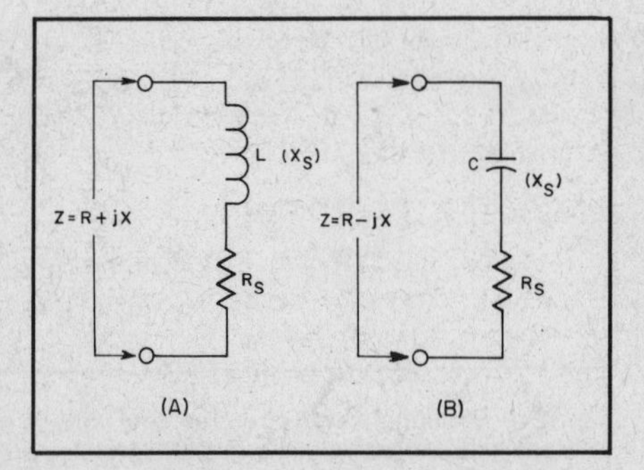


Fig. 14 — Series circuits which may be used to represent the input impedance of a length of transmission line.

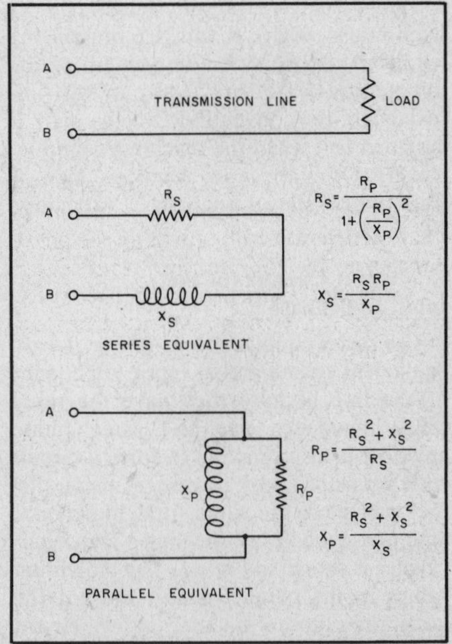


Fig. 15 — Input impedance of a line terminated in a pure resistance. The input impedance can be represented either by a resistance and reactance in series or a resistance and reactance in parallel. The relationships between the R and X values in the series and parallel equivalents are given by the formulas. X may be either inductive or capacitive, depending on the line length, $Z_o$, and the load.

remotely determining an antenna impedance, for example.

Named after its inventor, Phillip H. Smith, the Smith Chart was originally described in *Electronics* for January 1939. Smith Charts may be obtained at most university book stores. They may be ordered from Phillip H. Smith, Analog Instruments Co., P.O. Box 808, New Providence, NJ 07974. For 8-1/2 × 11-inch paper charts with normalized coordinates, request Form 82-BSPR. Smith Charts with 50-ohm coordinates (Form 5301-7569) are available from General Radio Co., West Concord, MA 01781. These charts also are available from ARRL Hq. (see Fig. 18.)

Although its appearance may at first seem somewhat formidable, the Smith Chart is really nothing more than a specialized type of graph, with curved, rather than rectangular, coordinate lines. The coordinate system consists simply of two families of circles — the resistance family and the reactance family. The *resistance circles* (Fig. 16) are centered on the *resistance axis* (the only straight line on the chart), and are tangent to the outer circle at the bottom of the chart. Each circle is assigned a value of resistance, which is indicated at the point where the circle crosses the resistance axis. All points along any one circle have the same resistance value.

The values assigned to these circles vary from zero at the top of the chart to infinity at the bottom, and actually represent a ratio with respect to the impedance value assigned to the center point of the chart, indicated 1.0. This center point is called *prime center*. If prime center is assigned a value of 100 ohms, then 200 ohms resistance is represented by the 2.0 circle, 50 ohms by the 0.5 circle, 20 ohms by the 0.2 circle, and so on. If a value of 50 is assigned to prime center, the 2.0 circle now represents 100 ohms, the 0.5 circle 25 ohms, and the 0.2 circle 10 ohms. In each case, it may be seen that the value on the chart is determined by dividing the actual resistance by the number assigned to prime center. This process if called *normalizing*. Conversely, values from the chart are converted back to actual resistance values by multiplying the chart value times the value assigned to prime center. This feature permits the use of the Smith Chart for any impedance values, and therefore with any type of uniform transmission line, whatever its impedance may be. As mentioned above, specialized versions of the Smith Chart may be obtained with a value of 50 ohms at prime center. These are intended for use with 50-ohm lines.

Now consider the *reactance circles* (Fig. 17) which appear as curved lines on the chart because only segments of the complete circles are drawn. These circles are tangent to the resistance axis, which itself is a member of the reactance family (with

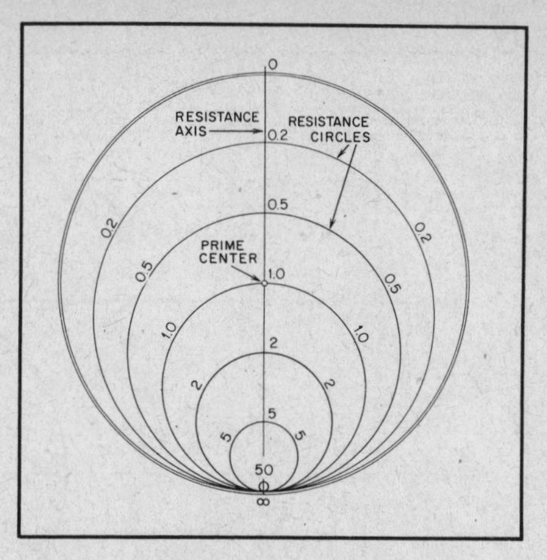


Fig. 16 — Resistance circles of the Smith Chart coordinate system.

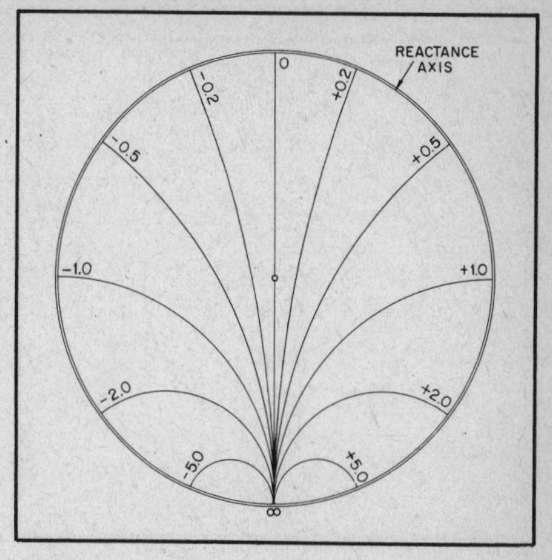


Fig. 17 — Reactance circles (segments) of the Smith Chart coordinate system.

a radius of infinity). The centers are displaced to the right or left on a line tangent to the bottom of the chart. The large outer circle bounding the coordinate portion of the chart is the reactance axis.

Each reactance circle segment is assigned a value of reactance, indicated near the point where the circle touches the reactance axis. All points along any one segment have the same reactance value. As with the resistance circles, the values assigned to each reactance circle are normalized with respect to the value assigned to prime center. Values to the right of the resistance axis are positive (inductive), and those to the left of the resistance axis are negative (capacitive).

When the resistance family and the reactance family of circles are combined, the coordinate system of the Smith Chart results, as shown in Fig. 18. Complex series impedance can be plotted on this coordinate system.

## IMPEDANCE PLOTTING

Suppose we have an impedance consisting of 50 ohms resistance and 100 ohms inductive reactance (Z = 50 + j100). If we assign a value of 100 ohms to prime center, we normalize the above impedance by dividing each component of the impedance by 100. The normalized impedance would then be 50/100 + j(100/100) = 0.5 + j1.0. This impedance would be plotted on the Smith Chart at the intersection of the 0.5 resistance circle and the +1.0 reactance circle, as indicated in Fig. 18. If a value of 50 ohms had been assigned to prime center, as for 50-ohm coaxial line, the same impedance would be plotted at the intersection of the 50/50 = 1.0 resistance circle, and the 100/50 = 2.0 positive reactance circle, or at 1 + j2 (also indicated in Fig. 18).

From these examples, it may be seen that the same impedance may be plotted at different points on the chart, depending upon the value assigned to prime center. It is customary when solving transmission-

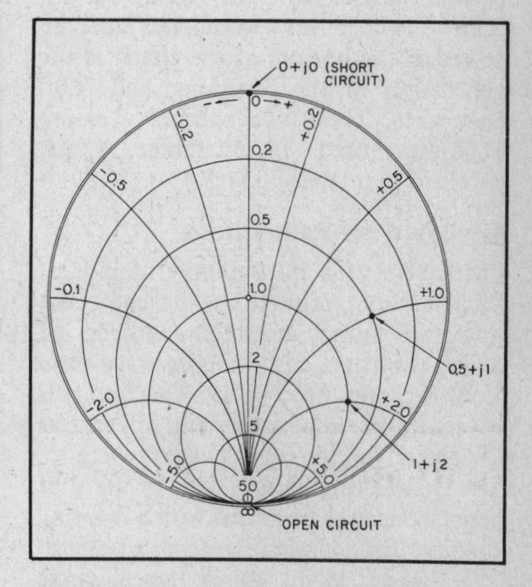


Fig. 18 — The complete coordinate system of the Smith Chart. For simplicity, only a few divisions are shown for the resistance and reactance values. Smith Chart forms are available from ARRL Hq. in two types, standard and expanded. The standard forms are of the type shown here, while the expanded forms are suitable only for work with SWR values below 1.6:1. At the time of this writing, five 8-1/2 × 11 inch Smith Chart forms are available for $1.

line problems to assign to prime center a value equal to the characteristic impedance, or $Z_0$, of the line being used. This value should always be recorded at the start of calculations, to avoid possible confusion later. In using the specialized charts with the value of 50 at prime center, it is, of course, not necessary to normalize impedances when working with 50-ohm line. The resistance and reactance values may be plotted directly.

### Short and Open Circuits

On the subject of plotting impedances, two special cases deserve consideration. These are short circuits and open circuits. A true short circuit has zero resistance and zero reactance, or 0 + j0. This impedance would be plotted at the top of the chart, at the intersection of the resistance and the

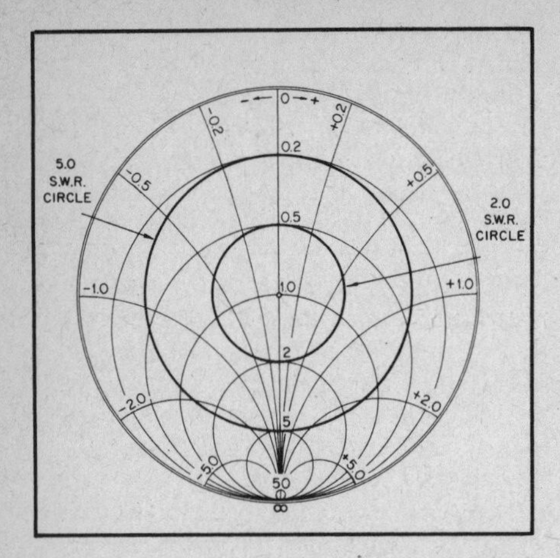


Fig. 19 — Smith Chart with SWR circles added

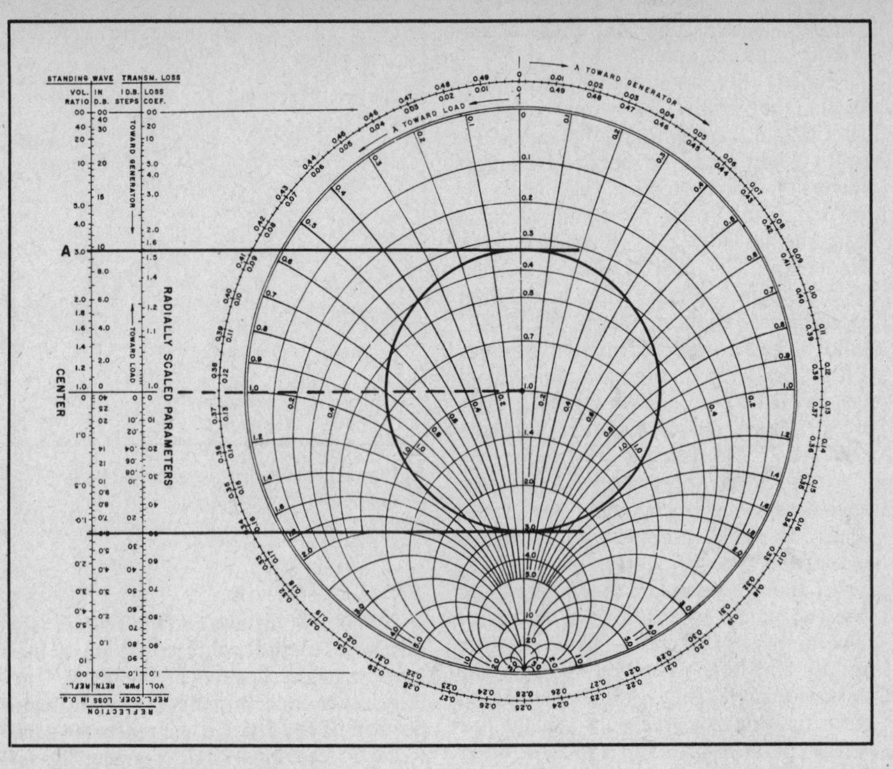


Fig. 20 — Example discussed in text.

reactance axes. An open circuit has infinite resistance, and would therefore be plotted at the bottom of the chart, at the intersection of the resistance and reactance axes. These two special cases are sometimes used in matching stubs, described in a later chapter.

### Standing-Wave-Ratio Circles

Members of a third family of circles, which are not printed on the chart but which are added during the process of solving problems, are ***standing-wave-ratio*** or ***SWR circles.*** See Fig. 19. This family is centered on prime center, and appears as concentric circles inside the reactance axis. During calculations, one or more of these circles may be added with a drawing compass. Each circle represents a value of SWR, every point on a given circle representing the same SWR. The SWR value for a given circle may be determined directly from the chart coordinate system, by reading the resistance value where the SWR circle crosses the resistance axis, below prime center. (The reading where the circle crosses the resistance axis above prime center indicates the inverse ratio.)

Consider the situation where a load mismatch in a length of line causes a 3-to-1 standing-wave ratio to exist. If we temporarily disregard line losses, we may state that the SWR remains constant throughout the entire length of this line. This is represented on the Smith Chart by drawing a 3:1 constant SWR circle (a circle with a radius of 3 on the resistance axis), as in Fig. 20. The design of the chart is such that any impedance encountered anywhere along the length of this mismatched line will fall on the SWR circle, and may be read from the coordinates merely by progressing around the SWR circle by an amount corresponding to the length of the line involved.

This brings into use the *wavelength scales*, which appear, in Fig. 20, near the perimeter of the Smith Chart. These scales are calibrated in terms of portions of an electrical wavelength along a transmission line. One scale, running counterclockwise, starts at the generator or input end of the line and progresses toward the load, while the other scale starts at the load and proceeds toward the generator in a clockwise direction. The complete circle represents one half wavelength. Progressing once around the perimeter of these scales corresponds to progressing along a transmission line for a half wavelength. Because impedances will repeat themselves every half wavelength along a piece of line, the chart may be used for any length of line by disregarding or subtracting from the line's total length an integral, or whole number, of half wavelengths.

Also shown in Fig. 20 is a means of transferring the radius of the SWR circle to the external scales of the chart, by drawing lines tangent to the circle. Instead, the radius of the SWR circles may be simply transferred to the external scale by placing the point of a drawing compass at the center, or 0, line and inscribing a short arc across the appropriate scale. It will be noted that when this is done in Fig. 20, the external STANDING-WAVE VOLTAGE-RATIO scale indicates the SWR to be 3.0 (at A) — our condition for initially drawing the circle on the chart (and the same as the SWR reading on the resistance axis).

### Solving Problems with the Smith Chart

Suppose we have a transmission line with a characteristic impedance of 50 ohms, and an electrical length of 0.3 wavelength. Also, suppose we terminate this line with an impedance having a resistive component of 25 ohms and an inductive reactance of 25 ohms ($Z = 25 + j25$), and desire to determine the input impedance to the line. Because the line is not terminated in its characteristic impedance, we know that standing waves will be present on the line, and that, therefore, the input impedance to the line will not be exactly 50 ohms. We proceed as follows: First, normalize the load impedance by dividing both the resistive and reactive components by 50 ($Z_0$ of the line being used). The normalized impedance in this case is $0.5 + j0.5$. This is plotted on the chart at the intersection of the 0.5 resistance and the +0.5 reactance circles, as in Fig. 21. Then draw a constant-SWR circle passing through this point. The radius of this circle may then be transferred to the external scales with the drawing compass. From the external SW-VR scale, it may be seen (at A) that the voltage ratio of 2.6 exists for this radius, indicating that our line is operating with an SWR of 2.6 to 1. This figure is converted to decibels in the adjacent scale, where 8.4 dB may be read (at B), indicating that the ratio of the voltage maximum to the voltage minimum along the line is 8.4 dB.

Next, with a straightedge, draw a radial line from prime center through the plotted point to intersect the wavelengths scale, and read a value from the wavelengths scale. Because we are starting from the load, we use the TOWARD-GENERATOR or

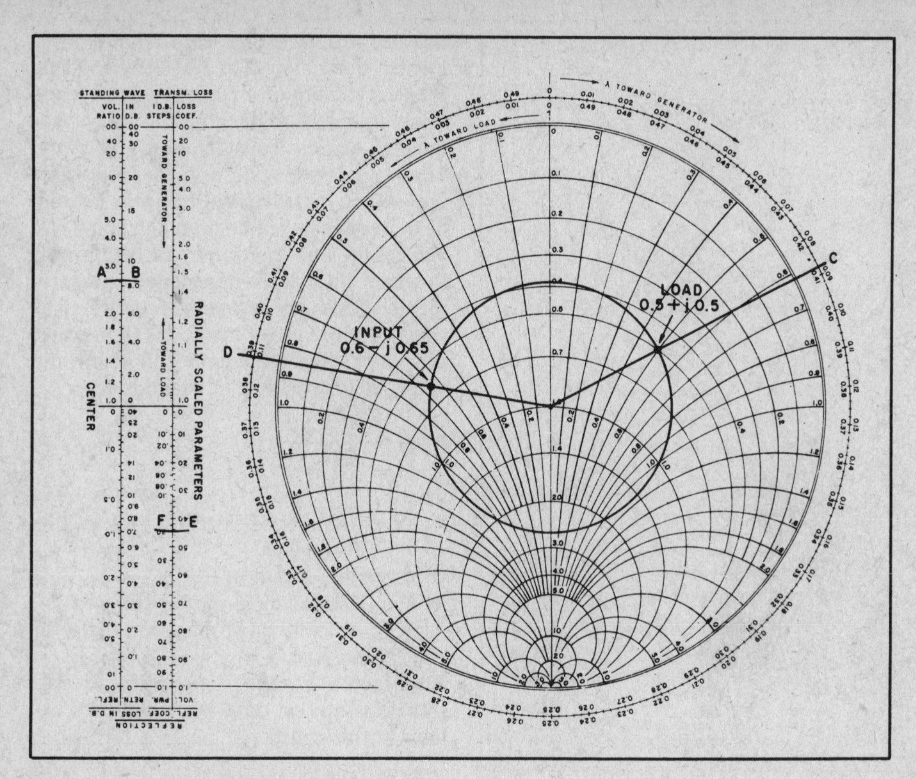


Fig. 21 — Example discussed in text.

outermost calibration, and read 0.088 wavelength (at C). To obtain the line input impedance, we merely find the point on the SWR circle that is 0.3 wavelength toward the generator from the plotted load impedance. This is accomplished by adding 0.3 (the length of the line in wavelengths) to the reference or starting point, 0.088; 0.3 + 0.088 = 0.388. Locate 0.388 on the TOWARD-GENERATOR scale (at D), and draw a second radial line from this point to prime center. The intersection of the new radial line with the SWR circle represents the line input impedance, in this case 0.6 − j0.65. To find the line impedance, multiply by 50 — the value assigned to prime center, which equals 30 — j32.5, or 30 ohms resistance and 32.5 ohms capacitive reactance. This is the impedance that a transmitter must match if such a system were a combination of antenna and transmission line, or is the impedance that would be measured on the impedance bridge if the measurement were taken at the line input.

In addition to the line input impedance and the SWR, the chart reveals several other operating characteristics of the above system of line and load, if a closer look is desired. For example, the voltage reflection coefficient, both magnitude and phase angle, for this particular load is given. The phase angle is read under the radial line drawn through the plot of the load impedance where the line intersects the ANGLE-OF-REFLECTION-COEFFICIENT scale. This scale is not included in Fig. 21, but will be found on the Smith Chart, just inside the wavelengths scales. In this example, the reading would be about 116.5 degrees. This indicates the angle by which the reflected voltage wave leads the incident wave at the load. It will be noted that angles on the left half, or capacitive-reactance side, of the chart are negative angles, a "negative" lead indicating that the reflected voltage wave actually lags the incident wave.

The magnitude of the voltage-reflection-coefficient may be read from the external REFLECTION-COEFFICIENT-VOLTAGE scale, and is seen to be approximately 0.44 (at E) for this example, meaning 44 percent of the incident voltage is reflected. Adjacent to this scale on the POWER calibration, it is noted (at F) that the power reflection coefficient is 0.20, indicating that 20 percent of the incident power would be reflected. (The amount of reflected power is proportional to the square of the reflected voltage.)

### Admittance Coordinates

Quite often it is desirable to convert impedance information to admittance data — conductance and susceptance. Working with admittances greatly simplifies determining the resultant when two complex impedances are connected in parallel, as in stub matching. The conductance values may be added directly, as may be the susceptance values, to arrive at the overall admittance for the parallel combination. This admittance may then be converted back to impedance data.

On the Smith Chart, the necessary conversion may be made very simply. The equivalent admittance of a plotted impedance value lies diametrically opposite the impedance point on the chart. In the foregoing example, where the normalized line input impedance is 0.6 — j0.65, the equivalent admittance lies at the intersection of the SWR circle and the extension of the straight line passing from point D to prime center. Although not shown in Fig. 21, the normalized admittance value may be read as 0.76 + j0.84 if the line is extended. (Capacitance is considered to be a positive susceptance and inductance a negative susceptance.) The admittance in siemens is determined by dividing the normalized values by the $Z_0$ of the line. For this example the admittance would be 0.76/50 + j0.84/50 = 0.0152 + j0.0168 siemen.

Of course admittance coordinates may be converted to impedance coordinates just as easily — by locating the point on the Smith Chart that is diametrically opposite the point representing the admittance coordinates, on the same SWR circle.

### Determining Antenna Impedances

To determine an antenna impedance from the Smith Chart, the procedure is similar to the previous example. The electrical length of the feed line must be known and the impedance value at the input end of the line must be determined through measurement, such as with an impedance-measuring bridge. In this case, the antenna is connected to the far end of the line and becomes the load for the line. Whether the antenna is intended purely for transmission of energy, or purely for reception makes no difference; the antenna is still the terminating or load impedance on the line as far as these measurements are concerned. The *input* or *generator* end of the line would be that end connected to the device for measurement of the impedance. In this type of problem, the measured impedance is plotted on the chart, and the TOWARD-LOAD wavelengths scale is used in conjunction with the electrical line length to determine the actual antenna impedance.

For example, assume we have a measured input impedance to a 50-ohm line of 70 −j25 ohms. The line is 2.35 wavelengths long, and is terminated in an antenna. We desire to determine the antenna feed impedance. Normalize the input impedance with respect to 50 ohms, which comes out 1.4 −j0.5, and plot this value on the chart. See Fig. 22. Draw a constant-SWR circle through the point, and transfer the radius to the external scales. The SWR of 1.7 may be read from the SW-VR scale (at A). Now draw a radial line from prime center through this plotted point to the wavelengths scale, and read a reference value, which is 0.195 (at

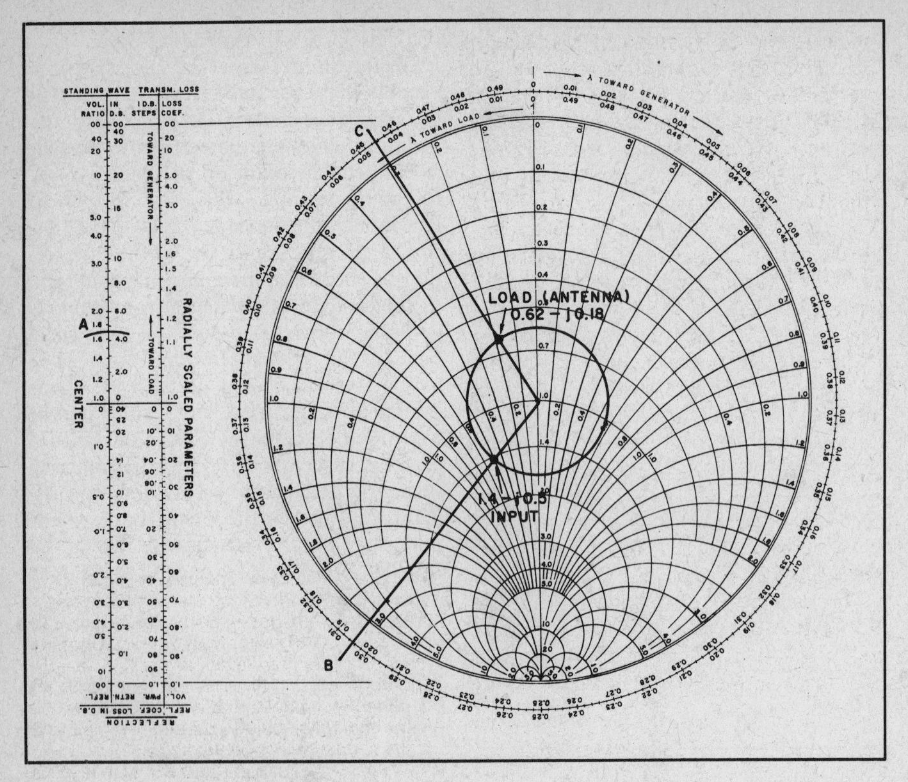


Fig. 22 — Example discussed in text.

B), on the TOWARD-LOAD scale. Remember, we are starting at the *generator* end of the transmission line.

To locate the load impedance on the SWR circle, we add the line length, 2.35 wavelengths to the reference value from the wavelengths scale, and locate the new value on the TOWARD-LOAD scale; 2.35 + 0.195 = 2.545. However, the calibrations extend only from 0 to 0.5, so we must subtract a whole number of half wavelengths from this value and use only the remaining value. In this situation, the largest integral number of half wavelengths that can be subtracted is 5, or 2.5 wavelengths. Thus, 2.545 − 2.5 = 0.045, and the 0.045 value is located on the TOWARD-LOAD scale (at C). A radial line is then drawn from this value to prime center, and the coordinates at the intersection of the second radial line and the SWR circle represent the load impedance. To read this value closely, some interpolation between the printed coordinate lines must be made, and the value of 0.62 − j0.18 is read. Multiplying by 50, the actual load or antenna impedance is 31 − j9 ohms, or 31 ohms resistance with 9 ohms capacitive reactance.

Problems may be entered on the chart in yet another manner. Suppose we have a length of 50-ohm line feeding a base-loaded resonant vertical ground-plane antenna which is shorter than a quarter wave. Further, suppose we have an SWR monitor in the line, and that it indicates an SWR of 1.7 to 1. The line is known to be 0.95 wavelength long. We desire to know both the input and the antenna impedances.

From that data given, we have no impedances to enter into the chart. We may, however, draw a circle representing the 1.7 SWR. We also know, from the definition of resonance, that the antenna presents a purely resistive load to the line; i.e. no reactive component. Thus, the antenna impedance must lie on the resistance axis. If we were to draw such an SWR circle and observe the chart with only the circle drawn, we would see two points which satisfy the resonance requirement for the load. These points are 0.59 + j0 and 1.7 + j0. Multiplying by 50, these values represent 29.5 and 85 ohms resistance. This may sound familiar, because, as was discussed earlier, when a line is terminated in a pure resistance, the SWR in the line equals $Z_R/Z_o$ or $Z_o/Z_R$, where $Z_R$ = load resistance and $Z_o$ = line impedance.

If we consider antenna fundamentals described earlier, we know that the theoretical impedance of a quarter-wave ground-plane antenna is approximately 36 ohms. We therefore can quite logically discard the 85-ohm impedance figure in favor of the 29.5-ohm value. This is then taken as the load-impedance value for the Smith Chart calculations. The line input impedance is found to be 0.64 − j0.21, or 32 − j10.5 ohms, after subtracting 0.5 wavelength from 0.95, and finding 0.45 wavelength on the TOWARD-GENERATOR scale. (The wavelength reference in this case is 0.)

### Determination of Line Length

In the example problems given so far in this section, the line length has conveniently been stated in wavelengths. The electrical length of a piece of line depends upon its physical length, the radio frequency under consideration, and the velocity of propagation in the line. If an impedance-measurement bridge is capable of quite reliable readings at high line-SWR values, the line length may be determined through line input-impedance measurements with short- or open-circuit terminations. A more direct method is to measure the line's physical length and apply the value to a formula. The formula is:

$$N = \frac{LF}{984K}$$

where

N = number of electrical wavelengths in the line,
L = line length in feet,
F = frequency in megahertz, and
K = velocity or propagation factor of the line.

The factor K may be obtained from transmission-line data tables that appear later in this chapter.

## ATTENUATION

The discussion in the preceding part of this chapter applies to all types of transmission lines, regardless of their physical construction. It is, however, based on the assumption that there is no power loss in the line. Every practical line will have some inherent loss, partly because of the resistance of the conductors, partly because power is consumed in every dielectric used for insulating the conductors, and partly because in many cases a small amount of power escapes from the line by radiation.

Losses in a line modify its characteristic impedance slightly, but usually not significantly. They will also affect the input impedance; in this case the theoretical values will be modified only slightly if the line is short and has only a small loss, but may be changed considerably if an appreciable proportion of the power input to the line is dissipated by the line itself. A large loss may exist because the line is long, because it has inherently high loss per unit length, because the standing-wave ratio is high, or because of a combination of two or all three of these factors.

The reflected power returning to the input terminals of the line is less when the line has losses than it would be if there were none. The overall effect is that the SWR changes along the line, being highest at the load and smallest at the input terminals. A long, high-loss line therefore tends to act, so far as its input impedance is concerned, as though the impedance match at the load end were better than is actually the case.

### Line Losses

The power lost in a transmission line is not directly proportional to the line length but varies logarithmically with the length. That is, if 10% of the input power is lost in a section of line of certain length, 10% of the remaining power will be lost in the next section of the same length, and so on. For this reason it is customary to express line losses in terms of decibels per unit length, since the decibel is a logarithmic unit. Calculations are very simple because the total loss in a line is found by multiplying the decibel loss per unit length by the total length of the line. Line loss is usually expressed in decibels per 100 feet. It is necessary to specify the frequency for which the loss applies, since the loss varies with frequency.

Conductor loss and dielectric loss both increase as the operating frequency is increased, but not in the same way. This, together with the fact that the relative amount of each type of loss depends on the actual construction of the line, makes it impossible to give a specific relationship between loss and frequency that will apply to all types of lines. Each line has to be considered individually. Actual loss values are given in a later section.

**Effect of SWR**

The power lost in a line is least when the line is terminated in a resistance equal to its characteristic impedance, and increases with an increase in the standing-wave ratio. This is because the effective values of both current and voltage become larger as the SWR becomes greater. The increase in effective current raises the ohmic losses in the conductors, and the increase in effective voltage increases the losses in the dielectric.

The increased loss caused by an SWR greater than 1 may or may not be serious. If the SWR at the load is not greater than 2, the *additional* loss caused by the standing waves, as compared with the loss when the line is perfectly matched, does not amount to more than about 1/2 dB even on very long lines. Since 1/2 dB is an undetectable change in signal strength, it can be said that from a practical standpoint an SWR of 2 or less is, so far as losses are concerned, every bit as good as a perfect match.

The effect of SWR on line loss is shown in Fig. 23. The horizontal axis is the attenuation, in decibels, of the line when perfectly matched. The vertical axis gives the *additional* attenuation, in decibels, caused by standing waves. For example, if the loss in a certain line is 4 dB when perfectly matched, an SWR of 3 on that same line will cause an additional loss of 1.1 dB, approximately. The total loss on the poorly matched line is therefore 4 + 1.1 = 5.1 dB. If the SWR were 10 instead of 3, the additional loss would be 4.3 dB, and the total loss 4 + 4.3 = 8.3 dB.

It is important to note that the curves in Fig. 23 that represent SWR are the values that exist *at the load*. In most cases of amateur operation, this will be at the antenna end of a length of transmission line. The SWR as measured at the *input* or transmitter end of the line will be less, depending on the line attenuation. It is not always convenient to measure SWR directly at the antenna. However, by using the graph shown in Fig. 24, the SWR at the load can be obtained by measuring it at the input to the transmission line and using the known (or estimated) loss of the transmission line. (See later section on testing coaxial cable under "Losses and Deterioration.") For example, if the SWR at the transmitter end of a line is measured as 3 to 1 and the line is known to have a total attenuation (under matched conditions) of 1 dB, the SWR at the load end of the line will be 4.5 to 1. From Fig. 23, the additional loss is nearly 1 dB because of the presence of the SWR. The total line loss in this case is 2 dB.

The equations relating input SWR, load SWR and total transmission-line loss are given below. Any one of these quantities may be calculated if the other two are known.

$$\text{SWR at load} = S_l = \frac{A + B}{A - B}$$

$$\text{SWR at input} = S_i = \frac{B + C}{B - C}$$

$$\text{Total loss} = L_t = 10 \log \left[ \frac{B^2 - C^2}{B(1 - C^2)} \right]$$

where

$$A = \frac{S_i + 1}{S_i - 1}$$

$$B = 10^{L_m/10}$$

$$C = \frac{S_l - 1}{S_l + 1}$$

$L_m$ = line loss in decibels when matched

It is of interest to note that when the line loss is high with perfect matching, the *additional* loss in decibels caused by the SWR tends to be constant regardless of the matched line loss. The reason for this is that the amount of power available to be reflected from the load is reduced, because relatively little power reaches the load in the first place. For example, if the line loss with perfect matching is 6 dB, only 25% of the power originally put into the line reaches the load. If the mismatch at the load (the SWR at the load) is 4 to 1, 36% of the power reaching the load will be reflected. Of the power originally put into the line, then, 0.25 × 0.36 = 0.09 or 9% will be reflected. This in turn will be attenuated 6 dB in traveling back to the input end of the line, so that only 0.09 × 0.25 = 0.0225 or slightly over 2% of the original power actually gets back to the input terminals. With such a small proportion of power returning to the input terminals the SWR measured at the *input*

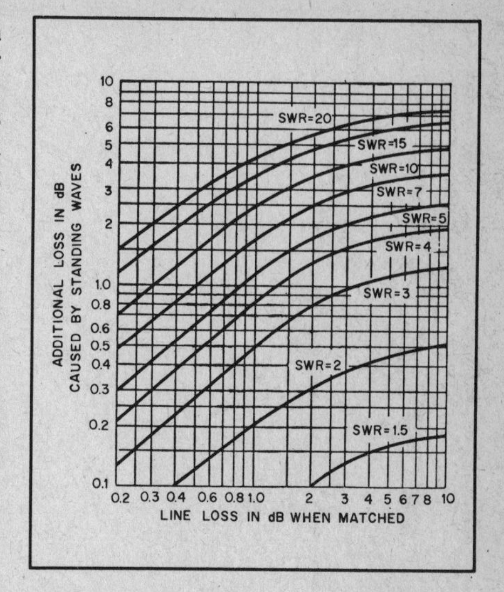


Fig. 23 — Increase in line loss because of standing waves (SWR measured at the load). To determine the total loss in decibels in a line having an SWR greater than 1, first determine the loss for the particular type of line, length and frequency, on the assumption that the line is perfectly matched (Table 1). Locate this point on the horizontal axis and move up to the curve corresponding to the actual SWR. The corresponding value on the vertical axis gives the additional loss in decibels caused by the standing waves.

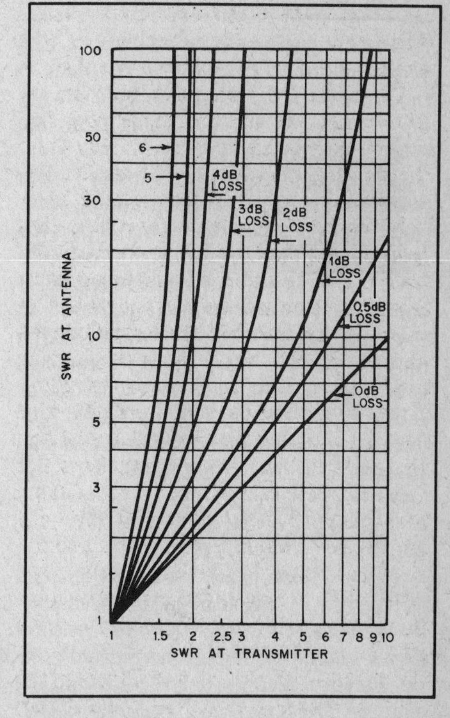


Fig. 24 — SWR at input end of transmission line vs. SWR at load end for various values of matched-line loss.

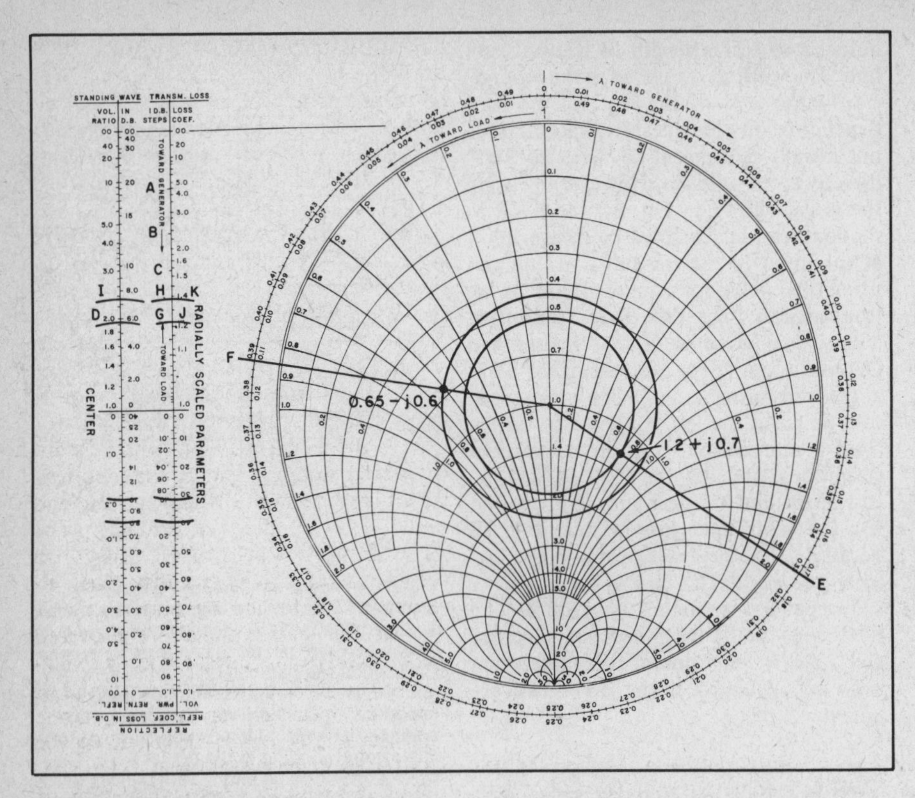


Fig. 25 — Example of Smith Chart calculations taking line losses into account.

end of the line would be only about 1.35 to 1 — although it is 4 to 1 at the load. In the presence of line losses the SWR always decreases along the line going from the load to the input end.

On lines having low losses when perfectly matched, a high standing-wave ratio may increase the power loss by a large factor. However, in this case the *total* loss may still be inconsequential in comparison with the power delivered to the load. An SWR of 10 on a line having only 0.3 dB loss when perfectly matched will cause an additional loss of 1 dB, as shown by the curves. This loss would produce a just-detectable difference in signal strength.

## LINE-LOSS CONSIDERATIONS WITH THE SMITH CHART

The problems presented earlier ignored attenuation, or line losses. Quite frequently it is not even necessary to consider losses when making calculations; any difference in readings obtained would be almost imperceptible on the Smith Chart. When the line losses become appreciable, as described above, loss considerations may be warranted in making Smith Chart calculations. This involves only one simple step, in addition to the procedures previously presented.

Because of line losses, the SWR does not remain constant throughout the length of the line, as just discussed. As a result, there is a decrease in SWR as one progresses away from the load. To truly represent this situation on the Smith Chart, instead of drawing a constant-SWR circle, it would be necessary to draw a spiral inward and clockwise from the load impedance toward the generator. The rate at which the curve spirals toward prime center is related to the attenuation in the line. Rather than drawing spiral curves, a simpler method is used in solving line-loss problems, by means of the external scale TRANSMISSION-LOSS, 1-DB STEPS in Fig. 25. Because this is only a relative scale, the decibel steps are not numbered.

If we start at the top end of this external scale and proceed in the direction indicated toward generator, the first dB step is seen to occur at a radius from center corresponding to an SWR of about 9 (at A); the second dB step falls at an SWR of about 4.5 (at B), the third at 3.0 (at C), and so forth, until the 15th dB step falls at an SWR of about 1.05 to 1. This means that a line terminated in a short or open circuit (infinite SWR) and having an attenuation of 15 dB, would exhibit an SWR of only 1.05 at its input. It will be noted that the dB steps near the lower end of the scale are very close together, and a line attenuation of 1 or 2 dB in this area will have only slight effect on the SWR. But near the upper end of the scale, 1- or 2-dB loss has considerable effect on the SWR.

In solving a problem using line-loss information, it is necessary only to modify the radius of the SWR circle by an amount indicated on the TRANSMISSION-LOSS, 1-DB-STEPS scale. This is accomplished by drawing a second SWR circle, of either greater or lesser radius than the first, as the case may be.

For example, assume that we have a 50-ohm line 0.282 wavelength long, with 1-dB inherent attenuation. The line input impedance is measured as 60 + j35 ohms. We desire to know the SWR at the input and at the load, and the load impedance. As before, we normalize the 60 + j35-ohm impedance, plot it on the chart, and draw a constant-SWR circle and a radial line through the point. In this case, the normalized impedance is 1.2 + j0.7. From Fig. 25, the SWR at the line input is seen to be 1.9 (at D), and the radial line is seen to cross the TOWARD-LOAD scale at 0.328 (at E). To the 0.328 we add the line length, 0.282, and arrive at a value of 0.610. To locate this point on the TOWARD-LOAD scale, first subtract 0.500, and locate 0.110 (at F); then draw a radial line from this point to prime center.

To account for line losses, transfer the radius of the SWR circle to the external 1-DB-STEPS scale. This radius will cross the external scale at G, the fifth decibel mark from the top. Since the line loss was given as 1 dB, we strike a new radius (at H), one "tick mark" higher (toward load) on the same scale. (This will be the fourth decibel tick mark from the top of the scale.) Now transfer this new radius back to the main chart, and scribe a new SWR circle of this radius. This new radius represents the SWR at the load, and is read as about 2.3 on the external SW-VR scale. At the intersection of the new circle and the load radial line, we read 0.65 − j0.6 as the normalized load impedance. Multiplying by 50, the actual load impedance is 32.5 − j30 ohms. The SWR in this problem was seen to increase from 1.9 at the line input to 2.3 (at I) at the load, with the 1-dB line loss taken into consideration.

In the example above, values were chosen to fall conveniently on or very near the "tick marks" on the 1-DB scale. Actually, it is a simple matter to interpolate between these marks when making a radius correction. When this is necessary, the relative distance between marks for each decibel step should be maintained while counting off the proper number of steps.

Adjacent to the 1-DB-STEPS scale lies a LOSS-COEFFICIENT scale. This scale provides a factor by which the matched-line loss in decibels should be multiplied to account for the increased losses in the line when standing waves are present. These added losses do not affect the standing-wave ratio or impedance calculations; they are merely the additional dielectric and copper losses of the line caused by the fact that the line conducts more average current and must withstand more average voltage in the presence of standing waves. In the above example and in Fig. 25, the loss coefficient at the input end is seen to be 1.21 (at J), and 1.39 (at K) at the load. As a good approximation, the loss coefficient may be averaged over the length of line under consideration; in this case, the average is 1.3. This means that the total losses in the line are 1.3 times the matched loss of the line (1 dB), or 1.3 dB, the same result that may be obtained from Fig. 23 for the data of the above example.

### Smith Chart Procedure Summary

To summarize briefly, any calculations made on the Smith Chart are performed in four basic steps, although not necessarily in the order listed.

1) Normalize and plot a line input (or load) impedance, and construct a constant-SWR circle.

2) Apply the line length to the wavelengths scales.

3) Determine attenuation or loss, if required, by means of a second SWR circle.

4) Read normalized load (or input) impedance, and convert to impedance in ohms.

The Smith Chart may be used for many types of problems other than those presented as examples here. The transformer action of a length of line — to transform a high impedance (with perhaps high reactance) to a purely resistive impedance of low value — was not mentioned. This is known as "tuning the line," for which the chart is very helpful, eliminating the need for cut-and-try procedures. The chart may also be used to calculate lengths for shorted or open matching stubs in a system, described later in this chapter. In fact, in any application where a transmission line is not perfectly matched, the Smith Chart can be of value.

## VOLTAGES AND CURRENTS ON LINES

The power reflected from a mismatched load does not represent an actual loss, except as it is attenuated in traveling back to the input end of the line. It merely represents power returned, and the actual effect is to reduce the power taken from the source. That is, it reduces the coupling between the power source and the line. This is easily overcome by readjusting the coupling until the actual power put into the line is the same as it would be with a matched load. In doing this, of course, the voltages and currents at loops along the line are increased.

As an example, suppose that a line having a characteristic impedance of 600 ohms is matched by a resistive load of 600 ohms and that 100 watts of power goes into the input terminals. The line simply looks like a 600-ohm resistance to the source of power. By Ohm's Law the current and voltage in such a matched line are

$$I = \sqrt{P/R}$$

$$E = \sqrt{PR}$$

Substituting 100 watts for P and 600 ohms

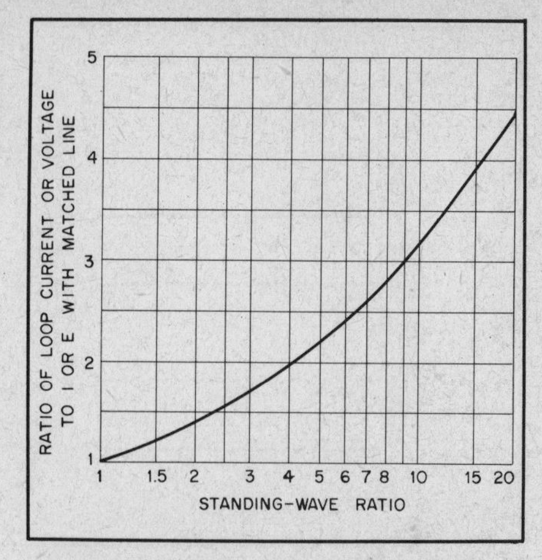


Fig. 26 — Increase in maximum value of current or voltage on a line with standing waves, as referred to the current or voltage on a perfectly matched line, for the same power delivered to the load. Voltage and current at minimum points are given by the reciprocals of the values along the vertical axis. The curve is plotted from the relationship, current (or voltage) ratio = $\sqrt{SWR}$.

for R, the current is 0.408 ampere and the potential is 245 volts. Assuming for the moment that the line has no losses, all the power will reach the load so the voltage and current at the load will be the same as at the input terminals.

Now suppose that the load is 60 ohms instead of 600 ohms. The SWR is 10, therefore. The reflection coefficient, or ratio of the reflected voltage or current to the voltage or current arriving at the load, is

$$\rho = \frac{SWR - 1}{SWR + 1}$$

In this case the reflection coefficient is (10 − 1)/(10 + 1) = 9/11 = 0.818, so that the reflected voltage and current are both equal to 81.8% of the incident voltage and current. The reflected *power* is proportional to the square of either the current or voltage, and so is equal to $(0.818)^2$ = 0.67 times the incident power, or 67 watts. Since we have assumed that the line has no losses, this amount of power arrives back at the input terminals and subtracts from the original 100 watts, leaving only 33 watts as the amount of power actually taken from the source.

In order to put 100 watts into the 60-ohm load the coupling to the source must be increased so that the incident power minus the reflected power equals 100 watts, and since the power absorbed by the load is only 33% of that reaching it, the incident power must equal 100/0.33 = 303 watts. In a perfectly matched line, the current and voltage with 303 watts input would be 0.71 ampere and 426 volts, respectively. The reflected current and voltage are 0.818 times these values, or 0.581 ampere and 348 volts. At current maxima or loops the current will therefore be 0.71 + 0.58 = 1.29 A, and at a minimum point will be 0.71 − 0.58 = 0.13 A. The voltage maxima and minima will be 426 + 348 = 774 volts and 426 − 348 = 78 volts. (Because of rounding off figures in the calculation process, the SWR does not work out to be exactly 10 in either the voltage or current case, but the error is very small.)

In the interests of simplicity this example has been based on a line with no losses, but the approximate effect of line attenuation could be included without much difficulty. If the matched-line loss were 3 dB, for instance, only half the input power would reach the load, so new values of current and voltage at the load would be computed accordingly. The reflected power would then be based on the attenuated figure, and then itself attenuated 3 dB to find the power arriving back at the input terminals. The overall result would be, as stated before, a reduction in the SWR at the input terminals as compared with that at the load, along with less actual power delivered to the load for the same power input to the line.

Fig. 26 shows the ratio of current or voltage at a loop, in the presence of standing waves, to the current or voltage that would exist with the same power in a perfectly matched line. Strictly speaking, the curve applies only near the load in the case of lines with appreciable losses. However, the curve shows the maximum possible value of current or voltage that can exist along the line whether there are line losses or not, and so is useful in determining whether or not a particular line can operate safely with a given SWR.

## SPECIAL CASES

Beside the primary purpose of transporting power from one point to another, transmission lines have properties that are useful in a variety of ways. One such special case is a line an exact multiple of one-quarter wavelength (90 degrees) long. As shown earlier, such a line will have a purely resistive input impedance when the termination is a pure resistance. Also, unterminated — i.e., short-circuited or open-circuited — lines can be used in place of conventional inductors and capacitors since such lines have an input impedance that is substantially a pure reactance when the line losses are low.

### The Half-Wavelength Line

When the line length is an even multiple of 90 degrees (that is, a multiple of a half wavelength), the input resistance is equal to the load resistance, regardless of the line $Z_0$. As a matter of fact, a line an *exact* multiple of a half wave in length (disregarding line losses) simply repeats, at its input or sending end, whatever impedance exists at its output or receiving end; it does not matter whether the impedance at the receiving end is resistive, reactive, or a combination of both. Sections of line having such length can be cut in or out without changing any of the operating conditions, at least when the losses in the line itself are negligible.

**Impedance Transformation with Quarter-Wave Lines**

The input impedance of a line an odd multiple of a quarter wavelength long is

$$Z_s = \frac{Z_o^2}{Z_L}$$

where $Z_s$ is the input impedance and $Z_L$ is the load impedance. If $Z_L$ is a pure resistance, $Z_s$ also will be a pure resistance. Rearranging this equation gives

$$Z_O = \sqrt{Z_s Z_L}$$

This means that if we have two values of impedance that we wish to "match," we can do so if we connect them together by a quarter-wave transmission line having a characteristic impedance equal to the square root of their product.

A quarter-wave line is, in effect, a transformer. It is frequently used as such in antenna work when it is desired, for example, to transform the impedance of an antenna to a new value that will match a given transmission line. This subject is considered in greater detail in a later chapter.

**Lines as Circuit Elements**

An open- or short-circuited line does not deliver any power to a load, and for that reason is not, strictly speaking a "transmission" line. However, the fact that a line of the proper length has inductive reactance makes it possible to substitute the line for a coil in an ordinary circuit. Likewise another line of appropriate length having capacitive reactance can be substituted for a capacitor.

Sections of lines used as circuit elements are usually a quarter wavelength or less long. The desired type of reactance (inductive or capacitive) or the desired type of resonance (series or parallel) is obtained by shorting or opening the far end of the line. The circuit equivalents of various types of line sections are shown in Fig. 27.

When a line section is used as a reactance, the amount of reactance is determined by the characteristic impedance and the electrical length of the line. The type of reactance exhibited at the input terminals of a line of given length depends on whether it is open- or short-circuited at the far end.

The equivalent "lumped" value for any "inductor" or "capacitor" may be determined with the aid of the Smith Chart. Line losses may be taken into account if desired, as explained in an earlier section.

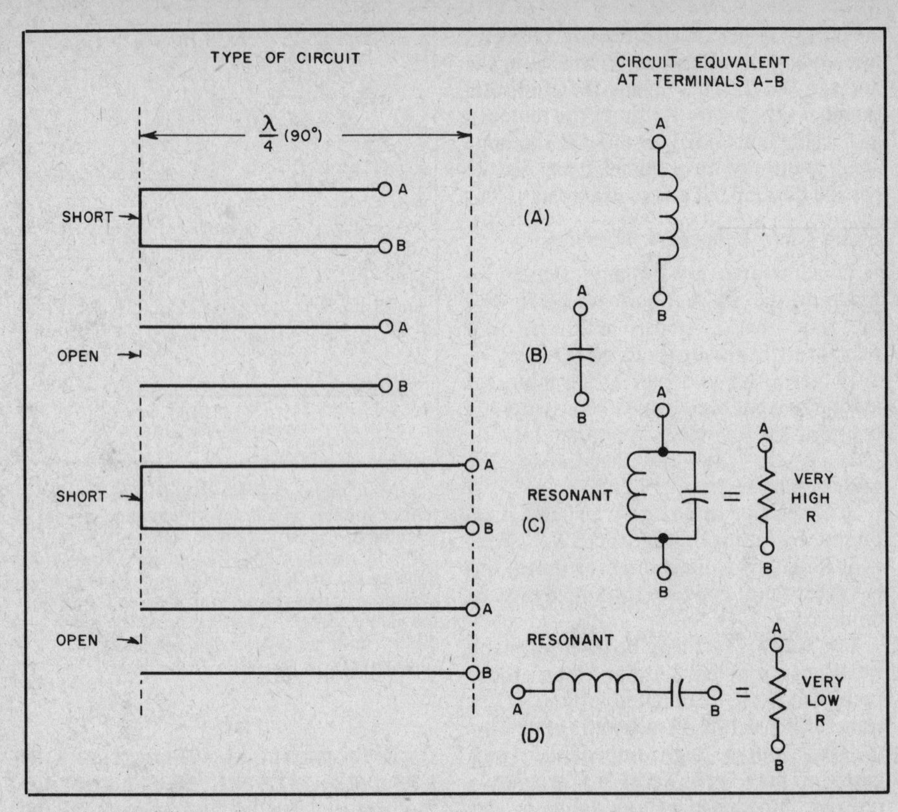


Fig. 27 — Lumped-constant circuit equivalents of open- and short-circuited transmission lines.

See Fig. 28. Remember that the right half of the Smith Chart coordinate system is used for impedances containing inductive reactances, and the left half for capacitive reactances. For example, a section of 600-ohm line 3/16-wavelength long (0.1875 λ) and short-circuited at the far end is represented by L1, drawn around a portion of the perimeter of the chart. The "load" is a short-circuit, 0 + j0 ohms, and the TOWARD GENERATOR wavelengths scale is used for marking off the line length. At A in Fig. 28 may be read the normalized impedance as seen looking into the length of line, 0 + j2.4. The reactance is therefore inductive, equal to 600 × 2.4 = 1440 ohms. The same line open-circuited (termination impedance = ∞, the point at the bottom of the chart) is represented by L2 in Fig. 28. At B the normalized line-input impedance may be read as 0 − j0.41; the reactance in this case is capacitive, 600 × 0.41 = 246 ohms. (Line losses are disregarded in these examples.) From Fig. 28 it is easy to visualize that if L1 were to be extended by a quarter wavelength, represented by L3, the line-input impedance would be identical to that obtained in the case represented by L2 alone. In the case of L2, the line is open-circuited at the far end, but in the case of L3 the line is terminated in a short. The added quarter wavelength of line for L3 provides the "transformer action" discussed in the previous section.

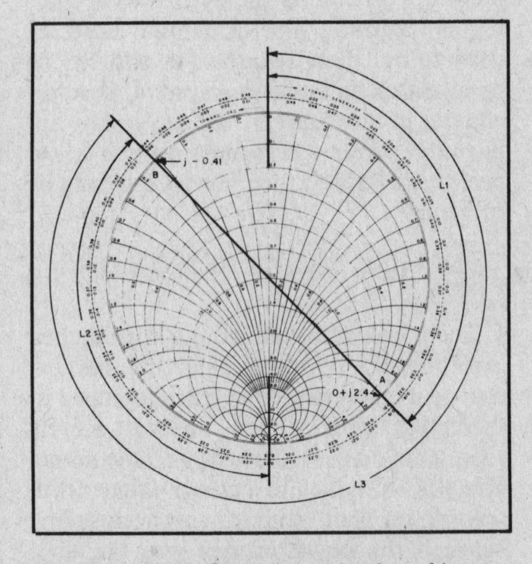


Fig. 28 — Smith Chart determination of input impedances for short- and open-circuited line sections, disregarding line losses.

**The equivalent inductance and capacitance as determined above can be found by substituting these values in the formulas relating inductance and capacitance to reactance, or by using the various charts and calculators available. The frequency corresponding to the line length in degrees must be used, of course. In this example, if the frequency is 14 MHz the equivalent inductance and capacitance in the two cases are 16.4 μH and 46. 2 pF, respectively. Note that when the line length is 45 degrees (0.125 wavelength) the reactance in either case is numerically equal to the characteristic impedance of the line. In using the Smith Chart it should be kept in mind that the**

electrical length of a line section depends on the frequency and velocity of propagation as well as on the actual physical length.

In the case of a line having no losses, and to a close approximation when the losses are small, the inductive reactance of a short-circuited line less than a quarter wave in length is

$$X_L \text{ (ohms)} = Z_o \tan \ell$$

where $\ell$ is the length of the line in electrical degrees and $Z_o$ is the characteristic impedance of the line. The capacitive reactance of an open-circuited line less than a quarter wave in length is

$$X_C \text{ (ohms)} = Z_o \cot \ell$$

At lengths of line that are exact multiples of a quarter wavelength, such lines have the properties of resonant circuits. At lengths where the input reactance passes through zero at the top of the Smith Chart, the line acts like a series-resonant circuit, as shown at D of Fig. 27. At lengths for which the reactances theoretically pass from "positive" to "negative" infinity at the bottom of the Smith Chart, the line simulates a parallel-resonant circuit, as shown at D of Fig. 27.

The effective Q of such linear resonant circuits is very high if the line losses, both in resistance and by radiation, are kept down. This can be done without much difficulty, particularly in coaxial lines, if air insulation is used between the conductors. Air-insulated open-wire lines are likewise very good at frequencies for which the conductor spacing is very small in terms of wavelength.

Applications of line sections as circuit elements in connection with antenna and transmission-line systems are discussed in a later chapter.

# Line Construction and Operating Characteristics

The two basic types of transmission lines, parallel-conductor and coaxial, can be constructed in a variety of forms. Both types can be divided into two classes: those in which the majority of the insulation between the conductors is air, only the minimum of solid dielectric necessary for mechanical support being used; and those in which the conductors are embedded and separated by a solid dielectric. The former class (air-insulated) has the lowest loss per unit length because there is no power loss in dry air so long as the voltage between conductors is below the value at which corona forms. At the maximum power permitted in amateur transmitters it is seldom necessary to consider corona unless the SWR on the line is very high.

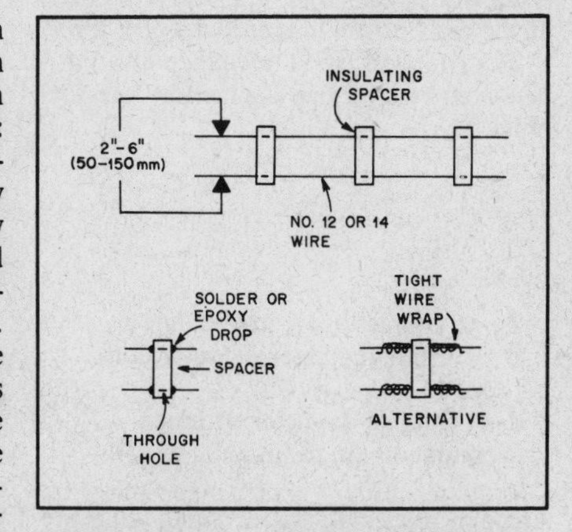


Fig. 29 — Typical open-wire line construction. The spacers may be held in place by using beads of solder or epoxy cement. Wire-Wrap can also be used as shown.

## AIR-INSULATED LINES

A typical type of construction used for parallel-conductor or "two-wire" air-insulated transmission lines is shown in Fig. 29. The two line wires are supported a fixed distance apart by means of insulating rods called spacers. Spacers may be made from material such as phenolic, polystyrene, plastic clothespins or plastic hair curlers. Materials commonly used in high quality spacers are isolantite, Lucite and polystyrene. The spacers used vary from 2 to 6 inches, the smaller spacings being desirable at the higher frequencies (28 MHz) so that radiation will be minimized. It is necessary to use the spacers at small enough intervals along the line to prevent the two wires from swinging appreciably with respect to each other in a wind. For amateur purposes lines using this construction ordinarily have no. 12 or no. 14 conductors, and the characteristic impedance is from 500 to 600 ohms. Although once in universal use, such lines have now been largely superseded by prefabricated lines.

Prefabricated open-wire lines (sold principally for television receiving applications) are available in nominal characteristic impedances of 450 and 300 ohms. The spacers, of low-loss material such as polystyrene, are molded on the conductors at relatively small intervals so there is no tendency for the conductors to swing with respect to each other. A conductor spacing of 1 inch (mm = in. × 25.4) is used in the "450-ohm" line and 1/2 inch in the "300-ohm" line. The conductor size is usually about no. 18. The impedances of such lines are somewhat lower than given by Fig. 30 for the same conductor size and spacing, because of the effect of the dielectric constant of the numerous spacers used. The attenuation is quite low and lines of this type are entirely satisfactory for transmitting applications at amateur powers.

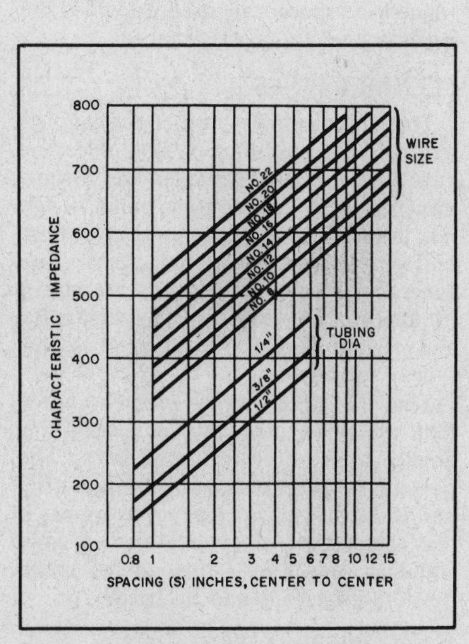


Fig. 30 — Characteristic impedance vs. conductor size and spacing for parallel-conductor lines.

When an air-insulated line having still lower characteristic impedance is needed, metal tubing having a diameter from 1/4 to 1/2 inch is frequently used. With the larger conductor diameter and relatively close spacing it is possible to build a line having a characterisic impedance as low as about 200 ohms. This type of construction is used principally for quarter-wave matching transformers at the higher frequencies.

### Characteristic Impedance

The characteristic impedance of an air-insulated parallel-conductor line, neglecting the effect of the insulating spacers, is given by:

$$Z_o = 276 \log \frac{2S}{d}$$

where

$Z_o$ = characteristic impedance
S = center-to-center distance between conductors
d = outer diameter of conductor (in same units as S)

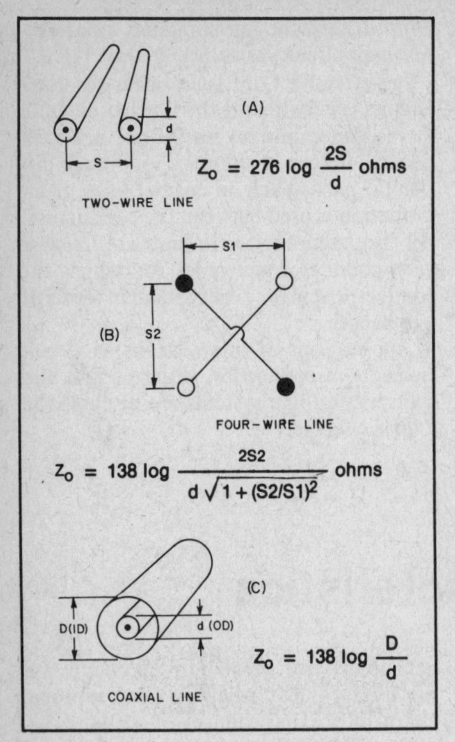


Fig. 31 — Construction of air-insulated transmission lines.

It does not matter what units are used for S and d so long as they are the same units. Both quantities may be measured in centimeters, inches, etc. Since it is necessary to have a table of common logarithms or a calculator to solve practical problems, the solution is given in graphical form in Fig. 30 for a number of common conductor sizes. Equations for determining the characteristic impedance of parallel-conductor lines and others are given in Fig. 31.

### Four-Wire Lines

Another type of parallel-conductor line that is useful in some special applications is the four-wire line (Fig. 31B). In cross-section, the conductors of the four-wire line are at the corners of a square, the spacings being of the same order as those used in two-wire lines. The conductors at opposite corners of the square are connected to operate in parallel. This type of line has a lower characteristic impedance than the simple two-wire type. Also, because of the more symmetrical construction it is better balanced, electrically, to ground and other objects that may be close to the line. The spacers for a four-wire line may be discs of insulating material, X-shaped members, etc.

### Coaxial Lines

In coaxial lines of the air-insulated type (Fig. 31C) a considerable proportion of the insulation between conductors may actually be a solid dielectric, because of the necessity for maintaining constant separation between the inner and outer conductors. This is particularly likely to be true in small-diameter lines. The inner conductor, usually a solid copper wire, is supported by insulating beads at the center of the copper-tubing outer conductor. The beads usually are isolantite and the wire is generally crimped on each side of each bead to prevent the beads from sliding. The material of which the beads are made, and the number of them per unit length of line, will affect the characteristic impedance of the line. The greater the number of beads in a given length, the lower the characteristic impedance compared with the value that would be obtained with air insulation only. The presence of the solid dielectric also increases the losses in the line. On the whole, however, a coaxial line of this type tends to have lower actual loss, at frequencies up to about 100 MHz, than any other line construction, provided the air inside the line can be kept dry. This usually means that air-tight seals must be used at the ends of the line and at every joint.

The characteristic impedance of an air-insulated coaxial line is given by the formula

$$Z_o = 138 \log \frac{D}{d}$$

where

$Z_o$ = characteristic impedance
D = inside diameter of outer conductor
d = Outside diameter of inner conductor (in same units as D)

Again it does not matter what units are used for D and d, so long as they are the same. Curves for typical conductor sizes are given in Fig. 32.

The formula and curves for coaxial lines are approximately correct for lines in which bead spacers are used, provided the beads are not too closely spaced.

## FLEXIBLE LINES

Transmission lines in which the conductors are separated by a flexible dielectric have a number of advantages over the air-insulated type. They are less bulky, weigh less in comparable types, maintain more uniform spacing between conductors, are generally easier to install, and are neater in appearance. Both parallel-conductor and coaxial lines are available with this type of insulation.

The chief disadvantage of such lines is that the power loss per unit length is greater than in air-insulated lines. The power loss causes heating of the dielectric, and if the heating is great enough — as it may be with high power and a high standing-wave ratio — the line may break down mechanically and electrically.

### Parallel-Conductor Lines

The construction of a number of types of flexible lines is shown in Fig. 33. In the

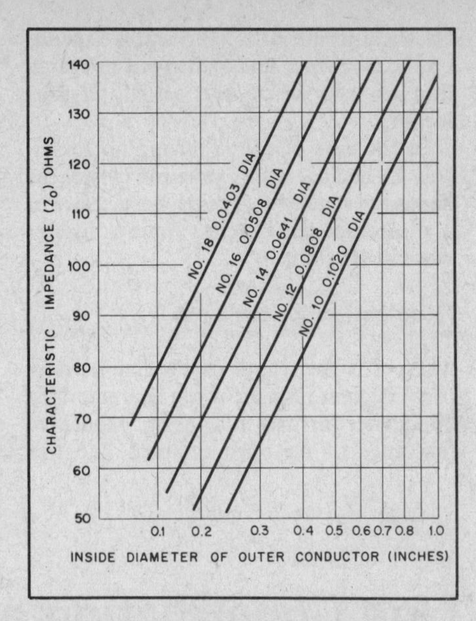


Fig. 32 — Characteristic impedance of typical air-insulated coaxial lines.

most common 300-ohm type (twin-lead) the conductors are stranded wire equivalent to no. 20 in cross-sectional area and are molded in the edges of a polyethylene ribbon about a half inch wide. The effective dielectric is partly solid and partly air. The presence of the solid dielectric lowers the characteristic impedance of the line as compared with the same conductors in air, the result being that the impedance is approximately 300 ohms. The fact that part of the field between the conductors exists outside the solid dielectric leads to an operating disadvantage in that dirt or moisture on the surface of the ribbon tends to change the characteristic impedance. The operation of the line is therefore affected by weather conditions. The effect will not be very serious in a line terminated in its characteristic impedance, but if there is a considerable standing-wave ratio a small change in $Z_o$ may cause wide fluctuations of the input impedance. Weather effects can be minimized by cleaning the line occasionally and giving it a thin coating of a water-repellent material such as silicone grease or automobile wax.

To overcome the effects of weather on the characteristic impedance and attenuation of ribbon-type line, another type of twin-lead is made using an air-core polyethylene tube with the conductors molded diametrically opposite each other in the walls. This increases the leakage path across the dielectric surface. Also, much of the electric field between the conductors is in the hollow center of the tube to make it watertight. This type of line is fairly impervious to weather effects. Care should be used when installing it, however, to make sure that any moisture that condenses on the inside with changes in temperature and humidity can drain out the bottom end of the tube and not be trapped in one section. This type of line is made in two conductor sizes (with different tube diameters), one for receiving applications and the other for transmitting.

The transmitting-type 75-ohm twin lead uses stranded conductors about equivalent to solid no. 12 wire, with quite close spacing between conductors. Because of the close spacing most of the field is confined to the solid dielectric, with very little existing in the surrounding air. This makes the 75-ohm line much less susceptible to weather effects than the 300-ohm ribbon type.

## COAXIAL CABLES

Coaxial cable is available in flexible and semiflexible formats. The fundamental design is the same in all types as shown in Fig. 33. The outer diameter varies from 0.060 to 3 inches (1.5 to 76 mm) and more. The larger the diameter the higher the power capability of the line, owing to the increase in dielectric thickness and conductor size. Generally, the losses decrease as the cable diameter increases, depending on the insulating material.

Some coaxial cables have stranded-wire center conductors while others employ a solid copper conductor. Similarly, the outer conductor (shield) may be a single layer of copper braid, a double layer of braid (more effective shielding), or solid aluminum (Hardline).

### Losses and Deterioration

The power capability and loss characteristics of coaxial cable depend largely on the dielectric material between the conductors. The commonly used cables and some of their properties are listed in Table 1. Fig. 34 is a graph showing the attenuation characteristics of the most popular lines. The outer insulating jacket (usually polyvinyl) of the cable is used solely as protection from dirt, moisture and chemicals. It has no electrical function. Exposure of the inner insulating material to moisture and chemicals will contaminate the dielectric over time and cause the cable to become lossy. The foam dielectric cables are less prone to contamination than are those having solid polyethylene insulation. Certain impregnated cables, such as Decibel Products *VB-8* and Times Wire & Cable Co. *Imperveon*, are immune to water and chemical damage, and may be buried if desired. They also have a self-healing property that is valuable when rodents chew into the line.

Coaxial cables that have been out of doors or buried should be checked for losses every two years. This can be accomplished by using the technique illustrated in Fig. 35. If the measured loss in watts equates to more than 1 dB over the rated loss per 100 feet, the cable should be replaced. The loss in decibels can be determined from

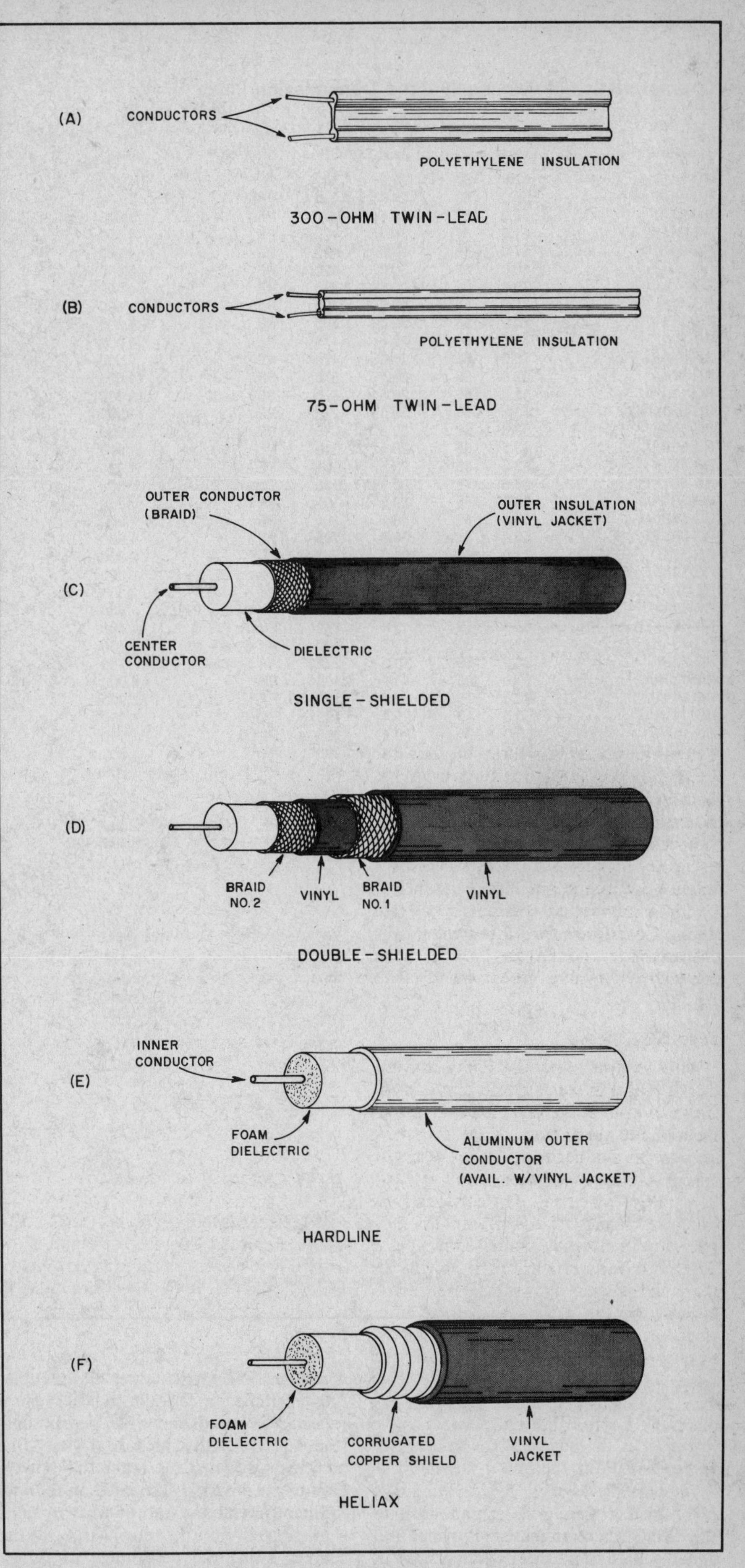


Fig. 33 — Construction of parallel-conductor and coaxial lines with solid dielectric. A common variation of the double-shielded design at D has the braids in continuous electrical contact.

**Table 1**

**Characteristics of Commonly Used Transmission Lines**

| *Type of line* | $Z_o$ *Ohms* | *Vel %* | *pF per foot* | *OD* | *Diel. Material* | **Max.** *Operating Volts (RMS)* |
|---|---|---|---|---|---|---|
| RG-8/U | 52.0 | 66 | 29.5 | .405 | PE | 4,000 |
| RG-8/U Foam | 50.0 | 80 | 25.4 | .405 | Foam PE | 1,500 |
| RG-8A/U | 52.0 | 66 | 29.5 | .405 | PE | 5,000 |
| RG-9/U | 51.0 | 66 | 30.0 | .420 | PE | 4,000 |
| RG-9A/U | 51.0 | 66 | 30.0 | .420 | PE | 4,000 |
| RG-9B/U | 50.0 | 66 | 30.8 | .420 | PE | 5,000 |
| RG-11/U | 75.0 | 66 | 20.6 | .405 | PE | 4,000 |
| RG-11/U Foam | 75.0 | 80 | 16.9 | .405 | Foam PE | 1,600 |
| RG-11A/U | 75.0 | 66 | 20.6 | .405 | PE | 5,000 |
| RG-12/U | 75.0 | 66 | 20.6 | .475 | PE | 4,000 |
| RG-12A/U | 75.0 | 66 | 20.6 | .475 | PE | 5,000 |
| RG-17/U | 52.0 | 66 | 29.5 | .870 | PE | 11,000 |
| RG-17A/U | 52.0 | 66 | 29.5 | .870 | PE | 11,000 |
| RG-55/U | 53.5 | 66 | 28.5 | .216 | PE | 1,900 |
| RG-55A/U | 50.0 | 66 | 30.8 | .216 | PE | 1,900 |
| RG-55B/U | 53.5 | 66 | 28.5 | .216 | PE | 1,900 |
| RG-58/U | 53.5 | 66 | 28.5 | .195 | PE | 1,900 |
| RG-58/U Foam | 53.5 | 79 | 28.5 | .195 | Foam PE | 600 |
| RG-58A/U | 53.5 | 66 | 28.5 | .195 | PE | 1,900 |
| RG-58B/U | 53.5 | 66 | 28.5 | .195 | PE | 1,900 |
| RG-58C/U | 50.0 | 66 | 30.8 | .195 | PE | 1,900 |
| RG-59/U | 73.0 | 66 | 21.0 | .242 | PE | 2,300 |
| RG-59/U Foam | 75.0 | 79 | 16.9 | .242 | Foam PE | 800 |
| RG-59A/U | 73.0 | 66 | 21.0 | .242 | PE | 2,300 |
| RG-62/U | 93.0 | 86 | 13.5 | .242 | Air Space PE | 750 |
| RG-62/U Foam | 95.0 | 79 | 13.4 | .242 | Foam PE | 700 |
| RG-62A/U | 93.0 | 86 | 13.5 | .242 | Air Space PE | 750 |
| RG-62B/U | 93.0 | 86 | 13.5 | .242 | Air Space PE | 750 |
| RG-133A/U | 95.0 | 66 | 16.2 | .405 | PE | 4,000 |
| RG-141/U | 50.0 | 70 | 29.4 | .190 | PTFE | 1,900 |
| RG-141A/U | 50.0 | 70 | 29.4 | .190 | PTFE | 1,900 |
| RG-142/U | 50.0 | 70 | 29.4 | .206 | PTFE | 1,900 |
| RG-142A/U | 50.0 | 70 | 29.4 | .206 | PTFE | 1,900 |
| RG-142B/U | 50.0 | 70 | 29.4 | .195 | PTFE | 1,900 |
| RG-174/U | 50.0 | 66 | 30.8 | .1 | PE | 1,500 |
| RG-213/U | 50.0 | 66 | 30.8 | .405 | PE | 5,000 |
| RG-215/U | 50.0 | 66 | 30.8 | .475 | PE | 5,000 |
| RG-216/U | 75.0 | 66 | 20.6 | .425 | PE | 5,000 |
| *Aluminum Jacket Foam Dielectric* | | | | | | |
| 1/2 inch | 50.0 | 81 | 25.0 | .5 | | 2,500 |
| 3/4 inch | 50.0 | 81 | 25.0 | .75 | | 4,000 |
| 7/8 inch | 50.0 | 81 | 25.0 | .875 | | 4,500 |
| 1/2 inch | 75.0 | 81 | 16.7 | .5 | | 2,500 |
| 3/4 inch | 75.0 | 81 | 16.7 | .75 | | 3,500 |
| 7/8 inch | 75.0 | 81 | 16.7 | .875 | | 4,000 |
| Open wire | — | 97 | — | — | | — |
| 75-ohm transmitting twin lead | 75.0 | 67 | 19.0 | — | | — |
| 300-ohm twin lead | 300.0 | 80 | 5.8 | — | | — |
| 300-ohm tubular | 300.0 | 77 | 4.6 | — | | — |
| *Open wire, TV type* | | | | | | |
| 1/2 inch | 300.0 | 95 | — | — | | — |
| 1 inch | 450.0 | 95 | — | — | | — |

| *Dielectric Designation* | *Name* | *Temperature Limits* |
|---|---|---|
| PE | Polyethylene | −65° to +80° C |
| Foam PE | Foamed Polyethylene | −65° to +80° C |
| PTFE | Polytetrafluoroethylene (Teflon) | −250° to +250° C |

$$dB = 10 \log \frac{P1}{P2}$$

where P1 is the power at the transmitter output and P2 is the power measured at $R_L$ of Fig. 35.

A lossless line will exhibit infinite VSWR when unterminated. Provided the signal source can operate safely into a severe mismatch, an SWR bridge or reflected-power meter can be used to determine the loss in an open-ended cable. For example, a VSWR of 3:1 or a reflected power reading of 25 percent corresponds to a 3-dB cable loss. Twenty-five percent is down 6 dB from full power. Half of the power is dissipated on the way to the open end and half of what remains is dissipated on the return trip to the source. Thus, the percentage of power reflected by an unterminated line, when converted to decibels, equals twice the attenuation. The graph and equation in Fig. 13 relate VSWR to forward and reflected power. The instruments available to most amateurs lose accuracy at VSWR values greater than about 5, so the method described here is useful principally as a go/no-go check on cables that are fairly long. For short, low-loss cables, only gross deterioration can be detected by the open-circuit VSWR test.

The pertinent characteristics of unmarked coaxial cables can be determined from the equations in Table 2. The most common impedance values are 50, 75 and 95 ohms. But, impedances from 25 to 125 ohms are available in special types of manufactured line. The 25-ohm cable (miniature) is used extensively in magnetic-core broadband transformers.

**Table 2**

**Coaxial Cable Formulas**

Eq. 1 — $C\ (\text{pF/ft}) = \dfrac{7.36\ \varepsilon}{\log (D/d)}$

Eq. 2 — $L\ (\mu\text{H/ft}) = 0.14 \log \dfrac{D}{d}$

Eq. 3 — $Z_o\ (\text{ohms}) = \sqrt{\dfrac{L}{C}} = \left(\dfrac{138}{\sqrt{\varepsilon}}\right)\left(\log \dfrac{D}{d}\right)$

Eq. 4 — Velocity % (ref. speed of light) $= \dfrac{100}{\sqrt{\varepsilon}}$

Eq. 5 — Time delay (ns/ft) $= 1.016 \sqrt{\varepsilon}$

Eq. 6 — $f\ (\text{cutoff/GHz}) = \dfrac{7.50}{\sqrt{\varepsilon}\,(D + d)}$

Eq. 7 — Ref. coeff. $= \rho = \dfrac{Z_L - Z_o}{Z_L + Z_o} = \dfrac{VSWR - 1}{VSWR + 1}$

Eq. 8 — $VSWR = \dfrac{1 + \rho}{1 - \rho}$

Eq. 9 — $V\ pk = \dfrac{(1.15\ Sd)(\log D/d)}{K}$

Eq. 10 — $A = \dfrac{0.435}{Z_o D}\left[\dfrac{D}{d}(K1 + K2)\right]\sqrt{f} + 2.78 \sqrt{\varepsilon}\,(P.F.)(f)$

A = atten. in dB/100 ft (30.5 m); d = OD of inner conductor; D = ID of outer conductor; S = max. voltage gradient of insulation in volts/mil; ε = dielectric constant; K = safety factor; K1 = strand factor; K2 = braid factor; f = freq. in MHz; P.F. = power factor. Obtain K1 and K2 data from mfr.

### Cable Capacitance

The capacitance between the conductors of coaxial cable will vary with the impedance and dielectric constant of the line. Hence, the lower the impedance the higher the capacitance per foot because the conductor spacing is decreased. The

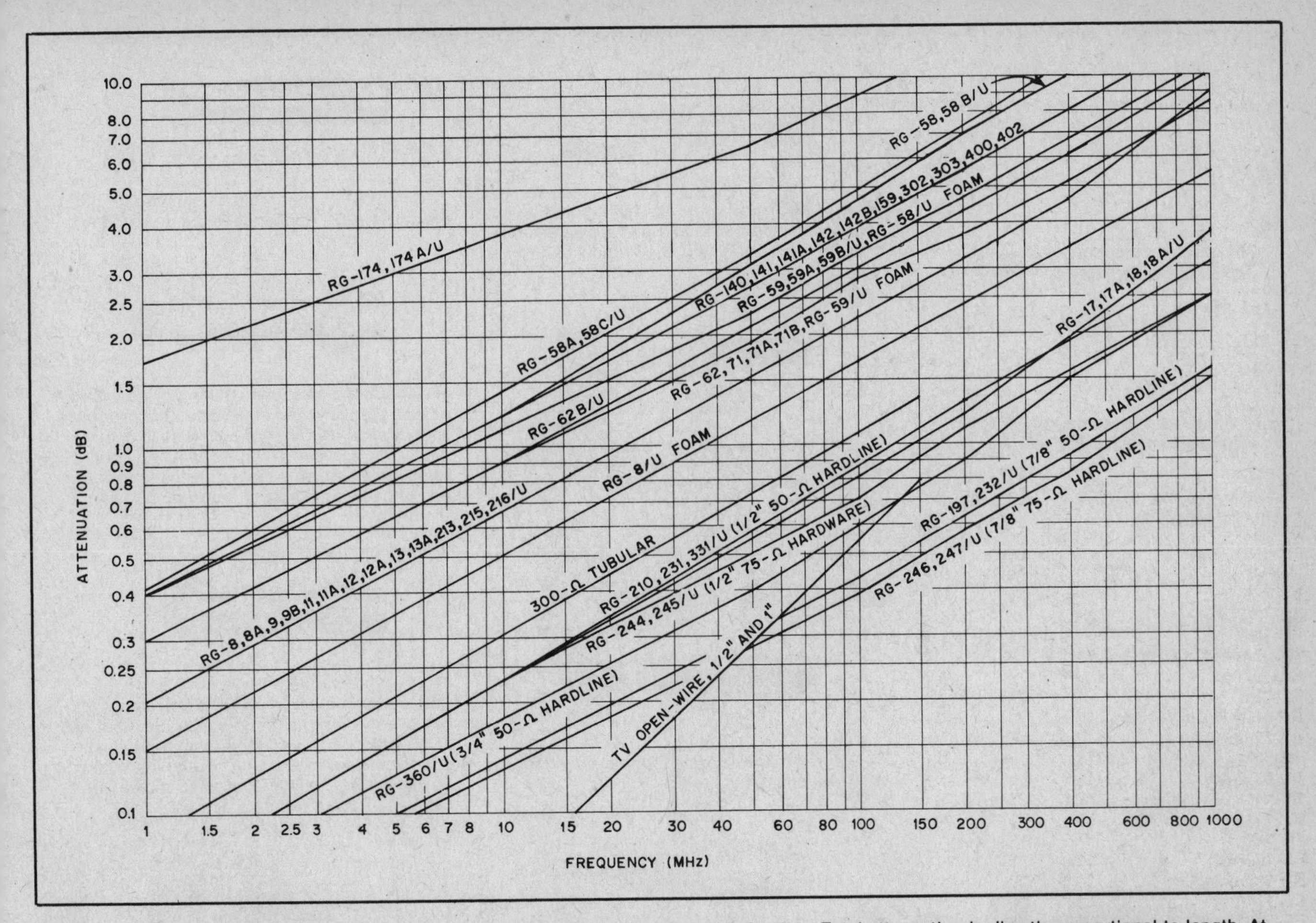


Fig. 34 — Nominal attenuation in decibels per 100 feet of various types of transmission line. Total attenuation is directly proportional to length. Attenuation will vary somewhat in actual cable samples, and generally increases with age in coaxial cables having a type I jacket. Cables grouped together in the above chart have approximately the same attenuation. Types having foam polyethylene dielectric have slightly lower loss than equivalent solid types, when not specifically shown above.

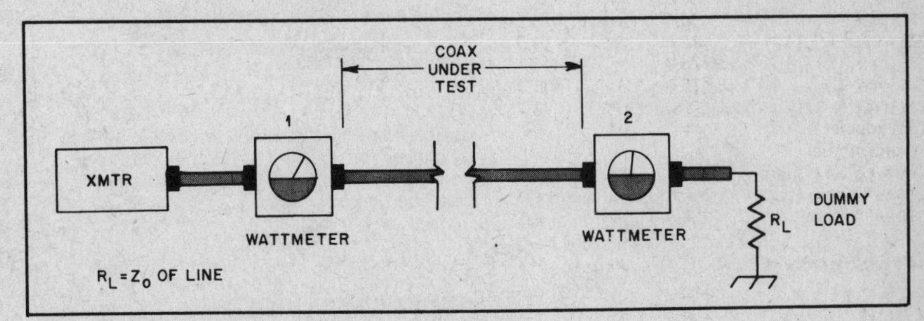


**Fig. 35 — Method for determining losses in transmission lines with an SWR of 1.**

capacitance also increases with dielectric constant.

### Voltage and Power Ratings

Selection of the correct coaxial cable for a particular application is not a casual matter. Not only is the loss of significance, but breakdown and heating (voltage and power) also need to be considered. If a cable were lossless, the power handling capability would be limited only by the breakdown voltage. RG-58/U, for example, can withstand an operating potential of 1900 V rms. In a 50-ohm system this equates to more than 72 kW. However, the current corresponding to this power level is 38 amperes, which would obviously melt the conductors in RG-58/U. In practical coaxial cables, the copper and dielectric losses, rather than breakdown voltage, limit the maximum power that can be accommodated. If 1000 W is applied to a cable having a loss of 3 dB, only 500 W is delivered to the load; the remaining 500 W must be dissipated in the cable. The dielectric and outer jacket are good thermal insulators, which prevent the conductors from efficiently transferring the heat to free air. As the frequency increases, the power capability of a cable decreases, because of increasing copper loss (skin effect) and dielectric loss. RG-58/U having a foam dielectric has a breakdown rating of only 600 V, yet it can handle substantially more power than its ordinary solid-dielectric counterpart because of the lower losses. Normally, the loss is not a matter of consequence (except as it affects power-handling capability) below 10 MHz in amateur applications, unless extremely long runs of cable are used. In general, full legal amateur power can be safely applied to inexpensive RG-58/U coax in the bands below 10 MHz. Cables of the RG-58/U family can withstand full amateur power through the vhf spectrum.

Excessive ac operating voltage in a coaxial cable can cause noise generation, dielectric damage and eventual breakdown between the conductors. Since the rms voltage is a function of the cable $Z_o$ and operating power in watts, it is a simple matter to determine the voltage in the line. For example, if the VSWR is unity, and the rf power applied to the cable is 800 W, and the cable impedance is 50 ohms, the developed voltage will be

$$E_{rms} = \sqrt{PR} = \sqrt{800 \times 50} = 200 \text{ V}$$

## BNC (UG-88/U) CONNECTORS

Connectors bearing suffix letters (UG-88C/U, etc.) differ slightly in internal construction; assembly and dimensions must be varied accordingly.

1) Cut end square and trim jacket 5/16″ for RG-58/U.

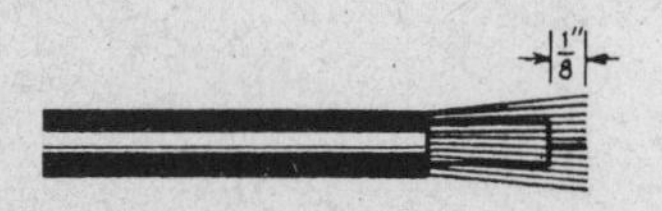


2) Fray shield and strip inner dielectric 1/8″. Tin center conductor.

3) Taper braid and slide nut (A), washer (B), gasket (C), and clamp (D), over braid. Clamp is inserted so that its inner shoulder fits squarely against end of cable jacket.

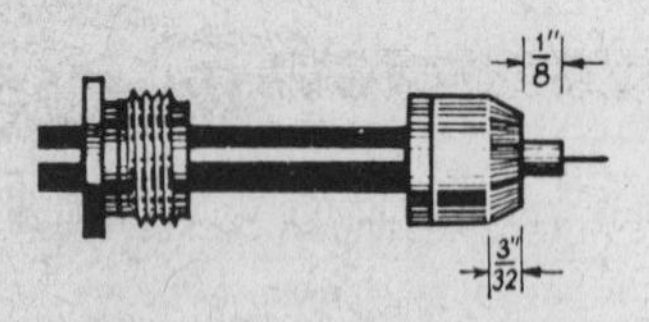


4) With clamp in place, comb out braid, fold back smooth as shown, and trim 3/32″ from end.

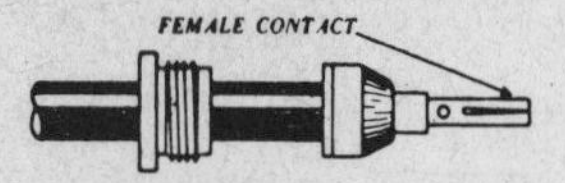


5) Tin center conductor of cable. Slip female contact in place and solder. Remove excess solder. Be sure cable dielectric is not heated excessively and swollen so as to prevent dielectric entering body.

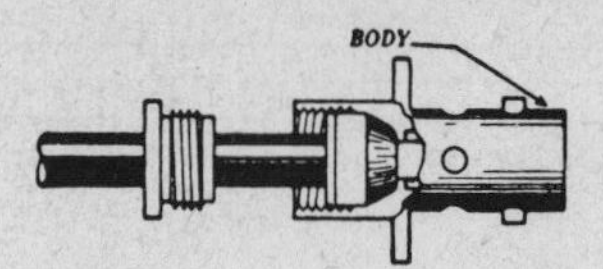


6) Push into body as far as it will go. Slide nut into body and screw into place with wrench until tight. Hold cable and shell rigidly and rotate nut.

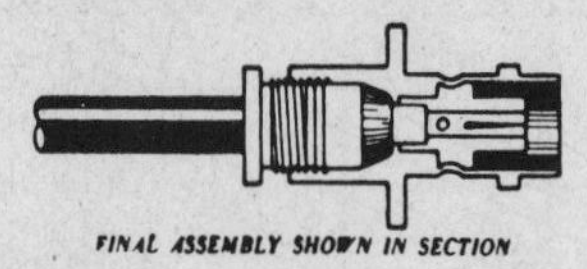


7) This assembly procedure applies to BNC jacks. The assembly for plugs is the same except for the use of male contacts and a plug body.

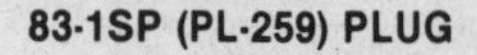

## 83-1SP (PL-259) PLUG

1) Cut end of cable even. Remove vinyl jacket 1-1/8″ — don't nick braid.

2) Bare 5/8″ of center conductor — don't nick conductor. Trim braided shield to 9/16″ and tin. Slide coupling ring on cable.

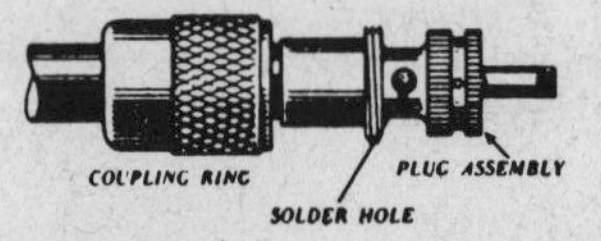


3) Screw the plug assembly on cable. Solder plug assembly to braid through solder holes. Solder conductor to contact sleeve. Screw coupling ring on assembly.

## N (UG-21/U) CONNECTORS

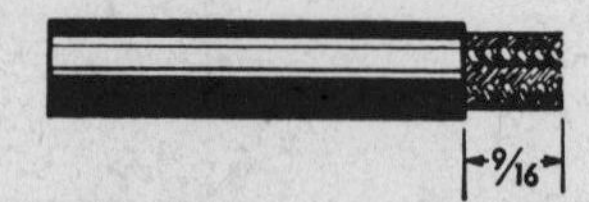


1) Remove 9/16″ of vinyl jacket. When using double-shielded cable remove 5/8″.

2) Comb out copper braid as shown. Cut off dielectric 7/32″ from end. Tin center conductor.

3) Taper braid as shown. Slide nut, washer and gasket over vinyl jacket. Slide clamp over braid with internal shoulder of clamp flush against end of vinyl jacket. When assembling connectors with gland, be sure knife-edge is toward end of cable and groove in gasket is toward the gland.

4) Smooth braid back over clamp and trim. Soft-solder contact to center conductor. Avoid use of excessive heat and solder. See that end of dielectric is clean. Contact must be flush against dielectric. Outside of contact must be free of solder. Female contact is shown; procedure is similar for male contact.

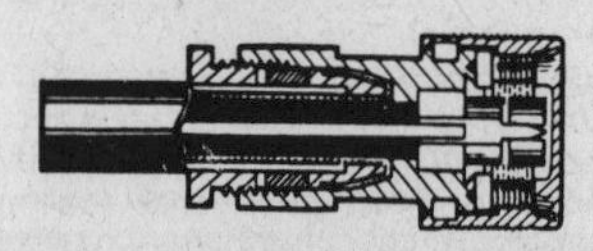

5) Slide body into place carefully so that contact enters hole in insulator (male contact shown). Face of dielectric must be flush against insulator. Slide completed assembly into body by pushing nut. When nut is in place, tighten with wrenches. In connectors with gland, knife edge should cut gasket in half by tightening sufficiently.

From Armed Services Index of R.F. Transmission Lines and Pittings, ASESA 49-2B.

Fig. 36 — Assembly instructions for BNC, N and UHF connectors.

If the VSWR is other than unity, this must be accounted for when computing the maximum voltage. This is done by multiplying the input voltage by the square root of the VSWR. Thus if the VSWR were 3:1, the maximum potential within the cable would be 346 volts, provided the line is at least an electrical quarter-wavelength long.

### SINGLE-WIRE LINE

There is one type of line, in addition to those already described, that deserves some mention since it is still used to a limited extent. This is the *single-wire line*, consisting simply of a single conductor running from the transmitter to the antenna. The "return" circuit for such a line is the earth; in fact, the second conductor of the line can be considered to be the image of the actual conductor in the same way

## 83 SERIES (SO-239) WITH HOODS

1) Cut end of cable even. Remove vinyl jacket to dimension appropriate for type of hood. Tin exposed braid.

2) Remove braid and dielectric to expose center conductor. Do not nick conductor.

3) Remove braid to expose dielectric to appropriate dimension. Tin center conductor. Soldering assembly depends on the hood used, as illustrated.

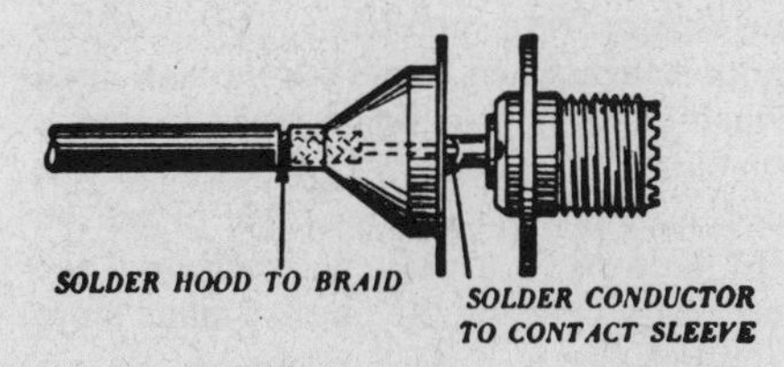


4) Slide hood over braid. Solder conductor to contact. Slide hood flush against receptacle and bolt both to chassis. Solder hood to braid as illustrated. Tape this junction if necessary. (For UG-177/U.)

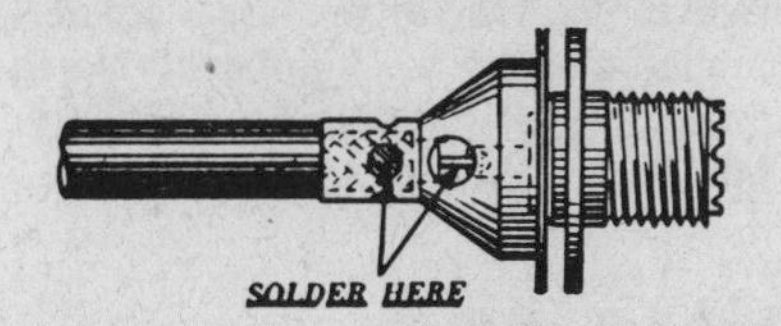


5) Slide hood over braid. Bring receptacle flush against hood. Solder hood to braid and conductor to contact sleeve through solder holes as illustrated. Tape junction if necessary. (For UG-372/U.)

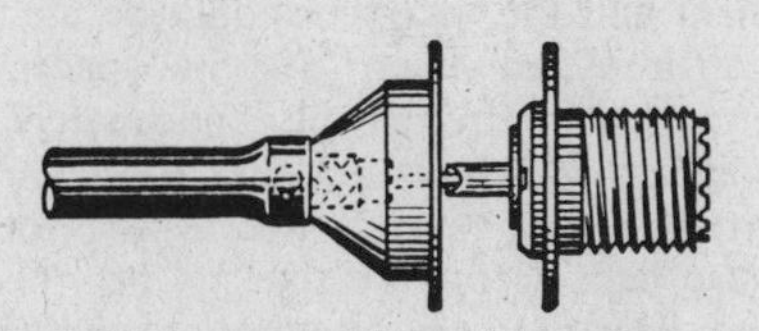

6) Slide hood over braid and force under vinyl. Place inner conductor in contact sleeve and solder. Push hood flush against receptacle. Solder hood to braid through solder holes. Tape junction if necessary. (For UG-106/U.)

## 83-1SP (PL-259) PLUG WITH ADAPTERS (UG-176/U OR UG-175/U)

1) Cut end of cable even. Remove vinyl jacket 3/4" — don't nick braid. Slide coupling ring and adapter on cable.

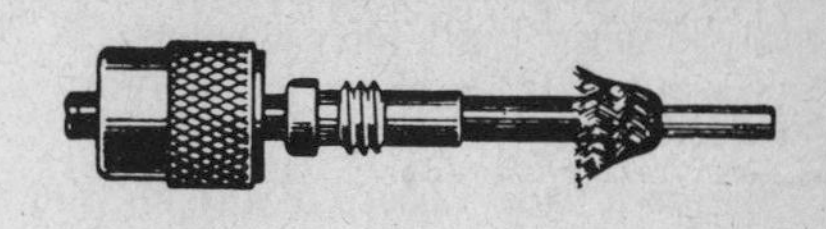

2) Fan braid slightly and fold back over cable.

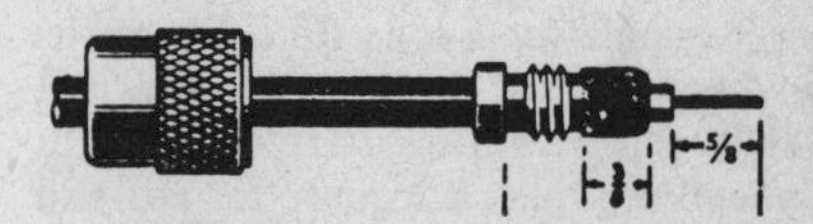

3) Position adapter to dimension shown. Press braid down over body of adapter and trim to 3/8". Bare 5/8" of conductor. Tin exposed center conductor.

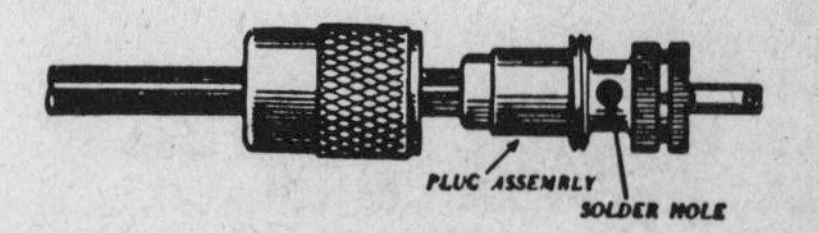


4) Screw the plug assembly on adapter. Solder braid to shell through solder holes. Solder conductor to contact sleeve.

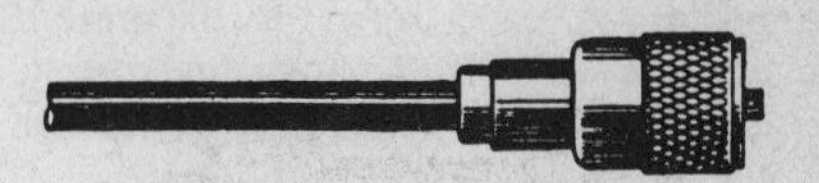

5) Screw coupling ring on plug assembly.

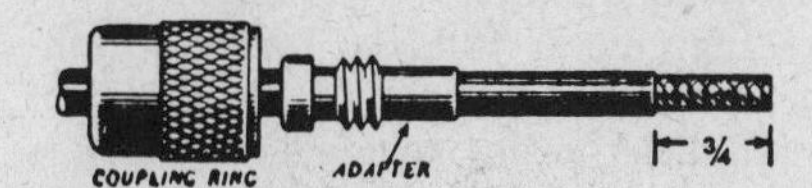

Fig. 37 — Complete electrical shield integrity in the UHF female connector requires that the shield be attached to the connector flange by means of a hood. The assembly is illustrated here. Additionally, cables in the RG-58, 59/U size group require a reducing adapter for use with the PL-259 connector. The lower portion of this figure shows the details for installing this component.

that an antenna strung above the earth has an image (see Chapter 2). The characteristic impedance of the single-wire line depends on the conductor size and the height of the wire above ground, ranging from 500 to 600 ohms for no. 12 or no. 14 conductors at heights of 10 to 30 feet. By connecting the line to the antenna at a point that represents a resistive impedance of 500 to 600 ohms the line can be matched and will operate without standing waves.

Although the single-wire line is very simple to install, it has at least two outstanding disadvantages. Since the return circuit is through the earth, the behavior of the system depends on the kind of ground over which the antenna and transmission lines are erected. In practice, it may not be possible to get the necessary good connection to actual ground that is required at the transmitter. Second, the line always radiates since there is no nearby second conductor to cancel the fields. The radiation will be minimum when the line is properly terminated because the line current is lowest under those conditions. However, the line is always a part of the radiating antenna system to a greater or lesser extent.

## ELECTRICAL LENGTH

Whenever reference is made to a line as being so many wavelengths (such as a "half wavelength" or "quarter wavelength") long, it is to be understood that the *electrical* length of the line is meant. The physical length corresponding to an electrical wavelength is given by

$$\text{length (feet)} = \frac{984}{f}V$$

where

f = frequency in megahertz
V = velocity factor

The *velocity factor* is the ratio of the actual velocity along the line to the velocity in free space. Values of V for several common types of lines are given in Table 1.

Because a quarter-wavelength line is frequently used as an impedance transformer, it is convenient to calculate the length of a quarter-wave line directly. The formula is

$$\text{length (feet)} = \frac{246}{f}V$$

## LINE INSTALLATION

One great advantage of coaxial line, particularly the flexible dielectric type, is that it can be installed with almost no regard for its surroundings. It requires no insulation, can be run on or in the ground or in piping, can be bent around corners with a reasonable radius, and can be "snaked" through places such as the space between walls where it would be impractical to use other types of lines. However, coaxial lines should always be operated in systems that permit a low standing-wave ratio, and precautions must be taken to prevent rf currents from flowing on the *outside* of the line. This point is discussed in a later chapter.

**Coaxial Fittings**

There is a wide variety of fittings and connectors designed to go with various sizes and types of solid-dielectric coaxial line. The "UHF" series of fittings is by far the most widely used type in the amateur field, largely because they have been available for a long time. These fittings, typified by the PL-259 plug and SO-239 chassis fitting (Armed Services numbers) are quite adequate for vhf and lower frequency applications, but are not weatherproof.

The "N" series fittings are designed to maintain constant impedance at cable joints. They are a bit harder to assemble than the "UHF" type, but are better for frequencies above 300 MHz or so. These fittings are weatherproof.

The "BNC" fittings are for small cable such as RG-58/U, RG-59/U and RG-62/U. They feature a bayonet-locking arrangement for quick connect and disconnect, and are weatherproof.

Methods of assembling the connectors to the cable are shown in Figs. 36 and 37. The most common or longest established connector in each series is illustrated. Several variations of each type exist. Assembly instructions for coaxial fittings not shown here are available from the manufacturers. One of the larger connector manufacturers is Amphenol North America, 33 East Franklin St., Danbury, CT 06810.

**Parallel-Wire Lines**

In installing a parallel-wire line, care must be used to prevent it from being affected by moisture, snow and ice. In home construction, only spacers that are impervious to moisture and are unaffected by sunlight and the weather should be used on air-insulated lines. Steatite spacers meet this requirement adequately, although they are somewhat heavy. The wider the line spacing the longer the leakage path across the spacers, but this cannot be carried too far without running into line radiation, particularly at the higher frequencies. Where an open-wire line must be anchored to a building or other structure, standoff insulators of a height comparable with the line spacing should be used if mounted in a spot that is open to the weather. Lead-in bushings for bringing the line into a building also should have a long leakage path.

The line should be kept away from other conductors, including downspouting, metal window frames, flashing, etc., by a distance equal to two or three times the line spacing. Conductors that are very close to the line will be coupled to it in greater or lesser degree, and the effect is that of placing an additional load across the line at the point where the coupling occurs. Reflections take place from this coupled "load," raising the standing-wave ratio. The effect is at its worst when one line wire is closer than the other to the external conductor. In such a case one wire carries a heavier load than the other, with the result that the line currents are no longer equal. The line then becomes "unbalanced."

Solid-dielectric two-wire lines have a relatively small external field because of the small spacing and can be mounted within a few inches of other conductors without much danger of coupling between the line and such conductors. Standoff insulators are available for supporting lines of this type when run along walls or similar structures.

Sharp bends should be avoided in any type of transmission line, because such bends cause a change in the characteristic impedance. The result is that reflections take place from each bend. This is of less importance when the SWR is high than when an attempt is being made to match the load to the line's characteristic impedance. It may be impossible to get the SWR down to a desired figure until the necessary bends in the line are made more gradual.

Chapter 4

# Coupling the Transmitter to the Line

In any system using a transmission line to feed the antenna, the load that the transmitter "sees" is the input impedance of the line. As shown earlier, this impedance is completely determined by the line length, the $Z_0$ of the line, and the impedance of the load (the antenna) at the output end of the line. The line length and $Z_0$ are generally matters of choice regardless of the type of antenna used. The antenna impedance, which may or may not be known accurately, is (with $Z_0$) the factor that determines the standing-wave ratio.

The SWR can be measured with relative ease, and from it the limits of variation in the line input impedance can be determined with little difficulty. It may be said, therefore, that the problem of transferring power from the transmitter to the line can be approached purely on the basis of the known $Z_0$ of the line and the maximum SWR that may be encountered.

Coupling systems that will deliver power into a flat line are readily designed. For all practical purposes the line can be considered to be flat if the SWR is no greater than about 1.5:1. That is, a coupling system designed to work into a pure resistance equal to the line $Z_0$ should have enough leeway to take care of the small variations in input impedance that will occur when the line length is changed, if the SWR is higher than 1:1 but no greater than 1.5:1.

So far as the transmitter itself is concerned, the requirements of present-day amateur operation almost invariably include complete shielding and provision for the use of low-pass filters to prevent harmonic interference with television reception. In almost all cases this means that the use of coaxial cable at the output of the transmitter is mandatory because it is inherently shielded. Current practice in transmitter design is to provide an output circuit that will work into a coaxial line of 50 to 75 ohms characteristic impedance.

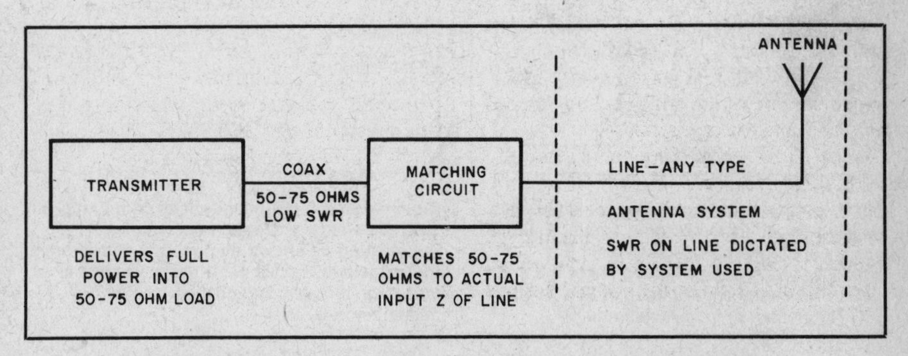


Fig. 1 — Essentials of a coupling system between transmitter and transmission line

This does not mean that a coaxial line must be used to feed the *antenna*; any type of line can be used.

If the input impedance of the transmission line that is to be connected to the transmitter differs appreciably from the impedance value that the transmitter output circuit is designed to operate, an impedance-matching network must be inserted between the transmitter and the line input terminals.

## THE MATCHING SYSTEM

The basic system is as shown in Fig. 1. Assuming that the transmitter is capable of delivering its rated power into a load on the order of 50 to 75 ohms, the coupling problem is one of designing a matching circuit that will transform the actual line input impedance into a resistance of 50 or 75 ohms. This resistance will be unbalanced; that is, one side will be grounded. The line to the antenna may be unbalanced (coaxial cable) or balanced (parallel-conductor line).

Many factors will influence the choice of line to be used with a given antenna. The shielding of coaxial cable offers advantages in incidental radiation and routing flexibility. Low loss and low cost should ensure that open wire line is around for a long time. Coaxial cable can perform acceptably even with significant VSWR. (Refer to information in Chapter 3.) A length of 100 feet of RG-8/U line has 1 dB loss at 30 MHz. If this line were used with a VSWR at the load of 5:1, the total line loss would be 2.1 dB. Open-wire line, in the same circumstances, has a matched loss of less than 0.1 dB. If this line was used with a VSWR of 20:1 the total loss would be less than 1 dB.

Several types of matching circuits are available. With some, such as an L network using fixed-value components, it is necessary to know the actual line-input impedance with fair accuracy in order to arrive at a proper design. This information is not essential with the circuits described in the next section, since they are capable of adjustment over a wide range.

## MATCHING WITH INDUCTIVE COUPLING

Inductively coupled matching circuits are shown in basic form in Fig. 2. R1 is the actual load resistance to which the power is to be delivered, and R2 is the resistance seen by the power source. R2 depends on the circuit design and adjustment; in general, the objective is to make it equal to 50 or 75 ohms. L1 and C1 form a resonant circuit capable of being tuned to the

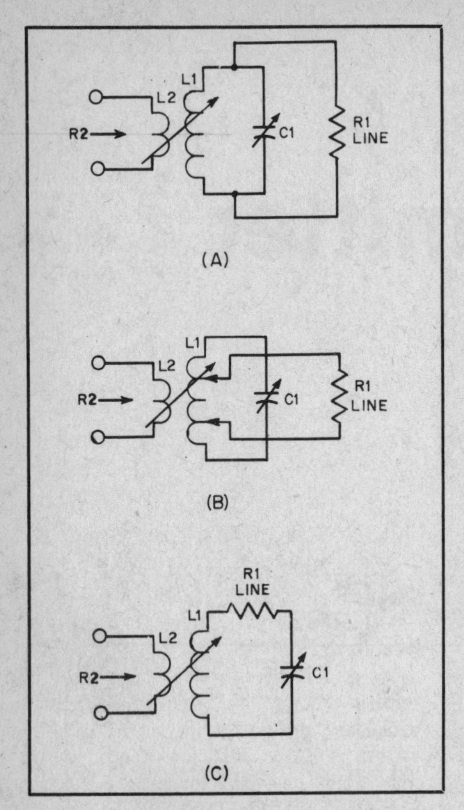


Fig. 2 — Circuit arrangements for inductively coupled impedance-matching circuit. A and B use a parallel-tuned coupling tank; B is equivalent to A when the taps are at the ends of L1. The series-tuned circuit at C is useful for very low values of load resistance, R1.

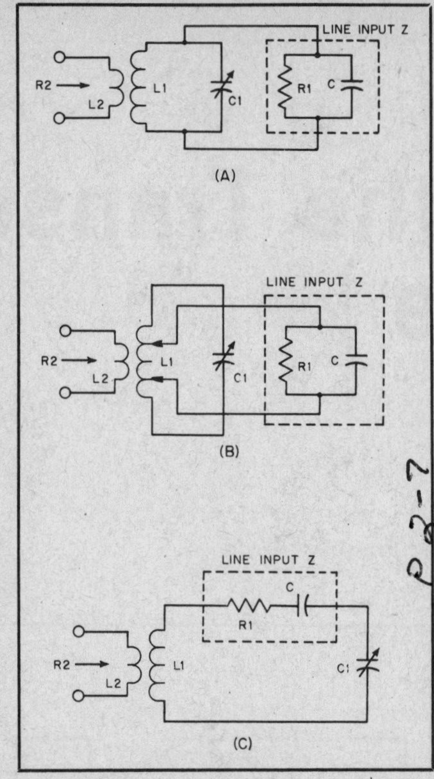

operating frequency. The coupling between L1 and L2 is adjustable.

The circuit formed by C1, L1 and L2 is equivalent to a transformer having a primary-to-secondary impedance ratio adjustable over wide limits. The resistance "coupled into" L2 from L1 depends on the effective Q of the circuit L1-C1-R1, the reactance of L2 at the operating frequency, and the coefficient of coupling, k, between the two coils. The approximate relationship is (assuming C1 is properly tuned)

$$R2 = k^2 X_{L2} Q$$

where $X_{L2}$ is the reactance of L2 at the operating frequency. The value of L2 is optimum when $X_{L2} = R2$, in which case the desired value of R2 is obtained when

$$k = \frac{1}{\sqrt{Q}}$$

This means that the desired value of R2 may be obtained by adjusting either the coupling, k, between the two coils, or by changing the Q of the circuit L1-C1-R1 — or, if necessary, by doing both. If the coupling is fixed, as is often the case, Q must be adjusted to attain a match. Note that increasing the value of Q is equivalent to tightening the coupling, and vice versa.

Fig. 3 — Line input impedances containing both resistance and reactance can be represented as shown enclosed in dotted lines, for capacitive reactance. If the reactance is inductive, a coil is substituted for the capacitance C.

If L2 does not have the optimum value, the match may still be obtained by adjusting k and Q, but one or the other — or both — must have a larger value than is needed when $X_{L2}$ is equal to R2. In general, it is desirable to use as low a value of Q as is practicable, since low Q values mean that the circuit requires little or no readjustment when shifting frequency within a band (provided R1 does not vary appreciably with frequency).

## Circuit Q

In Fig. 2A, Q is equal to R1 in ohms divided by the reactance of C1 in ohms, assuming L1-C1 is tuned to the operating frequency. This circuit is suitable for comparatively high values of R1 — from several hundred to several thousand ohms. In Fig. 2C, Q is equal to the reactance of C1 divided by the resistance of R1, L1-C1 again being tuned to the operating frequency. This circuit is suitable for low values of R1 — from a few ohms up to a hundred or so ohms. In Fig. 2B the Q depends on the placement of the taps on L1 as well as on the reactance of C1. This circuit is suitable for matching all values of R1 likely to be encountered in practice.

Note that to change Q in either A or C, Fig. 2, it is necessary to change the reactance of C1. Since the circuit is tuned essentially to resonance at the operating frequency, this means that the L/C ratio must be varied in order to change Q. In Fig. 2B a fixed L/C ratio may be used, since Q can be varied by changing the tap positions. The Q will increase as the taps are moved closer together, and will decrease as they are moved farther apart on L1.

## Reactive Loads — Series and Parallel Coupling

More often than not, the load represented by the input impedance of the transmission line is reactive as well as resistive. In such a case the load cannot be represented by a simple resistance, such as R1 in Fig. 2. As stated in an earlier chapter, we have the option of considering the load to be a resistance in parallel with a reactance, or as a resistance in series with a reactance. In Figs. 2A and B it is convenient to use the parallel equivalent of the line input impedance. The series equivalent is more suitable for Fig. 2C.

Thus in Fig. 3 at A and B the load might be represented by R1 in parallel with the capacitive reactance C, and in Fig. 3C by R1 in series with a capacitive reactance C. In A, the capacitance C is in parallel with C1 and so the total capacitance is the sum of the two. This is the effective capacitance that, with L1, tunes to the operating frequency. Obviously the setting of C1 will be at a lower value of capacitance with such a load than it would with a purely resistive load such as is shown in Fig. 2A.

In Fig. 3B the capacitance of C also increases the total capacitance effective in tuning the circuit. However, in this case the increase in effective tuning capacitance depends on the positions of the taps; if the taps are close together the effect of C on the tuning is relatively small, but it increases as the taps are moved farther apart.

In Fig. 3C, the capacitance C is in series with C1 and so the total capacitance is less than either. Hence the capacitance of C1 has to be increased in order to resonate the circuit, as compared with the purely resistive load shown in Fig. 2C.

If the reactive component of the load impedance is inductive, similar considerations apply. In such case an inductance would be substituted for the capacitance C shown in Fig. 3. The effect in Figs. 3A and B would be to decrease the effective inductance in the circuit, so C1 would require a larger value of capacitance in order to resonate the circuit to the operating frequency. In Fig. 3C the effective inductance would be increased, thus making it necessary to set C1 at a lower value of capacitance for resonating the circuit.

## Effect of Line Reactance on Circuit Q

The presence of reactance in the line input impedance can affect the Q of the matching circuit. If the reactance is capacitive, the Q will not change if resonance can be maintained by adjustment of C1 without changing either the value of L1 or the position of the taps in Fig. 3B (as compared with the Q when the load is purely resistive and has the same value of resistance, R1). If the load reactance is inductive the L/C ratio changes because the effective inductance in the circuit is changed and, in the ordinary case, L1 is not adjustable. This increases the Q in all three circuits of Fig. 3.

When the load has appreciable reactance it is not always possible to adjust the circuit to resonance by readjusting C1, as compared with the setting it would have with a purely resistive load. Such a situation may occur when the load reactance is low compared with the resistance in the parallel-equivalent circuit, or when the reactance is high compared with the resistance in the series-equivalent circuit. The very considerable detuning of the circuit that results is often accompanied by an increase in Q, sometimes to values that lead to excessively high circulating currents in the circuit. This causes the efficiency to suffer. (Ordinarily the power loss in matching circuits of this type is inconsequential, if the Q is below 10 and a good coil is used.) An unfavorable ratio of reactance to resistance in the input impedance of the line can exist if the SWR is high and the line length is near an odd multiple of one-eighth wavelength (45 degrees).

### Q of Line Input Impedance

The ratio between reactance and resistance in the equivalent input circuit — that is, the Q of the line input impedance — is a function of line length and SWR. There is no specific value of this Q of which it can be said that lower values are satisfactory while higher values are not. In part, the maximum tolerable value depends on the tuning range available in the matching circuit. If the tuning range is restricted (as it will be if the variable capacitor has relatively low maximum capacitance), compensating for the line input reactance by absorbing it in the matching circuit — that is, by retuning C1 in Fig. 3 — may not be possible. Also, if the Q of the matching circuit is low the effect of the line input reactance will be greater than it will when the matching-circuit Q is high.

As stated earlier, the optimum matching-circuit design is one in which the Q is low, i.e., a low reactance-to-resistance ratio.

### Compensating for Input Reactance

When the reactance/resistance ratio in the line input impedance is unfavorable it is advisable to take special steps to compensate for it. This can be done as shown in Fig. 4. Compensation consists of supplying external reactance of the same numerical value as the line reactance, but of the opposite kind. Thus in A, where the line input impedance is represented by resistance and capacitance in parallel, an inductance L having the same numerical value of reactance as C can be connected across the line terminals to "cancel out" the line reactance. (This is actually the same thing as tuning the line to resonance at the operating frequency.) Since the parallel combination of L and C is equivalent to an extremely high resistance at resonance, the input impedance of the line becomes a pure resistance having essentially the same resistance as R1 alone.

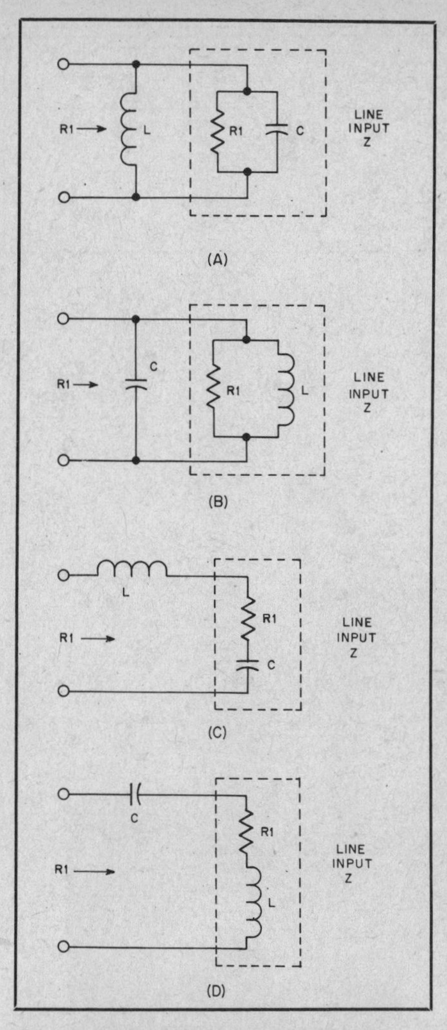


Fig. 4 — Compensating for reactance present in the line input impedance.

The case of an inductive line is shown at B. In this case the external reactance required is capacitive, of the same numerical value as the reactance of L. Where the series equivalent of the line input impedance is used the external reactance is connected in series, as shown at C and D in Fig. 4.

In general, these methods need not be used unless the matching circuit does not have sufficient range of adjustment to provide compensation for the line reactance as described earlier, or when such a large readjustment is required that the matching-circuit Q becomes undesirably high. The latter condition usually is accompanied by heating of the coil used in the matching network.

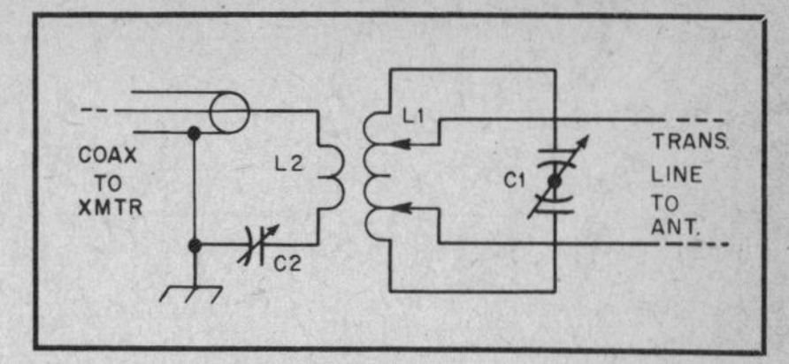


Fig. 5 — Using a variable capacitance, C2, as an alternative to variable mutual inductance between L1 and L2.

### Methods for Variable Coupling

The coupling between L1 and L2, Figs. 2 and 3, preferably should be adjustable. If the coupling is fixed, such as with a fixed-position link, the placement of the taps on L1 for proper matching becomes rather critical. The additional matching adjustment afforded by adjustable coupling between the coils facilitates the matching procedure considerably. L2 should be coupled to the center of L1 for the sake of maintaining balance, since the circuit is used with balanced lines.

If adjustable inductive coupling such as a swinging link is not feasible for mechanical reasons, an alternative is to use a variable capacitor in series with L2. This is shown in Fig. 5. Varying C2 changes the total reactance of the circuit formed by L2-C2, with much the same effect as varying the actual mutual inductance between L1 and L2. The capacitance of C2 should be such as to resonate with L2 at the lowest frequency in the band of operation. This calls for a fairly large value of capacitance at low frequencies (about 1000 pF at 3.5 MHz for 50-ohm line) if the reactance of L2 is equal to the line $Z_0$. To utilize a capacitor of more convenient size — maximum capacitance of perhaps 250-300 pF — a value of inductance may be used for L2 that will resonate at the lowest frequency with the maximum capacitance available.

On the higher frequency bands the problem of variable capacitors does not arise since a reactance of 50 to 75 ohms is within the range of conventional components.

### Circuit Balance

Fig. 5 shows C1 as a balanced or split-stator capacitor. This type of capacitor is desirable in a practical matching circuit to be used with a balanced line, since the two sections are symmetrical. The rotor assembly of the balanced capacitor may be grounded, if desired, or it may be left "floating" and the center of L1 may be grounded; or both may "float." Which method to use depends on considerations discussed later in connection with antenna

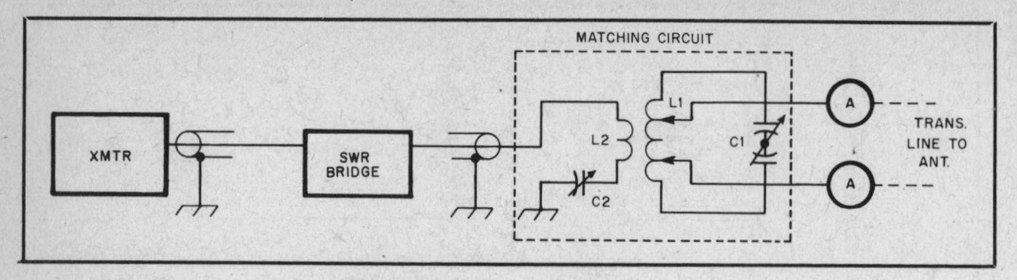

Fig. 6 — Adjustment setup using SWR indicator.

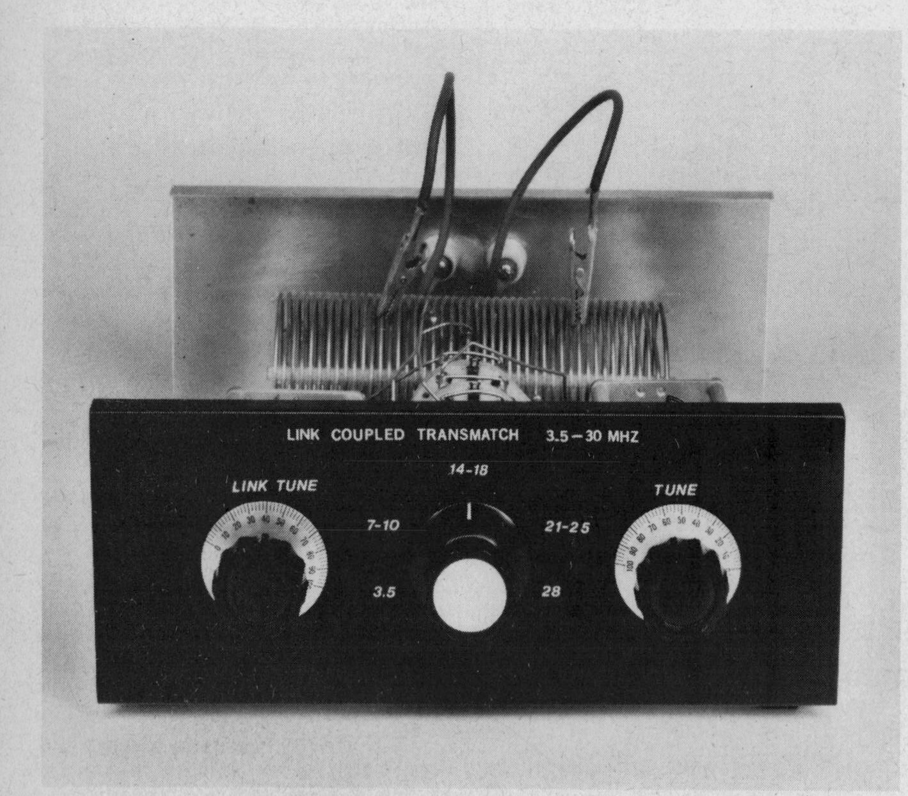

Fig. 7 — Exterior view of the band-switched link coupler. Alligator clips are used to select the proper tap positions of the coil.

currents on transmission lines. As an alternative to using a split-stator type of capacitor, a single-section capacitor may be used.

### Measurement of Line Input Current

The rf ammeters shown in Fig. 6 are not essential to the adjustment procedure but they, or some other form of output indicator, are useful accessories. In most cases the circuit adjustments that lead to a match as shown by the SWR indicator will also result in the most efficient power transfer to the transmission line. However, it is possible that a good match will be accompanied by excessive loss in the matching circuit. This is unlikely to happen if the steps described for obtaining a low Q are taken. If the settings are highly critical or it is impossible to obtain a match, the use of additional reactance compensation as described earlier is indicated.

Rf ammeters are useful for showing the comparative output obtained with various matching-network settings, and also for showing the improvement in output resulting from the use of reactance compensation when it seems to be required. Providing no basic circuit changes (such as grounding or ungrounding some part of the matching circuit) are made during such comparisons, the current shown by the ammeters will increase whenever the power put into the line is increased. Thus, the highest reading indicates the greatest transfer efficiency, assuming that the power input to the transmitter is kept constant.

If the line $Z_0$ is matched by the antenna, the current can be used to determine the actual power input to the line. The power at the input terminals is then equal to $I^2Z_0$ where I is the current and $Z_0$ is the characteristic impedance of the line. If there are standing waves on the line this relationship does not hold. In such a case the current that will flow into the line is determined by the line length, SWR, and whether the antenna impedance is higher or lower than the line impedance. Fig. 26 in Chapter 3 shows how the maximum current to be expected will vary with the standing-wave ratio. This information can be used in selecting the proper ammeter range.

Two ammeters, one in each line conductor, are shown in Fig. 6. The use of two instruments gives a check on the line balance, since the currents should be the same. However, a single meter can be switched from one conductor to the other. If only one instrument is used it is preferably left out of the circuit except when adjustments are being made, since it will add capacitance to the side in which it is inserted and thus cause some unbalance. This is particularly important when the instrument is mounted on a metal panel.

Since the resistive component of the input impedance of a line operating with an appreciable SWR is seldom known accurately, the rf current is of little value as a check on power input to such a line. However, it shows in a relative way the efficiency of the system as a whole. The set of coupling adjustments that results in the largest line current with the least final-amplifier plate current is the one that delivers the greatest power to the antenna with the lowest plate-power input.

For adjustment purposes, it is possible to substitute small flashlight lamps, shunted across a few inches of the line wires, for the rf ammeters. Their relative brightness shows when the current increases or decreases. They have the advantage of being inexpensive and of such small physical size that they do not unbalance the circuit.

## A LINK-COUPLED MATCHING NETWORK

Link coupling offers many advantages over other types of systems where a direct connection between the equipment and antenna is required. This is particularly true on 80 meters, where commercial broadcast stations often induce sufficient voltage to cause either rectification or front-end overload. Transceivers and receivers that show this tendency can usually be cured by using only magnetic coupling between the transceiver and antenna system. There is no direct connection and better isolation results along with the inherent band-pass characteristics of magnetically coupled tuned circuits.

Although link coupling can be used with either single-ended or balanced antenna systems, its most common application is with balanced feed. The model shown here is designed for 80- through 10-meter operation.

### The Circuit

The Transmatch or matching networks shown in Figs. 7 through 9 is a band-switched link coupler. L2 is the link and C1 is used to adjust the coupling. S1b selects the proper amount of link inductance for each band. L1 and L3 are located on each side of the link and are the coils to which the antenna is connected. Alligator clips are used to connect the antenna to the coil because antennas of different impedances must be connected at different points (taps) along the coil. Also, with most antennas it will be necessary to change taps for different bands of operation. C2 tunes L1 and L3 to resonance at the operating frequency.

Switch sections S1A and S1C select the amount of inductance necessary for each of the hf bands. The inductance of each of the coils has been optimized for antennas in the impedance range of roughly 20 to 600 ohms. Antennas that exhibit impedances well outside this range may require that some of the fixed connections to L1 and L3 be changed. Should this be necessary remember that the L1 and L3 sections must be kept symmetrical — the same number of turns on each coil.

### Construction

The unit is housed in a homemade aluminum enclosure that measures 9 × 8 × 3 1/2 inches. Any cabinet with similar dimensions that will accommodate the components may be used. L1, L2 and L3 are a one-piece assembly of B&W 3026 Miniductor stock. The individual coils are separated from each other by cutting two of the turns at the appropriate spots along the length of the coil. Then the inner ends of the outer sections are joined by a short wire that is run through the center of L2. Position the wire so that it will not come into contact with L2. Each of the fixed tap points on L1, L2 and L3 is located and lengths of hookup wire are attached. The coil is mounted in the enclosure and the connections between the coil and the bandswitch are made. Every other turn of L1 and L3 are pressed in toward the center of the coil to facilitate connection of the alligator clips.

As can be seen from the schematic, C2 must be isolated from ground. This can be accomplished by mounting the capacitor on steatite cones or other suitable insulating material. Make sure that the hole through the front panel for the shaft of C2 is large enough so the shaft does not come into contact with the chassis.

### Tune-up

The transmitter should be connected to the input of the Transmatch through some sort of instrument that will indicate SWR. S1 is set to the band of operation and the balanced line is connected to the in-

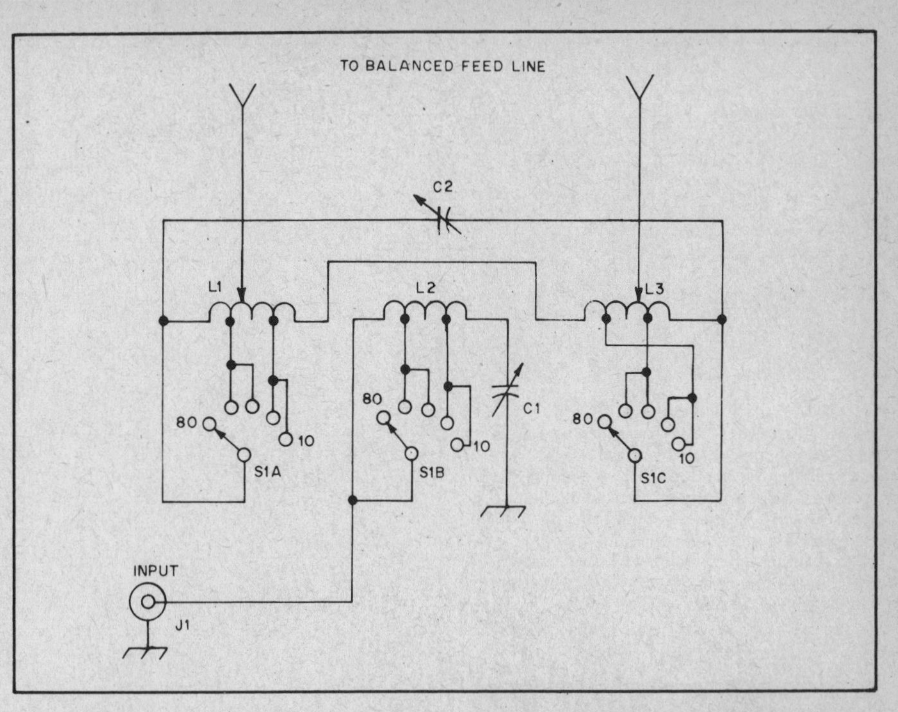


Fig. 8 — Schematic diagram of the link coupler. The connections marked as "to balanced feed line" are steatite feedthrough insulators. The arrows on the other ends of these connections are alligator clips.

C1 — 350 pF maximum, 0.0435 inch plate spacing or greater.
C2 — 100 pF maximum, 0.0435 inch plate spacing or greater.
J1 — Coaxial connector.
L1, L2, L3 — B&W 3026 Miniductor stock, 2-inch diameter, 8 turns per inch, no. 14 wire. Coils assembly consists of 48 turns, L1 and L3 are each 17 turns tapped at 8 and 11 turns from outside ends. L2 is 14 turns tapped at 8 and 12 turns from C1 end. See text for additional details.
S1 — 3-pole, 5-position ceramic rotary switch.

Fig. 9 — Interior view of the coupler showing the basic positions of the major components. Component placement is not critical, but the unit should be laid out for minimum lead lengths.

Fig. 10 — Exterior view of the SPC Transmatch. Radio Shack vernier drives are used for adjusting the tuning capacitors. A James Millen turns-counter drive is coupled to the rotary inductor. Green paint and green Dymo tape labels are used for panel decor. The cover is plain aluminum with a lightly grooved finish (sandpapered) which has been coated with clear lacquer. An aluminum foot holds the Transmatch at an easy access angle.

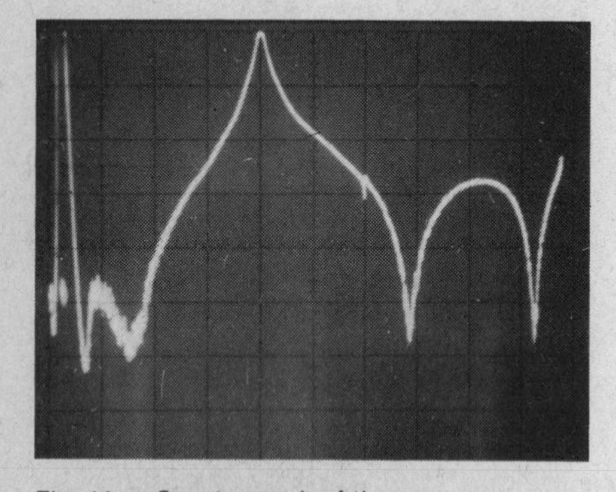

Fig. 11 — Spectrograph of the response characteristics of the SPC Transmatch looking into a 1000-ohm termination from a 50-ohm signal source. The scale divisions are 2 MHz horizontal and 10 dB vertical. The fundamental frequency is 8 MHz; the second harmonic is attenuated by 28 dB with these matching conditions.

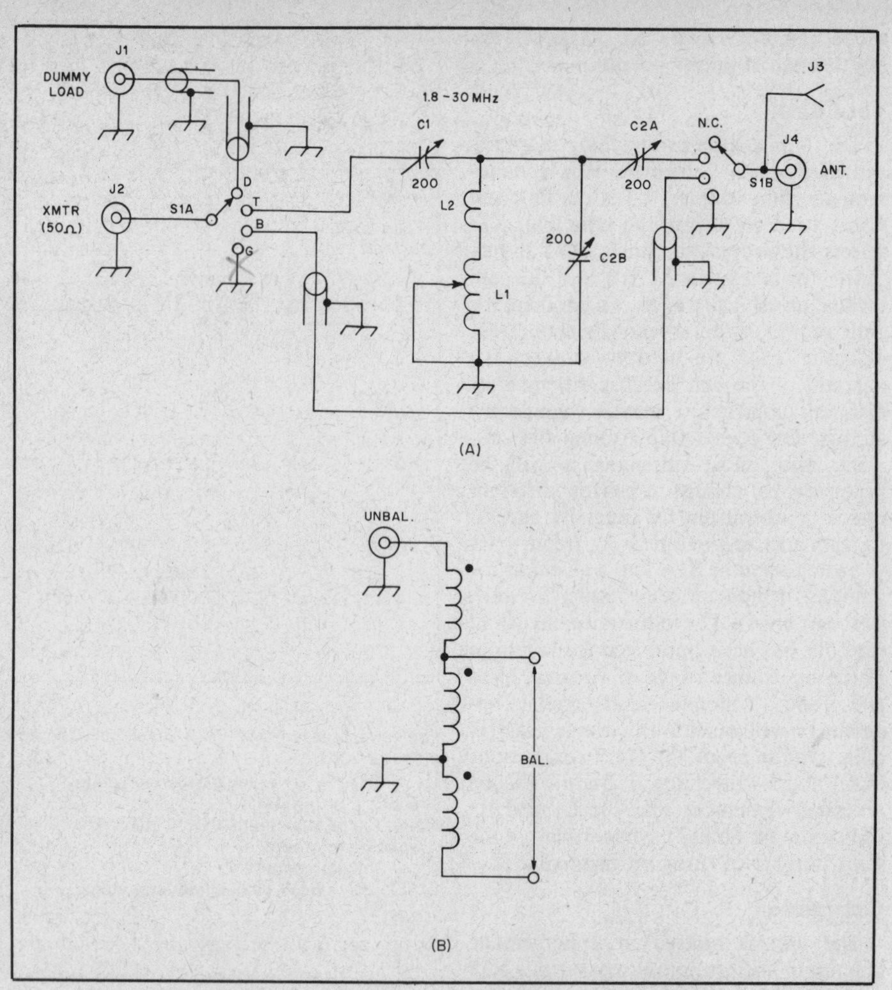


Fig. 12 — Schematic diagram of the SPC circuit. Capacitance is in picofarads.

C1 — 200-pF transmitting variable with plate spacing of 0.075 inch (2 mm) or greater. J.W. Miller Co. no. 2150 used here.

C2 — Dual-section variable. 200 pF per section. Same plate spacing as C1. J.W. Miller Co. no. 2151 used here. (Catalog no. 79, J.W. Miller Co., 19070 Reyes Ave., Compton, CA 90224.)

J1, J2, J4 — SO-239 style coaxial connector. J4 should have high-dielectric insulation if high-Z single-wire antennas are used at J3. Teflon insulation is recommended.

J3 — Ceramic feedthrough bushing.

L1 — Rotary inductor, 25μH min. inductance, E. F. Johnson 229-203 or equiv.

L2 — Three turns no. 8 copper wire, 1 inch (25 mm) ID × 1-1/2 inches (38 mm) long.

S1 — Large ceramic rotary wafer switch with heavy contacts. Two-pole, 4-position type. Surplus Centralab JV-9033 or equiv., two positions unused.

Z1 — Balun transformer. 12 turns no. 12 Formvar wire, trifilar, close-wound on 1-inch OD phenolic or PVC-tubing form.

sulators on the rear panel of the coupler. The alligator clips are attached to the mid points of coils L1 and L3 and power is applied. Adjust C1 and C2 for minimum reflected power. If a good match is not obtained, move the antenna tap points either closer to the ends or center of the coils. Again apply power and tune C1 and C2 until the best possible match is obtained. Continue moving the antenna taps until a 1-to-1 match is obtained.

The circuit described here is intended for power levels up to roughly 200 watts. Balance was checked by means of two rf ammeters, one in each leg of the feed line. Results showed the balance to be well within 1 dB.

## A TRANSMATCH FOR BALANCED OR UNBALANCED LINES

Most modern transmitters are designed to operate into loads of approximately 50 ohms. Solid-state transmitters produce progressively lower output power as the SWR on the transmission line increases, because of built-in SWR protection circuits. Therefore, it is useful to employ a matching network between the transmitter and the antenna feeder when antennas with complex impedances are used. One example of this need can be seen in the case of an 80-meter, coax-fed dipole antenna which has been cut for resonance at, say, 3.6 MHz. If this antenna were used in the 75-meter phone band, the SWR would be fairly high. A Transmatch could be used to give the transmitter a 50-ohm load, even though a significant mismatch was present at the antenna feed point. It is important to remember that the Transmatch will *not* correct the actual SWR condition; it only compensates for it as far as the transmitter is concerned. A Transmatch is useful also when employing a single-wire antenna for multiband use. By means of a balun at the Transmatch output it is possible to operate the transmitter into a balanced transmission line, such as a 300- or 600-ohm feed system of the type that would be used with a multiband tuned dipole, V beam or rhombic antenna.

A secondary benefit can be realized from a Transmatch. Under most conditions of normal application it will attenuate harmonics from the transmitter. The amount of attenuation depends on the circuit design and the loaded Q ($Q_L$) of the network after the impedance has been matched. The higher the $Q_L$, the greater the attenuation. The SPC Transmatch pictured in Figs. 10 and 13 was designed to provide high harmonic attenuation. See Fig. 11. The SPC (series-parallel capacitance) circuit maintains a band-pass response under load conditions of less than 25 ohms to more than 1000 ohms (from a 50-ohm transmitter). This is because a substantial amount of capacitance is always in parallel with the rotary inductor (C2B and L1 of Fig. 12).

The circuit of Fig. 12 operates from 1.8 to 30 MHz with the values shown. The SPC exhibits somewhat sharper tuning than other designs. This arises from the relatively high network Q, and is especially prominent at 40, 80 and 160 meters. For this reason there are vernier-drive dials on C1 and C2. They are also useful in logging the dial settings for changing bands or antennas.

### Construction

Figs. 10 and 13 show the structural details of the Transmatch. The cabinet is homemade from 16-gauge aluminum sheeting. L brackets are affixed to the right and left sides of the lower part of the cabinet to permit attachment of the U-shaped cover.

The conductors which join the components should be of heavy-gauge material to minimize stray inductance and heating. Wide strips of flashing copper are suitable for the conductor straps. The center conductor and insulation from RG-59/U polyfoam coaxial cable is used in this model for the wiring between the switch and the related components. The insulation is sufficient to prevent breakdown and arcing at 2 kW PEP input to the transmitter.

All leads should be kept as short as possible to help prevent degradation of the circuit Q. The stators of C1 and C2 should face toward the cabinet cover to minimize the stray capacitance between the capacitor plates and the bottom of the cabinet (important at the upper end of the Transmatch frequency range). Insulated ceramic shaft couplings are used between the vernier drives and C1 and C2, since the rotors of both capacitors are "floating" in this circuit. C1 and C2 are supported above the bottom plate on steatite cone insulators. S1 is attached to the rear apron of the cabinet by means of two metal standoff posts.

### Operation

The SPC Transmatch shown here is designed to handle the output from transmitters that operate up to 2 kW PEP. L2 improves the circuit Q at 10 and 15 meters. However, it may be omitted from the circuit if the rotary inductor (L1) has a tapered pitch at the minimum-inductance end. It may be necessary to omit L2 if the stray wiring inductance of the builder's version is high. Otherwise, it may be impossible to obtain a matched condition at 28 MHz with certain loads.

An SWR indicator is used between the transmitter and the Transmatch to show when a matched condition is attained. The builder may want to integrate an SWR meter in the Transmatch circuit between J2 and the arm of S1A (Fig 12A). If this is done there should be room for an edgewise panel meter above the vernier drive for C2.

Initial transmitter tuning should be done with a dummy load connected to J1, and with S1 in the D position. This will prevent interference that could otherwise occur if tuning is done "on the air." After the transmitter is properly tuned into the dummy load, unkey the transmitter and switch S1 to T (Transmatch). Never "hot switch" a Transmatch, as this can damage both transmitter and Transmatch. Set C1 and C2 at midrange. With a few watts of rf, adjust L1 for a decrease in reflected power. Then adjust C1 and C2 alternately for the lowest possible SWR. If the SWR cannot be reduced to 1:1, adjust L1 slightly and repeat the procedure. Finally, increase the transmitter power to maximum and touch up the Transmatch controls if necessary. When tuning, keep your transmissions brief and identify your station.

The air-wound balun of Fig. 12B can be used outboard from the Transmatch if a low-impedance balanced feeder is contemplated. Ferrite or powdered-iron core material is not used in the interest of avoiding TVI and harmonics which can result from core saturation. A detailed discussion of baluns can be found later in this chapter.

The B position of S1 permits switched-through operation when the Transmatch is not needed. The G position is used for grounding the antenna system, as necessary; a good-quality earth ground should be attached at all times to the Transmatch chassis.

### Final Comments

Surplus coils and capacitors are suitable for use in this circuit. L1 should have at least 25 $\mu$H of inductance, and the tuning capacitors need to have 150 pF or more of capacitance per section. Insertion loss through this Transmatch was measured at less than 0.5 dB at 600 watts of rf power on 7 MHz.

Fig. 13 — Interior view of the SPC Transmatch. L2 is mounted on the rear wall by means of two ceramic standoff insulators. C1 is on the left and C2 is at the right. The coaxial connectors, ground post and J3 are on the lower part of the rear panel.

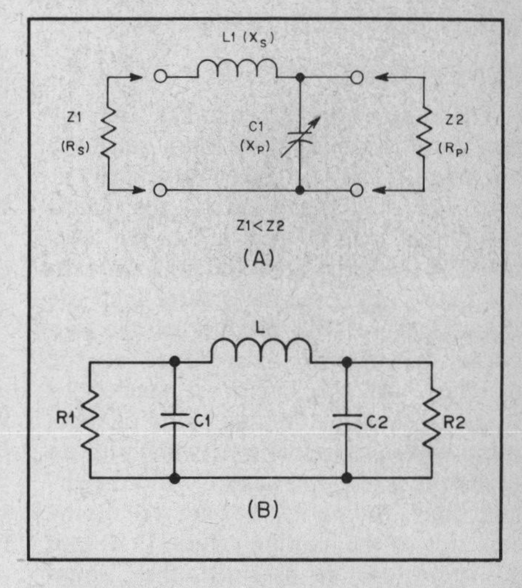


Fig. 14 — At A, the L matching network, consisting of L1 and C1, matches Z1 and Z2. The lower of the two impedances to be matched, Z1, must always be connected to the series-arm of the network and the higher impedance, Z2, to the shunt-arm side. The positions of the inductor and capacitor may be interchanged in the network. At B, the pi network, matching R1 to R2. The pi provides more flexibility than the L as a Transmatch circuit. See equations in the text for calculating component values.

## L NETWORKS

A comparatively simple but very useful matching circuit for unbalanced loads is the L network, Fig. 14A, L-network Transmatches are normally used for only a single band of operation, although multiband versions with switched or variable coil taps exist. To determine the range of circuit values for a matched condition, the input and load impedance values must be known or assumed. Otherwise a match may be found by trial and error.

In Fig. 14A, L1 is shown as the series reactance, $X_S$, and C1 as the shunt or parallel reactance, $X_P$. However, a capacitor may be used for the series reactance and an inductor for the shunt reactance, to satisfy mechanical or other considerations.

The ratio of the series reactance to the series resistance, $X_S/R_S$, is defined as the network Q. The four variables, $R_S$, $R_P$, $X_S$ and $X_P$, are related as given in the equations below. When any two values are known, the other two may be calculated.

$$Q = \sqrt{\frac{R_P}{R_S} - 1} = \frac{X_S}{R_S} = \frac{R_P}{X_P}$$

$$X_S = QR_S = \frac{QR_P}{1 + Q^2}$$

$$X_P = \frac{R_P}{Q} = \frac{R_P R_S}{X_S} = \frac{R_S^2 + X_S^2}{X_S}$$

$$R_S = \frac{R_P}{1 + Q^2} = \frac{X_S X_P}{R_P}$$

$$R_P = R_S(1 + Q^2) = QX_P$$

$$= \frac{R_S^2 + X_S^2}{R_S}$$

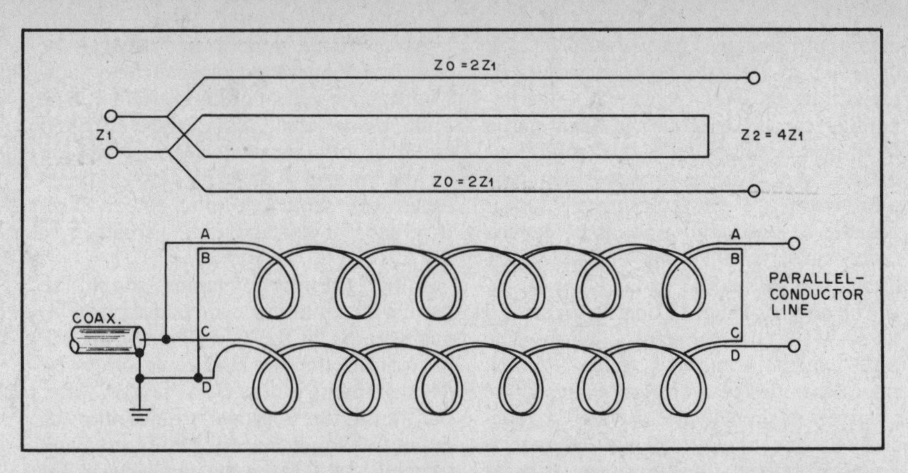


Fig. 15 — Baluns for matching between push-pull and single-ended circuits. The impedance ratio is 4:1 from the push-pull side to the unbalanced side. Coiling the lines (lower drawing) increases the frequency range over which satisfactory operation is obtained.

The reactance of loads that are not purely resistive may be taken into account and absorbed or compensated for in the reactances of the matching network. Inductive and capacitive reactance values may be converted to inductor and capacitor values for the operating frequency with standard reactance equations.

## PI NETWORKS

The pi network, shown in Fig. 14B, offers more flexibility than the L since the operating Q may be chosen practically at will. The only limitation on the circuit values that may be used is that the reactance of the series arm, the inductor L in the figure, must not be greater than the square root of the product of the two values of resistive impedance to be matched. As the circuit is applied to amateur use, this limiting value of reactance would represent a network with an undesirably low operating Q, and the circuit values ordinarily used are well on the safe side of the limiting values. If R1 and R2 are known or assumed, these equations may be used to determine the component values required for a match. The value of Q may be arbitrarily chosen, usually between 5 and 15 for most applications.

$$R1 > R2$$

$$X_{C1} = R1/Q$$

$$X_{C2} = R2 \sqrt{\frac{R1/R2}{Q^2 + 1 - R1/R2}}$$

$$X_L = \frac{Q \cdot R1 + R1 \cdot R2/X_{C2}}{Q^2 + 1}$$

The pi network may be used to match a low impedance to a rather high one, such as 50 ohms to several thousand ohms. Conversely, it may be used to match 50 ohms to a quite low value, such as 1 ohm or less. For Transmatch applications, C1 and C2 may be independently variable. L may be a roller inductor or a coil with switchable taps. Alternatively, a lead fitted with a suitable clip may be used to short out turns of a fixed inductor. In this way, a match may be obtained through trial and error. It will be possible to match two values of impedances with several different settings of L, C1 and C2. This results because the Q of the network is being changed. If a match is maintained with other adjustments, the Q of the circuit rises with increased capacitance at C1.

Quite often the load and source have reactive components along with resistance, but in most instances the pi network can still be used. The effect of these reactive components can be compensated for by changing one of the reactive elements in the matching network. For example, if some reactance was shunted across R2, the setting of C2 could be changed to compensate, whether that shunt reactance be inductive or capacitive.

## BALUNS

A centerfed antenna with open ends, of which the half-wave type is an example, is inherently a balanced radiator. When opened at the center and fed with a parallel-conductor line, this balance is maintained throughout the system, so long as the causes of feeder unbalance discussed in Chapter 5 are avoided.

If the antenna is fed at the center through a coaxial line, this balance is upset because one side of the radiator is connected to the shield while the other is connected to the inner conductor. On the side connected to the shield, a current can flow down over the *outside* of the coaxial line. The fields thus set up cannot be canceled by the fields from the inner conductor because the fields *inside* the line cannot escape through the shielding afforded by the outer conductor. Hence these "antenna" currents flowing on the outside of the line will be responsible for radiation.

Line radiation can be prevented by a number of devices whose purpose is to detune or decouple the line for "antenna" currents and thus greatly reduce their amplitude. Such devices generally are known as *baluns* (a contraction for "balanced to unbalanced").

The need for baluns arises in coupling a transmitter to a balanced transmission line, since the output circuits of most transmitters have one side grounded. (This type of output circuit is desirable for a number of reasons, including TVI reduction.) The most flexible type of balun for this purpose is the inductively coupled matching network described in a previous section in this chapter. This combines impedance matching with balanced-to-unbalanced operation, but has the disadvantage that it uses resonant circuits and thus can work over only a limited band of frequencies without readjustment. However, if a fixed impedance ratio in the balun can be tolerated, the coil balun described below can be used without adjustment over a frequency range of about 10:1 — 3 to 30 MHz, for example.

### Coil Baluns

The type of balun known as the "coil balun" is based on the principles of linear-transmission-line balun as shown in the upper drawing of Fig. 15. Two transmission lines of equal length having a characteristic impedance ($Z_o$) are connected in series at one end and in parallel at the other. At the series-connected end the lines are balanced to ground and will match an impedance equal to $2Z_o$. At the parallel-connected end the lines will be matched by an impedance equal to $Z_o/2$. One side may be connected to ground at the parallel-connected end, provided the two lines have a length such that, considering each line as a single wire, the balanced end is effectively decoupled from the parallel-connected end. This requires a length that is an odd multiple of a quarter wavelength.

A definite line length is required only for decoupling purposes, and so long as there is adequate decoupling the system will act as a 4:1 impedance transformer regardless of line length. If each line is wound into a coil, as in the lower drawing, the inductances so formed will act as choke coils and will tend to isolate the series-connected end from any ground connection that may be placed on the parallel-connected end. Balun coils made in this way will operate over a wide frequency range, since the choke inductance is not critical. The lower frequency limit is where the coils are no longer effective in isolating one end from the other; the length of line in each coil should be about equal to a quarter wavelength at the lowest frequency to be used.

The principal application of such coils is in going from a 300-ohm balanced line to a 75-ohm coaxial line. This requires that the $Z_0$ of the lines forming the coils be 150 ohms.

A balun of this type is simply a fixed-ratio transformer, when matched. It cannot compensate for inaccurate matching elsewhere in the system. With a "300-ohm" line on the balanced end, for example, a 75-ohm coax cable will not be matched unless the 300-ohm line actually is terminated in a 300-ohm load.

## TWO BROADBAND TOROIDAL BALUNS

Air-wound balun transformers are somewhat bulky when designed for operation in the 1.8- to 30-MHz range. A more compact broadband transformer can be realized by using toroidal ferrite core material as the foundation for bifilar-wound coil balun transformers. Two such baluns are described here.

In Fig. 16 at A, a 1:1 ratio balanced- to unbalanced-line transformer is shown. This transformer is useful in converting a 50-ohm balanced line condition to one that is 50 ohms, unbalanced. Similarly, the transformer will work between balanced and unbalanced 75-ohm impedances. A 4:1 ratio transformer is illustrated in Fig. 16 at B. This balun is useful for converting a 200-ohm balanced condition to one that is 50 ohms, unbalanced. In a like manner, the transformer can be used between a balanced 300-ohm point and a 75-ohm unbalanced line. Both balun transformers will handle 1000 watts of rf power and are designed to operate from 1.8 through 60 MHz.

Low-loss high-frequency ferrite core material is used for T1 and T2. The cores are made from Q2 material and are 0.5 inch thick, have an OD of 2.4 inches and the ID is 1.4 inches. The permeability rating of the cores is 40. Packaged 1-kilowatt balun kits, with winding instructions for 1:1 or 4:1 impedance transformation ratios, are available, but use a core of slightly different dimensions. Ferrite cores are available from several sources, including the following: Ferroxcube Corp. of America, 5083 Kings Hwy., Saugerties, NY 12477; Amidon Associates, 12033 Otsego St., North Hollywood, CA 91601; Palomar Engineers, P.O. Box 455, Escondido, CA 92025, and Radiokit, P.O. Box 411, Greenville, NH 03048.

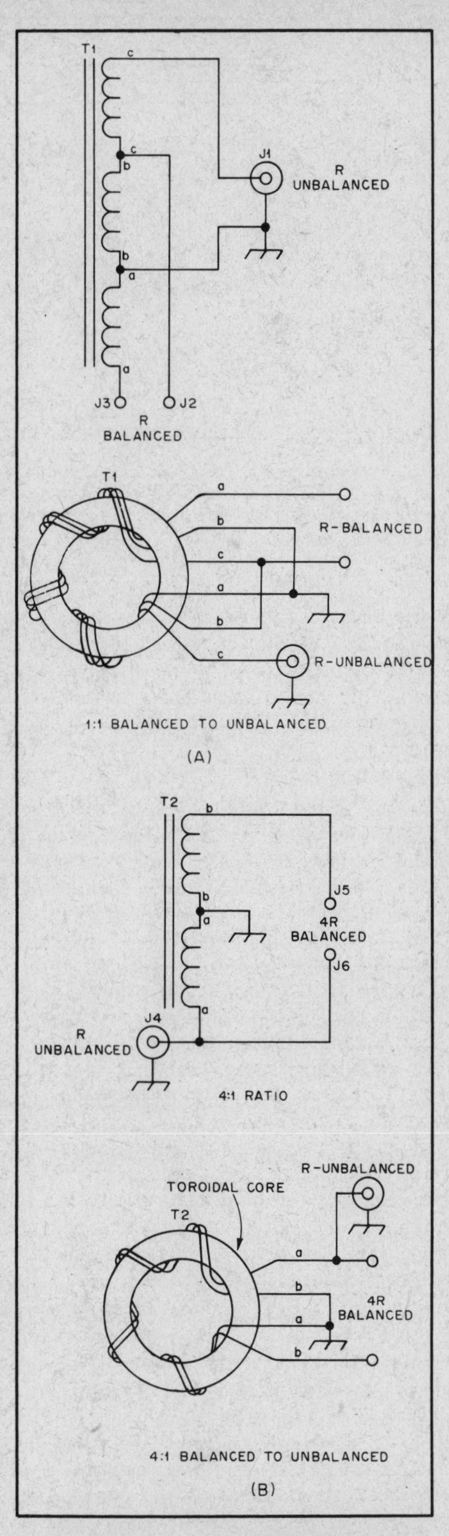


Fig. 16 — Schematic and pictorial representations of the balun transformers. T1 and T2 are wound on CF-123 toroid cores (see text). J1 and J4 are SO-239-type coax connectors, or similar. J2, J3, J5 and J6 are steatite feed-through bushings. The windings are labeled a, b and c to show the relationship between the pictorial and schematic illustrations.

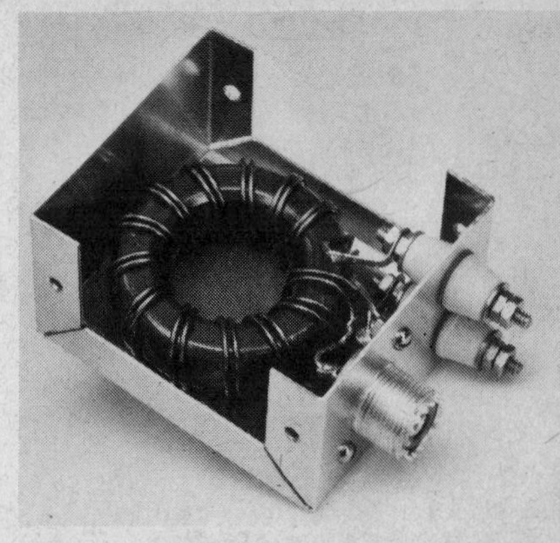

Fig. 17 — Layout of a kilowatt 4:1 toroidal balun transformer. Phenolic insulating board is mounted between the transformer and the Minibox wall to prevent short-circuiting. The board is held in place with epoxy cement. Cement is also used to secure the transformer to the board. For outdoor use, the Minibox cover can be installed, then sealed against the weather by applying epoxy cement along the seams of the box.

### Winding Information

The transformer shown in Fig. 16 at A has a trifilar winding consisting of 10 turns of no. 14 Formvar-insulated copper wire. A 10-turn bifilar winding of the same type of wire is used for the balun of Fig. 16 at B. If the cores have rough edges, they should be carefully sanded until smooth enough to prevent damage to the wire's Formvar insulation. The windings should be spaced around the entire core as shown in Fig. 17. Insulation can be used between the core material and the windings to increase the breakdown voltage of the balun.

## A 50- TO 75-OHM BROADBAND TRANSFORMER

Shown in Figs. 18 through 20 is a simple 50- to 75-ohm or 75- to 50-ohm transformer that is suitable for operation in the 2- to 30-MHz frequency range. A pair of these transformers is ideal for using 75-ohm CATV hardline in a 50-ohm system. In this application one transformer is used at each end of the cable run. At the antenna one transformer raises the 50-ohm impedance of the antenna to 75 ohms, thereby presenting a match to the 75-ohm cable. At the station end a transformer is used to step the 75-ohm line impedance down to 50 ohms.

The schematic diagram of the transformer is shown in Fig. 18, and the winding details are given in Fig. 19. C1 and C2 are compensating capacitors; the values shown were determined through swept return-loss measurements using a

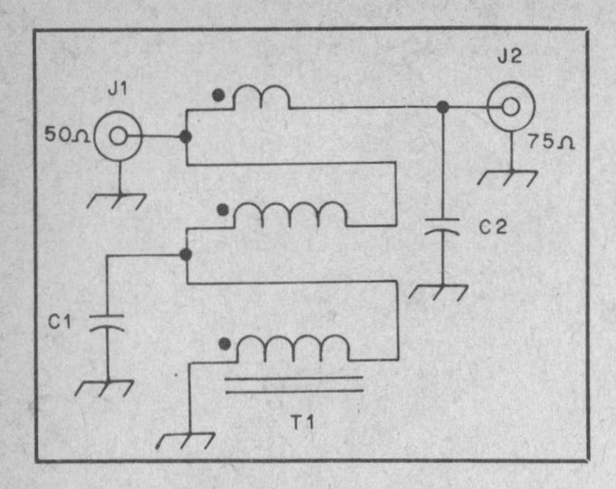

Fig. 18 — Schematic diagram of the 50- to 75-ohm transformer described in the text. C1 and C2 are compensating capacitors.
C1 — 100 pF silver mica.
C2 — 10 pF, silver mica.
J1, J2 — Coaxial connectors, builder's choice.
T1 — Transformer, 6 trifilar turns no. 14 enameled copper wire on an FT-200-61 (Q1 material, $\mu_i$ = 125) core. One winding has one-half the number of turns of the other two.

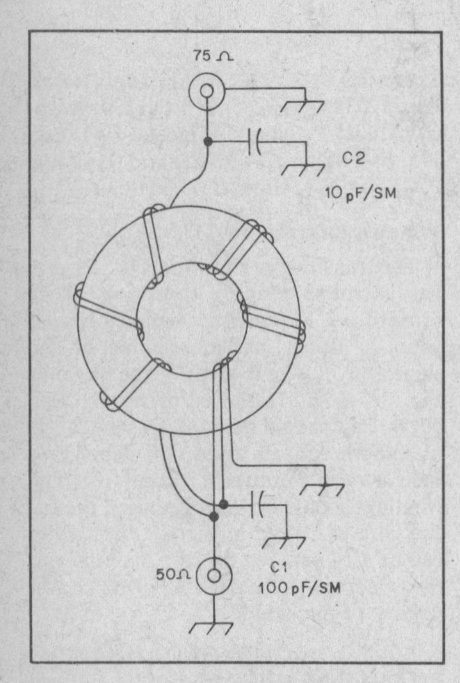

Fig. 19 — Pictorial drawing of the 50- to 75-ohm transformer showing details of the windings.

spectrum analyzer and tracking generator. The transformer consists of a trifilar winding of no. 14 enameled copper wire wound over an FT-200-61 (Q1 material) or equivalent core. As shown in Fig. 19, one winding has only half the number of turns of the other two. Care must be taken when connecting the loose ends so that the proper phasing of the turns is maintained. Improper phasing will become apparent when power is applied to the transformer.

If the core has sharp edges it is a good idea either to sand the edges until they are relatively smooth or wrap the core with

Fig. 20 — This is a photograph of a pair of the 50-to 75-ohm transformers. The units are identical.

tape. The one shown in the photograph was wrapped with ordinary vinyl electrical tape, although glass-cloth insulating tape would be better. The idea is to prevent chafing of the wire insulation.

### Construction

The easiest way to construct the transformer is to wind the three lengths of wire on the core at the same time. Different color wires will aid in identifying the ends of the windings. After all three windings are securely in place, the appropriate winding may be unwound three turns as shown in the diagram. This wire is the 75-ohm connection point. Connections at the 50-ohm end are a bit tricky, but if the information in Fig. 19 is followed carefully no problems should be encountered. Use the shortest connections possible, as long leads will degrade the high-frequency performance.

The balun is housed in a homemade aluminum enclosure measuring 3-1/2 × 3-3/4 × 1-1/4 inches. Any commercial cabinet of similar dimensions will work fine. In the unit shown in the photograph, several "blobs" of silicone seal (RTV) were used to hold the core in position. Alternatively, a piece of phenolic insulating material may be used between the core and the aluminum enclosure. Silicone seal is used to protect the inside of the unit from moisture. All joints and screw heads should receive a generous coating of RTV.

### Checkout

Checkout of the completed transformer or transformers is quite simple. If a 75-ohm dummy load is available connect it to the 75-ohm terminal of the transformer. Connect a transmitter and VSWR indicator (50 ohm) to the 50-ohm terminal of the transformer. Apply power (on each of the hf bands) and measure the VSWR looking into the transformer. Readings should be well under 1.3 to 1 on each of the bands. If a 75-ohm load is not available and two transformers have been constructed they may be checked out simultaneously as follows. Connect the 75-ohm terminals of both transformers together, either directly through a coaxial adaptor or through a length of 75-ohm cable. Attach a 50-ohm load to one of the 50-ohm terminals and connect a transmitter and VSWR indicator (50 ohm) to the remaining 50-ohm terminal. Apply power as outlined above and record the measurements. Readings should be under 1.3 to 1.

The transformers were checked in the ARRL laboratory under various mismatched conditions at the 1500-watt power level. No spurious signals (indicative of core saturation) could be found while viewing the lf, hf and vhf frequency range with a spectrum analyzer. A key-down, 1500-watt signal produced no noticeable core heating and only a slight increase in the temperature of the windings.

### Using the Baluns

For indoor applications, the transformers can be assembled open style, without benefit of a protective enclosure. For outdoor installations, such as at the antenna feed point, the balun should be encapsulated in epoxy resin or mounted in a suitable weatherproof enclosure. A Minibox, sealed against moisture, works nicely.

## BALUN TERMINATIONS

A word about baluns in Transmatches may be in order. Broadband transformers of the type found in many of the so-called Ultimate Transmatches are not suitable for use at high impedances. Disastrous results can be had when using these transformers with loads higher than, say, 300 ohms during high-power operation. The effectiveness of the transformer is questionable as well. At high peak rf voltages (high-Z load conditions such as 600-ohm feeders or an end-fed Hertz antenna) the core can saturate and the rf voltage can cause arcs between turns or between the winding and the core material. If a balanced-to-unbalanced transformation must be effected, try to keep the load impedance at 300 ohms or less. An airwound 1:1 balun with a trifilar winding is recommended over a transformer with ferrite or powdered-iron core material.

The principles on which baluns operate should make it obvious that the termination must be essentially a pure resistance in order for the proper impedance transformation to take place. If the termination is not resistive, the input impedance of each bifilar winding will depend on its electrical characteristics and the input impedance of the main transmission line; in other words, the impedance will vary just as it does with any transmission line, and the transformation ratio likewise will vary over wide limits.

Baluns alone are convenient as matching devices when the above condition can be met, since they require no adjustment. When used with a matching network as described earlier, however, the impedance-transformation ratio of a balun becomes of only secondary importance, and loads containing reactance may be tolerated so long as the losses in the balun itself do not become excessive.

## BIBLIOGRAPHY

Source material and more extended discussion of topics covered in this chapter can be found in the references given below and in the textbooks listed at the end of Chapter 2.


Belcher, D.K., "RF Matching Techniques, Design and Example," October 1972 *QST*, p. 24.

Dorbuck, T., "Matching-Network Design," March 1979 *QST*, p. 26.

Grammer, G., "Simplified Design of Impedance-Matching Networks," March, April, May 1957 *QST*.

Maxwell, M. W., "Another Look at Reflections," April, June, August, October 1973, April, December 1974 and August 1976 *QST*.

Pattison, B., "A Graphical Look at the L Network," March 1979 *QST*, p. 24.

## Chapter 5

# Coupling the Line to the Antenna

Throughout the discussion of transmission-line principles in Chapter 3, the operation of the line was described in terms of an abstract "load." This load had the electrical properties of resistance and, sometimes, reactance. It did not, however, have any physical attributes that associated it with a particular electrical device. That is, it could be anything at all that exhibits electrical resistance and/or reactance. The fact is that so far as the line is concerned, it does not matter what the load *is*, just as long as it will accept power.

Many amateurs make the mistake of confusing transmission lines with antennas, believing that because two identical antennas have different kinds of lines feeding them, or the same kind of line with different methods of coupling to the antenna, the "antennas" are different. There may be practical reasons why one system (including antenna, transmission line, and coupling method) may be preferred over another in a particular application. But to the transmission line, an antenna is just a load that terminates it, and the important thing is what that load looks like to the line in terms of resistance and reactance. *Any* kind of transmission line can be used with *any* kind of antenna, if the proper measures are taken to couple the two together.

### Frequency Range and SWR

Probably the principal factor that determines the way a transmission line is operated is the frequency range over which the antenna is to work. Very few types of antennas will present essentially the same load impedance to the line on harmonically related frequencies. As a result, the builder often is faced with choosing between (1) an antenna system that will permit operating the transmission line with a low standing-wave ratio, but is confined to one operating frequency or a narrow band of frequencies, and (2) a system that will permit operation in several harmonically related bands but with a large SWR on the line. (There are "multiband" systems which, in principle, make one antenna act as though it were a half-wave dipole on each of several amateur bands, by using "trap" circuits or multiple wires. Information on these is given in later chapters. Such an antenna can be assumed to be equivalent to a resonant half-wave dipole on each of the bands for which it is designed, and may be fed through a line, coaxial or otherwise, that has a $Z_0$ matching the antenna impedance, as described later in this chapter for simple dipoles.)

Methods of coupling the line to the antenna therefore divide, from a practical standpoint, into two classes. In the first, operation on several amateur bands is the prime consideration and the standing-wave ratio is secondary. The SWR is normally rather large and the input impedance of the line depends on the line length and the operating frequency.

In the second class, a conscious attempt is made, when necessary, to transform the antenna impedance to a value that matches the characteristic impedance of the line. When this is done the line operates with a very low standing-wave ratio and its input impedance is essentially a pure resistance, regardless of the line length. A transmission line can be considered to be "flat," within practical limits, if the SWR is not more than about 1.5 to 1.

### Losses

A principal reason for matching the antenna to the line impedance is that a flat line operates with the least power loss. While it is always desirable to reduce losses and thus increase efficiency, the effect of standing waves in this connection can be overemphasized. This is particularly true at the lower amateur frequencies, where the inherent loss in most types of lines is quite low even for runs that, in the average amateur installation, are rather long.

For example, 100 feet of 300-ohm receiving type twin-lead has a loss of only 0.18 dB at 3.5 MHz, as shown in Chapter 3. Even with an SWR as high as 10 to 1 the additional loss caused by standing waves is less than 0.7 dB. Since 1 dB represents the minimum detectable change in signal strength, it does not matter from this standpoint whether the line is flat or not. But at 144 MHz the loss in the same length of line perfectly matched is 2.8 dB, and an SWR of 10 to 1 that would mean an *additional* loss of 3.9 dB. At the higher frequency, then, it is worthwhile to match the antenna and line as closely as possible.

### Power Limitations

Another reason for matching is that certain types of lines, particularly those with solid dielectric, have definite voltage and current limitations. At the lower frequencies this is a far more compelling reason than power loss for at least approximate matching. Where the voltage and current must not exceed definite maximum values, the amount of power that the line can handle is inversely pro-

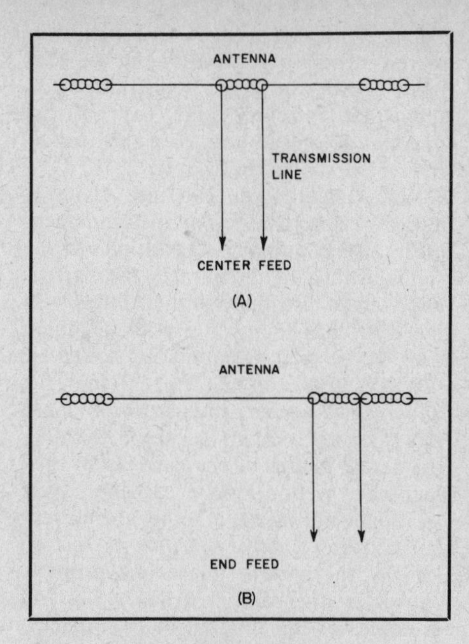


Fig. 1 — Center and end feed as used in simple antenna systems.

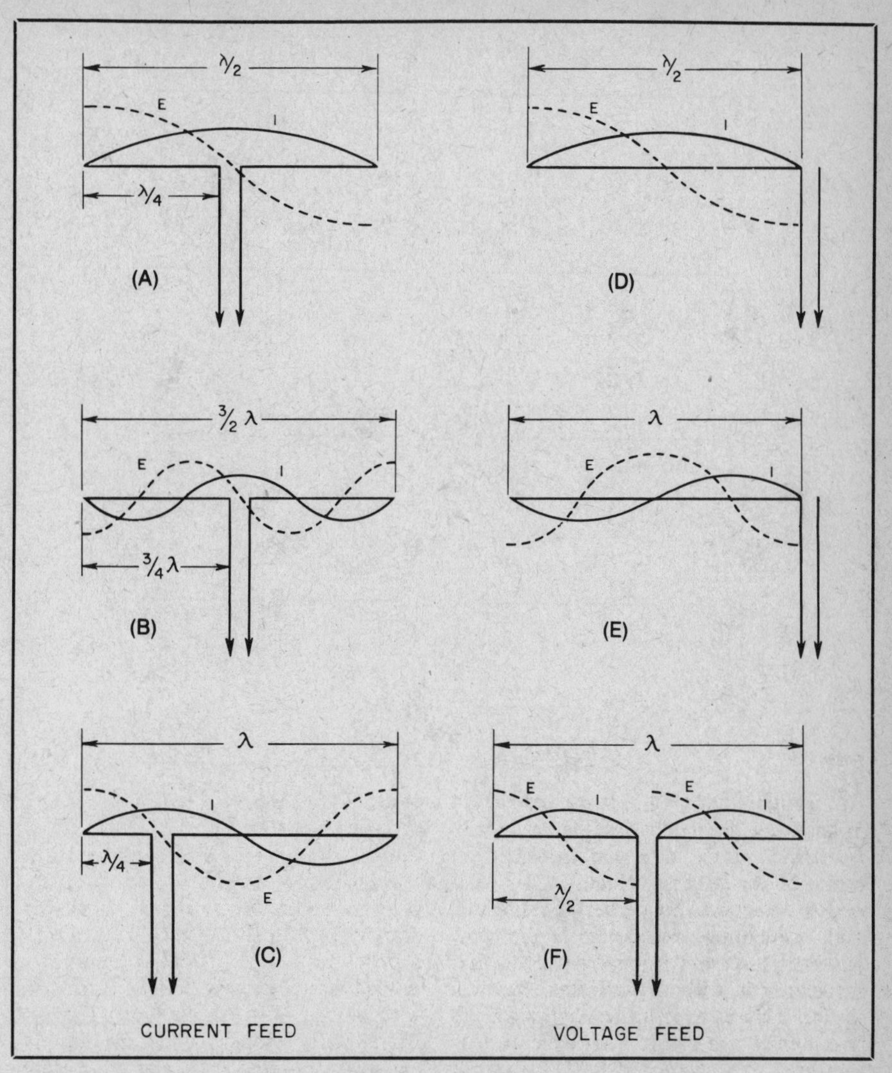


Fig. 2 — Current and voltage feed in antennas operated at the fundamental frequency, 2 times the fundamental, and 3 times the fundamental. The current and voltage distribution on the antenna are identical with both methods only at the fundamental frequency.

portional to the standing-wave ratio. If the safe rating on the 300-ohm line in the example above is 500 watts when perfectly matched, the line can handle only 50 watts with equal safety when the SWR is 10 to 1. Thus, despite the fact that the line losses are low enough to make no appreciable difference in the signal strength, the high SWR could be tolerated only with low-power transmitters.

### Line Radiation

Aside from power considerations, there is a more-or-less common belief that a flat line "does not radiate" while one with a high SWR does radiate. This impression is unjustified. It is true that the radiation from a parallel-conductor line increases with the current in the line, and that the effective line current increases with the SWR. However, the loss by radiation from a properly balanced line is so small (and is, furthermore, independent of the line length) that multiplying it several times still does not bring it out of the "negligible" classification.

Whenever a line radiates it is because of faulty installation (resulting in unbalance with parallel-conductor lines) or "antenna currents" on the line. Radiation from the latter cause can take place from either resonant or nonresonant lines, parallel-conductor or coaxial.

## UNMATCHED SYSTEMS

In many multiband systems or simple antennas where no attempt is made to match the antenna impedance to the characteristic impedance of the line, the customary practice is to connect the line either to the center of the antenna (center feed) as indicated in Fig. 1A, or to one end (end feed) as shown in Fig. 1B.

Because the line operates at a rather high standing-wave ratio, the best type to use is the open-wire line. Solid twin-lead of the 300-ohm receiving variety can also be used, but the power limitations discussed in the preceding section should be kept in mind. Although the manufacturers have placed no power rating on receiving-type 300-ohm line, it seems reasonable to make the assumption, based on the conductor size, that a current of 2 A can readily be carried by the line installed so that there is free air circulating about it. This corresponds to a power of 1200 watts in a matched 300-ohm line. When there are standing waves, the safe power can be found by dividing 1200 by the SWR. In a center-fed half-wave antenna, as in Fig. 1A, the SWR should not exceed about 5 to 1 (at the fundamental frequency), so receiving type 300-ohm twin-lead would appear to be safe for power outputs up to 250 watts or so.

Since there is little point in using a mismatched line to feed an antenna that is to operate on one amateur band only, the discussion to follow will be based on the assumption that the antenna is to be operated on its harmonics for multiband work.

### "Current" and "Voltage" Feed

Usual practice is to connect the transmission line to the antenna at a point where either a current or voltage loop occurs. If the feed point is at a current loop the antenna is said to be current fed; if at a voltage loop the antenna is voltage fed.

These terms should not be confused with center feed and end feed, because they do not necessarily have corresponding meanings. There is always a voltage loop at the end of a resonant antenna, no matter what the number of half wavelengths, so a resonant end-fed antenna is always voltage fed. This is illustrated at D and E in Fig. 2 for end-fed antennas a half wavelength long (antenna fundamental frequency) and one wavelength long (second harmonic). It would continue to be true for an end-fed antenna operated on any harmonic.

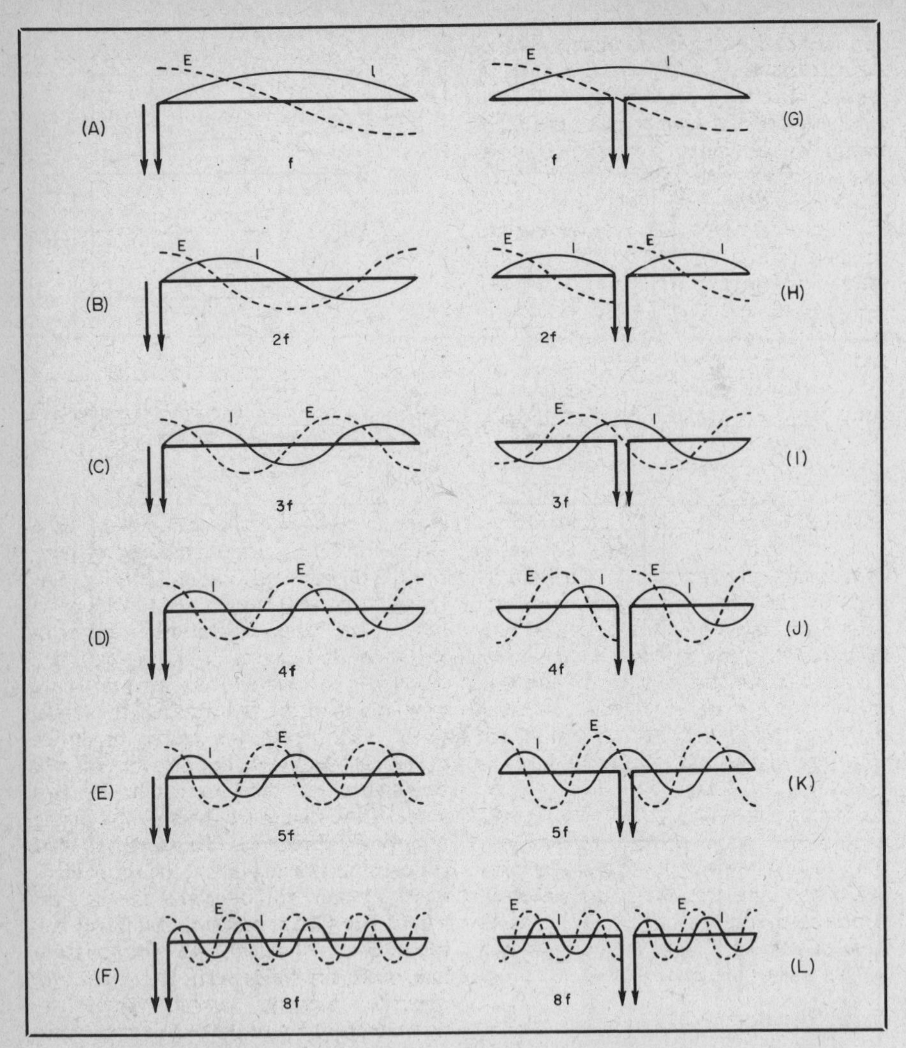

Fig. 3 — Current and voltage distribution at the fundamental frequency and various multiples, with both end feed and center feed. The distributions are the same with both types of feed only when the frequency is an odd multiple of the fundamental.

However, Fig. 2F shows voltage feed at the *center* of the antenna; in this case the antenna has a total length of two half wavelengths, each of which is voltage fed. Voltage feed is determined not by the physical position of the transmission line on the antenna, but by the fact that a voltage loop occurs on the antenna at the feed point. Since voltage loops always occur at integral multiples of a half wavelength from either end of a resonant antenna, feeding the antenna at any half-wavelength point constitutes voltage feed.

Typical cases of current feed are shown at A, B and C in Fig.2. The feed point is at a current loop, which always occurs at the midpoint of a half-wave section of the antenna. In order to feed at a current loop, the transmission line must be connected at a point that is an odd multiple of quarter wavelengths from either end of the resonant antenna. A center-fed antenna is also current fed *only* when the antenna length is an *odd* multiple of half wavelengths. Thus, the antenna in Fig. 2B is both center fed and current fed since it is three half wavelengths long. It would also be center fed and current fed if it were five, seven, etc., half wavelengths long.

To current feed a one-wavelength antenna, or any resonant antenna having a length that is an *even* multiple of one-half wavelength, it is necessary to shift the feed point from the center of the antenna (where a voltage loop always occurs in such a case) to the middle of one of the half-wave sections. This is indicated in Fig. 2C in the case of a one-wavelength antenna; current feed can be used if the line is connected to the antenna at a point 1/4 wavelength from either end.

### Operation on Harmonics

In the usual case of an antenna operated on several bands, the point at which the transmission line is attached is, of course, fixed. The antenna length is usually such that it is resonant at some frequency in the lowest frequency band to be used, and the transmission line is connected either to the center or the end. The current and voltage distribution along antennas fed at both points is shown in Fig. 3. With end feed, A to F inclusive, there is always a voltage loop at the feed point. Also, the current distribution is such that in every case the antenna operates as a true harmonic radiator of the type described in Chapter 2.

With center feed, the feed point is always at a current loop on the fundamental frequency and all *odd* multiples of the fundamental. In these cases the current and voltage distribution are identical with the distribution on an end-fed antenna. This can be seen by comparing A and G, C and I, and E and K, Fig. 3. (In I, the phase is reversed as compared with C, but this is merely for convenience in drawing; the actual phases of the currents in each half-wave section reverse each half cycle so it does not matter whether the current curve is drawn above or below the line, so long as the *relative* phases are properly shown in the same antenna.) On odd multiples of the fundamental frequency, therefore, the antenna operates as a true harmonic antenna.

On *even* multiples of the fundamental frequency the feed point with center feed is always at a voltage loop. This is shown at H, J and L in Fig. 3. Comparing B and H, it can be seen that the current distribution is different with center feed than with end feed. With center feed the currents in both half-wave sections of the antenna are in the same phase, but with end feed the current in one half-wave section is in reverse phase to the current in the other. This does not mean that one antenna is a better radiator than the other, but simply that the two will have different directional characteristics. The center-fed arrangement is commonly known as "two half-waves in phase," while the end-fed system is a "one-wavelength antenna" or "second-harmonic" antenna.

Similarly, the system at J has a different current and voltage distribution than the system at D, although both resonate at four times the fundamental frequency. A similar comparison can be made between F and L. The center-fed arrangement at J really consists of two one-wavelength antennas, while the arrangement at L has two 2-wavelength antennas. These have different directional characteristics than the 2-wavelength and 4-wavelength antennas (D and F) that resonate at the same multiple, respectively, of the fundamental frequency.

The reason for this difference between odd and even multiples of the fundamental frequency in the case of the center-fed antenna can be explained with the aid of Fig. 4. It will be recalled from Chapter 2 that the direction of current flow reverses in each half wavelength of wire. Also, in any transmission line the currents in the two wires always must be equal and flowing in opposite directions at any point along the line. Starting from the end of the antenna, the current must be flowing in one direction throughout the first half-

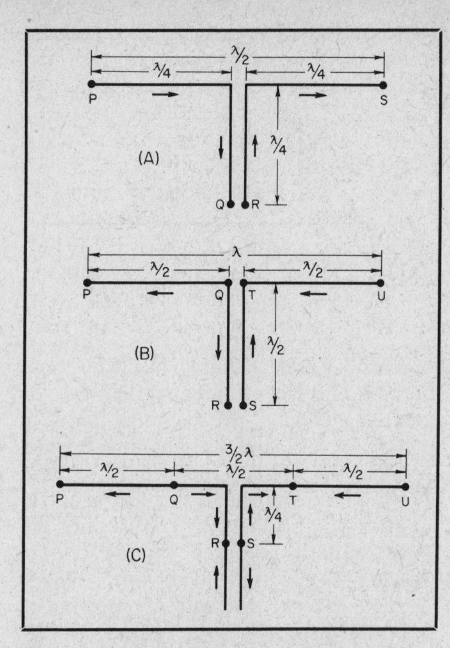


Fig. 4 — Showing how the type of feed changes from current to voltage, with a center-fed antenna, on twice the fundamental frequency, and back to current feed on three times the fundamental. The same change occurs between all even and odd frequency multiples.

wavelength section, whether this section is entirely antenna or partly antenna and partly one wire of the transmission line. Thus, in A, Fig. 4, the current flows in the same direction from P to Q, since this is all the same conductor. However, one quarter wave is in the antenna and one in the transmission line. The current in the other line wire, starting from R, must flow in the opposite direction in order to balance the current in the first wire, as shown by the arrow. And since the distance from R to S is 1/2 wavelength, the current continues to flow in the same direction all the way to S. The currents in the two halves of the *antenna* are therefore flowing in the same direction. Furthermore, the current is maximum 1/4 wavelength from the ends of the antenna, as previously explained, and so both the currents are maximum at the junction of the antenna and transmission line. This makes the current distribution along the length of the antenna exactly the same as with an end fed antenna.

Fig. 4B shows the case where the overall length of the antenna is one wavelength, making a half wave on each side. A half wavelength along the transmission line also is shown. If we assume that the current is flowing downward in the line conductor from Q to R, it must be flowing upward from S to T if the line currents are to balance. However, the distance from Q to P is 1/2 wavelength, and so the current in this section of the antenna must flow in the opposite direction to the currrent flowing in the section from Q to R. The current in section PQ is therefore flowing *away* from Q. Also, the current in section TU must be flowing in the opposite direction to the current in ST, and so is flowing *toward* T. The currents in the two half-wave sections of the antenna are therefore flowing in the same direction. That is, they are in the same phase.

With the above in mind, the direction of current flow in a 1-1/2 wavelength antenna, Fig. 4C, should be easy to follow. The center half-wave section QT corresponds to the half-wave antenna in A. The currents in the end sections, PQ and TU simply flow in the opposite direction to the current in QT. Thus the currents are out of phase in alternate half-wave sections.

The shift in voltage distribution between odd and even multiples of the fundamental frequency can be demonstrated by a similar method, making allowance for the fact that the voltage is maximum where the current is minimum, and vice versa. On all even multiples of the fundamental frequency there is a current minimum at the junction of the line and antenna, with center feed, because there is an integral number of half wavelengths in each side of the antenna. The voltage is maximum at the junction in such a case, and we have voltage feed. Where the multiple of the fundamental is odd, there is always a current maximum at the junction of the transmission line and antenna, as demonstrated by A and C in Fig. 4. At these points the voltage is minimum and we therefore have current feed.

## "Zepp" or End Feed

In the early days of short-wave communication an antenna consisting of a half-wave dipole, end-fed through a 1/4-wavelength transmission line, was developed as a trailing antenna for Zeppelin airships. In its use by amateurs, over the years, it has become popularly known as the "Zeppelin" or "Zepp" antenna. The term is now applied to practically any resonant antenna fed at the end by a two-wire transmission line.

The mechanism of end feed is perhaps somewhat difficult to visualize, since only one of the two wires of the transmission line is connected to the antenna while the other is simply left free. The difficulty lies in the natural tendency to think in terms of current flow in ordinary electrical circuits, where it is necessary to have a complete loop between both terminals of the power source before any current can flow at all. But as explained earlier, this limitation applies only to circuits in which the electromagnetic fields reach the most distant part of the circuit in a time interval that is negligible in comparison with the time of one cycle. When the circuit dimensions are comparable with the wavelength, no such complete loop is necessary. *The antenna itself is an example of an "open" circuit in which large currents can flow.*

One way of looking at end feed is to consider the entire length of wire, in-

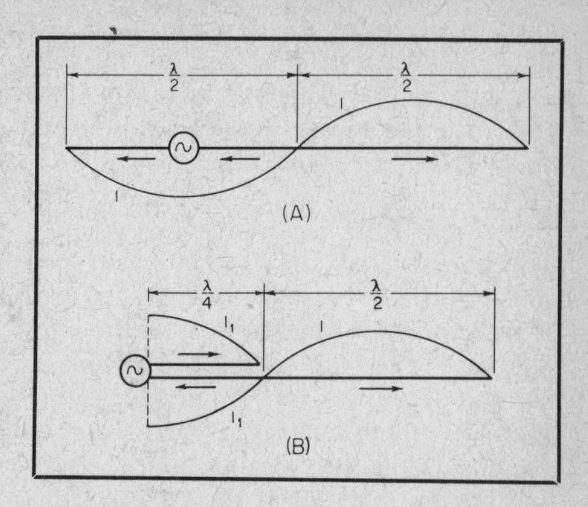


Fig. 5 — Folded-antenna analogy of transmission line for an end-fed antenna.

cluding both antenna and feeder, as a single unit. For example, suppose we have a wire one wavelength long, as in Fig. 5A, fed at a current loop by a source of rf power. The current distribution will be as shown by the curves, with the assumed directions indicated by the arrows. If we now fold back the 1/4-wavelength section to the left of the power source, as shown at B, the overall current distribution will be similar, but the currents in the two wires of the folded section will be flowing in opposite directions. The amplitudes of the currents at any point along the folded-back portion will be equal in the two wires. The folded section, therefore, has become a 1/4-wavelength transmission line, since the fields from the equal and opposite currents cancel. There is, however, nothing to prevent current from continuing to flow in the right-hand half-wavelength section, since there was current there before the left-hand section was folded.

This picture, although showing how power can flow from the transmission line to an antenna through end feed, lacks completeness. It does not take into account the fact that the current I1 in the transmission line is greatly different from the current I in the antenna. A more basic viewpoint is the one already mentioned in Chapter 2. The current is caused by electromagnetic fields traveling along the wire and simply constitutes a measurable manifestation of those fields; the current does not cause the fields. From this standpoint the transmission-line conductors merely serve as "guides" for the fields so the electromagnetic energy will go where we want it to go. When the energy reaches the end of the transmission line it meets another guide, in the form of the antenna, and continues along it. However, the antenna is a different form of guide; it has a single conductor while the line has two; it has no provision for preventing radiation while the line is designed for that very purpose. This is simply another way of saying that the impedance of the antenna differs from that of the transmission line, so there will be reflection when the energy traveling along the line arrives at the antenna. We are then back on familiar ground, in that we have a transmission line terminated in an impedance different from its characteristic impedance.

**Feeder Unbalance**

With end feed, the currents in the two line wires do not balance exactly and there is therefore some radiation from the line. The reason for this is that the current at the end of the free wire is zero (neglecting a small charging current in the insulator at the end) while the current does not go to zero at the junction of the "active" line wire and the antenna. This is because not all the energy going into the antenna is reflected back from the far end, some being radiated; hence the incident and reflected currents cannot completely cancel at a node.

In addition to this unavoidable line radiation a further unbalance will occur if the antenna is not exactly resonant at the operating frequency. If the frequency is too high (antenna too long) the current node does not occur at the junction of the antenna and "live" feeder, but moves out on the antenna. When the frequency is too low the node moves down the active feeder. Since the node on the free feeder has to occur at the end, either case is equivalent to shifting the position of the standing wave along one feeder wire but not the other. The further off resonance the antenna is operating, the greater the unbalance and the greater the line radiation. With center feed this unbalance does not occur, because the system is symmetrical with respect to the line.

To avoid line radiation it is always best to feed the antenna at its center of symmetry. In the case of simple antennas for operation in several bands, this means that center feed should be used. End feed is required only when the antenna is operated on an even harmonic to obtain a desired directional characteristic, and then only when it must be used on more than one band. For single-band operation it is always possible to feed an even-harmonic antenna at a current loop in one of the half-wave sections nearest the center.

**SWR with Wire Antennas**

When a line is connected to a single-wire antenna at a current loop the standing-wave ratio can be estimated with good-enough accuracy with the information found in Chapter 2. Although the actual value of the radiation resistance, as measured at a current loop, will vary with the height of the antenna above ground, the theoretical values will at least serve to establish whether the SWR will be high or low.

With center feed the line will connect to the antenna at a current loop on the fundamental frequency and all odd multiples, as shown by Fig. 3. At the fundamental frequency and usual antenna heights, the antenna resistance should lie between 50 and 100 ohms, so with a line having a characteristic impedance of 450 ohms the SWR will be $Z_0/R_L = 450/50 = 9$ to 1 as one limit and $450/100 = 4.5$ to 1 as the other. On the third harmonic the theoretical resistance is near 100 ohms, so the SWR should be about 4.5 to 1. For 300-ohm line the SWR can be expected to be between 3 and 6 on the antenna fundamental and about 3 to 1 on the third harmonic.

The impedances to be expected at voltage loops are less readily determined. Theoretical values are in the neighborhood of 5000 to 8000 ohms, depending on the antenna conductor size and the number of half wavelengths along the wire. Such experimental figures as are available indicate a lower order of resistance, with measurements and estimates running from 1000 to 5000 ohms. In any event, there will be some difference between end feed and center feed, since the current distribution on the antenna is different in these two cases at any given even multiple of the fundamental frequency. Also, the higher the multiple the lower the resistance at a voltage loop, so the SWR can be expected to decrease when an antenna is operated at a high multiple of its fundamental frequency. Assuming 4000 ohms for a wire antenna two half waves long, the SWR would be about 6 or 7 with a 600-ohm line and around 12 with a 300-ohm line. However, considerable variation is to be expected.

## ANTENNA CURRENTS ON TRANSMISSION LINES

In any discussion of transmission-line operation it is always assumed that the two conductors carry equal and opposite currents throughout their length. This is an ideal condition that may or may not be realized in practice. In the average case the chances are rather good that the currents will *not* be balanced unless special precautions are taken. Whether the line is matched or not has little to do with the situation.

Consider the half-wave antenna shown in Fig. 6 and assume that it is somehow fed by a source of power at its center, and that the instantaneous direction of current flow is as indicated by the arrows. In the neighborhood of the antenna is a group of conductors disposed in various ways with respect to the antenna itself. All of these conductors are in the field of the antenna and are therefore coupled to it. Consequently, when current flows in the antenna a voltage will be induced in each conductor. This causes a current flow determined by the induced voltage and the impedance of the conductor.

The degree of coupling depends on the position of the conductor with respect to the antenna, assuming that all the conductors in the figure are the same length. The coupling between the antenna and conductor IJ is greater than in any other case, because IJ is close to and parallel with the antenna. Ideally, the coupling between conductor GH and the antenna is zero, because the voltage induced by current flowing in the left-hand side of the antenna is exactly balanced by a voltage of opposite polarity induced by the current flowing in the right-hand side. This is because the two currents are flowing in opposite directions with respect to GH. Complete cancellation of the induced voltages can occur, of course, only if the currents in the two halves of the antenna are symmetrically distributed with respect to the center of the antenna, and also only if every point along GH is equidistant from any two points along the antenna that are likewise equidistant from the center. This cannot be true of any of the other conductors shown, so a finite voltage will be induced in any conductor in the vicinity of the antenna except one perpendicular to the antenna at its center.

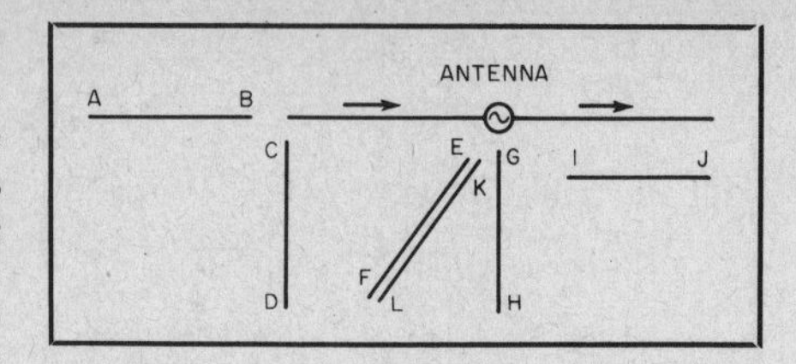


Fig. 6 — Coupling between antenna and conductors in the antenna field.

**Transmission Line in the Antenna Field**

Now consider the two conductors EF and KL, which are parallel and very close together. Except for the negligible spacing between them, the two conductors lie in the same position with respect to the antenna. Therefore, identical voltages will be induced in both, and the resulting currents will be flowing *in the same direction in both conductors.* It is only a short step to visualizing conductors EF and KL as the two conductors of a section of transmission line in the vicinity of the antenna. Because of coupling to the antenna, it is not only possible but *certain* that a voltage will be induced in the two conductors of the transmission line in parallel. The resulting current flow is in the same direction in both conductors, whereas the true transmission line currents are always flowing in opposite directions at each point along the line. These "parallel" currents are of the same nature as the current in the antenna itself, and hence are called "antenna" currents on the line. They are responsible for most of the radiation that takes place from transmission lines.

When there is an antenna current of appreciable amplitude on the line it will be found that not only are the line currents unbalanced but the apparent SWR is different in each conductor, and that the loops and nodes of current in one wire do

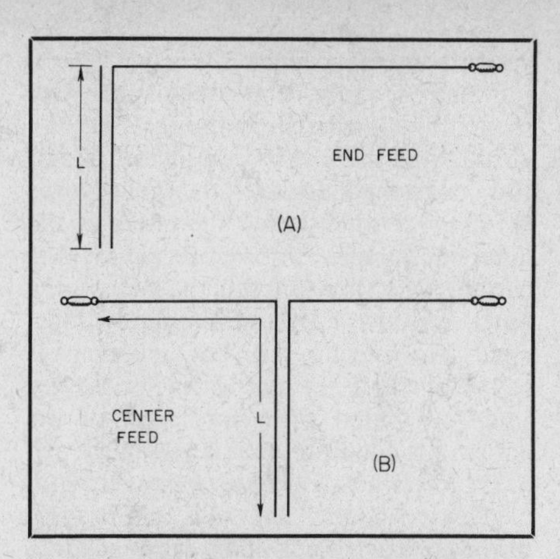


Fig. 7 — The important length for resonance to antenna currents coupled from the antenna to the line. In the center-fed system one side of the antenna is part of the "parallel"-resonant system.

not occur at corresponding points in the other wire. Under these conditions it is impossible to measure the true SWR.

It should be obvious from Fig. 6 that only in the case of a center-fed antenna can the coupling between the line and antenna be reduced to zero. There is always some such coupling when the antenna is end fed, so there is always the possibility that antenna currents of appreciable amplitude will exist on the line, contributing further to the inherent line unbalance in the end-fed arrangement. But the center-fed system also will have appreciable antenna-to-line coupling if the line is not brought off at right angles to the antenna for a distance of at least a half wavelength.

Antenna currents will be induced on lines of any type of construction. If the line is coax, the antenna current flows only on the *outside* of the outer conductor; no current is induced *inside* the line. However, an antenna current on the outside of coax is just as effective in causing radiation as a similar current induced in the two wires of a parallel-conductor line.

### Detuning the Line for Antenna Currents

The antenna current flowing on the line as a result of voltage induced from the antenna will be small if the overall circuit, considering the line simply as a single conductor, is not resonant at the operating frequency. The frequency (or frequencies) at which the system is resonant depends on the total length and whether the transmission line is grounded or not at the transmitter end.

If the line is connected to a coupling circuit that is not grounded, either directly or through a capacitance of more than a few picofarads, it is necessary to consider only the length of the antenna and line. In the end-fed arrangement, shown at A in Fig. 7, the line length, L, should not be an integral multiple or close to such a multiple of a half wavelength. In the center-fed system, Fig. 7B, the length of the line plus *one side* of the antenna should not be a multiple of a half wavelength. In this case the two halves of the antenna are simply in parallel so far as resonance for the induced "antenna" current on the line is concerned, because the line conductors themselves act in parallel. When the antenna is to be used in several bands, resonance of this type should be avoided at all frequencies to be used.

Transmission lines usually have bends, are at varying heights above ground, and so on, all of which will modify the resonant frequency. It is advisable to check the system for resonance at and near all operating frequencies before assuming that the line is safely detuned for antenna currents. This can be done by temporarily connecting the ends of the line together and coupling them through a small capacitance (not more than a few picofarads) to a resonance indicator such as a dip meter. Very short leads should be used between the meter and antenna. Fig. 8 shows the method. Once the resonance points are known it is a simple matter to prune the feeders to get as far away as possible from resonance at any frequency to be used.

Resonances in systems in which the coupling apparatus is grounded at the transmitter are not so easily predicted. The "ground" in such a case is usually the metal chassis of the transmitter itself, not actual ground. In the average amateur station it is not possible to get a connection to real ground without having a lead that is an appreciable fraction of wavelength long. At the higher frequencies, and particularly in the vhf region, the distance from the transmitter to ground may be one wavelength or more. Probably the best plan in such cases is to make the length L in Fig. 7 equal to a multiple of a half wavelength. If the transmitter has fairly large capacitance to ground, a system of this length will be effectively detuned for the fundamental and all even harmonics when grounded to the transmitter at the coupling apparatus. However, the resonance frequencies will depend on the arrangement and constants of the coupling system even in such a case, and preferably should be checked by means of the dip meter. If this test shows resonance at or near the operating frequency, alternative grounds (to a heating radiator, for example) should be tried until a combination is found that detunes the whole system.

It should be quite clear, from the mechanism that produces antenna currents on a transmission line, that such currents are entirely independent of the normal operation as a true transmission line. It does not matter whether the line is perfectly matched or is operated with a high standing-wave ratio. Nor does it matter what kind of line is used, air-insulated or solid-dielectric, parallel-conductor or

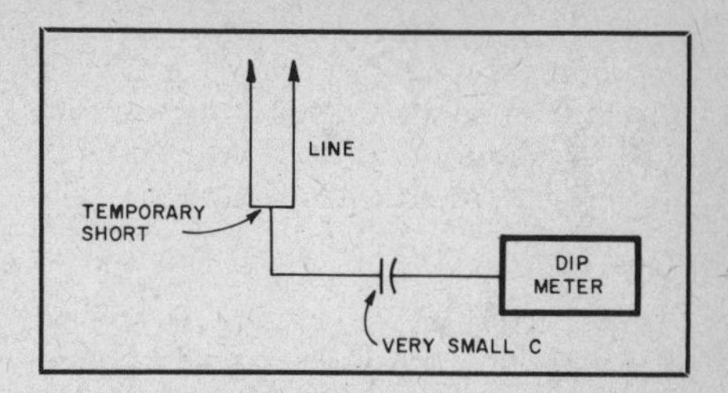


Fig. 8 — Using a dip meter to check resonance of the antenna system for antenna currents on the transmission line. The small capacitance may be a short length of wire connected to the feed line, coupled lightly to the dip-oscillator coil with a 1-turn loop.

coax. In every case, the antenna currents should be minimized by detuning the line if the line is to fulfill only its primary purpose of transferring power to the antenna.

### Other Causes of Unbalance

Unbalance in center-fed systems can arise even when the line is brought away at right angles to the antenna for a considerable distance. If both halves of the antenna are not symmetrically placed with respect to nearby conductors (such as power and telephone wires and downspouting) the antenna itself becomes unbalanced and the current distribution is different in the two halves. Because of this unbalance a voltage will be induced in the line even if the line is symmetrical with respect to the antenna.

## MATCHED LINES

Operating the transmission line at a low standing-wave ratio requires that the line be terminated, at its output end, in a resistive load matching the characteristic impedance of the line as closely as possible. The problem can be approached from two standpoints: (1) selecting a transmission line having a characteristic impedance that matches the antenna resistance at the point of connection; or (2) transforming the antenna resistance to a value that matches the $Z_0$ of the line selected.

The first approach is simple and direct, but its application is limited because the antenna impedance and line impedance are alike only in a few special cases. The second approach provides a good deal of freedom in that the antenna and line can be selected independently. Its disadvantage is that it is more complicated constructionally. Also, it sometimes calls for a somewhat tedious routine of measurement and adjustment before the desired match is achieved.

### Operating Considerations

As pointed out earlier in this chapter, most antenna systems show a marked change in resistance when going from the fundamental to multiples of the fundamental frequency. For this reason it is usually possible to match the line

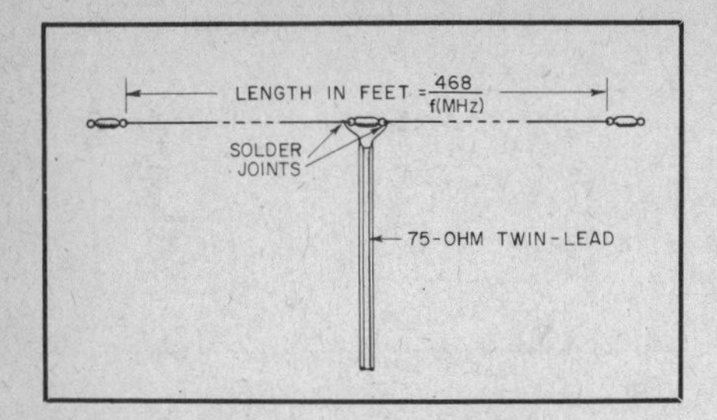


Fig. 9 — Half-wave dipole fed with 75-ohm twin-lead, giving a close match between antenna and line impedance. The leads in the "Y" from the end of the line to the ends of the center insulator should be as short as possible.

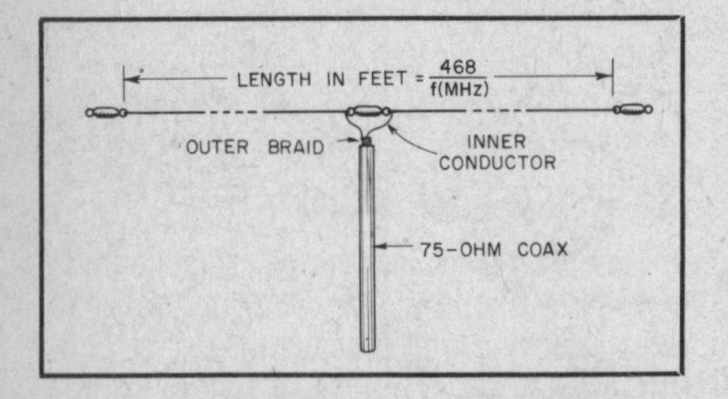


Fig. 10 — Half-wave antenna fed with 75-ohm coaxial cable. The outside of the outer conductor of the line may be grounded for lightning protection.

impedance only on one frequency. A matched antenna system is consequently a one-band affair, in most cases. It can, however, usually be operated over a fair frequency range in a given band. The frequency range over which the standing-wave ratio is low is determined by the impedance-vs.-frequency characteristic of the antenna. If the change in impedance is small for a given change in frequency, the SWR will be low over a fairly wide band of frequencies. However, if the impedance change is rapid (a sharply resonant or high-Q antenna — see discussion of Q later in this chapter) the SWR will also rise rapidly as the operating frequency is shifted to one side or the other of the frequency for which the antenna is resonant and for which the line is matched.

### Antenna Resonance

A point that needs emphasis in connection with matching the antenna to the line is that, with the exception of a few special cases discussed later in this chapter, the impedance at the point where the line is connected must be a *pure resistance*. This means that the antenna system must be resonant at the frequency for which the line is to be matched. (Some types of long-wire antennas are exceptions, in that their input impedances are resistive over a wide band of frequencies. Such systems are essentially nonresonant.) The higher the Q of the antenna system, the more essential it is that exact resonance be established before an attempt is made to match the line. This is particularly true of close-spaced parasitic arrays. With simple dipole and harmonic antennas, the tuning is not so critical and it is usually sufficient to cut the antenna to the length given by the appropriate formula in Chapter 2. The frequency should be selected to be at the center of the range of frequencies (which may be the entire width of an amateur band) over which the antenna is to be used.

## DIRECT MATCHING

As discussed in Chapter 2, the impedance at the center of a resonant half-wave antenna at heights of the order of 1/4 wavelength and more is resistive and is in the neighborhood of 70 ohms. This is fairly well matched by transmitting-type twin-lead having a characteristic impedance of 75 ohms. It is possible, therefore, to operate with a low SWR using the arrangement shown in Fig. 9. No precautions are necessary beyond those already described in connection with antenna-to-line coupling.

This system is badly mismatched on *even* multiples of the fundamental frequency, since the feed in such cases is at a high-impedance point. However, it is reasonably well matched at *odd* multiples of the fundamental. For example, an antenna resonant near the low-frequency end of the 7-MHz band will operate with a low SWR over the 21-MHz band (three times the fundamental).

The same method may be used to feed a harmonic antenna at any current loop along the wire. For lengths up to three or four wavelengths the SWR should not exceed 2 to 1 if the antenna is 1/4 or 1/2 wavelength above ground.

At the fundamental frequency the SWR should not exceed about 2 to 1 within a frequency range ±2% from the frequency of exact resonance. Such a variation corresponds approximately to the entire width of the 7-MHz band, if the antenna is resonant at the center of the band. A wire antenna is assumed. Antennas having a greater ratio of diameter to length will have a lower change in SWR with frequency.

### Coaxial Cable

Instead of using twin-lead as just described, the center of a half-wave dipole may be fed through 75-ohm coaxial cable such as RG-11/U, as shown in Fig. 10. Cable having an impedance of approximately 50 ohms, such as RG-8/U, also may be used, particularly in those cases where the antenna height is such as to lower the radiation resistance of the antenna, below 1/4 wavelength. (See Chapter 2.) The principle is exactly the same as with twin-lead, and the same remarks as to SWR apply. However, there is a considerable practical difference between the two types of line. With the parallel-conductor line the system is symmetrical, but with coaxial line it is inherently unbalanced.

Stated broadly, the unbalance with coaxial line is caused by the fact that the outside of the outer conductor is not coupled to the antenna in the same way as the inner conductor and the inside of the outer conductor. The overall result is that current will flow on the outside of the outer conductor in the simple arrangement shown in Fig. 10. The unbalance is rather small if the line diameter is very small compared with the length of the antenna, a condition that is met fairly well at the lower amateur frequencies. It is not negligible in the vhf and uhf range, however, nor should it be ignored at 28 MHz. The current that flows on the outside of the line because of this unbalance, it should be noted, does not arise from the same type of coupling as the "antenna" current previously discussed. The coupling pictured in Fig. 6 can still occur, *in addition*. However, the remedy is the same in both cases — the system must be detuned for currents on the outside of the line. This can be done by an actual resonance check using the method shown in Fig. 8.

### Balancing Devices

The unbalanced coupling described in the preceding paragraph can be nullified by the use of devices that prevent the unwanted current from flowing on the outside of the coaxial line. This may be done either by making the current cancel itself out or by choking it off. Devices of this type fall in a class of circuits usually termed *baluns*, a contraction for "balanced to unbalanced." The baluns described in Chapter 4 perform the same function, but the techniques described here are generally more suitable for mechanical reasons in coupling the line to the antenna.

The voltages at the antenna terminals in Fig. 10 are equal in amplitude with respect to ground but opposite in phase. Both these voltages act to cause a current to flow on the outside of the coax, and if the currents produced by both voltages were equal, the resultant current on the outside of the line would be zero since the currents are out of phase and would cancel each other. But since one antenna terminal is directly connected to the cable shield while the other is only weakly coupled to it, the voltage at the directly connected terminal produces a much larger current, and so there is relatively little cancellation.

The two currents could be made equal in amplitude by making a direct connection between the outside of the line and the antenna terminal that is connected to the inner conductor, but if it were done right at the antenna terminals the line and antenna would be short-circuited. If the connection is made through a conductor parallel to the line and a quarter wavelength long, as shown in Fig. 11A, the second conductor and the outside of the line act as a quarter-wave "insulator" for the normal voltage and current at the antenna terminals. (This is because a quarter-wave line short-circuited at the far end exhibits a very high resistive impedance, as explained in Chapter 3. On the other hand, any unbalanced current flowing on the outside of the line because of the direct connection between it and the antenna has a counterpart in an equal current flowing on the second conductor, because the latter is directly connected to the *other* antenna terminal. Where the two conductors are joined together at the bottom, the resultant of the two currents is zero, since they are of opposite phase. Thus, no current flows on the remainder of the transmission line.

Note that the length of the extra conductor has no particular bearing on its operation in balancing out the undesired current. The length is critical only in respect to preventing the normal operating of the antenna from being upset.

**Combined Balun and Matching Stub**

In certain antenna systems the balun length can be considerably shorter than a quarter-wavelength; the balun is, in fact, used as part of the matching system. This requires that the radiation resistance be fairly low as compared with the line $Z_0$ so that a match can be brought about by first shortening the antenna to make it have a capacitive reactance, and then using a shunt inductor across the antenna terminals to resonate the antenna and simultaneously raise the impedance to a value equal to the line $Z_0$. (See later sec-

than the length given by the formula. The shorting connection at the bottom may be installed permanently. With the dip meter coupled to the shorted end, check the frequency and cut off small lengths of the shield braid (cutting both lines equally) at the open ends until the stub is resonant at the desired frequency. In each case leave just enough inner conductor remaining to make a short connection to the antenna. After resonance has been established, solder the inner and outer conductors of the second piece of coax together and complete the connections indicated in Fig. 11A.

Another method is first to adjust the antenna length to the desired frequency, with the line and stub disconnected, then connect the balun and recheck the frequency. Its length may then be adjusted so that the overall system is again resonant at the desired frequency.

**Construction**

In constructing a balun of the type shown in Fig. 11A, the additional conductor and the line should be maintained parallel by suitable spacers. It is convenient to use a piece of coax for the second conductor; the inner conductor can simply be soldered to the outer conductor at both ends since it does not enter into the operation of the device. The two cables should be separated sufficiently so that the vinyl covering represents only a small proportion of the dielectric between them. Since the principal dielectric is air, the length of the quarter-wave section is based on a velocity factor of 0.95, approximately.

**Detuning Sleeves**

The detuning sleeve shown in Fig. 11B also is essentially an air-insulated quarter-wave line, but of the coaxial type, with the sleeve constituting the outer conductor and the outside of the coax line being the inner conductor. Because the impedance at the open end is very high, the unbalanced voltage on the coax line cannot cause much current to flow on the outside of the sleeve. Thus the sleeve acts like a choke coil in isolating the remainder of the line from the antenna. (The same viewpoint can be used in explaining the action of the quarter-wave arrangement shown at A, but is less easy to understand n the case of baluns less than 1/4 wavelength long.)

A sleeve of this type may be resonated y cutting a small longitudinal slot near he bottom, just large enough to take a ingle-turn loop which is, in turn, link-oupled to the dip meter. If the sleeve a little long to start with, a bit at a time an be cut off the top until the stub is esonant.

The diameter of the coaxial detuning eeve in B should be fairly large compared with the diameter of the cable it surunds. A diameter of two inches or so is tisfactory with half-inch cable. The sleeve should be symmetrically placed with respect to the center of the antenna so that it will be equally coupled to both sides. Otherwise a current will be induced from the antenna to the outside of the sleeve. This is particularly important at vhf and uhf.

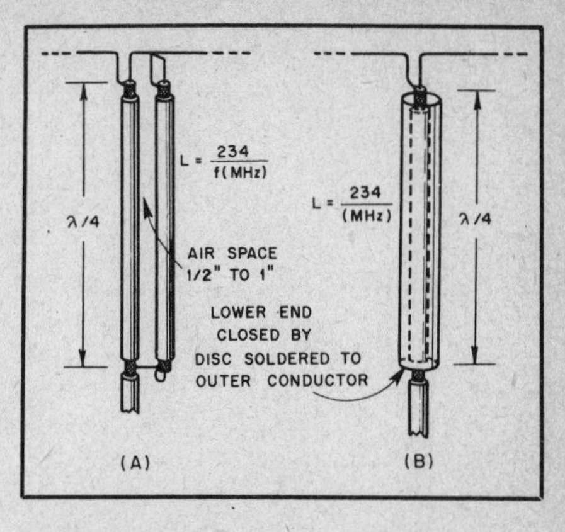

Fig. 11 — Methods of balancing the termination when a coaxial cable is connected to a balanced antenna.

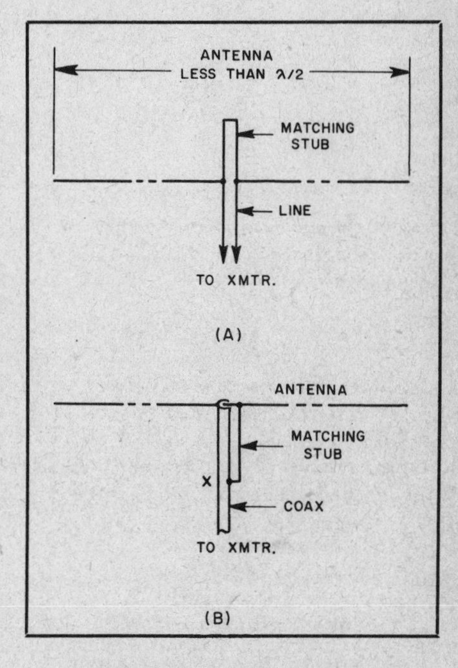

Fig. 12 — Combined matching stub and balun. A — basic arrangement; B — balun arrangement achieved by using a section of the outside of the coax feed line as one conductor of a matching stub.

In both the balancing methods shown in Fig. 11 the quarter-wave section should be cut to be resonant at exactly the same frequency as the antenna itself. These sections tend to have a beneficial effect on the impedance-frequency characteristic of the system, because their reactance varies in the opposite direction to that of the antenna. For instance, if the operating frequency is slightly below resonance the antenna has capacitive reactance, but the shorted quarter-wave sections or stubs have inductive reactance. Thus the reac-

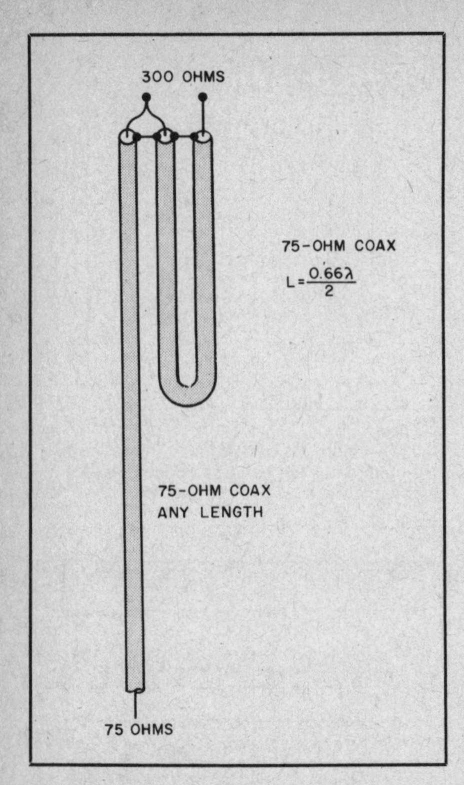


Fig. 13 — A balun that provides an impedance step-up ratio of 4:1.

Fig. 14 — An rf choke formed by coiling the feed line at the point of connection to the antenna. Electrical tape may be used to hold the coils in place.

tances tend to cancel, which prevents the impedance from changing rapidly and helps maintain a low standing-wave ratio on the line over a band of frequencies.

### Impedance Step-up Balun

A coax-line balun may also be constructed to give an impedance step-up ratio of 4:1. This form of balun is shown in Fig. 13. If 75-ohm line is used, as indicated, the balun will provide a match for a 300-ohm terminating impedance. The U-shaped section of line must be an electrical half wave in length, taking the velocity factor of the line into account. In most installations using this type of balun, it is customary to roll up the length of line represented by the U-shaped section into a coil of several inches in diameter. The coil turns may be bound together with electrical tape. Because of the bulk and weight of the balun, this type is seldom used with wire-line antennas suspended by insulators at the antenna ends. More commonly it is used with multielement antennas, where its weight may be supported by the boom of the antenna system.

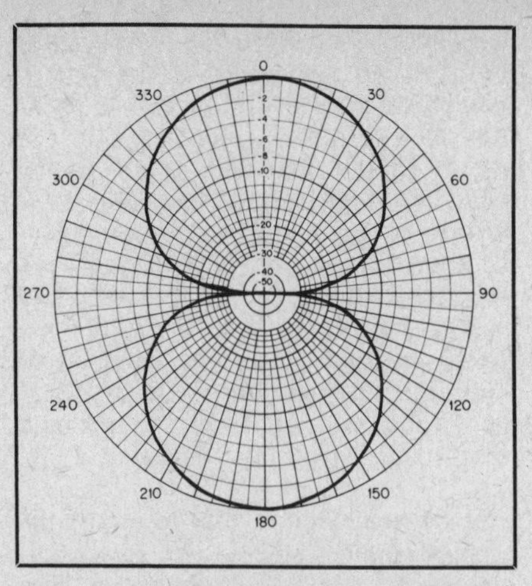


Fig. 15 — Classic response pattern of a half-wavelength dipole in free space. The concentric-circle scale is indicated in decibels down, relative to the response in a broadside direction from the axis of the dipole. The outer scale shows degrees of departure from one broadside direction. The axis of the conductor is common with the line between the 90° and 270° outer-scale markings.

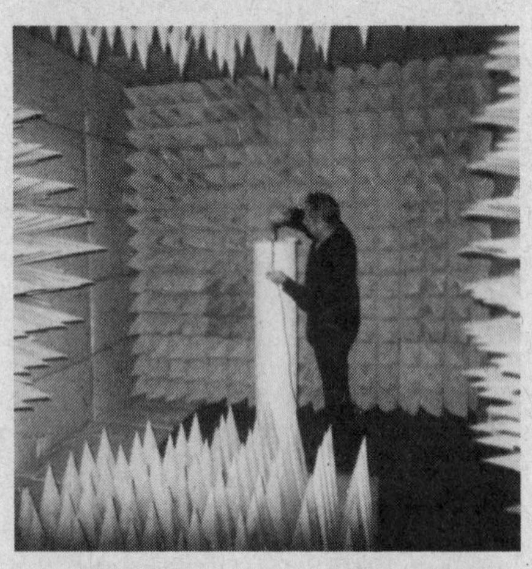

Fig. 16 — Positioning one of the test antennas on the rotatable Styrofoam support in the rf anechoic chamber.

### Coax-Line RF Choke

As was discussed earlier in this section, the unbalanced coupling that results from connecting coaxial line to a balanced antenna may be nullified by choking off the current from flowing on the outside of the feed line. A direct approach to this objective is shown in Fig. 14; where the line itself is formed into a coil at the antenna feed point. Ten turns of coax line coiled at a diameter of 6 inches has been found effective for the hf bands. The turns may be secured in a tight coil with electrical tape. This approach offers the advantage of requiring no pruning adjustments. The effectiveness of a choke of this sort decreases at the higher frequencies, however, because of the distributed capacitance among the turns.

### Pattern Distortion

Fig. 15 shows the classic "figure-eight" radiation pattern of a half-wave dipole in free space. This is actually an idealized pattern based on current flowing only in the antenna. As was stated earlier, current may flow on the outside of coaxial cable used to feed a dipole directly. That current will cause radiation and hence a distortion of the theoretical pattern.

Tests were made at the radio-frequency anechoic chamber at North Carolina State University to determine dipole patterns. An rf anechoic chamber is simply a room in which the walls, floor and ceiling are covered with a material that is designed to break up an electromagnetic wave and absorb its energy. An antenna placed in such a chamber can not "see," or be influenced by, any surface or objects that can reflect or reradiate electromagnetic energy. It is a simulation of "free space" right here on earth.

Two antennas were used for the tests. The source of rf was a half-wave balun-fed dipole, mounted horizontally at one end of the chamber. The type of balun used is shown in Fig. 11B. It was mounted at the same height as the receiving antenna and fed a few milliwatts of power at 1.6 GHz. The test antennas were then mounted, one at a time, horizontally, at the other end of the chamber, on a rotating support. The supports for both antennas were made of Styrofoam. The test antennas were then rotated a full 360 degrees. The received signal was carried to the receiver outside of the chamber on a coaxial feed line. The feed line dropped away from the antenna perpendicularly for a distance of about nine wavelengths. The chamber can be seen in Fig. 16.

Fig. 17 shows the pattern of the balun-fed antenna. The signal level in the nulls off the ends of the antenna is about 32 dB below the "broadside" signal level. Noise precluded indentifying nulls significantly deeper than that level with the setup used.

Fig. 18 shows the pattern of a dipole without the benefit of the balun. The peak amplitude of the signal is about 5 dB below that of the balun-fed antenna and one of the nulls, 30 degrees from broadside, is just as deep as was the null off the end of the balun-fed antenna. A couple of points about this trace need to be considered. First, the exact location of peaks and nulls is highly dependent on the relative location of the feed line as the antenna is rotated. In repeating the experiment with a different relative position of either, the pattern changes. Fig. 18 can only be considered as representative of how a half-wave dipole performs as a receiving antenna when used without a balun and when used with a long feed line. Second, the overall drop in signal level is not necessarily representative of what you should expect from the antenna in a

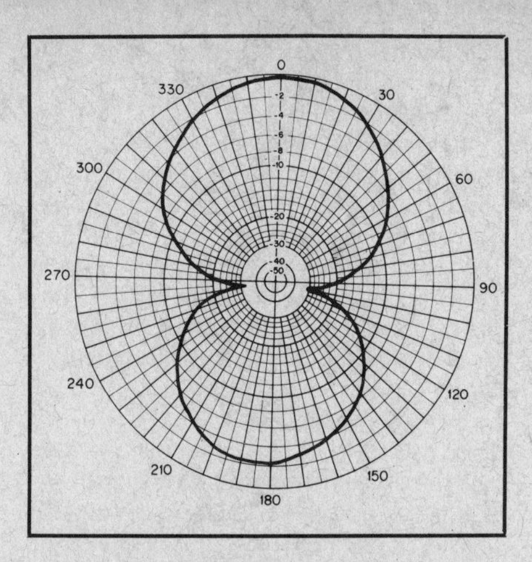


Fig. 17 — Response pattern of the balun-fed half-wavelength dipole in the rf anechoic chamber. The apparent front-to-back ratio exists because the antenna was not located at the exact center of the rotating support. This response and that of Fig. 18 are drawn to the same relative scale.

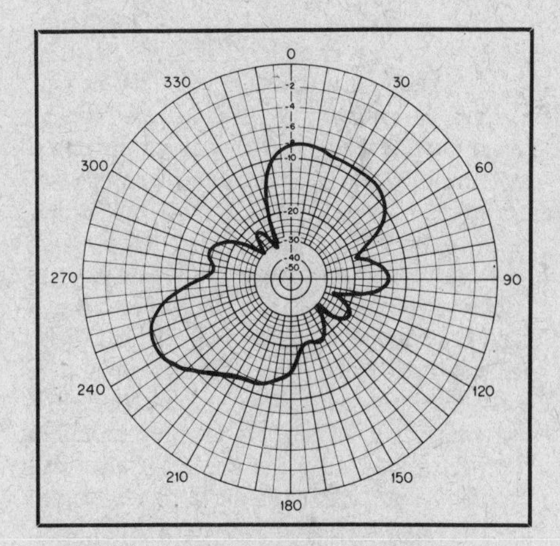


Fig. 18 — Response pattern of the half-wavelength dipole without a balun. The pattern changed significantly during tests if the coaxial feed line was relocated, no doubt caused by changes in the amplitude and phase of currents flowing on the outside of the line.

transmit application. Reciprocity notwithstanding, antenna currents flowing on the outside of the coax are, in general, lost to the receiver. These same currents, in the transmit mode, can radiate energy that effectively fills in the nulls noted here. The pattern of Fig. 17 is fully predictable and can be easily reproduced in a repeated experiment. That of Fig. 18 cannot.

A balun gives a predictable pattern. The biggest benefit that accrues from this feature is applicable to using a balanced element in a directional array.

The above should not necessarily be interpreted to mean that installing a balun on an 80-meter dipole is going to result in any detectable differences. Antennas interact with all kinds of reflecting and reradiating objects. Every piece of material in the vicinity of an antenna has an effect. The pattern of an 80-meter dipole might not look as bad as Fig. 18 does, but it probably doesn't look like Fig. 17, either. The majority of the variations between a real-world antenna pattern and an idealized pattern, at least with regard to simple antennas on the lower frequencies, will result from objects in the near field of the antenna. The additional variations introduced as a result of not using a balun in an application of a coaxial-fed balanced antenna will become most significant at higher frequencies with multielement antennas.

## QUARTER-WAVE TRANSFORMERS

The impedance-transforming properties of a quarter-wave transmission line can be used to good advantage in matching the antenna impedances to the characteristic impedance of the line. As described earlier, the input impedance of a quarter-wave line terminated in a resistive impedance $Z_R$ is

$$Z_S = \frac{Z_o^2}{Z_R}$$

Rearranging this equation gives

$$Z_o = \sqrt{Z_R Z_S}$$

This means that any value of load impedance $Z_R$ can be transformed into any desired value of impedance $Z_S$ at the input terminals of a quarter-wave line, provided the line can be constructed to have a characteristic impedance $Z_o$ equal to the square root of the product of the two impedances. The factor that limits the range of impedances that can be matched by this method is the range of values for $Z_o$ that is physically realizable. The latter range is approximately 50 to 600 ohms.

Practically any type of line can be used for the matching section, including both air-insulated and solid-dielectric lines. Such a matching arrangement is popularly known as the *"Q" matching system.*

One application of this type of matching section is in matching a half-wave antenna to a 600-ohm line, as shown in Fig. 19. Assuming that the antenna has a resistive impedance in the vicinity of 65 to 70 ohms, the required $Z_o$ of the matching section is approximately 200 ohms. A section of this type can be constructed of parallel tubing, from the data in Chapter 3.

The 1/4-wave transformer may be adjusted to resonance before being connected to the antenna by short-circuiting one end and coupling it inductively at that end to a dip meter. The length of the short-circuiting conductor lowers the frequency slightly, but this can be compensated for by adding half the length of the shorting bar to each conductor after resonating, measuring the shorting-bar length between the centers of the conductors.

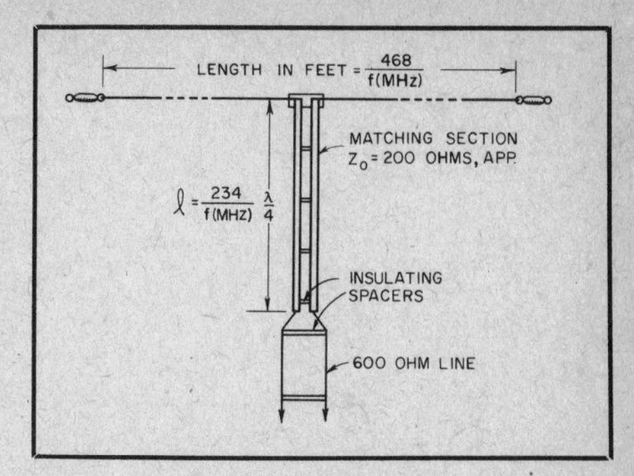


Fig. 19 — Matching a half-wave antenna to a 600-ohm line through a quarter-wave linear transformer. This arrangement is popularly known as the "Q" matching system.

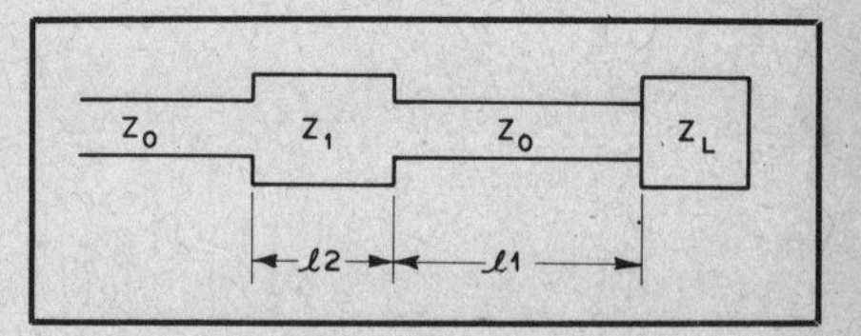


Fig. 20 — Series section transformer $Z_1$ for matching transmission-line $Z_o$ to load, $Z_L$.

### Driven Beam Elements

Another application for the quarter-wave "linear transformer" is in matching the very low antenna impedances encountered in close-spaced directional arrays to a transmission line having a characteristic impedance of 300 to 600 ohms. The observed impedances at the antenna feed point in such cases range from about 8 to 20 ohms. A matching section having a $Z_o$ of 75 ohms is useful with such arrays. The impedance at its input terminals will vary from approximately 700 ohms with an 8-ohm load to 280 ohms with a 20-ohm load.

Transmitting twin-lead is suitable for this application; such a short length is required that the loss in the matching section should not exceed about 0.6 dB even though the SWR in the matching section may be almost 10 to 1 in the extreme case.

### Series-Section Transformers

The series-section transformer has advantages over either stub tuning or the 1/4-λ transformer. The series-section transformer illustrated in Fig. 20 bears considerable resemblance to the 1/4-λ transformer. (Actually, the 1/4-λ tranformer is a special case of the series-section transformer.) The important differences are: first, that the matching section need not be located exactly at the load, second, that it may be less than a quarter wavelength long, and third, that there is great freedom in the choice of the characteristic impedance of the matching section.

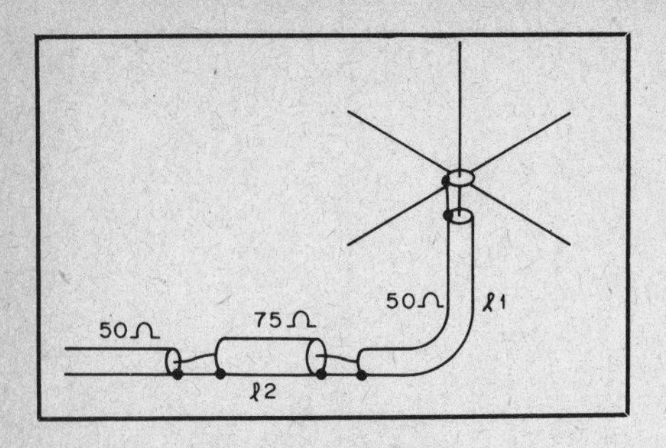


Fig. 21 — Example of series-section matching. A 38-Ω antenna is matched to 50-Ω coax by means of a length of 75-Ω cable.

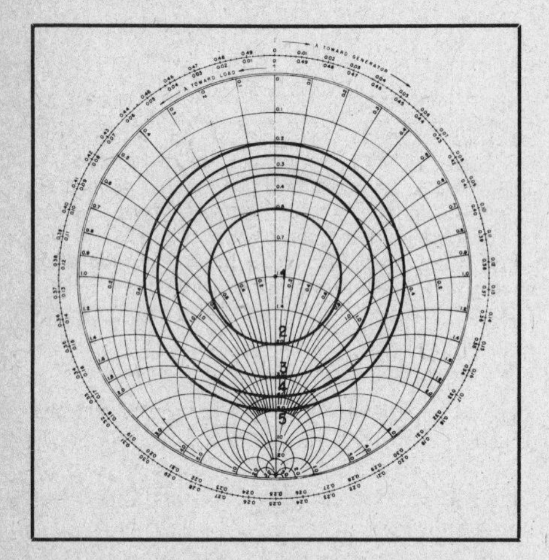

Fig. 22 — Constant-SWR circles for SWR = 2, 3, 4, and 5, showing impedance variation along 75-Ω line, normalized to 75Ω. Actual impedance is obtained by multiplying chart reading by 75 ohms.

In fact, the matching section can have *any* characteristic impedance that is not too close to that of the main line. Because of this freedom, it is almost always possible to find a length of commercially available line that will be suitable as a matching section. As an example, consider a 75-Ω line, a 300-Ω matching section, and a pure-resistance load. It can be shown that such a section may be used to match *any* resistance between 5 Ω and 1200 Ω to the main line.

The design of a series-section transformer consists of determining the length $\ell 2$ of the series or matching section and the distance $\ell 1$ from the load to the point where the section should be inserted into the main line. Three quantities must be known. These are the characteristic impedances of the main line and of the matching section, both assumed purely resistive, and the complex-load impedance. Either of two design methods may be used. One is algebraic, and the other is a graphic method using the Smith Chart. You can take your choice.

**Algebraic Design Method**

The two lengths $\ell 1$ and $\ell 2$ are to be determined from the characteristic impedances of the main line and the matching section, $Z_o$ and $Z_1$, respectively, and the load impedance $Z_L = R_L + jX_L$. The first step is to determine the normalized impedances.

$$n = \frac{Z_1}{Z_o}$$

$$r = \frac{R_L}{Z_o}$$

$$x = \frac{X_L}{Z_o}$$

Next, $\ell 2$ and $\ell 1$ are determined from the relations

$$\tan \ell 2 = B = \pm\sqrt{\frac{(r-1)^2 + x^2}{r\left(n - \frac{1}{n}\right)^2 - (r-1)^2 - x^2}}$$

$$\tan \ell 1 = A = \frac{\left(n - \frac{r}{n}\right)B + x}{r + xnB - 1}$$

Lengths $\ell 2$ and $\ell 1$ thus determined are electrical lengths in degrees. Actual lengths are obtained by dividing by 360° and multiplying by the wavelength measured along the line (main line or matching section, as the case may be), taking the velocity factor of the line into account.

The sign of B may be chosen either positive or negative, but the positive sign is preferred because it results in a shorter matching section. The sign of A may not be chosen but can turn out to be either positive or negative. If a negative sign occurs and an electronic calculator is then used to determine $\ell 1$, a negative electric length will result. If this happens, add 180°. The resultant electrical length will be correct both physically and mathematically.

In calculating B, if the quantity under the radical is negative, an imaginary value for B results. This would mean that $Z_1$, the impedance of the matching section, is too close to $Z_o$ and should be changed.

Limits on the characteristic impedance of $Z_1$ may be calculated in terms of the standing-wave ratio produced by the load on the main line without matching. For matching to occur, $Z_1$ should either be greater than $Z_o\sqrt{SWR}$ or less than $Z_o/\sqrt{SWR}$.

*An Example*

As an example, suppose we want to feed a 29-MHz ground plane vertical antenna with RG-58-type foam-dielectric coax (Fig. 21). We'll assume the antenna impedance to be 38 ohms, pure resistance, and use a length of RG-59/U foam-dielectric coax as the series section.

$Z_o$ is 50 ohms, $Z_1$ is 75 ohms, and both cables have a velocity factor of 0.79. (From above, $Z_1$ must have an impedance greater than 57.4 Ω or less than 43.6 Ω.) The design steps are as follows.

From the earlier equations, $n = 1.5$, $r = 0.76$, and $x = 0$.

Further, $B = 0.3500$ (positive sign chosen), $\ell 2 = 19.29°$ and $A = -1.4486$. Calculating $\ell 1$ yields $-55.38°$. Adding 180° to obtain a positive result gives $\ell 1 = 124.62°$.

To find the physical lengths $\ell 1'$ and $\ell 2'$ we first find the free-space wavelength.

$$\lambda_o = \frac{984}{f_{MHz}} \text{ feet}$$

and the transmission-line wavelength

$$\lambda = \lambda_o \times \text{velocity factor}$$

In the present case we find $\lambda = 26.81$ ft. Finally we have

$$\ell 1' = \frac{\ell 1 \times \lambda}{360} = 9.28 \text{ ft, and}$$

$$\ell 2' = \frac{\ell 2 \times \lambda}{360} = 1.44 \text{ ft}$$

This completes the calculations. Construction consists of cutting the main coax at a point 9.28 ft from the antenna and adding a 1.44-ft length of the 75-Ω cable.

**The Quarter-Wave Transformer**

The antenna in the preceding example could have been matched by a 1/4-λ transformer at the load. Such a transformer would have a characteristic impedance of 43.6 Ω. It is interesting to see what happens in the design of a series-section transformer if this value is chosen as the characteristic impedance of the series section.

Following the same steps as before, we find $n = 0.872$, $r = 0.76$, and $x = 0$.

From these values we find $B = \infty$ and $\ell 2 = 90°$. Further, $A = 0$ and $\ell 1 = 0°$. These results represent a quarter-wave section at the load, and indicate that, as stated earlier, the quarter-wave transformer is indeed a special case of the series-section transformer.

**Smith-Chart Solution**

A series-section transformer can be designed graphically with the aid of a Smith Chart, but this requires the use of the chart in its unfamiliar off-center mode. This mode is described in the next two paragraphs.

Fig. 22 shows the Smith Chart used in its familiar centered mode, with all impedances normalized to that of the transmission line, in this case 75 ohms, and all constant-SWR circles concentric with the normalized value $r = 1$ at the chart center. An actual impedance is recovered by multiplying a chart reading by the normalizing impedance of 75 ohms. If the actual (unnormalized) impedances represented by a constant-SWR circle in Fig. 22 are instead divided by a normalizing impedance of 300 ohms, a different picture results. A Smith Chart shows all possible impedances, and so a closed path such as a constant-SWR circle in Fig. 22 must again be represented by a closed path. In fact, it can be shown that the path remains a circle, but that the constant-SWR circles are no longer concentric. Fig. 23 shows the circles that result when the impedances along a mismatched 75-Ω line are normalized by dividing by 300 ohms instead of 75. The constant-SWR circles still surround the point corresponding to the characteristic impedance of the line ($r = 0.25$) but are no longer concentric with it. Note that the normalized impedances read from corresponding points on Figs. 22 and 23 are different but that the actual, unnormalized, impedances are exactly the same.

Let's turn now to the example shown in Fig. 24. A complex load of $Z_L = 600 + j900$ ohms is to be fed with 300-Ω line, and a 75-Ω series section is to be used. These characteristic impedances agree with those used in Fig. 23, and thus Fig. 23 can be used to find the impedance variation along the 75-Ω series section. In particular, the constant-SWR circle which passes through the chart center, SWR = 4 in this case, passes through all the impedances (normalized to 300 ohms) which the 75-Ω series section is able to match to the 300-Ω main line. The length $\ell 1$ of 300-Ω line has the job of transforming the load impedance to some impedance on this matching circle.

Fig. 25 shows the whole process more clearly, with all impedances normalized to 300 Ω. Here the normalized load impedance $Z_L = 2 + j3$ is shown at R, and the matching circle appears centered on the real axis and passing through the points $r = 1$ and $r = n^2 = 0.0625$. A constant-SWR circle is drawn from R to an intersection with the matching circle at Q or Q′ and the corresponding length $\ell 1$ (or $\ell 1'$) can be read directly from the Smith Chart.

Although the impedance locus from Q to P is shown in Fig. 25, the length $\ell 2$ cannot be determined directly from this chart. This is because the matching circle is not concentric with the chart center, as it must be if the length indications on the periphery of the Smith Chart are to be used. This problem is overcome by forming Fig. 26, which is the same as Fig. 25 except that all impedances have been divided by $n = 0.25$, resulting in a Smith Chart normalized to 75 ohms instead of 300. The matching circle and the chart center are now concentric, and the series-section length $\ell 2$, the distance between Q and P, can be taken directly from the chart.

In fact it is not necessary to construct the entire impedance locus shown in Fig. 26. It is sufficient to plot $Z_Q/n$ ($Z_Q$ is read from Fig. 25) and $Z_P/n = 1/n$, connect

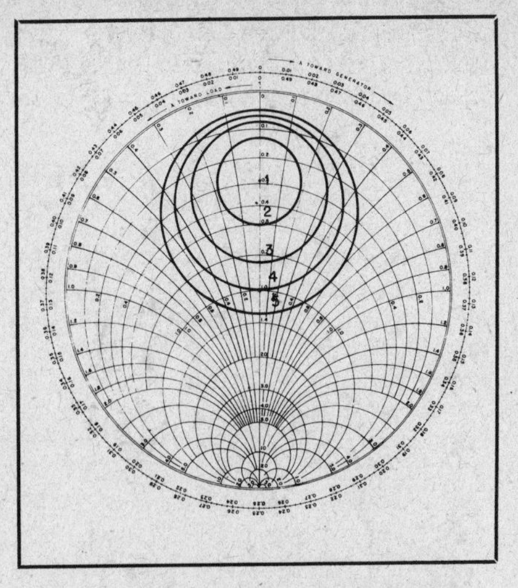


Fig. 23 — Paths of constant SWR for SWR = 2, 3, 4 and 5, showing impedance variation along 75-Ω line, normalized to 300 Ω. Normalized impedances differ from those in Fig. 22, but actual impedances are obtained by multiplying chart readings by 300 ohms and are the same as those corresponding in Fig. 22. Paths remain circles but are no longer concentric. One, the matching circle, SWR = 4 in this case, passes through the chart center and is thus the locus of all impedances which can be matched to a 300-Ω line.

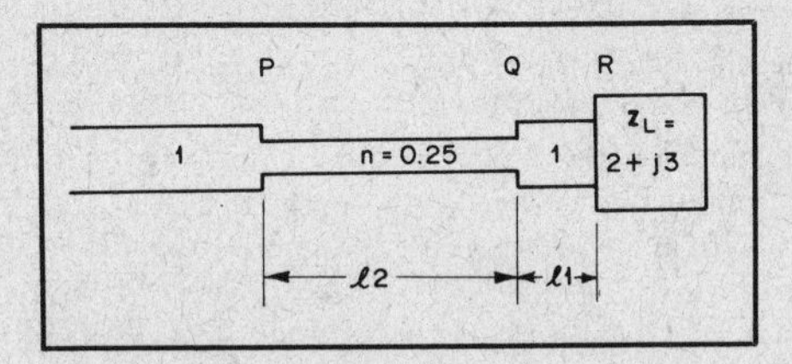


Fig. 24 — Example for solution by Smith Chart. All impedances are normalized to 300 ohms.

them by a circular arc centered on the chart center, and to determine the arc length $\ell 2$ from the Smith Chart.

The steps necessary to design a series-section transformer by means of the Smith Chart can now be listed:

1) Normalize all impedances by dividing by the characteristic impedance of the main line.

2) On a Smith Chart plot the normalized load impedance $Z_L$ at R and construct the matching circle so that its center is on the real axis and it passes through the points $r = 1$ and $r = n^2$.

3) Construct a constant-SWR circle centered on the chart center through point R. This circle should intersect the matching circle at two points. One of these points, normally the one resulting in the shorter clockwise distance along the matching circle to the chart center, is chosen as point Q, and the clockwise distance from R to Q is read from the chart and taken to be $\ell 1$.

4) Read the impedance $Z_Q$ from the chart, calculate $Z_Q/n$ and plot it as point Q on a second Smith Chart. Also plot $r = 1/n$ as point P.

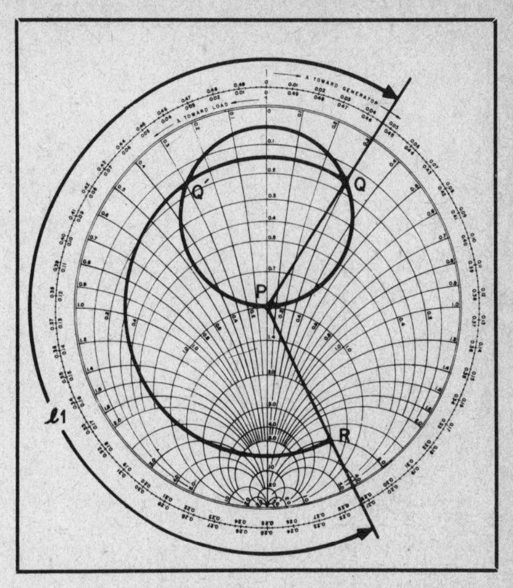


Fig. 25 — Smith Chart representation of example shown in Fig. 24. Impedance locus always has clockwise direction from load to generator, first along constant-SWR circle from load at R to intersection with matching circle at Q or Q′, then along matching circle to chart center at P. Length $\ell 1$ can be determined directly from chart.

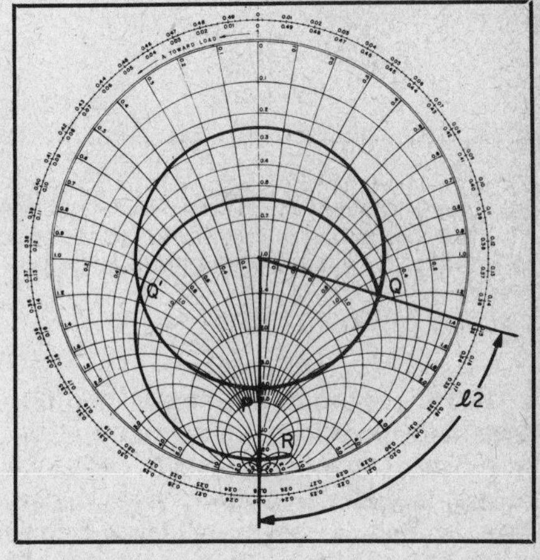


Fig. 26 — Same impedance locus as shown in Fig. 25 except normalized to 75 ohms instead of 300. The matching circle is now concentric with chart center, and $\ell 2$ can be determined directly from chart. In this example, $\ell 1 = 0.332\ \lambda$ and $\ell 2 = 0.102\ \lambda$.

5) On this second chart construct a circular arc, centered on the chart center, clockwise from Q to P. The length of this arc, read from the chart, represents $\ell 2$. The design of the transformer is now complete.

The Smith Chart construction shows that two design solutions are usually possible, corresponding to the two intersections of the load constant-SWR circle with the matching circle, and also corresponding to positive and negative values of the square-root radical in the equation given earlier for B. It may happen, however, that the load circle misses the matching circle completely, in which case no solution is possible. The cure is to enlarge the matching circle by choosing a series section whose impedance departs

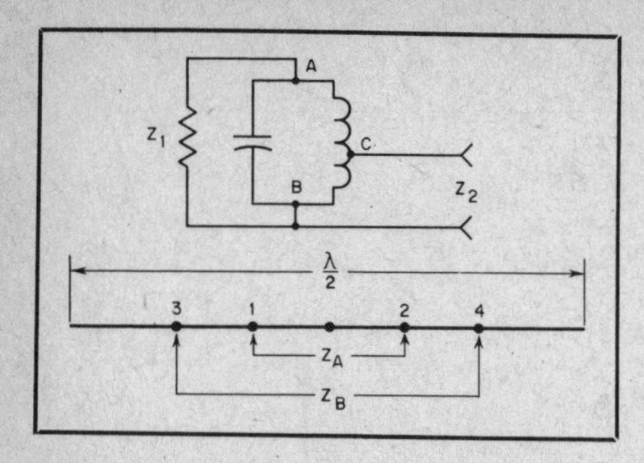


Fig. 27 — Impedance transformation with a resonant circuit, together with antenna analogy.

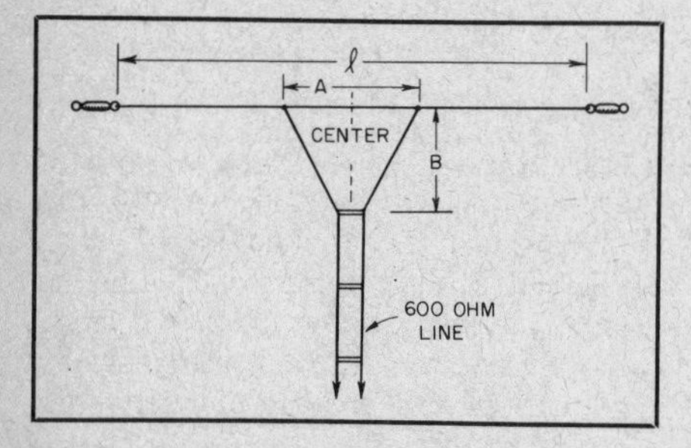


Fig. 28 — The "delta" matching system.

more from that of the main line.

A final possibility is that, rather than intersecting the matching circle, the load circle is tangent to it. There is then but one solution — that of the 1/4-λ transformer.

## DELTA MATCHING

Among the properties of a coil-and-capacitor resonant circuit is that of transforming impedances. If a resistive impedance, $Z_1$ in Fig. 27, is connected across the outer terminals AB of a resonant LC circuit, the impedance $Z_2$ as viewed looking into another pair of terminals such as BC will also be resistive, but will have a different value depending on the mutual coupling between the parts of the coil associated with each pair of terminals. $Z_2$ will be less than $Z_1$ in the circuit shown. Of course this relationship will be reversed if $Z_1$ is connected across terminals BC and $Z_2$ is viewed from terminals AB.

A resonant antenna has properties similar to those of a tuned circuit. The impedance presented between any two points symmetrically placed with respect to the center of a half-wave antenna will depend on the distance between the points. The greater the separation, the higher the value of impedance, up to the limiting value that exists between the open ends of the antenna. This is also suggested in Fig. 27. The impedance $Z_A$ between terminals 1 and 2 is lower than the impedance $Z_B$ between terminals 3 and 4. Both impedances, however, are purely resistive if the antenna is resonant.

This principle in used in the *delta* matching system shown in Fig. 28. The center impedance of a half-wave dipole is too low to be matched directly by any practicable type of air-insulated parallel-conductor line. However, it is possible to find, between two points, a value of impedance that can be matched to such a line when a "fanned" section or delta is used to couple the line and antenna. The antenna length $\ell$, should be based on the formula in Chapter 2, using the appropriate factor for the length/diameter ratio. The ends of the delta or "Y" should be attached at points equidistant from the center of the antenna. When so connected, the terminating impedance for the line will be essentially purely resistive.

Based on experimental data for the case of a simple half-wave antenna coupled to a 600-ohm line, the total distance, A, between the ends of the delta should be 0.120 λ for frequencies below 30 MHz, and 0.115 λ for frequencies above 30 MHz. The length of the delta, distance B, should be 0.150 λ. These values are based on a wavelength in air, and on the assumption that the center impedance of the antenna is approximately 70 ohms. The dimensions will require modifications if the actual impedance is very much different.

The delta match can be used for matching the driven element of a directive array to a transmission line, but if the impedance of the element is low — as is frequently the case — the proper dimensions for A and B must be found by experimentation.

The delta match is somewhat awkward to adjust when the proper dimensions are unknown, because both the length and width of the delta must be varied. An additional disadvantage is that there is always some radiation from the delta. This is because the conductors are not close enough together to meet the requirement (for negligible radiation) that the spacing should be very small in comparison with the wavelength.

## FOLDED DIPOLES

Basic information on the folded dipole antenna appears in Chapter 2. The two-wire system of Chapter 2 is an especially useful one because the input impedance is so close to 300 ohms that it can be fed directly with 300-ohm twin-lead or with open line without any other matching arrangement, and the line will operate with a low standing-wave ratio. The antenna itself can be built like an open-wire line; that is, the two conductors can be held apart by regular feeder spreaders. TV "ladder" line is quite suitable. It is also possible to use 300-ohm line for the antenna, in addition to using it for the transmission line. Additional construction information is contained in Chapter 13. Since the antenna section does not operate as a transmission line, but simply as two wires in parallel, the velocity factor of twin-lead can be ignored in computing the antenna

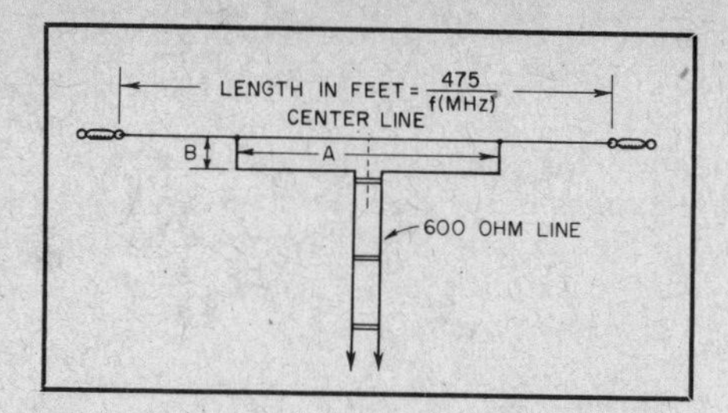


Fig. 29 — The T matching system, applied to a half-wave antenna and 600-ohm line.

length. The reactance of the folded-dipole antenna varies less rapidly with frequency changes away from resonance than single-wire antenna. Therefore it is possible to operate over a wide band of frequencies, while maintaining a low SWR on the line, than with a simple dipole. This is partly explained by the fact that the two conductors in parallel form a single conductor of greater effective diameter.

For reasons described in Chapter 2, a folded dipole will not accept power at twice the fundamental frequency. However, the current distribution is correct for harmonic operation on odd multiples of the fundamental. Because the radiation resistance is not greatly different for a three-half-wave antenna and a single half wave, a folded dipole can be operated on its third harmonic with a low SWR in a 300-ohm line. A 7-MHz folded dipole, consequently, can be used for the 21-MHz band as well.

### Spacing Adjustment of Multi- and Unequal-Conductor Dipoles

Chapter 2 shows how a wide range of impedance step-up ratios is available by varying the size of the conductors and/or using more than two. Because the relatively large effective thickness of the antenna reduces the rate of change of reactance with frequency, the tuning becomes relatively broad. It is a good idea, however, to check the resonant frequency with a dip meter in making length adjustments. The transmission line should be disconnected and the antenna terminals temporarily short-circuited when this check is being made.

As shown by the charts in Chapter 2, there are two special cases where the impedance ratio of the folded dipole is independent of the spacing between conductors. These are for a ratio of 4:1 with the two-conductor dipole and a ratio of 9:1 in the three-conductor case. In all other cases the impedance ratio can be varied by adjustment of the spacing. The adjustment range is quite limited when ratios near 4 and 9, respectively, are used, but increases with the departure in either direction from these "fixed" values. This offers a means for final adjustment of the match to the transmission line when the antenna resistance is known approximately but not exactly.

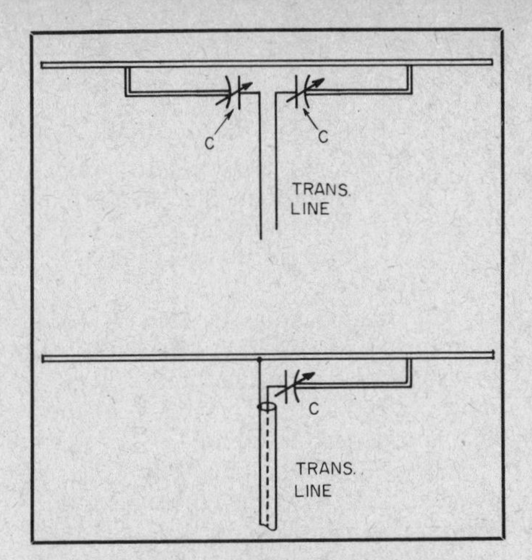


Fig. 30 — Series capacitors for tuning out residual reactance with the T and gamma matching systems. A maximum capacitance of 150 pF in each capacitor should provide sufficient adjustment range, in the average case, for 14-MHz operation. Proportionately smaller capacitance values can be used on higher frequency bands. Receiving-type plate spacing will be satisfactory for power levels up to a few hundred watts.

If a suitable match cannot be obtained by adjustment of spacing, there is no alternative but to change the ratio of conductor diameters. The impedance ratio decreases with an increase in spacing, and vice versa. Hence, if a match cannot be brought about by changing the spacing, such a change will at least indicate whether the ratio of $d_2/d_1$ should be increased or decreased.

## THE T AND GAMMA

The "T" matching system shown in Fig. 29 has a considerable resemblance to the folded dipole; in fact, if the distance A is extended to the full length of the antenna, the system becomes an ordinary folded dipole. The T has considerable flexibility in impedance ratio and is more convenient, constructionally, than the folded dipole when used with the driven element of a rotatable parasitic array. Since it is a symmetrical system it is inherently balanced, and so is well suited to use with parallel-conductor transmission lines. If coaxial line is used, some form of balun, as described earlier, should be installed. Alternatively, the gamma form described below can be used with unbalanced lines.

The current flowing at the input terminals of the T consists of the normal antenna current divided between the radiator and the T conductors in a way that depends on their relative diameters and the spacing between them, with a superimposed transmission-line current flowing in each half of the T and its associated section of the antenna. Each such T conductor and the associated antenna conductor can be looked upon as a section of transmission line shorted at the end. Since it is shorter than 1/4 wavelength it has inductive reactance; as a consequence, if the antenna itself is exactly resonant at the operating frequency, the input impedance of the T will show inductive reactance as well as resistance. The reactance must be tuned out if a good match to the transmission line is to be secured. This can be done either by shortening the antenna to obtain a value of capacitive reactance that will reflect through the matching system to cancel the inductive reactance at the input terminals, or by inserting a capacitance of the proper value in series at the input terminals as shown in Fig. 30, upper drawing.

A theoretical analysis has shown that the part of the impedance step-up arising from the spacing and ratio of conductor diameters is approximately the same as given for the folded dipole in Chapter 2. The actual impedance ratio is, however, considerably modified by the length A of the matching section (Fig. 29). The trends can be stated as follows:

1) The input impedance increases as the distance A is made larger, but not indefinitely. There is in general a distance A that will give a maximum value of input impedance, after which further increase in A will cause the impedance to decrease.

2) The distance A at which the input impedance reaches a maximum is smaller as $d_2/d_1$ (using the notation of Chapter 2) is made larger, and becomes smaller as the spacing between the conductors is increased.

3) The maximum impedance values occur in the region where A is 40 to 60 percent of the antenna length in the average case.

4) Higher values of input inpedance can be realized when the antenna is shortened to cancel the inductive reactance of the matching section.

### Simple Dipole Matching

For a dipole having an approximate impedance of 70 ohms, the T matching-section dimensions for matching a 600-ohm line are given by the following formulas:

$$\text{A (feet)} = \frac{180.5}{f(\text{MHz})}$$

$$\text{B (inches)} = \frac{114}{f(\text{MHz})}$$

These formulas apply for wire antennas with the matching section made of the same size wire. With an antenna element of different impedance, or for matching a line having a $Z_0$ other than 600 ohms, the matching-section dimensions can be determined experimentally.

### The Gamma

The gamma arrangement shown in Fig. 31 is an unbalanced version of the T, suitable for use with coaxial lines. Except for the fact that the matching section is connected between the center and one side of the antenna, the remarks above about the behavior of the T apply equally well. The inherent reactance of the matching section can be canceled either by shortening the antenna appropriately or by using the resonant length and installing a capacitor C, as shown in the lower drawing of Fig. 30.

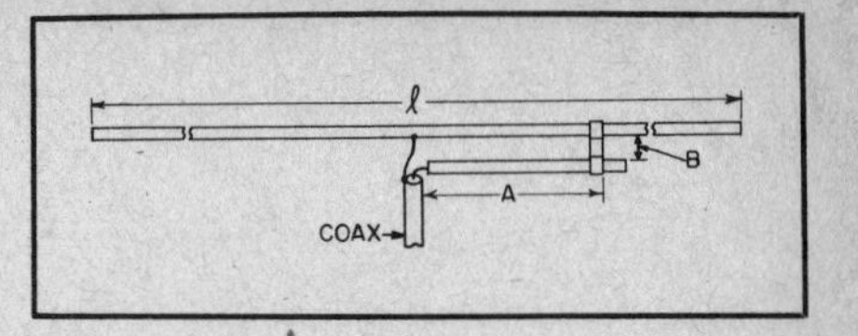


Fig. 31 — The gamma match, as used with tubing elements. The transmission line may be either 52-ohm or 75-ohm coax.

The gamma match has been widely used for matching coaxial cable to all-metal parastic beams for a number of years. Because it is well suited to "plumber's delight" construction, where all the metal parts are electrically and mechanically connected, it has become quite popular for amateur arrays.

Because of the many variable factors — driven-element length, gamma rod length, rod diameter, spacing between rod and driven element, and value of series capacitors — a number of combinations will provide the desired match. The task of finding a proper combination can be a tedious one, however, as the settings are interrelated. A few "rules of thumb" have evolved that provide a starting point for the various factors. For matching a multielement array made of aluminum tubing to 52-ohm line, the length of the rod should be 0.04 to 0.05 λ, its diameter 1/3 to 1/2 that of the driven element, and its spacing (center to center from the driven element), approximately 0.007 λ. The capacitance value should be approximately 7 pF per meter of wavelength, i.e., about 140 pF for 20-meter operation. The exact gamma dimensions and value for the capacitor will depend on the radiation resistance of the driven element, and whether or not it is resonant. These starting-point dimensions are for an array having a feed-point impedance of about 25 ohms, with the driven element shortened approximately 3% from resonance.

### Calculating Gamma Dimensions

D. H. Healey, W3PG, has developed a method of determining by calculations whether or not a particular set of parameters for a gamma match will be suitable for obtaining the desired impedance transformation. (See bibliography at the end of this chapter.) The procedure uses mathematical equations and the Smith Chart (see Chapter 3), and consists of the following basic steps.

1) Find the impedance step-up ratio for the gamma rod and element diameters and spacing. (See Chapter 2; use the rod

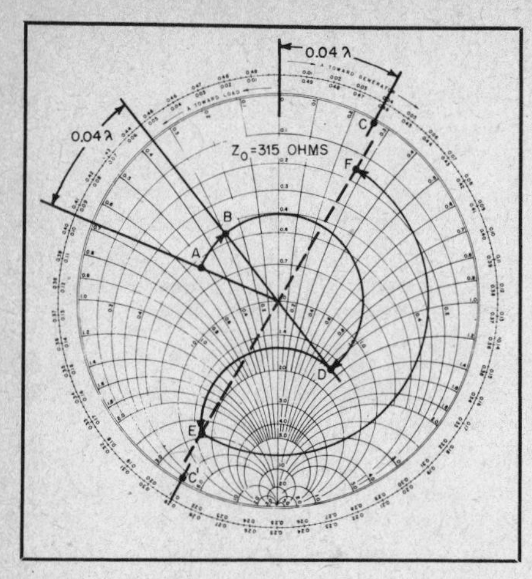

Fig. 32 — Smith Chart calculation of gamma dimensions. See text.

diameter as $d_1$ and the element diameter as $d_2$.)

2) Determine the $Z_o$ of the "transmission line" formed by the gamma rod and the element, considering them as two parallel conductors. Use the equation

$$Z_o = 276 \log_{10} \frac{2S}{\sqrt{d_1 d_2}} \text{ ohms}$$

where the terms are the same as in Chapter 2.

3) Assign (or assume) a length for the gamma rod, expressed in electrical degrees. Call this angle $\theta$.

4) Determine the increased impedance of the driven element over its center-point impedance, caused by its being fed off center. Use the equation

$$Z_2 = \frac{Z_1}{\cos^2\theta}$$

where $Z_2$ is the impedance at the tap point and $Z_1$ is the complex impedance at the center of the element.

5) Determine the "load" impedance at the antenna end of the gamma "transmission line." This is the resultant value of Step 1 above multiplied by the value for $Z_2$, taken as R + jX from Step 4. Normalize this impedance value to the $Z_o$ of the gamma "transmission line" determined in Step 2. Plot this normalized impedance on the Smith Chart.

6) Using the TOWARD GENERATOR wavelengths scale of the Smith Chart, take the "transmission line" length (rod length) into account and determine the normalized input impedance to this line. This impedance represents the portion of the total impedance at the gamma feed point which arises from the antenna alone.

7) In shunt with the impedance from Step 6 is an inductive reactance caused by the short-circuit termination on the gamma "transmission line" itself. Determine the normalized value of this inductance either from the Smith Chart (taking 0 + j0 as the load and the rod length into account on the TOWARD GENERATOR wavelengths scale) or from the equation

$$X_p = j \tan \theta \text{ ohms}$$

Also, Fig. 37 of a later section may be used.

8) Invert the line input impedance (from Step 6) to obtain the equivalent admittance, G + jB. This may be done by locating the point on the chart which is diametrically opposite that for the plot of the impedance. (Remember that inductance is considered to be a negative susceptance, and capacitance a positive susceptance.) Similarly, invert the inductance value from Step 7. (This susceptance will simply be the reciprocal of the reactance.)

9) Add the two parallel susceptance components from Step 8, taking algebraic signs into account. Plot the new admittance on the Smith Chart, G (from Step 8) + jB (from this step).

10) Invert the admittance of Step 9 to impedance by locating the diametrically opposite point on the Smith Chart. Convert the normalized resistance and reactance components to ohms by multiplying each by the line $Z_o$ (from Step 2). This impedance is that which terminates the transmission line with no gamma capacitor. A capacitor having the reactance of the X component of the impedance should be used to cancel the inductance, leaving a purely resistive line termination. If the dimensions were properly chosen, this value will be near the $Z_o$ of the coaxial feed line.

As an example, assume a 20-meter Yagi beam is to be matched to 50-ohm line. The driven element is 1-1/2 inches in diameter, and the gamma rod is a length of 1/2-inch tubing, spaced 6 inches from the element (center to center). Initially, the rod length is adjusted to 0.04 λ, or 14.4° (33 inches). The driven element has been shortened by 3% from its resonant length.

Following Step 1, from Chapter 2, the impedance step-up ratio is 6.4.

From the equation of Step 2, the $Z_o$ of the transmission line is 315 ohms.

From Step 3, $\theta$ = 14.4°.

For Step 4, assume the antenna has a radiation resistance of 25 ohms and a capacitive reactance component of 25 ohms (about the reactance which would result from the 3% shortening). The overall impedance of the driven element is therefore 25 – j25 ohms. Using this value for $Z_1$ in the equation of Step 4, $Z_2$ is determined to be 26.6 – j26.6.

In Step 5, the value obtained above for $Z_2$, 26.6, – j26.6, is multiplied by the step-up ratio (Step 1), 6.4. The resultant impedance is 170 – j170 ohms. Normalized to the $Z_o$ of the gamma "transmission line," 315 ohms, this impedance is 0.54 – j0.54. This value is plotted on the Smith Chart, shown at point A of Fig. 32.

Taking the line length into account (0.04 λ) as directed in Step 6, the normalized line input impedance is found to be 0.44 – j0.29, as shown at point B of Fig. 32.

From the equation in Step 7, $X_p$ is found to be j0.257. This same value may be determined from the Smith Chart, as shown at point C, or from Fig. 37 for a matching-section length of 14.4°.

Point D is found on the Smith Chart as directed in Step 8, and represents a normalized admittance of 1.6 + j1.07 siemens. The inductance from Step 7, above, inverted to susceptance, is –j1/0.257 = –j3.89. (This same value may be read diametrically opposite point C in Fig. 32, at point C'.)

Proceeding as indicated in Step 9, the admittance components of the parallel combination are 1.6 + j1.07 – j3.89 = 1.6 – j2.82. This admittance is plotted as shown at point E in Fig. 32.

Inverting the above admittance to its equivalent impedance, point F of Fig. 32, the normalized value of 0.16 + j0.27 is read. Multiplying each value by 315 (from Step 2), the input impedance to the gamma section is found to be 50.4 + j85 ohms.

Thus, a series capacitor having a reactance of 85 ohms is required to cancel the inductance in the gamma section (from the standard reactance equation the required capacitance is 134 pF), and a very good match is provided for 50-ohm line.

### Adjustment

After installation of the antenna, the proper constants for the T and gamma must be determined experimentally. The use of the variable series capacitors, as shown in Fig. 30, is recommended for ease of adjustment. With a trial position of the tap or taps on the antenna, measure the SWR on the transmission line and adjust C (both capacitors simultaneously in the case of the T) for minimum SWR. If it is not close to 1 to 1, try another tap position and repeat. It may be necessary to try another size of conductor for the matching section if satisfactory results cannot be secured. Changing the spacing will show which direction to go in this respect, just as in the case of the folded dipole discussed in the preceding section.

## THE OMEGA MATCH

The omega match is a slightly modified form of the gamma match. In addition to the series capacitor, a shunt capacitor is used to aid in canceling a portion of the inductive reactance introduced by the gamma section. This is shown in Fig. 33. C1 is the usual series capacitor. The addition of C2 makes it possible to use a shorter gamma rod, or makes it easier to

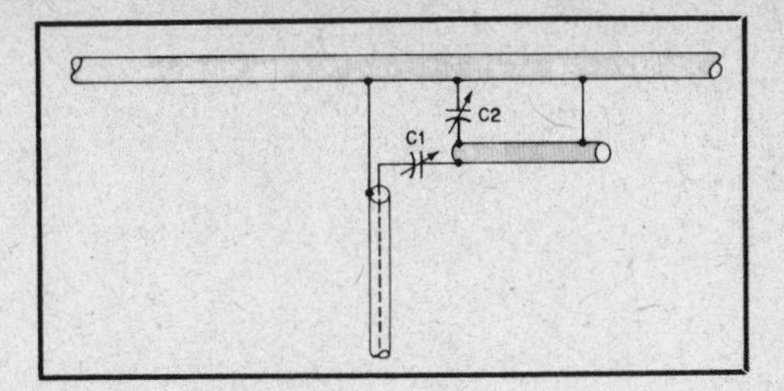


Fig. 33 — The omega match.

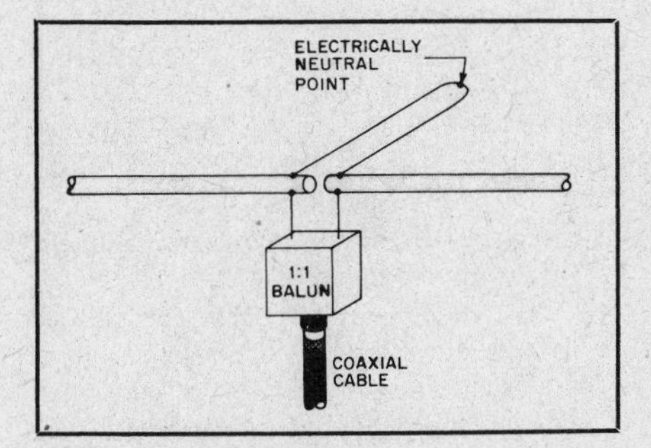


Fig. 34 — The hairpin match.

obtain the desired match when the driven element is resonant. (The effect of the shunt capacitor may be taken into account when calculating gamma dimensions, as in the foregoing section, during the performance of Step 9.) During adjustment, C2 will serve primarily to determine the resistive component of the load as seen by the coax line, and C1 serves to cancel any reactance.

## THE HAIRPIN AND BETA MATCHES

The usual form of the *hairpin match* is shown in Fig. 34. Basically, the hairpin is a form of an L-matching network. Because it is somewhat easier to adjust for the desired terminating impedance than the gamma match, it is preferred by many amateurs. Its disadvantages, compared with the gamma, are that it must be fed with a balanced line (a balun may be used with a coax feeder, as shown in Fig. 34), and the driven element must be split at the center. This latter requirement complicates the mechanical mounting arrangement for the element, by ruling out "plumber's delight" construction.

As indicated in Fig. 34, the center point of the hairpin is electrically neutral. As such, it may be grounded or connected to the remainder of the antenna structure. The hairpin itself is usually secured by attaching this neutral point to the boom of the antenna array. The *beta match* is electrically identical to the hairpin match, the difference being in the mechanical construction of the matching section. With the beta match, the conductors of the matching section straddle the boom, one conductor being located on either side, and the electrically neutral point consists of a sliding or adjustable shorting clamp placed around the boom and the two matching-section conductors.

The electrical operation of the hairpin match has been treated extensively by Gooch, Gardner and Roberts (see bibliography at the end of this chapter). The antenna is matched to the transmission line by forming an equivalent parallel-resonant circuit in which the antenna resistance appears in series with the capacitance. The impedance of this type parallel-resonant circuit varies almost inversely with the series antenna resistance, and therefore can cause a very small antenna resistance to appear as a very large resistance at the terminals of the resonant circuit. The values of inductance and capacitance are chosen to transform the antenna resistance to a resistance value equal to the characteristic impedance of the transmission line. The capacitive portion of this circuit is produced by slightly shortening the antenna driven element. For a given frequency the impedance of a shortened half-wave element appears as the antenna resistance and a capacitance in series, as indicated schematically in Fig. 35B. The inductive portion of the resonant circuit at C is a hairpin of heavy wire or small tubing which is connected across the driven-element center terminals. The diagram of C is redrawn in D to show the circuit in conventional L-network form. $R_A$, the radiation resistance, is a smaller value than $R_{IN}$, the impedance of the feed line. (In L-network matching, the higher of the two impedances to be matched is connected to the shunt-arm side of the network, and the lower impedance to the series-arm side.)

Instead of using a separate hairpin matching section and a balun, as shown in Fig. 34, a combined matching stub and balun may be used with coaxial line. This is shown in Fig. 12B. The principles of operation are discussed in a later section of this chapter titled "Stubs on Coaxial Lines."

If the approximate radiation resistance of the antenna system is known, Figs. 36 and 37 may be used to gain an idea of the hairpin dimensions necessary for the desired match. The curves of Fig. 36 were obtained from design equations for L-network matching. Fig. 37 is based on the equation, $X_p = j \tan \theta$, which gives the inductive reactance as normalized to the $Z_o$ of the hairpin, looking at it as a short-circuit-terminated length of transmission line. For example, if an antenna-system impedance of 20 ohms is to be matched to 52-ohm line, Fig. 36 indicates that the inductive reactance required for the hairpin is 41 ohms. If the hairpin is constructed of quarter-inch tubing spaced 1-1/2 inches, its characteristic impedance is 300 ohms (from Chapter 3.) Normalizing the required 41-ohm reactance to this impedance, 41/300 = 0.137. Entering the chart of Fig. 37 with this value, 0.137, on the scale at the bottom, it may be seen that the hairpin length should be 7.8 electrical degrees, or 7.8/360 wavelength. For purposes of these calculations, taking a 97.5% velocity factor into account, the wavelength in inches is $11,500/f_{MHz}$. If the antenna is to be used on 20 meters, the required hairpin length is 7.8/360 × 11,500/14 = 17.8 inches. The length of the hairpin affects primarily the resistive component of the terminating impedance as seen by the feed line. Greater resistances are obtained with longer hair-

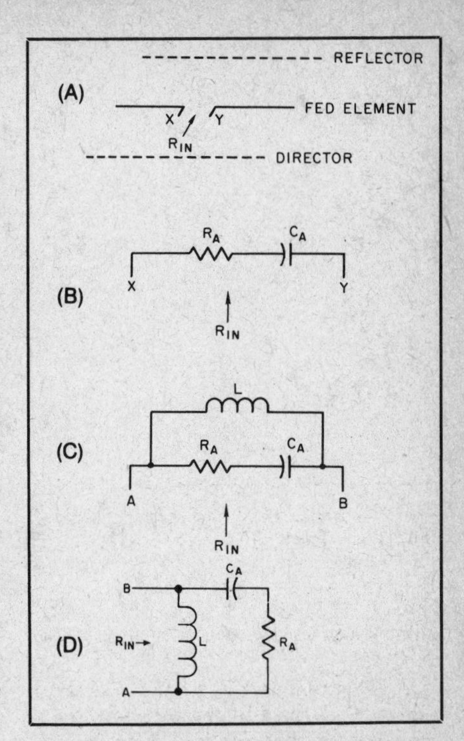


Fig. 35 — For the Yagi antenna shown at A, the input impedance at its operating frequency is represented at B, if the driven element is shorter than its resonant length. By adding an inductor, as shown at C, a low value of $R_A$ is made to appear as a higher impedance at terminals AB. At D, the diagram of C is redrawn in the usual L-network configuration.

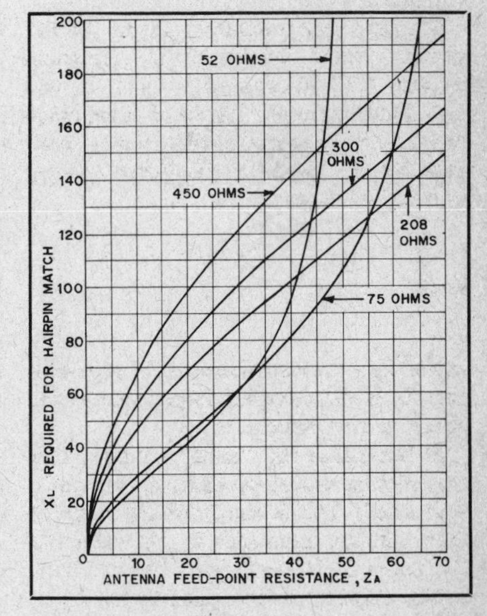


Fig. 36 — Reactance required to match various antenna resistances to common line or balun impedance.

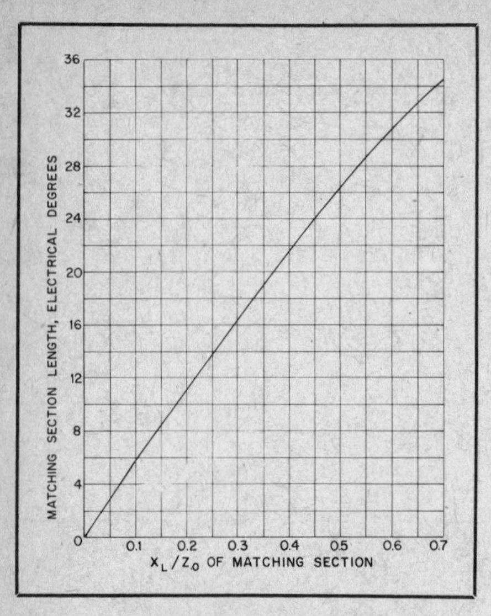


Fig. 37 — Inductive reactance (normalized to $Z_o$ of matching section), scale at bottom, versus required matching-section length, scale at left. To determine the length in wavelengths, divide the number of electrical degrees by 360. For open-wire line, a velocity factor of 97.5% should be taken into account when determining the electrical length.

pin sections, and smaller resistances with shorter sections. Reactance at the feed-point terminals is tuned out by adjusting the length of the driven element, as necessary. If a fixed-length hairpin section is in use, a small range of adjustment may be made in the effective value of the inductance by spreading or squeezing together the conductors of the hairpin. Spreading the conductors apart will have the same effect as lengthening the hairpin, while placing them closer together will effectively shorten it.

## MATCHING STUBS

As explained in Chapter 3, a mismatch-terminated transmission line less than 1/4 wavelength long has an input impedance that is both resistive and reactive. The equivalent circuit of the line input impedance can be formed either of resistance and reactance in series or resistance and reactance is parallel. Depending on the line length, the series-resistance component, $R_S$, can have any value between the terminating resistance, $Z_R$ (when the line has zero length) and $Z_o^2/Z_R$ (when the line is exactly 1/4 wave long). The same thing is true of $R_P$, the parallel-resistance component. $R_S$ and $R_P$ do not have the same values at the same line length, however, other than zero and 1/4 wavelength. With either equivalent there is some line length that will give a value of $R_S$ or $R_P$ equal to the characteristic impedance of the line. However, there will always be reactance along with the resistance. But if provision is made for canceling or "tuning out" this reactive part of the input impedance, only the resistance will remain. Since this resistance is equal to the $Z_o$ of the transmission line, the section from the reactance-cancellation point back to the generator will be properly matched.

Tuning out the reactance in the equivalent series circuit requires that a reactance of the same value as $X_S$, but of opposite kind, be inserted in series with the line. Tuning out the reactance in the equivalent parallel circuit requires that a reactance of the same value as $X_P$ but of the opposite kind be connected across the line. In practice it is convenient to use the parallel-equivalent circuit. The transmission line is simply connected to the load (which of course is usually a resonant antenna) and then a reactance of the proper value is connected across the line at the proper distance from the load. From this point back to the transmitter there are no standing waves on the line.

A convenient type of reactance to use is a section of transmission line less than one-quarter wavelength long, either open-circuited or short-circuited, depending on whether capacitive reactance or inductive reactance is called for. Reactances formed from sections of transmission line are called *matching stubs,* and are designated as *open* or *closed* depending on whether the free end is open- or short-circuited. The two types of matching stubs are shown in the sketches of Fig. 38.

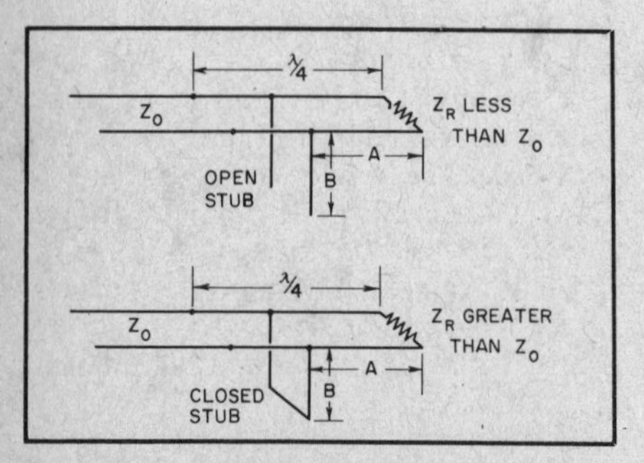


Fig. 38 — Use of open or closed stubs for canceling the parallel reactive component of input impedance.

The distance from the load to the stub (dimension A in Fig. 38) and the length of the stub, B, depend on the characteristic impedances of the line and stub and on the ratio of $Z_R$ to $Z_o$. Since the ratio of $Z_R$ to $Z_o$ is also the standing-wave ratio in the absence of matching, the dimensions are a function of the standing-wave ratio. If the line and stub have the same $Z_o$, dimensions A and B are dependent on the standing-wave ratio only. Consequently, if the standing-wave ratio can be measured before the stub is installed, the stub can be properly located and its length determined even though the actual value of load impedance is not known.

Typical applications of matching stubs are shown in Fig. 39, where open-wire line is being used. From inspection of these drawings it will be recognized that when an antenna is fed at a current loop, as in Fig. 39A, $Z_R$ is less than $Z_o$ (in the average case) and therefore an open stub is called for, installed within the first quarter wavelength of line measured from the antenna. Voltage feed, as at B, corresponds to $Z_R$ greater than $Z_o$ and therefore requires a closed stub.

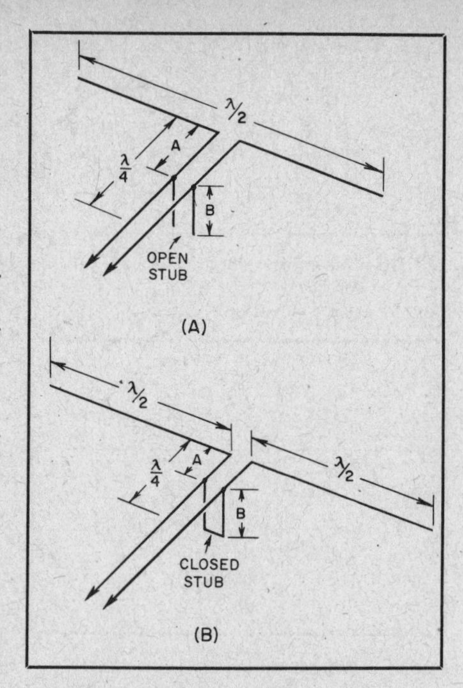


Fig. 39 — Application of matching stubs to common types of antennas.

The Smith Chart may be used to determine the length of the stub and its distance from the load (see later example). If the load is a pure resistance and the characteristic impedances of the line and stub are identical, the lengths may be determined by equations. For the closed stub when $Z_R$ is greater than $Z_o$, they are

$$\tan A = \sqrt{SWR} \text{ and } \cot B = \frac{SWR - 1}{\sqrt{SWR}}$$

For the open stub when $Z_R$ is less than $Z_o$

$$\cot A = \sqrt{SWR} \text{ and } \tan B = \frac{SWR - 1}{\sqrt{SWR}}$$

In these equations the lengths A and B are the distance from the stub to the load and the length of the stub, respectively, as shown in Fig. 39. These lengths are expressed in electrical degrees, equal to 360 times the lengths in wavelengths.

In using the Smith Chart or the above equations it must be remembered that the wavelength along the line is not the same as in free space. If an open-wire line is used the velocity factor of 0.975 will apply. When solid-dielectric line is used the free-space wavelength as given by the curves must be multiplied by the appropriate velocity factor to obtain the actual length of A and B (see Chapter 3.)

Although the equations above do not apply when the characteristic impedances of the line and stub are not the same, this does not mean that the line cannot be matched under such conditions. The stub can have any desired characteristic impedance if its length is chosen so that it has the proper value of reactance. Using the Smith Chart, the correct lengths can be determined without difficulty for dissimilar types of line.

In using matching stubs it should be noted that the length and location of the stub should be based on the standing-wave ratio *at the load.* If the line is long and has fairly high losses, measuring the SWR at the input end will not give the true value at the load. This point was discussed in Chapter 3 in the section on attenuation.

### Reactive Loads

In this discussion of matching stubs it has been assumed that the load is a pure resistance. This is the most desirable condition, since the antenna that represents the load preferably should be tuned to resonance before any attempt is made to match the line. Nevertheless, matching stubs can be used even when the load is considerably reactive. A reactive load simply means that the loops and nodes of the standing waves of voltage and current along the line do not occur at integral multiples of 1/4 wavelength from the load. To use the equations above it is necessary to find a point along the line at which a current loop or node occurs. Then the first set of equations gives the stub length and distance *toward the transmitter* from a current *loop*. The second set gives the stub length and distance *toward the transmitter* from a current *node*.

### Stubs on Coaxial Lines

The principles outlined in the preceding section apply also to coaxial lines. The coaxial cases corresponding to the open-wire cases shown in Fig. 38 are given in Fig. 40. The equations given earlier may be used to determine the dimensions A and B. In a practical installation the junction of the transmission line and stub would be a T connector.

A special case of the use of a coaxial matching stub in which the stub is associated with the transmission line in such a way as to form a balun has been described earlier in this chapter (Fig. 12). The principles used are those just described. The antenna is shortened to introduce just enough reactance at its input terminals to permit the matching stub to be connected at that point, rather than at some other point along the transmission line as in the general cases discussed here. To use this method the antenna resistance must be lower than the $Z_0$ of the main transmission line, since the resistance is transformed to a higher value. In beam antennas this will nearly always be the case.

### Matching Sections

If the two antenna systems in Fig. 39 are redrawn in somewhat different fashion, as shown in Fig. 41, a system results that differs in no consequential way from the matching stubs described previously, but in which the stub formed by A and B together is called a "quarter-wave matching section." The justification for this is that a quarter-wave section of line is similar to a resonant circuit, as described earlier in this chapter, and it is therefore possible to use it to transform impedances by tapping at the appropriate point along the line.

Earlier equations give design data for matching sections, A being the distance from the antenna to the point at which the line is connected, and A + B being the total length of the matching section. The equations apply only in the case where the characteristic impedance of the matching section and transmission line are the same. Equations are available for the case where the matching section has a different $Z_0$ than the line, but are somewhat complicated and will not be given here, since it is generally impossible to make the line and matching section similar in construction.

### Adjustment

In the experimental adjustment of any type of matched line it is necessary to measure the standing-wave ratio with fair accuracy in order to tell when the adjustments are being made in the proper direction. In the case of matching stubs, experience has shown that experimental adjustment is unnecessary, from a practical standpoint, if the SWR is first measured with the stub not connected to the transmission line, and the stub is then installed according to the design data.

## DESIGNING STUB MATCHES WITH THE SMITH CHART

Fig. 40A shows the case of a line terminated in a load impedance less than the characteristic impedance of the line, calling for an open (capacitive) stub for impedance matching. As an example, suppose that the antenna is a close-spaced array fed by a 52-ohm line, and that the standing-wave ratio has been determined to be 3.1:1. From this information, a constant-SWR circle may be drawn on the Smith Chart. Its radius is such that it intersects the lower portion of the resistance axis at the SWR value, 3.1, as shown in Fig. 42.

Since the stubs of Fig. 40 are connected in parallel with the transmission line, determining the design of the matching arrangement is simplified if Smith Chart values are dealt with as admittances, rather than impedances. (An admittance is simply the reciprocal of the associated impedance.) This leaves less chance for errors in making calculations, by eliminating the need for making series-equivalent to parallel-equivalent circuit conversions and back, or else for using

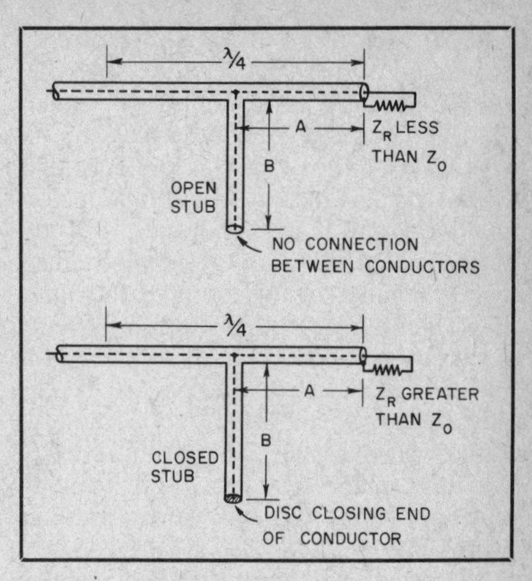


Fig. 40 — Open and closed stubs on coaxial lines.

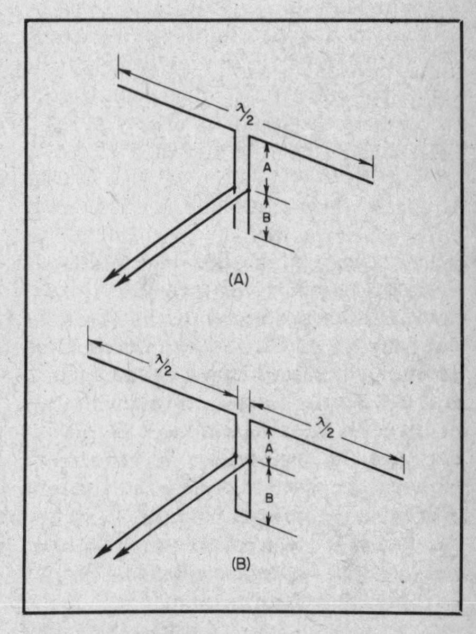


Fig. 41 — Application of matching sections to common antenna types.

complicated equations for determining the resultant value of two complex impedances connected in parallel.

A complex impedance, Z, is equal to R + jX, as described in Chapter 3. The equivalent admittance, Y, is equal to G − jB, where G is the conductance component and B the susceptance. (Inductance is taken as negative susceptance, and capacitance as positive.) Conductance and susceptance values are plotted and handled on the Smith Chart in the same manner as are resistance and reactance. Because of the way in which the Smith Chart is designed, the coordinates for an admittance will be located at a point which is diametrically opposite the plot for its impedance counterpart — on the same SWR circle, but on the opposite side of prime center.

Assuming that the close-spaced array of the foregoing example has been resonated at the operating frequency, it will present a purely resistive termination for the load end of the 52-ohm line. From earlier information of this chapter, it is known that the impedance of the antenna equals $Z_0$/SWR = 52/3.1 ≈ 16.8 ohms. If this value were to be plotted as an impedance on the Smith Chart, it would first be normalized (16.8/52 = 0.32) and then plotted as 0.32 + j0. Although not necessary for the solution of this example, this value is plotted at point A in Fig. 42. What is necessary is a plot of the admittance for the antenna as a load. This is the reciprocal of the impedance; 1/16.8 ohms equals 0.060 siemen. To plot this point it is first normalized by ***multiplying*** the conductance and susceptance values by the $Z_0$ of the line. Thus, (0.060 + j0) × 52 = 3.1 + j0. This admittance value is shown plotted at point B in Fig. 42. It may be seen that points A and B are diametrically opposite each other on the chart.

Actually, for the solution of this example, it wasn't necessary to compute the values for either point A or point B as in the above paragraph, for they were both determined from the known SWR value of 3.1. As may be seen in Fig. 42, the points are located on the constant-SWR circle which was already drawn, at the two places where it intersects the resistance axis. The plotted value for point A, 0.32, is simply the reciprocal of the value for point B, 3.1. However, an understanding of the relationship between impedance and admittance is easier to gain with simple examples such as this.

In stub matching, the stub is to be connected at a point in the line where the conductive component equals the $Z_0$ of the line. Point B represents the admittance of the load, which is the antenna. Various admittances will be encountered along the line, when moving in a direction indicated by the TOWARD GENERATOR wavelengths scale, but all admittance plots must fall on the constant-SWR circle. Moving clockwise around the SWR circle from point B, it is seen that the line input conductance will be 1.0 (normalized $Z_0$ of the line) at point C, 0.082 λ toward the transmitter from the antenna. Thus, the stub should be connected at this location on the line.

The normalized admittance at point C, the point representing the location of the stub, is 1 − j1.2 siemens, having an inductive susceptance component. A capacitive susceptance having a normalized value of +j1.2 siemens is required across the line at the point of stub connection, to cancel the inductance. This capacitance is to be obtained from the stub section itself; the problem now is to determine how long the stub should be. This is done by first plotting the susceptance required for cancellation, 0 + j1.2, on the chart (point D in Fig. 42). This point represents the input admittance as seen looking into the stub. The "load" or termination for the stub section is found by moving in the TOWARD LOAD direction around the chart,

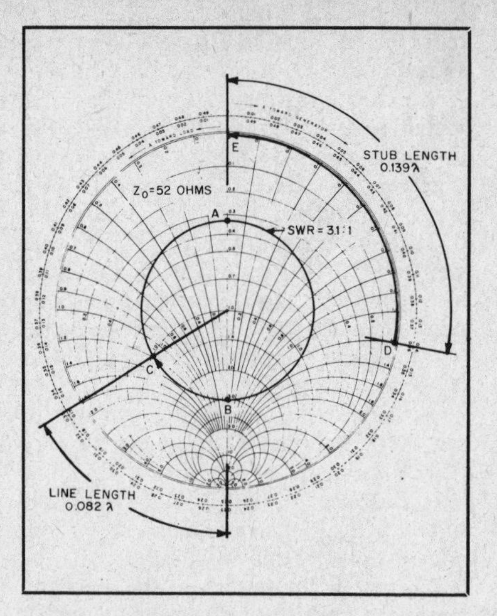


Fig. 42 — Smith Chart method of determining the dimensions for stub matching.

and will appear at the closest point on the resistance/conductance axis, either at the top or the bottom of the chart. Moving counter-clockwise from point D, this is located at E, at the top of the chart, 0.139 λ away. From this we know the required stub length. The "load" at the far end of the stub, from Fig. 40A, should be an open circuit. This load, as represented on the Smith Chart, has a normalized admittance of 0 + j0 siemen, which is equivalent to an open circuit.

When the stub, having an input admittance of 0 + j1.2 siemens, is connected in parallel with the line at a point 0.082 λ from the load, where the line input admittance is 1.0 − j1.2, the resultant admittance is the sum of the individual admittances. The conductance components are added directly, as are the susceptance components. In this case, 1.0 − j1.2 + j1.2 = 1.0 + j0 siemen. Thus, the line from the point of stub connection to the transmitter will be terminated in a load which offers a perfect match. When determining the physical line lengths for stub matching, it is important to remember that the velocity factor for the type of line in use must be considered.

## MATCHING WITH LUMPED CONSTANTS

It was pointed out earlier that the purpose of a matching stub is to cancel the reactive component of line impedance at the point of connection. In other words, the stub is simply a reactance of the proper kind and value shunted across the line. It does not matter what physical shape this reactance takes. It can be a section of transmission line or a "lumped" inductance or capacitance, as desired. In the above example with the Smith Chart solution, a capacitive reactance was required. A capacitor having the same value of reactance can be used just as well. There are cases where, from an installation standpoint, it may be considerably more convenient to use a capacitor in place of a stub. This is particularly true when open-wire feeders are used. If a variable capacitor is used, it becomes possible to adjust the capacitance to the exact value required.

The proper value of reactance may be determined from Smith Chart information. In the previous example, the required susceptance, ***normalized,*** was +j1.2 siemens. This is converted into actual siemens by dividing by the line $Z_0$; 1.2/52 = 0.023 siemen, capacitance. The required capacitive reactance is the reciprocal of this latter value, 1/0.023 = 43.5 ohms. If the frequency is 14.2 MHz, for instance, 43.5 ohms corresponds to a capacitance of 258 pF. A 325-pF variable capacitor connected across the line 0.082 wavelength from the antenna terminals would provide ample adjustment range. The rms voltage across the capacitor is $E = \sqrt{P \bullet Z_0}$ and for 500 watts, for example, would be $E = \sqrt{500 \times 52} = 161$ volts. The peak voltage is 1.41 times the rms value, or 227 volts.

## FLEXIBLE SECTIONS FOR ROTATABLE ARRAYS

When open-wire transmission line is used there is likely to be trouble with shorting or grounding of feeders in rotatable arrays unless some special precautions are taken. Usually some form of insulated flexible line is connected between the antenna and a stationary support at the top of the tower or mast on which the antenna is mounted.

Such a flexible section can take several forms, and it can be made to do double duty. Probably the most satisfactory system for arrays that are not designed to be fed with coaxial line, is to use a flexible section of coax with coaxial baluns at both ends. The outer conductor of the coax may be grounded to the tower or to the beam antenna framework, wherever it is advantageous to do so. Such a flexible section is shown in Fig. 43. If the coaxial section is made any multiple of a half wave in electrical length, the impedance of the array will be repeated at the bottom of the flexible section.

Another method is to use twin-lead for the flexible section. The 300-ohm tubular type designed for transmitting applications is recommended. Here, again, half-wave sections repeat the antenna feed impedance at the bottom end. The twin-lead section may also be made an odd multiple of a quarter wavelength, in which case it will act as a Q section, giving an impedance step-down between a 450-ohm line and an antenna impedance of 200 ohms.

## GROUND-PLANE ANTENNAS

The same principles discussed earlier

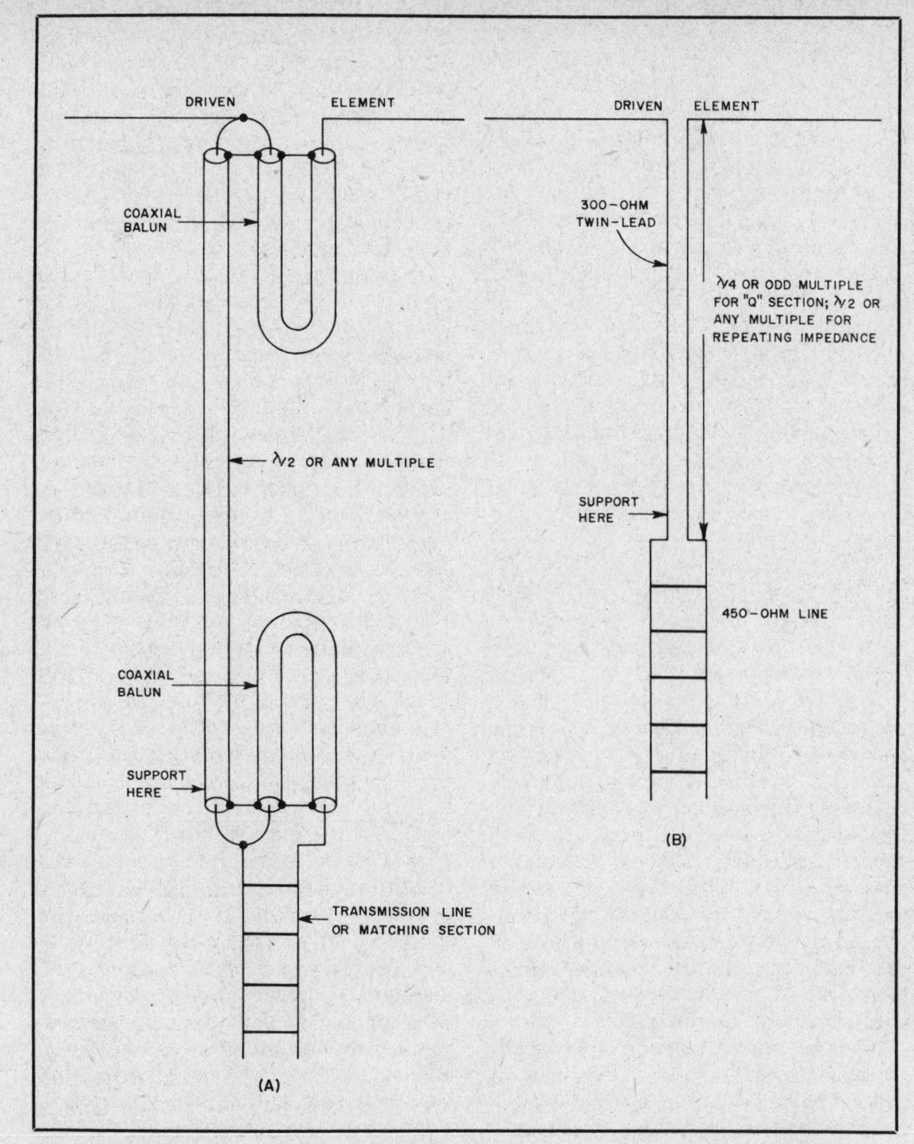


Fig. 43 — Flexible sections for rotatable arrays. Coax may be used, as at A. If the coax section is any multiple of a half wavelength, the antenna impedance will be repeated at the bottom end. Twin-lead may be used either as a Q section or as an impedance repeater, as shown in B.

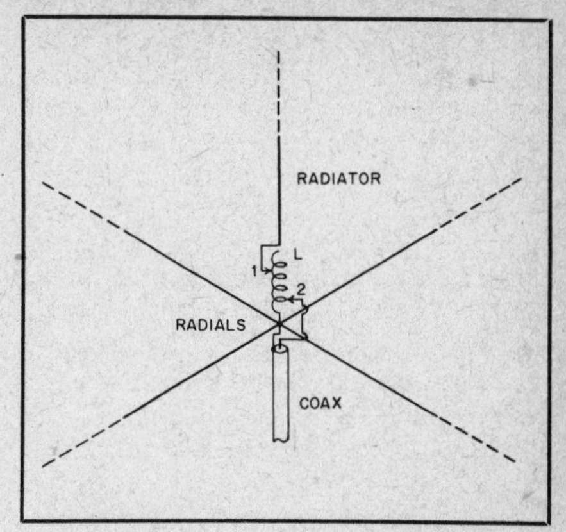


Fig. 44 — Matching to ground-plane antenna by tapped coil. This requires that the antenna (but not radials) be shorter than the resonant length.

also apply to an unsymmetrical system such as the grounded antenna or the ground-plane antenna. In the case of the quarter-wave ground-plane antenna a straightforward design procedure for matching is possible because the radiation resistance is essentially independent of the physical height of the system (provided the radiator is reasonably clear of other conductors in the vicinity) and there is no ground-connection resistance to be included in the total resistance to be matched.

The ground-plane antenna lends itself well to direct connection to coaxial line, so this type of line is nearly always used. Several matching methods are available. If the antenna length can be adjusted to resonance, the stub matching system previously described is convenient.

A second method of matching, particularly convenient for small antennas (28 MHz and higher frequencies) mounted on top of a supporting mast or pole, requires shortening the antenna to the high-frequency side of resonance so that it shows a particular value of capacitive reactance at its base. The antenna terminals are then shunted by an inductive reactance, which may have the physical form either of a coil or a closed stub, to restore resonance and simultaneously transform the radiation resistance to the proper value for matching the transmission line. This concept is the same as for the hairpin match, described in detail earlier.

## Tapped-Coil Matching

The matching arrangement shown in Fig. 44 is a more general form of the method just mentioned, in that it does not require adjusting the radiator height to an exact value. The radiator must be shortened so that the system will show capacitive reactance, but any convenient amount of shortening can be used. This system is particularly useful on lower frequencies where it may not be possible to obtain a height approximating a quarter wavelength.

The antenna impedance is matched to the characteristic impedance of the line by adjusting the taps on L. As a preliminary adjustment, before attaching the line tap (2), the radiator tap (1) may be set for resonance at the operating frequency as indicated by a dip meter coupled to L. The line tap (2) is then moved along the coil to find the point that gives minimum standing-wave ratio as indicated by an SWR indicator. To bring the SWR down to 1 to 1 it will usually be necessary to make a small readjustment of the radiator tap (1) and perhaps further "touch up" the line tap (2), since the adjustments interact to some extent.

This method is equivalent to tapping down on a parallel-resonant circuit to match a low value of resistance to a higher value connected across the whole circuit. The antenna impedance can be represented by a capacitance in parallel with a resistance which is much higher than the actual radiation resistance. The transformation of resistance comes about by using the parallel equivalent of the radiation resistance and capacitive reactance in series, using the relationship given in Chapter 3.

## Matching by Length Adjustment

Still another method of matching may be used when the antenna length is not fixed by other considerations. As shown in Chapter 2 under "Grounded Antennas," the radiation resistance as measured at the base of a ground-plane antenna increases with the antenna height, and it is possible to choose a height such that the base radiation resistance will equal the $Z_0$ of the transmission line to be used. The heights of most interest are a little over

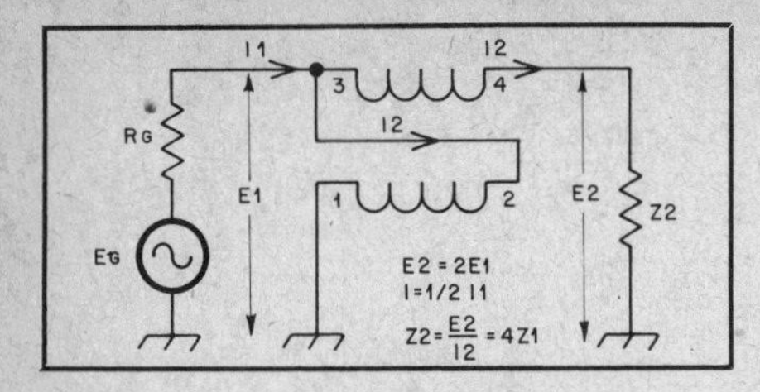


Fig. 45 — Four-to-one broadband bifilar transformer. Upper winding can be tapped at appropriate points to obtain other ratios such as 1.5:1, 2:1, and 3:1.

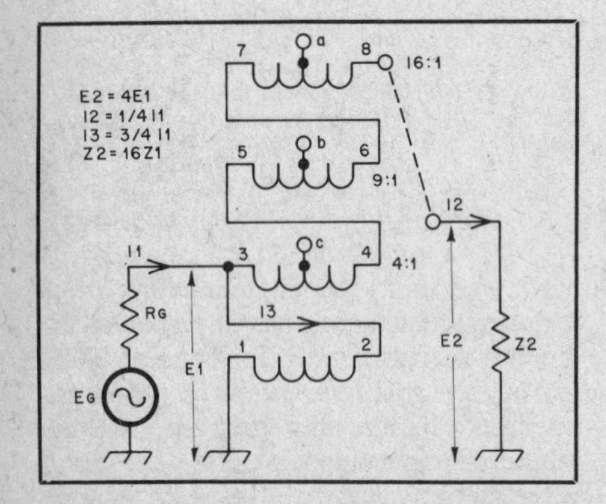


Fig. 46 — Four-winding, broadband, variable impedance transformer. Connections a, b, and c can be placed at appropriate points to yield various ratios from 1.5:1 to 16:1.

Fig. 47 — A 4-winding, wideband transformer (with front cover removed) with connections made for matching ratios of 4:1, 6:1, 9:1 and 16:1. The 6:1 ratio is the top coaxial connector and, from left to right, 16:1, 9:1, and 4:1 are the others. There are 10 (quadrifilar) turns of no. 14 enameled wire on a Q1, 2.5 inch OD ferrite core.

100 degrees (0.28 wavelength), where the resistance is approximately 52 ohms, and about 113 degrees (0.32 wavelength), where the resistance is 75 ohms, to match the two common types of coaxial line. These heights are quite practicable for ground-plane antennas for 14 MHz and higher frequencies. The lengths (heights) in degrees as given above do not require any correction for length/diameter ratio; i.e., they are free-space lengths.

Since the antenna is not resonant at these lengths, its input impedance will be reactive as well as resistive. The reactance must be tuned out in order to make the line see a purely resistive load equal to its characteristic impedance. This can be done with a series capacitor of the proper value. The approximate value of capacitive reactance required, for antennas of typical length/diameter ratio, is about 100 ohms for the 52-ohm case and about 200 ohms for the 75-ohm case. The corresponding capacitance values for the frequency in question can be determined from appropriate charts or by equation. Variable capacitors of sufficient range should be used.

In an analysis by Robert Stephens, W3MIR, the impedance and reactance versus height data of Chapter 2 for ground-plane antennas has been converted to conductance and susceptance information. For parallel-equivalent matching, an input conductance of 1/52 siemen is needed. For an antenna length-to-diameter ratio of approximately 1000, there are two heights for which the conductance is this value — one at 0.234 $\lambda$ and the other at 0.255$\lambda$. At the shorter height, the susceptance is positive (capacitive), 0.0178 siemen, and at the longer it is negative, 0.0126 siemen. As far as radiation is concerned, one is as good as the other and the choice becomes the one of the simpler mechanical approach for the particular antenna. If the antenna is made 0.234 wavelength high, its capacitive reactance may be canceled with a shunt inductor having a reactance of 56 ohms at the operating frequency, resulting in a 52-ohm termination for the feed line. If the antenna is 0.255 wavelength, a 52-ohm match may be obtained with a shunt capacitor of 79 ohms. Similarly, a match may be obtained for 75-ohm coax with a height of 0.23 wavelength and a shunt inductor of 56 ohms, or with a height of 0.26 wavelength and a shunt capacitor of 78 ohms. These reactance values may be obtained with lumped constants or with stubs, as described earlier. As mentioned above, these heights do not require any correction factor; they are free-space lengths.

The adjustment of systems like these requires only that the capacitance or inductance be varied until the lowest possible SWR is obtained. If the lengths mentioned above are used, the SWR should be close enough to 1 to 1 to make a fine adjustment of the length unnecessary.

**Broadband Matching Transformers**

Broadband transformers have been used widely because of their inherent bandwidth ratios (as high as 20,000:1) from a few tens of kilohertz to over a thousand megahertz. This is possible because of the transmission-line nature of the windings. The interwinding capacitance is a component of the characteristic impedance and therefore, unlike the conventional transformer, forms no resonances which seriously limit the bandwidth. At low frequencies, where interwinding capacitances can be neglected, these transformers are similar in operation to the conventional transformer. The main difference (and a very important one from a power standpoint) is that the windings tend to cancel out the induced flux in the core. Thus, high-permeability ferrite cores, which are not only highly nonlinear but also suffer serious damage even at flux levels as low as 200 to 500 gauss, can be used. This greatly extends the low-frequency range of performance. Since higher permeability also permits fewer turns at the lower frequencies, high-frequency performance is also improved since the upper cutoff is determined mainly from transmission-line considerations. At the high-frequency cutoff, the effect of the core is negligible.

Bifilar matching transformers lend themselves to unbalanced operation. That is, both input and output terminals can have a common ground connection. This eliminates the third magnetizing winding required in balanced-to-unbalanced (balun) operation. (See Chapter 4 for a discussion of baluns.) By adding third and fourth windings, as well as tapping windings at appropriate points, various combinations of broadband matching can be obtained. Fig. 45 shows a 4:1 unbalanced-to-unbalanced configuration. No. 14 wire can be used and it will easily handle 1000 watts of power. By tapping at points 1/4, 1/2 and 3/4 of the way along the top winding, ratios of approximately 1.5:1, 2:1 and 3:1 can also be obtained. It should be noted that one of the wires should be covered with vinyl electrical tape in order to prevent voltage breakdown between the windings. This is necessary when a step-up ratio is used at high power to match antennas with impedances greater than 50 ohms.

Fig. 46 shows a transformer with four windings, permitting wide-band matching ratios as high as 16:1. Fig. 47 shows a four-winding transformer with taps at 4:1, 6:1, 9:1, and 16:1. In tracing the current flow in the windings when using the 16:1 tap, one sees that the top three windings carry the same current. The bottom winding, in order to maintain the proper potentials, sustains a current three times greater. The bottom current cancels out the core flux caused by the other three windings. If this transformer is used to match into low impedances, such as 3 to 4 ohms, the current in the bottom winding can be as high as 15 amperes if the high side of the transformer is fed with 50-ohm cable handling a kilowatt of power. If one needs a 16:1 match like this at high power, then cascading two 4:1 transformers is recommended. In this case, the transformer at the lowest impedance side only requires each winding to handle 7.5 A. Thus, even no. 14 wire would suffice in this application.

The popular cores used in these applications are 2.5 inch OD ferrites of Q1 and Q2 material and powdered iron cores, of 2-inch OD. The permeabilities of these cores, $\mu$, are nominally 125, 40 and 10 respectively. Powdered-iron cores of permeabilities 8 and 25 are also available.

In all cases these cores can be made to operate over the 10- to 160-meter bands with full power capability and very low loss. The main difference in their design is that lower permeability cores require more turns at the lower frequencies. For example, Q1 material required 10 turns to cover the 160-meter band. Q2 required 12 turns, and powdered iron ($\mu$ = 10) required 14 turns. Since the more common powdered-iron core is generally smaller in diameter and requires more turns because of lower permeability, higher ratios are sometimes difficult to obtain because of physical limitations. When you are working with low impedance levels, unwanted parasitic inductances come into play, particularly on 14 MHz and above. In this case lead lengths should be kept to a minimum.

## BANDWIDTH AND ANTENNA Q

Although more properly a subject for discussion in connection with antenna fundamentals, the bandwidth of the antenna is considered here because as a practical matter the change in antenna impedance with frequency is reflected as a change in the standing-wave ratio on the transmission line. Thus, when an antenna is matched to the line at one frequency — usually in the center of the band of frequencies over which the antenna is to be used — a shift in the operating frequency will be accompanied by a change in the SWR. This would not occur if the antenna impedance were purely resistive and constant regardless of frequency, but unfortunately no practical antennas are that "flat."

In the frequency region around resonance the resistance change is fairly small and, by itself, would not affect the SWR enough to matter, practically. The principal cause of the change in SWR is the change in the reactive component of the antenna impedance when the frequency is varied. If the reactance changes rapidly with frequency the SWR will rise rapidly off resonance, but if the rate of reactance change is small the shift in SWR likewise will be small. Hence an antenna that has a relatively slow rate of reactance change will cover a wider-frequency band, for a given value of SWR at the band limits (such as 2 to 1 or 3 to 1), than one having a relatively rapid rate of reactance change.

In the region around the resonant frequency of the antenna the impedance as measured at a current loop varies with frequency in essentially the same way as the impedance of a series-resonant circuit using lumped constants. It is therefore possible to define a quantity Q for the antenna in the same way as Q is defined in a series-resonant circuit. The Q of the antenna is a measure of the antenna's selectivity, just as the Q of an ordinary circuit is a measure of its selectivity.

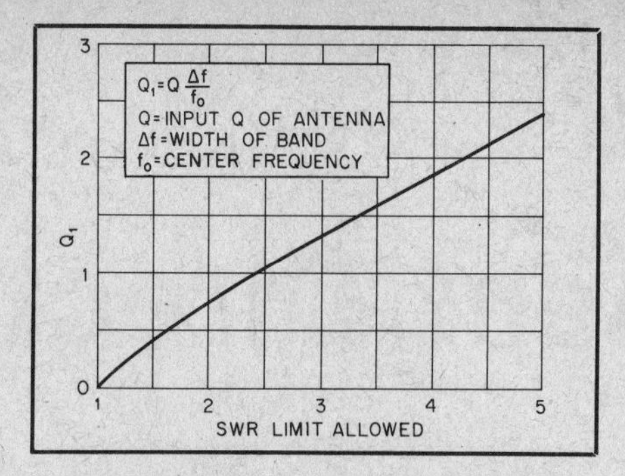


Fig. 48 — Bandwidth in terms of SWR limits, as a function of antenna is Q. The inset formula gives an effective Q. Q1 is determined by the fractional band (Δf/fo) and the actual antenna Q as defined in the text.

The Q of the antenna can be found by measuring its input resistance and reactance at some frequency close to the resonant frequency (at exact resonance the antenna is purely resistive and there is no reactive component). Then, for frequency changes of less than 5 percent from the exact resonant frequency, the antenna Q is given with sufficient accuracy by the following formula:

$$Q = \frac{X}{R} \cdot \frac{1}{2n}$$

where X and R are the measured reactance and resistance and n is the percentage difference, expressed as a decimal, between the antenna resonant frequency and the frequency at which X and R were measured. For example, if the frequency used for the measurement differs from the resonant frequency by 2 percent, n = 0.02.

For an ordinary half-wave dipole, the approximate Q values vary from about 14 for a length/diameter ratio of 25,000, to about 8 for a ratio of 1250. In parasitic arrays with close spacing between elements the input Q may be well over 50, depending on the spacing and tuning (See parasitic arrays in Chapter 6).

### SWR vs. Q

If the Q of the antenna is known, the variation in SWR over the operating band can be determined from Fig. 48. It is assumed that the antenna is matched to the line at the center frequency of the band. Conversely, if a limit is set on the SWR, the width of the band that can be covered can be found from Fig. 48. As an example, suppose that a dipole having a Q of 15 (more or less typical of a wire antenna) is to be used over the 3.5-4 MHz band and that it is matched with a 1-to-1 SWR at the band center. Then Δf/fo = 0.5/3.75 = 0.133 and Q1 = 15 × 0.133 = 2. The SWR that can be expected at the band edges, 3.5 and 4 MHz, is shown by the chart to be a bit over 4 to 1. If it should be decided arbitrarily that no more than a 2-to-1 standing-wave ratio is allowable, Q1 is 0.75 and from the formula in Fig. 48 the total bandwidth is found to be 187.5 kHz.

### Effect of Matching Network

The measurement of resistance and reactance to determine Q should be made at the input terminals of the matching network, if one is required. The selectivity of the matching network has just as much effect on the bandwidth, in terms of SWR on the line, as the selectivity of the antenna itself. Where the greatest possible bandwidth is wanted a low-Q matching network must be used. This is not always controllable, particularly when the antenna resistance differs considerably from the $Z_0$ of the line to which it is to be matched. A large impedance ratio usually means that large values of reactance must be used in the matching section; in other words, the Q of the matching section in such cases tends to be higher than desirable. Simple systems having direction matching, such as a dipole fed with 75-ohm line or a folded dipole matched to the line, will have the greatest bandwidth, other things being equal, because no matching network is required.

## BIBLIOGRAPHY

Source material and more extended discussion of topics covered in this chapter can be found in the references given below and in the textbooks listed at the end of Chapter 2.


Eggers, B.A., "An Analysis of the Balun, " April 1980 *QST*.

Geiser, D., "Resistive Impedance Matching with Quarter-Wave Lines," February 1964 *QST*.

Gooch, J. D., Gardner, O. E. and Roberts, G. L., "The Hairpin Match," April 1962 *QST*.

Grammer, G., "Simplified Design of Impedance-Matching Networks," March, April and May 1957 *QST*.

Healey, D., "An Examination of the Gamma Match," April 1969 *QST*.

Kraus, J. D. and Sturgeon, S.S., "The T-Matched Antenna," September 1940 *QST*.

Maxwell, M. W., "Another Look at Reflections," *QST*, April, June, August and October 1973; April and December 1974; and August 1976.

Regier, F. A., "Series-Section Transmission-Line Impedance Matching," July 1978 *QST*.

Sevick, J., "Simple Broadband Matching Networks," January 1976 *QST*.

Stephens, R. E., "Admittance Matching the Ground-Plane Antenna to Coaxial Transmission Line," Technical Correspondence, April 1973 *QST*.

# Chapter 6

# Multielement Directive Arrays

The gain and directivity offered by an array of elements represents a worthwhile improvement both in transmitting and receiving. Power gain in an antenna is the same as an equivalent increase in the transmitter power. But, unlike increasing the power of one's own transmitter, it works equally well on signals received from the favored direction. In addition, the directivity reduces the strength of signals coming from the directions not favored, and so helps discriminate against a good deal of interference.

One common method of securing gain and directivity is to combine the radiation from a group of half-wave dipoles in such a way as to concentrate it in a desired direction. The way in which such combinations affect the directivity has been explained in Chapter 2. A few words of additional explanation may help make it clear how power gain is achieved.

In Fig. 1, imagine that the four circles, A, B, C and D, represent four dipoles so far separated from each other that the coupling between them is negligible. The point P is supposed to be so far away from the dipoles that the distance from P to each one is exactly the same (obviously P would have to be much farther away than it is shown in this drawing). Under these conditions the fields from all the dipoles will add up at P if all four are fed rf currents in the same phase.

Let us say that a certain current, I, in dipole A will produce a certain value of field strength, E, at the distant point P. The same current in any of the other dipoles will produce the same field at P. Thus, if only dipoles A and B are operating, each with a current I, the field at P will be 2E. With A, B and C operating, the field will be 3E, and with all four operating with the same I, the field will be 4E. Since the power received at P is proportional to the square of the field strength, the relative power received at P is 1, 4, 9 and 16, depending on whether one, two, three or four dipoles are operating.

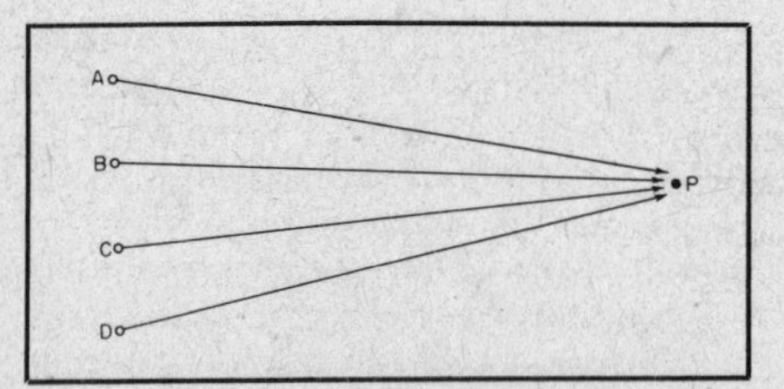

Fig. 1 — Fields from separate antennas combine at a distant point, P, to produce a field strength that exceeds the field produced by the same power in a single antenna

**Table 1**
**Comparison of Dipoles with Negligible Coupling (See Fig. 1)**

| *Dipoles* | *Relative Output Power* | *Relative Input Power* | *Power Gain* | *Gain in dB* |
|---|---|---|---|---|
| A only | 1 | 1 | 1 | 0 |
| A and B | 4 | 2 | 2 | 3 |
| A, B and C | 9 | 3 | 3 | 4.8 |
| A, B, C and D | 16 | 4 | 4 | 6 |

Now, since all four dipoles are alike and there is no coupling between them, the same power must be put into each in order to cause the current I to flow. For two dipoles the relative power input is 2, for three dipoles it is 3, for four dipoles 4, and so on. The gain in each case is the relative received (or output) power divided by the relative input power. Thus we have the results shown in Table 1. The power gain is directly proportional to the number of elements used.

It is well to have clearly in mind the conditions under which this relationship is true:

1) The fields from the separate antenna elements must be in phase at the receiving point.

2) The currents in all elements must be identical.

3) The elements must be separated in such a way that the current induced in one by another is negligible, i.e., the radiation resistance of each element must be the same as it would have been had the other elements not been there.

Very few antenna arrays meet all these conditions exactly. However, it may be said that the power gain of a directive array consisting of dipole elements in which optimum values of element spacing are used is approximately proportional to the number of elements. It is possible, though, for an estimate based on this rule to be in error by a factor of two or more.

## DEFINITIONS

The "element" in a multielement directive array is usually a half-wave dipole or a quarter-wave vertical element above ground. The length is not always an exact electrical half or quarter wavelength, because in some types of arrays it is desirable that the element show either inductive or capacitive reactance. However, the departure in length from resonance is ordinarily small (not more than 5%, in the usual case) and so has no appreciable effect on the radiating properties of the element.

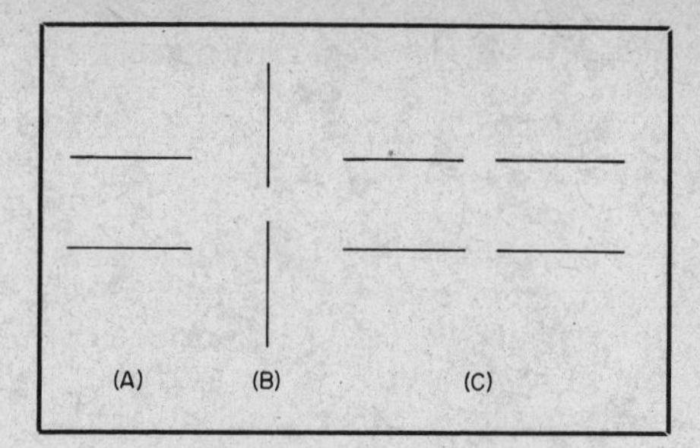


Fig. 2 — Parallel (A) and collinear (B) antenna elements. The array shown at C combines both parallel and collinear elements.

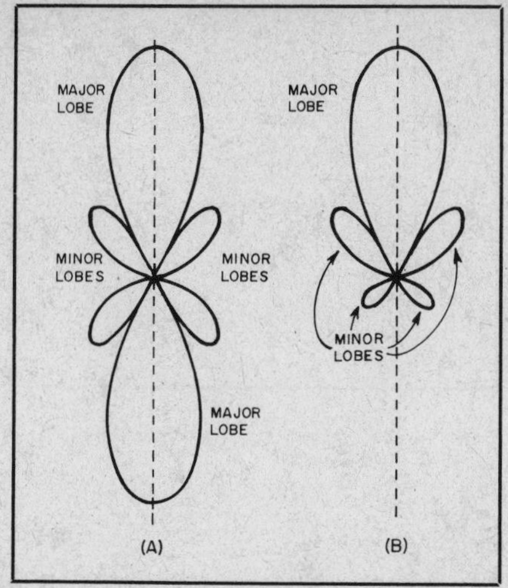


Fig. 4 — Typical bidirectional pattern (A) and unidirectional directive pattern (B). These drawings also illustrate the application of the terms "major" and "minor" to the pattern lobes.

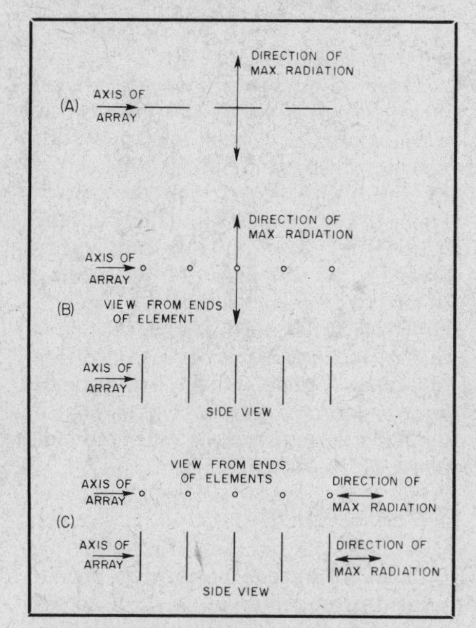


Fig. 3 — Representative broadside arrays are shown at A and B, the first with collinear elements, the second with parallel elements. An end-fire array is shown at C. Practical arrays may combine both broadside and end-fire directivity, including both parallel and collinear elements.

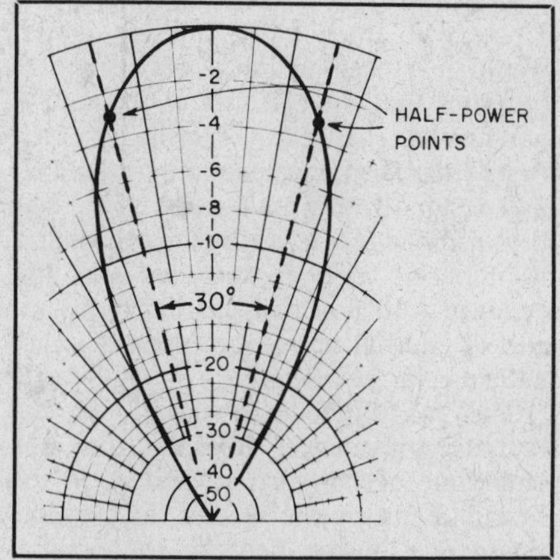


Fig. 5 — The width of a beam is the angular distance between the directions at which the received or transmitted power is one half the maximum power (–3dB).

Antenna elements in multielement arrays of the type considered in this chapter are always either parallel, as at A in Fig. 2, or collinear (end-to-end), Fig. 2B. Fig. 2C shows an array combining both parallel and collinear elements. The elements can be either horizontal or vertical, depending on whether horizontal or vertical polarization is desired. Except for space communications, there is seldom any reason for mixing polarization, so arrays are customarily constructed with all elements similarly polarized.

A *driven element* is one supplied power from the transmitter, usually through a transmission line. A *parasitic element* is one that obtains power solely through coupling to another element in the array because of its proximity to such an element.

A *driven array* is one in which all the elements are driven elements. A *parasitic array* is one in which one or more of the elements are parasitic elements. At least one element in a parasitic array has to be a driven element, since it is necessary to introduce power into the array.

A *broadside array* is one in which the principal direction of radiation is perpendicular to the axis of the array and to the plane containing the elements. An *end-fire array* is one in which the principal direction of radiation coincides with the direction of the array axis. These definitions are illustrated by Fig. 3.

A *bidirectional array* is one that radiates equally well in either direction along the line of maximum radiation. A bidirectional pattern is shown in Fig. 4A. A *unidirectional array* is one that has only one principal direction of radiation, as illustrated by the pattern in Fig. 4B.

The *major lobes* of the directive pattern are those in which the radiation is maximum. Lobes of lesser radiation intensity are called *minor lobes*. The *beamwidth* of a directive antenna is the width, in degrees, of the major lobe between the two directions at which the relative radiated power is equal to one half its value at the peak of the lobe. At these "half-power points" the field intensity is equal to 0.707 times its maximum value, or down 3 dB from maximum. Fig. 5 is an example of a lobe having a beamwidth of 30 degrees.

Unless specified otherwise, the term "gain" as used in this chapter is the power gain over a half-wave dipole of the same orientation and height as the array under discussion, and having the same power input. Gain may either be measured experimentally or determined by calculation. Experimental measurement is difficult and often subject to considerable error, since in addition to the normal errors in measurement (the accuracy of simple rf measuring equipment is relatively poor, and even high-quality instruments suffer in accuracy compared with their low-frequency and dc counterparts) the accuracy depends considerably on conditions — the antenna site, including height, terrain characteristics, and surroundings — under which the measurements are made. Calculations are frequently based on the measured or theoretical directive patterns (see Chapter 2) of the antenna. An approximate formula often used is

$$G = 10 \log \frac{25,000}{\theta_H \theta_V}$$

where

$G$ = decibel gain over a dipole in its favored direction
$\theta_H$ = horizontal half-power beamwidth in degrees
$\theta_V$ = vertical half-power beamwidth in degrees.

This formula, strictly speaking, applies only to antennas having approximately equal and narrow E and H-plane (see Chapter 2) beamwidths — up to about 20 degrees — and no large minor lobes. The error may be considerable when the formula is applied to simple directive antennas having relatively large beamwidths. The error is in the direction of making the calculated gain larger than the actual gain.

*Front-to-back ratio* is the ratio of the power radiated in the favored direction to the power radiated in the opposite direction.

### Phase

The term "phase" has the same meaning when used in connection with the currents flowing in antenna elements as it does in ordinary circuit work. For example, two currents are in phase when they reach their maximum values, flowing in the same direction, at the same instant.

The direction of current flow depends on the way in which power is applied to the element.

This is illustrated in Fig. 6. Assume that by some means an identical voltage is applied to each of the dipoles at the end marked A. Assume also that the instantaneous polarity of the voltage is such that the current is flowing away from the point at which the voltage is applied. The arrows show the assumed current directions. Then the currents in elements 1 and 2 are completely in phase, since they are flowing in the same direction in space and are caused by the same voltage. However, the current in element 3 is flowing in the *opposite* direction in space because the voltage is applied to the opposite end of the element. The current in element 3 is therefore 180 degrees out of phase with the currents in elements 1 and 2.

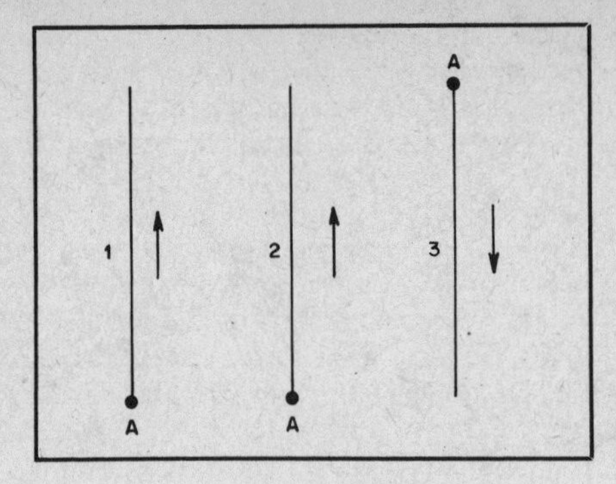


Fig. 6 — Illustrating phasing of currents in antenna elements.

The phasing of driven elements depends on the direction of the element, the phase of the applied voltage, and the point at which the voltage is applied. In most systems used by amateurs the voltages applied to the elements are practically always exactly in or exactly out of phase with each other. Also, the axes of the elements are nearly always in the same direction, since parallel or collinear elements are invariably used. The currents in driven elements in such systems are therefore always either exactly in or out of phase with the currents in other elements.

It is possible to use phase differences of less than 180 degrees in driven arrays. One important case is where the voltage applied to one set of elements differs by 90 degrees from the voltage applied to another set. However, making provision for proper phasing in such systems is considerably more of a problem than in the case of simple 0- or 180-degree phasing.

In parasitic arrays the phase of the currents in the parasitic elements depends on the spacing and tuning, as described later.

### Ground Effects

The effect of the ground is the same with a directive antenna as it is with a simple dipole antenna. The reflection factors discussed in Chapter 2 may therefore be applied to the vertical pattern of an array, subject to the same modifications mentioned in that chapter. In cases where the array elements are not all at the same height, the reflection factor for the *mean* height of the array must be used. The mean height is the average of the heights measured from the ground to the centers of the lowest and highest elements.

## MUTUAL IMPEDANCE

Consider two half-wave dipoles that are fairly close to each other. When power is applied to one and current flows, a voltage will be induced in the second by the electromagnetic field, and current will flow in it as well. The current in antenna no. 2 will in turn induce a voltage in antenna no. 1, causing a current to flow in the latter. The total current in no. 1 is then the sum (taking phase into account) of the original current and the induced current.

If the voltage applied to antenna no. 1 has not changed, the fact that the amplitude of the current flowing is different, with antenna no. 2 present, than it would have been had no. 2 not been there indicates that the presence of the second antenna has changed the impedance of the first. This effect is called mutual coupling, and results in a mutual impedance. The actual impedance of an antenna element is the sum of its self-impedance (the impedance with no other antennas present) and its mutual impedance with all other antennas in the vicinity.

The magnitude and nature of the mutual impedance depends on the amplitude of the current induced in the first antenna by the second, and on the phase relationship between the original and induced currents. The amplitude and phase of the induced current depend on the spacing between the antennas and whether or not the second antenna is tuned to resonance.

### Amplitude of Induced Current

The induced current will be largest when the two antennas are close together and are parallel. Under these conditions the voltage induced in the second antenna by the first, and in the first by the second, has its greatest value and causes the largest current flow. The coupling decreases as the parallel antennas are moved farther apart.

The coupling between collinear antennas is comparatively small, and so the mutual impedance between such antennas is likewise small. It is not negligible, however.

### Phase Relationships

When the separation between the two antennas is an appreciable fraction of a wavelength, a measurable period of time elapses before the field from antenna no. 1 reaches antenna no. 2. There is a similar time lapse before the field set up by the current in no. 2 gets back to induce a current in no. 1. Hence the current induced in no. 1 by no. 2 will have a phase relationship with the original current in no. 1 that depends on the spacing between the two antennas.

The induced current can range all the way from being completely in phase with the original current to being completely out of phase with it. In the first case the total current is larger than the original current and the antenna impedance is reduced. In the second, the total current is smaller and the impedance is increased. At intermediate phase relationships the impedance will be lowered or raised depending on whether the induced current is mostly in or mostly out of phase with the original current.

Except in the special cases when the induced current is exactly in or out of phase with the original current, the induced current causes the phase of the total current to shift with respect to the applied voltage. The mutual impedance, in other words, has both resistive and reactive components. Consequently, the presence of a second antenna nearby may cause the impedance of an antenna to be reactive — that is, the antenna will be detuned from resonance — even though its self-impedance is entirely resistive. The amount of detuning depends on the magnitude and phase of the induced current.

### Tuning Conditions

A third factor that affects the impedance of antenna no. 1 when no. 2 is present is the tuning of no. 2. If no. 2 is not exactly resonant the current that flows in it as a result of the induced voltage will either lead or lag the phase it would have had if the antenna were resonant. This causes an additional phase advance or delay that affects the phase of the current induced back in no. 1. Such a phase lag has an effect similar to a change in the spacing between self-resonant antennas. However, a change in tuning is not exactly equivalent to a change in spacing because the two methods do not have the same effect on the amplitude of the induced current.

## MUTUAL IMPEDANCE AND GAIN

The mutual impedance between antennas is important because it determines the amount of current that will flow for a given amount of power supplied. It must be remembered that it is the *current* that determines the field strength from the antenna. Other things being equal, if the mutual impedance between two antennas is such that the currents are greater for the same total power than would be the case if the two antennas were not coupled, the power gain will be greater than that shown in Table 1. On the other hand, if the mutual impedance is such as to reduce the current, the gain will be less than if the antennas were not coupled.

The calculation of mutual impedance

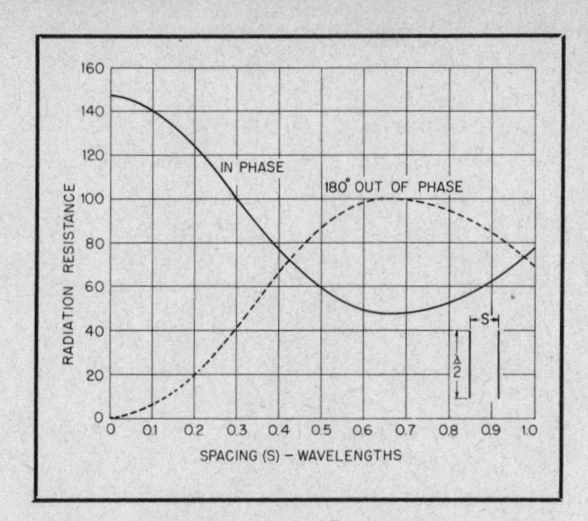


Fig. 7 — Radiation resistance measured at the center of one element as a function of the spacing between two parallel half-wave self-resonant antenna elements. For ground-mounted quarter-wave vertical elements, divide these resistances by two.

between antennas is a difficult problem, but data are available for several special cases. Two simple but important ones are shown in Figs. 7 and 8. These graphs do not show the mutual impedance, but instead show a more useful quantity — the radiation resistance measured at the center of an antenna as it is affected by the spacing between two antennas.

As shown by the solid curve in Fig. 7, the radiation resistance at the center of either dipole, when the two are self-resonant, parallel, and operated in phase, decreases rapidly as the spacing between them is increased until the spacing is about 0.7 wavelength. This is a broadside array. The maximum gain is secured from a pair of such elements when the spacing is in this region, because the current is larger for the same power and the fields from the two arrive in phase at a distant point placed on a line perpendicular to the line joining the two antennas.

The broken curve in Fig. 7, representing two antennas operated 180 degrees out of phase (end-fire), cannot be interpreted quite so simply. The radiation resistance decreases with decreasing spacing in this case. However, the fields from the two antennas add up in phase at a distant point in the favored direction only when the spacing is one-half wavelength (in the range of spacings considered). At smaller spacings the fields become increasingly out of phase, so the total field is less than the simple sum of the two. The latter factor decreases the gain at the same time that the reduction in radiation resistance is increasing it. As shown later in this chapter, the gain goes through a maximum when the spacing is in the region of 1/8 wavelength.

The curve for two collinear elements in phase, Fig. 8, shows that the radiation

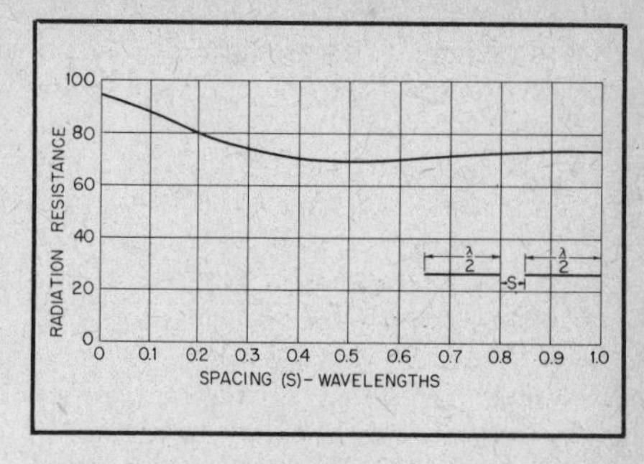


Fig. 8 — Radiation resistance measured at the center of one element as a function of the spacing between the ends of two collinear self-resonant half-wave antenna elements operated in phase.

resistance decreases and goes through a broad minimum in the region of 0.3- to 0.5-wavelength spacing between the adjacent ends of the antennas. Since the minimum is not significantly less than the radiation resistance of an isolated antenna, the gain will not exceed the gain calculated on the basis of uncoupled antennas. That is, the best that two collinear elements will give, even with the optimum spacing, is a power gain of about 2 (3 dB). When the separation between the ends is very small — the usual method of operation — the gain is reduced.

# Driven Arrays

Driven arrays may be either broadside or end-fire, and may consist of collinear elements, parallel elements, or a combination of both. The number of elements that it is practical to use depends on the frequency and the space available for the antenna. Fairly elaborate arrays, using as many as 16 or even 32 elements, can be installed in a rather small space when the operating frequency is in the vhf range, and more at uhf. At lower frequencies the construction of antennas with a large number of elements would be impractical for most amateurs.

Of course the simplest of driven arrays is one with just two elements. If the elements are collinear, they are always fed in phase. The effects of mutual coupling are not great, as illustrated in Fig. 8. Therefore, feeding power to each element in the presence of the other presents no significant problems. This may not be the case when the elements are parallel to each other, but because the combination of spacing and phasing arrangements is infinite, the number of possible radiation patterns is endless. This is illustrated in Fig. 9. When the elements are fed in phase, a broadside pattern always results. At spacings of less than 5/8 wavelength with the elements fed 180° out of phase, an end-fire pattern always results. With intermediate amounts of phase difference, the results cannot be so simply stated. Patterns evolve which are not symmetrical in all four quadrants.

In the plots of Fig. 9 the elements are assumed to have the same length. Further, equal currents are assumed to be flowing in each element, a condition that may be difficult to obtain in practice.

Because of the effects of mutual coupling between the two driven elements, greater or lesser currents will flow in each with changes in spacing and phasing, as described in the previous section. This, in turn, affects the gain of the array in a way which cannot be shown merely by plotting the *shapes* of the patterns, as has been done in Fig. 9. This gain information is also shown, adjacent to the pattern for each combination of spacing and phasing, referenced to a single element. For example, a pair of elements fed in phase at a spacing of a half wavelength will have a gain in the direction of maximum radiation of 4.0 dB over a single element.

It is characteristic of broadside arrays that the power gain is proportional to the length of the array but is substantially independent of the number of elements used, provided the optimum element spacing is not exceeded. This means, for example, that a 5-element array and a 6-element array will have the same gain, provided the elements in both are spaced so that the overall array length is the same. Although this principle is seldom used for the purpose of reducing the number of elements, because of complications introduced in feeding power to each element in the proper phase, it does illustrate the fact that there is nothing to be gained by increasing the number of elements if the space occupied by the antenna is not increased proportionally.

Generally speaking, the maximum gain in the smallest linear dimensions will result when the antenna combines both broadside and end-fire directivity and uses both parallel and collinear elements. In this way the antenna is spread over a greater volume of space, which has the same effect as extending its length to much greater extent in one linear direction.

### FEEDING DRIVEN ARRAYS

Not the least of the problems encountered in constructing multielement driven arrays is that of supplying the required amount of power to each element and making sure that the currents in the elements are in the proper phase. The directive patterns given in this chapter are based on the assumption that each element carries the same current and that the phasing is exact. If the element currents differ, or if the phasing is not proper, the actual

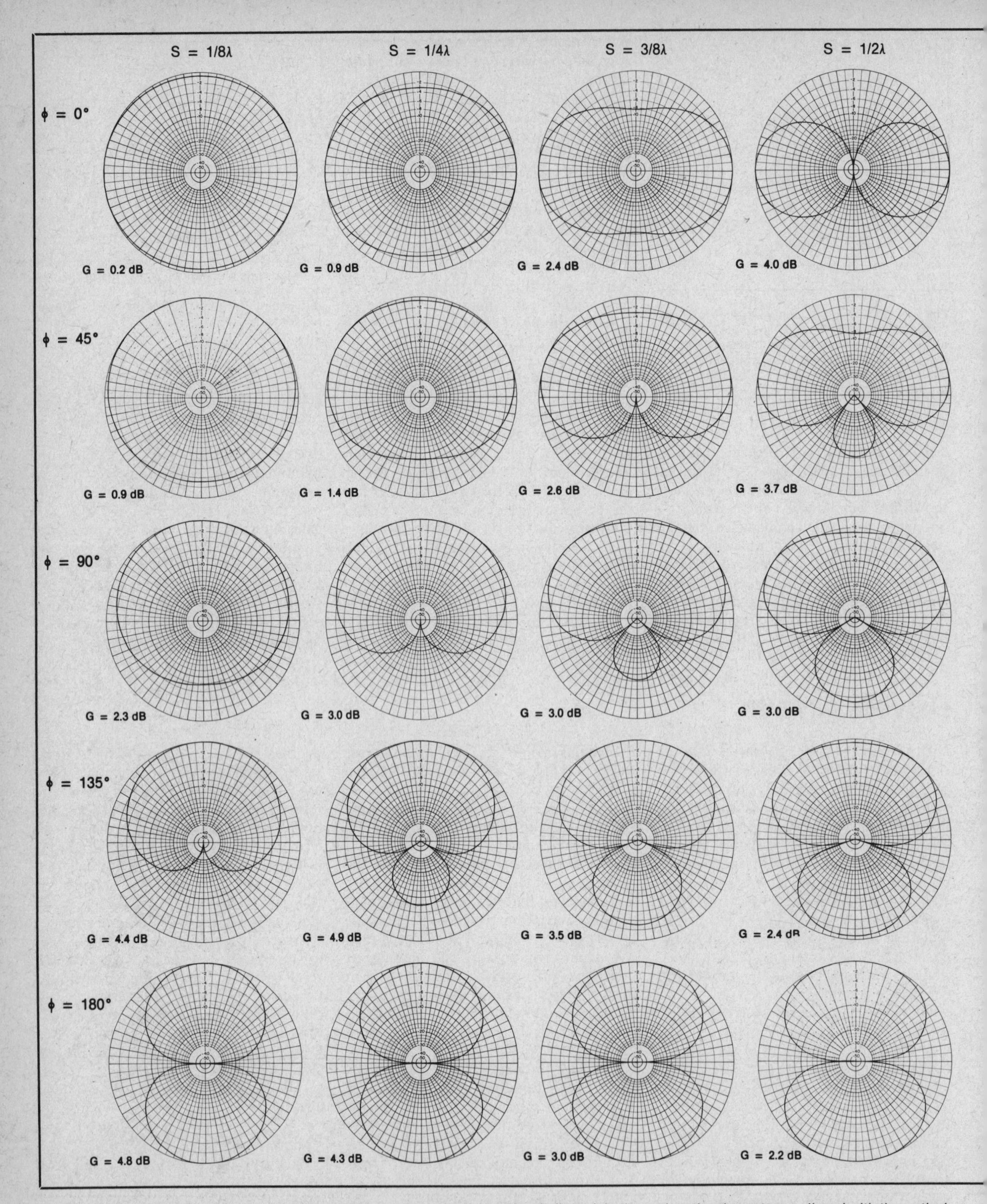


Fig. 9 — H-plane patterns of two parallel driven elements, spaced and phased as indicated. In these plots the elements are aligned with the vertical axis, and the uppermost element is the one of lagging phase at angles other than 0°. The two elements are assumed to be the same length, with exactly equal currents. The gain figure associated with each pattern indicates that of the array over a single element. The plots may be interpreted

directive patterns will not be quite like those shown. Small departures will not greatly affect the gain, but may increase the beamwidth and introduce minor lobes — or emphasize those that exist already.

If the directive properties of beam antennas are to be fully realized, care must be used to prevent antenna currents from flowing on transmission lines (see Chapter 5) used as interconnections between elements, as well as on the main transmission line. If radiation takes place

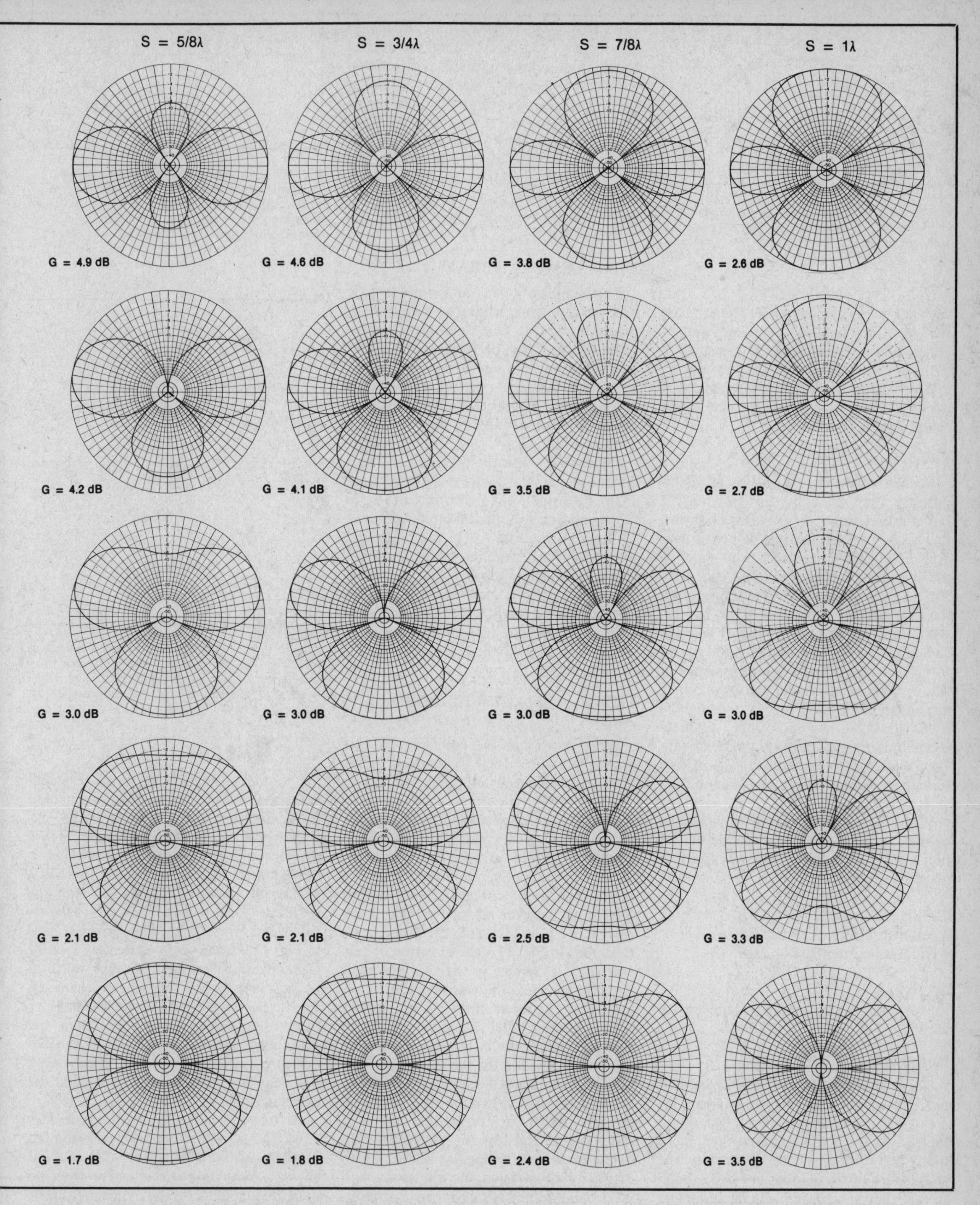


as the horizontal or azimuthal pattern of two vertical elements at a 0° elevation angle, or the free-space pattern of two horizontal elements when viewed on end, with one element above the other.

from these lines or if signals can be picked up on them, the directive effects may be masked by such stray radiation or pickup. Although this may not greatly affect the gain either in transmission or reception, received signals coming from undesired directions will not be suppressed to the extent that is possible with a well-designed system.

## COLLINEAR ARRAYS

Collinear arrays are always operated

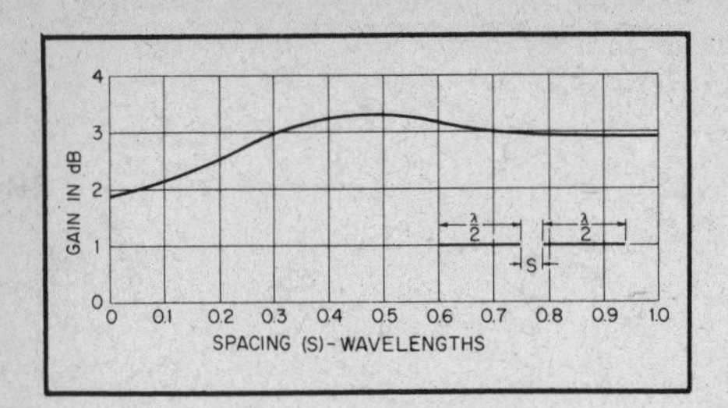


Fig. 10 — Gain of two collinear half-wave elements as a function of spacing between the adjacent ends.

with the elements in phase. (If alternate elements in such an array are out of phase, the system simply becomes a harmonic-type antenna.) A collinear array is a broadside radiator, the direction of maximum radiation being at right angles to the line of the antenna.

### Power Gain

Because of the nature of the mutual impedance between collinear elements the radiation resistance is increased as shown in Fig. 8. For this reason the power gain does not increase in direct proportion to the number of elements. The gain with two elements, as the spacing between them is varied, is shown by Fig. 10. Although the gain is greatest when the end-to-end spacing is in the region of 0.3 to 0.5 wavelength, the use of spacings of this order is inconvenient constructionally and introduces problems in feeding the two elements. As a result, collinear elements are almost always operated with their ends quite close together — in wire antennas, usually with just a strain insulator between.

With very small spacing between the ends of adjacent elements the theoretical power gain of collinear arrays is approximately as follows:

2 collinear elements — 1.9 dB
3 collinear elements — 3.2 dB
4 collinear elements — 4.3 dB

More than four elements are rarely used.

### Directivity

The directivity of a collinear array, in a plane containing the axis of the array, increases with its length. Small secondary lobes appear in the pattern when more than two elements are used, but the amplitudes of these lobes are low enough so that they are not important. In a plane at right angles to the array the directive diagram is a circle, no matter what the number of elements. Collinear operation, therefore, affects only E-plane directivity, the plane containing the antenna. At right angles to the wire the pattern is the same as that of the half-wave elements of which it is composed.

When a collinear array is mounted with the elements vertical, the antenna radiates equally well in all geographical directions. An array of such "stacked" collinear elements tends to confine the radiation to low vertical angles.

If a collinear array is mounted horizontally, the directive pattern in the vertical plane at right angles to the array is the same as the vertical pattern of a simple half-wave antenna at the same height (Chapter 2).

## TWO-ELEMENT ARRAY

The simplest and most popular collinear array is one using two elements, as shown in Fig. 11. This system is commonly known as "two half-waves in phase," and the manner in which the desired current distribution is secured has been described in Chapter 5. The directive pattern in a plane containing the wire axis is shown in Fig. 12.

Depending on the conductor size, height, and similar factors, the impedance at the feed point can be expected to be in the range from about 4000 to 6000 ohms, for wire antennas. If the elements are made of tubing having a low length/diameter ratio, values as low as 1000 ohms are representative. The system can be fed through an open-wire tuned line with negligible loss for ordinary line lengths, or a matching section may be used if desired.

## THREE- AND FOUR-ELEMENT ARRAY

When more than two collinear elements are used it is necessary to connect "phasing" stubs between adjacent elements in order to bring the currents in all elements in phase. It will be recalled from Chapter 2 that in a long wire the direction of current flow reverses in each half-wave section. Consequently, collinear elements cannot simply be connected end to end; there must be some means for making the current flow in the same direction in all elements. In Fig. 13A the direction of current flow is correct in the two left-hand elements because the transmission line is connected between them. The phasing stub between the second and third elements makes the instantaneous current direction correct in the third element. This stub may be looked upon simply as the alternate half-wave section of a long-wire antenna folded back on itself to cancel its radiation. In Fig. 13A the part to the right of the transmission line has a total length of three half wavelengths, the center half wave being folded back to form a quarter-wave phase-reversing stub. No data are available on the impedance at the feed point in this arrangement, but various considerations indicate that it should be over 1000 ohms.

An alternative method of feeding three collinear elements is shown in Fig. 13B. In this case power is applied at the center of the middle element and phase-reversing stubs are used between this element and both of the outer elements. The impedance at the feed point in this case is somewhat over 300 ohms and provides a close match to 300-ohm line. The SWR will be less than 2 to 1 when 600-ohm line is used. Center feed of this type is somewhat preferable to the arrangement in Fig. 13A because the system as a whole is balanced. This assures more uniform power distribution among the elements. In A, the right-hand element is likely to receive somewhat less power than the other two because a portion of the fed power is radiated by the middle element before it can reach the one located at the extreme right.

A four-element array is shown in Fig. 13C. The system is symmetrical when fed between the two center elements as shown. As in the three-element case, no data are available on the impedance at the feed point. However, the SWR with a 600-ohm line should not be much over 2 to 1. Fig. 14 shows the directive pattern of a four-element array. The sharpness of the three-element pattern is intermediate between Figs. 12 and 14, with a small minor lobe at right angles to the array axis.

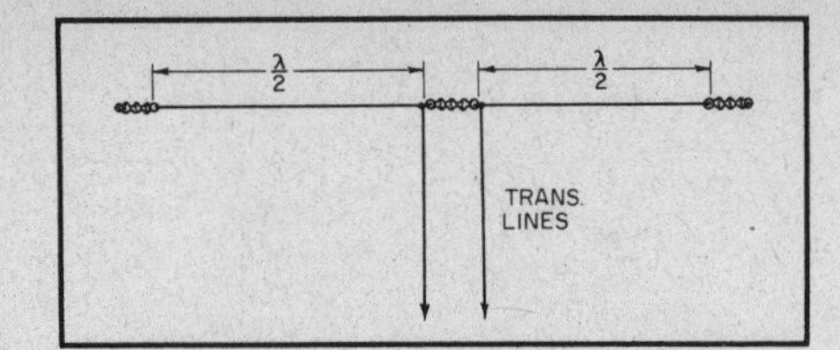


Fig. 11 — A 2-element collinear array ("two half-waves in phase"). The transmission line shown would operate as a tuned line. A matching section can be substituted and a nonresonant line used if desired.

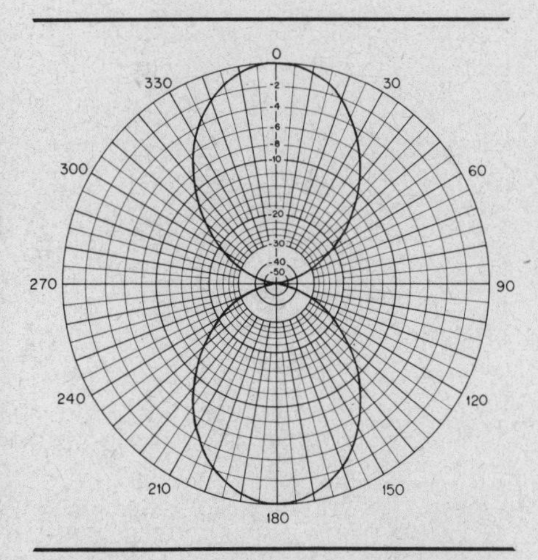

Fig. 12 — Free-space E-plane directive diagram for a two-element collinear array. The axis of the elements lies along the 90°-270° line. This is the horizontal pattern at low wave angles when the array is horizontal.

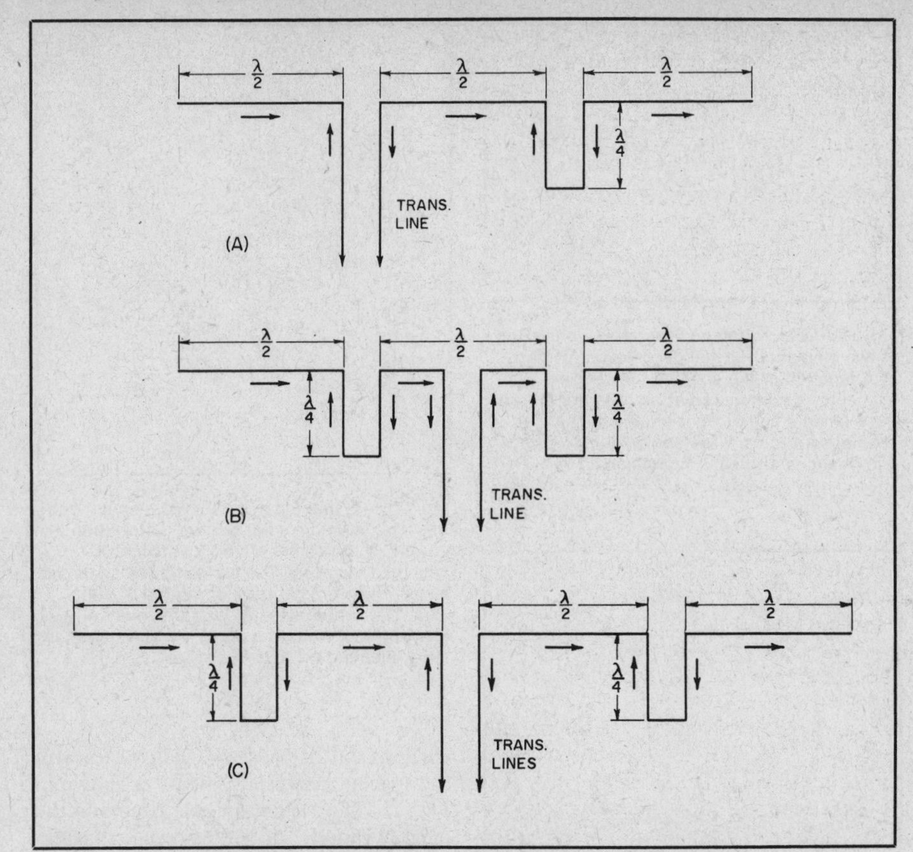

Fig. 13 — Three- and four-element collinear arrays. Alternative methods of feeding a 3-element array are shown at A and B. These drawings also show the current distribution on the antenna elements and phasing stubs. A matched transmission line can be substituted for the tuned line by using a suitable matching section.

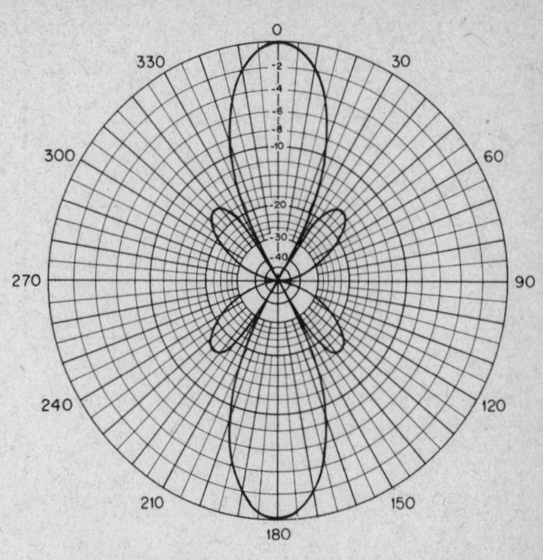

Fig. 14 — E-plane pattern for a 4-element collinear array. The axis of the elements lies along the 90°-270° line.

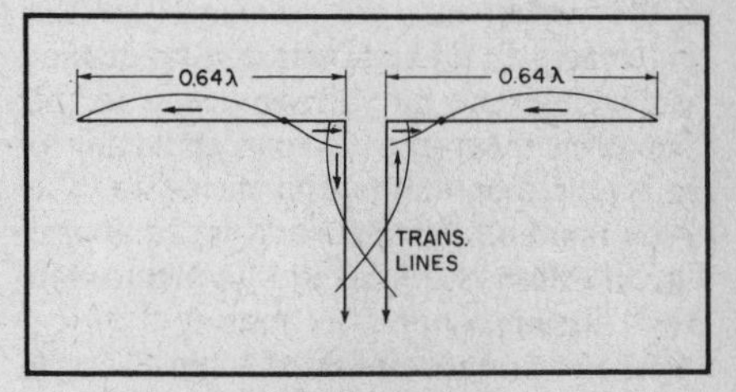

Fig. 15 — The extended double Zepp. This system gives somewhat more gain than two half-wave collinear elements.

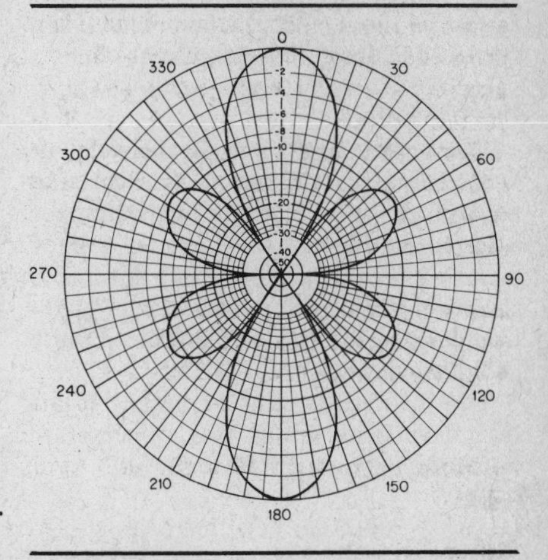

Fig. 16 — E-plane pattern for the extended double Zepp. This is also the horizontal directional pattern when the elements are horizontal. The axis of the elements lies along the 90°-270° line.

Collinear arrays can be extended to more than four elements. However, the simple two-element collinear array is the type most used, since it lends itself well to multiband operation. More than two collinear elements are seldom used because more gain can be obtained from other types of arrays.

### Adjustment

In any of the collinear systems described the lengths of the radiating elements in feet can be found from the formula 468/f(MHz). The lengths of the phasing stubs can be found from the formulas given in Chapter 5 for the type of line used. If the stub is open-wire line (500 to 600 ohms impedance) it is satisfactory to use a velocity factor of 0.975 in the formula for a quarter-wave line. On-the-ground adjustment is, in general, an unnecessary refinement. If desired, however, the following procedure may be used when the system has more than two elements.

Disconnect all stubs and all elements except those directly connected to the transmission line (in the case of feed such as is shown in Fig. 13B leave only the center element connected to the line). Adjust the elements to resonance, using the still-connected element. When the proper length is determined, cut all other elements to the same length. Make the phasing stubs slightly long and use a shorting bar to adjust their length. Connect the elements to the stubs and adjust the stubs to resonance, as indicated by maximum current in the shorting bars or by the standing-wave ratio on the transmission line. If more than three or four elements are used it is best to add elements two at a time (one at each end of the array), resonating the system each time before a new pair is added.

## THE EXTENDED DOUBLE ZEPP

An expedient that may be adopted to obtain the higher gain that goes with wider spacing in a simple system of two collinear elements is to make the elements somewhat longer than 1/2 wavelength. As shown in Fig. 15, this increases the spacing between the two in-phase half-wave sections at the ends of the wires. The section in the center carries a current of opposite phase, but if this section is short the current will be small because it represents only the outer ends of a half-wave section. Because of the small current and short length the radiation from the center is small. The optimum length for each element is 0.64 wavelength. At greater lengths the system tends to act as a long-wire antenna and the gain decreases.

This system is known as the "extended double Zepp." The gain over a half-wave dipole is approximately 3 dB, as compared with slightly less than 2 dB for two collinear dipoles. The directional pattern in the plane containing the axis of the antenna is shown in Fig. 16. As in the case of all other collinear arrays, the free-space pattern in the plane at right angles to the

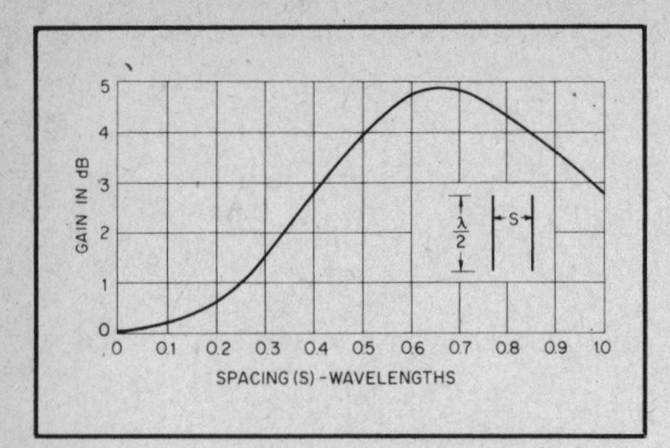


Fig. 17 — Power gain as a function of the spacing between two parallel elements operated in phase (broadside).

antenna elements is the same as that of a half-wave antenna, i.e., is circular.

## BROADSIDE ARRAYS WITH PARALLEL ELEMENTS

To obtain broadside directivity with parallel elements the currents in the elements must all be in phase. At a distant point lying on a line perpendicular to the axis of the array and also perpendicular to the plane containing the elements, the fields from all elements add up in phase. The situation is similar to that pictured in Fig. 1 in this chapter.

Broadside arrays of this type theoretically can have any number of elements. However, practical limitations of construction and available space usually limit the number of broadside parallel elements to two, in the amateur bands below 30 MHz, when horizontal polarization is used. More than four such elements seldom are used even at vhf.

### Power Gain

The power gain of a parallel-element broadside array depends on the spacing between elements as well as on the number of elements. The way in which the gain of a two-element array varies with spacing is shown in Fig. 17. The greatest gain is obtained when the spacing is in the vicinity of 0.7 wavelength.

The theoretical gains of broadside arrays having more than two elements are approximately as follows:

| *No. of Parallel Elements* | *dB Gain with 1/2-Wave Spacing* | *dB Gain with 3/4-Wave Spacing* |
|---|---|---|
| 3 | 5 | 7 |
| 4 | 6 | 8.5 |
| 5 | 7 | 10 |
| 6 | 8 | 11 |

The elements must, of course, all lie in the same plane and all be fed in phase.

### Directivity

The sharpness of the directive pattern depends on spacing between elements and on the number of elements. Larger element spacing will sharpen the main lobe, for a given number of elements. The two-element array has no minor lobes when the spacing is 1/2 wavelength, but small minor lobes appear at greater spacings, as indicated in Fig. 9. When three or more elements are used the pattern always has minor lobes.

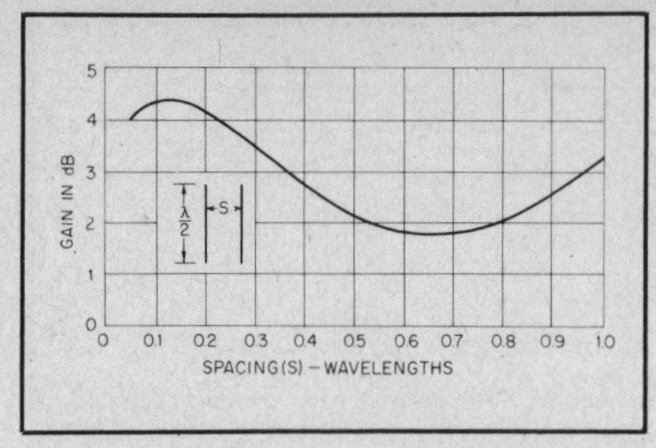


Fig. 18 — Gain of an end-fire array consisting of two elements fed 180 degrees out of phase, as a function of the spacing between elements. Maximum radiation is in the plane of the elements and at right angles to them at spacings up to 0.5 wavelength, but the direction changes at greater spacings.

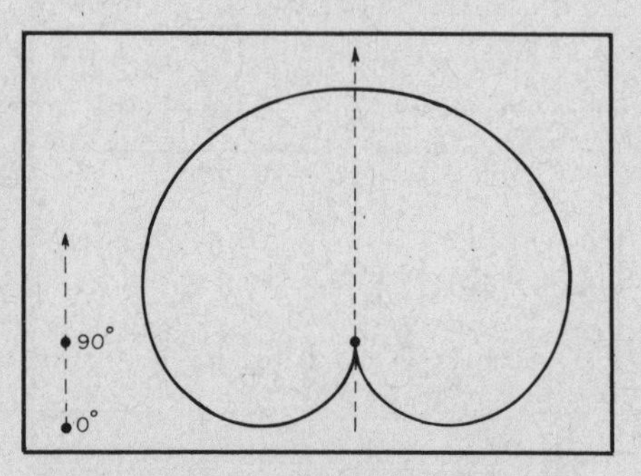


Fig. 19 — Representative H-plane pattern for a 2-element end-fire array with 90-degree phasing. The elements lie along a vertical axis, with the uppermost element the one of lagging phase.

## END-FIRE ARRAYS

The term "end-fire" covers a number of different methods of operation, all having in common the fact that the maximum radiation takes place along the array axis, and that the array consists of a number of parallel elements in one plane. End-fire arrays can be either bidirectional or unidirectional. In the bidirectional type commonly used by amateurs there are only two elements, and these are operated with currents 180 degrees out of phase. Even though adjustment tends to be complicated, unidirectional end-fire driven arrays have also seen some amateur use, primarily as a pair of phased, ground-mounted quarter-wave vertical elements. Horizontally polarized unidirectional end-fire arrays see little amateur use except in logarithmic periodic arrays (described later in this chapter). Instead, horizontally polarized unidirectional arrays usually have parasitic elements (also described later in this chapter).

## TWO-ELEMENT ARRAY

In the two-element array with equal currents out of phase the gain varies with the spacing between elements as shown in Fig. 18. The maximum gain occurs in the neighborhood of 1/8-wave spacing. Below 0.05-wave spacing the gain decreases rapidly, since the system is approaching the spacing used for nonradiating transmission lines.

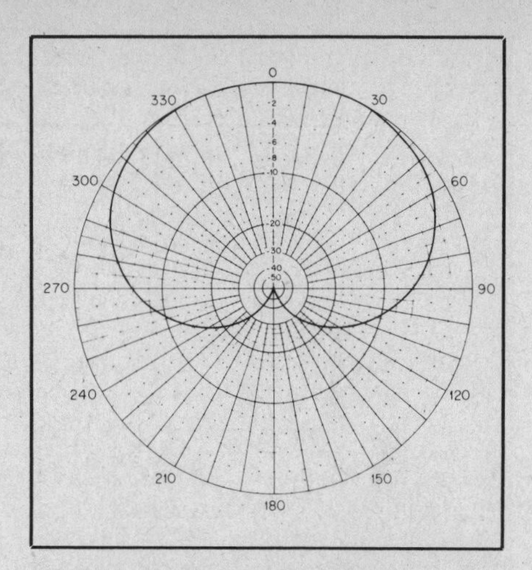


Fig. 20 — H-plane pattern for a 3-element end-fire array with binomial current distribution (the current in the center element is twice that in each end element). The elements are spaced a quarter wavelength apart along the 0°-180° axis. The center element lags the lower element by 90°, while the upper element lags the lower element by 180° in phase.

The radiation resistance at the feed point of either element is very low at the spacings giving the greatest gain, as shown by Fig. 7. The spacings most frequently used are 1/8 and 1/4 wavelength, at which the resistances of center-fed half-wave elements are about 8 and 32 ohms, respectively. When the spacing is 1/8 wavelength it is advisable to use good-sized conductors — preferably tubing — for the elements because with the radiation resistance so low the heat losses in the conductor can represent an appreciable portion of the power supplied to the antenna. Excessive conductor losses will mean that the theoretical gain cannot be realized.

## UNIDIRECTIONAL END-FIRE ARRAYS

Two parallel elements spaced 1/4 wavelength apart and fed equal currents 90 degrees out of phase will have a directional pattern, in the plane at right angles to the plane of the array, as represented in Fig. 19. The maximum radiation is in the direction from the element in which the current leads to the element in which the current lags. In the opposite direction the fields from the two elements cancel.

When the currents in the elements are neither in phase nor 180 degrees out of phase the radiation resistances of the elements are not equal. This complicates the problem of feeding equal currents to the elements. If the currents are not equal, one or more minor lobes will appear in the pattern and decrease the front-to-back ratio. The adjustment process is likely to be tedious and requires field-strength measurements in order to get the best performance.

More than two elements can be used in a unidirectional end-fire array. The requirement for unidirectivity is that there must be a progressive phase shift in the element currents equal to the spacing, in electrical degrees, between the elements. The amplitudes of the currents in the various elements also must be properly related. This requires "binomial" current distribution — i.e., the ratios of the currents in the elements must be proportional to the coefficients of the binomial series. In the case of three elements, this requires that the current in the center element be twice that in the two outside elements, for 90-degree (quarter-wave) spacing and element current phasing. This antenna has an overall length of 1/2 wavelength. The directive diagram is shown in Fig. 20.

## COMBINATION DRIVEN ARRAYS

Broadside, end-fire and collinear elements can readily be combined to increase gain and directivity, and this is in fact usually done when more than two elements are used in an array. Combinations of this type give more gain, in a given amount of space, than plain arrays of the types just described. The combinations that can be worked out are almost endless, but in this section we shall describe only a few of the simpler types.

The accurate calculation of the power gain of a multielement array requires a knowledge of the mutual impedances between all elements. For approximate purposes it is sufficient to assume that each *set* (collinear, broadside, end-fire) will have the gains as given earlier, and then simply add up the gains for the combination. This neglects the effects of cross-coupling between sets of elements. However, the array configurations are such that the mutual impedances from cross-coupling should be relatively small, particularly when the spacings are 1/4 wavelength or more, so that the estimated gain should be reasonably close to the actual gain.

### FOUR-ELEMENT END-FIRE AND COLLINEAR ARRAY

The array shown in Fig. 21 combines collinear in-phase elements with parallel out-of-phase elements to give both broadside and end-fire directivity. It is popularly known as a "two-section W8JK" or "two-section flat-top beam." The approximate gain calculated as described above is 6.2 dB with 1/8-wave spacing and 5.7 dB with 1/4-wave spacing. Directive patterns are given in Figs. 22 and 23.

The impedance between elements at the point where the phasing line is connected is of the order of several thousand ohms. The SWR with an unmatched line consequently is quite high, and this system should be constructed with open-wire line (500 or 600 ohms) if the line is to be resonant. To use a matched line a closed stub 3/16 wavelength long can be connected at the transmission-line junction shown in Fig. 21, and the transmission line itself can then be tapped on this matching section at the point resulting in the lowest line SWR. This point can be determined by trial.

With 1/4-wave spacing the SWR on a 600-ohm line is estimated to be in the vicinity of 3 or 4 to 1.

This type of antenna can be operated on two bands having a frequency ratio of 2 to 1, if a resonant feed line is used. For example, if designed for 28 MHz with 1/4-wave spacing between elements it can be operated on 14 MHz as a simple end-fire array having 1/8-wave spacing.

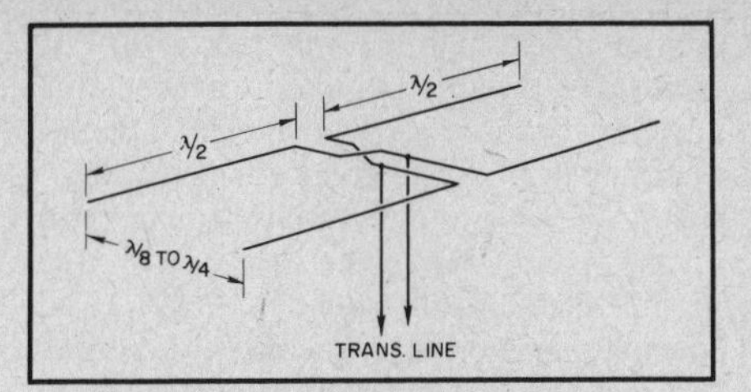


Fig. 21 — A 4-element array combining collinear broadside elements and parallel end-fire elements, popularly known as the W8JK array.

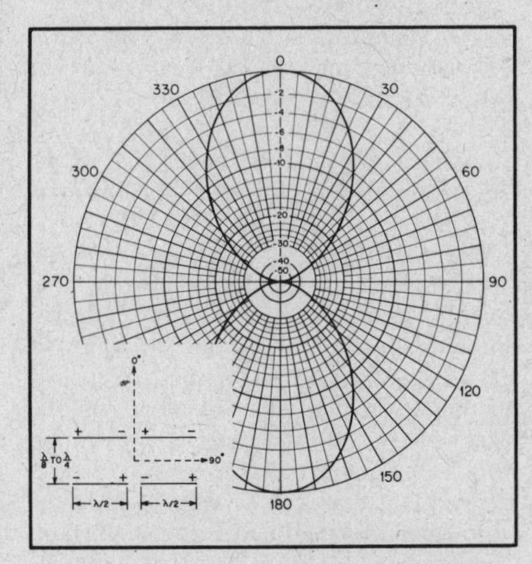


Fig. 22 — E-plane pattern for the antenna shown in Fig. 21. The elements are parallel to the 90°-270° line in this diagram. Less than a 1° change in half-power beamwidth results when the spacing is changed from 1/8 to 1/4 wavelength.

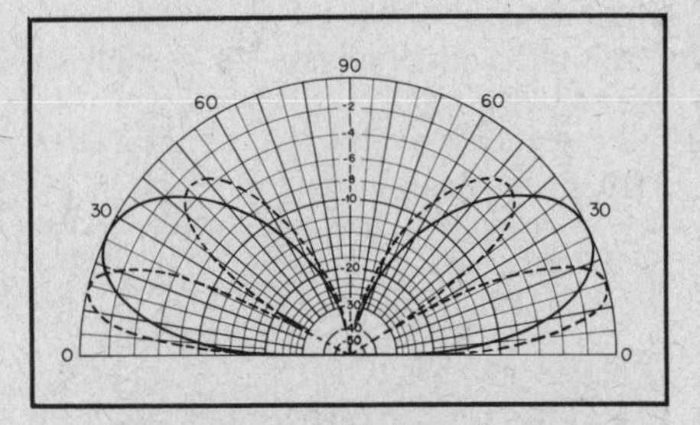


Fig. 23 — Vertical pattern for the 4-element antenna of Fig. 21 when mounted horizontally. Solid curve, height 1/2 wavelength; broken curve, height 1 wavelengfth. Fig. 22 gives the horizontal pattern.

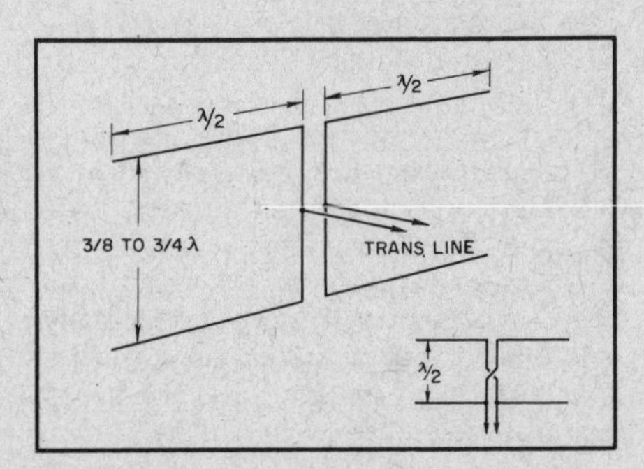


Fig. 24 — Four-element broadside array ("lazy-H") using collinear and parallel elements.

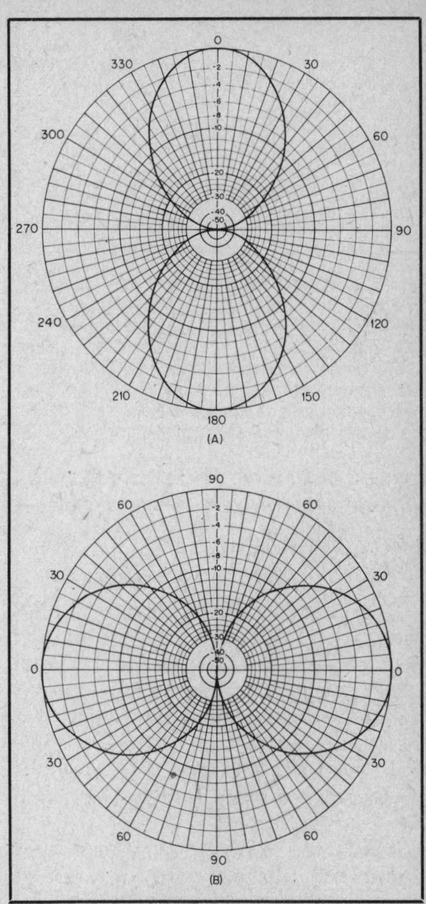


Fig. 25 — Free-space directive diagrams of the 4-element antenna shown in Fig. 24. At A is the E-plane pattern, the horizontal directive pattern at low wave angles when the antenna is mounted with the elements horizontal. The axis of the elements lies along the 90°-270° line. At B is the H-plane pattern, viewed as if one set of elements is above the other from the ends of the elements.

### FOUR-ELEMENT BROADSIDE ARRAY

The 4-element array shown in Fig. 24 is

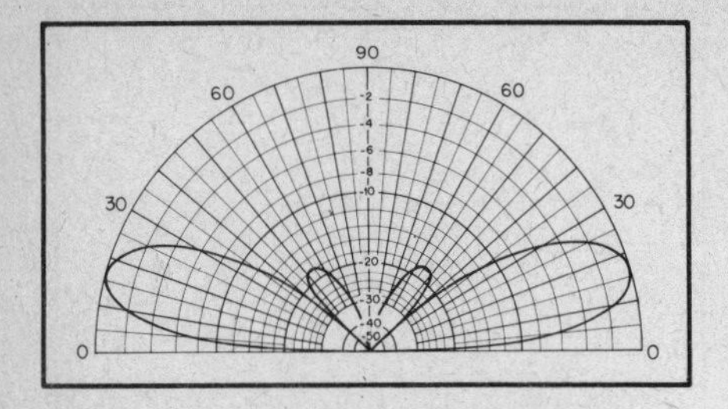


Fig. 26 — Vertical pattern of the 4-element broadside antenna of Fig. 24, when mounted with the elements horizontal and the lower set 1/2 wavelength above ground. "Stacked" arrays of this type give best results when the lowest elements are at least 1/2 wavelength high. The gain is reduced and the wave angle raised if the lowest elements are close to ground.

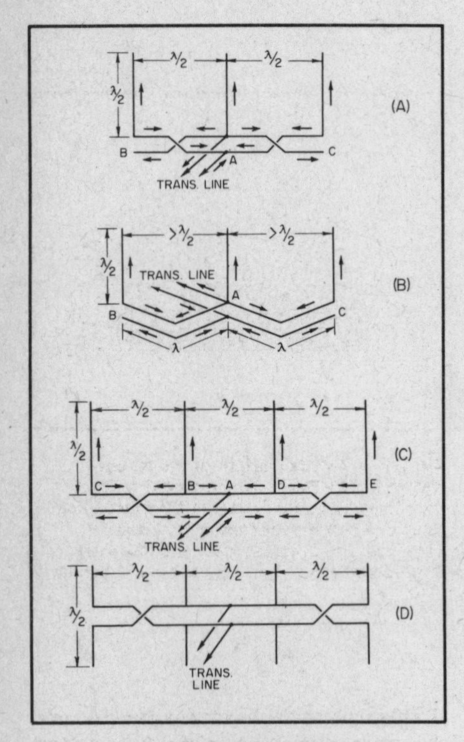


Fig. 27 — Methods of feeding 3- and 4-element broadside arrays with parallel elements.

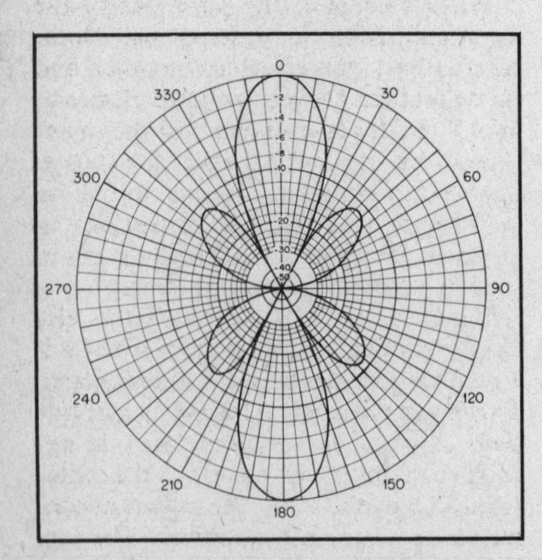


Fig. 28 — Free-space H-plane pattern of a 4-element broadside array using parallel elements. This is also the horizontal directive pattern at low wave angles for a vertically polarized array. The axis of the elements lies along the 90°-270° line.

commonly known as the "lazy-H." It consists of a set of two collinear elements and a set of two parallel elements, all operated in phase to give broadside directivity. The gain and directivity will depend on the spacing, as in the case of a simple parallel-element broadside array. The spacing may be chosen between the limits shown on the drawing, but spacings below 3/8 wavelength are not worthwhile because the gain is small. Estimated gains are as follows

3/8-wave spacing — 4.4 dB
1/2-wave spacing — 5.9 dB
5/8-wave spacing — 6.7 dB
3/4-wave spacing — 6.6 dB

Half-wave spacing is generally used. Directive patterns for this spacing are given in Figs. 25 and 26.

With half-wave spacing between parallel elements the impedance at the junction of the phasing line and transmission line is resistive and is in the vicinity of 100 ohms. With larger or smaller spacing the impedance at this junction will be reactive as well as resistive. Matching stubs are recommended in cases where a nonresonant line is to be used. They may be calculated and adjusted as described in Chapter 5.

The system shown in Fig. 24 may be used on two bands having a 2-to-1 frequency relationship. It should be designed for the higher of the two frequencies, using 3/4-wave spacing between parallel elements. It will then operate on the lower frequency as a simple broadside array with 3/8-wave spacing.

An alternative method of feeding is shown in the small diagram in Fig. 24. In this case the elements and the phasing line must be adjusted exactly to an electrical half wavelength. The impedance at the feed point will be resistive and of the order of 2000 ohms.

## OTHER FORMS OF MULTIELEMENT DRIVEN ARRAYS

For those who have the available room, multielement arrays based on the broadside concept have something to offer. The antennas are large but of simple design and noncritical dimensions; they are also very economical in terms of gain per unit of cost.

Three- and 4-element arrays are shown in Fig. 27. In the 3-element array with half-wave spacing, A, the array is fed at the center. This is the most desirable point in that it tends to keep the power distribution between elements uniform. However, the transmission line could be connected at either point B or C of Fig. 27A, with only slight skewing of the radiation pattern.

When the spacing is greater than 1/2 wavelength, the phasing lines must be one wavelength long and are not transposed between elements. This is shown at B in Fig. 27. With this arrangement, any element spacing up to one wavelength can be used, if the phasing lines can be folded as suggested in the drawing.

The 4-element array at C is fed at the center of the system to make the power distribution between elements as uniform as possible. However, the transmission line could be connected at either point B, C, D or E. In such case the section of phasing line between B and D must be transposed in order to make the currents flow in the same direction in all elements. The 4-element array at C and the 3-element array at B have approximately the same gain when the element spacing in the latter is 3/4 wavelength.

An alternative feeding method is shown at D of Fig. 27. This system can also be applied to the 3-element arrays, and will result in better symmetry in any case. It is only necessary to move the phasing line to the center of each element, making connection to both sides of the line instead of one only.

The free-space pattern for a 4-element array with half-wave spacing is shown in Fig. 28. This is also approximately the pattern for a 3-element array with 3/4-wavelength spacing.

Larger arrays can be designed and constructed by following the phasing principles shown in the drawings. No accurate figures are available for the impedances at the various feed points indicated in Fig. 27. It can be estimated to be in the vicinity of 1000 ohms when the feed point is at a junction between the phasing line and a half-wave element, becoming smaller as the number of elements in the array is increased. When the feed point is midway between end-fed elements as in Fig. 27C, the impedance of a 4-element array as seen by the transmission line is in the vicinity of 200 to 300 ohms, with 600-ohm open-wire phasing lines. The impedance at the feed point with the antenna shown at D should be about 1500 ohms.

### FOUR-ELEMENT BROADSIDE AND END-FIRE ARRAY

The array shown in Fig. 29 combines parallel elements in broadside and end-fire directivity. Approximate gains can be calculated by adding the values from Figs. 17 and 18 for the element spacings used. The smallest (physically) array — 3/8-wave spacing between broadside and 1/8-wave spacing between end-fire elements — has an estimated gain of 6.8 dB and the largest — 3/4- and 1/4-wave spacing, respectively — about 8.5 dB. The optimum element spacings are 5/8 wave broadside and 1/8 wave end-fire, giving an overall gain estimated at 9.3 dB. Directive patterns are given in Figs. 30 and 31.

The impedance at the feed point will not be purely resistive unless the element lengths are correct and the phasing lines are exactly a half wavelength long. (This

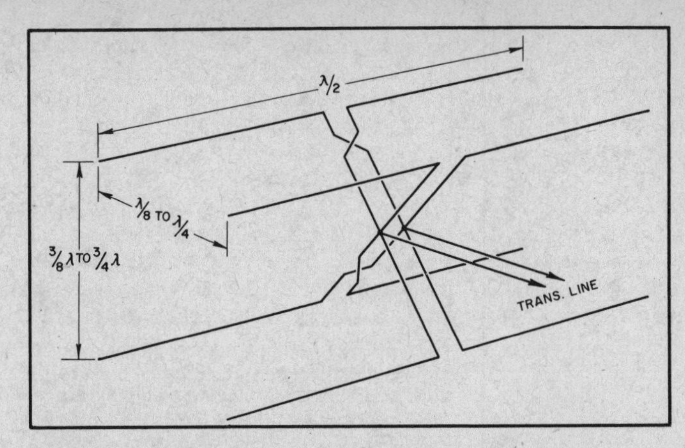


Fig. 29 — Four-element array combining both broadside and end-fire elements.

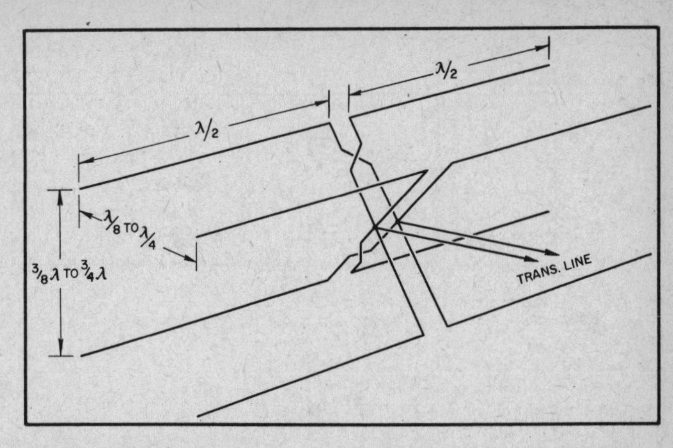


Fig. 32 — Eight-element driven array combining collinear and parallel elements for broadside and end-fire directivity.

requires somewhat less than half-wave spacing between broadside elements.) In this case the impedance at the junction is estimated to be over 10,000 ohms. With other element spacings the impedance at the junction will be reactive as well as resistive, but in any event the standing-wave ratio will be quite large. An open-wire line can be used as a resonant line, or a matching section may be used for nonresonant operation.

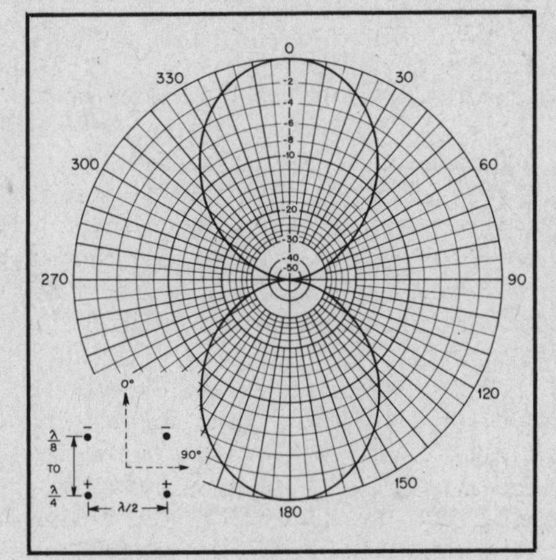


Fig. 30 — Free-space H-plane pattern of the 4-element antenna shown in Fig. 29.

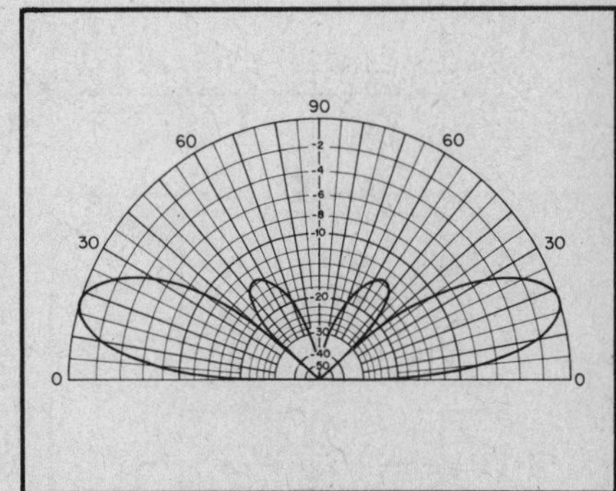


Fig. 31 — Vertical pattern of the antenna shown in Fig. 29 at a mean height of 3/4 wavelength (lowest elements 1/2 wave above ground) when the antenna is horizontally polarized. For optimum gain and low wave angle the mean height should be at least 3/4 wavelength.

## EIGHT-ELEMENT DRIVEN ARRAY

The array shown in Fig. 32 is a combination of collinear and parallel elements in broadside and end-fire directivity. The gain can be calculated as described earlier, using Figs. 10, 17 and 18. Common practice is to use half-wave spacing for the parallel broadside elements and 1/8-wave spacing for the end-fire elements. This gives an estimated gain of about 10 dB. Directive patterns for an array using these spacings are similar to those of Figs. 30 and 31, being somewhat sharper.

Although even approximate figures are not available, the SWR with this arrangement will be high. Matching stubs are recommended for making the line nonresonant. Their position and length can be determined as described in Chapter 5.

This system can be used on two bands related in frequency by a 2-to-1 ratio, providing it is designed for the higher of the two with 3/4-wave spacing between the parallel broadside elements and 1/4-wave spacing between the end-fire elements. On the lower frequency it will then operate as a four-element antenna of the type shown in Fig. 29 with 3/8-wave broadside spacing and 1/8-wave end-fire spacing. For two-band operation a resonant transmission line must be used.

## OTHER DRIVEN SYSTEMS

Two other types of driven antennas are worthy of mention, although their use by amateurs has been rather limited. The Sterba array, shown at A in Fig. 33, is a broadside radiator consisting of both collinear and parallel elements with 1/2-wave spacing between the latter. Its distinctive feature is the method of closing the ends of the system. For direct current and low-frequency ac, the system forms a closed loop, which is advantageous in that heating currents can be sent through the wires to melt the ice that forms in cold climates. There is comparatively little radiation from the vertical connecting wires at the ends because the currents are relatively small and are flowing in opposite directions with respect to the center (the voltage loop is marked with a dot in this drawing).

The system obviously can be extended as far as desired. The approximate gain is the sum of the gains of one set of collinear elements and one set of broadside elements, counting the two 1/4-wave sections at the ends as one element. The antenna shown, for example, is about equivalent to one set of four collinear elements and one set of two parallel broadside elements, so the total gain is approximately 4.3 + 4.0 = 8.3 dB. Horizontal polarization is the only practicable type at the lower frequencies, and the lower set of elements should be at least 1/2 wavelength above ground for best results.

When feeding at the point shown the impedance is of the order of 600 ohms. Alternatively, this point can be closed and the system fed between any two elements, as at X. In this case a point near the center should be chosen so that the power distribution between elements will be as uniform as possible. The impedance at any such point will be 1000 ohms or less in systems with six or more elements.

The Bruce array is shown at B in Fig. 33. It consists simply of a single wire folded so that the vertical sections carry large currents in phase while the horizontal sections carry small currents flowing in opposite directions with respect to the center (indicated by the dot). The radiation consequently is vertically polarized. The gain is proportional to the length of the array but is somewhat smaller, because of the short radiating elements, than is obtainable from a broadside array of half-wave parallel elements of the same overall

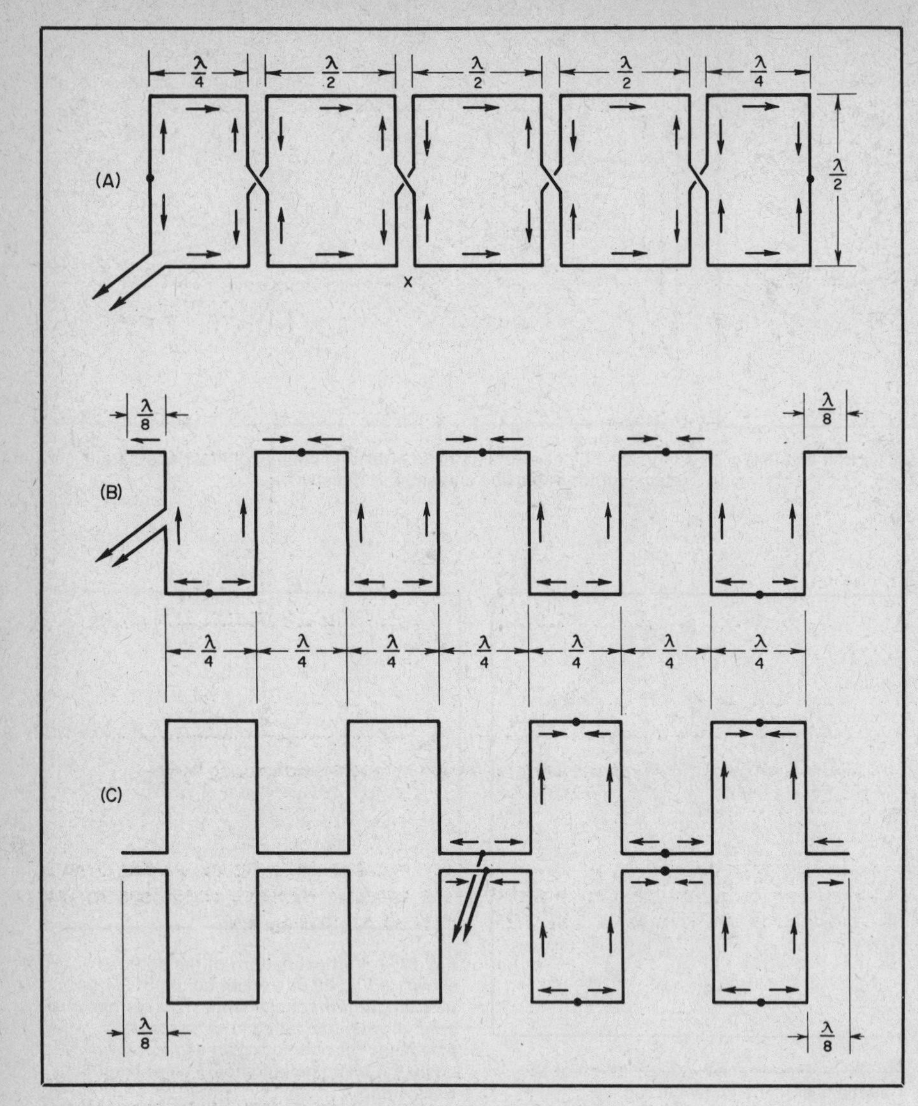


Fig. 33 — The Sterba array (A) and two forms of the Bruce array (B and C).

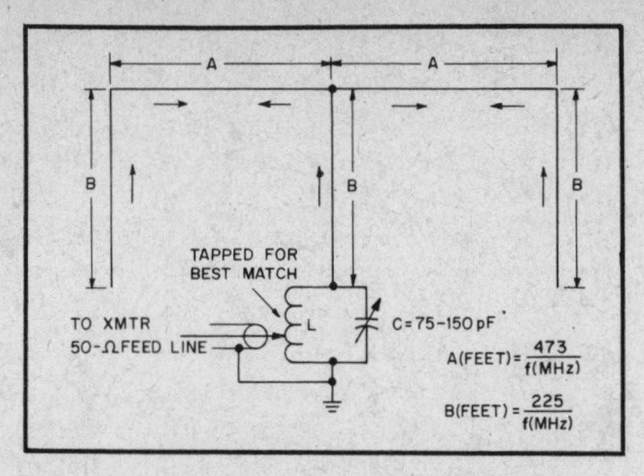


Fig. 34 — The bobtail curtain is an excellent low-angle radiator having broadside bidirectional characteristics. Current distribution is represented by the arrows. Dimensions A and B (in feet) can be determined by the formulas.

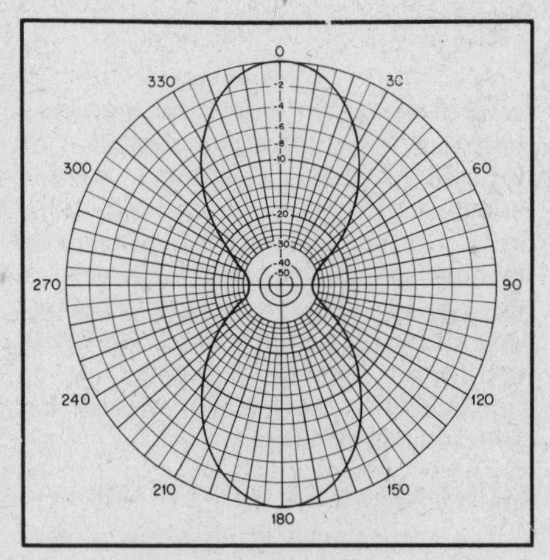


Fig. 35 — Horizontal directive diagram of a bobtail curtain antenna. The array lies along the 90°-270° axis. The three broadside elements are assumed to be spaced 173° apart with 2.5° phase lead at the end elements, and a current ratio of 1:2:1. The theoretical gain is 4.9 dB over a single vertical element.

length. The array should be two or more wavelengths long to secure a worthwhile gain. The system can be fed at any current loop; these occur at the centers of the vertical wires.

Another form of the Bruce array is shown at C. Because the radiating elements have twice the height, the gain is increased. The system can be fed at the center of any of the connecting lines.

## BOBTAIL CURTAIN

The antenna system of Fig. 34 uses the principles of cophased verticals to produce a broadside, bidirectional pattern providing approximately 5 dB of gain over a single element. The antenna performs as three in-phase top-fed vertical radiators approximately 1/4 wavelength in height and spaced approximately a half wavelength. It is most effective for low-angle signals and makes an excellent long-distance antenna for either 3.5 or 7 MHz.

The three vertical sections are the actual radiating components, but only the center element is fed directly. The two horizontal parts, A, act as phasing lines and contribute very little to the radiation pattern. Because the current in the center element must be divided between the end sections, the current distribution approaches a binomial 1:2:1 ratio. The radiation pattern is shown in Fig. 35.

The vertical elements should be as vertical as possible. The height for the horizontal portion should be slightly greater than B, as shown in Fig. 34. The tuning network is resonant at the operating frequency. The L/C ratio should be fairly low to provide good loading characteristics. As a starting point, a maximum value of 75 to 150 pF is recommended, and the inductor value is determined by C and the operating frequency. The network is first tuned to resonance and then the tap point is adjusted for the best match. A slight readjustment of C may be necessary. A link coil consisting of a few turns can also be used to feed the antenna.

## CHECKING PHASING

In the antenna diagrams earlier in this chapter the relative direction of current flow in the various antenna elements and connecting lines was shown by arrows. In laying out any antenna system it is necessary to know that the phasing lines are properly connected; otherwise the antenna may have entirely different characteristics than anticipated. The phasing may be checked either on the basis of current direction or polarity of voltages. There are two rules to remember:

1) In every half-wave section of wire, starting from an open end, the current directions reverse. In terms of voltage, the polarity reverses at each half-wave point, starting from an open end.

2) Currents in transmission lines always must flow in opposite directions in adjacent wires. In terms of voltage, polarities always must be opposite.

Examples of the use of current direction and voltage polarity are given at A and B, respectively, in Fig. 36. The half-wave points in the system are marked by the small circles. When current in one section flows toward a circle, the current in the next section must also flow toward it, and vice versa. In the 4-element antenna shown at A, the current in the upper right-hand element cannot flow toward the transmission line, because then the current in the right-hand section of the phasing line would have to flow upward and thus would be flowing in the same direction as the current in the left-hand wire. The phasing line would simply act like two wires in parallel in such a case.

C shows the effect of transposing the phasing line. This transposition reverses the direction of current flow in the lower pair of elements, as compared with A, and thus changes the array from a combination collinear and end-fire arrangement into a collinear-broadside array.

The drawing at D shows what happens when the transmission line is connected at the center of a section of phasing line. Viewed from the main transmission line the two parts of the phasing line are simply in parallel, so the half wavelength is measured from the antenna element along the upper section of phasing line and thence along the transmission line. The distance from the lower elements is measured in the same way. Obviously the two sections of phasing line should be the same length. If they are not, the current distribution becomes quite complicated; the element currents are neither in phase nor 180 degrees out of phase, and the elements at opposite ends of the lines do not receive the same power. To change the element current phasing at D into the phasing at A, simply transpose the wires in one section of the phasing line; this reverses the direction of current flow in the antenna elements connected to that section of phasing line.

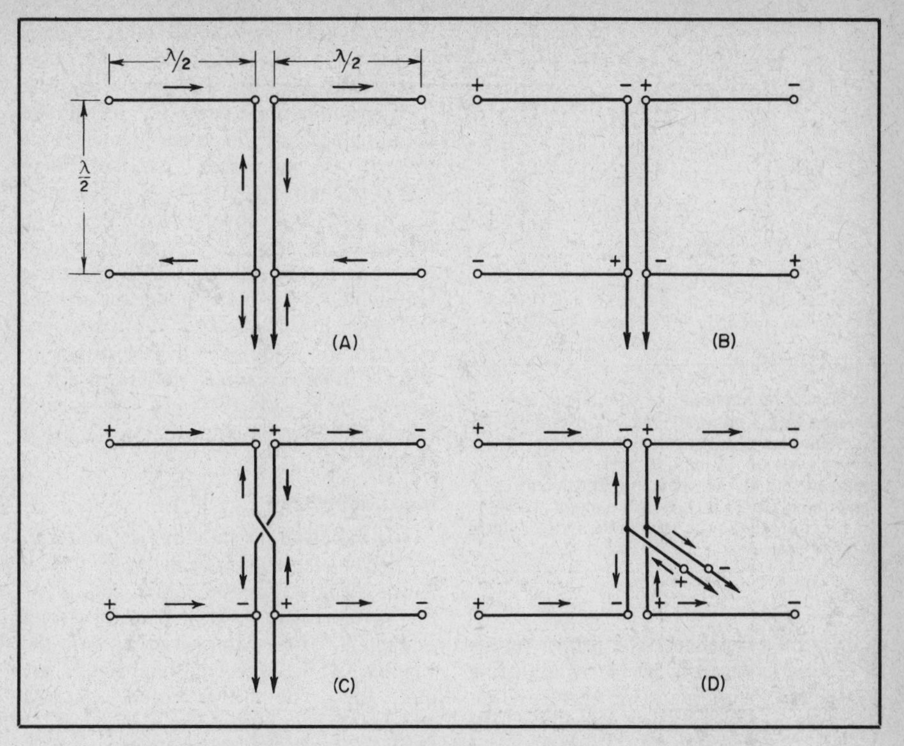

Fig. 36 — Methods of checking the phase of currents in elements and phasing lines.

# Parasitic Arrays

Multielement arrays containing parasitic elements are called "parasitic" arrays even though at least one and sometimes more than one of the elements is driven. A parasitic element obtains its power through electromagnetic coupling with a driven element, as contrasted with receiving it by direct connection to the power source. A parasitic array with linear (dipole-type) elements is frequently called a Yagi or Yagi-Uda antenna, after the inventors.

As explained earlier in this chapter in the section on mutual impedance, the amplitude and phase of the current induced in an antenna element will depend on its tuning and the spacing between it and the driven element to which it is coupled. The fact that the relative phases of the currents in driven and parasitic elements can be adjusted is very advantageous. For example, the spacing and tuning can be adjusted to approximate the conditions that exist when two driven elements 1/4 wavelength apart are operated with a phase difference of 90 degrees (which gives a unidirectional pattern as shown in Fig. 19). However, complete cancellation of radiation in the rear direction is not possible when a parasitic element is used. This is because it is usually not possible to make amplitude and phase *both* reach desired values simultaneously. Nevertheless, a properly designed parasitic array can be adjusted to have a large front-to-back ratio.

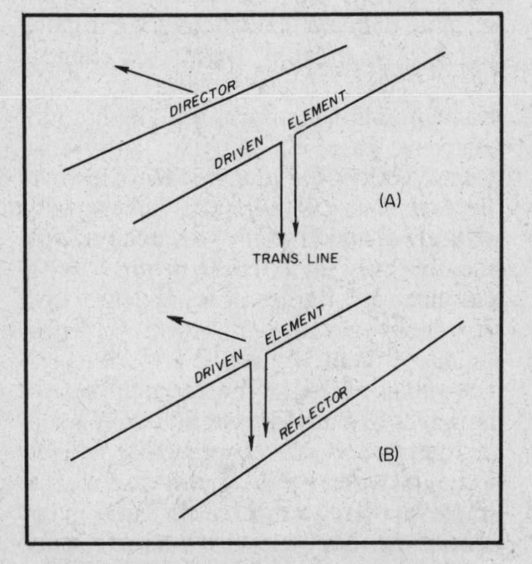

Fig. 37 — Antenna systems using a single parasitic element. In A the parasitic element acts as a director, in B as a reflector. The arrows show the direction in which maximum radiation takes place.

The substantially unidirectional characteristic and relatively simple electrical configuration of an array using parasitic elements make it especially useful for antenna systems that are to be rotated to aim the beam in any desired direction.

### REFLECTORS AND DIRECTORS

Although there are special cases where a parasitic array will have a bidirectional (but usually not symmetrical) pattern, in most applications the pattern tends to be unidirectional. A parasitic element is called a director when it makes the radiation maximum along the perpendicular line from the driven to the parasitic element, as shown at A in Fig. 37. When the maximum radiation is in the opposite direction — that is, from the parasitic element through the driven element as at B — the parasitic element is called a reflector.

Whether the parasitic element operates as a director or reflector is determined by the relative phases of the currents in the driven and parasitic elements. At the element spacings commonly used (1/4 wavelength or less) the current in the parasitic element will be in the right phase to make the element act as a reflector when its tuning is adjusted to the low-frequency side of resonance (inductive reactance). The parasitic element will act as a director when its tuning is adjusted to the high-frequency side of resonance

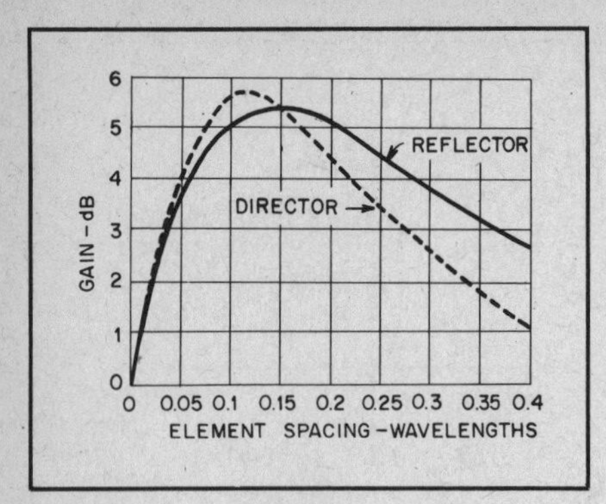


Fig. 38 — The maximum possible gain obtainable with a parasitic element over a half-wave antenna alone, assuming that the parasitic element tuning is adjusted for greatest gain at each spacing. These curves assume no ohmic losses in the elements. In practical antennas the gain is less, particularly at close spacings.

(capacitive reactance). The proper tuning is ordinarily accomplished by adjusting the lengths of the parasitic elements, but the elements can be "loaded" at the center with lumped inductance or capacitance to achieve the same purpose. If the parasitic element is self-resonant the element spacing determines whether it will act as a reflector or director.

## THE 2-ELEMENT BEAM

The maximum gain theoretically obtainable with a single parasitic element, as a function of the spacing, is shown in Fig. 38 (from analysis by G. H. Brown). The two curves show the greatest gain to be expected when the element is tuned for optimum performance either as a director or reflector. This shift from director to reflector, with the corresponding shift in direction as shown in Fig. 38, is accomplished simply by tuning the parasitic element — usually, in practice, by changing its length.

With the parasitic element tuned to act as a director, maximum gain is secured when the spacing is approximately 0.1 wavelength. When the parasitic element is tuned to work as a reflector, the spacing that gives maximum gain is about 0.15 wavelength, with a fairly broad peak. The director will give slightly more gain than the reflector, but the difference is less than 1/2 dB.

In only two cases are the gains shown in Fig. 38 secured when the parasitic element is self-resonant. These occur at 0.1- and 0.25-wavelength spacing, with the parasitic element acting as director and reflector, respectively. For reflector operation, it is necessary to tune the parasitic element to a lower frequency than resonance to secure maximum gain at all spacings less than 0.25 wavelength, while at greater spacings the reverse is true. The closer the spacing the greater the detuning required. On the other hand, the director must be detuned toward a higher frequency (that is, its length must be made less than the self-resonant length) at spacings greater than 0.1 wavelength in order to secure maximum gain. The amount of detuning necessary becomes greater as the spacing is increased. At less than 0.1-wavelength spacing the director must be tuned to a lower frequency than resonance to secure the maximum gains indicated by the curve. (Generally these requirements for maximum gain are not followed in practice for close-spaced directors or wide-spaced reflectors; instead, forward gain is sacrificed for a higher front-to-back ratio or greater bandwidth, as discussed in a later section.)

### Input Impedance

The radiation resistance at the center of the driven element varies as shown in Fig. 39 for the spacings and tuning conditions that give the gains indicated by the curves of Fig. 38. These values, especially in the vicinity of 0.1-wavelength spacing, are quite low. The curves coincide at 0.1 wavelength, both showing a value of 14 ohms.

The low radiation resistance at the spacings giving highest gain tends to reduce the radiation efficiency. This is because, with a fixed loss resistance, more of the power supplied to the antenna is lost in heat and less is radiated, as the radiation resistance approaches the loss resistance in magnitude.

The loss resistance can be decreased by using low-resistance conductors for the antenna elements. This means, principally, large-diameter conductors — usually tubing of aluminum, copper or copper-plated steel. Such conductors have mechanical advantages as well, in that it is relatively easy to provide adjustable sliding sections for changing length, while the fact that they can be largely self-supporting makes them well adapted for rotatable antenna construction. With half-inch or larger tubing the loss resistance in any 2-element antenna should be small.

With low radiation resistance the standing waves of both current and voltage on the antenna reach considerably higher maximum values than is the case with a simple dipole. For this reason losses in insulators at the ends of the elements become more serious. The use of tubing rather than wire helps reduce the end voltage, and furthermore the tubing does not require support at the ends, thus eliminating the insulators.

The mutual impedance between two parallel antenna elements contains reactance as well as resistance, so that the presence of a director or reflector near the driven element affects not only the radiation resistance of the driven element but also introduces a reactive component (assuming resonance if the parasitic element were not there). In other words, the

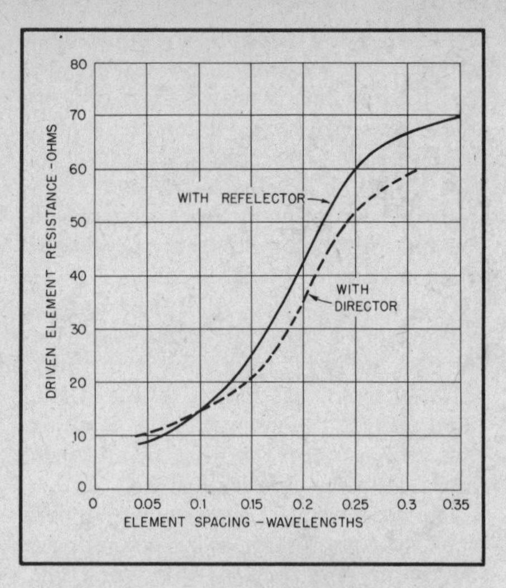


Fig. 39 — Radiation resistance at center of driven element as a function of element spacing, when the parasitic element is adjusted for the gains given in Fig. 38.

parasitic element detunes the driven element. The degree of detuning depends on the spacing and tuning of the parasitic element, and also on the length/diameter ratios of the elements.

With the parasitic element tuned for maximum gain, the effect of the coupled reactance is to make the driven element "look" more inductive with the parasitic element tuned as a reflector than it does when the parasitic element is tuned as a director. That is, the driven element should be slightly longer, if the parasitic element is a director, than when the parasitic element is a reflector. These remarks apply to spacings between about 0.1 and 0.25 wavelength, but are not necessarily true for other spacings.

### Self-Resonant Parasitic Elements

The special case of the self-resonant parasitic element is of interest, since it gives a good idea of the performance as a whole of 2-element systems, even though the results can be modified by detuning the parasitic element. Fig. 40 shows gain and radiation resistance as a function of the element spacing for this case. Relative field strength in the direction A of the small drawing is indicated by curve A; similarly for curve B.The front-to-back ratio at any spacing is the difference between the values given by curves A and B at that spacing. Whether the parasitic element is functioning principally as a director or reflector is determined by whether curve A or curve B is on top; it can be seen that the principal function shifts at about 0.14-wavelength spacing. That is, at closer spacings the parasitic element is principally a director, while at greater spacings it is chiefly a reflector. At 0.14 wavelength the radiation is the same in both directions; in other words, the antenna is bidirectional with a theoretical gain of about 4 dB.

The front-to-back ratios that can be

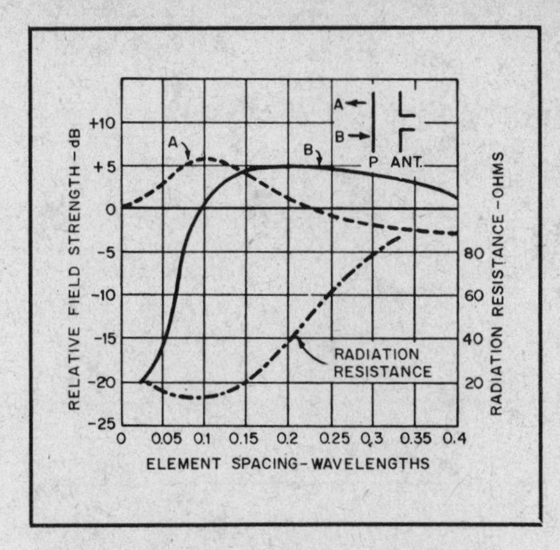


Fig. 40 — Theoretical gain of a 2-element parasitic array over a half-wave dipole as a function of element spacing when the parasitic element is self-resonant.

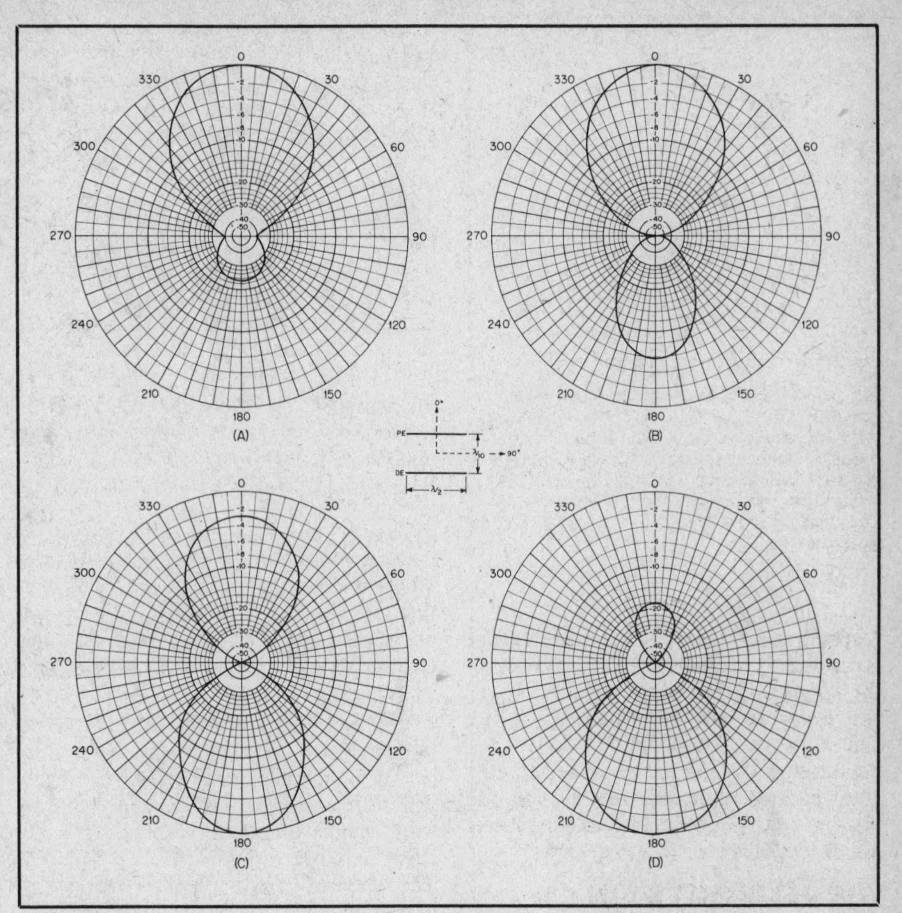

Fig. 41 — Experimentally determined horizontal directive patterns of horizontally polarized 2-element parasitic arrays at a height of 1-1/4 wavelengths. These patterns are for a wave angle of 12 degrees. The spacing between elements is 0.1 wavelength. The patterns should not be compared for gain, since they are plotted on a relative basis with the peak response for each arbitrarily plotted at 0 dB. The curves represent the following conditions, approximately:

A — Parasitic element tuned for maximum gain as a director.
B — Parasitic element self-resonant.
C — Parasitic element tuned for maximum gain as a reflector.
D — Parasitic element tuned for maximum front-to-back ratio as a reflector.

secured with the parasitic element self-resonant are not very great except in the case of extremely close spacings. Spacings of the order of 0.05 wavelength are not very practicable with outdoor construction since it is difficult to make the elements sufficiently stable mechanically. Ordinary practice is to use spacings of at least 0.1 wavelength and tune the parasitic element for greatest attenuation in the backward direction.

The radiation resistance increases rapidly for spacings greater than 0.15 wavelength, while the gain, with the parasitic element acting as a reflector, decreases quite slowly. If front-to-back ratio is not an important consideration, a spacing as great as 0.25 wavelength can be used without much reduction in gain. At this spacing the radiation resistance approaches that of a half-wave antenna alone. Spacings of this order are particularly suited to antennas using wire elements, such as multielement arrays consisting of combinations of collinear and broadside elements.

### Front-to-Back Ratio

The tuning conditions that give maximum gain forward do not give maximum signal reduction, or attenuation to the rear. It is necessary to sacrifice some gain to get the highest front-to-back ratio. The reduction in backward response is brought about by adjustment of the tuning or length of the parasitic element. With a reflector, the length must be made slightly greater than that which gives maximum gain, at spacings up to 0.25 wavelength. The director must be shortened somewhat to achieve the same end, with spacings of 0.1 wavelength and more. The tuning condition, or element length, which gives maximum attenuation to the rear is considerably more critical than that for maximum gain, so that a good front-to-back ratio can be secured without sacrificing more than a small part of the gain.

For the sake of good reception, general practice is to adjust for maximum front-to-back ratio rather than for maximum gain. Larger front-to-back ratios can be secured with the parasitic element operated as a director rather than as a reflector. With the optimum director spacing of 0.1 wavelength, the front-to-back ratio with the director tuning adjusted for maximum gain is only 5.5 dB (the back radiation is equal to that from the driven element alone). By proper director tuning, however, the ratio can be increased to 17 dB; the gain in the desired direction is in this case 4.5 dB, or 1 dB less than the maximum obtainable.

### Directional Patterns

The directional patterns obtained with two-element arrays will vary considerably with the tuning and spacing of the parasitic element. Typical patterns are shown in Figs. 41 and 42, for four cases where the parasitic element tuning or length is approximately adjusted for optimum gain as a director, for self-resonance, for optimum gain as a reflector, and for optimum front-to-back ratio as a reflector. Over this range of adjustment the width of the main beam does not change significantly. These patterns are based on experimental measurements.

### Bandwidth

The bandwidth of the antenna can be specified in various ways, such as the width of the band over which the gain is higher than some stated figure, the band over which at least a given front-to-back ratio is obtained, or the band over which the standing-wave ratio on the transmission line can be maintained below a chosen value. The latter is probably the most useful, since the SWR not only determines the percentage power loss in the transmission line but also affects the coupling between the transmitter and the line.

The bandwidth from this latter standpoint depends on the Q of the antenna (see Chapter 5). The Q of close-spaced parasitic arrays is quite high, with the result that the frequency range over which the SWR will stay below a specified maxi-

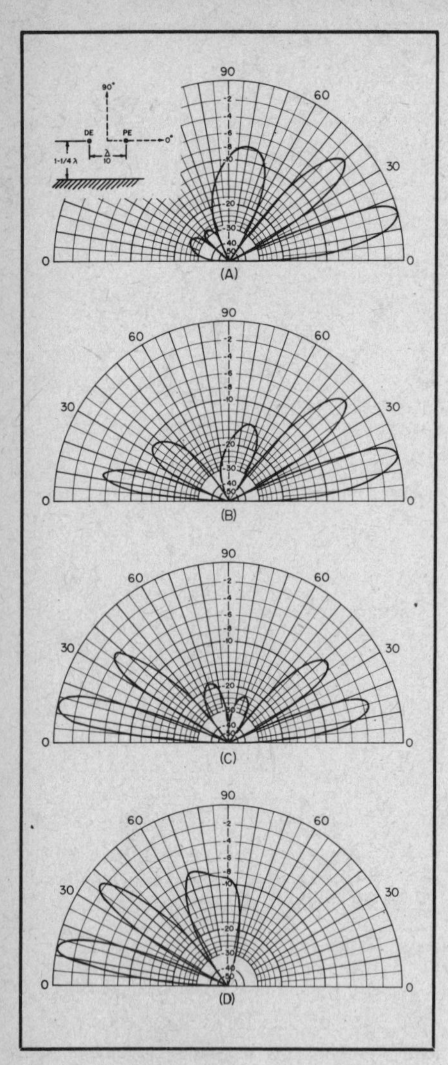


Fig. 42 — Vertical patterns of a horizontally polarized 2-element array under the conditions given in Fig. 41. These patterns are in the vertical plane at right angles to the antenna elements. Patterns of Figs. 41 and 42 are based on measurements by J. L. Gillson, W3GAU.)

**Table 2**

**Feed Impedance and Front-to-Back Ratio of a Fed Dipole with One Director**

**Data based on measurements by J. P. Shanklin.**

| *Element Spacing* | *Fed Dipole Length* | *Director Length* | *Input Resistance at Band Center* | *Q* | *Front-to-Back Ratio at Band Center* |
|---|---|---|---|---|---|
| 0.050 λ | 0.509 λ | 0.484 λ | 13.2 ohms | 53.2 | 20.0 (26 dB) |
| 0.075 | 0.504 | 0.476 | 24.4 | 29.4 | 8.3 (18 dB) |
| 0.100 | 0.504 | 0.469 | 28.1 | 20.0 | 4.3 (12.7 dB) |

mum value is relatively narrow. Data for a driven element and close-spaced director from experimental measurements are given in Table 2. The antenna with 0.075-wavelength spacing will, through a suitable matching device, operate with an SWR of less than 3 to 1 over a band having a width equal to about 3 percent of the center frequency (this corresponds to the width of the 14-MHz band for a 14-MHz antenna, for example) and maintain a front-to-back ratio of approximately 10 dB or better over this band. At greater element spacings than those shown in the table the Q is smaller and the bandwidth consequently greater, but the front-to-back ratio is smaller. This is to be expected from the trend shown by the curves of Fig. 40. The gain is practically constant at about 5 dB for all spacings shown in Table 2.

The same series of experimental measurements showed that with the parasitic element tuned as a reflector for maximum front-to-back ratio the optimum spacing was 0.2 wavelength. The maximum front-to-back ratio was determined to be 16 dB. In both the director and reflector cases the front-to-back ratio decreased rather rapidly as the operating frequency was moved away from the frequency for which the system was tuned. With the reflector at 0.2-wavelength spacing and tuned for maximum front-to-back ratio the input resistance was found to be 72 ohms and the Q of the antenna was 4.7.

The antenna elements used in these measurements had a length/diameter ratio of 330. A smaller length/diameter ratio will decrease the rate of reactance change with length and hence decrease the Q, while a larger ratio will increase the Q. The use of fairly thick elements is desirable when maximum bandwidth is sought.

## THE 3-ELEMENT BEAM

It is readily possible to use more than one parasitic element in conjunction with a single driven element. With two parasitic elements the optimum gain and directivity result when one is used as a reflector and the second as a director. Such an antenna is shown in Fig. 43.

As the number of parasitic elements is increased, the problem of determining the optimum element spacings and lengths to meet given specifications — i.e., maximum gain, maximum front-to-back ratio, maximum bandwidth, and so on — becomes extremely tedious because of the large number of variables. In general, it can be said that when one of these quantities — gain, front-to-back ratio, or bandwidth — is maximized the other two cannot be. Also, if it is desired to design the antenna to have a specific input impedance for matching a transmission line, the other three cannot be maximized.

### Power Gain

A theoretical investigation of the 3-element case (director, driven element and reflector) has indicated a maximum gain of slightly more than 7 dB (Uda and Mushiake). A number of experimental investigations has shown that the optimum spacing between the driven element and reflector is in the region of 0.15 to 0.25 wavelength, with 0.2 wavelength representing probably the best overall choice.

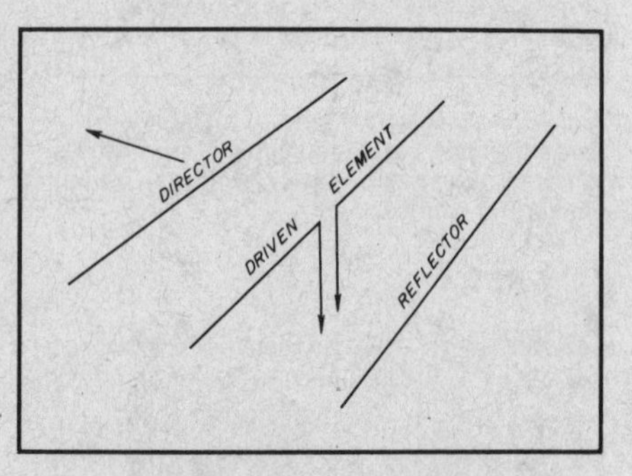


Fig. 43 — Antenna system using a driven element and two parasitic elements, one as a reflector and one as a director.

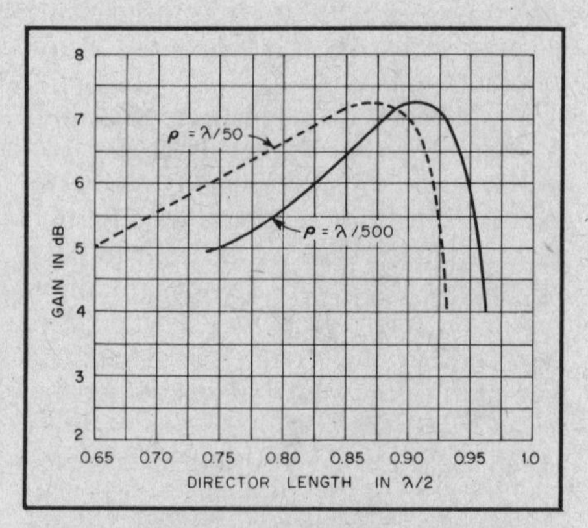


Fig. 44 — Gain of a 3-element Yagi over a dipole as a function of the director length for 0.2-wavelength spacing between driven element and director and the same spacing between driven element and reflector. These curves show how the element thickness affects the optimum length; ρ is the element radius expressed as a fraction of the wavelength. λ/500 corresponds to an element radius of approximately 1/2 inch at 50 MHz. Where the fractional-wavelength radius is smaller, as on the lower frequencies, the optimum director length will be somewhat greater.

With 0.2-wavelength reflector spacing, Fig. 44 shows the variation in gain with director length, with the director also spaced 0.2 wavelength from the driven element, and Fig. 45 shows the gain variation with direction spacing. It is obvious that the director spacing is not especially critical, and that the overall length of the array (boom length in the case of a rotatable antenna) can be anywhere between 0.35 and 0.45 wavelength with no appreciable difference in gain.

Wide spacing of both elements is desirable not only because it results in high gain but also because adjustment of tuning or element length is less critical and the input resistance of the driven element is higher than with close spacing. The latter feature improves the efficiency of the antenna and makes a greater bandwidth possible. However, a total antenna length, director to reflector, of more than 0.3 wavelength at frequencies of the order of 14 MHz introduces considerable difficulty from a constructional standpoint, so lengths of 0.25 to 0.3 wavelength are frequently used for this band, even though they are less than optimum.

In general, the gain of the antenna drops off less rapidly when the reflector length is increased beyond the optimum value than it does for a corresponding decrease below the optimum value. The opposite is true of a director, as shown by Fig. 44. It is therefore advisable to err, if necessary, on the long side for a reflector and on the short side for a director. This also tends to make the antenna performance less dependent on the exact frequency at which it is operated, because an increase above the design frequency has the same effect as increasing the length of both parasitic elements, while a decrease in frequency has the same effect as shortening both elements. By making the director slightly short and the reflector slightly long, there will be a greater spread between the upper and lower frequencies at which the gain starts to show a rapid decrease.

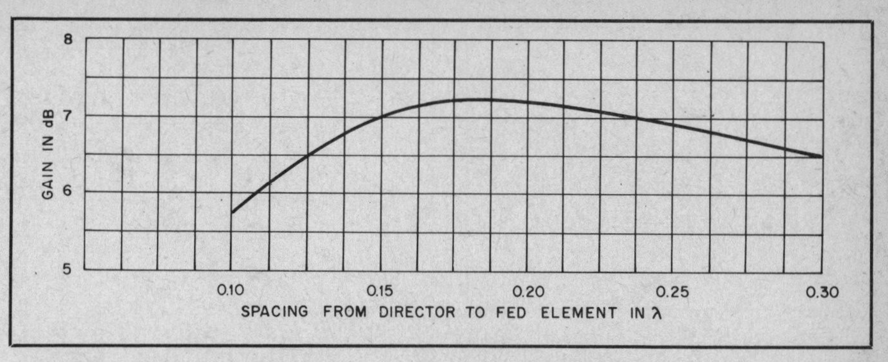


Fig. 45 — Gain of 3-element Yagi versus director spacing, the reflector spacing being fixed at 0.2 wavelength. (Curves of Figs. 44 and 45 are from work of Carl Greenblum.)

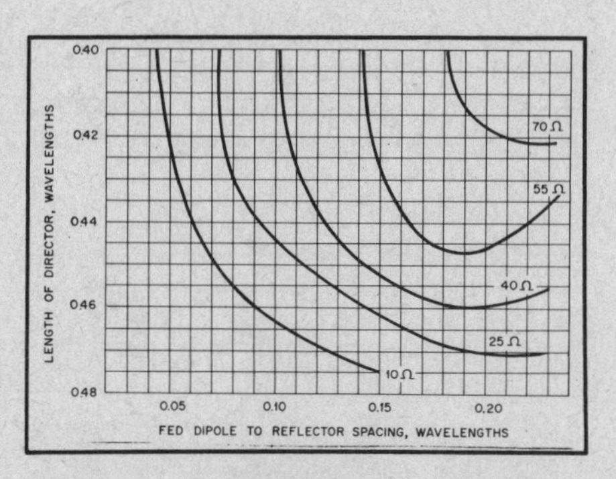


Fig. 46 — Resonant resistance of fed dipole in a 3-element parasitic antenna, overall length 0.3 wavelength. (Based on measurements by J. P. Shanklin.)

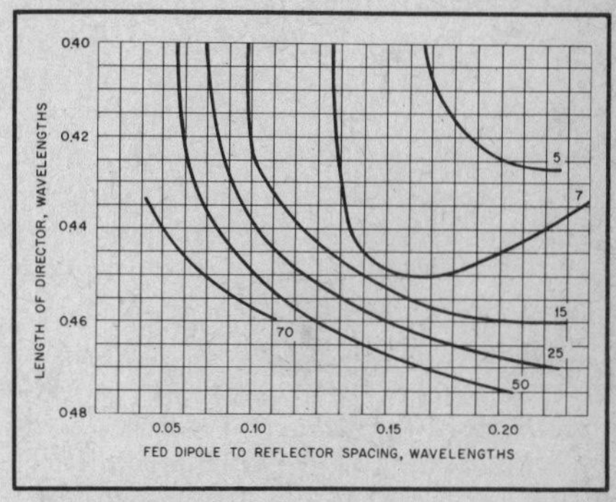


Fig. 47 — Q of input impedance of fed dipole in a 3-element parasitic antenna, overall length 0.3 wavelength. (J. P. Shanklin)

### Input Impedance

The radiation resistance as measured at the center of the driven element of a 3-element array can vary over a fairly wide range since it is a function of the spacing and tuning of the parasitic elements. There are, however, certain fairly well-defined trends: (1) The resistance tends to reach a minimum at the parasitic-element tuning condition that gives maximum gain, becoming larger as the element is detuned in either direction — that is, made longer or shorter. (2) The resistance tends to be lower, the closer the spacing between the parasitic and driven elements. Values of the order of 10 ohms are typical with a 3-element beam having 0.1-wavelength director spacing, when the director length is adjusted for maximum gain. This can be raised considerably — to 50 ohms or more — by sufficient change in director length at a sacrifice of gain. The minimum value of resistance increases with increased director spacing, and is of the order of 30 ohms at a spacing of 0.25 wavelength.

As in the case of the 2-element beam, tuning and spacing of the parasitic elements affect the reactance of the driven element; that is, a change in the spacing or length of the parasitic elements will tend to change the resonant frequency of the driven element. It is generally found, however, that the resonant length of the driven element with the parasitic elements properly tuned does not differ greatly from its resonant length with the parasitic elements removed. This is because the two parasitic elements reflect opposite kinds of reactance into the driven element and hence tend to cancel each other's effects in this respect.

Fig. 46 shows the input resistance of 3-element arrays having an overall length (director to reflector) of 0.3 wavelength. The curves give resistance contours as a function of the spacing between the driven element and the reflector and the length of the director. So long as the reflector length was in the optimum region for a good front-to-back ratio, as described in the next section, small changes in reflector length were found to have only a comparatively small effect on the input resistance. In using Fig. 46, it is to be understood that the spacing between the director and driven element is equal to the difference between 0.3 wavelength and the selected driven-element-to-reflector spacing, since the length of the array was held constant. The elements used in obtaining the data in Fig. 46 had a length/diameter ratio of 330.

### Front-to-Back Ratio

The element lengths and spacings are more critical when a high front-to-back ratio is the objective than when the antenna is designed for maximum gain. Some gain must be sacrificed for the sake of a good front-to-back ratio, just as in the case of the 2-element array. In general, a high front-to-back ratio requires fairly close spacing between the director and driven element, but considerably larger spacings are optimum for the reflector.

The front-to-back ratio will change more rapidly than the gain when the operating frequency differs from that for which the antenna was adjusted.

The front-to-back ratio tends to decrease with increased spacing among the elements. However, with a director spacing of about 0.2 wavelength, it is possible to secure a very good front-to-*side* ratio, which may be a useful feature in some locations. As in the case of front-to-back ratio, the reflector spacing has a considerably lesser effect.

### Bandwidth

The bandwidth with respect to input impedance, as evidenced by the change in standing-wave ratio over a band of frequencies, will in general be smaller the smaller the input resistance. This in turn becomes smaller when the spacing between elements is decreased. Hence, close spacings are usually associated with small bandwidths, especially when the element lengths are adjusted for maximum gain.

Fig. 47 shows how the Q of a 3-element antenna having a total length of 0.3

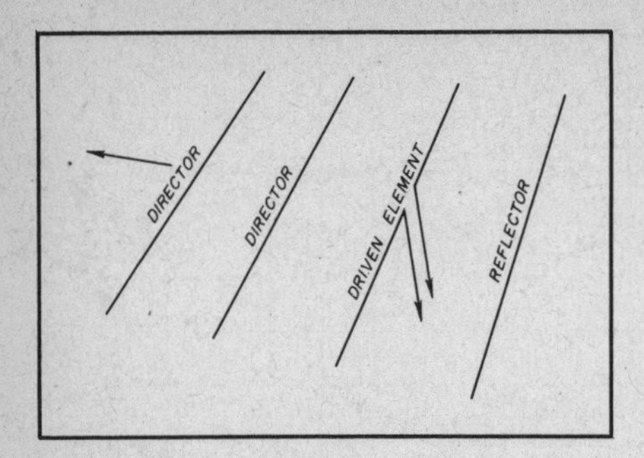


Fig. 48 — A 4-element antenna system, using two directors and one reflector in conjunction with a driven element.

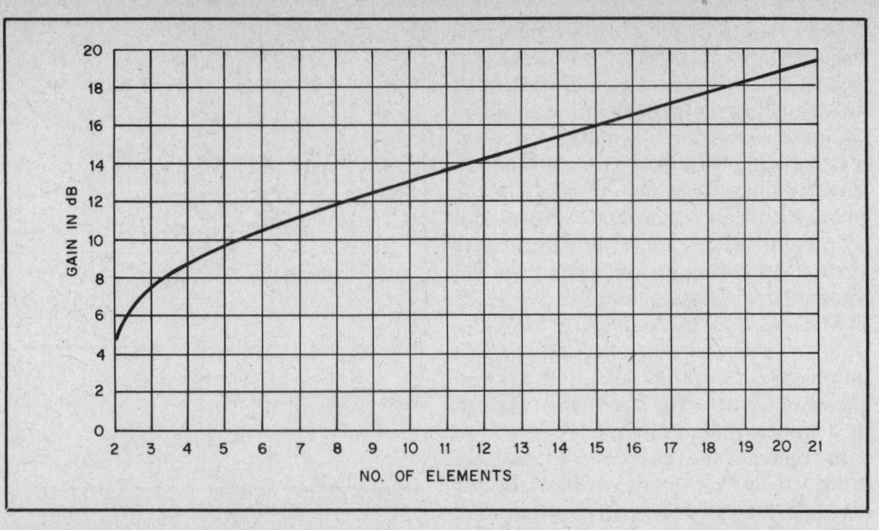


Fig. 49 — Gain in decibels over a half-wave dipole vs. the number of elements of the Yagi array, assuming the array length is as given in Fig. 50. (C. Greenblum)

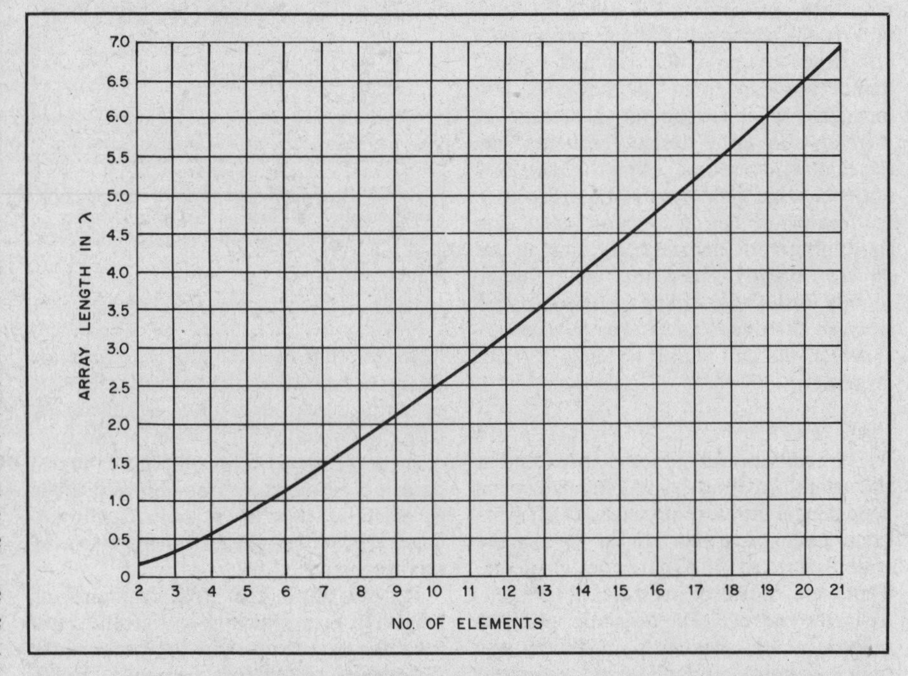


Fig. 50 — Optimum length of Yagi antenna as a function of number of elements. (C. Greenblum)

wavelength varies as a function of spacing and director length. The data are for an element length/diameter ratio of 330, but should hold sufficiently well for ratios between 200 and 400. From the standpoint of impedance bandwidth, the upper right-hand region of the chart is best, since this region is associated with low values of Q.

In these measurements it was found that the length of the reflector for optimum front-to-back ratio did not vary over much of a range. In the "low-Q" region of Fig. 47 it was 0.51 wavelength, increasing to 0.525 wavelength in the "high-Q" region. The proper driven-element length was found to be 0.49 wavelength (at the center of the band) for all conditions.

Similar tests were made on antennas having overall lengths of 0.2 and 0.4 wavelength. The conclusion was (1) that the smaller length would give high front-to-back ratios but with high Q values and consequently small bandwidth; (2) with 0.4-wavelength overall the Q values were low enough for good bandwidth but the front-to-back ratio was smaller.

The Q values given by the chart can be used as described in Chapter 5 to find the bandwidth over which the SWR will not exceed a specified value.

The low values of radiation resistance are accompanied by a high degree of selectivity in the antenna; that is, its impedance is constant over only a small frequency range. These changes in impedance make it troublesome to couple power from the transmitter to the line. Such difficulties can be reduced by using wider spacing — in particular, using spacings of the order of 0.2 wavelength or more.

### Directive Patterns

Directive patterns for 3-element arrays, based on experimental measurements made by J. L. Gillson, W3GAU, at vhf, show that the beam is somewhat sharper, as is to be expected, when the parasitic-element tuning is adjusted for maximum gain. Increasing the height of the antenna will of course lower the wave angle since the shape and amplitude of the vertical lobes are determined by the ground-reflection factors given in Chapter 2, as well as by the free-space pattern of the antenna itself.

## FOUR-ELEMENT ARRAYS

Parasitic arrays having a driven element and three parasitic elements — reflector and two directors — are frequently used at the higher frequencies, 28 MHz and up. This type of antenna is shown in Fig. 48.

Close spacing is undesirable in a four-element antenna because of the low radiation resistance. An optimum design, based on an experimental determination at 50 MHz, uses the following spacings:

Driven element to reflector — 0.2 wavelength
Driven element to first director — 0.2 wavelength
First director to second director — 0.25 wavelength

Using a length/diameter ratio of about 100 for the elements, the element lengths for maximum gain were found to be

Reflector — 0.51 wavelength
Driven element — 0.47 wavelength
First director — 0.45 wavelength
Second director — 0.44 wavelength

The input resistance with the above spacings and dimensions was of the order of 30 ohms, and the antenna gave useful gain over a total bandwidth equal to about 4 percent of the center frequency.

## LONG YAGIS

Parasitic arrays are not limited as to the number of elements that can be used, although it is hardly practical to use more than four at frequencies below 30 MHz. However, on the vhf bands an array that is long in terms of wavelength is often of practicable physical size. Several independent investigations of the properties of multielement Yagi antennas have shown that in a general way the gain of the antenna expressed as a power ratio is proportional to the length of the array, provided the number, lengths and spacings of the elements are chosen properly.

The results of one such study are shown in terms of the number of elements in the antenna in Figs. 49 and 50. In every case the antenna consists of a driven element, one reflector and a series of directors properly spaced and tuned. Thus if the antenna is to have a gain of 12 dB, Fig. 49 shows that 8 elements — driven, reflector, and six directors — will be required, and Fig. 50 shows that for such an 8-element antenna the array length required is 1.75 wavelength.

Table 3 shows the optimum element spacings determined from investigations by C. Greenblum. There is a fair amount of latitude in the placement of the elements along the length of the array, although the optimum tuning of the element will vary somewhat with the exact spacing chosen. Within the spacing ranges shown, the gain will not vary more than 1 dB provided the director lengths are suitably adjusted.

The optimum director lengths are in general greater, the closer the particular director is to the driven element, but the length does not uniformly decrease with increasing distance from the driven element. Fig. 51 shows the experimentally determined lengths for various element diameters, based on cylindrical elements supported by mounting through a cylindrical metal boom two or three times the element diameter. The curves probably would not be useful for other shapes.

In another study of long Yagi antennas at vhf, J. A. Kmosko, W2NLY, and H. G. Johnson, W6QKI, reached essentially the same general conclusions concerning the relationship between overall antenna length and power gain, although their gain figures differ from those of Greenblum. The comparison is shown in Fig. 52. The Kmosko-Johnson results are based on a somewhat different element spacing and a construction in which thin director elements are supported above the metal boom rather than running through it. In their optimum design the first director is spaced 0.1 wavelength from the driven element. The next two directors are slightly over 0.1 wavelength apart, the fourth director is approximately 0.2 wavelength from the third, and succeeding directors are spaced 0.4 wavelength apart. The Kmosko-Johnson figures are based on a simplified method of computing gain from the beamwidth of the antenna pattern, the beamwidths having been measured experimentally.

**Table 3**

**Optimum Element Spacings for Multielement Yagi Arrays**

| *No. Elements* | *R-DE* | *DE-$D_1$* | *$D_1$-$D_2$* | *$D_2$-$D_3$* | *$D_3$-$D_4$* | *$D_4$-$D_5$* | *$D_5$-$D_6$* |
|---|---|---|---|---|---|---|---|
| 2 | 0.15λ-0.2λ | | | | | | |
| 2 | | 0.07λ-0.11λ | | | | | |
| 3 | 0.16-0.23 | 0.16-0.19 | | | | | |
| 4 | 0.18-0.22 | 0.13-0.17 | 0.14λ-0.18λ | | | | |
| 5 | 0.18-0.22 | 0.14-0.17 | 0.15-0.20 | 0.17λ-0.23λ | | | |
| 6 | 0.16-0.20 | 0.14-0.17 | 0.16-0.25 | 0.22-0.30 | 0.25λ-0.32λ | | |
| 8 | 0.16-0.20 | 0.14-0.16 | 0.18-0.25 | 0.25-0.35 | 0.27-0.32 | 0.27λ-0.33λ | 0.30λ-0.40λ |
| 8 to N | 0.16-0.20 | 0.14-0.16 | 0.18-0.25 | 0.25-0.35 | 0.27-0.32 | 0.27-0.33 | 0.35-0.42 |

DE — Driven Element; R — Reflector; D — Director; N — any number; director spacings beyond $D_6$ should be 0.35-0.42λ.

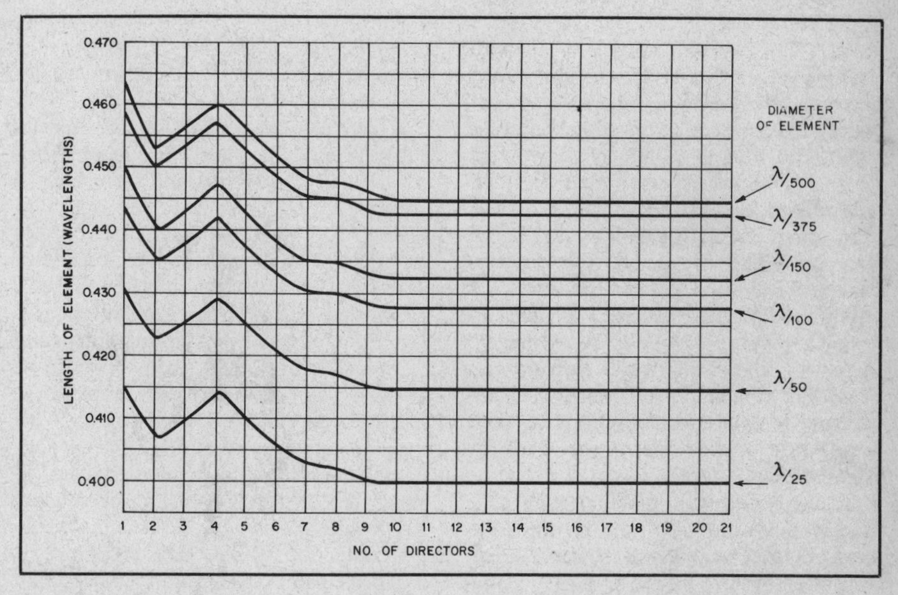


Fig. 51 — Length of director vs. its position in the array, for various element thicknesses.

The Greenblum data is from experimental measurement of gain.

Experimental gain figures based on measurements made at 3.3-centimeter wavelength by H. W. Ehrenspeck and H. Poehler, shown by a third curve in Fig. 52, indicate lower gain for a given antenna length but confirm the gain-vs.-length trend. These measurements were made over a large ground plane using elements of the order of one-quarter wavelength high. The general conclusions of this study were (1) that the reflector spacing and tuning is independent of the other antenna dimensions, the optimum fed-element to reflector spacing being in the neighborhood of 0.25 wavelength but not critical; (2) for a given antenna length the gain is practically independent of the number of directors provided the director-to-director spacing does not exceed 0.4 wavelength; (3) that the optimum director tuning differs with different director spacings but that for constant spacings all directors can be similarly tuned; and (4) a slight improvement in gain results from using an extra director spaced about 0.1 wavelength from the driven element.

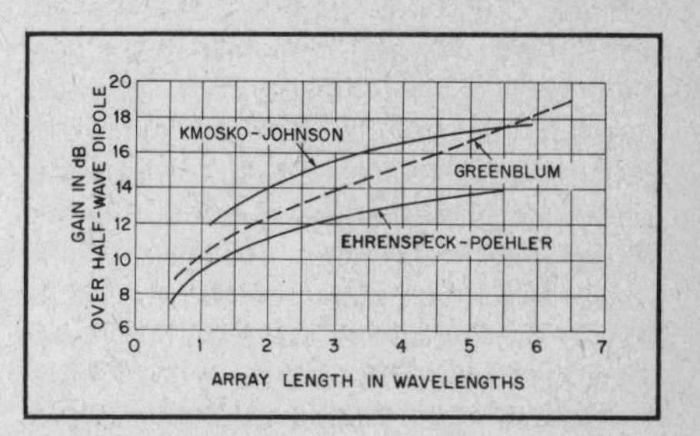


Fig. 52 — Gain of long Yagi antennas as a function of overall length. The antenna consists of a driven element, a single reflector spaced approximately one-fourth wavelength from the driven element, and a series of directors spaced as described in the text. The three curves represent the results of three independent studies.

The agreement between these three sets of measurements is not as close as might be wished, which simply confirms the difficulty of determining optimum design where a multiplicity of elements is used, and of measuring gain with a degree of accuracy that will permit reconciliation of the results obtained by various observers. There is, however, agreement on the

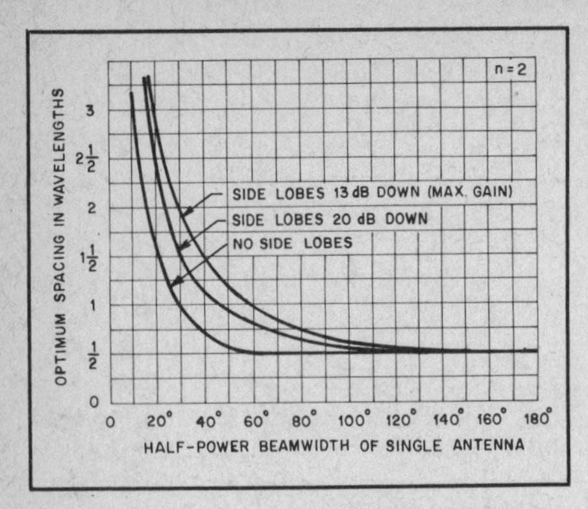


Fig. 53 — Optimum stacking spacing for two antennas. The spacing for no side lobes, especially for small beamwidths, may result in almost no gain improvement with stacking.

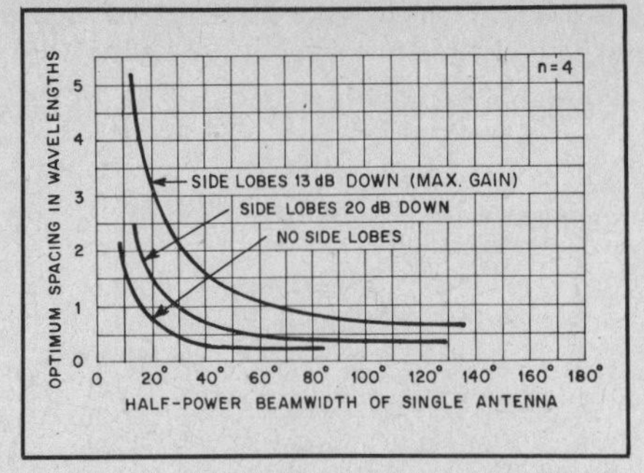


Fig. 54 — Optimum stacking space for four antennas. Spacings less than 1/2 wavelength are physically possible only for shortened dipoles in the case of collinear elements or for stacking in the plane perpendicular to the plane of polarization.

general principle that length is of greater importance than the number of elements, within the limit of a maximum element spacing of 0.4 wavelength.

It is an interesting fact that the feed-point impedance and bandwidth of long Yagis depends almost entirely on the two or three parasitic elements closest to the driven element, presumably because those farther from the driven element are relatively loosely coupled to it. In this respect, therefore, the information already given in connection with three-element arrays is quite applicable.

## STACKED YAGIS

Parasitic arrays can be stacked either in broadside or collinear fashion for additional directivity and gain. The increase in gain that can be realized is dependent on the spacing between the individual arrays. It is assumed, of course, that all the individual arrays making up the stacked system are identical, and that in the case of broadside stacking the corresponding elements are parallel and lie in planes perpendicular to the axis of the individual arrays. In collinear stacking, it is assumed that the corresponding elements are collinear and all elements of the individual arrays lie in the same plane. In both cases the driven elements must be fed in phase.

The decrease in beamwidth of the main radiation lobe that accompanies stacking is, in the general case, accompanied by a splitting off of one or more sets of side lobes. These will have an amplitude depending on the shape of the directive pattern of the unit array, the number of unit arrays, and their spacing. An optimum spacing is one which gives as much gain as possible on the condition that these side lobes do not exceed some specified amplitude relative to the main lobe. Fig. 53 shows the optimum spacings for three such conditions (no side lobes, side lobes down 20 dB and side lobes down 13 dB) as a function of the half-power beamwidth of the unit array, from calculations by H. W. Kaspar, K2GAL, Maximum gain occurs when the side lobes are approximately 13 dB down, as indicated in the figure.

A single 3-element array will have a half-power beamwidth (free space) of approximately 75 degrees, and from Fig. 53 it can be determined that the optimum spacing for maximum gain will be 3/4 wavelength. Measurements by Greenblum have shown that the stacking gain that can be realized with two such Yagi antennas is approximately 3 dB, remaining practically constant at spacings from 3/4 to 2 wavelengths, but decreasing rapidly at spacings less than 3/4 wavelength. With spacings less than about 1/2 wavelength stacking does not give enough gain to make the construction of a stacked array worthwhile.

If reduction of side-lobe amplitude is the principal consideration rather than gain, smaller spacings are optimum, as shown by the curves. A similar set of curves for four stacked unit arrays is given in Fig. 54.

The spacings in Fig. 53 and 54 are measured between the array centers. When the unit arrays are stacked in a collinear arrangement, a spacing of less than 1/2 wavelength is physically impossible with full-size elements, since at 1/2-wavelength spacing the ends of the collinear elements will be practically touching.

## FEEDING AND ADJUSTING

The problems of matching and adjusting parasitic arrays for maximum performance are the same in principle as with other antenna systems. Adjustment of element lengths for optimum performance usually necessitates measurements of relative field strength. However, the experience of a great many amateurs who have followed the rather laborious procedure (adjusting each element a little at a time, and measuring the relative field after each such change) has accumulated a large amount of data on optimum lengths. Depending on the objective in designing the antenna — i.e., maximum gain, maximum front-to-back ratio, etc. — it is possible to predetermine the actual element lengths for a given center frequency and thus avoid the necessity for such adjustments. Charts giving proper element lengths for 3-element beams are shown in Chapter 9 for the maximum-gain condition, and data for maximum front-to-back ratio were given earlier in this chapter. The principal adjustment that actually needs to be made is to match the antenna to the transmission line so the standing-wave ratio is minimized.

## METHODS OF FEED

The driven element in a parasitic array is a load for the transmission line in the same way that a driven element in any antenna system is such a load. It differs from the load presented by a simple dipole only in that the resistance may be quite low, especially if close spacings are used between elements, and the rate of change of reactance as the operating frequency is moved away from the design frequency may be greater. With low input resistance, a fairly large impedance step-up is required for matching practicable lines, and the amount of mismatch will increase more rapidly than with a simple dipole when the applied frequency is varied from that at which the line is matched.

Practically any of the matching systems detailed in Chapter 5 are applicable. The folded or multiconductor dipole used as the driven element furnishes a useful method of transforming a low antenna resistance to a value suitable for matching a transmission line. Design details are given in Chapter 5.

The choice of a matching system is affected by constructional considerations, since parasitic arrays are usually built to be rotated in operation. The T match and gamma match are favorites with many amateurs because they fit in well, constructionally, when the driven element is made of tubing. Another matching system suitable for continuously rotatable antennas uses two large inductively coupled loops, one fixed and one rotatable. Open-wire line is preferred for such coupling, however, so this method is seldom used these days. Because of the attendant problems with installation of open-wire lines, most amateurs prefer coaxial lines for carrying rf power up a tower or other support and use rotators with mechanical stops to avoid entangling the feed line around the tower.

### Broadening the Response

It has already been pointed out that the tuning conditions giving maximum gain with parasitic elements are not highly critical. However the varying amounts of reactance coupled into the driven element, as well as the fact that the radiation resistance at the center of the driven element is often very low, cause the impedance to change rapidly when the applied frequency is varied above or below the design frequency.

This impedance change can be made less rapid by using fairly wide spacing between elements, as already mentioned. It is also beneficial to use elements having a fairly low ratio of length to diameter because, as explained in Chapter 2, the impedance change with frequency is reduced when the antenna conductor has a large diameter.

The use of a folded-dipole driven element is beneficial in broadening the frequency characteristic of the antenna because of the smaller effective length/diameter ratio that results from using two or more conductors instead of one and the shorted-transmission-line effect of the conductors.

## ADJUSTING PARASITIC ARRAYS

There are two separate processes in adjusting an array with parasitic elements. One is the determination of the optimum element lengths, depending on whether maximum gain or maximum front-to-back ratio is desired. The other is matching the antenna to the transmission line. The second is usually dependent on the first, and the results observed on adjusting the element tuning may well be meaningless unless the line is equally well matched under all tuning conditions.

As stated earlier, the element tuning for maximum gain is not excessively critical, and the dimensions given by the following formulas have been found to work well in practice for 3-element antennas:

$$\text{Driven element: Length (ft)} = \frac{475}{f(\text{MHz})}$$

$$\text{Director: Length (ft)} = \frac{455}{f(\text{MHz})}$$

$$\text{Reflector: Length (ft)} = \frac{500}{f(\text{MHz})}$$

These are average lengths determined experimentally for elements having a length/diameter ratio of 200 to 400, and with element spacings from 0.1 to 0.2 wavelength.

Many amateurs have found that very satisfactory results are secured simply by cutting the elements to the lengths given by these formulas. It has been a rather common experience that, after a considerable amount of time has been spent in trying all possible adjustments, the dimensions finally determined to be optimum are very close to those given by the formulas above or the charts contained in Chapter 9 (the difference between the charts and the formulas above amount to only about one percent in most cases), and the actual difference in gain is negligible. It appears safe to say, therefore, that in the average case there is probably little to be realized, in the way of increased gain, by spending much time in adjusting element lengths. The front-to-back ratio can often be improved, however, since it depends much more on the exact element tuning. In general, the reflector tuning is the more critical.

If the array is put up by formula, the only adjustment that need be made is to match the driven element to the transmission line. The adjustment procedure for each type of matching arrangement is described in Chapter 5.

### Test Setup

The only practicable method of adjusting parasitic element lengths for best performance is to measure the field strength from the antenna as adjustments are made. Measurements on a relative basis are entirely satisfactory for the purpose of determining the operating conditions that result in the maximum output or greatest front-to-back ratio. For this purpose the measuring equipment does not need to be calibrated; the only requirement is that it indicate whether the signal is stronger or weaker.

If the help of a nearby amateur owning a receiver with an S meter can be enlisted, the S-meter indications can be used to indicate the relative field strength. A few precautions must be taken if this method is to be reliable. The receiving antenna must have the same polarization as the transmitting antenna under test (that is usually horizontal) and should be reasonably high above its surroundings. The receiving system should be checked for pickup on the transmission line to make sure that the indications given by the receiver are caused entirely by energy picked up by the receiving antenna itself. This can be checked by temporarily disconnecting the line from the antenna (but leaving it in place) and observing the signal strength on the S meter. If the reading is not several S units below the reading with the antenna connected, the readings cannot be relied upon when adjusting the transmitting antenna for maximum gain. In checking the front-to-back ratio, the stray pickup at the receiving installation must be well below the smallest signal received via the antenna, if the adjustments are to mean anything at all.

Another method of checking field strength is to use a field-strength indicator of the diode-detector type. The preferable method of using such an indicator is to connect it to a dipole antenna mounted some distance away and at a height at least equal to that of the transmitting antenna. There should be no obstructions between the two antennas, and both should have the same polarization. The receiving dipole need not be a half wave long, although that length is desirable because it will increase the ratio of energy picked up on the antenna to energy picked up by stray means. To prevent coupling effects the distance between the two antennas should be at least three wavelengths. At shorter distances the mutual impedance may be large enough to cause the receiving antenna to tend to become part of the transmitting system, which can lead to false results. A recommended type of indicating system is shown in Fig. 55. The transmission line should drop vertically down to the indicator, to avoid stray pickup. This pickup can be checked as described in the preceding paragraph. If the distance between the two antennas is such that greater sensitivity is needed a reflector may be placed 1/4 wavelength behind the receiving dipole.

### Adjustment Procedure

It is advisable first to set the element lengths to those given by the formulas and then match the driven element to the transmission line obtaining as low an SWR as possible. In subsequent adjustments a close watch should be kept on the SWR, and the transmitter power input should be maintained at exactly the same figure throughout. If the SWR changes enough to affect the coupling at the transmitter when an adjustment is made, but not enough to raise the line loss significantly, readjust the coupling to bring the input back to the same value. If the line loss increases more than a fraction of a decibel, *rematch at the driven element*. If this is not done, the results may be entirely misleading; it is absolutely necessary to maintain constant power input to the driven element if adjustment of director or reflectors is to give meaningful results.

The experience of most amateurs in adjusting parasitic arrays indicates that there is not a great deal of preference in the order in which elements are tuned, but that there is slightly less interaction if the director is first adjusted to give maximum gain and the reflector is then adjusted to give either maximum gain or maximum front-to-back ratio, whichever is desired. After the second parasitic element has been adjusted, go back and check the tuning of the first to make sure that it has not been thrown out of adjustment by the mutual coupling. If there are three parasitic elements, the other two should be checked each time an appreciable change is made in one. The actual lengths should not be very far from those given by the formulas when the optimum settings are finally determined. As already pointed out, the reflector length may be somewhat greater when adjusted to give maximum front-to-back ratio.

Radiation from the transmission line must be eliminated, or at least reduced to a very low value compared with the radiation from the antenna itself, if errors are to be avoided. Conditions are usually favorable to low line radiation in horizontally polarized rotatable parasitic arrays because the line is usually symmetrical

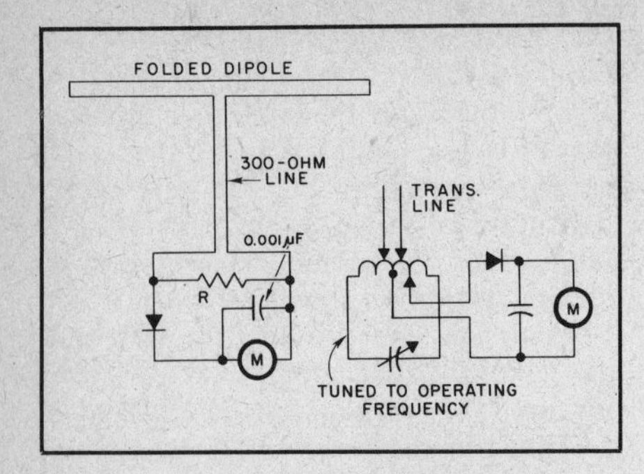


Fig. 55 — Field-strength measurement setup. The folded dipole should be at least as high as the antenna under test and should be three or more wavelengths away. R should be a 300-ohm composition resistor to provide a proper load for the line, so that a line of any desired length can be used. If the sensitivity is not high enough with this arrangement the alternative connections at the right will result in increased meter readings. The taps are adjusted for maximum reading, keeping the transmission-line taps spaced equally on either side of the coil center tap. The indicating meter, M, may be either a microammeter or 0-1 milliammeter.

with respect to the antenna and is brought away perpendicular to it, at least for a half wavelength or so. Nevertheless the line radiation can be appreciable unless the line is detuned as described in Chapter 5. With coaxial line some method of line balancing at the antenna always should be incorporated to avoid "skewing" the beam pattern or lowering the front-to-back ratio.

After arriving at the optimum adjustments at the frequency for which the antenna was designed, the performance should be checked over a frequency range either side of the design frequency to observe the sharpness of response. If the field strength falls off rapidly with frequency, it may be desirable to shorten the director a bit to increase the gain at frequencies above resonance and lengthen the reflector slightly to increase it at frequencies below resonance. Do not confuse the change in SWR with the change in antenna gain. The antenna itself may give good gain over a considerable frequency range, but the SWR may vary between wide limits in this range. To check the *antenna* behavior, keep the power input to the transmission line constant and rematch the driven element to the line, as suggested above, whenever the line losses increase appreciably. If such rematching is found necessary over the band of frequencies to be used, it may be advisable to retune the system to give a higher input resistance and thus decrease the selectivity, even though some gain is sacrificed in so doing.

**Adjustment by Reception**

As an alternative to applying power to the array and checking the field strength, it is possible to adjust the array by measuring received signal strength. It is impracticable to do this on distant signals because of fading. The most reliable method is to erect a temporary antenna of the type recommended for field-strength measurements (Fig. 55) and excite it from a low-power oscillator. The same precautions with respect to distance between the two antennas apply.

In this method, as in the one where the transmitting antenna is excited, it is necessary to minimize line radiation and pickup if the results are to be reliable. The same tests may be applied. However, it is less easy to keep the SWR under control. In the receiving case the SWR on the transmission line depends on the load presented by the *receiver* to the line. Under most conditions the SWR will be reasonably constant over an amateur band, although its value may not be known. However, the energy transfer from the antenna to the line depends on the mismatch between the driven element and the line. There is no convenient way to check this in the receiving case. About all that can be done is to apply power to the array after a set of tuning conditions has been reached, and then rematch at the driven element if necessary. After rematching, the measurement will have to be repeated. Thus double checking is necessary if the results are to be comparable with those obtained by the field-strength method.

## THE QUAD ANTENNA

In this chapter it has been assumed that the various antenna arrays have been assemblies of linear half-wave (or approximately half-wave) dipole elements. However, other element forms may be used according to the same basic principles. For example, loops of various types may be combined into directive arrays. A popular type of parasitic array using loops is the quad antenna, in which loops having a perimeter of one wavelength are used in much the same way as dipole elements in the Yagi antenna.

The quad antenna was originally designed by Moore, W9LZX, in the late 1940s. Since its inception, there has been extensive controversy whether the quad is a better performer than a Yagi. This argument continues, but over the years several facts have become apparent. For example, Lindsay (W7ZQ, ex-WØHTH) has made many comparisons between quads and Yagis. His data show that the quad has a gain of approximately 2 decibels over a Yagi for the same array length. Another argument that has existed is that for a given array height, the quad has a lower angle of radiation than a Yagi. Even among authorities there is disagreement on this point. However, the H-plane pattern of a quad is slightly greater than that of a Yagi at the half-power points. This means that the quad covers a wider area in the vertical plane.

The full-wave loop has been discussed

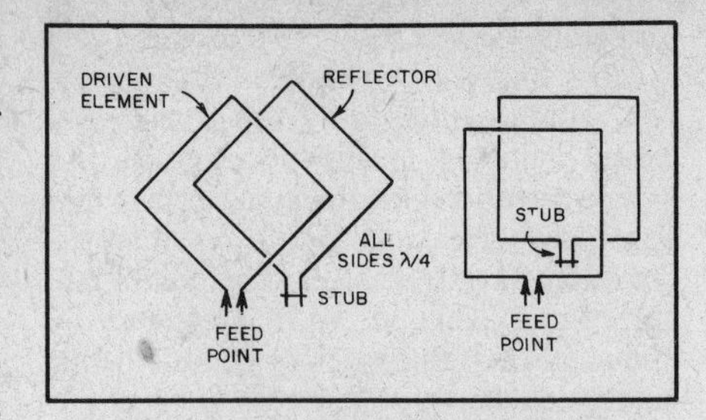


Fig. 56 — The basic quad antenna, with driven loop and reflector loop. The loops are electrically one wavelength in circumference (1/4 wavelength on a side). Both configurations shown give horizontal polarization; for vertical polarization, the driven element should be fed at one of the side corners in the arrangement at the left, or at the center of a vertical side in the "square" quad at the right.

in Chapter 2. Two such loops, one as a driven element and one as a reflector, are shown in Fig. 56. This is the original version of the quad; in subsequent development, loops tuned as directors have been added in front of the driven element. The square loops may be mounted either with the corners lying on horizontal and vertical lines, as shown at the left, or with two sides horizontal and two vertical (right). The feed points shown for these two cases will result in horizontal polarization, which is commonly used.

The parasitic element is tuned in much the same way as the parasitic element in a Yagi antenna. That is, the parasitic loop is tuned to a lower frequency than the operating frequency when the element is to act as a reflector, and to a higher frequency when it acts as a director. Fig. 56 shows the parasitic element with an adjustable tuning stub, a convenient method of tuning since the resonant frequency can be changed simply by changing the position of the shorting bar on the stub. In practice, it has been found that the length around the loop should be approximately 3 percent greater than the self-resonant length if the element is a reflector, and about 3 percent shorter than the self-resonant length if the parasitic element is a director. Approximate formulas for the loop lengths in feet are

$$\text{Driven element} = \frac{1005}{f(\text{MHz})}$$

$$\text{Reflector} = \frac{1030}{f(\text{MHz})}$$

$$\text{Director} = \frac{975}{f(\text{MHz})}$$

for quad antennas intended for operation below 30 MHz. At vhf, where the ratio of loop circumference to conductor diameter is usually relatively small, the circumference must be increased in comparison to the wavelength. For example, a one-wavelength loop constructed of quarter-inch tubing for 144 MHz should have a circumference about 4.5 percent greater than the wavelength in free space, as compared to the approximately 2 percent increase in the formula above for the driven element.

In any case, on-the-ground adjustment is required if optimum results are to be secured, especially with respect to front-to-back ratio. The method of adjustment parallels that outlined previously for the Yagi antenna.

Element spacings of the order of 0.14 to 0.2 wavelength are generally used, the smaller spacings being employed in antennas having more than two elements, where the structural support for elements with larger spacings tends to become difficult. The feed-point impedances of antennas having element spacings of this order have been found to be in the 40- to 60-ohm range, so the driven element can be fed through coaxial cable with only a small mismatch. For spacings of the order of 0.25 wavelength (physically feasible for two elements, or for several elements at 28 MHz) the impedance more closely approximates the impedance of a driven loop alone (see Chapter 2) — that is, 80 or 90 ohms.

The feed methods described in Chapter 5 can be used, just as in the case of the Yagi.

**Directive Patterns and Gain**

The small gain of a one-wavelength loop over a half-wave dipole carries over into arrays of loops. That is, if a quad parasitic array and a Yagi having the same overall length (boom length) are compared, the quad will have approximately 2 dB greater gain than the Yagi, as mentioned earlier. This assumes that both antennas have the optimum number of elements for the antenna length; the number of elements is not necessarily the same in both when the antennas are long.

## THE LOG-PERIODIC DIPOLE ARRAY

The log-periodic dipole array (LPDA) consists of a system of driven elements, but not all elements in the system are active on a single frequency of operation. Depending upon its design parameters, the LPDA can be operated over a range of frequencies having a ratio of 2:1 or higher, and over this range its electrical characteristics — gain, feed-point impedance, front-to-back ratio, etc. — will remain more or less constant. This is not true of any of the types of antennas discussed earlier in this chapter, for either the gain factor or the front-to-back ratio, or both, deteriorate rapidly as the frequency of operation departs from the design frequency of the array. And because the antenna designs discussed earlier are based upon resonant elements, off-resonance operation introduces reactance which causes the SWR in the feeder system to increase.

As may be seen in Fig. 57, the log-periodic array consists of several dipole elements which are each of different lengths and different relative spacings. A distributive type of feeder system is used to excite the individual elements. The element lengths and relative spacings, beginning from the feed point for the array, are seen to increase smoothly in dimension, being greater for each element than for the previous element in the array. It is this feature upon which the design of the LPDA is based, and which permits changes in frequency to be made without greatly affecting the electrical operation. With changes in operating frequency, there is a smooth transition along the array of the elements which comprise the active region. The following information was provided by Peter Rhodes, K4EWG.

A good LPDA may be designed for any band, hf to uhf, and can be built to meet the amateur's requirements at nominal cost: high forward gain, good front-to-back ratio, low VSWR, and a boom length equivalent to a full sized three-element Yagi. The LPDA exhibits a relatively low SWR (usually not greater than 2 to 1) over a wide band of frequencies. A well-designed LPDA can yield a 1.3-to-1 SWR over a 1.8-to-1 frequency range with a typical directivity of 9.5 dB. (Directivity is the ratio of maximum radiation intensity in the forward direction to the average radiation intensity from the array. Assuming no resistive losses in the antenna system, 9.5 dB directivity equates to 9.5 dB gain over an isotropic radiator or approximately 7.4 dB gain over a half-wave dipole.

**Basic Theory**

The LPDA is frequency independent in that the electrical properties such as the mean resistance level, $R_o$, characteristic impedance of the feed line, $Z_o$, and driving-point admittance, $Y_o$, vary periodically with the logarithm of the frequency. As the frequency f1 is shifted to another frequency f2 within the pass-band of the antenna, the relationship is

$$f2 = f1/\tau \qquad \text{(Eq. 1)}$$

where

$\tau$ = a design parameter, a constant; $\tau < 1.0$. Also

$f3 = f1/\tau^2$
$f4 = f1/\tau^3$
.
.
.
$f_n = f1/\tau^{n-1}$
$n = 1, 2, 3, \ldots n$
f1 = lowest frequency
$f_n$ = highest frequency

The design parameter $\tau$ is a geometric constant near 1.0 that is used to determine the element lengths, $\ell$, and element spacings, d, as shown in Fig. 57. That is,

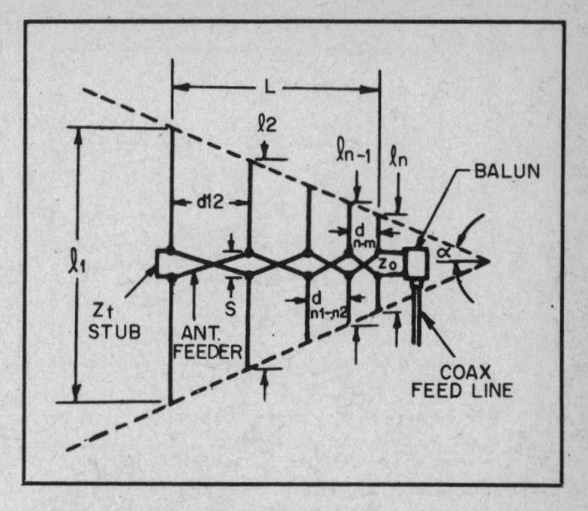


Fig. 57 — Schematic diagram of log-periodic dipole array, with some of the design parameters indicated. Design factors are:

$$\tau = \frac{\ell_n}{\ell_{n-1}} = \frac{d_{n,n-1}}{d_{n-2,n-1}}$$

$$\sigma = \frac{d_{n,n-1}}{2\ell_{n-1}}$$

$$h_n = \frac{\ell_n}{2},$$

where

$\ell$ = el. length
h = el. half length
d = element spacing
$\tau$ = design constant
$\sigma$ = relative spacing constant
S = feeder spacing
$Z_o$ = char. impedance of antenna feeder

$\ell 2 = \tau \ell 1$
$\ell 3 = \tau \ell 2$
.
.
.

$$\ell_n = \tau \ell_{(n-1)} \qquad \text{(Eq. 2)}$$

where

$\ell_n$ = shortest element length, and

$d23 = \tau d12$
$d34 = \tau d23$
.
.
.

$$d_{n-1,n} = \tau d_{n-2,n-1} \qquad \text{(Eq. 3)}$$

where

d23 = spacing between elements 2 and 3.

Each element is driven with a phase shift of 180° by switching or alternating element connections, as shown in Fig. 57. The dipoles near the input, being nearly out of phase and close together, nearly cancel each other's radiation. As the element spacing, d, expands there comes a point along the array where the phase delay in the transmission line combined with the 180° switch gives a total of 360°. This puts the radiated fields from the two dipoles in phase in a direction toward the apex. Hence, a lobe coming off the apex results.

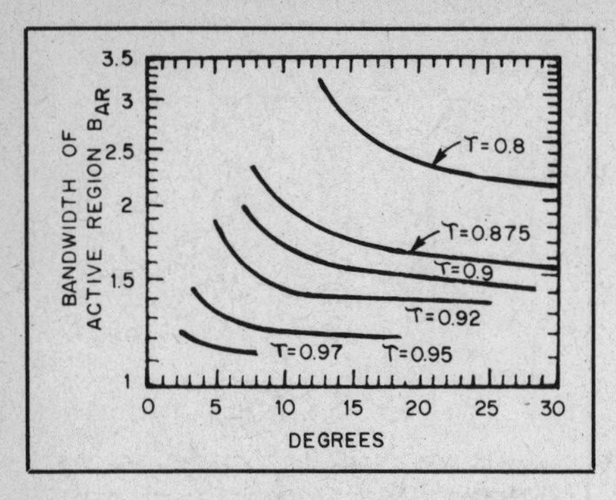

Fig. 58 — Design graph showing the relationships between the bandwidth of the active region, ∝ and τ.

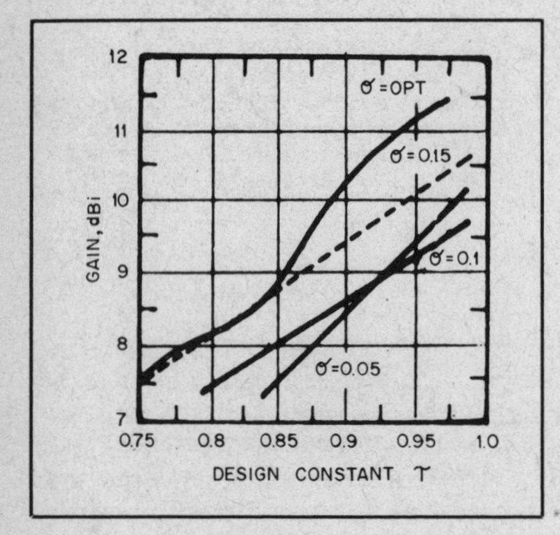

Fig. 59 — Showing the relationship between gain, τ and σ. Gain over a dipole may be obtained by subtracting 2.14 from the gain over isotropic values indicated.

This phase relationship exists in a set of dipoles known as the "active region." If we assume that an LPDA is designed for a given frequency range, then that design must include an active region of dipoles for the highest and lowest design frequency. It has a bandwidth which we shall call $B_{ar}$ (bandwidth of the active region).

Assume for the moment that we have a 12-element LPDA. Currents flowing in the elements are both real and imaginary, the real current flowing in the resistive component of the impedance of a particular dipole, and the imaginary flowing in the reactive component. Assume that the operating frequency is such that element number 6 is near to being half-wave resonant. The imaginary parts of the currents in shorter elements 7 to 12 are capacitive, while those in longer elements 1 to 5 are inductive. The capacitive current components in shorter elements 9 and 10 exceed the conductive components; hence, these elements receive little power from the feeder and act as parasitic directors. The inductive current components in longer elements 4 and 5 are dominant and they act like parasitic reflectors. Elements 6, 7 and 8 receive most of their power from the feeder and act like driven elements. The amplitudes of the currents in the remaining elements are small and they may be ignored as primary contributors to the radiation field. Hence, we have a generalized Yagi array with seven elements comprising the active region. It should be noted that this active region is for a specific set of design parameters ($\tau = 0.93$, $\sigma = 0.175$). The number of elements making up the active region will vary with τ and σ. Adding additional elements on either side of the active region cannot significantly modify the circuit or field properties of the array.

This active region determines the basic design parameters for the array, and sets the bandwidth for the structure, $B_s$. That is, for a design-frequency coverage of bandwidth B, there exists an associated bandwidth of the active region such that

$$B_s = B \times B_{ar} \quad \text{(Eq. 4)}$$

where

$$B = \text{operating bandwidth} = \frac{f_n}{f1} \quad \text{(Eq. 5)}$$

f1 = lowest freq., MHz
$f_n$ = highest freq., MHz

$B_{ar}$ varies with τ and ∝ as shown in Fig. 58. Element lengths which fall outside $B_{ar}$ play an insignificant role in the operation of the array. The gain of an LPDA is determined by the design parameter τ and the relative element spacing constant σ. There exists an optimum value for σ, $\sigma_{opt}$, for each τ in the range $0.8 \leqslant \tau < 1.0$, for which the gain is maximum; however, the increase in gain achieved by using $\sigma_{opt}$ and τ near 1.0 (i.e., $\tau = 0.98$) is only 3 dB when compared with the minimum σ ($\sigma_{min} = 0.05$) and $\tau = 0.9$, shown in Fig. 59.

An increase in τ means more elements and optimum σ means a long boom. A high-gain (8.5 dBi) LPDA can be designed in the hf region with $\tau = 0.9$ and $\sigma = 0.05$. The relationship of τ, σ and ∝ is as follows:

$$\sigma = (1/4)(1 - \tau) \cot \propto \quad \text{(Eq. 6)}$$

where
∝ = 1/2 the apex angle
τ = design constant
σ = relative spacing constant

$$\text{also } \sigma = \frac{d_{n,n-1}}{2\ell_{n-1}} \quad \text{(Eq. 7)}$$

$$\sigma_{opt} = 0.258\tau - 0.066 \quad \text{(Eq. 8)}$$

The method of feeding the antenna is rather simple. As shown in Fig. 57, a balanced feeder is required for each element, and all adjacent elements are fed with a 180° phase shift by alternating element connections. In this section the term *antenna feeder* is defined as that line which connects each adjacent element. The *feed line* is that line between antenna and transmitter. The characteristic impedance of the antenna feeder, $Z_o$, must be determined so that the feed-line impedance and type of balun can be determined. The antenna-feeder impedance $Z_o$ depends on the mean radiation resistance level $R_o$ (required input impedance of the active region elements — see Fig. 60) and average characteristic impedance of a dipole, $Z_{av}$. ($Z_{av}$ is a function of element radius a and the resonant element half length, where $h = \lambda/4$. See Fig. 61). The relationship is as follows:

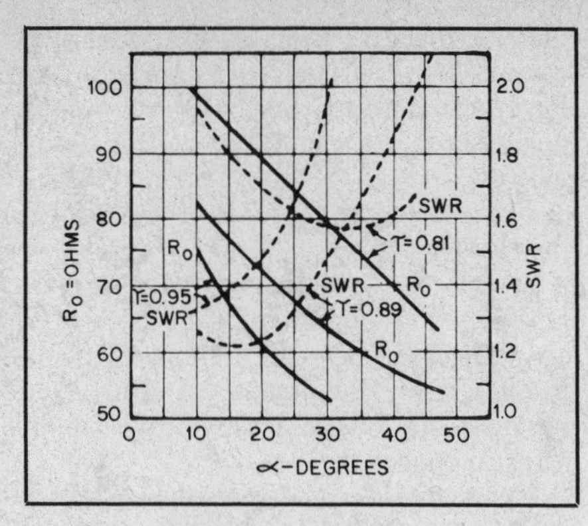

Fig. 60 — Showing various design parameters versus mean resistance level.

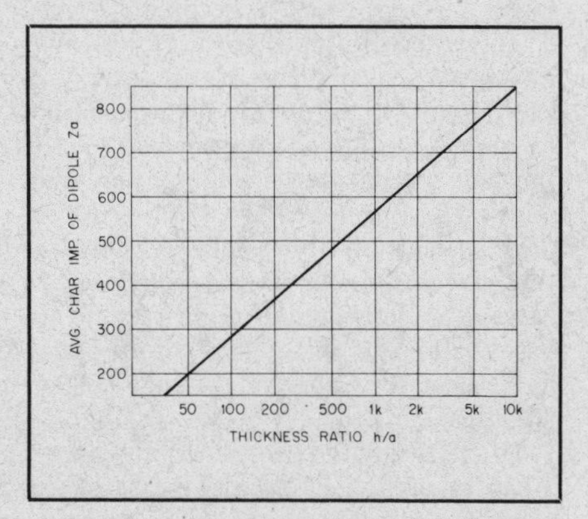

Fig. 61 — Average characteristic impedance of dipole versus thickness ratio. (See Eq. 10.)

$$Z_o = \frac{R_o^2}{8\sigma' Z_{av}} + R_o \sqrt{\left(\frac{R_o}{8\sigma' Z_{av}}\right)^2 + 1} \quad \text{(Eq. 9)}$$

where
$Z_o$ = characteristic impedance of feeder
$R_o$ = mean radiation resistance level of required input impedance of the active region
$Z_{av}$ = average characteristic impedance of a dipole

$$= 120 \left(\ln \frac{h}{a} - 2.55\right) \quad \text{(Eq. 10)}$$

h = el. half length
a = radius of el.

$$\sigma' = \text{mean spacing factor} = \frac{\sigma}{\sqrt{\tau}} \quad \text{(Eq. 11)}$$

From Fig. 60 we can see that $R_o$ decreases with increasing $\tau$ and increasing $\alpha$. Also the VSWR with respect to $R_o$ has a minimum value of about 1.1 to 1 at $\sigma$ optimum, and a value of 1.8 to 1 at $\sigma = 0.05$. These SWR values are acceptable when using standard RG-8/U 52-ohm and RG-11/U 72-ohm coax for the feed line. However, a one-to-one VSWR match can be obtained at the transmitter end using a coax-to-coax Transmatch. A Transmatch will enable the transmitter low-pass filter to see a 52-ohm load on each frequency within the array passband. The Transmatch also eliminates possible harmonic radiation caused by the frequency-independent nature of the array.

Once the value of $Z_o$ has been determined for each band within the array passband, the balun and feed line may be chosen. That is, if $Z_o = 100$ ohms, a good choice for the balun would be 1 to 1 balanced to unbalanced, and 72-ohm coax feed line. If $Z_o = 220$ ohms, choose a 4 to 1 balun and 52-ohm coax feed line, and so on. The balun may be omitted if the array is to be fed with an open-wire feed line.

The terminating impedance, $Z_t$, may be omitted. However, if it is used, it should have a length no longer than $\lambda_{max}/8$. The terminating impedance tends to increase the front-to-back ratio for the lowest frequency used. For hf-band operation a 6-inch shorting jumper wire may be used for $Z_t$. When $Z_t$ is simply a short-circuit jumper the longest element behaves as a passive reflector. It also might be noted that one could increase the front-to-back ratio on the lowest frequency by moving the passive reflector (no. 1 element) a distance of 0.15 to 0.25 $\lambda$ behind element no. 2, as would be done in the case of an ordinary Yagi parasitic reflector. This of course would necessitate lengthening the boom. The front-to-back ratio increases somewhat as the frequency increases. This is because more of the shorter inside elements form the active region, and the longer elements become additional reflectors.

### Design Procedure

A systematic step-by-step design procedure of the LPDA follows. This procedure may be used for designing any LPDA for any desired bandwidth.

1) Decide on an operating bandwidth B between f1, lowest frequency and $f_n$, highest frequency, using Eq. 5.

2) Choose $\tau$ and $\sigma$ to give a desired gain (Fig. 59).

$0.8 \leqslant \tau \leqslant 0.98$

$0.05 \leqslant \sigma \leqslant \sigma_{opt}$

The value of $\sigma_{opt}$ may be determined from Eq. 8.

3) Determine the apex half-angle $\alpha$

$$\cot \alpha = \frac{4\sigma}{1 - \tau}$$

4) Determine the bandwidth of the active group $B_{ar}$ from Fig. 58.

5) Determine the structure (array) bandwidth $B_s$ from Eq. 4.

6) Determine the boom length, L, number of elements, N, and longest element length, $\ell 1$.

$$L = \left[\frac{1}{4}\left(1 - \frac{1}{B_s}\right) \cot \alpha\right] \lambda_{max} \qquad \text{(Eq. 12)}$$

$$N = 1 + \frac{\log B_s}{\log \frac{1}{\tau}} \qquad \text{(Eq. 13)}$$

$$\ell 1 = \frac{492}{f1}$$

where $\lambda_{max}$ = longest free-space wavelength = 984/f1. Examine L, N and $\ell 1$ and determine whether or not the array size is acceptable for your needs. If the array is too large, increase $\alpha$ by 5° and repeat steps 2 through 6.

7) Determine the terminating stub $Z_t$. (Note: For hf arrays short out the longest element with a 6-inch jumper. For vhf and uhf arrays use:

$Z_t = \lambda_{max}/8$.

8) Once the final values of $\tau$ and $\sigma$ are found, the characteristic impedance of the feeder $Z_o$ must be determined so the type of balun and feed line can be found. Use Eq. 9. Determine $R_o$ from Fig. 60, $Z_{av}$ from Fig. 61 and $\sigma^1$ from Eq. 11. Note: Values for h/a, $Z_{av}$ and $Z_o$ must be determined for each amateur band within the array passband. Choose the element half-length h nearest h = $\lambda$/4, at the center frequency of each amateur band. Once $Z_o$ is found for each band, choose whatever combination of balun and feed line will give the lowest SWR on each band.

9) Solve for the remaining element lengths from Eq. 2.

10) Determine the element spacing d12 from

$$d12 = 1/2\,(\ell 1 - \ell 2) \cot \alpha \qquad \text{(Eq. 14)}$$

and the remaining element-to-element spacings from Eq. 3.

This completes the design. Construction information for an array designed by this procedure is contained in Chapter 9. The measured radiation pattern for a 12-element LPDA is shown in Fig. 62.

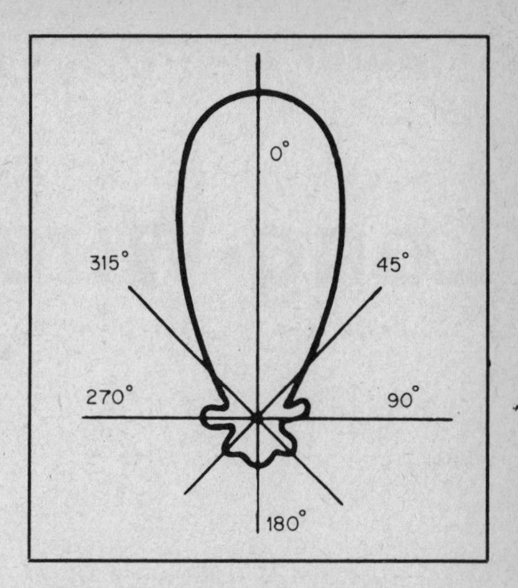


Fig. 62 — Measured radiation pattern for the lowest frequency band (14 MHz) of a 12-element 13-30 MHz log periodic dipole array. For its Design parameters, $\tau = 0.9$ and $\sigma = 0.05$. The measured front-to-back ratio is 14.4 dB at 14 MHz, and increases to 21 dB at 28 MHz.

## LOG-YAG ARRAYS

This section has dealt with the basic LPDA system. However, there are several high-gain array possibilities using this type of antenna as a basis. Tilting the elements toward the apex will increase the gain 3 to 5 dB. Adding parasitic directors and a reflector will increase both gain and front-to-back ratio for a specific frequency within the passband. The LPDA-Yagi combination is very simple. Use the LPDA design procedures within the set of driven elements, and place parasitic elements at normal Yagi spacings from the LPDA end elements. Use standard Yagi design procedures for the parasitic elements. An example of a single-band high-gain LPDA-Yagi would be a 2- or 3-element LPDA for 21.0 to 21.45 MHz with the addition of two or three parasitic directors and one parasitic reflector. The combinations are endless.

## BIBLIOGRAPHY

Source material and more extended discussion of topics covered in this chapter can be found in these references.


Brown, "Directional Antennas." *Proc. I.R.E.*, January 1937.

Carrel, "The Design of Log-Periodic Dipole Antennas." *1961 I.R.E. International Convention Record*, Part 1, Antennas and Propagation, pp. 61-75; also Ph.D. thesis, Univ. of Illinois, Urbana, 1961.

Carter, "Circuit Relations in Radiating Systems and Applications to Antenna Problems." *Proc. I.R.E.*, June 1932.

Ehrenspek and Poehler, "A New Method of Obtaining Maximum Gain from Yagi Antennas." *I.R.E. Transactions on Antennas and Propagation*, October 1959.

Greenblum, "Notes on the Development of Yagi Arrays." *QST*, Part 1, August 1956; Part 2, September 1956.

Isbell, "Log-Periodic Dipole Arrays." *I.R.E. Transactions on Antennas and Propagation*, Vol. AP-8, No. 3, May 1960, pp. 260-267.

Kasper, "Array Design with Optimum Antenna Spacing." *QST*, November 1960.

King, Mack and Sandler, *Arrays of Cylindrical Dipoles*. London: Cambridge Univ. Press, 1968.

Kmosko and Johnson, "*Long* Long Yagis." *QST*, January 1956.

Laport, *Radio Antenna Engineering*. New York: McGraw-Hill Book Co., 1952.

Lawson, "Simple Arrays or Vertical Antenna Elements." *QST*, May 1971.

Rhodes, "The Log-Periodic Dipole Array." *QST*, November 1973.

Rumsey, *Frequency Independent Antennas*. New York: Academic Press, 1966, pp. 71-78.

Southworth, "Certain Factors Affecting the Gain of Directive Antennas." *Proc. I.R.E.*, September 1930.

Terman, *Radio Engineering*. New York: McGraw-Hill Book Co.

Uda and Mushiake, *Yagi-Uda Antenna*, Sendai, Japan: Sasaki Publishing Co.

Viezbicke, P. P., "Yagi Antenna Design," *NBS Technical Note 688*, National Bureau of Standards, Boulder, Colo., December 1976.

# Chapter 7

# Long-Wire Antennas

The power gain and directive characteristics of the harmonic wires (which are "long" in terms of wavelength) described in Chapter 2 make them useful for long-distance transmission and reception on the higher frequencies. In addition, long wires can be combined to form antennas of various shapes that will increase the gain and directivity over a single wire. The term "long wire," as used in this chapter, means any such configuration, not just a straight-wire antenna.

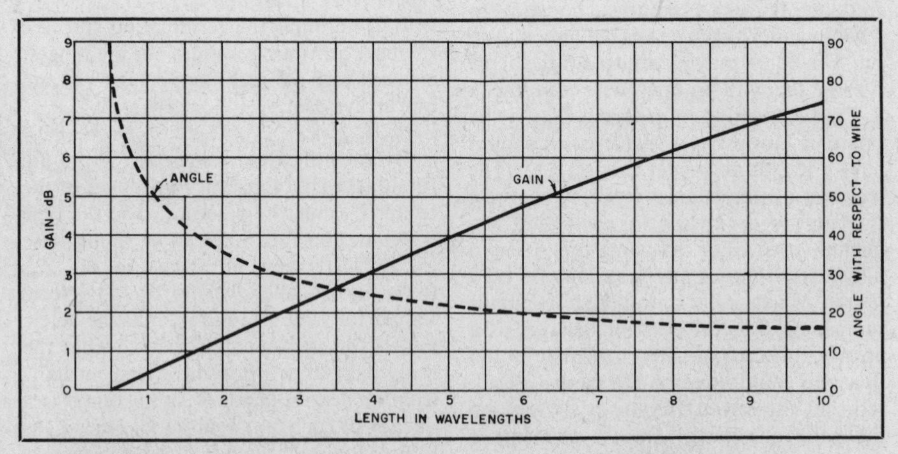


Fig. 1 — Theoretical gain of a long-wire antenna over a dipole as a function of wire length. The angle, with respect to the wire, at which the radiation intensity is maximum also is shown.

**Long Wires vs. Multielement Arrays**

In general, the gain obtained with long-wire antennas is not as great, when the space available for the antenna is limited, as can be secured from the multielement arrays in Chapter 6. To offset this, the long-wire antenna has advantages of its own. The construction of long-wire antennas is simple both electrically and mechanically, and there are no especially critical dimensions or adjustments. The long-wire antenna will work well and give satisfactory gain and directivity over a 2-to-1 frequency range; in addition, it will accept power and radiate it well on any frequency for which its overall length is not less than about a half wavelength. Since a wire is not "long," even at 28 MHz, unless its length is at least equal to a half wavelength on 3.5 MHz, any long-wire can be used on all amateur bands that are useful for long-distance communication.

As between two directive antennas having the same theoretical gain, one a multielement array and the other a long-wire antenna, many amateurs have found that the long-wire antenna seems more effective in reception. One possible explanation is that there is a diversity effect with a long-wire antenna because it is spread out over a large distance, rather than being concentrated in a small space; this may raise the average level of received energy for ionospheric propagation. Another factor is that long-wire antennas have directive patterns that are sharp in both the horizontal and vertical planes, and tend to concentrate the radiation at the low vertical angles that are most useful at the higher frequencies. This is not true of some types of multielement arrays.

**General Characteristics of Long-Wire Antennas**

Whether the long-wire antenna is a single wire running in one direction or is formed into a V, rhombic, or some other configuration, there are certain general principles that apply and some performance features that are common to all types. The first of these is that the power gain of a long-wire antenna as compared with a half-wave dipole is not considerable until the antenna is really "long" — long, that is, when the lengths are measured in wavelengths, rather than in a specific number of feet. The reason for this is that the fields radiated by elementary lengths of wire along the antenna do not combine, at a distance, in as simple a fashion as the fields from half-wave dipoles used as described in Chapter 6. There is no point in space, for example, where the distant fields from all points along the wire are exactly in phase (as they are, in the optimum direction, in the case of two or more collinear or broadside dipoles when fed with in-phase currents). Consequently, the field strength at a distance always is less than would be obtained if the same length of wire were cut up into properly phased and separately driven dipoles. As the wire is made longer the fields combine to form an increasingly intense main lobe, but this lobe does not develop appreciably until the wire is several wavelengths long. This is indicated by the curve showing gain in Fig. 1. The longer the antenna, the sharper the lobe becomes, and since it is really a cone of radiation about the wire in free space, it becomes sharper in all planes. Also, the greater the length, the smaller the angle with the wire at which the maximum radiation occurs.

### Directivity

Because many points along a long wire are carrying currents in different phase (usually with different current amplitude as well) the field pattern at a distance becomes more complex as the wire is made longer. This complexity is manifested in a series of minor lobes, the number of which increases with the wire length. The intensity of radiation from the minor lobes is frequently as great as, and sometimes greater than, the radiation from a half-wave dipole. The energy radiated in the minor lobes is not available to improve the gain in the major lobe, which is another reason why a long-wire antenna must be long to give appreciable gain.

Driven and parasitic arrays of the simple types described in Chapter 6 do not have minor lobes of any great consequence. For that reason they frequently seem to have better directivity than long-wire antennas, because their response in directions other than that at which the antenna is aimed is well down. This will be so even if a multielement array and a long-wire antenna have the same actual gain in the favored direction. For amateur work, particularly with directive antennas that cannot be rotated, the minor lobes of a long-wire antenna have some advantages. In most directions the antenna will be as good as a half-wave dipole, and in addition will give high gain in the most favored direction; thus a long-wire antenna (depending on the design) frequently is a good all-around radiator in addition to being a good directive antenna.

In the discussion of directive patterns of long-wire antennas in this chapter, it should be kept in mind that the radiation patterns of resonant long wires are based on the assumption that each half-wave section of wire carries a current of the same amplitude. As pointed out in Chapter 2, this is not exactly true, since energy is radiated as it travels along the wire. For this reason it is to be anticipated that, although the theoretical pattern is bidirectional and identical in both directions, actually the radiation (and reception) will be best in one direction. This effect becomes more marked as the antenna is made longer.

### Wave Angles

The wave angle at which maximum radiation takes place from a long-wire antenna depends largely on the same factors that operate in the case of simple dipoles and multielement antennas. That is, the directive pattern in the presence of ground is found by adding the free-space vertical-plane pattern of the antenna to the ground-reflection factors for the particular antenna height used. These factors

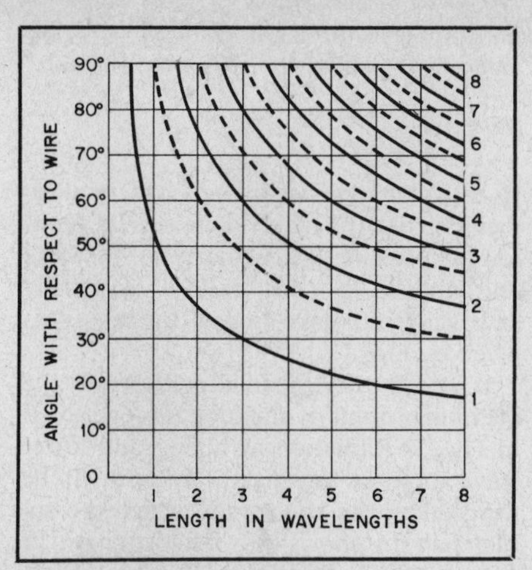


Fig. 2 — Angles at which radiation from long wires is maximum (solid curves) and zero (broken curves). The major lobe, no. 1, has the power gains given by Fig. 1. Secondary lobes have smaller amplitude, but the maxima may exceed the radiation intensity from a half-wave dipole.

are discussed in Chapter 2.

As mentioned a few paragraphs ago, the free-space radiation pattern of a long-wire antenna has a major lobe that forms a cone around the wire. The angle at which maximum radiation takes place becomes smaller, with respect to the wire, as the wire length is increased, also shown in Fig. 1. For this reason a long-wire antenna is primarily a low-angle radiator when installed horizontally above the ground. Its performance in this respect is improved by selecting a height that also tends to concentrate the radiation at low wave angles. This is also discussed in Chapter 2.

Antenna systems formed from ordinary horizontal dipoles that are not stacked have, in most cases, a rather broad vertical pattern; the wave angle at which the radiation is maximum therefore depends chiefly on the antenna height. However, with a long-wire antenna the wave angle at which the major lobe is maximum can never be higher than the angle at which the first null occurs (see Fig. 2), even if the antenna height is very low. (The efficiency may be less at very small heights, partly because the pattern is affected in such a way as to put a greater proportion of the total power into the minor lobe.) The result is that when radiation at wave angles below 15 or 20 degrees is under consideration, a long-wire antenna is less sensitive to height than are the multi-element arrays or a simple dipole. To assure good results, however, the antenna should have a height equivalent to at least a half wavelength at 14 MHz — that is, a minimum height of about 30 feet. Great heights will give a worthwhile improvement at wave angles below 10 degrees.

With an antenna of fixed physical length and height, both length and height increase, in terms of wavelength, as the frequency is increased. The overall effect is that both the antenna and the ground reflections tend to keep the system operating at high effectiveness throughout the frequency range. At low frequencies the wave angle is raised, but high wave angles are useful at 3.5 and 7 MHz. At high frequencies the converse is true. Good all-around performance usually results on all bands when the antenna is designed to be optimum in the 14-MHz band.

### Calculating Length

In this chapter, lengths are always discussed in terms of wavelengths. There can obviously be nothing very critical about wire lengths in an antenna system that will work over a frequency range including several amateur bands. The antenna characteristics change very slowly with length, except when the wires are short (i.e., around one wavelength), and there is no need to try to establish exact resonance at a particular frequency.

The formula for harmonic wires given in Chapter 2 is quite satisfactory for determining the lengths of any of the antenna systems to be described. For convenience, the formula is repeated here in the following form:

$$\text{Length (feet)} = \frac{984\,(N - 0.025)}{\text{Freq. (MHz)}}$$

where N is the number of wavelengths in the antenna. In cases where exact resonance is desired for some reason (for obtaining a resistive load for a transmission line at a particular frequency, for example) it is best established by trimming the wire length until measurement of the resonant frequency shows it to be correct.

## LONG SINGLE WIRES

In Fig. 1 the solid curve shows that the gain in decibels of a long wire increases almost linearly with the length of the antenna. The gain does not become appreciable until the antenna is about four wavelengths long, where it is equivalent to doubling the transmitter power (3 dB). The actual gain over a half-wave dipole when the antenna is at a practical height above ground will depend on the way in which the radiation resistance of the long-wire antenna and the comparison dipole are affected by the height. The exact way in which the radiation resistance of a long wire varies with height depends on its length. In general, the resistance does not fluctuate as much, in terms of percentage, as does the resistance of a half-wave antenna. This is particularly true at heights from one-half wavelength up.

The nulls bounding the lobes in the directive pattern of a long wire are fairly sharp and are frequently somewhat obscured, in practice, by irregularities in the pattern. The locations of nulls and maxima for antennas up to eight wavelengths long are shown in Fig. 2.

### Orientation

The broken curve of Fig. 1 shows the angle with the wire at which the radiation intensity is maximum. As shown in Chapter 2, there are two main lobes to the directive patterns of long-wire antennas; each makes the same angle with respect to the wire. The solid pattern, considered in free space, is the hollow cone formed by rotating the wire on its axis.

When the antenna is mounted horizontally above the ground, the situation depicted in Fig. 3 exists. Only one of the two lobes is considered in this drawing, and its lower half is cut off by the ground. The maximum intensity of radiation in the remaining half occurs through the broken-line semicircle; that is, the angle B (between the wire direction and the line marked *wave direction*) is the angle given by Fig. 1 for the particular antenna length used.

In the practical case, there will be some wave angle (A) that is optimum for the frequency and the distance between the transmitter and receiver. Then for that wave angle the wire direction and the optimum geographical direction of transmission are related by the angle C. If the wave angle is very low, B and C will be practically equal. But as the wave angle becomes higher the angle C becomes smaller; in other words, the best direction of transmission and the direction of the wire more nearly coincide. They coincide exactly when C is zero; that is, when the wave angle is the same as the angle given by Fig. 1.

The maximum radiation from the antenna can be aligned with a particular geographical direction at a given wave angle by means of the following formula:

$$\cos C = \frac{\cos B}{\cos A}$$

In most amateur work the chief requirement is that the wave angle should be as low as possible, particularly at 14 MHz and above. In such case it is usually satisfactory to make angle C the same as is given by Fig. 1.

It should be borne in mind that only the *maximum* point of the lobe is represented in Fig. 3. Radiation at higher and lower wave angles in any given direction will be proportional to the way in which the actual pattern shows the field strength to vary as compared with the maximum point of the lobe.

### Tilted Wires

Fig. 3 shows that when the wave angle is equal to the angle which the maximum intensity of the lobe makes with the wire, the best transmitting or receiving direction is that of the wire itself. If the wave angle is less than the lobe angle, the best direction can be made to coincide with the direction of the wire by tilting the wire enough to make the lobe and wave angle coincide. This is shown in Fig. 4, for the case of a one-wavelength antenna tilted so that the maximum radiation from one lobe is horizontal to the left, and from the other is horizontal to the right (zero wave angle). The solid pattern can be visualized by imagining the plane diagram rotating about the antenna as an axis.

Since the antenna is neither vertical nor horizontal in this case, the radiation is part horizontally polarized and part vertically polarized. Computing the effect of the ground becomes complicated, because the horizontal and vertical components must be handled separately. In general, the directive pattern at any given wave angle becomes unsymmetrical when the antenna is tilted. For small amounts of tilt (less than the amount that directs the lobe angle horizontally) and for low wave angles the effect is to shift the optimum direction closer to the line of the antenna. This is true in the direction in which the antenna slopes downward. In the opposite direction the low-angle radiation is reduced.

### Feeding Long Wires

It has been pointed out in Chapter 5 that a harmonic antenna can be fed only at the end or at a current loop. Since a current loop changes to a node when the antenna is operated at any even multiple of the frequency for which it is designed, a long-wire antenna will operate as a true long wire *on all bands* only when it is fed at the end.

A common method of feeding is to use a resonant open-wire line, as described in Chapter 5. This system will work on all bands down to the one, if any, at which the antenna is only a half wave long. Any convenient line length can be used if the transmitter is matched to the line input impedance by the methods described in Chapter 4.

Two arrangements for using nonresonant lines are given in Fig. 5. The one at A

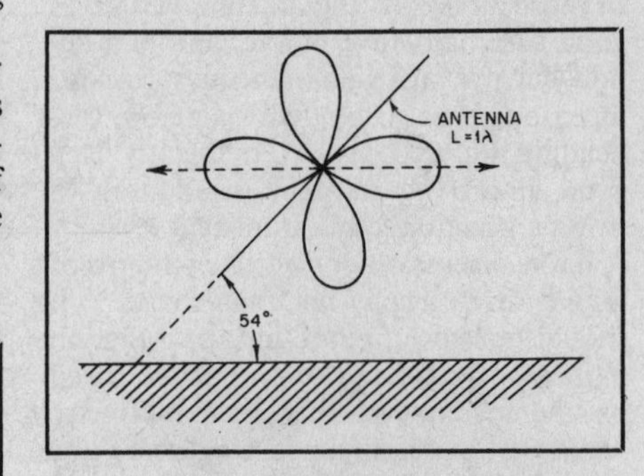


Fig. 4 — Alignment of lobes for horizontal transmission by tilting a long wire in the vertical plane.

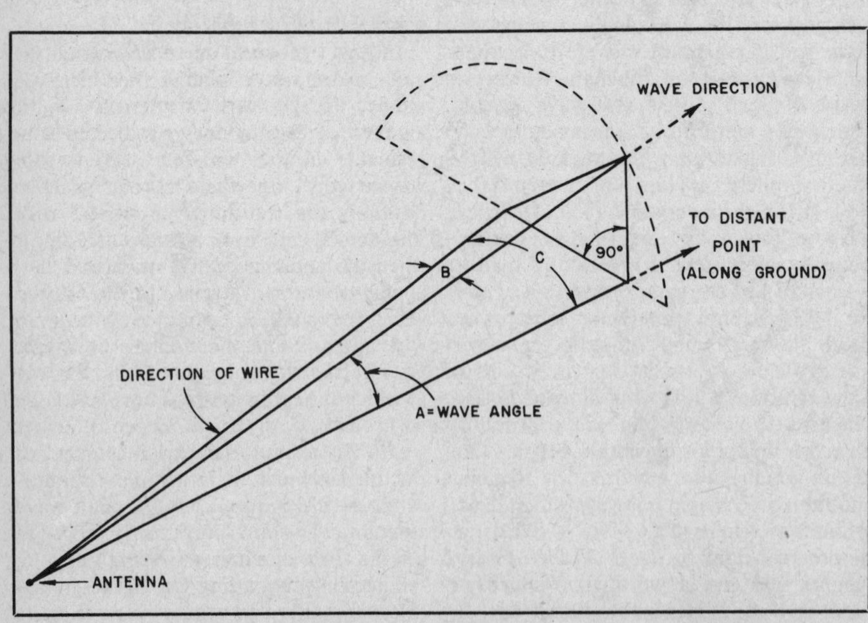


Fig. 3 — This drawing shows how the hollow cone of radiated energy from a long wire (broken-line arc) results in different wave angles (A) for various angles between the direction of the wire and the direction to the distant point (C).

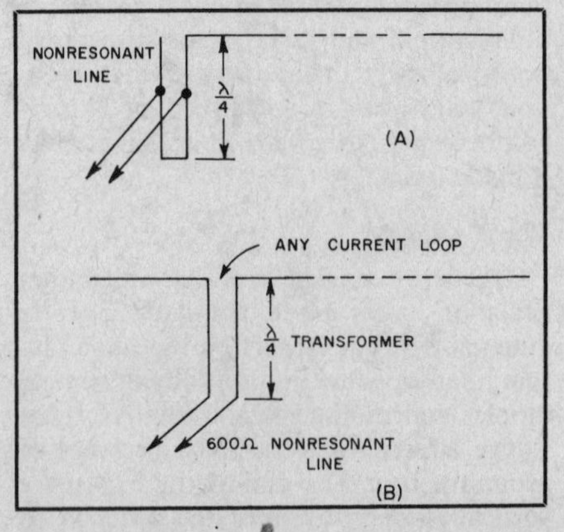


Fig. 5 — Methods of feeding long single-wire antennas.

is useful for one band only since the matching section must be a quarter wave long, approximately, unless a different matching section is used for each band. In B, the Q-section impedance should be adjusted to match the antenna to the line as described in Chapter 5, using the value of radiation resistance given in Chapter 2, Fig. 20. This method is best suited to working with a 600-ohm transmission line. Although it will work as designed only on one band, the antenna can be used on other bands by treating the line and matching transformer as a resonant line. In such case, as mentioned earlier, the antenna will not radiate as a true long wire on even multiples of the frequency for which the matching system is designed.

The end-fed arrangement, although the most convenient when tuned feeders are used, suffers the disadvantage that there is likely to be a considerable antenna current on the line, as described in Chapter 5. In addition, the antenna reactance changes rapidly with frequency for the reasons outlined in Chapter 2. Consequently, when the wire is several wavelengths long a relatively small change in frequency — a fraction of the width of a band — may require major changes in the adjustment of the transmitter-to-line coupling apparatus. Also, the line becomes unbalanced at all frequencies between those at which the antenna is exactly resonant. This leads to a considerable amount of radiation from the line. The unbalance can be overcome by using two wires in one of the arrangements described in succeeding sections.

## COMBINATIONS OF RESONANT LONG WIRES

The directivity and gain of long wires may be increased by using two wires so placed in relation to each other as to make the fields from both combine to produce the greatest possible field strength at a distant point. The principle is similar to that used in forming the multielement arrays described in Chapter 6. The maximum radiation from a long wire occurs at an angle of less than 90 degrees with respect to the wire, however, so different physical relationships must be used.

### Parallel Wires

One possible method of using two (or more) long wires is to place them in parallel, with a spacing of 1/2 wavelength or so, and feed the two in phase. In the direction of the wires the fields will add up in phase. However, since the wave angle is greatest in the direction of the wire, as shown by Fig. 3, this method will result in rather high-angle radiation unless the wires are several wavelengths long. The wave angle can be lowered, for a given antenna length, by tilting the wires as described earlier. With a parallel arrangement of this sort the gain should be about 3 dB over a single wire of the same length, at spacings in the vicinity of 1/2 wavelength.

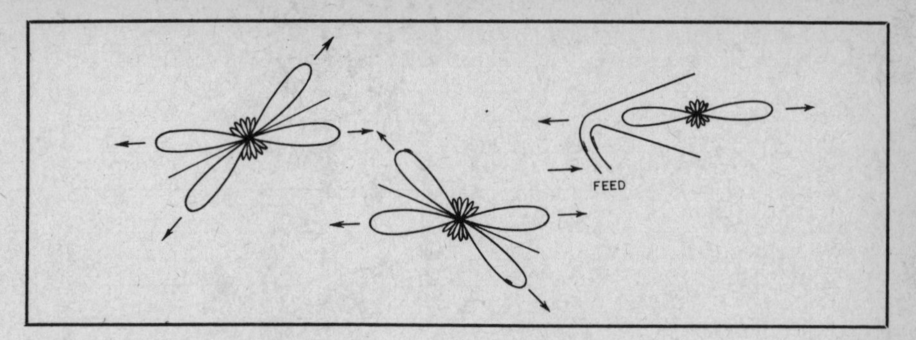


Fig. 6 — Two long wires and their respective patterns are shown at the left. If these two wires are combined to form a V with an angle that is twice that of the major lobes of the wires and with the wires excited out of phase, the radiation along the bisector of the V adds and the radiation in the other directions tends to cancel.

### THE V ANTENNA

Instead of using two long wires parallel to each other, they may be placed in the form of a horizontal V, with the angle at the apex of the V equal to twice the angle given by Fig. 1 for the particular length of wire used. The currents in the two wires should be out of phase. Under these conditions the plane directive patterns of the individual wires combine as is indicated in Fig. 6. Along a line in the plane of the antenna and bisecting the V the fields from the individual wires reinforce each other at a distant point. The other pair of lobes in the plane pattern is more or less eliminated, so that the pattern becomes essentially bidirectional.

The directional pattern of an antenna of this type is sharper in both the horizontal and vertical planes than the patterns of the individual wires composing it. Maximum radiation in both planes is along the line bisecting the V. There are minor lobes in both the horizontal and vertical patterns but if the legs are long in terms of wavelength the amplitude of the minor lobes is small. When the antenna is mounted horizontally above the ground, the wave angle at which the radiation from the major lobe is maximum is determined by the height, but cannot exceed the angle values shown in Fig. 1 for the leg length used. Only the minor lobes give high-angle radiation.

The gain and directivity of a V depend on the length of the legs. An approximate idea of the gain for the V antenna may be obtained by adding 3 dB to the gain value from Fig. 1 for the corresponding leg length. The actual gain will be modified by the mutual impedance between the sides of the V, and will be somewhat higher than indicated by the values determined as above, especially at the longer leg lengths. With 8-wavelength legs, the gain is approximately 4 dB greater than that indicated for a single wire in Fig. 1.

### Lobe Alignment

It is possible to align the lobes from the individual wires with a particular wave angle by the method described in connection with Fig. 3. At very low wave angles the change in the apex angle is extremely small; for example, if the desired wave angle is 5 degrees the apex angles of twice the value given in Fig. 1 will not be reduced more than a degree or so, even at the longest leg lengths which might be used.

When the legs are long, alignment does not necessarily mean that the greatest signal strength will be secured at the wave angle for which the apex angle is chosen. It must be remembered that the polarization of the radiated field is the same as that of a plane containing the wire. As illustrated by the diagram of Fig. 3, at any wave angle other than zero the plane containing the wire and passing through the desired wave angle is not horizontal. In the limiting case where the wave angle and the angle of maximum radiation from the wire are the same the plane is vertical, and the radiation at that wave angle is vertically polarized. At in-between angles the polarization consists of both horizontal and vertical components.

When two wires are combined into a V the polarization planes have opposite slopes. In the plane bisecting the V, this makes the horizontally polarized components of the two fields add together numerically, but the vertically polarized components are out of phase and cancel completely. As the wave angle is increased the horizontally polarized components become smaller, so the intensity of horizontally polarized radiation decreases. On the other hand, the vertically polarized components become more intense but always cancel each other. The overall result is that although alignment for a given wave angle will increase the useful radiation at that angle, the wave angle at which maximum radiation occurs (in the direction of the line bisecting the V) is always below the wave angle for which the wires are aligned. As shown by Fig. 7, the difference between the apex angles required for optimum alignment of the lobes at wave angles of zero and 15 degrees is rather small, even when the legs

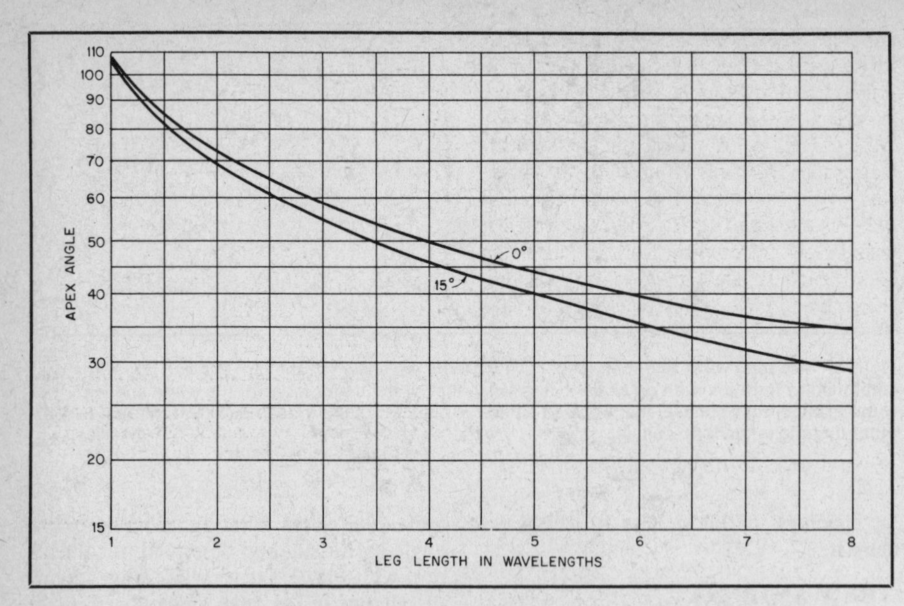


Fig. 7 — Apex angle of V antenna for alignment of main lobe at different wave angles, as a function of leg length in wavelengths.

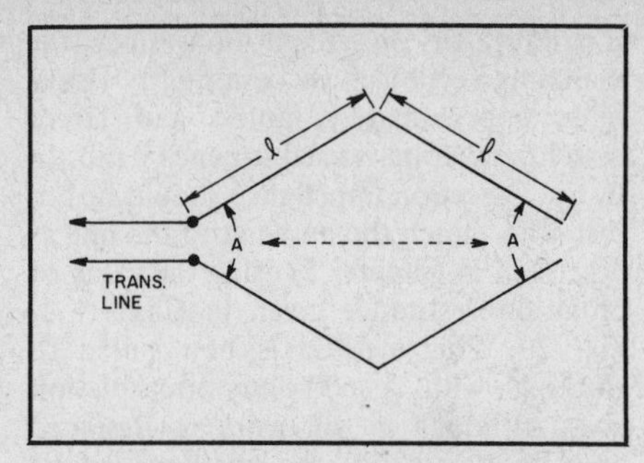


Fig. 8 — The resonant rhombic or diamond-shaped antenna. All legs are the same length, and opposite angles of the diamond are equal.

are many wavelengths long.

For long-distance transmission and reception the lowest possible wave angle usually is the best. Consequently, it is good practice to choose an apex angle between the limits represented by the two curves in Fig. 7. The actual wave angle at which the radiation is maximum will depend on the shape of the vertical pattern and the height of the antenna above ground.

When the leg length is small, there is some advantage in reducing the apex angle of the V because this changes the mutual impedance in such a way as to increase the gain of the antenna. For example, the optimum apex angle in the case of 1-$\lambda$ legs is 90 degrees.

### Multiband Design

When a V antenna is used over a range of frequencies — such as 14 to 28 MHz — its characteristics over the frequency range will not change greatly if the legs are sufficiently long at the lowest frequency. The apex angle, at zero wave angle, for a 5-wavelength V (each leg approximately 350 feet long at 14 MHz) is 44 degrees. At 21 MHz, where the legs are 7.5 wavelengths long, the optimum angle is 36 degrees, and at 28 MHz where the leg length is 10 wavelengths it is 32 degrees. Such an antenna will operate well on all three frequencies if the apex angle is about 35 degrees. From Fig. 7, a 35-degree apex angle with a 5-wavelength V will align the lobes at a wave angle of something over 15 degrees, but this is not too high when it is kept in mind that the maximum radiation actually will be at a lower angle. At 28 MHz the apex angle is a little large, but the chief effect will be a small reduction in gain and a slight broadening of the horizontal pattern, together with a tendency to reduce the wave angle at which the radiation is maximum. The same antenna can be used at 3.5 and 7 MHz, and on these bands the fact that the wave angle is raised is of less consequence, since high wave angles are useful. The gain will be small, however, because the legs are not long at these frequencies.

### Other V Combinations

The gain can be increased about 3 dB by stacking two Vs one above the other, a half wavelength apart, and feeding them in phase with each other. This will result in a lowered angle of radiation. The bottom V should be at least a quarter wavelength above the ground, and preferably a half wavelength.

Two V antennas can be broadsided to form a W, giving an additional 3-dB gain. However, two transmission lines are required and this, plus the fact that five poles are needed to support the system, renders it normally impractical for the amateur.

The V antenna can be made unidirectional by using another V placed an odd multiple of a quarter wavelength in back of the first and exciting the two with a phase difference of 90 degrees. The system will be unidirectional in the direction of the antenna with the lagging current. However, the V reflector is not normally employed by amateurs at low frequencies because it restricts the use to one band and requires a fairly elaborate supporting structure. Stacked Vs with driven reflectors could, however, be built for the 200- to 500-MHz region without much difficulty. The overall gain for such an antenna (two stacked Vs, each with a V reflector) is about 9 dB greater than the gains given in Fig. 1.

### Feeding the V

The V antenna is most conveniently fed by tuned feeders, since they permit multiband operation. Although the length of the wires in a V beam is not at all critical, it is important that both wires be of the same electrical length. If it is desired to use a nonresonant line, probably the most appropriate matching system is that using a stub or quarter-wave matching section. The adjustment is described in Chapter 5.

## THE RESONANT RHOMBIC ANTENNA

The diamond-shaped or rhombic antenna shown in Fig. 8 can be looked upon as two acute-angle Vs placed end-to-end. This arrangement has two advantages over the simple V that have caused it to be favored by amateurs. For the same total wire length it gives somewhat greater gain than the V; a rhombic 4 wavelengths on a leg, for example, has a gain of better than 1 dB over a V antenna with 8 wavelengths on a leg. And the directional pattern of the rhombic is less affected by frequency than the V when the antenna is used over a wide frequency range. This is because a change in frequency causes the major lobe from one leg to shift in one direction while the lobe from the opposite leg shifts the other way. This tends to make the optimum direction stay the same over a considerable frequency range. The disadvantage of the rhombic as compared with the V is that one additional support is required.

The same factors that govern the design of the V antenna apply in the case of the resonant rhombic. The angle A in the drawing is the same as that for a V having a leg length equal to $\ell$. If it is desired to align the lobes from individual wires with the wave angle, the curves of Fig. 7 may be used, again using the length of one leg in taking the data from the curves. The diamond-shaped antenna also can be operated as a nonresonant antenna, as described later in this chapter, and much of the discussion in that section applies to the resonant rhombic as well.

The direction of maximum radiation with a resonant rhombic is given by the arrows in Fig. 8, i.e., the antenna is bidirectional. There are minor lobes in other directions, their number and intensity depending on the leg length. When used at

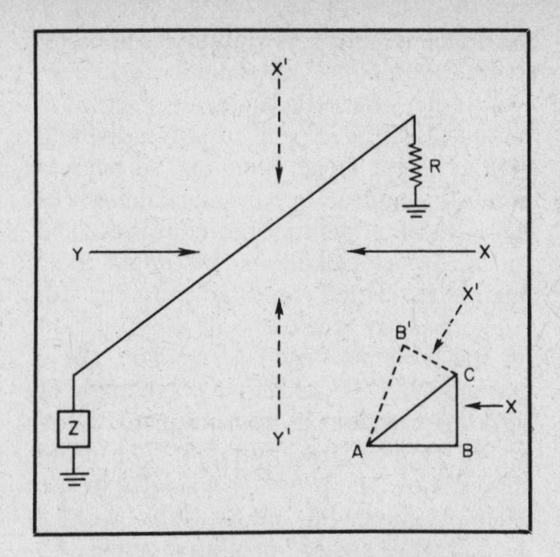


Fig. 9 — Nonresonant long-wire antenna.

frequencies below the vhf region the rhombic antenna is always mounted with the plane containing the wires horizontal. The polarization in this plane, and also in the perpendicular plane that bisects the rhombic, is horizontal. At 144 MHz and above, the dimensions are such that the antenna can be mounted with the plane containing the wires vertical if vertical polarization is desired.

When the rhombic antenna is to be used on several amateur bands it is advisable to choose the apex angle, A, on the basis of the leg length in wavelengths at 14 MHz. This point is covered in more detail in connection both with the V and with the nonresonant rhombic. Although the gain on higher frequency bands will not be quite as favorable as if the antenna had been designed for the higher frequencies, the system will radiate well at the low angles that are necessary at such frequencies. At frequencies below the design frequency the greater apex angle of the rhombic (as compared with a V of the same total length) is more favorable to good radiation than in the case of the V.

The resonant rhombic antenna can be fed in the same way as the V. Resonant feeders are necessary if the antenna is to be used in several amateur bands.

## NONRESONANT LONG-WIRE ANTENNAS

All the antenna systems we have considered so far have been based on resonant operation, that is, with standing waves of current and voltage along the wire. Although most antenna designs are based on using resonant wires, resonance is by no means a necessary condition for the wire to radiate and intercept electromagnetic waves.

In Fig. 9 let us suppose that the wire is parallel with the ground (horizontal) and is terminated by a load Z equal to its characteristic impedance, $Z_0$. The load Z can represent a receiver matched to the line. The resistor R is also equal to the $Z_0$ of the wire. A wave coming from the direction X will strike the wire first at its far end and sweep across the wire at an angle until it reaches the end at which Z is connected. In doing so it will induce voltages in the antenna and currents will flow as a result. The current flowing toward Z is the useful output of the antenna, while the current flowing toward R will be absorbed in R. The same thing is true of a wave coming from the direction X′. In such an antenna there are no standing waves, because all received power is absorbed at either end.

The greatest possible power will be delivered to the load Z when the individual currents induced as the wave sweeps across the wire all combine properly on reaching the load. The currents will reach Z in optimum phase when the time required for a current to flow from the far end of the antenna to Z is exactly one-half cycle longer than the time taken by the wave to sweep over the antenna. Since a half cycle is equivalent to a half wavelength greater than the distance traversed by the wave from the instant it strikes the far end of the antenna to the instant that it reaches the near end. This is shown by the small drawing, where AC represents the antenna, BC is a line perpendicular to the wave direction, and AB is the distance traveled by the wave in sweeping past AC. AB must be one-half wavelength shorter than AC. Similarly, AB′ must be the same length as AB for a wave arriving from X′.

A wave arriving at the antenna from the opposite direction Y (or Y′), will similarly result in the largest possible current at the *far* end. However, the far end is terminated in R, which is equal to Z, so all the power delivered to R by the wave arriving from Y will be absorbed in R. The current traveling to Z will produce a signal in Z in proportion to its amplitude. If the antenna length is such that all the individual currents arrive at Z in such phase as to add up to zero, there will be no current through Z. At other lengths the resultant current may reach appreciable values. The lengths that give zero amplitude are those which are odd multiples of 1/4 wavelength, beginning at 3/4 wavelength. The response from the Y direction is greatest when the antenna is any even multiple of 1/2 wavelength long; the higher the multiple, the smaller the response.

### Directional Characteristics

The explanation above considers the phase but not the relative amplitudes of the individual currents reaching the load. When the appropriate correction is made, the angle with the wire at which radiation or response is maximum is given by the curve of Fig. 10. The response drops off gradually on either side of the maximum point, resulting in lobes in the directive pattern much like those for harmonic antennas, except that the system is substantially unidirectional. Typical patterns are shown in Fig. 11. When the antenna length is 3/2 wavelength or greater there are also angles at which secondary maxima occur; these secondary maxima (minor lobes) have their peaks approximately at angles for which the length AB, Fig. 9, is less than AC by any odd multiples of one-half wavelength. When AB is shorter than AC by an *even* multiple of a half wavelength, the induced currents cancel each other completely at Z, and in such cases there is a null for waves arriving in the direction perpendicular to BC.

The antenna of Fig. 9 responds to horizontally polarized signals when mounted horizontally. If the wire lies in a plane that is vertical with respect to the earth it responds to vertically polarized signals. By reciprocity, the characteristics for transmitting are the same as for receiving. For average conductor diameters and heights above ground, 20 or 30 feet,

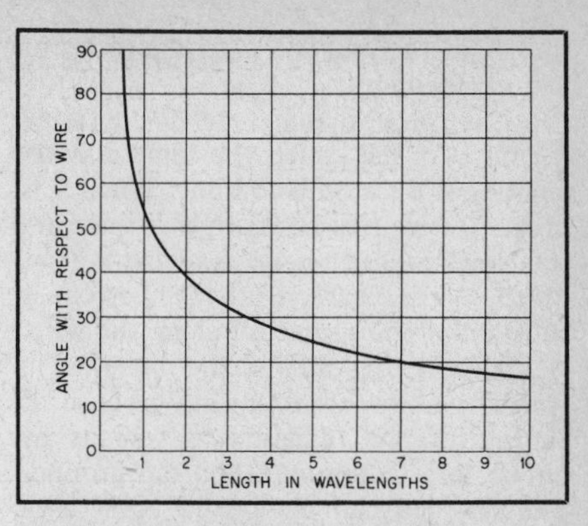


Fig. 10 — Angle with respect to wire axis at which the radiation from a nonresonant long-wire antenna is maximum.

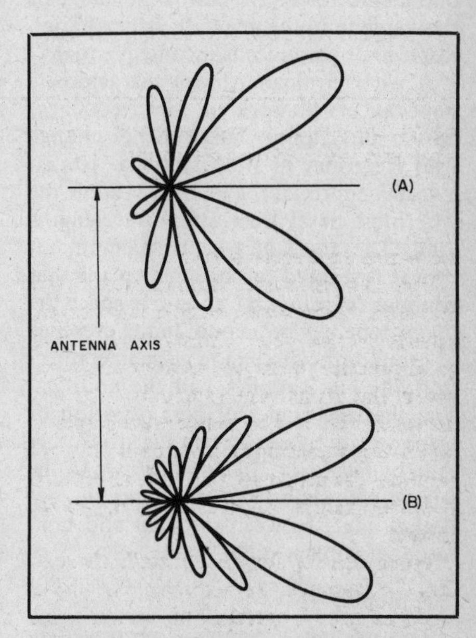


Fig. 11 — Typical radiation patterns (cross section of solid figure) for terminated long wires. At A, the length is two wavelengths, at B, four wavelengths, both for an idealized case in which there is no decrease of current along the wire. In practice, the pattern is somewhat distorted by wire attenuation.

the $Z_0$ of the antenna is of the order of 500 to 600 ohms.

It is apparent that an antenna operating in this way has much the same characteristics as a transmission line. When it is properly terminated at both ends there are traveling waves, but no standing waves, on the wire. Consequently the current is substantially the same all along the wire. Actually, it decreases in the direction in which the current is flowing because of energy loss by radiation as well as by ohmic loss in the wire and the ground. The antenna can be looked upon as a transmission line terminated in its characteristic impedance, but having such wide spacing between conductors (the second conductor in this case is the image of the antenna in the ground) that radiation losses are by no means inconsequential.

A wire terminated in its characteristic impedance will work on any frequency, but its directional characteristics change with frequency as shown by Fig. 10. To give any appreciable gain over a dipole the wire must be at least a few wavelengths long. The angle at which maximum response is secured can be in any plane that contains the wire axis, so in free space the major lobe will be a cone. In the presence of ground, the discussion given in connection with Fig. 3 applies, with the modification that the angles of best radiation or response are those given in Fig. 10, rather than by Figs. 1 or 2. As comparison of the curves will show, the difference in optimum angle between resonant and nonresonant wires is quite small.

### The Sloping V

The sloping V antenna, portrayed in Fig. 12, is a nonresonant system. Even though it is simple to construct and offers multiband operation, it has not seen much use by amateurs. Only a single support is required, and the antenna should provide several decibels of gain over a frequency ratio of 3 to 1 or greater.

For satisfactory performance, the leg length, $\ell$, should be a minimum of one wavelength at the lowest operating frequency. The height of the support may be 1/2 to 3/4 of the leg length. The feed-point impedance of the sloping V is in the order of 600 ohms. Therefore open wire may be used for the feeder or, alternatively, a coaxial transmission line and a step-up transformer balun at the apex of the V may be used.

The terminating resistors should be noninductive with a value of 300 ohms and a dissipation rating equal to one half the transmitter output power. The grounded end of the resistors should be connected to a good rf ground, such as a radial system extending beneath the wires of the V. A single ground stake at each termination point will likely be insufficient; a pair of wires, one running from each termination point to the base of the support will probably prove superior.

By using the data presented earlier in this chapter, it should be possible to calculate the apex angle and support height for optimum lobe alignment from the two wires at a given frequency. Ground reflections will complicate the calculations, however, as both vertical and horizontal polarization components are present. Dimensions that have proved useful for point-to-point communications work on frequencies from 14 to 30 MHz are a support height of 60 ft, a leg length of 100 ft and an apex angle of 36° (m = ft × 0.3048).

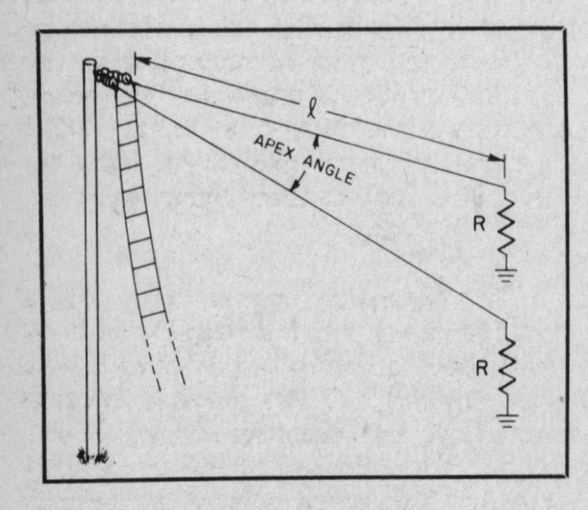


Fig. 12 — The sloping V antenna.

## THE NONRESONANT RHOMBIC ANTENNA

The highest development of the long-wire antenna is the nonresonant rhombic, shown schematically in Fig. 13. It consists of four conductors joined to form a diamond, or rhombus. All sides of the antenna have the same length and the opposite corner angles are equal. The antenna can be considered as being made up of two V antennas placed end to end and terminated by a noninductive resistor to produce a unidirectional pattern. The terminating resistor is connected between the far ends of the two sides, and is made approximately equal to the characteristic impedance of the antenna as a unit. The rhombic may be constructed either horizontally or vertically, but practically always is horizontal at frequencies below 54 MHz, since the pole height required is considerably less. Also, horizontal polarization is equally, if not more, satisfactory at these frequencies.

The basic principle of combining lobes of maximum radiation from the four individual wires constituting the rhombus or diamond is the same in either the nonresonant type shown in Fig. 13, or the resonant type described earlier in this chapter. The included angles should differ slightly because of the differences between resonant and nonresonant wires, as just described, but the differences are almost negligible.

### Tilt Angle

It is a matter of custom, in dealing with the nonresonant or terminated rhombic, to talk about the "tilt angle" ($\phi$ in Fig. 13), rather than the angle of maximum radiation with respect to an individual wire. The tilt angle is simply 90 degrees minus the angle of maximum radiation. In the case of a rhombic antenna designed for zero wave angle the tilt angle is 90 degrees minus the values given in Fig. 10.

Fig. 14 shows the tilt angle as a function of the antenna leg length. The curve marked "0°" is used for a wave angle of zero degrees; that is, maximum radiation in the plane of the antenna. The other curves show the proper tilt angles to use when aligning the major lobe with a desired wave angle. For a wave angle of 5 degrees the difference in tilt angle is less than one degree for the range of lengths shown. Just as in the case of the resonant V and resonant rhombic, alignment of the wave angle and lobes always results in still greater radiation at a lower wave angle, and for the same reason, but also results in the greatest possible radiation at the desired wave angle.

The broken curve marked "optimum length" shows the leg length at which maximum gain is secured at a chosen wave angle. Increasing the leg length beyond the optimum will result in lessened gain, and for that reason the curves do not extend beyond the optimum length. Note that the optimum length becomes greater as the desired wave angle is smaller. Leg lengths over 6 λ are not recommended because the directive pattern becomes so sharp that the antenna performance is highly variable with small changes in the angle, both horizontal and vertical, at which an incoming wave reaches the antenna. Since these angles vary to some extent in ionospheric propagation, it does not pay to attempt to use too great a degree of directivity.

### Multiband Design

When a rhombic antenna is to be used over a considerable frequency range it is worth paying some attention to the effect of the tilt angle on the gain and directive pattern at various frequencies. For example, suppose the antenna is to be used at frequencies up to and including the 28 MHz band, and that the leg length is to be 6 wavelengths on the latter frequency. For

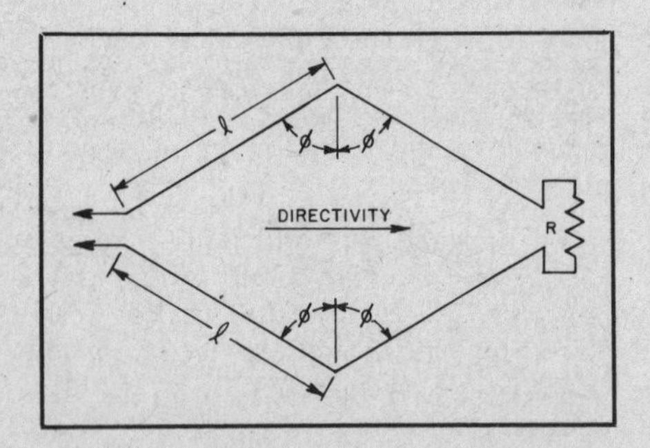


Fig. 13 — The nonresonant rhombic antenna.

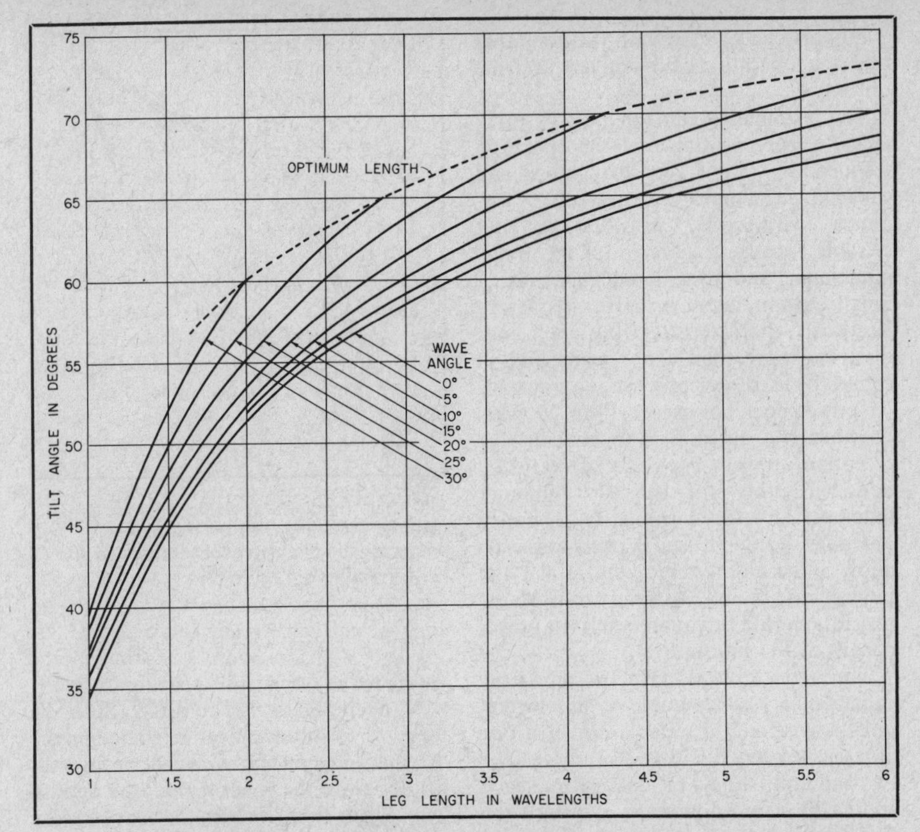


Fig. 14 — Rhombic-antenna design chart. For any given leg length, the curves show the proper tilt angle to give maximum radiation at the selected wave angle. The broken curve marked "optimum length" shows the leg length that gives the maximum possible output at the selected wave angle. The optimum length as given by the curves should be multiplied by 0.74 to obtain the leg length for which the wave angle and main lobe are aligned (see text, "Alignment of Lobes," p. 7-10).

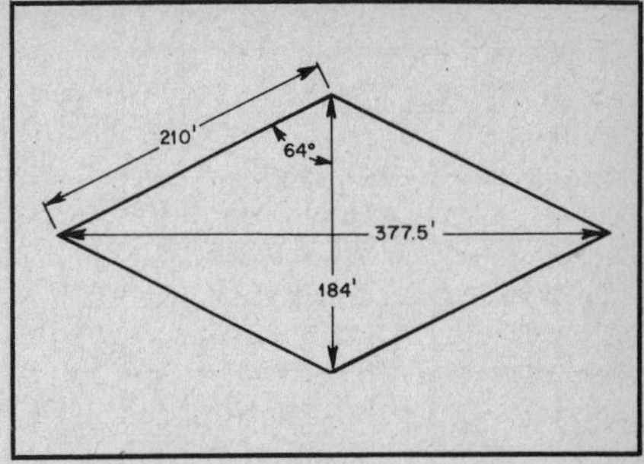


Fig. 16 — Rhombic antenna dimensions for a compromise design between 20- and 10-meter requirements, as discussed in the text. The leg length is 6 λ on 10 meters, 3 λ on 20.

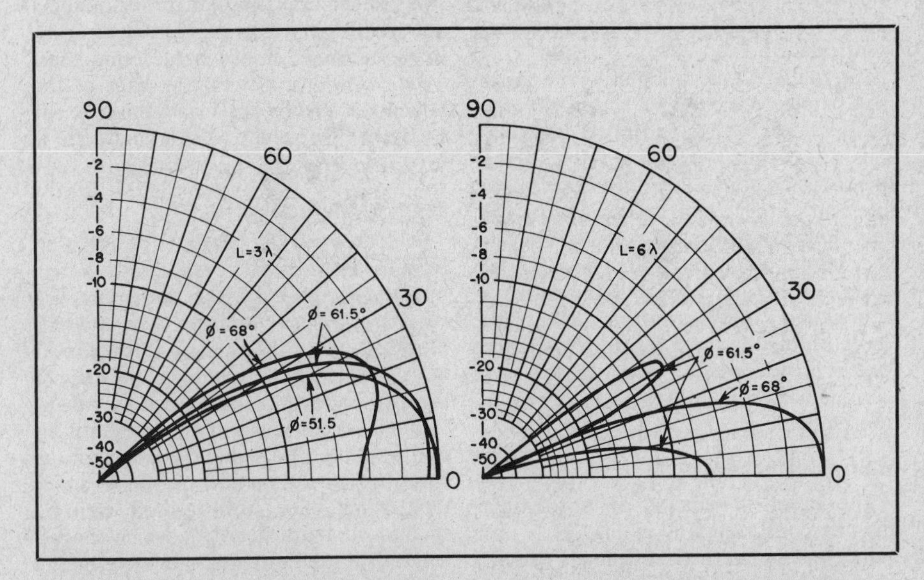


Fig. 15 — These drawings show the effect of tilt angle on the free-space vertical pattern of a nonresonant rhombic antenna having a leg length of 3 wavelengths at one frequency and 6 wavelengths at twice the frequency. These patterns apply only in the direction of the antenna axis. Minor lobes above 30° are not shown.

zero wave angle the optimum tilt angle is 68 degrees, and the calculated free-space directive pattern in the vertical plane bisecting the antenna is shown in the right-hand drawing of Fig. 15. At 14 MHz this same antenna has a leg length of three wavelengths, which calls for a tilt angle of 58.5 degrees for maximum radiation at zero wave angle. The calculated patterns for tilt angles of 58.5 and 68 degrees are shown in the left-hand drawing in Fig. 15, and it is seen that if the optimum tilt for 28-MHz operation is used the gain will be reduced and the wave angle raised at 14 MHz. In an attempt at a compromise, we might select a wave angle of 15 degrees, rather than zero, for 14 MHz since, as shown by Fig. 14, the tilt angle is larger and thus more nearly coincides with the tilt angle for zero wave angle on 28 MHz. From the chart, the tilt angle for 3 wavelengths on a leg and a 15-degree wave angle is 61.5 degrees. The patterns with this tilt angle are shown in Fig. 15 for both the 14- and 28-MHz cases. The effect at 28 MHz is to decrease the gain at zero wave angle by more than 6 dB and to split the radiation in the vertical plane into two lobes, one of which is at a wave angle too high to be useful at this frequency.

Inasmuch as the gain increases with the leg length in wavelengths, it is probably better to favor the lower frequency in choosing the tilt angle. In the present example, the best compromise probably would be to split the difference between the optimum tilt angle for the 15-degree wave angle at 14 MHz and that for zero wave angle at 28 MHz; that is, use a tilt angle of about 64 degrees. Design dimensions for such an antenna are given in Fig. 16.

The patterns of Fig. 15 are in the vertical plane through the center of the antenna only. In vertical planes making an angle with the antenna axis, the patterns may differ considerably. The effect of a tilt angle that is smaller than the optimum is to broaden the horizontal pattern, so at 28 MHz the antenna in the example would be less directive in the horizontal plane than would be the case if it were designed for optimum performance at that frequency. It should also be noted that the patterns given in Fig. 15 are free-space patterns and must be multiplied by the ground-reflection factors for the actual antenna height used, if the actual vertical patterns are to be determined. (Also see later discussion on lobe alignment.)

### Power Gain

The theoretical power gain of a nonresonant rhombic antenna over a dipole, both in free space, is given by the curve of Fig. 17. This curve is for zero wave angle and includes an allowance of 3 dB for power dissipated in the terminating resistor. The actual gain of an antenna mounted horizontally above the ground,

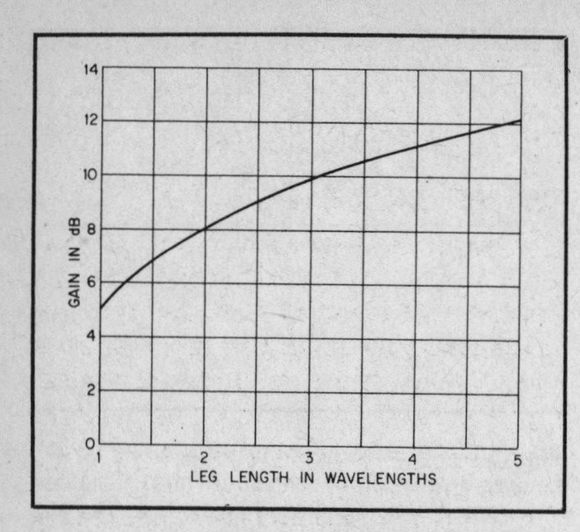


Fig. 17 — Theoretical gain of a nonresonant rhombic antenna over a half-wave dipole in free space. This curve includes an allowance of 3 dB for loss in the terminating resistor.

as compared with a dipole at the same height, can be expected to vary a bit either way from the figures given by the curve. The power lost in the terminating resistor is probably less than 3 dB in the average installation, since more than half of the power input is radiated before the end of the antenna is reached. However, there is also more power loss in the wire and in the ground under the antenna than in the case of a simple dipole, so the 3 dB figure is probably a representative estimate of *overall* loss.

**Termination**

Although there is no marked difference in the gain obtainable with resonant and nonresonant rhombics of comparable design, the nonresonant antenna has the advantage that over a wide frequency range it presents an essentially resistive and constant load to the transmitter coupling apparatus. In addition, nonresonant operation makes the antenna substantially unidirectional, while the unterminated or resonant rhombic is always bidirectional, although not symmetrically so. In a sense, it can be considered that the power dissipated in the terminating resistor is simply power that would have been radiated in the other direction had the resistor not been there, so the fact that some of the power (about one-third) is used up in heating the resistor does not mean an actual loss in the desired direction.

The characteristic impedance of an ordinary rhombic antenna, looking into the input end, is in the order of 700 to 800 ohms when properly terminated in a resistance at the far end. The terminating resistance required to bring about the matching condition usually is slightly higher than the input impedance because of the loss of energy through radiation by the time the far end is reached. The correct value usually will be found to be of the order of 800 ohms, and should be determined experimentally if the flattest possible antenna is desired. However, for average work a noninductive resistance of 800 ohms can be used with the assurance that the operation will not be far from optimum.

The terminating resistor must be practically a pure resistance at the operating frequencies; that is, its inductance and capacitance should be negligible. Ordinary wire-wound resistors are not suitable because they have far too much inductance and distributed capacitance. Small carbon resistors have satisfactory electrical characteristics but will not dissipate more than a few watts and so cannot be used, except when the transmitter power does not exceed 10 or 20 watts or when the antenna is to be used for reception only. The special resistors designed either for use as "dummy" antennas or for terminating rhombic antennas should be used in other cases. To allow a factor of safety, the total rated power dissipation of the resistor or resistors should be equal to half the power output of the transmitter.

To reduce the effects of stray capacitance it is desirable to use several units, say three, in series even when one alone will safely dissipate the power. The two end units should be identical and each should have one fourth to one third the total resistance, with the center unit making up the difference. The units should be installed in a weatherproof housing at the end of the antenna to protect them and to permit mounting without mechanical strain. The connecting leads should be short so that little extraneous inductance is introduced.

Alternatively, the terminating resistance may be placed at the end of an 800-ohm line connected to the end of the antenna. This will permit placing the resistors and their housing at a point convenient for adjustment rather than at the top of the pole. Resistance wire may be used for this line, so that a portion of the power will be dissipated before it reaches the resistive termination, thus permitting the use of lower wattage lumped resistors. The line length is not critical, since it operates without standing waves and hence is nonresonant.

**Multiwire Rhombics**

The input impedance of a rhombic antenna constructed as in Fig. 13 is not quite constant as the frequency is varied. This is because the varying separation between the wires causes the characteristic impedance of the antenna to vary along its length. The variation in $Z_0$ can be minimized by a conductor arrangement that increases the capacitance per unit length in proportion to the separation between the wires.

The method of accomplishing this is shown in Fig. 18. Three conductors are used, joined together at the ends but with increasing separation as the junction between legs is approached. As used in commercial installations having legs several wavelengths long, the spacing between wires at the center is 3 to 4 feet. Since all three wires should have the same length, the top and bottom wires will be slightly farther from the support than the middle wire. Using three wires in this way reduces the $Z_0$ of the antenna to approximately 600 ohms, thus providing a better match for a practicable open-wire line, in addition to smoothing out the impedance variations over the frequency range.

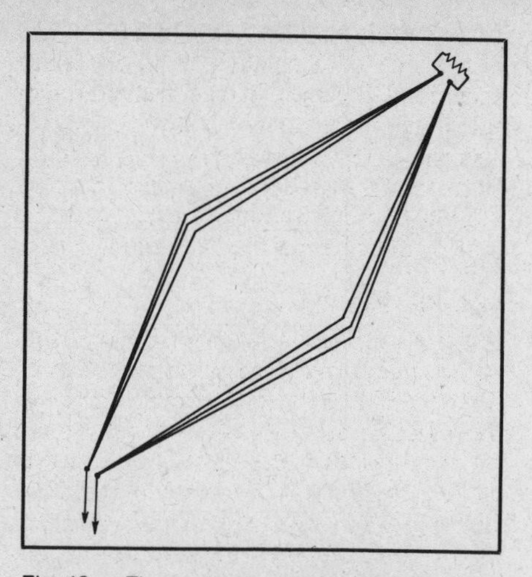

Fig. 18 — Three-wire rhombic antenna. Use of multiple wires improves the impedance characteristic of a nonresonant rhombic and increases the gain somewhat.

A similar effect, although not quite so good, is obtained by using two wires instead of three. It has been found that, with the 3-wire system, the gain of the antenna is slightly greater (in the order of 1 dB) than when only a single conductor is used.

**Front-to-Back Ratio**

It is theoretically possible to obtain an infinite front-to-back ratio with a terminated rhombic antenna, and in practice very large values can actually be secured. However, when the antenna is terminated in its characteristic impedance the infinite front-to-back ratio can be secured only at frequencies for which the leg length is an odd multiple of a quarter wavelength, as described in the section on nonresonant long wires. The front-to-back ratio is smallest at frequencies for which the leg length is a multiple of a half wavelength.

When the leg length is not an odd multiple of a quarter wave at the frequency under consideration, the front-to-back ratio can be made very high by slightly decreasing the value of terminating resistance. This permits a small reflection from the far end of the antenna which cancels out, at the input end, the residual response. With large antennas the front-to-back ratio may be made very large over the whole frequency range by experimental adjustment of the terminating resistance. Modification of the terminating resistance can result in a splitting of the back null into two nulls, one on either side of a small lobe in the back direction. Changes in the value of terminating resistance thus permit "steering" the back null over a small horizontal range so that signals coming from a particular spot not exactly to the rear of the antenna may be minimized.

### Ground Effects

Reflections from the ground play exactly the same part in determining the vertical directive pattern of a horizontal rhombic antenna that they play with other horizontal antennas. Consequently, if a low wave angle is desired it is necessary to make the height great enough to bring the reflection factor into the higher range of values given by the charts in Chapter 2.

### Alignment of Lobes, Wave Angle, and Ground Reflections

When maximum antenna response is desired at a particular wave angle (or maximum radiation is desired at that angle) the major lobe of the antenna cannot only be aligned with the wave angle as previously described but also with a maximum in the ground-reflection factor. When this is done it is no longer possible to consider the antenna height independently of other aspects of rhombic design. The wave angle, leg length, and height become mutually dependent.

This method of design is of particular value when the antenna is built to be used over fixed transmission distances for which the optimum wave angle is known. It has had wide application in commercial work with nonresonant rhombic antennas, but seems less desirable for amateur use where, for the long-distance work for which rhombic antennas are built, the lowest wave angle that can be obtained is the most desirable. Alignment of all three factors is limited in application because it leads to impracticable heights and leg lengths for small wave angles. Consequently, when a fairly broad range of low wave angles is the objective, it is more satisfactory to design for a low wave angle and simply make the antenna as high as possible.

Fig. 19 shows the lowest height at which ground reflections make the radiation maximum at a desired wave angle. It can be used in conjunction with Fig. 14 for complete alignment of the antenna. For example, if the desired wave angle is 20 degrees, Fig. 19 shows that the height must be 0.75 wavelength. From Fig. 14, the optimum leg length is 4.2 wavelengths and the tilt angle is just under 70 degrees. A rhombic antenna so designed will have the maximum possible output that can be obtained at a wave angle of 20 degrees; no other set of dimensions will be as good. However, it will have still greater output at some angle lower than 20 degrees, for the reasons given earlier. When it is desired to make the maximum output of the antenna occur at the 20-degree wave angle, it may be accomplished by using the same height and tilt angle, but with the leg length reduced by 26 percent. Thus for such alignment the leg length should be $4.2 \times 0.74 = 3.1$ wavelengths. The output at the 20-degree wave angle will be smaller than with 4.2-wavelength legs, however, despite the fact that the smaller antenna has its maximum radiation at 20 degrees. The reduction in gain is about 1.5 dB.

### Methods of Feed

If the broad frequency characteristic of the rhombic antenna is to be utilized fully, the feeder system must be similarly broad. Open-wire transmission line of the same characteristic impedance as that shown at the antenna input terminals, or approximately 700 to 800 ohms, may be used. Data for the construction of such lines is given in Chapter 3. While the usual matching stub can be used to provide an impedance transformation to more satisfactory line impedances, this limits the operation of the antenna to a comparatively narrow range of frequencies centering about that for which the stub is adjusted. Probably a more satisfactory arrangement would be to use a coaxial transmission line and a broadband transformer balun at the antenna feed point.

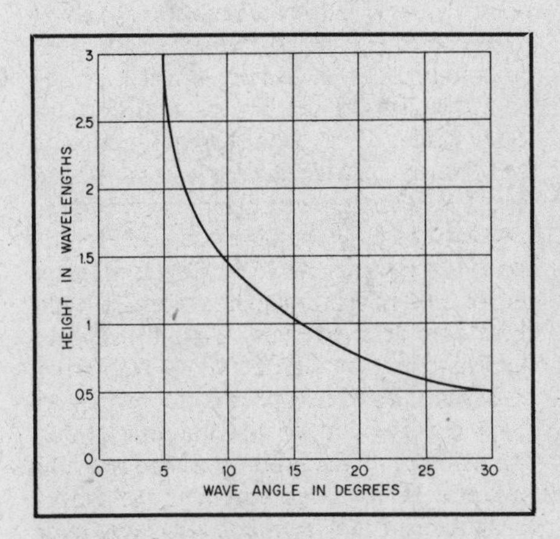


Fig. 19 — Antenna height to be used for securing maximum radiation at a desired wave angle. This curve applies to any type of horizontal antenna.

## WAVE ANTENNAS

### THE BEVERAGE ANTENNA

Perhaps the best known type of wave antenna is the Beverage. Many 160-meter enthusiasts have used Beverage antennas to enhance the effective signal-to-noise ratio while attempting to extract weak signals from the sometimes high levels of atmospheric noise and interference. Alternative antenna systems have been developed and used over the years, such as loops and long spans of unterminated wire on or slightly above the ground, but nothing seems to surpass the Beverage antenna for 160-meter weak-signal reception.

Although some radio amateurs have reported improved reception from Beverage antennas at 3.5 and even 7 MHz, the suggested upper frequency limit is 2.0 MHz. Occasional improved reception at hf may result from propagation conditions at a given time. However, because the incoming sky waves above medium frequency arrive at moderate and high angles and because the polarity changes at random during reflection from the ionosphere, the Beverage is not suited to effective use in that part of the spectrum. The wave antenna is responsive mostly to incoming waves of low angle — those that tend to follow the contour of the earth and maintain a constant polarization.

Fig. 20 shows the simplest form of Beverage antenna. It consists simply of a wire, at least one wavelength long, stretched in the direction of the transmitting station. Assume that the transmitting station is east of the receiving station, and that the receiver is placed at the west end of the antenna, as shown. The traveling wave from the transmitting station moves from east toward the west at the velocity of light. As the wave moves along the antenna, it induces currents in the wire that travel in both directions. The current that travels east moves against the motion of the wave and builds down to practically zero if the antenna is one wavelength long. The currents that travel west,

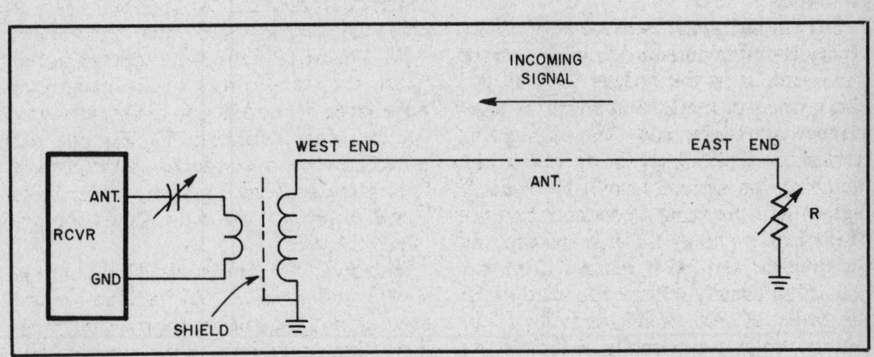


Fig. 20 — The simplest Beverage antenna.

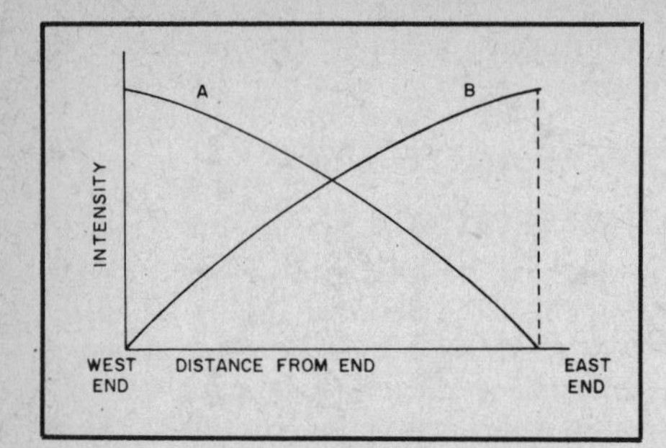


Fig. 21 — Curve A shows how the current increments add in phase at the west end of the antenna. Curve B illustrates how the static and interference add at the east end of the antenna (see text).

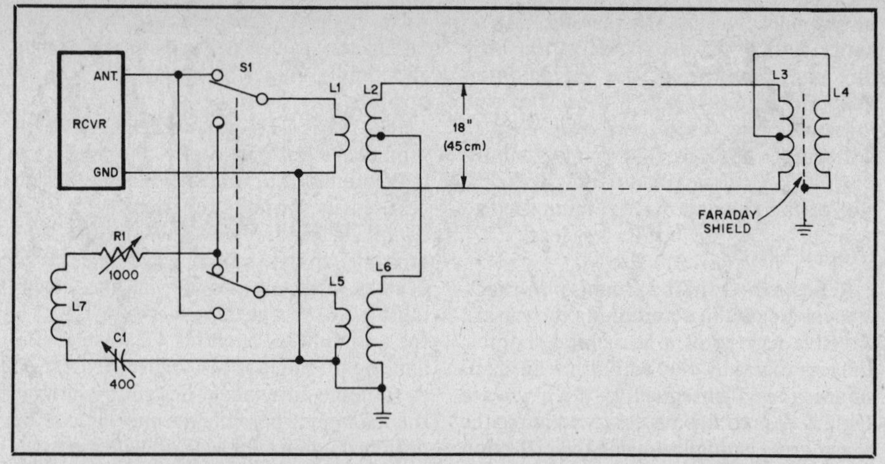


Fig. 24 — The two-wire Beverage antenna permits direction switching (S1) and null steering (C1 and R1). Optimum height is 10 feet above ground.

C1 — 400-pF variable capacitor.
L1, L5 — 6 turns no. 14 enam. wire, 2.5-in. dia, close wound. Install inside L2 and L6 near the center of the windings.
L2, L6 — 56 turns no. 14 enam. wire, close wound on 3.5-in. dia form, center tapped.
L3 — 60 turns no. 26 enam. wire, 4-in. dia, 4 in. long, center tapped. Coil is wrapped with single layer of metal foil that has 7/8-in. split full length to prevent shorted-turn effect. Shield is grounded.
L4 — 40 turns no. 26 enam. wire centered over L3.
L7 — 40 turns no. 24 enam. wire, 1-in. dia, 1-in. long (250 μH).
R1 — 1000-ohm linear-taper carbon control.
S1 — Dpdt switch.

however, travel along the wire with practically the velocity of light, and, therefore, move along with the wave in space. The current increments all add up in phase at the west end, producing a strong signal, as shown by curve A in Fig. 21. In a like manner, static or interference originating in the west will build up to a maximum at the east end of the antenna, as shown by curve B in Fig. 21.

If the east end of the antenna were open or grounded, all of the energy represented by curve B would be reflected and would travel back over the antenna to the west end, where part of the energy would pass to ground through the receiver and part could be reflected again, depending on the impedance of the receiver input circuit. The horizontal directive pattern would be bidirectional, as shown in Fig. 22. The reception from the west is not as good as from the east, as some of the energy is lost because of attenuation in the wire and some reradiation as the reflected wave travels back from east to west.

To make the antenna unidirectional, it is necessary to eliminate the reflections at the end farthest from the receiver. This is accomplished simply by terminating the antenna with a noninductive resistance at the far end. If this resistance is made equal to the surge impedance of the wire, it absorbs all of the energy and prevents reflections. The pattern becomes unidirectional, as shown in Fig. 23.

The value of the surge impedance depends on the size, number and height of the wire above ground, but is independent of the length of the wire. For practical contruction with no. 12 copper wire, the surge impedance lies between 200 and 400 ohms.

A two-wire Beverage offers several advantages. This flexible antenna allows the receiver and terminating resistor to be placed at either end of the antenna. If the terminating resistance is replaced with an adjustable-impedance termination, the rejection null of the antenna can be steered to minimize interference and noise.

Fig. 24 shows a two-wire Beverage antenna. When S1 is in the position shown, maximum reception is from the

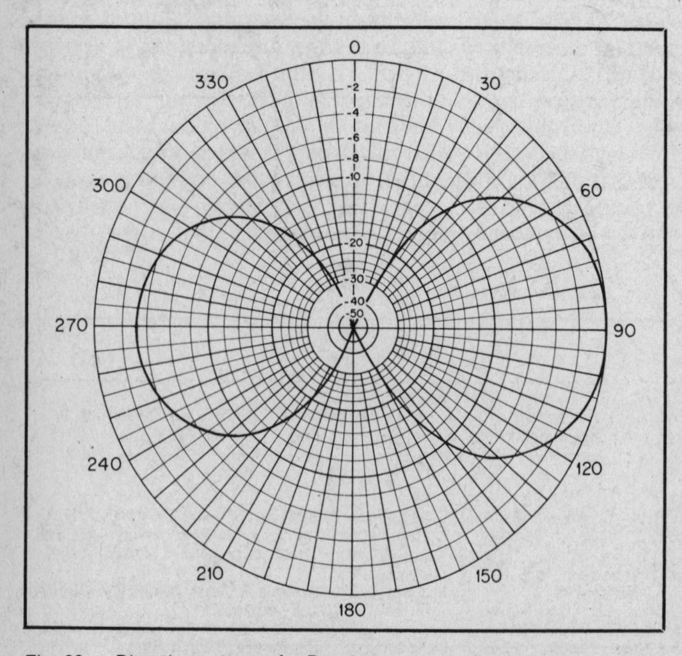


Fig. 22 — Directive pattern of a Beverage antenna that is one wavelength long. It does not have a terminating resistor.

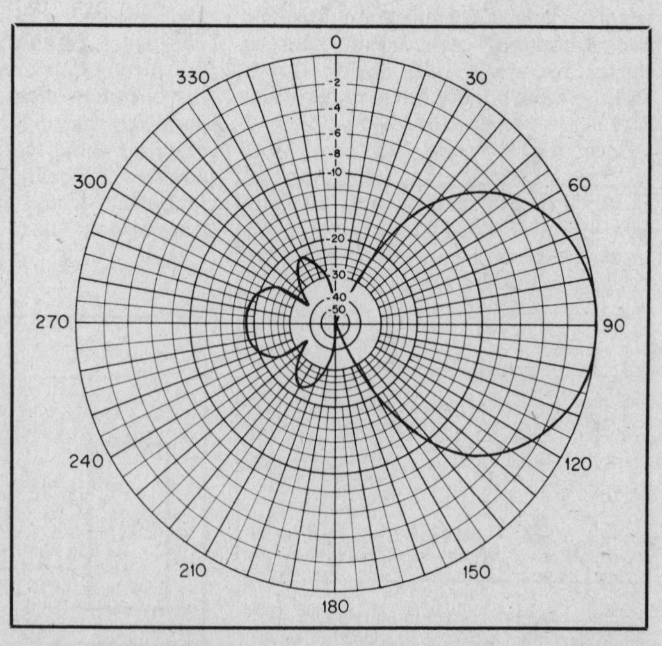


Fig. 23 — Directive pattern for a 1-wavelength Beverage. The antenna has been terminated properly.

direction away from the receiver end of the antenna. With S1 in the other position, the directive pattern is switched 180°. By adjusting R1 and C1 the null position can be moved.

Beverage antennas work best over ground having poor conductivity. A good rf ground is required, however, for the terminating resistance. A single ground rod may represent no ground at all. A quality ground system contains a substantial number of buried radial wires. If an extensive ground arrangement isn't practical, use as much wire as possible, even if some of the radials are quite short. Sufficient wire should be used to ensure that the ground resistance is as low as possible.

Optimum length for a two-wire Beverage antenna is 1 to 2 wavelengths; for a single-wire version it is 1 to 3 wavelengths. For 160 meters, the optimum height is 10 to 20 feet, with only a small advantage for the higher end of the range.

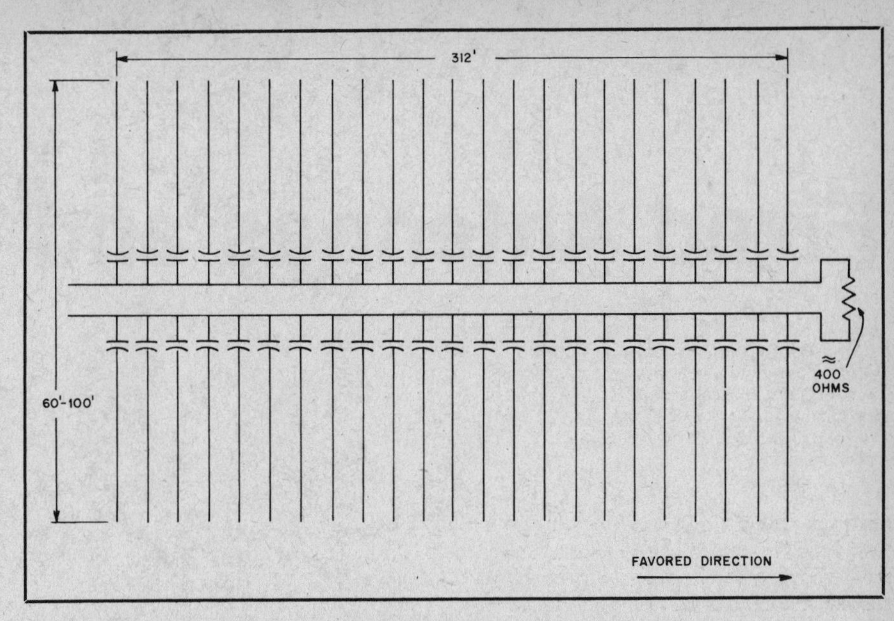


Fig. 25 — The fishbone antenna provides higher gain per acre than does a rhombic. It is essentially a wave antenna which evolved from the Beverage.

## FISHBONE ANTENNAS

Another type of wave antenna is the fishbone, illustrated in Fig. 25. Its impedance is approximately 400 ohms. The antenna is formed of closely spaced elements that are lightly coupled, capacitively, to a long, terminated transmission line. The capacitors are chosen to have a value that will keep the velocity of propagation of the line more than 90% of that in air. The elements are usually spaced approximately 0.1 wavelength or slightly more so that an average of 7 or more elements is used for each full wavelength of transmission-line length. This antenna obtains low-angle response primarily as a function of its height, and therefore, is generally installed 60 to 120 feet above ground. If the antenna is to be used for transmission, the capacitors should be of the transmitting type, as they will be required to handle substantial current.

The English HAD fishbone antenna, shown in its two-bay form in Fig. 26, is of less complicated design than the one just described. It may be used singly, of course, and may be fed with 600-ohm open-wire line. Installation and operational charcteristics are similar to the standard fishbone antenna.

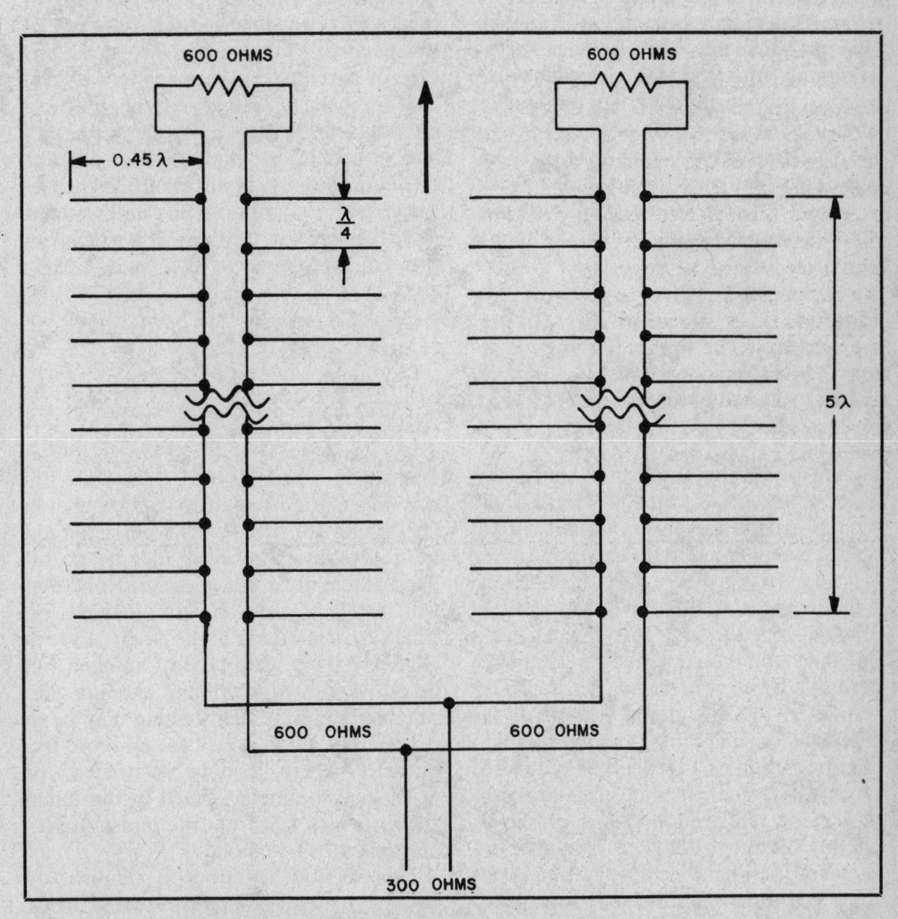


Fig. 26 — The English HAD fishbone antenna is a simplified version of the standard fishbone. It may be used as a single-bay antenna fed with 600-ohm open-wire line.

## BIBLIOGRAPHY

Source material and more extended discussion of topics covered in this chapter can be found in the references given below.


Bailey, A., Dean, S. W. and Wintringham, W. T. "The Receiving System for Long-Wave Transatlantic Radio Telephony." *The Bell System Technical Journal,* April 1929.

Belrose, J. S. "Beverage Antennas for Amateur Communications." Technical Correspondence, *QST,* September 1981.

Beverage, H. H. "Antennas," *RCA Review,* July 1939.

Beverage, H. H., and DeMaw, D. "The Classic Beverage Antenna Revisited." *QST,* January 1982.

Booth, B. "Weak-Signal Reception on 160 — Some Antenna Notes." *QST,* June 1977.

Bruce, E. "Developments in Short-Wave Directive Antennas." *Proc. I.R.E.,* August 1931.

Bruce, E., Beck, A. C. and Lowry, L. R. "Horizontal Rhombic Antennas." *Proc. I.R.E.,* January 1935.

Carter, P. S., Hansell, C. W. and Lindenblad, N. E. "Development of Directive Transmitting Antennas by R.C.A. Communications." *Proc. I.R.E.,* October 1931.

Harper, A. E., *Rhombic Antenna Design* (New York: D. Van Nostrand Co., Inc.)

Laport, E. A. "Design Data for Horizontal Rhombic Antennas." *RCA Review,* March 1952.

Misek, V. A. *The Beverage Antenna Handbook* (Wason Rd., Hudson, NH: W1WCR, 1977).

# Chapter 8

# Fixed Antennas and Supports

Fixed antennas are generally less expensive, less complicated and easier to erect than rotatable ones. These features make them highly attractive for any station, and probably the best choice for the beginning amateur. Fixed antennas can be made highly directive, but most of the ones described in this chapter are of the general-purpose type and are not specifically designed for directional work.

An antenna can be designed to operate on a single amateur band, on several bands or over a large, continuous portion of the hf spectrum. Multiband antennas are usually a compromise between performance and convenience on at least one of the bands. Given sufficient space, any type of antenna can be made to work on any frequency. However, the practical considerations of cost, support structures and propagation characteristics dictate different antenna types for different frequencies. Economical all-band systems are treated in this chapter, as are ones optimized for the 1.8-, 3.5- and 7-MHz bands. Antennas designed especially for the bands between 14 and 30 MHz are covered in the chapters entitled "Multielement Directive Arrays" and "Rotatable Antennas and Supports."

Some of the multibanding schemes presented in this chapter also have the effect of shortening an antenna, although this is not the primary objective. A separate chapter, "HF Antennas for Restricted Space," addresses the subject of miniature antennas.

## Multiband Antennas

For operation in a number of bands such as those between 3.5 and 30 MHz it would be impractical, for most amateurs, to put up a separate antenna for each band. Nor is it necessary; a dipole, cut for the lowest frequency band to be used, can be operated readily on higher frequencies if one is willing to accept the fact that such harmonic-type operation leads to a change in the directional pattern of the antenna (see Chapter 2), and if one is willing to use "tuned" feeders. A center-fed single-wire antenna can be made to accept power and radiate it with high efficiency on any frequency higher than its fundamental resonant frequency and, with some reduction in efficiency and bandwidth, on frequencies as low as one half the fundamental.

In fact, it is not necessary for an antenna to be a full half-wavelength long at the lowest frequency. It has been determined that an antenna can be considerably shorter than a half wavelength, as much as one-quarter wavelength, and still be a very efficient radiator at the lowest frequency.

In addition, methods have been devised for making a single antenna structure operate on a number of bands while still offering a good match to a transmission line, usually of the coaxial type. It should be understood, however, that a "multiband antenna" is not *necessarily* one that will match a given line on all bands on which it is intended to be used. Even a relatively short whip type of antenna can be operated as a multiband antenna with suitable loading. Such loading may be in the form of a coil at its base on those frequencies where loading is needed, or which may be incorporated in the tuned feeders which run from the transmitter to the base of the antenna.

This section describes a number of systems that can be used on two or more bands. Beam antennas are treated separately in later chapters.

### DIRECTLY FED ANTENNAS

The simplest multiband antenna is a random length of no. 12 or no. 14 wire. Power can be fed to the wire on practically any frequency by one or the other of the methods shown in Fig. 1. If the wire is made either 67 or 135 feet long, it can also be fed through a tuned circuit, as in

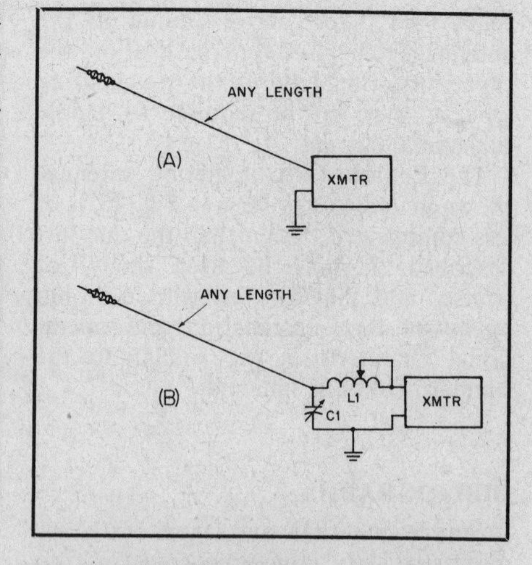


Fig. 1 — (A) Random-length wire driven directly from the pi-network output of a transmitter. (B) L network for use in cases where sufficient loading cannot be obtained with (A). C1 should have about the same plate spacing as the final tank capacitor in a vacuum-tube type of transmitter; a maximum capacitance of 100 pF is sufficient if L1 is 20 to 25 μH. A suitable coil would consist of 30 turns of no. 12 wire, 2-1/2 inches in diameter, 6 turns per inch. Bare wire should be used so the tap can be placed as required for loading transmitter.

Fig. 2. It is advantageous to use an SWR bridge or other indicator in the coax line at the point marked "X."

If a 28- or 50-MHz rotary beam has been installed, in many cases it will be possible to use the beam feed line as an antenna on the lower frequencies. Connecting the two wires of the feeder together at the station end will give a random-length wire that can be conveniently coupled to the transmitter as in Fig. 1. The rotary system at the far end will serve only to "end load" the wire and will not have much other effect.

One disadvantage of all such directly fed systems is that part of the antenna is practically within the station, and there is a good chance that trouble with rf feedback will be encountered. The rf within the station can often be minimized by choosing a length of wire so that a *current loop* occurs at or near the transmitter. This means using a wire length of a quarter wavelength (65 feet at 80 meters, 33 feet at 40 meters), or an odd multiple of a quarter wavelength (3/4 wavelength is 195 feet at 80 meters, 100 feet at 40 meters). Obviously, this can be done for only one band in the case of even harmonically related bands, since the wire length that gives a current loop at the transmitter will give a voltage loop at two (or four) times that frequency.

When one is operating with a random-wire antenna, as in Figs. 1 and 2, it is wise to try different types of grounds on the various bands, to see which will give the best results. In many cases it will be satisfactory to return to the transmitter chassis for the ground, or directly to a convenient water pipe. If neither of these works well (or the water pipe is not available), a length of no. 12 or no. 14 wire (approximately 1/4 wavelength long) can often be used to good advantage. Connect the wire at the point in the circuit that is shown grounded, and run it out and down the side of the house, or support it a few feet above the ground if the station is on the first floor or in the basement. It should not be connected to actual ground at any point.

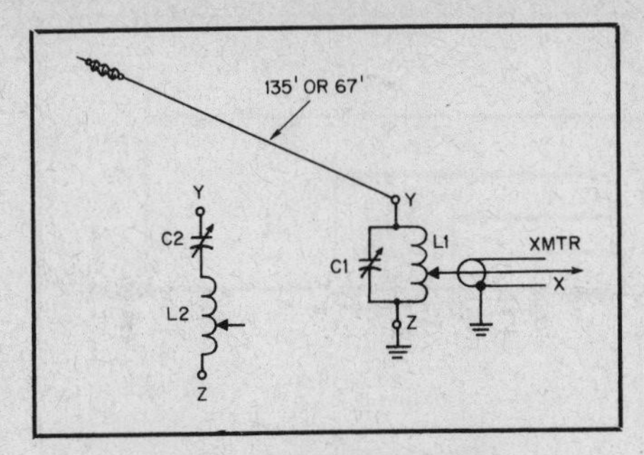


Fig. 2 — If the antenna length is 135 feet, a parallel-tuned coupling circuit can be used on each amateur band from 3.5 through 30 MHz, with the possible exception of the WARC 30-, 17- and 12-meter bands. C1 should duplicate the final-tank tuning capacitor and L1 should have the same dimensions as the final-tank inductor on the band being used. If the wire is 67 feet long, series tuning can be used on 3.5 MHz as shown at the left; parallel tuning will be required on 7 MHz and higher frequency bands. C2 and L2 will in general duplicate the final tank tuning capacitor and inductor, the same as with parallel tuning. The L network shown in Fig. 1B is also suitable for these antenna lengths.

## END-FED ANTENNAS

When a straight-wire antenna is fed at one end by a two-wire line, the length of the antenna portion becomes critical if radiation from the line is to be held to a minimum. Such an antenna system for multiband operation is the "end-fed" or "Zepp-fed" antenna shown in Fig. 3. The antenna length is made a half wavelength at the lowest operating frequency. The feeder length can be anything that is convenient, but feeder lengths that are multiples of a quarter wavelength generally give trouble with parallel currents and radiation from the feeder portion of the system. The feeder can be an open-wire line of no. 14 solid copper wire spaced 4 or 6 inches with ceramic or plastic spacers.

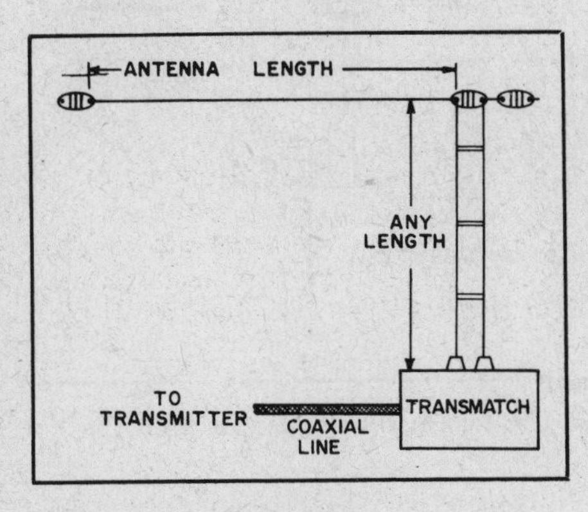


Fig. 3 — An end-fed Zepp antenna for multiband use.

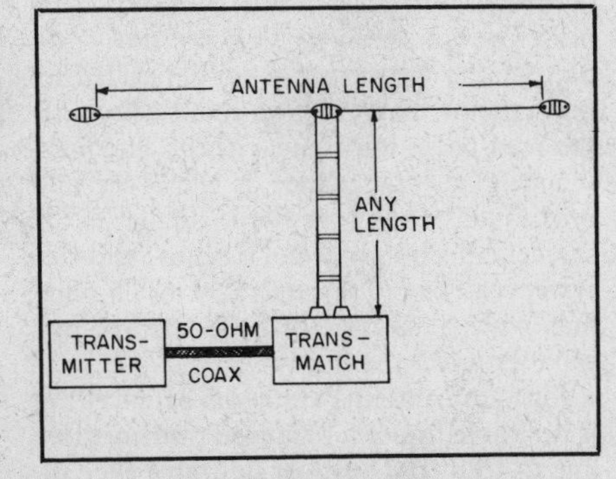


Fig. 4 — A center-fed antenna system for multiband use.

Open-wire TV line (not the type with a solid web of dielectric) is a convenient type to use. This type of line is available in approximately 300- and 450-ohm characteristic impedances.

If one has room for only a 67-foot flat top and yet wants to operate in the 3.5-MHz band, the two feeder wires can be tied together at the transmitter end and the entire system treated as a random-length wire fed directly, as in Fig. 1.

The simplest insurance against parallel currents that could cause feed-line radiation is to use a feeder length that is not a multiple of a quarter wavelength. A Transmatch can be used to provide multiband coverage with an end-fed antenna with any length of open-wire feed line, as shown in Fig. 3

## CENTER-FED ANTENNAS

The simplest and most flexible (and also least expensive) all-band antennas are those using open-wire parallel-conductor feeders to the center of the antenna, as in Fig. 4. Because each half of the flat top is the same length, the feeder currents will be balanced at all frequencies unless, of course, unbalance is introduced by one half of the antenna being closer to ground (or a grounded object) than the other. For best results and to maintain feed-current balance, the feeder should run away at right angles to the antenna, preferably for at least a quarter wavelength.

Center feed is not only more desirable than end feed because of inherently better balance, but generally also results in a lower standing-wave ratio on the transmission line, provided a parallel-conductor line having a characteristic impedance of 450 to 600 ohms is used. TV-type open-wire line is satisfactory for all but possibly high-power installations (over 500 watts), where heavier wire and wider spacing is desirable to handle the larger currents and voltages.

The length of the antenna is not critical, nor is the length of the line. As mentioned earlier, the length of the antenna can be considerably less than one-half wavelength and still be very effective. If the overall length is at least one-quarter wavelength at the lowest frequency, a quite usable system will result. The only difficulty that may exist with this type of system is the matter of coupling the antenna-system load to the transmitter. Most modern transmitters are designed to work into a 50-ohm coaxial load. With this type of antenna system a coupling network (a Transmatch) is required.

### Feed-Line Radiation

The preceding sections have pointed out means of reducing or eliminating feed-line radiation. However, it should be emphasized that any radiation from a transmission line is not "lost" energy and is not necessarily harmful. Whether or not feed-line radiation is important depends entirely on the antenna system being used. For example, feed-line radiation is *not* desirable when a directive array is being used. Such radiation can distort the desired pattern of such an array, producing responses in unwanted directions. In other words, one wants radiation *only* from the directive array.

On the other hand, in the case of a

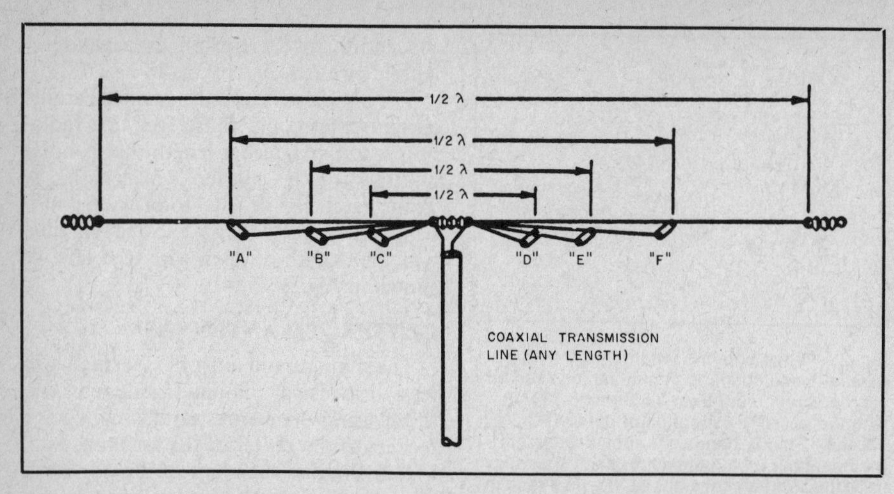


Fig. 5 — Multiband antenna using paralleled dipoles all connected to a common low-impedance transmission line. The half-wave dimensions may be either for the centers of the various bands or selected to fit favorite frequencies in each band. Length of half wave in feet is 468/frequency in MHz.

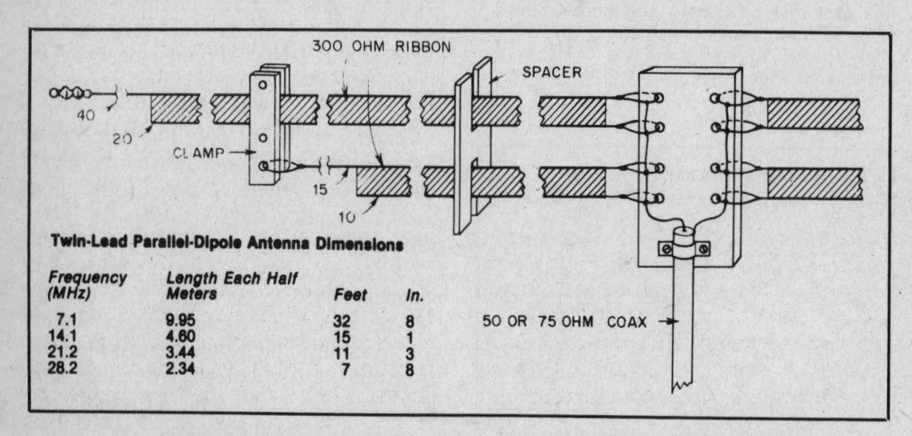


**Twin-Lead Parallel-Dipole Antenna Dimensions**

| *Frequency (MHz)* | *Length Each Half Meters* | *Feet* | *In.* |
|---|---|---|---|
| 7.1 | 9.95 | 32 | 8 |
| 14.1 | 4.60 | 15 | 1 |
| 21.2 | 3.44 | 11 | 3 |
| 28.2 | 2.34 | 7 | 8 |

Fig. 6 — Sketch showing how the twin-lead multiple-dipole antenna system is assembled. The excess wire and insulation are stripped away.

multiband dipole where general coverage is desired, if the feed line happens to radiate, such energy could actually have a desirable effect. Antenna purists may dispute such a premise, but from a practical standpoint where one is not concerned with a directive pattern, much time and labor can be saved by ignoring possible transmission-line radiation.

## MULTIPLE-DIPOLE ANTENNAS

The antenna system shown in Fig. 5 consists of a group of center-fed dipoles all connected in parallel at the point where the transmission line joins them. One such dipole is used for each band on which it is desired to work, and as many as four have been used, as indicated in the sketch. It is not generally necessary to provide a separate dipole for the 21-MHz band since a 7-MHz dipole works satisfactorily as a third-harmonic antenna on this band. A dipole cut for the cw portion of the 7-MHz band will resonate in the phone segment of the 21-MHz band because the end effect operates only on the outer quarter-wave sections of the harmonic antenna.

Although there is some interaction between the dipoles it has been found in practice that the ones that are not resonant at the frequency actually applied to the antenna have only a small effect on the feed-point impedance of the "active" dipole. This impedance is therefore approximately that of a single dipole, or in the neighborhood of 60-70 ohms, and the system can be fed through a 50- or 75-ohm line with a satisfactorily low standing-wave ratio on the line.

Since the antenna system is balanced, it is desirable to use a balanced transmission line to feed it. The most desirable type of line is 75-ohm transmitting twin-lead. However, either 52-ohm or 75-ohm coaxial line can be used; coax line introduces some unbalance, but this is tolerable on the lower frequencies.

The separation between the dipoles for the various frequencies does not seem to be especially critical. One set of wires can be suspended from the next larger set, using insulating spreaders (of the type used for feeder spreaders) to give a separation of a few inches.

An interesting method of construction used successfully by ON4UF is shown in Fig. 6. The antenna has four dipoles (for 7, 14, 21 and 28 MHz) constructed from 300-ohm ribbon transmission line. A single length of ribbon makes two dipoles. Thus, two lengths, as shown in the sketch, serve to make dipoles for four bands. Ribbon with copper-clad steel conductors (Amphenol type 14-022) should be used because all of the weight, including that of the feed line, must be supported by the uppermost wire.

Two pieces of ribbon are first cut to a length suitable for the two halves of the longest dipole. Then one of the conductors in each piece is cut to proper length for the next band higher in frequency. The excess wire and insulation is stripped away. A second pair of lengths is prepared in the same manner, except that the lengths are appropriate for the next two higher frequency bands.

A piece of thick polystyrene sheet drilled with holes for anchoring each wire serves as the central insulator. The shorter pair of dipoles is suspended the width of the ribbon below the longer pair by clamps also made of poly sheet. Intermediate spacers are made by sawing slots in pieces of poly sheet so that they will fit the ribbon snugly.

The multiple-dipole principle can also be applied to vertical antennas. Parallel or fanned quarter-wavelength elements of wire or tubing can be worked against ground or tuned radials from a common feed point.

## TRAP DIPOLES

By using tuned circuits of appropriate design strategically placed in a dipole, the antenna can be made to show what is essentially fundamental resonance at a number of different frequencies. The general principle is illustrated by Fig. 7. The two inner lengths of wire, X, together form a simple dipole resonant at the highest band desired, say 14 MHz. The tuned circuits L1-C1 are also resonant at this frequency, and when connected as shown offer a very high impedance to rf current of that frequency which may be flowing in the section X-X. Effectively, therefore, these two tuned circuits act as insulators for the inner dipole, and the outer sections beyond L1-C1 are inactive.

However, on the next lower frequency band, say 7 MHz, L1-C1 shows an inductive reactance and is the electrical equivalent of a coil, Fig. 7B. If the two sections marked Y are now added and their length adjusted so that, together with the loading coils represented by the inductive reactance of L1-C1, the system out to the ends of the Y sections is resonant at 7 MHz. This part of the antenna is equivalent to a loaded dipole on 7 MHz and will exhibit about the same impedance at the feed point as a simple dipole for that band. The tuned circuit L2-C2 is resonant at 7 MHz and acts as a high im-

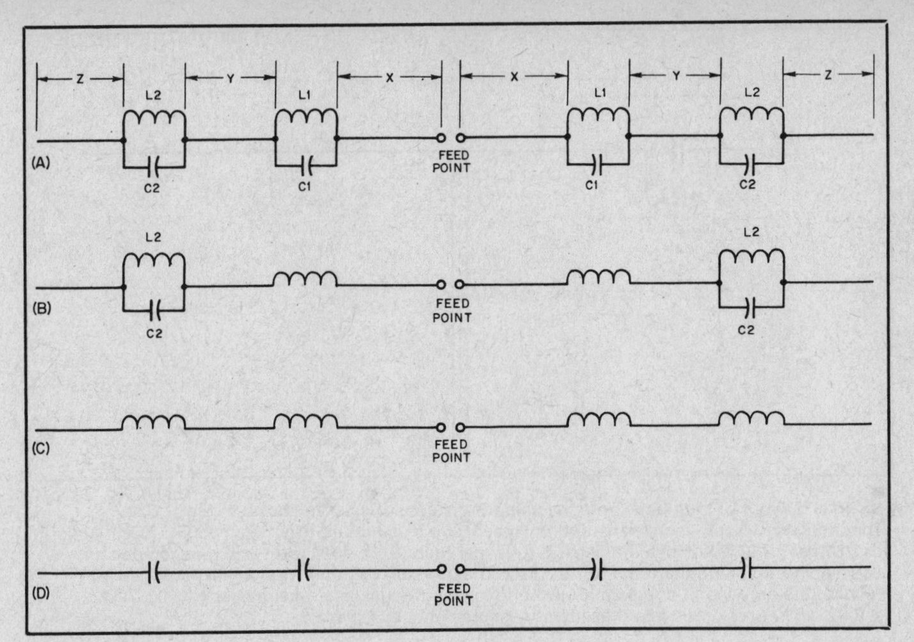


Fig. 7 — Development of the "trap" dipole for operation on fundamental-type resonance in several bands.

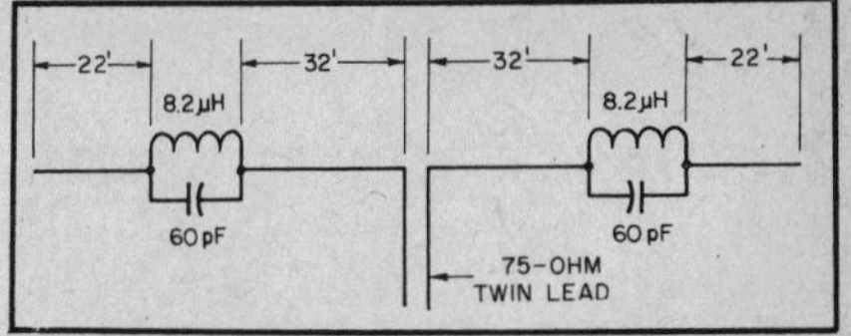


Fig. 8 — Five-band (3.5, 7, 14, 21 and 28 MHz) trap dipole for operation with 75-ohm feeder at low SWR (W3DZZ). The balanced (parallel-conductor) line indicated is desirable, but 75-ohm coax can be substituted with some sacrifice of symmetry in the system. Dimensions given are for resonance (lowest SWR) at 3750, 7200, 14,150, and 29,500 kHz. Resonance is very broad on the 21-MHz band, with SWR less than 2:1 throughout the band.

pedance for this frequency, so the 7-MHz dipole is in turn insulated, for all practical purposes, from the remaining outer parts of the antenna.

Carrying the same reasoning one step further, L2-C2 shows inductive reactance on the next lower frequency band, 3.5 MHz, and is equivalent to a coil on that band, Fig. 7C. The length of the added sections, Z-Z is adjusted so that, together with the two sets of equivalent loading coils now indicated in C, the whole system is resonant as a loaded dipole on 3.5 MHz. The reactance of the parallel LC circuit is given by

$$\frac{-X_L X_C}{X_L - X_C}$$

and the loading effect of this reactance can be determined from a loaded dipole chart in Chapter 10. A single transmission line having a characteristic impedance of the same order as the feed-point impedance of a simple dipole can be connected at the center of the antenna and will be satisfactorily matched on all three bands, and so will operate at a low SWR on all three. A line of 75-ohm impedance is satisfactory; coax may be used, but twin-lead will maintain better balance in the system since the antenna itself is symmetrical.

### Trap Losses

Since the tuned circuits have some inherent losses the efficiency of this system depends on the Qs of the tuned circuits. Low-loss (high-Q) coils should be used, and the capacitor losses likewise should be kept as low as possible. With tuned circuits that are good in this respect — comparable with the low-loss components used in transmitter tank circuits, for example — the reduction in efficiency compared with the efficiency of a simple dipole is small, but tuned circuits of low unloaded Q can lose an appreciable portion of the power supplied to the antenna.

The above commentary applies to traps assembled from conventional components. The important function of a trap is to provide a high isolating impedance, and this impedance is directly proportional to Q. Unfortunately, high Q restricts the antenna bandwidth, because the traps provide maximum isolation only at resonance. A new type of trap described by Gary O'Neil, N3GO, in October 1981 *Ham Radio* achieves high impedance with low Q, effectively overcoming the bandwidth problem. The N3GO trap is fabricated from a single length of coaxial cable. The cable is wound around a form as a single-layer coil, and the shield becomes the trap inductor. The capacitance between the center conductor and shield resonates the trap. At each end of the coil the center conductor and shield are separated. At the "inside" end of the trap, nearer the antenna feed point, the shield is connected to the antenna wire. At the outside end, the center conductor is attached to the outside antenna wire. The center conductor from the inside end is joined to the shield from the outside end to complete the trap. Constructed in this way, the trap provides high isolation over a greater bandwidth than is possible with conventional traps.

### Dimensions

The lengths of the added antenna sections, Y and Z in the example, must in general be determined experimentally. The length required for resonance in a given band depends on the length/diameter ratio of the antenna conductor and on the L/C ratio of the trap acting as a loading coil. The effective reactance of an LC circuit on half the frequency to which it is resonant is equal to 2/3 the reactance of the inductance at the resonant frequency. For example, if L1-C1 of Fig. 7 resonates at 14 MHz and L1 has a reactance of 300 ohms at 14 MHz, the inductive reactance of the circuit at 7 MHz will be equal to 2/3 × 300 = 200 ohms. The added antenna section, Y, would have to be cut to the proper length to resonate at 7 MHz with this amount of loading. Since any reasonable L/C ratio can be used in the trap without affecting its performance materially at its resonant frequency, the L/C ratio can be varied to control the added antenna length required. The added section will be shorter with high-L trap circuits and longer with high-C traps.

### Higher Frequencies

On bands higher than that for which the inner dipole is resonant, all traps in the system show capacitive reactance. Thus at such frequencies the antenna has the equivalent circuit shown at D in Fig. 7. The capacitive reactances have the effect of raising the resonant frequency of the system as compared with a simple dipole of the same overall length.

This effect is greatest near the resonant frequency of the inner dipole X-X and becomes less marked as the frequency is increased, since the capacitive reactance decreases with increasing frequency. The system therefore can be used on higher frequency bands as a harmonic-type antenna, but obtaining resonance with low impedance will require careful balancing of the trap L/C ratios and the lengths of the various antenna sections.

### Five-Band Antenna

One such system has been worked out by W3DZZ for the five pre-WARC amateur bands from 3.5 to 30 MHz. Dimensions are given in Fig. 8. Only one set of traps is used, resonant at 7 MHz to isolate the inner (7-MHz) dipole from the outer sections, which cause the overall

Fig. 9 — Easily constructed trap for wire antennas (W2CYK). The ceramic insulator is 4-1/4 inches long (Birnbach 668). The clamps are small service connectors available from electrical supply and hardware stores (Burndy KS90 Servits).

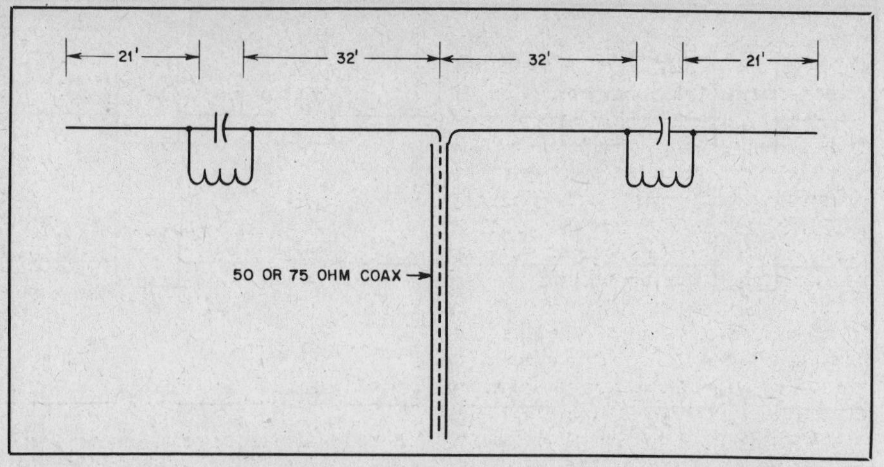


Fig. 10 — Layout of multiband antenna using traps constructed as shown in Fig. 9. The capacitors are 100 pF each, transmitting type, 5000-volt dc rating (Centralab 850SL-100N). Coils are 9 turns no. 12, 2-1/2 inch diameter, 6 turns per inch (B&W 3029) with end turns spread as necessary to resonate the traps to 7200 kHz. These traps, with the wire dimensions shown, resonate the antenna at approximately the following frequencies on each band: 3900, 7250, 14,100, 21,500 and 29,900 kHz (based on measurements by W9YJH).

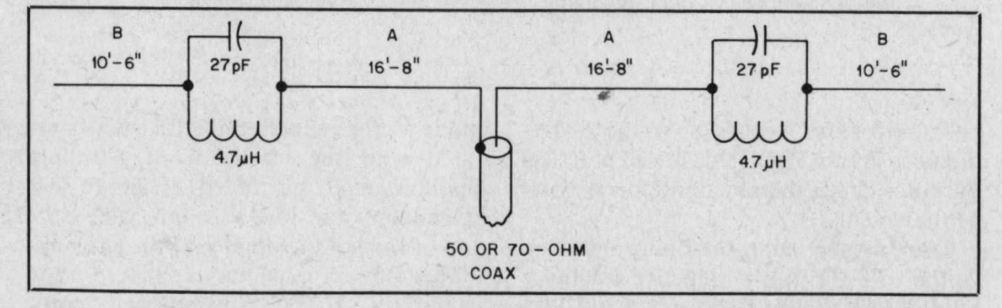


Fig. 11 — Sketch showing dimensions of a trap dipole covering the 40- through 10-meter bands (K2GU).

system to be resonant in the 3.5-MHz band. On 14, 21 and 28 MHz the antenna works on the capacitive-reactance principle just outlined. Using 75-ohm twin-lead as a feeder, the SWR with this antenna was under 2 to 1 throughout the three high-frequency bands, and the SWR was comparable with that obtained with similarly fed simple dipoles on 7 and 3.5 MHz.

### Trap Construction

Traps frequently are built with coaxial aluminum tubes (usually with polystyrene tubing between them for insulation) for the capacitor, with the coil either self-supporting or wound on a form of larger diameter than the tubular capacitor. The coil is then mounted coaxially with the capacitor to form a unit assembly that can be supported at each end by the antenna wires. In another type of trap devised by Lattin (see bibliography at the end of this chapter), the coil is supported *inside* an aluminum tube, and the trap capacitor is obtained in the form of capacitance between the coil and the outer tube. This type of trap is inherently weatherproof.

A simpler type of trap, easily assembled from readily available components, is shown in Fig. 9. A small transmitting-type ceramic capacitor is used, together with a length of commercially available coil material, these being supported by an ordinary antenna strain insulator. The circuit constants and antenna dimensions differ slightly from those of Fig. 8, in order to bring the antenna resonance points closer to the centers of the various phone bands. Construction data are given in Fig. 10. If a 10-turn length of inductor is used, a half turn from each end may be used to slip through the anchor holes in the insulator to act as leads.

The components used in these traps are sufficiently weatherproof in themselves so that no additional treatment for this purpose has been found to be necessary. However, if it is desired to protect them from the accumulation of snow or ice a plastic cover can be made by cutting two discs of polystyrene slightly larger in diameter than the coil, drilling at the center to pass the antenna wires, and cementing a plastic cylinder on the edges of the discs. The cylinder can be made by wrapping two turns or so of 0.02-inch poly or Lucite sheet around the discs, if no suitable ready-made tubing is available. Plastic drinking glasses and soft 2-liter soft-drink bottles are easily adaptable for use as trap covers.

The construction of traps assembled entirely from coaxial cable is treated by R. Johns, W3JIP, in May 1981 *QST* and G. O'Neil, N3GO, in October 1981 *Ham Radio*.

### Four-Band Trap Dipole

In case there is not enough room available for erecting the 100-odd-foot length required for the five-band antennas just described, Fig. 11 shows a four-band dipole operating on the same principle that requires only half the linear space. The trap construction is the same as shown in Fig. 9. With the dimensions given in Fig. 11 the resonance points are 7200, 14,100, 21,150 and 28,400 kHz. The capacitors are 27-pF transmitting-type ceramic (Centralab type 857). The inductors are 9 turns of no. 12, 2-1/2 inches in diameter; 6 turns per inch (B&W 3029), adjusted so that the trap resonates at 14,100 kHz before installation in the antenna.

## VERTICAL ANTENNAS

There are two basic types of vertical antennas; either type can be used in multiband configurations. The first is the ground-mounted vertical and the second, the ground plane. These antennas are described in detail in Chapter 2.

The efficiency of any ground-mounted vertical depends a great deal on earth losses. As pointed out in Chapter 2, these losses can be reduced or eliminated with an adequate radial system. Considerable experimentation has been conducted on this subject by Sevick, and several important results were obtained. It was determined that a radial system consisting of 40 to 50 radials, two-tenths wavelength long, would reduce the earth losses to about 2 ohms when a quarter-wave radiator was being used. These radials should be on the earth's surface, or if buried, placed not more than an inch or so below ground.

Otherwise, the rf current would have to travel through the lossy earth before reaching the radials. In a multiband vertical system, the radials should be 0.2 wavelength long for the lowest band, i.e., 55 feet long for 80-meter operation. Any wire size may be used for the radials. The radials should fan out in a circle, radiating from the base of the antenna. A metal plate, such as a piece of sheet copper, can be used at the center connection.

The other common type of vertical is the ground-plane antenna. Normally, this antenna is mounted above ground with the radials fanning out from the base of the antenna. The vertical portion of the antenna is usually an electrical quarter wavelength, as is each of the radials. In this type of antenna, the system of radials acts somewhat like an rf choke, to prevent rf currents from flowing in the supporting structure, so the number of radials is not as important a factor as it is with a ground-mounted vertical system. From a practical standpoint, the customary number of radials is four or five. In a multiband configuration, quarter-wave radials are required for each band of operation with the ground-plane antenna. This is not so with the ground-mounted antenna, where the ground plane is relied upon to provide an image of the radiating section. In the latter case, as long as the ground-screen radials are approximately 0.2 wavelength long at the lowest frequency, this length will be more than adequate for the higher bands.

**Short Vertical Antennas**

A short vertical antenna can be operated on several bands by loading it at the base, the general arrangement being similar to Figs. 1 and 2. That is, for multiband work the vertical can be handled by the same methods that are used for random-length wires.

A vertical antenna should not be longer than about 3/4 wavelength at the highest frequency to be used, however, if low-angle radiation is wanted. If the antenna is to be used on 28 MHz and lower frequencies, therefore, it should not be more than approximately 25 feet high, and the shortest possible ground lead should be used. If the base of the antenna is well above actual ground, the ground lead should run to the nearest water or heating pipe.

Another method of feeding is shown in Fig. 12. L1 is a loading coil of adjustable inductance so the antenna can be tuned to resonate on the desired band. It is tapped for adjustment of tuning, and a second tap permits using the coil as a transformer for matching a coax line to the transmitter. C1 is not strictly necessary, but may be helpful on the lower frequencies, 3.5 and 7 MHz, if the antenna is quite short. In that case C1 makes it possible to tune to resonance with a coil of reasonable dimensions at L1. C1 may also be useful

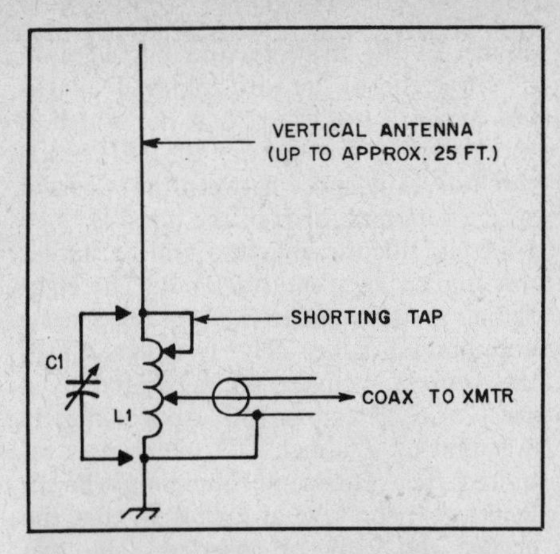


Fig. 12 — Multiband vertical antenna system using base loading for resonating on 10 to 80 meters. L1 should be wound with bare wire so it can be tapped at every turn, using no. 12 wire. A convenient size is 2-1/2 inches in diameter, 6 turns per inch (such as B&W 3029). Number of turns required depends on antenna and ground lead length, more turns being required as the antenna and ground lead are made shorter. For a 25-foot antenna and a ground lead of the order of 5 feet, L1 should have about 30 turns. The use of C1 is explained in the text. The smallest capacitance that will permit matching the coax cable should be used; a maximum capacitance of 100 to 150 pF will be sufficient in any case.

on other bands as well, if the system cannot be matched to the feed line with a coil alone.

The coil and capacitor should preferably be installed at the base of the antenna, but if this cannot be done a wire can be run from the antenna base to the nearest convenient location for mounting L1 and C1. The extra wire will of course be a part of the antenna, and since it may have to run through unfavorable surroundings it is best to avoid its use if at all possible.

This system is best adjusted with the help of an SWR indicator. Connect the coax line across a few turns of L1 and take trial positions of the shorting tap until the SWR reaches its lowest value. Then vary the line tap similarly; this should bring the SWR down to a low value. Small adjustments of both taps then should reduce the SWR to close to 1 to 1. If not, try adding C1 and go through the same procedure, varying C1 each time a tap position is changed.

**Trap Verticals**

The trap principle described in Fig. 7 for center-fed dipoles also can be used for vertical antennas. There are two principal differences: Only one half of the dipole is used, the ground connection taking the place of the missing half, and the feed-point impedance is one half the feed-point impedance of a dipole. Thus it is in the vicinity of 30 ohms (plus the ground-connection resistance), so 52-ohm cable should be used since it is the commonly available type that comes closest to matching.

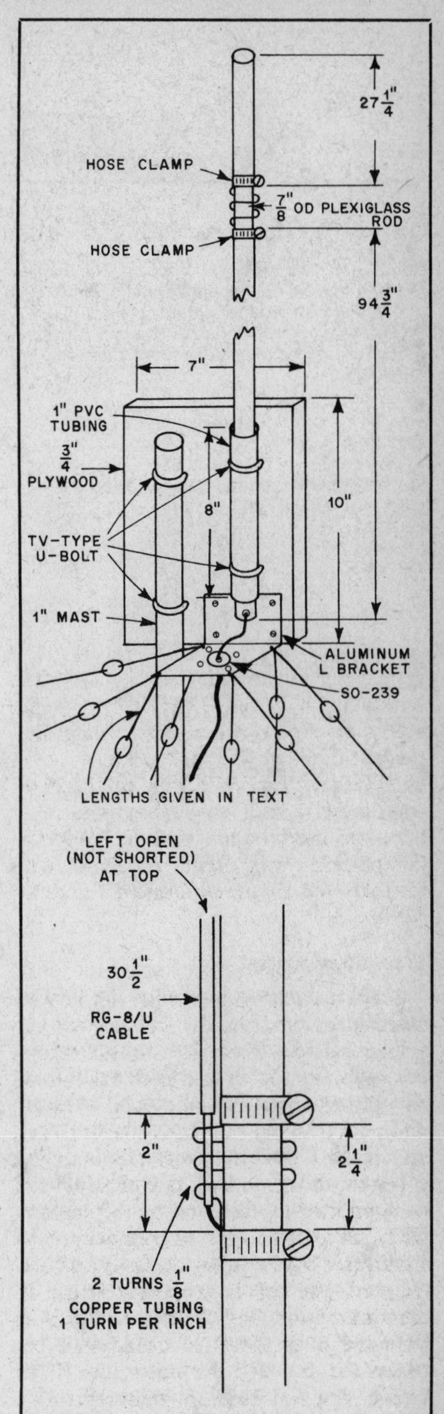


Fig. 13 — Constructional details of the 10- and 15-meter band antenna system.

As in the case of any vertical antenna, a good ground is essential, and the ground lead should be short. Some amateurs have reported successfully using a ground plane dimensioned for the lowest frequency to be used; for example, if the lowest frequency is 7 MHz, the ground-plane radials can be approximately 34 feet long.

## A TRAP VERTICAL FOR 10 AND 15 METERS

Simple antennas covering the upper hf bands can be quite compact and inexpensive. The two-band vertical ground plane

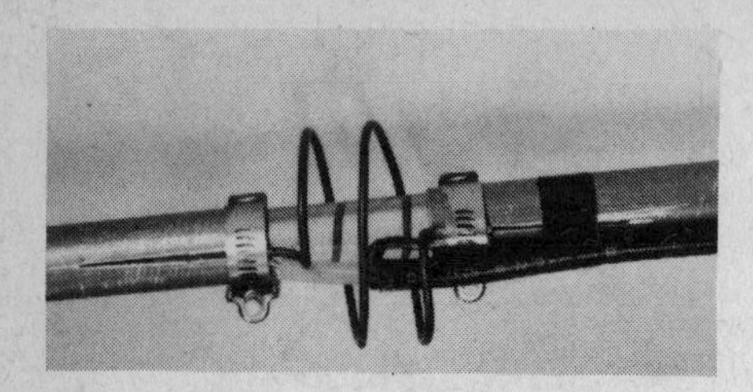

Fig. 14 — A close-up view of a trap. The leads from the coaxial-cable capacitor should be soldered directly to the pigtails of the coil. These connections should be coated with varnish after they have been secured under the hose clamps.

Fig. 15 — The base assembly of the 10- and 15-m vertical. The SO-239 coaxial connector and hood can be seen in the center of the aluminum L bracket. The U bolts are TV-type antenna hardware. The plywood should be coated with varnish or similar material.

described here is highly effective for long-distance communication when installed in the clear.

Figs. 13, 14 and 15 show the important assembly details. The vertical section of the antenna is mounted on a 3/4-inch thick piece of plywood board that measures 7 × 10 inches. Several coats of exterior varnish or similar material will help protect the wood from inclement weather. Both the mast and the radiator are mounted on the piece of wood by means of TV U-bolt hardware. The vertical is electrically isolated from the wood with a piece of 1-inch diameter PVC tubing. A piece approximately 8 inches long is required, and it is of the schedule-80 variety. To prepare the tubing it must be slit along the entire length on one side. A hacksaw will work quite well. The PVC fits rather snugly on the aluminum tubing and will have to be "persuaded" with the aid of a hammer. The mast is mounted directly on the wood with no insulation. An SO-239 coaxial connector and four solder lugs are mounted on an L-shaped bracket made from a piece of aluminum sheet. A short length of test probe wire, or inner conductor of RG-58/U cable, is soldered to the inner terminal of the connector. A UG-106/U connector hood is then slid over the wire and onto the coaxial connector. The hood and connector are bolted to the aluminum bracket. Two wood screws are used to secure the aluminum bracket to the plywood as shown in the drawing and photograph. The free end of the wire coming from the connector is soldered to a lug which is mounted on the bottom of the vertical radiator. Any space between the wire and where it passes through the hood is filled with GE silicone glue and seal or similar material to keep moisture out. The eight radials are soldered to the four lugs on the aluminum bracket. The two sections of the vertical member are separated by a piece of clear acrylic rod. Approximately 8 inches of 7/8-inch OD material is required. The aluminum tubing must be slit lengthwise for several inches so that the acrylic rod may be inserted. The two pieces of aluminum tubing are separated by 2-1/4 inches.

The trap capacitor is made from RG-8/U coaxial cable and is 30.5 inches long. RG-8/U cable has 29.5 pF of capacitance per foot and RG-58/U has 28.5 pF per foot. RG-8/U cable is recommended over RG-58/U because of its higher breakdown-voltage characteristic. The braid should be pulled back 2 inches on one end of the cable, and the center conductor soldered to one end of the coil. Solder the braid to the other end of the coil. Compression type hose clamps are placed over the capacitor/coil leads and put in position at the edges of the aluminum tubing. When tightened securely, the clamps serve a two-fold purpose — they keep the trap in contact with the vertical members and prevent the aluminum tubing from slipping off the acrylic rod. The coaxial-cable capacitor runs upward along the top section of the antenna. This is the side of the antenna to which the braid of the capacitor is connected. A cork or plastic cap should be placed in the very top of the antenna to keep moisture out.

### Installation and Operation

The antenna may be mounted in position using a TV-type tripod, chimney, wall or vent mount. Alternatively, a telescoping mast or ordinary steel TV masting may be used, in which case the radials may be used as guys for the structure.The 10-meter radials are 8 feet 5 inches long, and the 15-meter radials are 11 feet 7 inches.

Any length of 50-ohm cable may be used to feed the antenna. The SWR at resonance should be on the order of 1.2 to 1.5 to 1 on both bands. The reason the SWR is not 1 is that the feedpoint resistance is something other than 50 ohms — closer to 35 or 40 ohms.

## ADAPTING MANUFACTURED TRAP VERTICALS TO THE WARC BANDS

The frequency coverage of a multiband vertical antenna can be modified simply by altering the lengths of the tubing sections and/or adding a trap. Several companies manufacture trap verticals covering 40, 20, 15 and 10 meters. Many amateurs

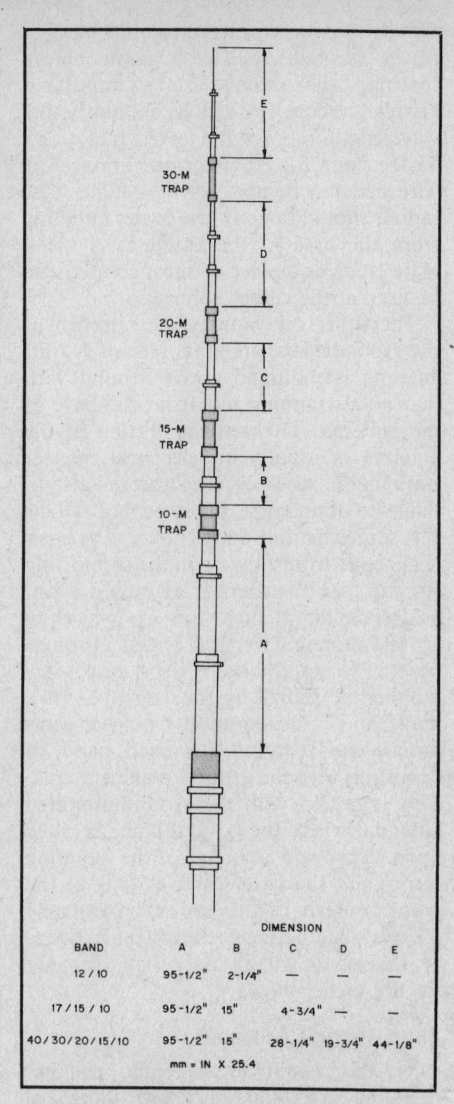


| BAND | DIMENSION A | B | C | D | E |
|---|---|---|---|---|---|
| 12 / 10 | 95-1/2" | 2-1/4" | — | — | — |
| 17 / 15 / 10 | 95-1/2" | 15" | 4-3/4" | — | — |
| 40 / 30 / 20 / 15 / 10 | 95-1/2" | 15" | 28-1/4" | 19-3/4" | 44-1/8" |

mm = IN X 25.4

Fig. 16 — Modified dimensions for the ATV-series Cushcraft vertical antennas for some frequency combinations that include the WARC bands. The 30-meter trap inductor consists of 20 turns of no. 16 enameled wire close-wound on a 5/8-in. dia Plexiglas rod. The capacitor is a 29-3/4-in. length of RG-58/U cable.

roof-mount these antennas, because an effective ground radial system isn't practical, to keep children away from the antenna, or to clear metal-frame buildings. On the three highest-frequency bands, the tubing and radial lengths are convenient for rooftop installations, but 40 meters sometimes presents problems. Prudence dictates erecting an antenna with the assumption that it will fall down. When the antenna falls, it and the radial system must clear any nearby power lines. Where this consideration rules out 40-meter operation, careful measurement may show that 30-meter dimensions will allow adequate safety. The antenna is resonated by pruning the tubing above the 20-meter trap and installing tuned radials.

Several new frequency combinations are possible. The simpler ones, 12/10, 17/15/10, and 40/30/20/15/10 meters,

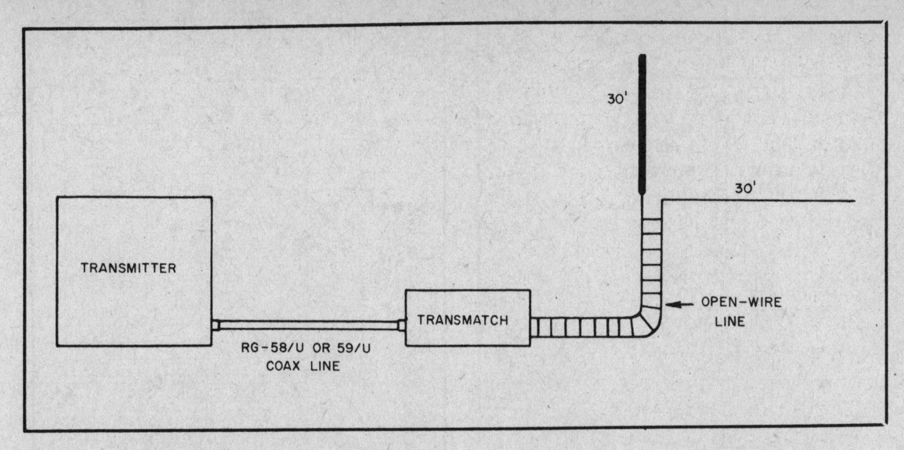


Fig. 17 — Vertical and horizontal conductors combined. This system can be used on all bands from 3.5 to 28 MHz with good results.

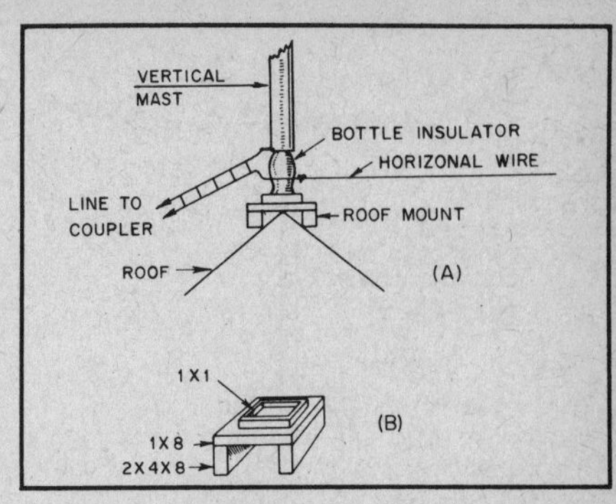


Fig. 18 — One method of mounting the vertical section on a rooftop. The mounting base dimensions can be adjusted to fit the pitch of the roof. The 1 × 1 pieces should fit snugly around the bottom of the bottle to keep it from shifting position.

are shown in Fig. 16 applied to the popular ATV series of trap verticals manufactured by Cushcraft. Operation in the 30-meter band requires an additional trap — use Fig. 14 as a guide for constructing this component.

### Combining Vertical and Horizontal Conductors

The performance of vertical antennas such as just described depends a great deal on the ground system. You have no way of knowing whether or not you have a "good" ground, in the rf sense. If you can eliminate the ground connection as a part of the antenna system, it simplifies things. Fig. 17 shows how it can be done. Instead of a ground, the system is completed by a wire — preferably, but not necessarily, horizontal — of the same length as the antenna. This makes a balanced system somewhat like the center-fed dipole.

It is desirable that the length of each conductor be on the order of 30 feet, as shown in the drawing, if the 3.5-MHz band is to be used. At 7 MHz, this length doesn't really represent a compromise, since it is almost a half wavelength overall on that band. Because the shape of the antenna differs from that of a regular half-wave dipole, the radiation characteristics will be different, but the efficiency will be high on 7 MHz and higher frequencies. Although the radiating part is only about a quarter wavelength at 3.5 MHz the efficiency on this band, too, will be higher than it would be with a grounded system. If one is not interested in 3.5 MHz and can't use the dimensions shown, the lengths can be reduced. Fifteen feet in both the vertical and horizontal conductors will not do too badly on 7 MHz and will not be greatly handicapped, as compared with a halfwave dipole, on 14 MHz and higher.

The vertical part can be mounted in a number of ways. However, if it can be put on the roof of your house, the extra height will be worthwhile. Fig. 18 suggests a simple base mount using a soft-drink bottle as an insulator. Get one with a neck diameter that will fit into the tubing used for the vertical part of the antenna. To help prevent possible breakage, put a piece of some elastic material such as rubber sheet around the bottle where the tubing rests on it.

The wire conductor doesn't actually have to be horizontal. It can be at practically any angle that will let it run off in a straight line to a point where it can be secured. Use an insulator at this point, of course.

TV ladder line should be used for the feeder in this system. On most bands the standing-wave ratio will be high, and you will lose a good deal of power in the line if you try to use coax, or even 300-ohm twin-lead. This system can be tuned up by using an SWR indicator in the coax line between the transmitter and a Transmatch.

### HARMONIC RADIATION FROM MULTIBAND ANTENNAS

Since a multiband antenna is intentionally designed for operation on a number of different frequencies, any harmonics or spurious frequencies that happen to coincide with one of the antenna's resonant frequencies will be radiated with very little, if any, attenuation. Particular care should be exercised, therefore, to prevent such harmonics from reaching the antenna.

Multiband antennas using tuned feeders have a certain inherent amount of built-in protection against such radiation since it is nearly always necessary to use a tuned coupling circuit between the transmitter and the feeder. This adds considerable selectivity to the system and helps to discriminate against all frequencies other than the desired one.

Multiple dipoles and trap antennas do not have this feature, since the objective in design is to make the antenna show as nearly as possible the same resistive impedance in all the amateur bands the antenna is intended to cover. It is advisable to conduct tests with other amateur stations to determine whether harmonics of the transmitting frequency can be heard at a distance of, say, a mile or so. If they can, more selectivity should be added to the system since a harmonic that is heard locally, even if weak, may be quite strong at a distance because of propagation vagaries.

## Antennas for 3.5 and 7 MHz

Multiband antennas constructed as described earlier in this chapter obviously will be useful on 3.5 and 7 MHz, and, in fact, the end-fed and center-fed antennas are quite widely used for 3.5- and 7-MHz operation. The center-fed system is better because it is inherently balanced on both bands and there is less chance for feeder radiation and rf feedback troubles, but either system will give a good account of itself. On these frequencies the height of the antenna is not too important, and anything over 35 feet will work well for average operation. This section is concerned principally with antennas designed for use on one band only.

### HALF-WAVELENGTH ANTENNAS

An untuned or "flat" feed line is a logical choice on any band because the losses are low, but it generally limits the

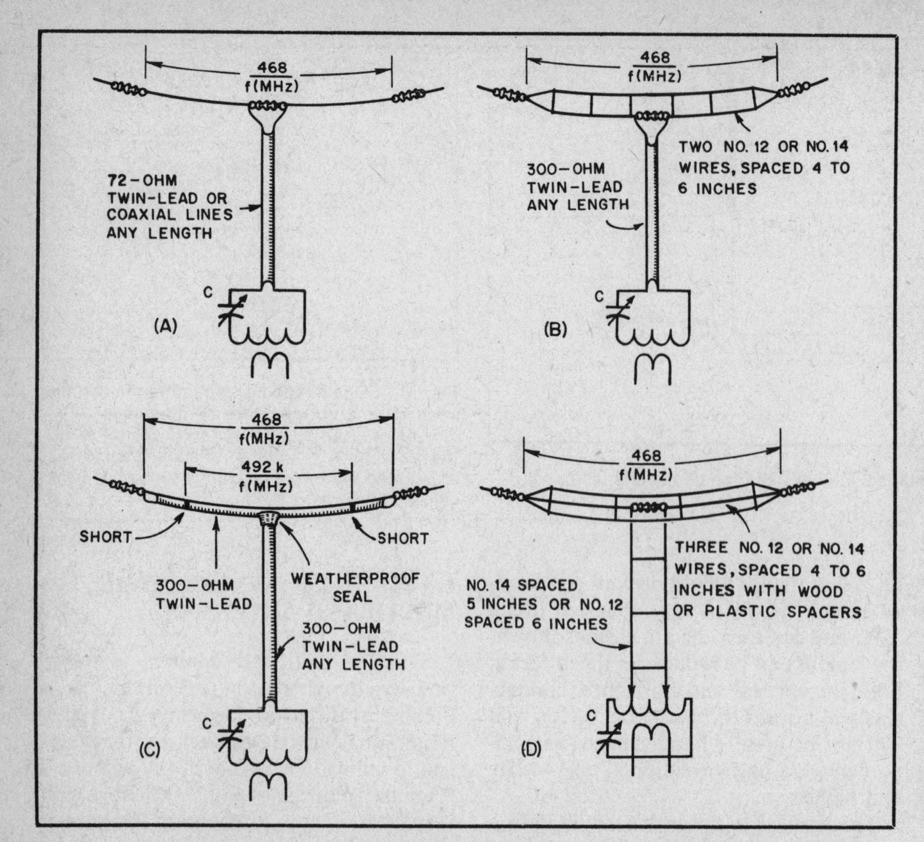


Fig. 19 — Half-wavelength antennas for single-band operation. The multiwire types shown in B, C and D offer a better match to the feeder over a somewhat wider range of frequencies but otherwise the performances are identical. At C, K = velocity factor for the twin-lead. The feeder should run away from the antenna at a right angle for as great a distance as possible. In the coupling circuits shown, tuned circuits should resonate to the operating frequency. In the series-tuned circuits of A, B, and C, high L and low C are recommended, and in D the inductance and capacitance should be similar to the output-amplifier tank, with the feeders tapped across at least 1/2 the coil. The tapped-coil matching circuit or the Transmatch, both shown in Chapter 6, can be substituted in each case.

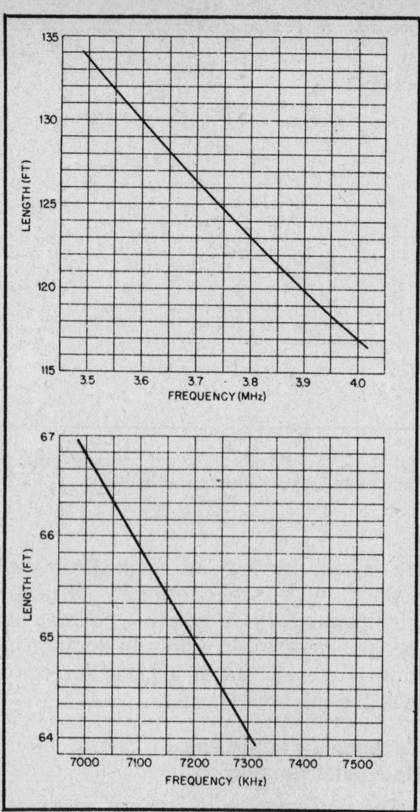


Fig. 20 — These charts can be used to determine the length of a half-wave antenna of wire.

use of the antenna to one band. Where only single-band operation is wanted, the half-wave antenna fed with untuned line is one of the most popular systems on the 3.5- and 7-MHz bands. If the antenna is a single-wire affair, its impedance is in the vicinity of 60 ohms. The most logical way to feed the antenna is with 72-ohm twin-lead or 50- or 72-ohm coaxial line. The heavy-duty twin-lead and the coaxial line present support problems, but these can be overcome by using a small auxiliary pole to take the weight of the line. The line should come away from the antenna at right angles, and it can be of any length.

A "folded dipole" shows an impedance of 300 ohms, and so it can be fed directly with any length of 300-ohm TV line. The line should come away from the antenna at as close to a right angle as possible. The folded dipole can be made of ordinary wire spaced by lightweight wooden or plastic spacers, 4 or 6 inches long, or a piece of 300-ohm TV line can be used for the folded dipole.

A folded dipole can be fed with a 600-ohm open-wire line with only a 2-to-1 SWR, but a nearly perfect match can be obtained with 600-ohm open line and a three-wire dipole.

The three types of half-wavelength antennas just discussed are shown in Fig. 19. One advantage of the two- and three-wire antennas over the single wire is that they offer a better match over a band. This is particularly important if full coverage of the 3.5-MHz band is contemplated.

While there are many other methods of matching lines to half-wavelength antennas, the three mentioned are the most practical ones. It is possible, for example, to use a quarter-wavelength transformer of 150-ohm twin-lead to match a single-wire half-wavelength antenna to 300-ohm feed line. But if 300-ohm feed line is to be used, a folded dipole offers an excellent match without the necessity for a matching section.

The formula shown above each antenna in Fig. 19 can be used to compute the length at any frequency, or the length can be obtained directly from the charts in Fig. 20.

**Inverted-V Dipole**

The halves of a dipole may be sloped to form an inverted V, as shown in Fig. 21. This has the advantages of requiring only a single high support and less horizontal space. Many amateurs have reported that the dipole in this form is more effective than a horizontal antenna, especially for frequencies of 7 MHz and lower.

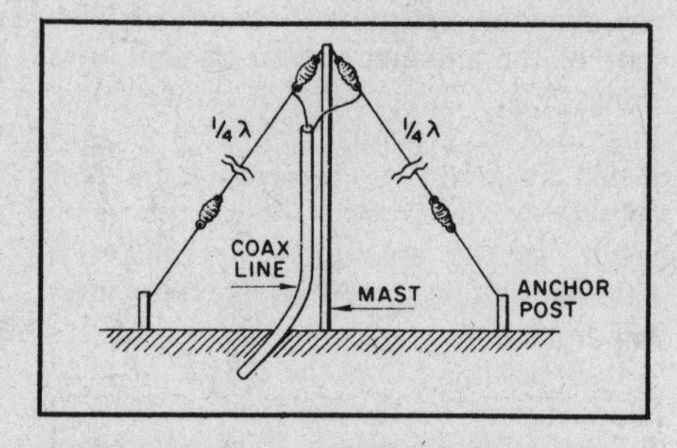


Fig. 21 — The inverted-V dipole. The length and apex angle should be adjusted as described in the text.

Sloping of the wires results in a decrease in the resonant frequency and a decrease in feed-point impedance and bandwidth. Thus, for the same frequency, the length of the dipole must be decreased somewhat. The angle at the apex is not critical, although it should probably be made no smaller than 90 degrees. Because of the lower impedance, a 50-ohm line should be used, and the usual procedure is to adjust the angle for lowest SWR while keeping the dipole resonant by adjustment of length. Bandwidth may be increased by using multiconductor elements, such as the cage configuration.

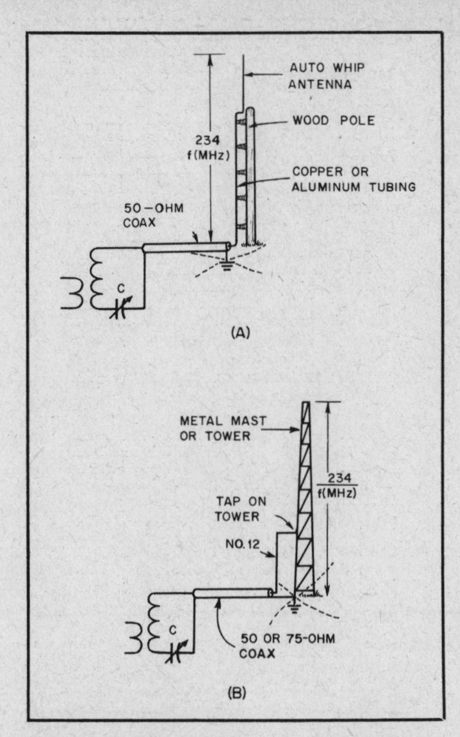


Fig. 22 — Vertical antennas are effective for 3.5- or 7-MHz work. The quarter-wavelength antenna shown at A is fed directly with 50-ohm coaxial line, and the resulting standing-wave ratio is usually less than 1.5 to 1, depending on the ground resistance. If a grounded antenna is used as at B, the antenna can be shunt-fed with either 50- or 75-ohm coaxial line. The tap for best match and the value of C will have to be found by experiment; the line running up the side of the antenna should be spaced 6 to 12 inches from the antenna. The length (height) of the antenna can be computed from the formula, or it can be obtained from Fig. 20 by using just one half the length indicated in the chart. For example, at 3.6 MHz, the length is 130'/2 = 65'.

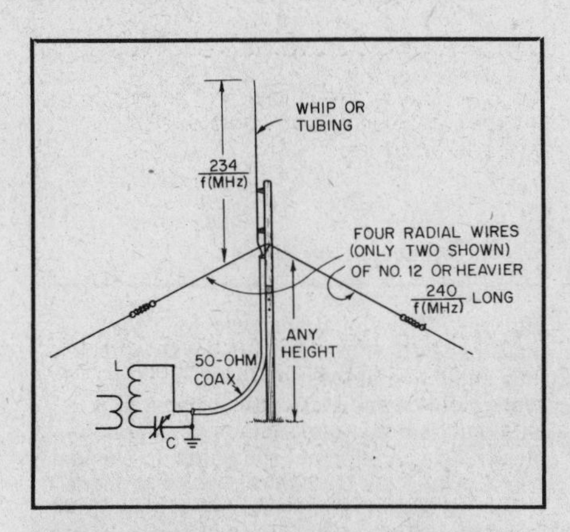


Fig. 23 — A ground-plane antenna is effective for DX work on 7 MHz. Although its base can be any height above ground, losses in the ground underneath will be reduced by keeping the bottom of the antenna and the ground plane as high above ground as possible. Feeding the antenna directly with 50-ohm coaxial cable will result in a low standing-wave ratio. The length of the vertical radiator can be computed from the formula, or it can be obtained from Fig. 20 by using just one half the length indicated in the chart. The radial wires are 2.5% longer. For example, at 7.1 MHz, the radiator is 65' 11"/2 ≈ 33'; the radials are 1.025 x 33 = 33'10".

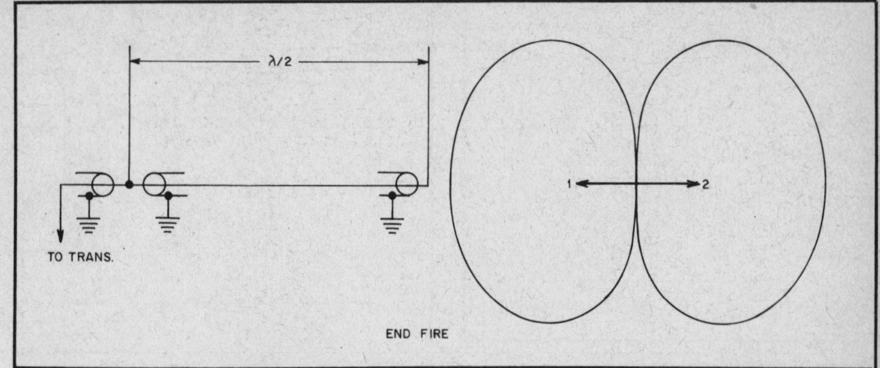


Fig. 24 — Pattern for two 1/4-λ verticals spaced one-half wavelength apart fed 180 degrees out of phase. The arrow represents the axis of the elements.

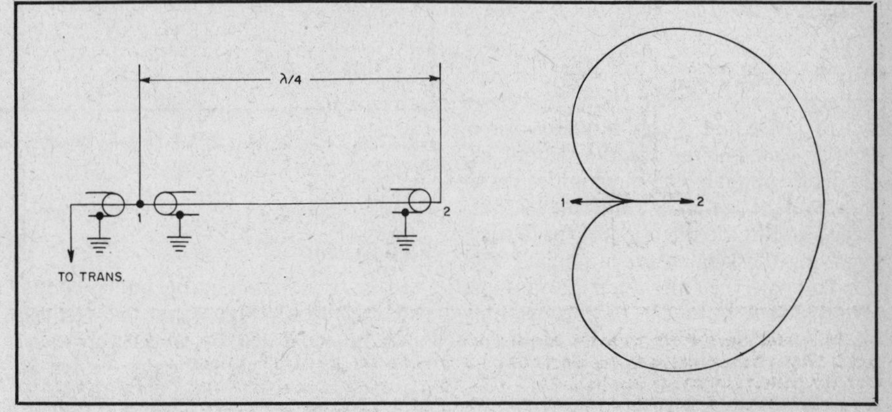


Fig. 25 — Pattern for two 1/4-λ verticals spaced 1/4 wavelength apart and fed 90 degrees out of phase. The arrow represents the axis of the elements, with the element on the right being the one of lagging phase.

## VERTICAL ANTENNAS

For 3.5-MHz work, the vertical can be a quarter wavelength long (if one can get the height), or it can be something less than this and "top-loaded." The bottom of the antenna has only to clear the ground by inches. Probably the cheapest construction of a quarter-wavelength vertical involves running copper or aluminum wire alongside a wooden mast. A metal tower can also be used as a radiator. If the tower is grounded, the antenna can be "shunt-fed," as shown in B of Fig. 22. The "gamma" matching system may also be used. A good ground system is essential in feeding a quarter-wavelength vertical antenna. The ground can be either a convenient water-pipe system or a number of radial wires extending out from the base of the antenna for about a quarter wavelength.

### The Ground Plane

The size of a ground-plane antenna makes it a little impractical for 3.5-MHz work, but it can be used at 7 MHz to good advantage, particularly for DX work. This type of antenna can be placed higher above ground than an ordinary vertical without decreasing the low-angle radiation. The vertical member can be a length of self-supporting tubing at the top of a short mast, and the radials can be lengths of wire used also to support the mast. The radials do not have to be exactly horizontal, as shown in Fig. 23.

The ground-plane antenna can be fed directly with 50-ohm cable, although the resulting SWR on the line will not be as low as it will if the antenna is designed with a stub matching section. However, the additional loss caused by an SWR as high as 2 to 1 will be inappreciable even in cable runs of several hundred feet when the frequency is as low as 7 MHz.

## PHASED VERTICALS

Two or more vertical antennas spaced a half wavelength apart can be operated as a single antenna system to obtain additional gain and a directional pattern. The following design for 40-meter phased verticals is contributed by Gary Elliott, KH6HCM/W7UXP. An 80-meter version can be constructed by proper scaling. There are practical ways that verticals for 40 meters can be combined, end-fire and broadside. In the broadside configuration, the two verticals are fed in phase, producing a figure-eight pattern that is broadside to the plane of the verticals. In an end-fire arrangement, the two verticals are fed out of phase, and a figure-eight pattern is obtained that is in line with the

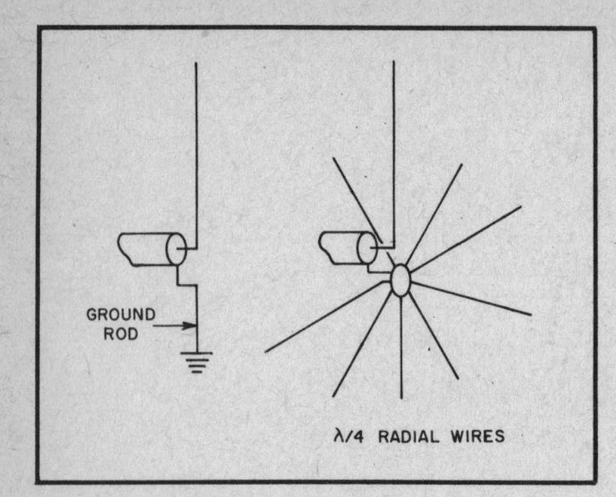


Fig. 26 — An 8- to 10-ft. ground rod may provide a satisfactory ground system in marshy or beach areas, but in most locations a system of radial wires will be necessary.

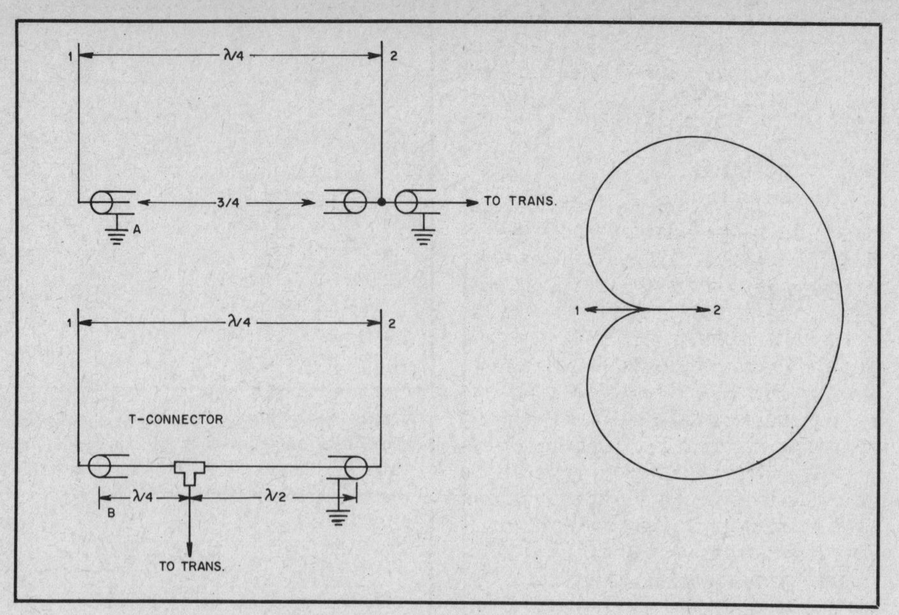


Fig. 27 — Two methods of feeding the phased verticals.

two antennas, Fig. 24. However, an end-fire pair of verticals can be fed 90 degrees out of phase and spaced a quarter wavelength apart, and the resulting pattern will be unidirectional. The direction of maximum radiation is in line with the two verticals, and in the direction of the vertical receiving the lagging excitation; see Fig. 25. This pattern, with theoretically infinite front-to-back ratio, can be obtained only if the elements are excited by equal currents. This condition is difficult to realize in practice because the phases of the *induced* currents cause the feed impedances to be grossly unequal.

## Construction

Physically, each vertical is constructed of telescoping aluminum tubing that starts off at 1-1/2-inch dia and tapers down to 1/4-inch dia at the top. The length of each vertical is 32 feet. Each vertical is supported on two standoff insulators set on a 2 × 4, 6 feet long and strapped to a fence. An alternative method of mounting would be a 2 × 4 about 8 feet long and set about 2 feet in the ground.

Originally each vertical element was 32 feet, 6 inches long, 234/f (MHz). After one vertical was mounted on the 2 × 4 it was raised into position and the resonant frequency was checked with an antenna noise bridge. It was found that the vertical resonated too low in frequency, about 6.9 MHz. This was to be expected as the fundamental equation for the quarter-wave vertical, 234/f, is only reasonably correct for very small-diameter tubing or antenna wire. When larger diameter tubing (1-1/4 inch and larger) is used, the physical length will be shorter than this, as described in Chapter 2. Using the antenna noise bridge, an inch at a time was cut off the top until the resonant frequency was 7100 kHz. This resulted in 6 inches being cut off, thus making the vertical exactly 32 feet long.

The ground system is very important in the operation of a vertical. The two usual methods of obtaining a ground system with verticals are shown in Fig. 26.

## Feed System

In order to obtain the unidirectional pattern shown in Fig. 25, the two verticals must be separated by a quarter wavelength, and one vertical must be fed 90 degrees behind the other. Two suggested feed methods are shown in Fig. 27. An electrical section of line cannot be used by itself to connect the two verticals together to obtain the 90-degree lag because of the velocity factor of RG-8/U. The length of an electrical quarter wavelength of transmission line is based on the calculation:

$$\frac{246 \times 0.66}{7.1\ \text{MHz}} = 22'\ 10''$$

(Further information concerning velocity factor and transmission lines can be found in Chapter 3 in the section on electrical length.) Obviously, 22 feet, 10 inches of coax cannot be used, as the verticals are spaced 34.6 feet apart. This is overcome and a 90-degree lag is still obtained by using a 3/4-wavelength section of transmission line between the two verticals, Fig. 27A. The SWR is less than 1.25 to 1 across the entire band, using 52-ohm coax and no matching network.

## PHASED HORIZONTAL ARRAYS

Phased arrays with horizontal elements can be used to advantage at 7 MHz, if they can be placed at least 40 feet above ground. Any of the usual combinations will be effective. If a bidirectional characteristic is desired, the W8JK type of array, shown at A in Fig. 28, is a good one. If a unidirectional characteristic is required, two elements can be mounted about 20 feet apart and provision included for tuning one of the elements as either a director or reflector, as shown in Fig. 28B.

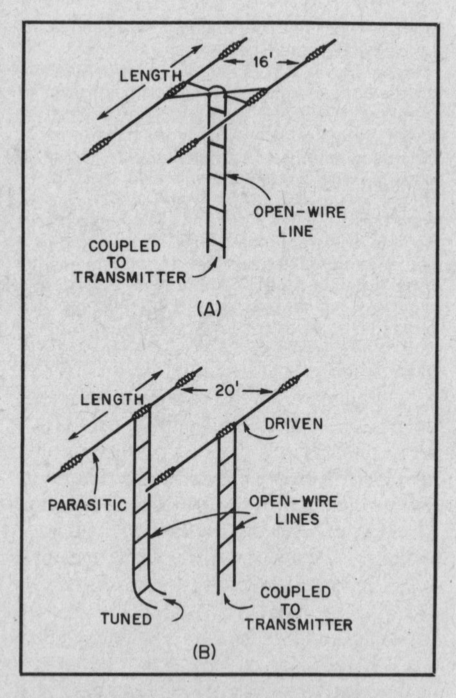


Fig. 28 — Directional antennas for 7 MHz. To realize any advantage from these antennas, they should be at least 40 feet high. The system at A is bidirectional, and that at B is unidirectional in a direction depending upon the tuning conditions of the parasitic element. The length of the elements in either antenna should be exactly the same, but any length from 60 to 150 feet can be used. If the length of the antenna at A is between 60 and 80 feet, the antenna will be bidirectional along the same line on both 7 and 14 MHz. The system at B can be made to work on 7 and 14 MHz in the same way, by keeping the length between 60 and 80 feet.

The parasitic element is tuned at the end of its feed line with a series- or parallel-tuned circuit (whichever would normally be required to couple power into the line), and the proper tuning condition can be found by using the system for receiving and listening to distant stations along the line of maximum radiation of the antenna. Tuning the feeder to the parasitic element will peak up the signal.

## 40-METER LOOP

An effective but simple 40-meter antenna that has a theoretical gain of approximately 2 dB over a dipole is a full-wave, closed loop. A full-wavelength closed loop need not be square. It can be trapezoidal, rectangular, circular, or some distorted configuration in between those shapes. For best results, however, the builder should attempt to make the loop as square as possible. The more rectangular the shape the greater the cancellation of energy in the system, and the less effective it will be. The effect is similar to that of a dipole, its effectiveness becoming impaired as the ends of the dipole are brought closer and closer together. The practical limit can be seen in the inverted-V antenna, where a 90-degree apex angle between the legs is the minimum value ordinarily used. Angles that are less than 90 degrees cause serious cancellations of the rf energy.

The loop can be fed in the center of one of the vertical sides if vertical polarization is desired. For horizontal polarization it is necessary to feed either of the horizontal sides at the center.

Optimum directivity occurs at right angles to the plane of the loop, or in more simple terms, broadside from the loop. Therefore, one should try to hang the system from available supports which will enable the antenna to radiate the maximum amount in some favored direction.

Just how the wire is erected will depend on what is available in one's yard. Trees are always handy for supporting antennas, and in many instances the house is high enough to be included in the lineup of solid objects from which to hang a radiator. If only one supporting structure is available it should be a simple matter to put up an A frame or pipe mast to use as a second support. (Also, tower owners see Fig. 29 inset.)

The overall length of the wire used in a loop is determined in feet from the formula 1005/f(MHz). Hence, for operation at 7125 kHz the overall wire length will be 141 feet. The matching transformer, an electrical quarter wavelength of 75-ohm coax cable, can be computed by dividing 246 by the operating frequency in MHz, then multiplying that number by the velocity factor of the cable being used. Thus, for operation at 7125 kHz, 246/7.125 MHz = 34.53 feet. If coax with solid polyethylene insulation is used a velocity factor of 0.66 must be employed. Foam-polyethylene coax has a velocity factor of 0.80. Assuming RG-59/U is used, the length of the matching transformer becomes 34.53 (feet) × 0.66 = 22.79 feet, or 22 feet, 9-1/2 inches.

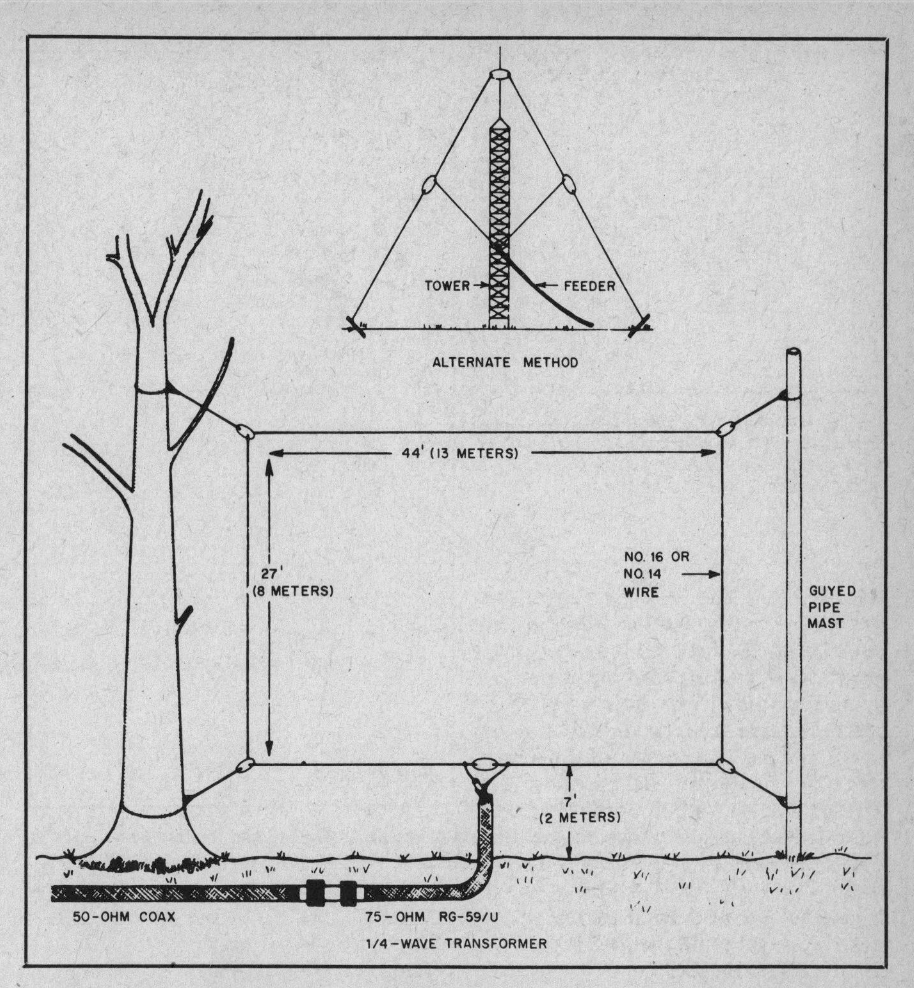


Fig. 29 — Details of the full-wave loop. The dimensions given are for operation at the low end of 40 meters (7050 kHz). The height above ground was 7 feet in this instance, though improved performance should result if the builder can install the loop higher above ground without sacrificing length on the vertical sides. The inset illustrates how a single supporting structure can be used to hold the loop in a diamond-shaped configuration. Feeding the diamond at the lower tip provides radiation in the horizontal plane. Feeding the system at either side will result in vertical polarization of the radiated signal.

This same loop antenna may be used on the 20- and 15-meter bands, although its pattern will be somewhat different than on its fundamental frequency. Also, a slight mismatch will occur, but this can be overcome by a simple matching network. When the loop is mounted in a vertical plane, it tends to favor low-angle signals. If a high-angle system is desired, say for 80 meters, the full-wave loop can be mounted in a horizontal plane, 30 or more feet above ground. This arrangement will direct most of the energy virtually straight up, providing optimum sky-wave coverage on a short-haul basis.

## 40-METER "SLOPER" SYSTEM

One of the more popular antennas for 3.5 and 7 MHz is the sloping dipole. David Pietraszewski, K1WA, has made an extensive study of sloping dipoles at different heights with reflectors at the 3-GHz frequency range. From his experiments, he developed the novel 40-meter antenna system described here. With several sloping dipoles supported by a single mast and a switching network, an antenna with directional characteristics and forward gain can be simply constructed. This 40-meter system uses several "slopers" equally spaced around a common center support. Each dipole is cut to a half wavelength and fed at the center with 52-ohm coax. The length of each feed line is 36 feet. This length is just over 3/8 λ, which provides a useful quality. All of the feed lines go to a common point on the support (tower) where the switching takes place. At 7 MHz, the 36-foot length of coax looks inductive to the antenna when the end at the switching box is open circuited. This has the effect of adding inductance at the center of the sloping dipole element, which electrically lengthens the element. The 36-foot length of feed line serves to increase the length of the element about 5%. This makes any unused element appear to be a reflector.

The array is simple and effective. By selecting one of the slopers through a relay box located at the tower, the system becomes a parasitic array which can be electrically rotated. All but the driven element of the array become reflectors.

The basic physical layout is shown in Fig. 30. The height of the support point

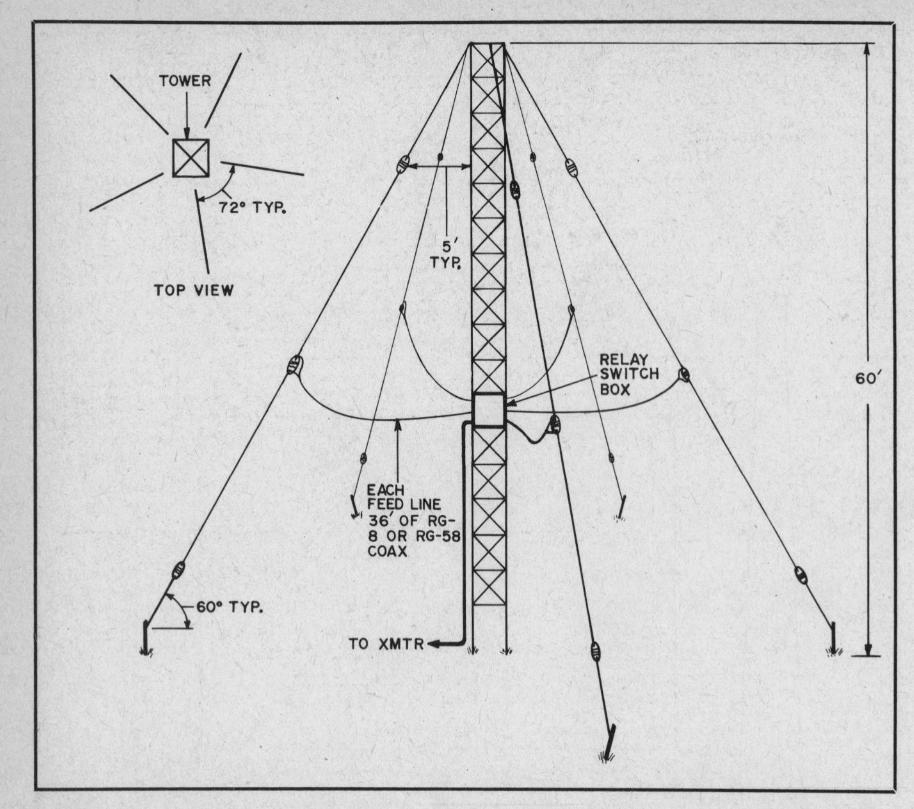


Fig. 30 — Five sloping dipoles suspended from one support. Directivity and forward gain can be obtained from this simple array. The top view shows how the elements should be spaced around the support.

Fig. 32 — The basic materials required for the sloper system. Control box at left and relay box at right.

Fig. 31 — Inside view of relay box. Four relays provide control over five antennas. See text. The relays pictured here are Potter and Brumfield type MR11D.

should be about 60 feet, but can be less and still give reasonable results. The upper portion of the sloper is 5 feet from the tower, suspended by rope. The wire makes an angle of 60 degrees with the ground. In Fig. 31, the switch box is shown containing all the necessary relays required to select the proper feed line for the desired direction. One feed line is selected at a time and opens the feed lines of those remaining. In this way the array is electrically rotated. These relays are controlled from inside the shack with an appropriate power supply and rotary switch. For safety reasons and simplicity, 12-volt dc relays are used. The control line consists of a five conductor cable, one wire used as a common connection; the others go to the four relays. By using diodes in series with the relays and a dual-polarity power supply, the number of control wires can be reduced, as shown in Fig. 33B.

Measurements indicate that this sloper array provides up to 20 dB front-to-back ratio and forward gain of about 4 dB. If one direction is the only concern, the switching system can be eliminated and the reflectors should be cut 5 percent longer than the resonant frequency. The one feature which is worth noting is the good front-to-back ratio. By arranging the system properly, a null can be placed in an unwanted direction, thus making it an effective receiving antenna. In the tests conducted with this antenna, the number of reflectors used were as few as one and as many as five. The optimum combination appeared to occur with four reflectors and one driven element. No tests were conducted with more than five reflectors. This same array can be scaled to 80 meters for similar results. The basic materials required for the sloper system are shown in Fig. 32.

## THE QUARTER-WAVELENGTH "HALF-SLOPER"

Perhaps one of the easiest antennas to install is the quarter-wavelength sloper. A sloping half-wavelength dipole is known among radio amateurs as a "full sloper" or "sloper." If only one half of it is used it becomes a "half-sloper." The performance of the two types of sloping antennas is similar: They exhibit some directivity in the direction of the slope and radiate energy at low angles respective to the horizon. The wave polarization is vertical.

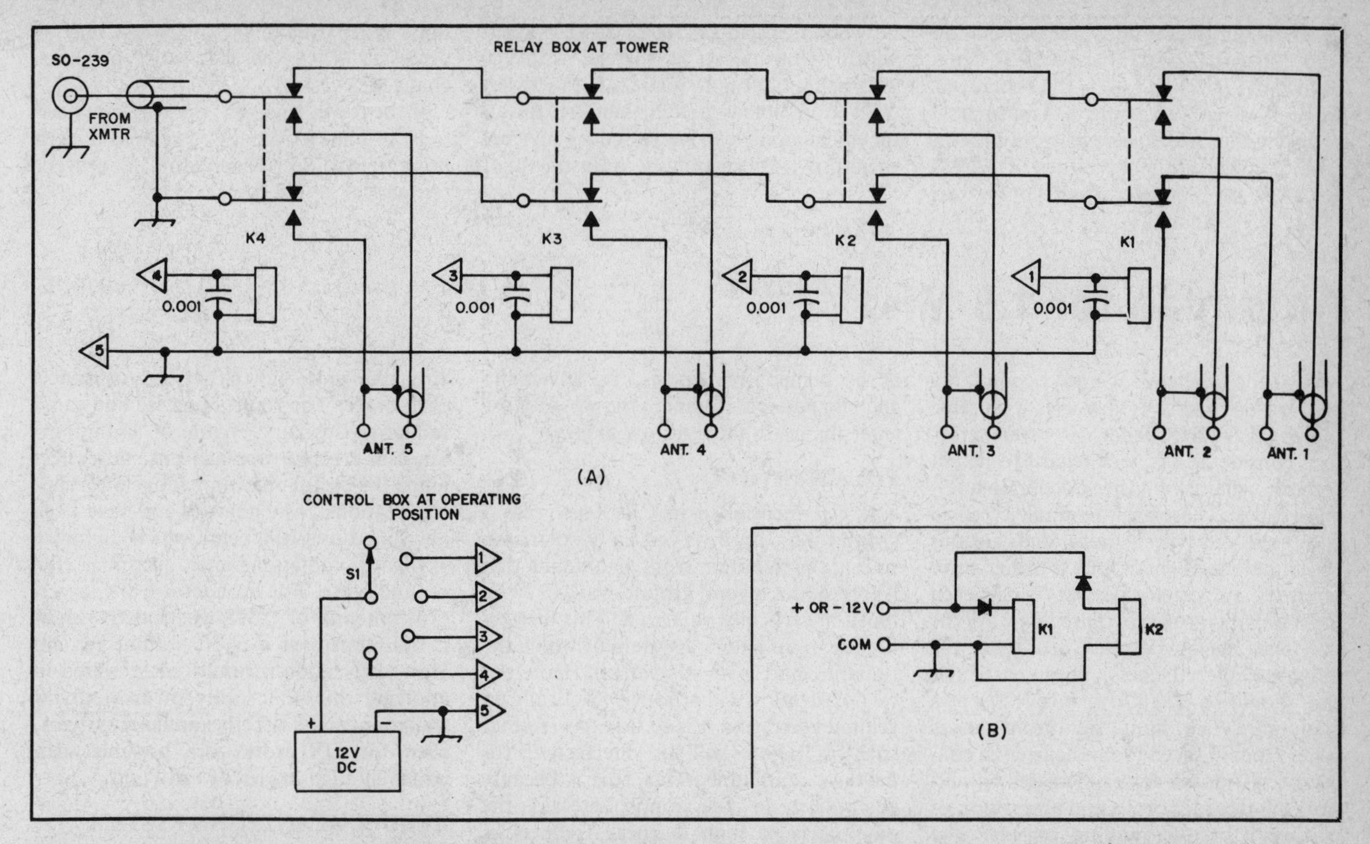


Fig. 33 — Schematic diagram for sloper control system. All relays are 12-volt dc, dpdt with 8-A contact ratings. In A, the basic layout, excluding control cable and antennas. Note that the braid of the coax is also open-circuited when not in use. Each relay is bypassed with 0.001-μF capacitors. The power supply is a low-current type. In B, diodes are used to reduce the number of control wires when using dc relays. See text.

The amount of directivity will range from 3 to 6 dB, depending upon the individual installation, and will be observed in the slope direction.

The advantage of the half-sloper over the full sloper is that the current portion of the antenna is higher. Also, only half as much wire is required to build the antenna for a given amateur band. The disadvantage of the half-sloper is that it is sometimes impossible to obtain a low value of VSWR when using coaxial-cable feed. This perplexing phenomenon is brought about by the manner in which the antenna is installed. Factors that affect the feed impedance are tower height, height of the attachment point, enclosed angle between the sloper and the tower and what is mounted atop the tower (hf or vhf beams). Also the quality of the ground under the tower (ground conductivity, radials, etc.) has a marked effect on the antenna performance. The final VSWR can vary (after optimization) from 1:1 to as high as 6:1. Generally speaking, the closer the low end of the slope wire is to ground, the more difficult it will be to obtain a good match.

**Basic Recommendations**

This excellent DX type of antenna is usually installed on a metal supporting structure such as a mast or tower. The support needs to be grounded at the lower end, preferably to a buried or on-ground radial system. If a nonconductive support is used, the outside of the coax braid becomes the return circuit and should be grounded at the base of the support. As a starting point one can attach the sloper so that the feed point is approximately 1/4 wavelength above ground. If the tower is not high enough to permit this, the antenna should be affixed as high on the supporting structure as possible. It is suggested that the amateur start with an enclosed angle of approximately 45 degrees, as indicated in Fig. 34. The wire may be cut in accordance with L(ft) = 260/f(MHz). This will allow sufficient extra length for pruning the wire for the lowest indicated VSWR.

A metal tower or mast becomes an operating part of the half-sloper system. In effect it and the slope wire function somewhat like an inverted-V antenna. In other words, the tower operates as the missing half of the dipole, hence its height and the top loading (beams) play a significant role.

The 50-ohm transmission line can be taped to the tower leg at frequent intervals to make it secure. The best method is to bring it to earth level, then route it to the operating position along the surface of the ground if it can't be buried. This will ensure adequate rf decoupling, which will help prevent rf energy from affecting the equipment in the station. Rotator cable and other feed lines on the tower or mast should be treated in a similar manner.

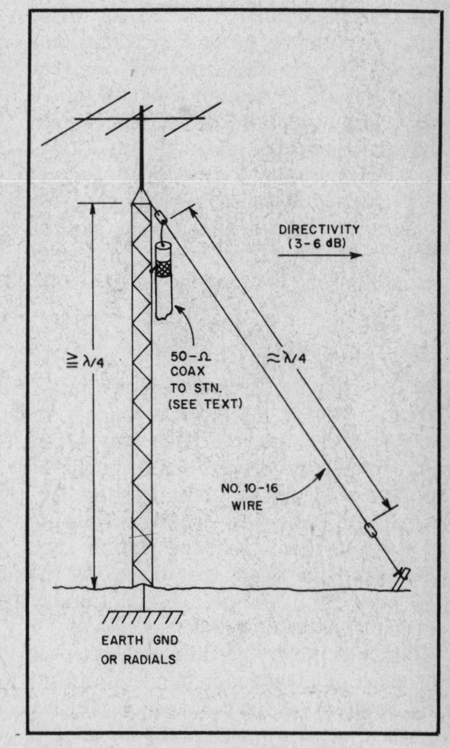


Fig. 34 — The quarter-wave half-sloper antenna.

Adjustment of the half-sloper is done with a VSWR indicator in the 50-ohm transmission line. A compromise can be found between the enclosed angle and wire length, providing the lowest VSWR attainable in the center of the chosen part of an amateur band. If the VSWR "bottoms out" at 2:1 or lower, the system will work fine without using a Transmatch, provided the transmitter can work into the load. Typical optimum values of VSWR for 40- or 80-meter half-slopers are between 1.3:1 and 2:1. A 100-kHz bandwidth is normal on 80 meters, with 200 kHz being typical at 7 MHz. If the lowest VSWR possible is greater than 2:1, the attachment point can be raised or lowered to improve the match. Readjustment of the wire length and enclosed angle may be necessary when the feed-point height is changed.

If the tower is guyed, the wire will need to be insulated from the tower and broken up with additional insulators to prevent resonance.

# Antennas for 1.8 MHz

Any of the antennas or arrays commonly used on the higher frequencies is suitable for the 1.8-MHz band. However, practical considerations with regard to height and size usually limit the selection to a few basic types. These are the dipole, vertical wire, end-fed wire, loop, and various combinations of these four. Further compromises are often necessary since even these antennas are still quite large. As the size and height decrease, so does the radiating effectiveness, and particular care should be taken to reduce losses to a minimum. The most significant losses result from induced ground currents, conductor resistance, losses in matching networks and loading coils and absorption of rf energy by surrounding objects. The type of antenna installation finally selected is often dictated by those losses most easily eliminated. For example, vertical antennas are usually considered the most desirable ones to use on 1.8 MHz, but if a suitable ground system is not feasible the ground losses will be very high. In such a case, an ordinary dipole may give superior performance even though the angle of maximum radiation is further from optimum than that of a vertical. Some experimentation is often necessary to find the best system, and the purpose of this section is to aid the reader in selecting the best one for his particular station.

### PROPAGATION ON 1.8 MHz

While important, propagation characteristics on 1.8 MHz are secondary to system losses since the latter may offset any attempt to optimize for the angle of radiation. Generally speaking, the 1.8-MHz band has similar properties to those of the a-m broadcast band (550 to 1600 kHz) but with greater significance of the sky wave. In this respect, it is not unlike the higher amateur frequencies such as 3.5 MHz, and most nighttime contacts over distances of a few hundred miles on 160 m will be by sky-wave propagation. During the daytime, absorption of the sky wave in the D region is almost complete but reliable communication is still possible by means of the ground wave.

With respect to sky-wave transmission, 160-meter waves entering the ionosphere, even vertically, are reflected to earth, so that there is no such phenomenon as skip distance on these frequencies. However, as at higher frequencies, to cover the greatest possible distance, the waves must enter the ionosphere at low angles.

### POLARIZATION

It was mentioned in Chapter 1 that a ground wave must be vertically polarized, so that the radiation from an antenna that is to produce a good ground wave likewise must be vertically polarized. This dictates the use of an antenna system of which the radiating part is mostly vertical. Horizontal polarization will produce practically no ground wave, and it is to be expected that such radiation will be ineffective for daytime communication. This is because absorption in the ionosphere in the daytime is so high at these frequencies that the reflected wave is too weak to be useful. At night a horizontal antenna will give better results than it will during the day. Ionospheric conditions permit the reflected wave to return to earth with less attenuation.

Some confusion over the term *ground wave* exists, since there are a number of propagation modes that go by this name. Here, only the type that travels over and near a conducting surface will be considered. If the surface is flat and has a very high conductivity, the attenuation of the wave follows a simple inverse-distance law. That is, every time the distance is doubled, the field strength drops by 6 dB. This law also holds for spherical surfaces for some distance, and then the field strength drops quite rapidly. For the earth, the break point is approximately 100 miles (160 km).

The conductivity of the surface is an important factor in ground-wave propagation. For example, sea water can almost be considered a perfect conductor for this purpose, at frequencies well into the hf range. However, there may be as much as an additional 20 dB of attenuation for a 1- to 10-mile path over poor-conducting earth, compared to an equivalent path over the sea. The conductivity of sea water is roughly 400 times as great as good-conducting land (agricultural regions) and 4000 times better than poor land (cities and industrial areas).

After sundown, the propagation depends upon both the ground wave and sky wave. At the limit of the ground-wave region, the two may have equal field strengths and may either aid or cancel each other. The result is severe and rapid fading in this zone. While of less importance in amateur applications, this effect limits the useful nighttime range of broadcast stations. Antenna designs have been developed over the years which minimize sky-wave radiation and maximize the ground wave. For broadcast work, a vertical antenna of 0.528-wavelength height is optimum over a good ground system. However, caution should be exercised in applying this philosophy to amateur installations since effective antenna systems, even for DX work, are possible with relatively high angles of radiation.

### REDUCING ANTENNA SYSTEM LOSSES

As the length of an antenna becomes small compared with the wavelength being used, the radiation resistance, Ra, drops to a very low value, as discussed in Chapter 2. The various losses can be represented by a resistance, $R_L$, in series with Ra. $R_L$ may be larger than Ra in practical cases. Therefore, in an antenna system with high losses, most of the applied power is dissipated in the loss resistance and very little is radiated in Ra. Since Ra is mostly dependent upon antenna construction, efforts to reduce the loss resistance will normally not affect the radiation resistance. Efficiency can be improved significantly by keeping the loss resistance as low as possible.

The simplest losses to reduce are the conductor losses. Since electrically short antennas, such as dipoles and end-fed wires, exhibit a large series capacitive reactance, a loading coil is commonly used to tune out the reactance. If not part of the radiating system, the coil should have as high a Q as possible. Incorporating the loading coil into the radiating system not only simplifies loading-coil construction, but may actually increase the efficiency by redistributing the current in the antenna. Such loading coils are designed for low loss, rather than high Q. The reason for this is that one of the parameters resulting in low coil Q is radiation. But the latter is exactly the desired result in an antenna system. The radiation from a coil increases as its length-to-diameter ratio increases. In some instances, the entire antenna may consist of a single coil. A helically wound vertical is an example of this type. In any of the loading coils that are part of the radiating system, the conductor diameter should be as large as possible, and very close spacing between turns should be avoided.

The effect of the earth on antenna loss can best be seen by examination of Fig. 35A. If a vertical radiator that is short compared with a wavelength is placed over a ground plane, the antenna current will consist of two components. Part of the current flows through Cw, which is the capacitance of the vertical to the radial wires, and part flows through Ce, the capacitance of the vertical to the earth. For a small number of radials, Ce will be much greater than Cw, and most of the current will flow through the circuit consisting of Ce and Re (the earth resistance). Power will be dissipated in Re which will not contribute to the radiation. The solution to the problem is to increase the number of radials. This will increase Cw, but, of more importance, will reduce Re by providing more return paths. Theory and experiments have shown that the ideal radial system with a 0.528-λ vertical consists of approximately 120 radials, each a half-wave long. If fewer radials are used (12), little is to be gained by running them out so far. The converse is also true. If space restricts the length of the radials, increasing the number much over 12 will have little effect for an antenna of this height. Since the current is greatest near the base of the antenna, a ground screen will also help if only a few radials are used.

Another method to reduce the ground currents is shown in Fig. 35B. By raising the antenna and ground plane off the earth, Cw stays the same in value but Ce is considerably reduced (such a system is sometimes called a counterpoise). This decreased influence of the earth is also the reason why as few as three radials are sufficient for hf and vhf ground-plane antennas that are several feet above the earth.

The simple lumped antenna-capacitance analysis is a good approximation to actual operation if the vertical is electrically short, but analysis becomes more complicated for greater antenna lengths. For instance, the maximum ground loss for the 0.528-wave broadcast vertical mentioned earlier occurs at a point 0.35 wavelength away from the base.

Location of the antenna is perhaps more critical with regard to receiving applications than transmitting ones. Sources of strong local noise, such as TV sets and power lines, can cause considerable difficulty on 1.8 MHz. However, the proximity of rf-absorbing objects such as steel buildings may cut down on transmitting efficiency also. Since most installations are tailored to the space available, little can be done about the problem except to see that the other losses are kept to a minimum.

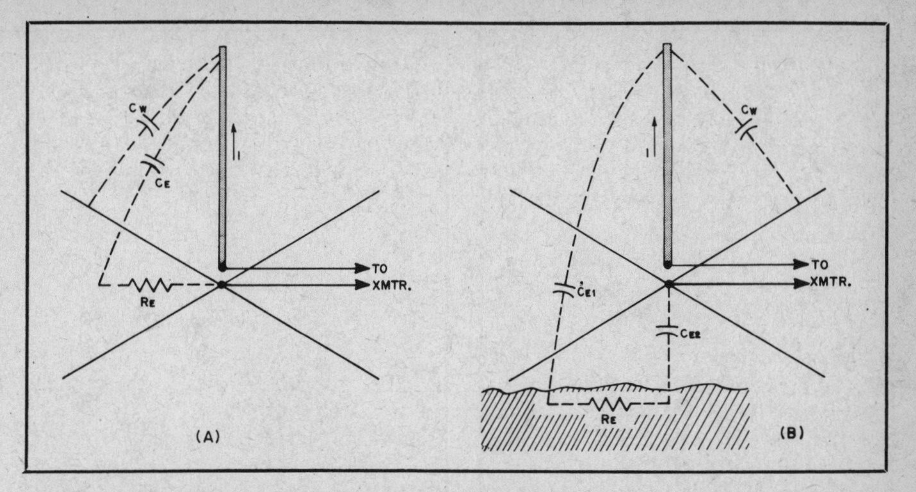


Fig. 35 — Drawing showing how earth currents affect the losses in a shortened-vertical antenna system. In A, the current through the combination of Ce and Re may be appreciable if Ce is much greater than Cw, the capacitance of the vertical to the ground wires. This ratio can be improved (up to a point) by using more radials. By raising the entire antenna system off the ground, Ce (which consists of the series combination of $Ce_1$ and $Ce_2$) is decreased while $C_W$ stays the same. The radial system shown at B is sometimes called a counterpoise.

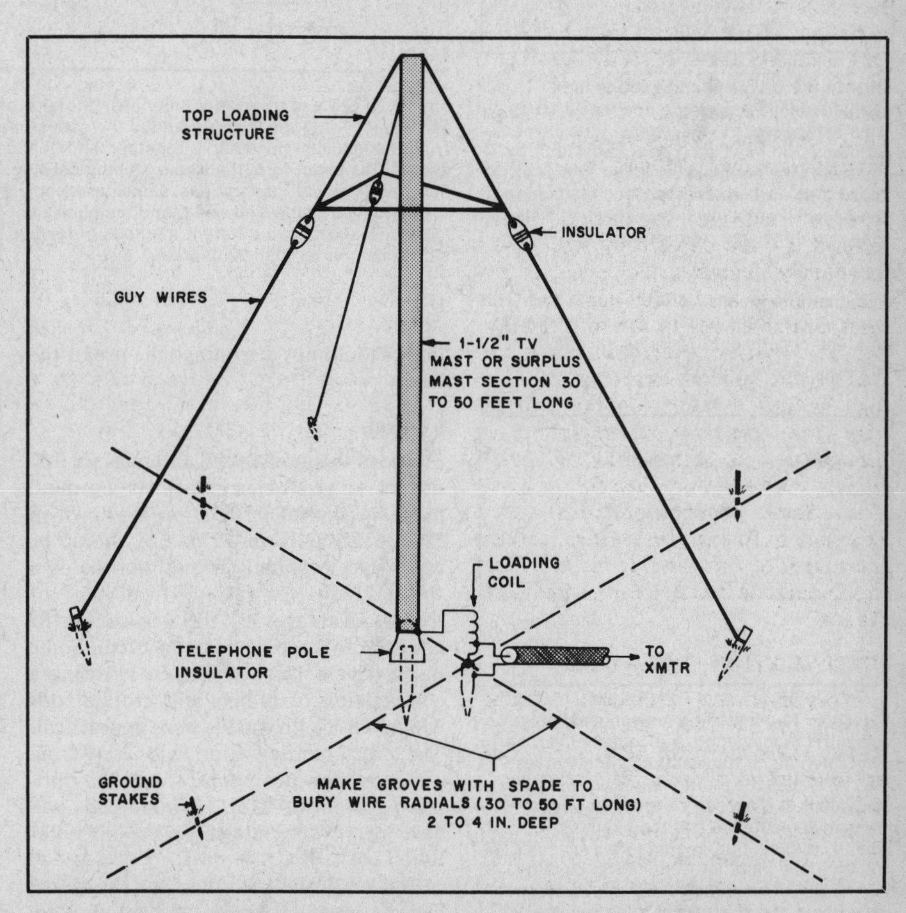


Fig. 36 — Physical layout of a typical vertical antenna suitable for 1.8-MHz operation. Without the top-loading structure or capacitance hat, the radiation resistance would be approximately 1 ohm for a 30-foot vertical, and 3.3 ohms for a 50-foot height. The loading inductance for the 30-foot vertical is approximately 400 μH. Once the antenna is approximately tuned to resonance with the base loading coil, a suitable tap near the low end of the coil can be found which will give the best match for the transmitter. The radiation resistance can be increased by the use of a top-loading structure consisting of the guy wires (broken up near the top by insulators) which are connected by a horizontal wire, as shown. The radial system consists of wires buried a few inches underground.

## ANTENNA TYPES

A common misconception is that antennas for 160 meters have to be much larger, higher and more elaborate than those for the higher bands. When one considers that even the gigantic antennas used for

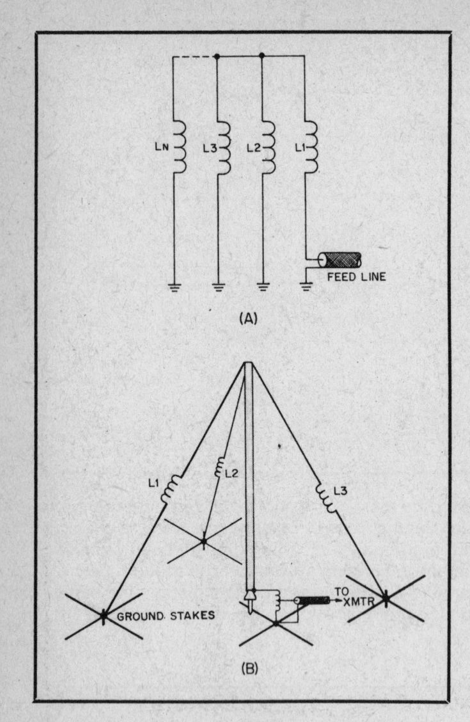


Fig. 37 — Possible configurations for a multiple-tuned vertical antenna for 1.8 MHz. Used extensively in vlf systems, little experimentation has been performed with it by amateurs. The principle is similar to that of the folded dipole where an impedance transformation occurs from a lower to higher value simplifying matching. The ratio is equal to $N^2$ where N is the number of elements. In the system shown at B, the step-up ratio would be 16, since the total number of elements is four. The exact values of the loading inductors should be found experimentally, being such that the current in each leg is the same.

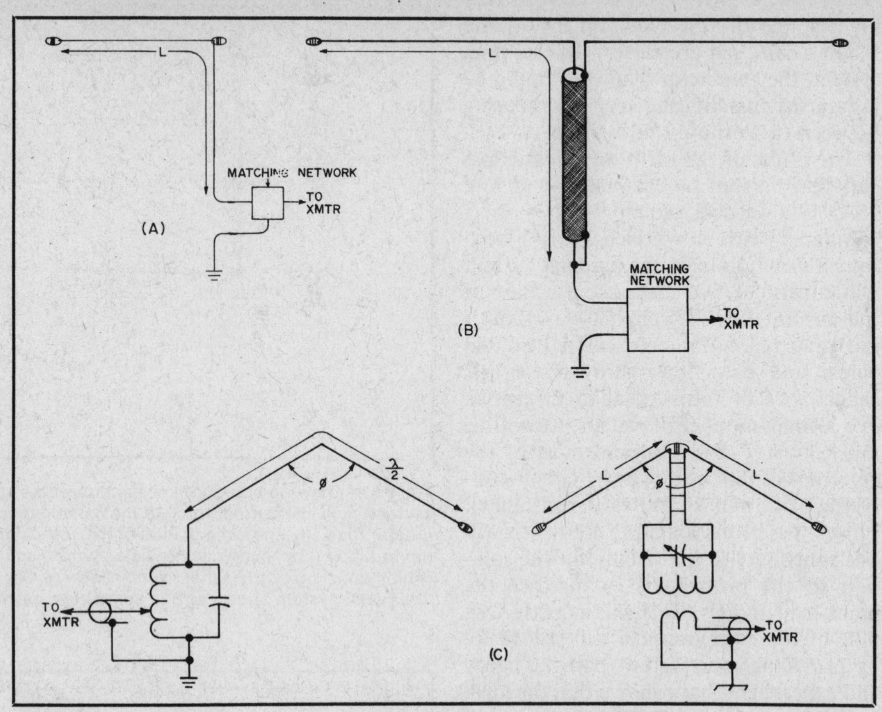


Fig. 39 — L, T and inverted-V antennas. The type of matching network suitable for the L antenna will depend upon the length L and is the same for a straight horizontal antenna (see Fig. 38). By tying the feed-line conductors together, an hf-band dipole can be used as a T antenna for 160 meters. The exact form that the matching network will take depends on the lengths of both the horizontal and vertical portions. Considering only the length of one leg should be sufficient for the majority of cases, however, and the equivalent L-antenna network can be used. The arrangement at C shows two different methods of feeding an inverted V. In either case, the apex angle, $\phi$, should be greater than 90 degrees.

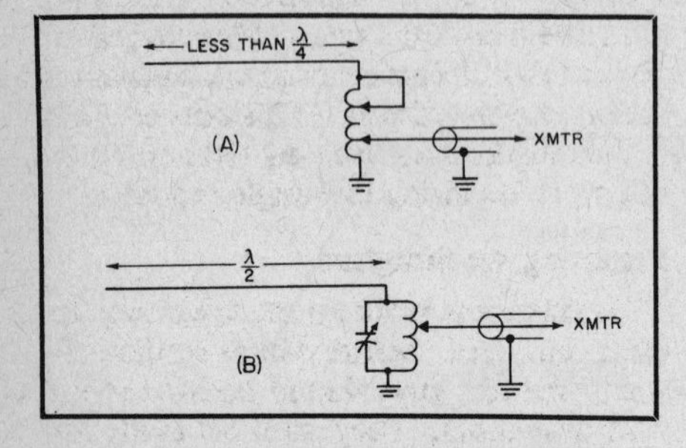


Fig. 38 — Two matching networks suitable for use with random-length horizontal (or vertical) wire antennas. If the electrical length is less than 1/4 wave, the input impedance will be equivalent to a resistance in series with a capacitive reactance and the circuit at A should be used. For lengths in the vicinity of 1/2 wave, the input impedance will be fairly high and may have reactances that are either inductive or capacitive. For this case, the parallel-tuned circuit in B should be used.

vlf work have radiating efficiencies of approximately one percent, it is not surprising that many contacts on 160 meters can be made with little more than a piece of wire a few feet off the ground, or even from mobile installations. While it is true that a larger and more sophisticated system may perform better than a smaller one, the point here is that space restrictions should not discourage the use of the band.

## Verticals

One of the most useful antennas for 160 meters is a vertical radiator over a ground plane. A typical installation is shown in Fig. 36. Some form of loading should be used, since economics would not justify a full-sized quarter-wave vertical. One of the disadvantages of the vertical is the necessity for a good ground system. Some improvement has been noted by using a combination of radials and ground rods where full-length radials were impractical. The exact configuration will vary from one installation to another, and the optimum placement of the ground rods will have to be determined by experimentation.

For verticals less than an electrical quarter wave in height, the input reactance without loading will be capacitive. A simple series loading coil should be used to tune this reactance out and the coil may be the only matching network necessary.

Normally, matching to the feed line or transmitter can be accomplished with simple networks or a Transmatch. However, a unique method that also improves the radiating efficiency is used with certain vlf antennas. The technique, called multiple tuning, is illustrated in Fig. 37A. A series of verticals is fed through a common flattop structure with one of the "downleads" also acting as a feed point. If the verticals are closely spaced (in comparison with a wavelength), the entire system can be considered to be one vertical with N times the current of one of the downleads taken alone. The result is that the radiation resistance is $N^2 \times Ra$, where Ra is the radiation resistance of a single vertical. This is the same principle as acquiring an impedance step-up in a multiconductor folded dipole. If the ground losses are also considered, the effective loss resistance (Rg) would also be transformed by the same amount. However, since the current distribution in the ground is usually improved by using this method, the ratio of $Ra/R_L$ is also improved. The disadvantages of the system are increased complexity and difficulty of adjustment. While little if any use of this principle has been applied to amateur systems for 1.8 MHz, it offers some interesting possibilities where a good ground system is impractical. The construction approach shown at B of Fig. 37 may be used for the erection of an experimental antenna of this type.

## Horizontal Antennas

In cases where a good ground system is not practical and when most of the operation will rely on sky-wave propagation, horizontal antennas can be used (see Fig. 38). The relative simplicity of construction of an end-fed wire antenna makes it an attractive one for portable operation or where supporting structures are without much height.

As is the case with electrically short verticals, the input impedance of horizontal end-fed antennas less than a quarter wave in length can be considered to be a resistance in series with a capacitive reactance. Matching networks for the end-fed wire are identical to those used for verticals.

Balanced center-fed antennas are also useful, even though they may be electrically short for 160 meters and at heights typical of those used at the higher bands. For example, an 80-meter doublet fed with open-wire line may also be used on 160 meters with the appropriate matching network at the transmitter. Care should be taken to preserve the balanced configuration of the doublet in matching to this type. If one side of the feed line is connected to ground, part of the return circuit may be through the power line. This increases the chances of interference from appliances such as TV sets and fluorescent lamps on the same circuit. Also, since there are usually connections on the power service that are not soldered, rectification may take place. The result may be mixing of local broadcast stations with products on 160 meters. Filtering will not eliminate the problem because the products are in the same band with the desired signals. Problems of this type are usually less severe as the electrical length of the doublet approaches a half wave.

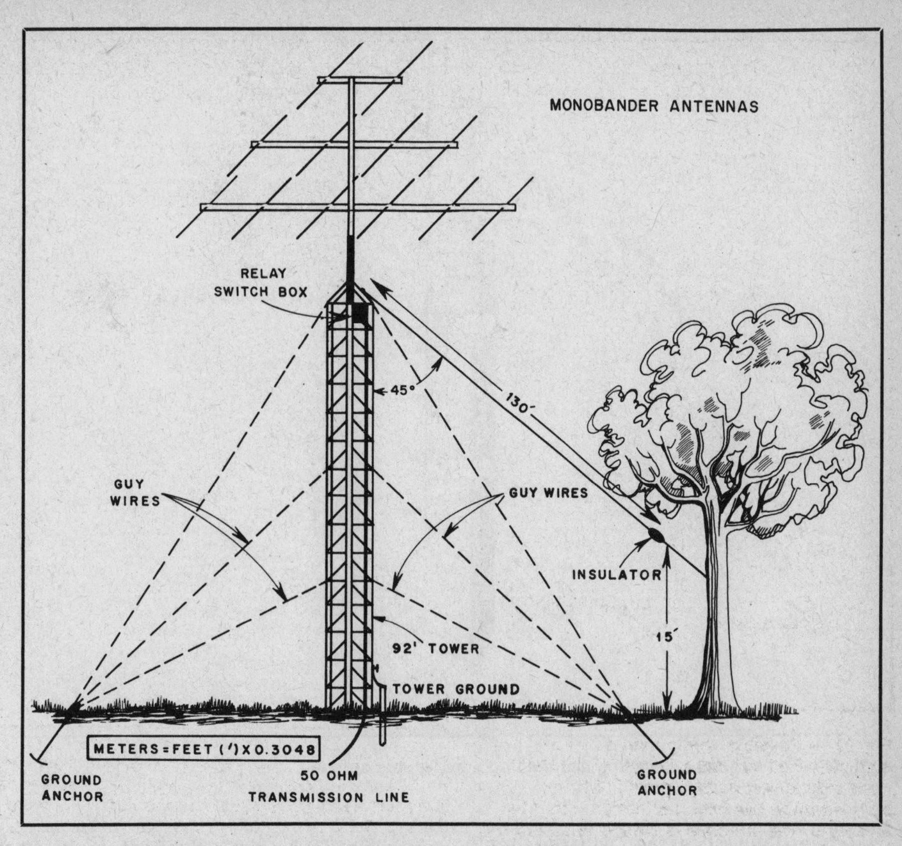


Fig. 40 — The W1CF half-sloper is arranged in this manner. Three monoband antennas atop the tower provide some capacitive loading.

#### Combinations of Vertical and Horizontal Antennas

The L and T antennas are the most common examples where combinations of horizontal and vertical radiators can be used to advantage. Various types are shown in Fig. 39. Here, the philosophy is usually to run the vertical portion up as high as possible with the horizontal part merely acting as a top-loading structure. Such a system can be considered to be equivalent to a vertical, and performance should be improved by the use of a ground system. Running the horizontal portion out to great distances may or may not improve the performance, unless the height is also increased.

A dipole fed with coaxial cable for a higher frequency band can be used as a T antenna by tying the feed-line conductors together at the transmitter. This will also work with dipoles fed with open-wire line; however, they may work just as well (or better) by using them in the more conventional manner discussed earlier. The inverted-V antenna has also given good results on 160 meters. While the center of the antenna should be as high as possible, the total angle of the V should not be less than 90 degrees at the apex. This will be determined by the height of the apex and how high the ends of the antenna are located above the ground. For angles less than 90 degrees, the radiation efficiency drops very rapidly. The drawings in Fig. 38 show some of the configurations of the various antenna types discussed.

### 160-METER ANTENNA SYSTEMS USING TOWERS

An existing metallic tower used to support hf or vhf beam antennas can be pressed into services as an integral part of a 160-meter radiating system. The quarter-wave sloper, or half-sloper discussed in the sections devoted to 3.5 and 7 MHz, will also perform well on 1.8 MHz. Those prominent 160-meter operators who have had success with this antenna suggest a minimum tower height of 50 feet. Dana Atchley, W1CF, uses the configuration sketched in Fig. 40. He reports that the uninsulated guy wires act as an effective counterpoise for the sloping wire. At Fig. 41 is the feed system used by Doug DeMaw, W1FB, on a 50-foot self-supporting tower. The ground for the W1FB system is provided by buried radials connected to the tower base.

A tower can also be used as a true vertical antenna, provided a good ground system is available. The shunt-fed tower is at its best on 160, where a full quarter-wavelength vertical antenna is rarely possible. Almost any tower height can be used. If the beam structure provides some top loading, so much the better — but anything can be made to radiate, if it is fed properly. A self-supporting, aluminum, crank-up, tilt-over tower is used at W5RTQ with a TH6DXX tribander mounted at 70 feet. Measurements showed that the entire structure has about the same properties as a 125-foot vertical. It thus works quite well on 160 and 80 in DX work requiring low-angle radiation.

#### Preparing the Structure

Usually some work on the tower system must be done before shunt-feeding is tried. Metallic guys should be broken up with insulators. They can be made to simulate top loading, if needed, by judicious placement of the first insulators. Don't overdo it; there is no need to "tune the radiator to resonance" in this way. If the tower is fastened to a house at a point more than about one-fourth of the height of the tower, it may be desirable to insulate the tower from the building. Plexiglas sheet, 1/4-inch or more thick, can be bent to any desired shape for this purpose, if it is heated in an oven and bent while hot.

All cables should be taped tightly to the tower, preferably on the inside, and run down to the ground level. It is not necessary to bond shielded cables to the tower electrically, but there should be no exceptions to the down-to-the-ground rule.

Though the effects of ground losses on feed-point impedance are less severe with

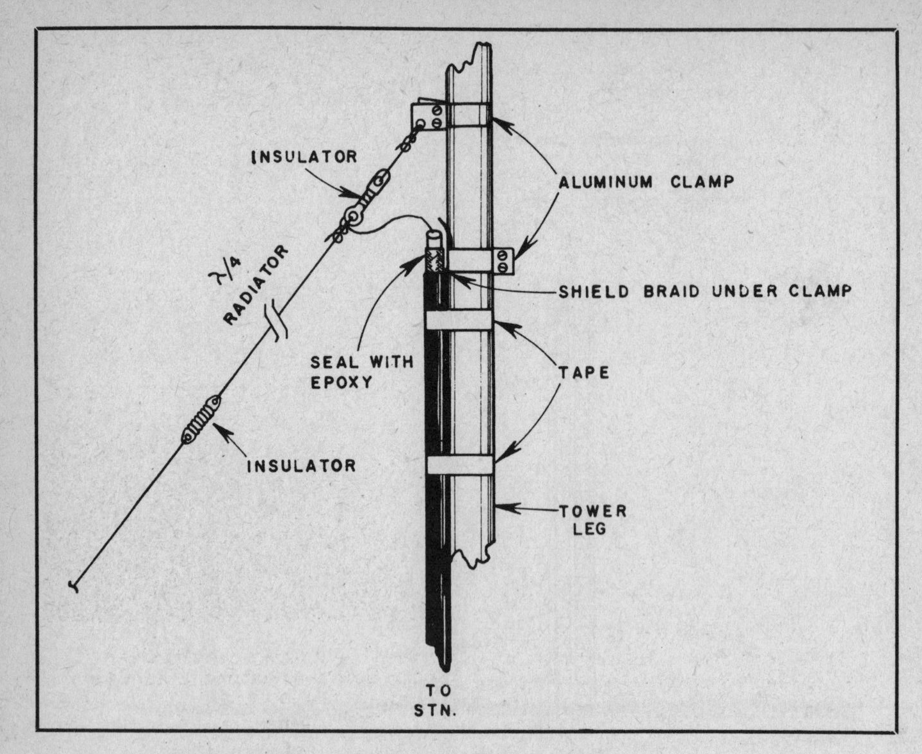


Fig. 41 — A method of installing and feeding a half-sloper antenna.

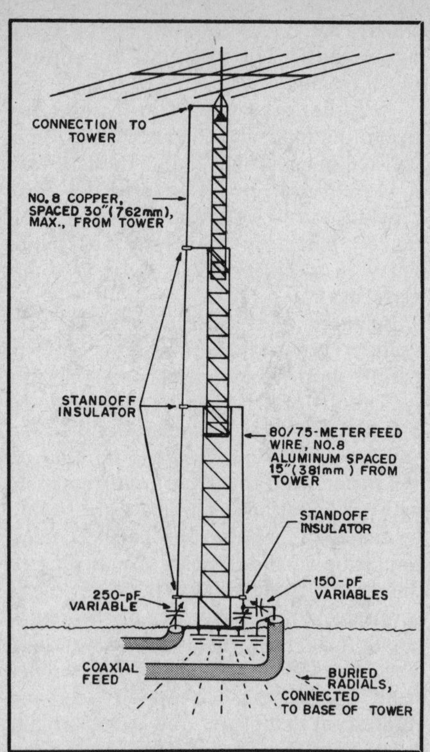


Fig. 42 — Principal details of the shunt-fed tower at W5RTQ. The 160-meter feed, left side, connects to the top of the tower through a horizontal arm of 1-inch diameter aluminum tubing. The other arms have stand-off insulators at their outer ends, made of 1-foot lengths of plastic water pipe. The connection for 80/75, right, is made similarly, at 28 feet, but two variable capacitors are used, to permit adjustment of matching with large changes in frequency.

the shunt-fed vertical than with the simple quarter-wave antenna, a good system of buried radials is very desirable. The ideal would be 120 radials, each 250 feet long, but fewer and shorter ones must often suffice. You can sneak them around corners of houses, along fences or sidewalks, wherever they can be put a few inches under the surface, or even on the earth surface. Aluminum clothesline wire is used extensively at W5RTQ, and it stands up well. Neoprene-covered aluminum wire may be safer in highly acid soils. Contact with the soil is not important. Deep-driven ground rods, and connection to underground copper water pipes, are good, if usable.

### Installing the Shunt Feed

Principal details of the shunt-fed tower for 80 and 160 meters are shown in Fig. 42. Rigid rod or tubing can be used for the feed portion, but heavy gauge aluminum or copper wire is easier to work with. Flexible stranded no. 8 copper wire is used for the 160-meter feed at W5RTQ, because when the tower is cranked down, the feed wire must come down with it. Connection is made at the top, 68 feet, through a 4-foot length of aluminum tubing clamped to the top of the tower, horizontally. The wire is clamped to the tubing at the outer end, and runs down vertically through standoff insulators. These are made by fitting 12-inch lengths of PVC plastic water pipe over 3-foot lengths of aluminum tubing. These are clamped to the tower at 15- to 20-foot intervals, with the bottom clamp about 3 feet above ground. The lengths given allow for adjustment of the tower-to-wire spacing over a range of about 12 to 36 inches, for impedance matching.

The gamma-match capacitor for 160 is a 250-pF variable with about 1/16-inch plate spacing, which is adequate for power levels up to about 200 watts.

### Tuning Procedure

The 160-meter feed wire should be connected to the top of a structure 75 feet tall or less. Mount the standoff insulators so as to have a spacing of about 24 inches between wire and tower. Pull the wire taut and clamp it in place at the bottom insulator. Leave a little slack below to permit adjustment of the wire spacing, if necessary.

Adjust the series capacitor in the 160-meter line for minimum reflected power, as indicated on an SWR meter connected between the coax and the connector on the capacitor housing. Make this adjustment at a frequency near the middle of your expected operating range. If a high SWR is indicated, try moving the wire closer to the tower. Just the lower part of the wire need be moved for an indication as to whether reduced spacing is needed. If the SWR drops, move all insulators closer to the tower, and try again. If the SWR goes up, increase the spacing. There will be a practical range of about 12 to 36 inches. If going down to 12 inches does not give a low SWR, try connecting the top a bit farther down the tower. If wide spacing does not make it, the omega match shown for 80-meter work should be tried. No adjustment of spacing is needed with the latter arrangement, which may be necessary with short towers or installations having little or no top loading.

The two-capacitor arrangement is also useful for working in more than one 25-kHz segment of the 160-meter band. Tune up on the highest frequency, say 1990 kHz, using the single capacitor, making the settings of wire spacing and connection point permanent for this frequency. To move to the lower frequency, say 1810 kHz, connect the second capacitor into the circuit and adjust it for the new frequency. Switching the second capacitor in and out then allows changing from one segment to the other, with no more than a slight retuning of the first capacitor.

### A Different Approach

Fig. 43 shows the method used by Doug DeMaw, W1FB, to gamma-match his self-supporting 50-foot tower. A wire cage simulates a gamma rod of the proper diameter. The tuning capacitor is fashioned from telescoping sections of 1-1/2-and 1-1/4-inch aluminum tubing with polyethylene sheet serving as the dielectric. This capacitor is more than adequate for power levels of 100 watts.

## A RECEIVING LOOP FOR 160 METERS

Small shielded loop antennas can be

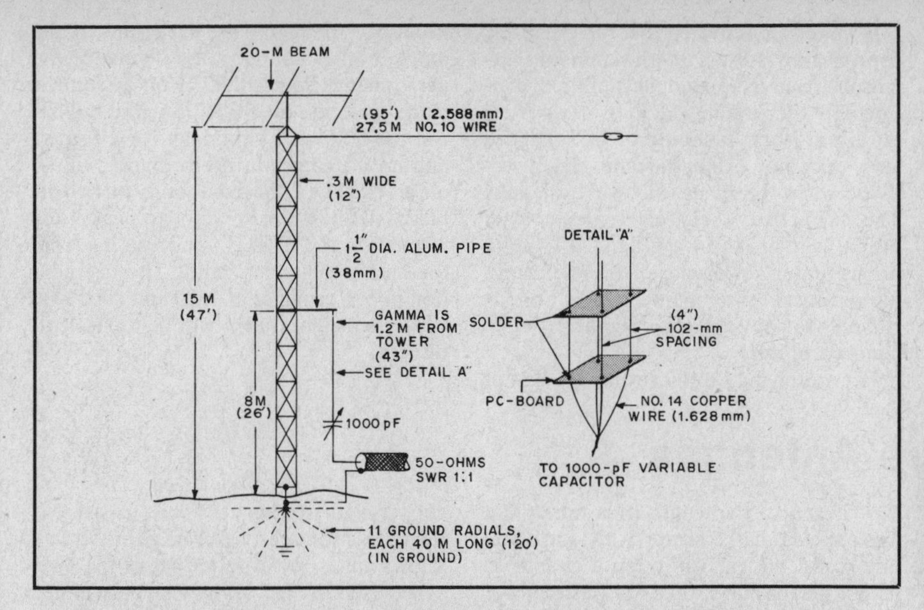


Fig. 43 — Details and dimensions in English and metric for gamma feeding a 50-foot tower as a 160-meter quarter-wavelength vertical antenna. The rotator cable and coaxial feed line for the 20-meter beam is taped to the tower legs and run into the shack from ground level. No rf decoupling networks are necessary.

Fig. 44 — The 160-meter shielded loop. Bamboo cross arms are used to support the antenna.

used to improve reception under certain conditions, especially at the lower amateur frequencies. The foregoing is particularly true when high levels of man-made noise are prevalent, when the second-harmonic energy from a nearby broadcast station falls in the 160-meter band, or when interference exists from some other amateur station in the immediate area. A properly constructed and tuned small loop will exhibit approximately 30 dB of front-to-side response, the maximum response being at right angles to the plane of the loop. Therefore, noise and interference can be reduced significantly or completely nulled out, by rotating the loop so that it is sideways to the interference-causing source. Generally speaking, small shielded loops are far less responsive to man-made noise than are the larger antennas used for transmitting and receiving. But a trade-off in performance must be accepted when using the loop, for the strength of received signals will be 10 or 15 dB less than when using a full-size resonant antenna. This condition is not a handicap on 160 or 80 meters, provided the station receiver has normal sensitivity and overall gain. Because a front-to-side ratio of 30 dB may be expected, a shielded loop can be used to eliminate a variety of receiving problems if made rotatable, as shown in Fig. 44.

To obtain the sharp bidirectional pattern of a small loop, the overall length of the conductor must not exceed 0.1 wavelength. The loop of Fig. 45 has a conductor length of 20 feet. At 1.810 MHz, 20 feet is 0.036 wavelength. With this style of loop, 0.036 wavelength is the maximum practical dimension if one is to tune the element to resonance. This limitation results from the distributed capacitance between the shield and inner conductor of the loop. RG-59/U was used for the loop element in this example. The capacitance per foot for this cable is 21 pF, resulting in a total distributed capacitance of 420 pF. An additional 100 pF was needed to resonate the loop at 1.810 MHz. Therefore, the approximate inductance of the loop is 15 μH. The effect of the capacitance becomes less pronounced at the higher end of the hf spectrum, provided the same percentage of a wavelength is used in computing the conductor length. The ratio between the distributed capacitance and the lumped capacitance used at the feed point becomes greater at resonance. These facts should be contemplated when scaling the loop to those bands above 160 meters.

There will not be a major difference in the construction requirements of the loop if coaxial cables other than RG-59/U are used. The line impedance is not significant with respect to the loop element. Various types of coaxial line exhibit different amounts of capacitance per foot, however, thereby requiring more or less capacitance across the feed point to establish resonance.

Shielded loops are not affected noticeably by nearby objects, and therefore they can be installed indoors or out after being tuned to resonance. Moving them from one place to another does not affect the tuning significantly.

In the model shown here it can be seen that a supporting structure was fashioned from bamboo poles. The X frame is held together at the center by means of two U bolts. The loop element is taped to the cross arms to form a square. It is likely that one could use metal cross arms without degrading the antenna performance. Alternatively, wood can be used for the supporting frame.

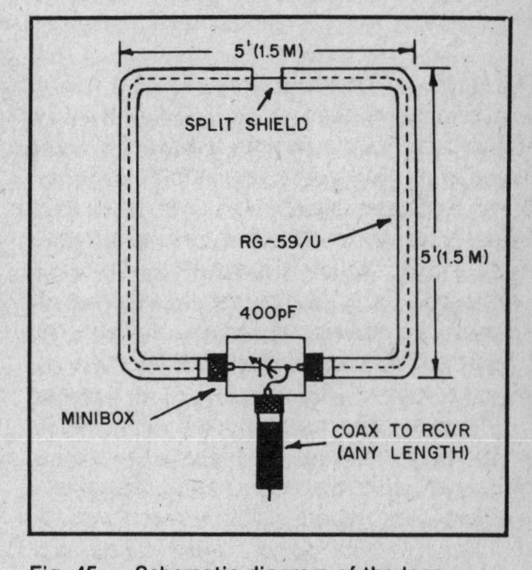


Fig. 45 — Schematic diagram of the loop antenna. The dimensions are not critical provided overall length of the loop element does not exceed approximately 0.1 wavelength. Small loops which are one half or less the size of this one will prove useful where limited space is a consideration.

A Minibox was used at the feed point of the loop to contain the resonating variable capacitor. In this model a 50- to 400-pF compression trimmer is used to establish resonance. It is necessary to weatherproof the box for outdoor installations.

The shield braid of the loop coax is removed for a length of one inch directly opposite the feed point. The exposed areas should be treated with a sealing compound once this is done.

In operation this receiving loop has been very effective in nulling out second-harmonic energy from local broadcast stations. During DX and contest operation on 160 meters it helped prevent receiver overloading from nearby 160-meter stations that share the band. The marked reduction in response to noise has made the loop a valuable station accessory when receiving weak signals. It is not used all of the time, but is available when needed by connecting it to the receiver through an antenna-selector switch. Reception of European stations with the loop has been possible from New England at times when other antennas were totally ineffective because of noise.

It was also discovered that the effects of approaching storms (with attendant atmospheric noise) could be nullified considerably by rotating the loop away from the storm front. It should be said that the loop does not exhibit meaningful directivity when receiving sky-wave signals. The directivity characteristics relate primarily to ground-wave signals. This is a bonus feature in disguise, for when nulling out local noise or interference, one is still able to copy sky-wave signals from all compass points!

For receiving applications it is not necessary to match the feed line to the loop, though doing so may enhance the performance somewhat. If no attempt is made to secure an SWR of 1, the builder can use 50- or 75-ohm coax for a feeder, and no difference in performance will be observed. The Q of this loop is sufficiently low to allow the operator to peak it for resonance at 1900 kHz and use it across the entire 160-meter band. The degradation in performance at 1800 and 2000 kHz will be so slight that it will be difficult to discern.

# Construction of Wire Antennas

Although wire antennas are relatively simple, they can constitute a potential hazard unless properly constructed. Antennas should *never* be run under or over public-utility (telephone or power) lines. Several amateurs have lost their lives by failing to observe this precaution.

## ANTENNA MATERIALS

The rf resistance of copper wire increases as the size of the wire decreases. However, in most types of antennas that are commonly constructed of wire, the rf resistance, even for quite small sizes of wire, will not be so high, compared to the radiation resistance, that the efficiency of the antenna will suffer greatly. Wire sizes as small as no. 30, or even smaller, have been used quite successfully in the construction of "invisible" antennas in areas where there is local objection to the erection of more conventional types. In most cases, the selection of wire for an antenna will be based primarily on the physical properties of the wire, since the suspension of wire from elevated supports places a strain on the wire[1].

### WIRE TYPES

Wire having an enamel-type coating is preferable to bare wire, since the coating resists oxidation and corrosion. Several types of wire having this type of coating are available, depending on the strength needed. "Soft-drawn" or annealed copper wire is easiest to handle but, unfortunately, is subject to considerable stretch under stress. It should therefore be avoided, except for applications where the wire will be under little or no tension, or where some change in length can be tolerated. (For instance, the length of a horizontal antenna fed at the center with open-wire line is not critical, although a change in length may require some readjustment of coupling to the transmitter.)

"Hard-drawn" copper wire or copper-clad steel, especially the latter, is harder to handle, because it has a tendency to spiral when it is unrolled. However, these types are mandatory for applications where significant stretch cannot be tolerated. Care should be exercised in using this wire to make sure that kinks do not develop that may cause the wire to break at far under normal stress. After the coil has been unwound, it is advisable to suspend the wire a few feet above ground for a day or two before making use of it. The wire should not be recoiled before installing.

The size of the wire to be selected, and the choice between hard-drawn and copper-clad, will depend on the length of the unsupported span, the amount of sag that can be tolerated, the stability of the supports under wind pressure, and whether or not an unsupported transmission line is to be suspended from the span.

### Wire Tension

Table 1 shows the maximum rated working tensions of hard-drawn and copper-clad steel wire of various sizes. If the tension on a wire can be adjusted to a known value, the expected sag of the wire, as depicted in Fig. 46, may be determined in advance of installation with the aid of Table 1 and the nomograph of Fig. 47. Even though there may be no convenient method of determining the tension in pounds, calculation of the expected sag for practicable working tensions is often desirable. If the calculated sag is greater than allowable it may be reduced by any one or a combination of the following:

1) Providing additional supports, thereby decreasing the span,

2) Increasing the tension in the wire if less than recommended,

3) Decreasing the size of the wire.

Conversely, if the sag in a wire of a particular installation is measured, the tension can be determined by reversing the procedure.

### Instructions for Using the Nomograph

1) From Table 1, find the weight (pounds/1000 feet) for the particular wire size and material to be used.

2) Draw a line from the value obtained above, plotted on the weight axis, to the desired span (feet) on the span axis, Fig. 47.

**Table 1**
**Stressed Antenna Wire**

| *American Wire Gauge* | *Recommended Tension[1] (pounds)* | | *Weight (pounds per 1000 feet)* | |
|---|---|---|---|---|
| | *Copper-clad steel[2]* | *Hard-drawn copper* | *Copper-clad steel[2]* | *Hard-drawn Copper* |
| 4 | 495 | 214 | 115.8 | 126 |
| 6 | 310 | 130 | 72.9 | 79.5 |
| 8 | 195 | 84 | 45.5 | 50 |
| 10 | 120 | 52 | 28.8 | 31.4 |
| 12 | 75 | 32 | 18.1 | 19.8 |
| 14 | 50 | 20 | 11.4 | 12.4 |
| 16 | 31 | 13 | 7.1 | 7.8 |
| 18 | 19 | 8 | 4.5 | 4.9 |
| 20 | 12 | 5 | 2.8 | 3.1 |

[1]Approximately one-tenth the breaking load. Might be increased 50 percent if end supports are firm and there is no danger of ice loading.
[2]"Copperweld," 40 percent copper.

[1]The National Electric Code of the National Fire Protection Association contains a section on amateur stations in which a number of recommendations are made concerning minimum size of antenna wire and the manner of bringing the transmission line into the station. The code in itself does not have the force of law, but it is frequently made a part of local building regulations, which are enforceable. The provisions of the code may also be written into, or referred to, in fire and liability insurance documents. A copy of this code may be obtained from National Fire Protection Association, Batterymarch Park, Quincy, MA 02269 (price $8.25 at this writing).

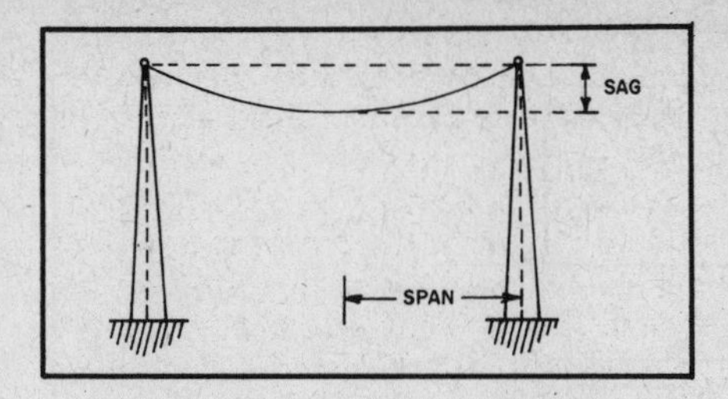


Fig. 46 — The span and sag of a long-wire antenna.

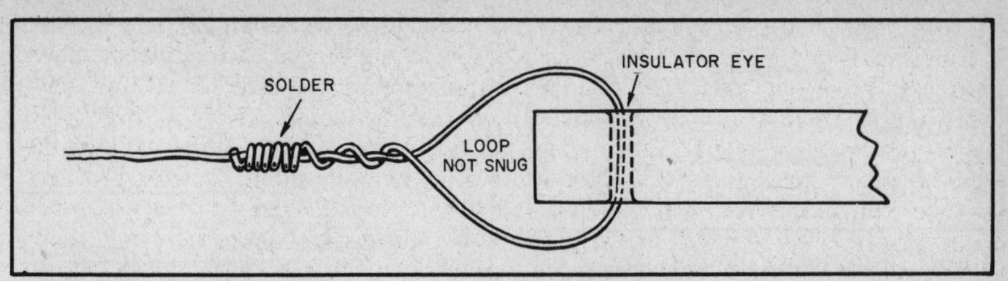


Fig. 49 — In fastening antenna wire to an insulator, the wire loop should not be made too snug. After completion, solder should be flowed into the turns. When the joint has cooled completely, it should be sprayed with acrylic.

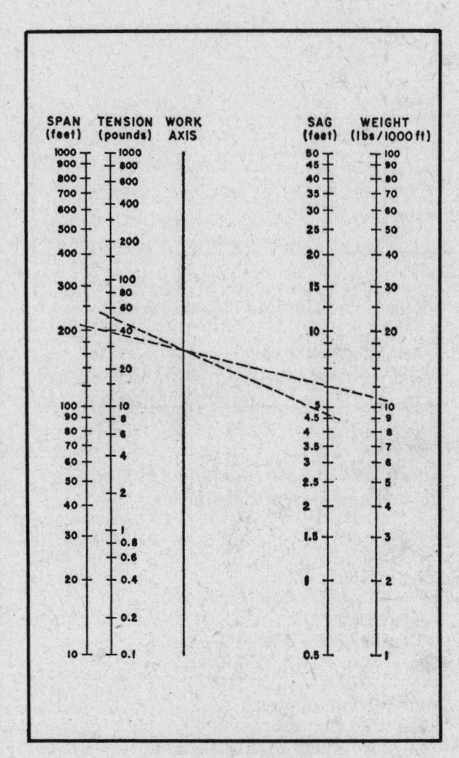


Fig. 47 — Nomograph for determining wire sag. (K1AFR)

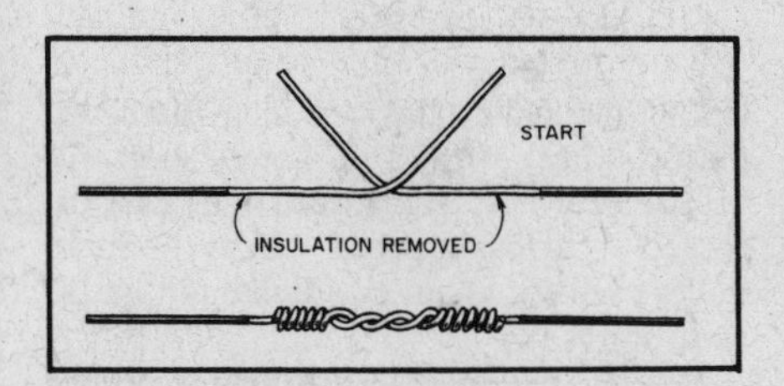


Fig. 48 — Correct method of splicing antenna wire.

3) Choose an operating tension level (pounds) consistent with the values presented in Table 1 (preferably less than the recommended wire tension).

4) Construct a line from the tension value chosen, plotted on the tension axis, through the crossover point of the work axis and the original line constructed from step 2, above, and continue this new line to the sag axis.

5) Read the sag (feet) on the sag axis. Example:

Weight = 11 pounds/1000 feet.
Span = 210 feet.
Tension = 50 pounds.

Answer:

Sag = 4.7 feet.

Of course, these calculations do not take the weight of a feed line into account, if it is supported by the antenna wire.

### Wire Splicing

Wire antennas should preferably be made with unbroken lengths of wire. In instances where this is not feasible, wire sections should be spliced as shown in Fig. 48. The enamel insulation should be removed for a distance of about 6 inches from the end of each section by scraping with a knife or rubbing with sandpaper until the copper underneath is bright. The turns of wire should be brought up tight around the standing part of the wire by twisting with broad-nose pliers.

The crevices formed by the wire should be completely filled with solder. Since most antenna soldering must be done outdoors, the ordinary soldering iron or gun may not provide sufficient heat, and the use of a propane torch may become desirable. The joint should be heated sufficiently so that the solder will flow freely into the joint when the source of heat is removed momentarily. After the joint has cooled completely, it should be wiped clean with a cloth, and then sprayed generously with acrylic to discourage corrosion.

## ANTENNA INSULATION

To prevent loss of power, the antenna should be well insulated from ground, particularly at the outer end or ends, since these points are always at a comparatively high rf potential. If an antenna is to be installed indoors (in an attic, for instance) the antenna may be suspended directly from the wood rafters without additional insulation, if the wood is permanently dry. When the antenna is located outside, where it is exposed to wet weather, however, much greater care should be given to the selection of proper insulators.

### Insulator Leakage

The insulators should be of material that will not absorb moisture. Most insulators designed specifically for antenna use are made of glass or glazed porcelain. Aside from this, the length of an insulator in proportion to its surface area is indicative of its comparative insulating ability. A long thin insulator will have less leakage than a short thick insulator. Some antenna insulators are deeply ribbed to increase the surface leakage path without increasing the physical length of the insulator. Shorter insulators can be used at low-potential points, such as at the center of a dipole. However, if such an antenna is to be fed with open-wire line and used on several bands, the center insulator should be the same as those used at the ends, because high rf potential will exist across the center insulator on some bands.

### Insulator Stress

As with the antenna wire, the insulator must have sufficient physical strength to sustain the stress without danger of breakage. Long elastic bands or lengths of nylon fishing line provide long leakage paths and make satisfactory insulators within their limits to resist mechanical strain. They are often used in antennas of the "invisible" type mentioned earlier.

For low-power work with short antennas not subject to appreciable stress, almost any small glass or glazed-porcelain insulator will do. Homemade insulators of Lucite rod or sheet will also be satisfactory. More care is required in the selection of insulators for longer spans and higher transmitter power.

For the same material, the breaking tension of an insulator will be proportional to its cross-sectional area. It should be remembered, however, that the wire hole at the end of the insulator decreases the effective cross-sectional area. For this reason, insulators designed to carry heavy strains are fitted with heavy metal end caps, the eyes being formed in the metal cap, rather than in the insulating material itself. The following stress ratings of several antenna insulators made by E. F. Johnson are typical:

5/8 inch square by 4 inches long — 400 lb.

1 inch diameter by 7 or 12 inches long — 800 lb.

1-1/2 inches diameter by 8, 12 or 20 inches long, with special metal end caps — 5000 lb.

These are rated breaking tensions. The actual working tensions should be limited to

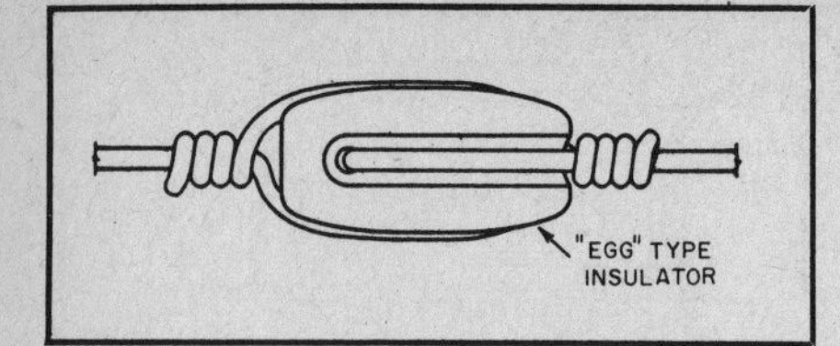


Fig. 50 — Conventional manner of fastening to a strain insulator. This method decreases the leakage path and increases capacitance, as discussed in the text.

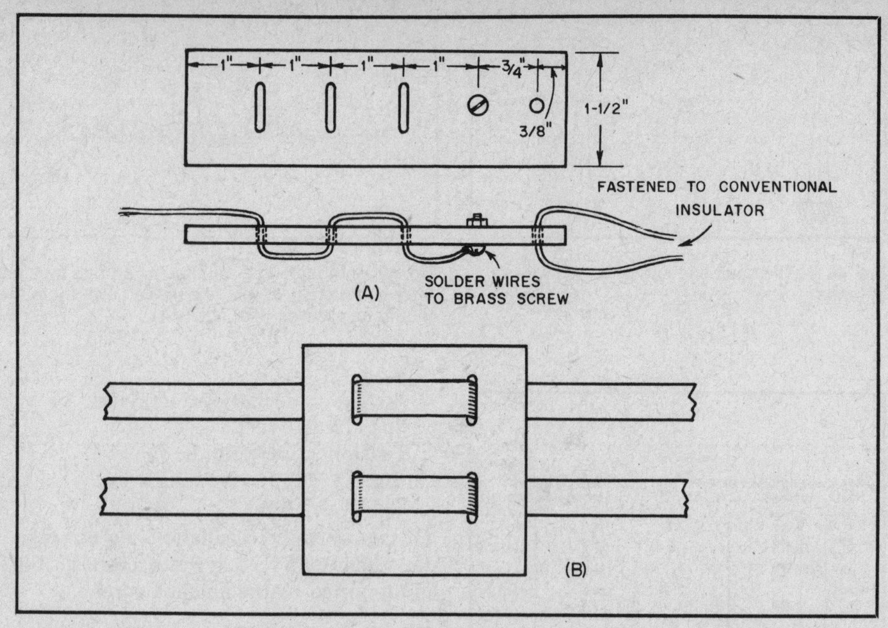


Fig. 51 — A — Insulator for ends of folded dipoles, or multiple dipoles made of 300-ohm ribbon. B — A method of suspending one ribbon dipole from another in a multiband dipole system.

not more than 25% of the breaking rating.

The antenna wire should be attached to the insulators as shown in Fig. 49. Care should be taken to avoid sharp angular bends in the wire in looping it through the insulator eye. The loop should be generous enough in size that it will not bind the end of the insulator tightly. If the length of the antenna is critical, the length should be measured to the outward end of the loop, where it passes through the eye of the insulator. The soldering should be done as described earlier for the wire splice.

### Strain Insulators

Strain insulators have their holes at right angles, since they are designed to be connected as shown in Fig. 50. It can be seen that this arrangement places the insulating material under compression, rather than tension. An insulator connected this way can withstand much greater stress. Furthermore, if the insulator should break, the wire will not collapse, since the two wire loops are interlocked. Because the wire is wrapped around the insulator, however, the leakage path is reduced drastically, and the capacitance between the wire loops provides an additional leakage path. For this reason the use of the strain insulator is usually confined to such applications as breaking up resonances in guy wire, where high levels of stress prevail, and where the rf insulation is of less importance. Such insulators might, however, be suitable for use at low-potential points on the antenna, such as at the centers of dipoles. These insulators may also be fastened in the conventional manner if the wire will not be under sufficient tension to break the eyes out.

### Insulators for Ribbon-Line Antennas

Fig. 51A shows the sketch of an insulator designed to be used at the ends of a folded dipole, or a multiple dipole made of ribbon line. It should be made approximately as shown, out of Lucite or bakelite about 1/4-inch thick. The advantage of this arrangement is that the strain of the antenna is shared by the conductors and the plastic webbing of the ribbon, which adds considerable strength. After soldering, the screw should be sprayed with acrylic.

Fig. 51B shows a similar arrangement for suspending one dipole from another in a multiple-dipole system. If better insulation is desired, these insulators can be wired to a conventional insulator.

## PULLEYS AND HALYARDS

Pulleys and halyards commonly used to raise and lower the antenna are also items that must be capable of taking the same strain as the antenna wire and insulators. Unfortunately little specific information on the stress ratings of most pulleys is available. Several types of pulleys are readily available at almost any hardware store. Among these are small galvanized pulleys designed for awnings, and several styles and sizes of clothesline pulleys. In judging the stress that any pulley might handle, particular attention should be paid to the diameter of the shaft, how securely the shaft is fitted into the sheath, and the size and material of which the frame is made. Heavier and stronger pulleys are those used in marine work.

Another important factor to be considered in the selection of a pulley is its ability to resist corrosion. Galvanized awning pulleys are probably the most susceptible to corrosion. While the frame or sheath usually stands up well, these pulleys usually fail at the shaft, which eventually rusts out, allowing the grooved wheel to break away under tension.

Most good-quality clothesline pulleys are made of alloys which do not corrode readily. Since they are designed to carry at least 50 feet of line loaded with wet clothing in stiff winds, they should be adequate for normal spans of 100 to 150 feet between stable supports. One type of clothesline pulley has a 4-inch diameter plastic wheel with a 1/4-inch shaft in bronze bearings. The sheath is of cast or forged corrosion-proof alloy. Such pulleys sell for about two dollars in hardware stores.

**Table 2**

**Approximate Safe Working Tension (lb) for Various Halyard Materials**

Manila Rope
1/4"—120 3/8"—270 1/2"—530 5/8"—800
Polypropylene Rope
1/4"—270 3/8"—530 1/2"—840
Nylon Rope
1/4"—300 3/8"—660 1/2—1140
7 × 11 Galvanized Sash Cord
1/16"—30 1/8"—125 3/16"—250 1/4"—450
High-Strength Stranded
Galvanized Guy Wire
1/8"—400 3/16"—700 1/4"—1200
Rayon-filled Plastic Clothesline
7/32"—60 to 70

Marine pulleys have good weather-resisting qualities, since they are usually made of bronze, but they are comparatively expensive and are not designed to carry heavy loads. For extremely long spans, the wood-sheathed pulleys used in "block and tackle" devices, and for sail hoisting should fill the requirements.

### Halyards

Table 2 shows recommended maximum tensions for various sizes and types of line and rope suitable for hoisting halyards. Probably the best type for general amateur use for spans up to 150 or 200 feet is 1/4-inch nylon rope. It is somewhat more expensive than ordinary rope of the same size, but it weathers much better.

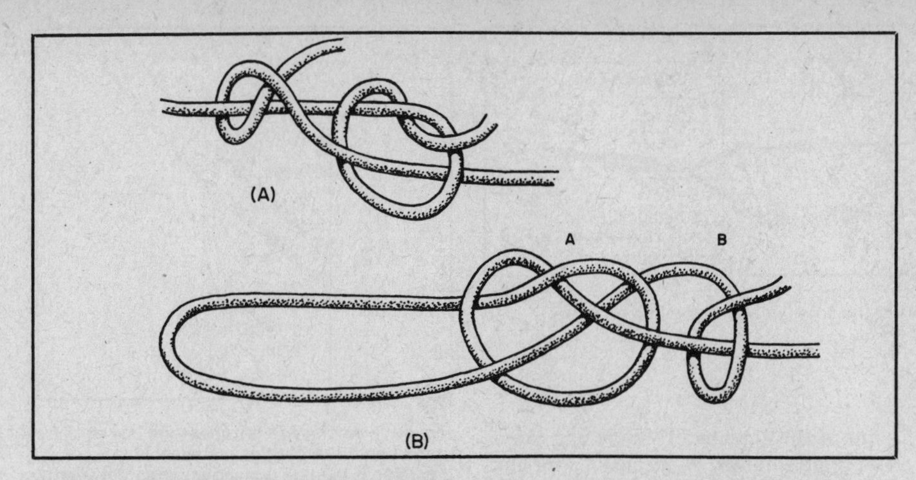


Fig. 52 — This is one type of knot that will hold with smooth rope, such as nylon. A shows the knot for splicing two ends. B shows the use of a smilar knot in forming a loop, as might be needed for attaching an insulator to a halyard. Knot A is first formed loosely 10 or 12 inches from the end of the rope; then the end is passed through the eye of the insulator and knot A. Knot B is then formed and both knots pulled tight. (K7HDB)

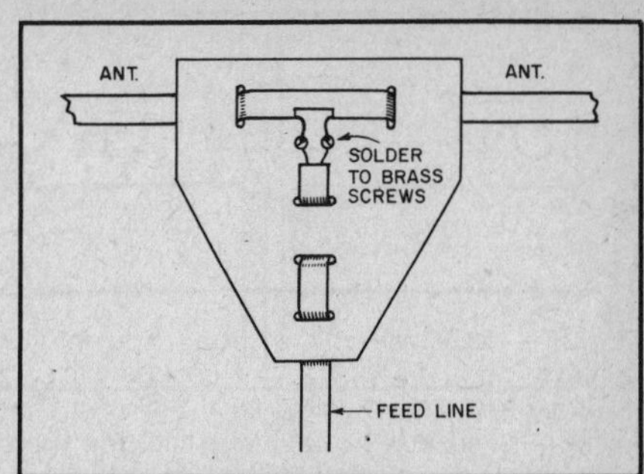


Fig. 54 — Strain reliever for conductors of 300-ohm ribbon line in a folded dipole. The piece can be made from 1/4-inch Lucite sheet.

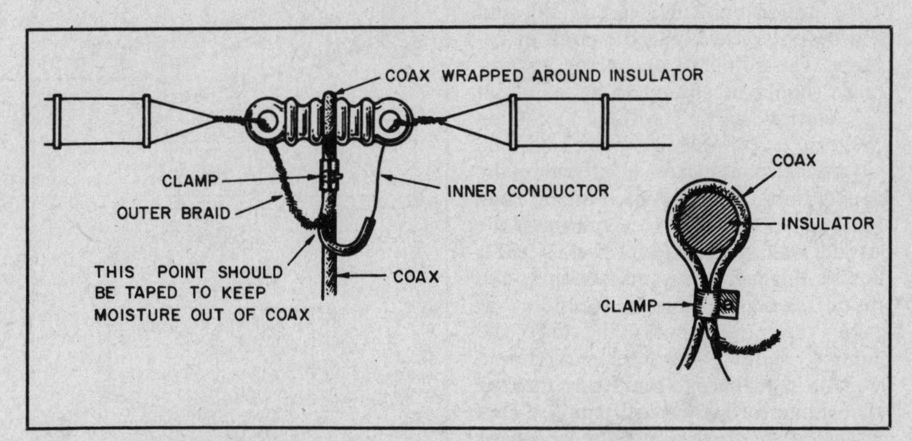


Fig. 53 — Method of relieving strain on conductors of coaxial cable in feeding a dipole. Six or eight wraps of solid copper bus wire may be used in place of the clamp.

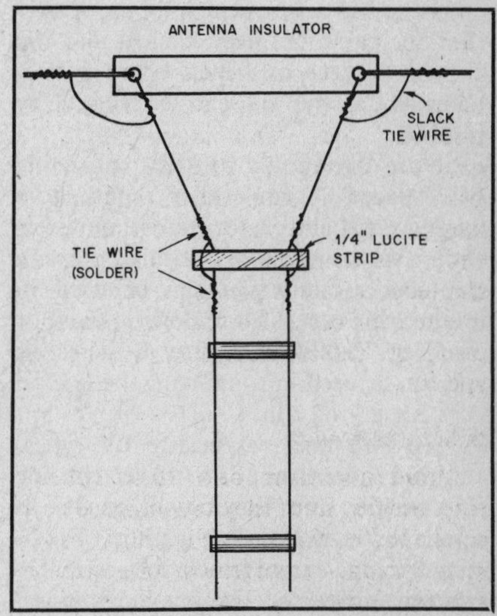


Fig. 55 — Method of connecting open-wire line to center antenna insulator. The Lucite strip keeps the feed-line conductors from pulling away from the spreaders when TV open-wire line is used.

Furthermore, it has a certain amount of elasticity to accommodate gusts of wind, and is particularly recommended for antennas using trees as supports.

Most types of synthetic rope are slippery, and some types of knots ordinarily used for rope will not hold well. Fig. 52 shows a knot that should hold well, even with nylon rope or plastic line.

For exceptionally long spans, stranded galvanized steel sash cord is suitable. Cable advertised as "wire rope" usually does not weather well. A convenience in antenna hoisting (usually a necessity with metal halyards) is the boat winch sold at marinas, and also at such places as Sears.

## INSTALLING TRANSMISSION LINES

In connecting coaxial cable or 300-ohm ribbon line to a dipole that does not have a support at the center, it is essential that the conductors of the line be relieved of the weight of the cable or ribbon. Fig. 53 shows a method of accomplishing this with coaxial cable. The cable is looped around the center antenna insulator, and clamped before making connections to the antenna. In Fig. 54, the weight of the ribbon line is removed from the conductor by threading the line through a sheet of insulating material. The sheet is suspended from the antenna by threading the antenna through the sheet. This arrangement is particularly suited to folded dipoles made of 300-ohm ribbon.

In connecting an open-wire line to an antenna, the conductors of the line should be anchored to the insulator by threading them through the eyes of the insulator two or three times, and twisting the wire back on itself before soldering fast. A slack tie wire should then be used between the feeder conductor and the antenna, as shown in Fig. 55. (The tie wires may be extensions of the line conductors themselves.)

When using TV-type open-wire line, the tendency of the line to twist and short out close to the antenna can be counteracted by making the center insulator of the antenna longer than the spacing of the line, as shown in Fig. 55. In this case, a heavier spreader insulator should be added just below the antenna insulator to prevent side stress from pulling the conductors away from the light plastic feeder spreaders.

### Running Line from Antenna to Station

Coaxial cable requires no particular care in running from the antenna to the station entrance, except to protect it from mechanical damage. If the antenna is not supported at the center, the line should be fastened to a post more than head high located under the center of the antenna, allowing enough slack between the post and the antenna to take care of any movement of the antenna in the wind. If the antenna feed point is supported by a tower or mast, the cable can be taped at intervals to the mast, or to one leg of the tower.

If desired, coaxial cable can be buried a few inches in the ground in making the run from the antenna to the station. A deep slit can be cut by pushing a square-

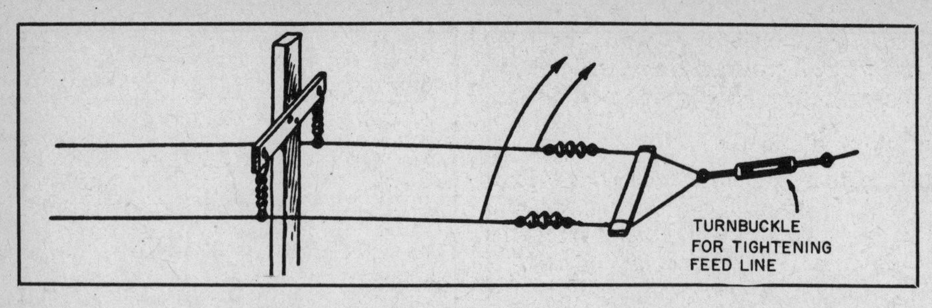


Fig. 56 — A support for open-wire line. The support at the antenna end of the line must be sufficiently rigid to stand the tension of the line.

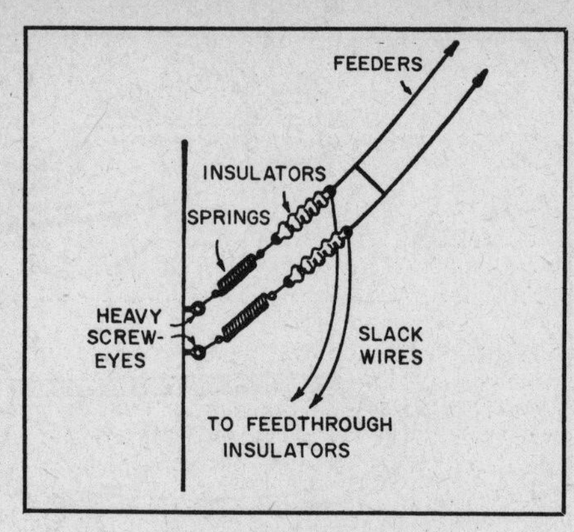


Fig. 58 — Anchorage for open-wire line at the station end. The springs are especially desirable if the line is not supported between the antenna and the anchoring point.

end spade full depth into the ground, and moving the handle back and forth to widen the slit before removing the spade. After the cable has been pushed into the slit with a piece of 1-inch board 3 or 4 inches wide, the slit can be closed by tamping.

Ribbon line should be kept reasonably well spaced from other conductors running parallel to it for more than a few feet. TV-type standoff insulators with strap-clamp mountings can be used in running this type of line down a mast or tower leg. Similar insulators of the screw type can be used in supporting the line on poles for a long run.

Open-wire lines, especially TV types, require frequent supports to keep the line from twisting and shorting out, as well as to relieve the strain. One method of supporting a long run of heavy open-wire line is shown in Fig. 56. The line must be anchored securely at a point under the feed point of the antenna. TV-type line can be supported similarly by means of wire links fastened to the insulators. Fig. 57 shows a method of supporting an open-wire line from a tower.

Fig. 57 — A board fitted with standoff insulators and clamped to the tower with U bolts keeps open-wire line suitably spaced from a tower. (W4NML)

To keep the line clear of pedestrians and vehicles, it is usually desirable to anchor the feed line at the eave or rafter line of the station building (see Fig. 58), and then drop it vertically to the point of entrance. The points of anchorage and entrance should be chosen so as to permit the vertical drop without crossing windows.

If the station is located in a room on the ground floor, one way of bringing coax transmission line in is to go through the outside wall below floor level, feed it through the basement, and then up to the station through a hole in the floor. In making the entrance hole in the side of the building, suitable measurements should be made in advance to make sure that the hole will go through the sill 2 or 3 inches above the foundation line (and between joists if the bore is parallel to the joists). The line should be allowed to sag below the entrance-hole level to allow rain water to drip off, and not follow the line into the building.

Open-wire line can be fed in a similar manner, although it will require a separate hole for each conductor. The hole should be insulated with lengths of polystyrene or Lucite tubing, and should be drilled with a slight downward slant toward the outside of the building to prevent rain seepage. With TV-type line, it will be necessary to remove a few of the spreader insulators, cut the line before passing through the holes (allowing enough length to reach the inside), and splice the remainder on the inside.

If the station is located above ground level, or there is other objection to the procedure described above, entrance can be made at a window, using the arrangement shown in Fig. 59. An Amphenol type 83-1F (UG-363/U) connector can be used as shown in Fig. 60; ceramic feedthrough insulators can be used for open-wire line. Ribbon line can be run through clearance holes in the panel, and secured by a winding of tape on either side of the panel, or by cutting the retaining rings and insulators from a pair of TV standoff

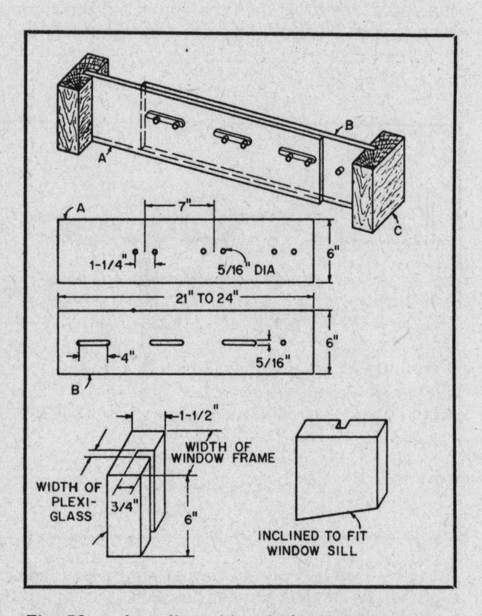


Fig. 59 — An adjustable window lead-in panel made of two sheets of Lucite. A feedthrough connector for coax line can be made as shown in Fig. 60. Ceramic feedthrough insulators are suitable for open-wire line. (W1RVE)

insulators, and clamping one on each side of the panel.

## LIGHTNING PROTECTION

Two or three types of lightning arresters for coaxial cable are available on the market. These are designed to join two lengths of coax cable. If the antenna feed point is at the top of a well-grounded tower, the arrester can be fastened securely to the top of the tower for grounding purposes. A short length of cable, terminated in a coaxial plug, is then run from the antenna feed point to one receptacle of the arrester, while the transmission line is run from the other arrester receptacle to the station. Such arresters may also be placed at the entrance point to the station, if a suitable ground connection is

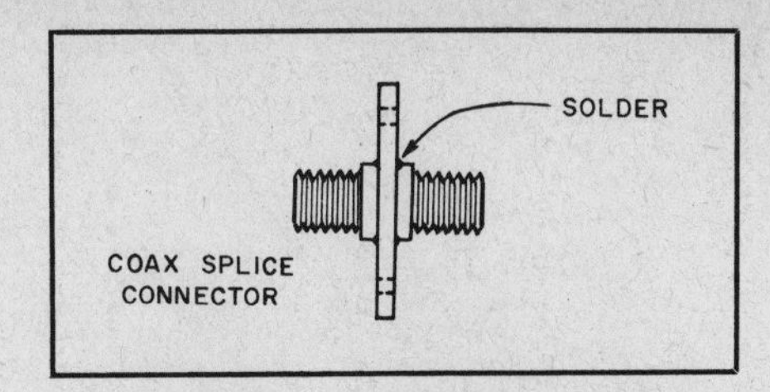


Fig. 60 — Feedthrough connector for coax line. An Amphenol 83-1J (PL-258) connector, the type used to splice sections of coax line together, is soldered into a hole cut in a brass mounting flange. Amphenol bulk adapter 83-1F may be used instead.

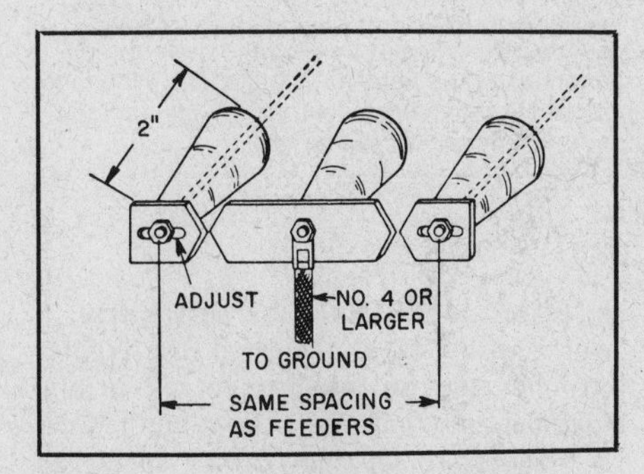


Fig. 61 — A simple lightning arrester for open-wire line made from three standoff or feed-through insulators and sections of 1/8 × 1/2-inch brass or copper strap. It should be installed in the line at the point where the line enters the station. The heavy ground lead should be as short and direct as possible. The gap setting should be adjusted to the minimum width that will prohibit arcing when the transmitter is operated.

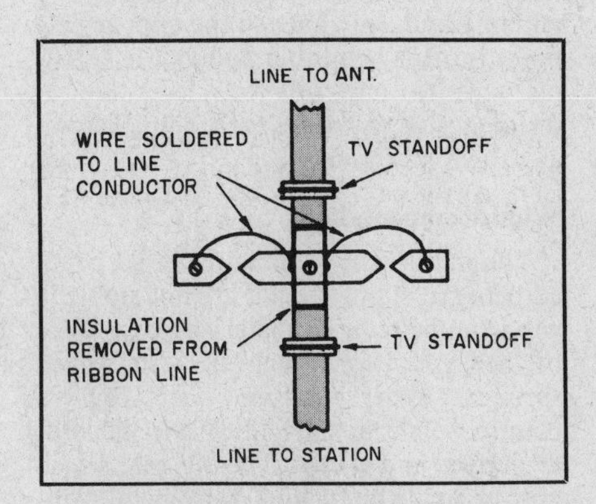


Fig. 62 — The lightning arrester of Fig. 61 may be used with 300-ohm ribbon line in the manner shown here. The TV standoffs support the line an inch or so away from the grounded center member of the arrester.

available at that point (or arresters may be placed at both points for added insurance).

The construction of a homemade arrester for open-wire line is shown in Fig. 61. This type of arrester can be adapted to ribbon line, as shown in Fig. 62. The two TV standoff insulators should elevate the ribbon line an inch or so away from the center member of the arrester. Sufficient insulation should be removed from the line where it crosses the arrester to permit soldering the arrester connecting leads.

### Lightning Grounds

Lightning-ground connecting leads should be of conductor size equivalent to at least no. 10 wire. The no. 8 aluminum wire used for TV-antenna grounds is satisfactory. Copper braid 3/4-inch wide (Belden 8662-10) is also suitable. The conductor should run in a straight line to the grounding point. The ground connection may be made to a water-piping system, the grounded metal frame of a building, or to one or more 5/8-inch ground rods driven to a depth of at least 8 feet.

## SUPPORTS FOR WIRE ANTENNAS

A prime consideration in the selection of a support for an antenna is that of structural safety. Building regulations in many localities require that a permit be secured in advance of the erection of structures of certain types, often including antenna poles or towers. In general, localities having such requirements also will have building safety codes that must be observed. Such regulations may govern the method and materials used in construction of, for example, a self-supporting tower. Checking with your local government building department before putting up a tower may save a good deal of difficulty later, since a tower would have to be taken down or modified if not approved by the building inspector on safety grounds.

Municipalities have the right and duty to enforce any reasonable regulations having to do with the safety of life or property. The courts generally have recognized, however, that municipal authority does not extend to esthetic questions; i.e., the fact that someone may object to the mere presence of a pole or tower, or an antenna structure, because in his opinion it detracts from the beauty of the neighborhood, is not grounds for refusing to issue a permit for a safe structure to be erected.

Even where such regulations do not exist or are not enforced, the amateur should be careful to select a type of support and a location for it that will minimize the chances of collapse and, if collapse does occur, will minimize the chances that someone will be injured or property damaged. A single injury can be far more costly than the price of a more rugged support.

### TREES AS ANTENNA SUPPORTS

From the beginning of Amateur Radio, trees have been used widely for supporting wire antennas. Trees cost nothing, of course, and will often provide a means of supporting a wire antenna at considerable height. However, as an antenna support, a tree is highly unstable in the presence of wind, unless the tree is a very large one

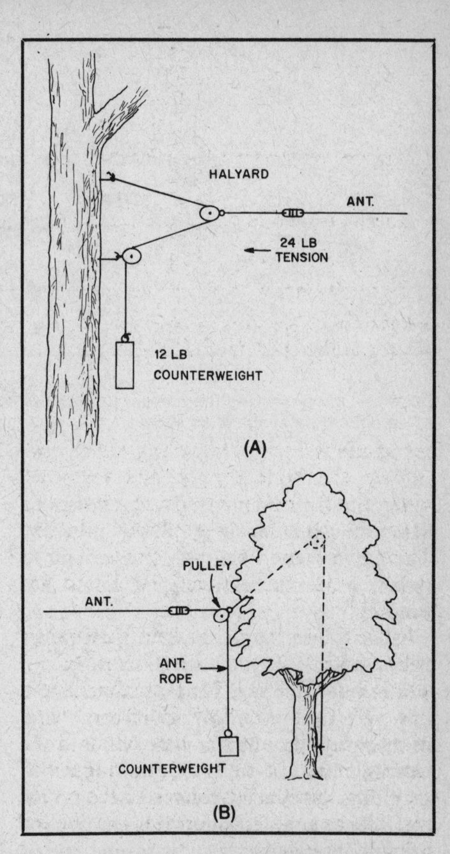


Fig. 63 — Methods of counterweighting to minimize antenna movement. The method at A limits the fall of the counterweight should the antenna break. It also has a 2 to 1 mechanical advantage, as indicated. The method at B has the disadvantage that the point of support in the tree must be much higher than the end of the antenna.

and the antenna is suspended from a point well down on the tree trunk. As a result, the antenna must be constructed much more sturdily than would be necessary with stable supports. Even with rugged construction, it is unlikely that an antenna suspended from a tree, or between trees, will stand up indefinitely, and occasional repair or replacement usually must be expected.

There are two general methods of securing a pulley to a tree. If the tree can be climbed safely to the desired level, a pulley can be wired to the trunk of the tree, as shown in Fig. 63. If, after passing the halyard through the pulley, both ends of the halyard are simply brought back down to ground along the trunk of the tree, there may be difficulty in bringing the antenna end of the halyard out where it will be clear of branches. To avoid this, one end of the halyard can be tied temporarily to the tree at the pulley level, while the remainder of the halyard is coiled up, and the coil thrown out horizontally from this level, in the direction in which the antenna will run. It may help to have the antenna end of the halyard weighted. Then, after attaching the antenna to the halyard, the other end is untied from the tree, passed through the

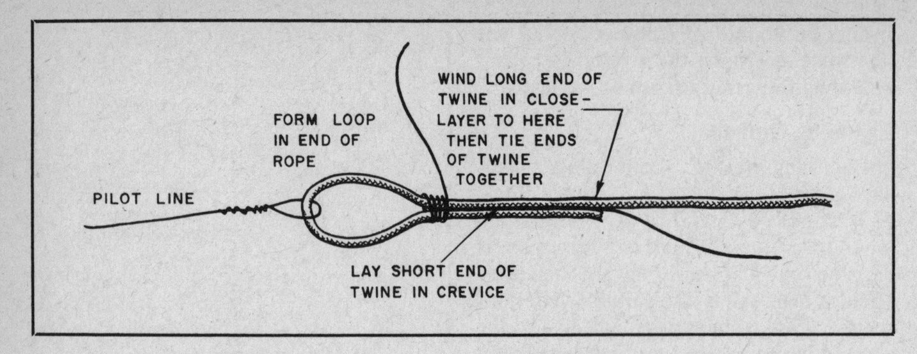


Fig. 64 — In connecting the halyard to the pilot line, a large knot that might snag in the crotch of a tree should be avoided, as shown.

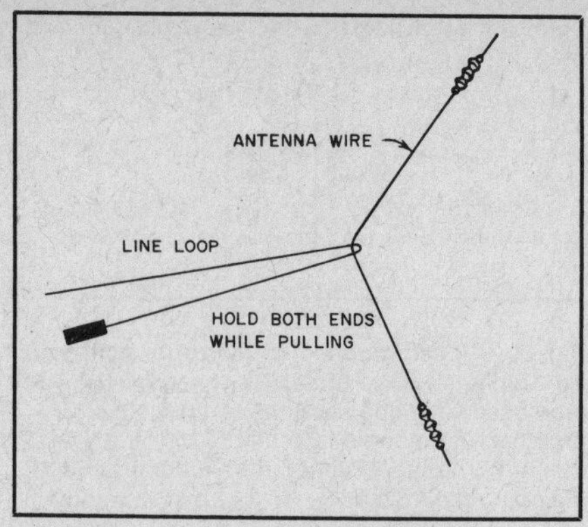


Fig. 65 — A weighted line thrown over the antenna can be used to pull the antenna to one side to avoid overhanging obstructions, such as branches of trees in the path of the antenna, as the antenna is pulled up. When the obstruction has been cleared, the line can be removed by releasing one end.

pulley, and brought to ground along the tree trunk in as straight a line as possible. The halyard need be only long enough to reach the ground after the antenna has been hauled up, since additional rope can be tied to the halyard when it becomes necessary to lower the antenna.

The other method consists of passing a line over the tree from ground level, and using this line to haul a pulley up into the tree and hold it there. Several ingenious methods have been used to accomplish this. The simplest method employs a weighted pilot line, such as fishing line or mason's chalk line. By grasping the line about two feet from the weight, the weight is swung back and forth, pendulum style, and then heaved with an underhand motion in the direction of the tree top. Several trials may be necessary to determine the optimum size of the weight for the line selected, the distance between the weight and the hand before throwing, and the point in the arc of the swing where the line released. The weight, however, must be sufficently large to assure that it will carry the pilot line back to ground after passing over the tree. Flipping the end of the line up and down so as to put a traveling wave on the line often helps to induce the weight to drop down if the weight is marginal. The higher the tree, the lighter the weight and the pilot line must be. A glove should be worn on the throwing hand, because a line running swiftly through the bare hand can cause a severe burn.

If there is a clear line of sight between ground and a particularly desirable crotch in the tree, it may be possible to hit the crotch eventually after a sufficient number of tries. Otherwise, it is best to try to heave the pilot line completely over the tree, as close to the center line of the tree as possible. If it is necessary to retrieve the line and start over again, the line should be drawn back very slowly; otherwise the swinging weight may wrap the line around a small limb, making retrieval impossible.

Stretching the line out in a straight line on the ground before throwing may help to keep the line from snarling, but it places extra drag on the line, and the line may snag on obstructions overhanging the line when it is thrown. Another method is to make a stationary reel by driving eight nails, arranged in a circle, through a 1-inch board. After winding the line around the circle formed by the nails, the line should reel off readily when the weighted end of the line is thrown. The board should be tilted at approximately right angles to the path of the throw.

Other devices that have been used successfully to pass a pilot line over a tree are the bow and arrow with heavy thread tied to the arrow, and the short casting rod and spinning reel used by fishermen. The Wrist Rocket slingshot made from surgical rubber tubing and a metal frame has proved highly effective as an antenna launching device. Still another method that has been used where sufficient space is available is to fly a kite. After the kite has reached sufficient altitude, simply walk around the tree until the kite string lines up with the center of the tree. Then pay out string until the kite falls to the earth. This method has been used successfully to pass a line over a patch of woods between two higher supports, which would have been impossible using any other method.

The pilot line can be used to pull successively heavier lines over the tree until one of adequate size to take the strain of the antenna has been reached. This line is then used to haul a pulley up into the tree after the antenna halyard has been threaded through the pulley. The line that holds the pulley must be capable of withstanding considerable chafing where it passes through the crotch, and at points where lower branches may rub against the standing part. For this reason, it may be advisable to use galvanized sash cord or stranded guy wire for raising the pulley.

Especially with larger sizes of line or cable, care must be taken when splicing the pilot line to the heavier line to use a splice that will minimize the chances that the splice cannot be coaxed through the tree crotch. One type of splice is shown in Fig. 64.

The crotch in which the line first comes to rest may not be sufficiently strong to stand up under the tension of the antenna. However, if the line has been passed over, or close to, the center line of the tree, it will usually break through the lighter crotches and finally come to rest in one sufficiently strong lower down on the tree.

Needless to say, any of the suggested methods should be used with due respect to persons or property in the immediate vicinity. A child's sponge-rubber ball (baseball size) makes a safe weight for heaving a heavy thread line or fishing line.

If the antenna wire becomes snagged in lower branches of the tree when the wire is pulled up, or if branches of other trees in the vicinity interfere with raising the antenna, a weighted line thrown over the antenna and slid along to the appropriate point is often helpful in pulling the antenna wire to one side to clear the interference as the antenna is being raised, as shown in Fig. 65.

### Wind Compensation

The movement of an antenna suspended between supports that are not stable in wind can be reduced materially by the use of heavy springs, such as screen-door springs under tension, or by a counterweight at the end of one halyard, as shown in Fig. 63. The weight, which may be made up of junk-yard metal, window sash weights, or a galvanized pail filled with sand or stone, should be adjusted experimentally for best results under existing conditions. Fig. 66 shows a convenient way of fastening the counterweight to the halyard. It avoids the necessity for untying a knot in the halyard which may have hardened under tension and exposure to the weather.

## TREES AS SUPPORTS FOR VERTICAL WIRE ANTENNAS

Trees can often be used to support vertical as well as horizontal antennas. If the tree is a tall one with overhanging branches, the scheme of Fig. 67 may be used. The top end of the antenna is secured to a halyard passed over the limb, brought back to ground level, and fastened to the trunk of the tree.

## MASTS

Where suitable trees are not available, or a more stable form of support is desired, masts are suitable for wire antennas of reasonable span length. At one time, most amateur masts were constructed of lumber, but the TV industry has brought out metal masting that is inexpensive and much more durable than wood. However, there are some applications where wood is necessary or desirable.

### The "A-Frame" Mast

A light and relatively inexpensive mast is shown in Fig. 68. In lengths up to 40 feet it is very easy to erect and will stand without difficulty the pull of ordinary wire antenna systems. The lumber used is 2 × 2 straight-grained pine (which many lumber yards know as hemlock) or even fir stock. The uprights can be as long as 22 feet each (for a mast slightly over 40 feet high) and the crosspieces are cut to fit. Four pieces of 2 × 2, 22 feet long, will provide enough and to spare. The only other materials required are five 1/4-inch carriage bolts 5-1/2 inches long, a few spikes, about 300 feet of stranded or solid galvanized iron wire for the guys or stays, enough glazed-porcelain compression insulators ("eggs") to break up the guys into sections, and the usual pulley and halyard rope. If the strain insulators are put in every 20 feet, approximately 15 of them will be enough.

After selecting and purchasing the lumber — which should be straight-grained and knot-free — three sawhorses or boxes should be set up and the mast assembled in the manner indicated in Fig. 69. At this stage it is a good plan to give the mast two coats of "outside-white" house paint or latex.

After the second coat of paint is dry, attach the guys and rig the pulley for the antenna halyard. The pulley anchorage should be at the point where the top stays are attached so that the back stay will assume the greater part of the load tension. It is better to use wire wrapped around the mast with a small through-bolt to prevent sliding down, than to use eye bolts.

If the mast is to stand on the ground, a couple of stakes should be driven to keep the bottom from slipping. At this point the mast may be "walked up" by a helper. If it is to go on a roof, first stand it up against the side of the building and then hoist it, from the roof, keeping it vertical. The whole assembly is light enough for two men to perform the complete operation — lifting the mast, carrying it to its permanent berth, and fastening the guys with the mast vertical all the while. It is therefore entirely practicable to put up

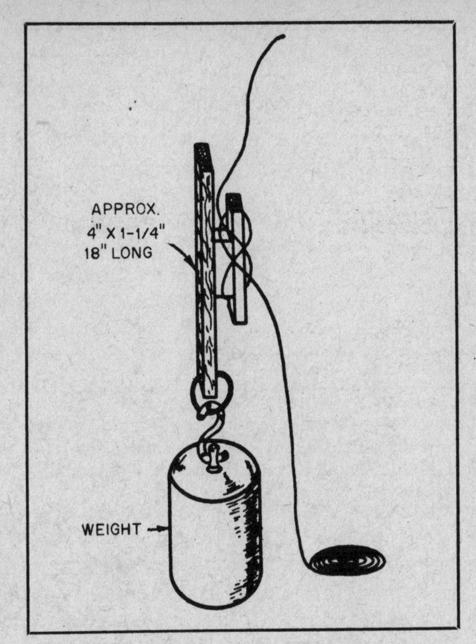


Fig. 66 — The cleat avoids the necessity of having to untie a knot that may have been weather-hardened.

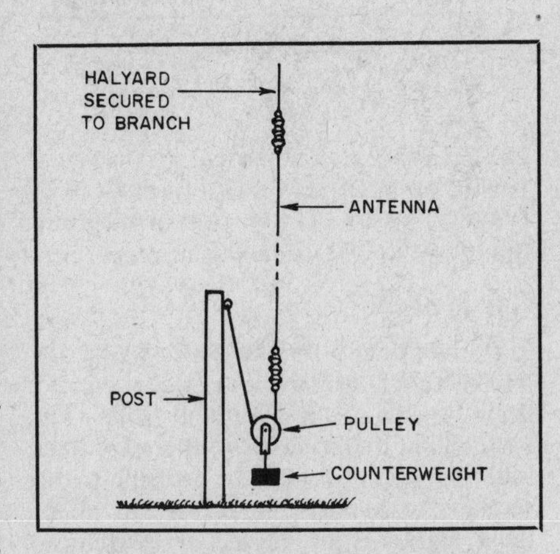


Fig. 67 — Counterweight for a vertical antenna suspended from an overhanging tree branch.

this kind of mast on a small flat area of roof that would prohibit the erection of one that had to be raised to the vertical in its final location.

### TV Masting

TV masting is available in 5- and 10-foot lengths, 1-1/4 inches in diameter, in both steel and aluminum. These sections are crimped at one end to permit sections to be joined together. However, a form that will usually be found more convenient is the telescoping TV mast available from many electronic supply houses. The masts may be obtained with three, four or five 10-foot sections, and come complete with guying rings and a means of locking the sections in place after they have been extended. These masts are stronger than the nontelescoping type because the top section is 1-1/4 inches in diameter, and the diameter increases toward the bottom section which is 2-1/2 inches in diameter in the 50-foot mast.

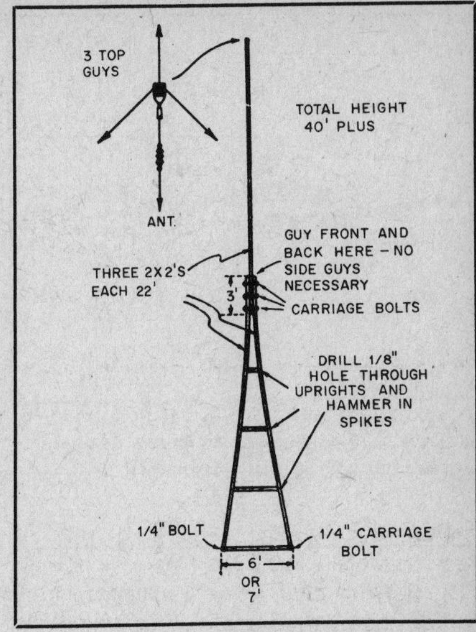


Fig. 68 — The "A-frame" mast, lightweight and easily constructed and erected.

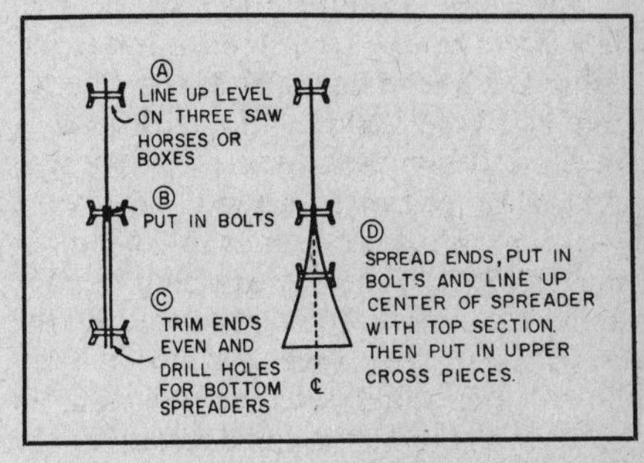


Fig. 69 — Method of assembling the "A-frame" mast.

Guy rings are provided at 10-foot intervals, but guys may not be required in all points. Guys at the top are essential and at least one other set near the center of the mast will usually be found necessary to keep the mast from bowing. If the mast has any tendency to whip in the wind, or to bow under the stress of the antenna, additional guys should be added at the obvious points.

### Mast Guying

Three guy wires in each set will usually be adequate for a mast. These should be spaced equally around the mast. The number of sets of guys will depend on the height of the mast, its natural sturdiness, and the required antenna tension. A 30-foot mast will usually require two sets of guys, while a 50-foot mast will need at least three sets. One guy of the top set should be run in a direction directly opposite to the direction in which the antenna will run, the other two being spaced 120 degrees with respect to the first, as shown in the inset, Fig. 68.

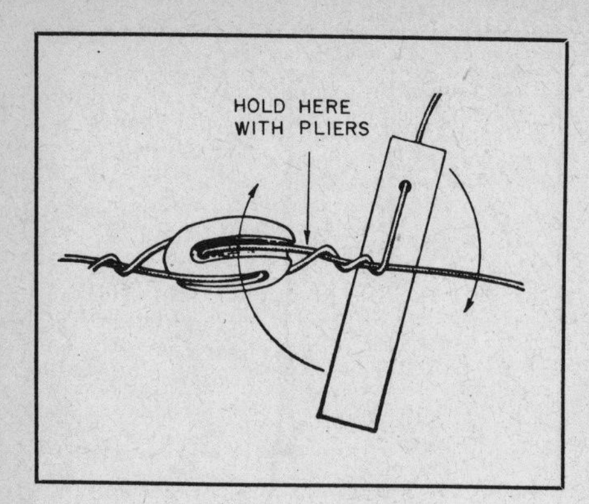

Fig. 70 — Simple lever for twisting solid guy wires in attaching strain insulators.

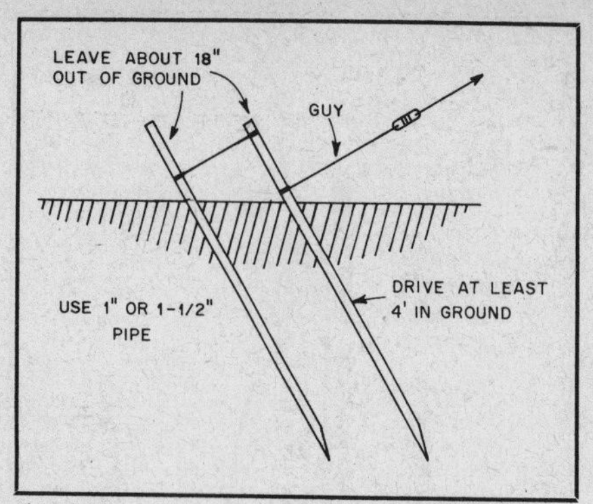

Fig. 72 — Driven guy anchors. One pipe will usually be sufficient for a small mast. For added strength, a second pipe may be added, as shown.

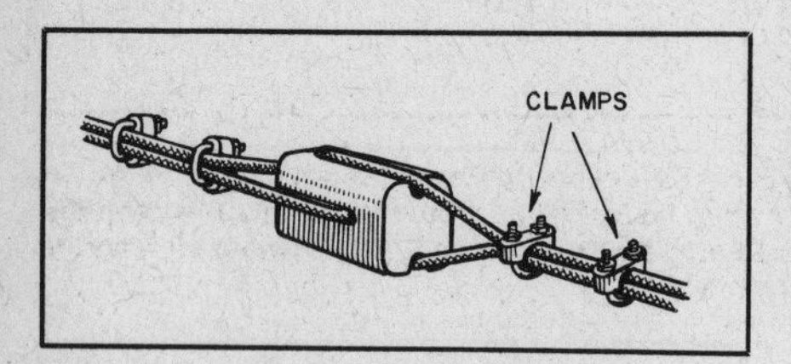

Fig. 71 — Stranded guy wire should be attached to strain insulators by means of standard cable clamps to fit the size of wire used.

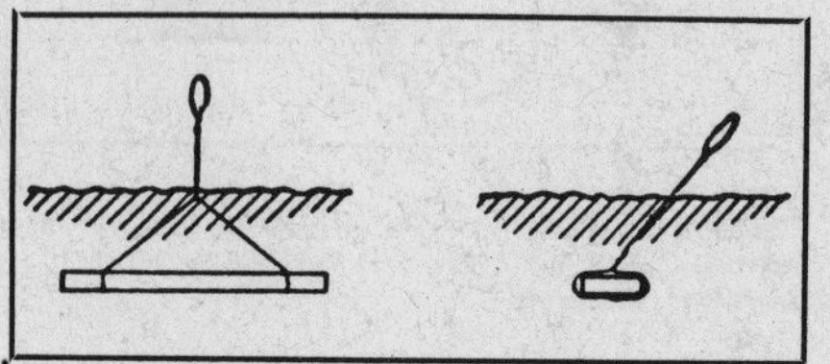
Fig. 73 — Buried "dead-man" type of guy anchor (see text).

The general rule is that the top guys should be anchored at distances from the base of the mast equal to not less than 60% of the height of the mast. At the 60% distance, the stress on the guy wire opposite the antenna will be approximately twice the tension on the antenna. As the distance between the guy anchorage and the base of the mast is decreased, the tension on the rear guy in proportion to the tension on the antenna rises rapidly, the extra tension resulting in additional compression on the mast, which increases the tendency for the mast to buckle.

The function of additional sets of guys is to correct for any tendency that the mast may have to bow or buckle under the compression imposed by the top guys. To avoid possible mechanical resonance in the mast that might cause the mast to have a tendency to vibrate, the sets of guys should not be spaced equally on the mast. A second set of guys should be placed at approximately 60% of the distance between the ground and the top of the mast, while a third set should be placed at about 60% of the distance between the ground and the second set.

The additional set of guys should be anchored at distances from the base of the mast equal to not less than 60% of the distance between ground and the points of attachment on the mast. In practice, the same anchors are usually used for all sets of guys, which means that this requirement is met automatically if the top set has been anchored at the correct distance.

To avoid electrical resonances which might cause distortion of the normal radiation pattern of the antenna, it is advisable to break each guy into sections of 19 to 20 feet by the insertion of strain insulators (see Figs. 70 and 71).

### Guy Material

Within their stress ratings, any of the halyard materials listed in Table 2 may be used for the construction of guys. The nonmetalic materials have the advantage that they do not have to be broken up into sections to avoid resonances, but all of these materials are subject to stretching, which may cause mechanical problems in permanent installations. At rated working load tension, dry manila rope stretches about 5 percent, while nylon rope stretches about 20 percent.

The antenna wire listed in Table 1 is also suitable for guys, particularly the copper-clad steel types. Solid galvanized steel wire is also used widely for making guys. This wire has approximately twice the tension ratings of similar sizes of copper-clad wire, but it is more susceptible to corrosion. Stranded galvanized wire sold for guying TV masts is also suitable for light-duty applications, but is susceptible to corrosion.

### Guy Anchors

Figs. 72 and 73 show two different styles of guy anchors. In Fig. 72, one or more pipes are driven into the earth at right angles to the guy wire. If a single pipe proves to be inadequate, another pipe can be added in tandem, as shown. Steel fence posts may be used in the same manner. Fig. 73 shows a "dead-man" type of anchor. The buried anchor may consist of one or more pipes 5 or 6 feet long, or scrap automobile parts, such as bumpers or wheels. The anchors should be buried 3 or 4 feet in the ground. Some tower manufacturers make heavy auger-type anchors that screw into the earth. These anchors are usually heavier than required for guying a mast, although they may be more convenient to install. Trees and buildings may also be used as guy anchorages if they are located appropriately. Care should be exercised, however, to make sure that the tree is of adequate size, or that the fastening to a building can be made sufficiently secure.

### Guy Tension

Most troubles encountered in mast guying are a result of pulling the guy wires too tight. Guy-wire tension should never be more than is necessary to correct for obvious bowing or movement under wind pressure. In most cases, the tension needed will not require the use of turnbuckles, with the possible exception of the guy opposite the antenna. If any great difficulty is experienced in eliminating bowing from the mast, the antenna tension should be reduced.

## ERECTING A MAST OR OTHER TYPE OF SUPPORT

The erection of a mast of 30 feet or less can usually be done by simply "walking" the mast up after blocking the bottom end securely so that it can neither slip along the ground or upend when the mast is raised. An assistant should be stationed at each guy wire, and in the last stages of raising, some assistance may be desirable by pulling on the proper guy wire. Further assistance may be gained by using the halyards in the same manner. As the mast is raised, it may be helpful to follow the under side of the mast with a scissors rest (Fig. 74), should a pause in the hoisting become necessary. The rest may also be used to assist in the raising, if an assistant is used on each leg.

As the mast nears the vertical position, those holding the guy wires should be ready to make the guys fast temporarily to prevent the mast from falling in one direction or another. The guys can then be adjusted, one at a time, until the mast is perfectly straight.

For a mast over 30 feet, a "gin" of some form may be required, as shown in Fig. 74. Several turns of rope are wound around a point on the mast above center. The ends of the rope are then brought together and passed over the limb of a tree. The rope should be pulled as the mast is walked up to keep the mast from bending at the center. If a tree is not available, a post, such as a 2 × 4, temporarily erected and guyed, can be used. After the mast has been erected, the assisting rope can be removed by walking one end around the mast (inside the guy wires).

Much sturdier supports are telephone

poles and towers. Towers are discussed in Chapter 9 in reference to rotatable antennas. Such supports may require no guying, but they are not often used solely for the support of wire antennas because of their relatively high cost. However, for antenna heights in excess of 50 feet, they are usually the most practical form of support.

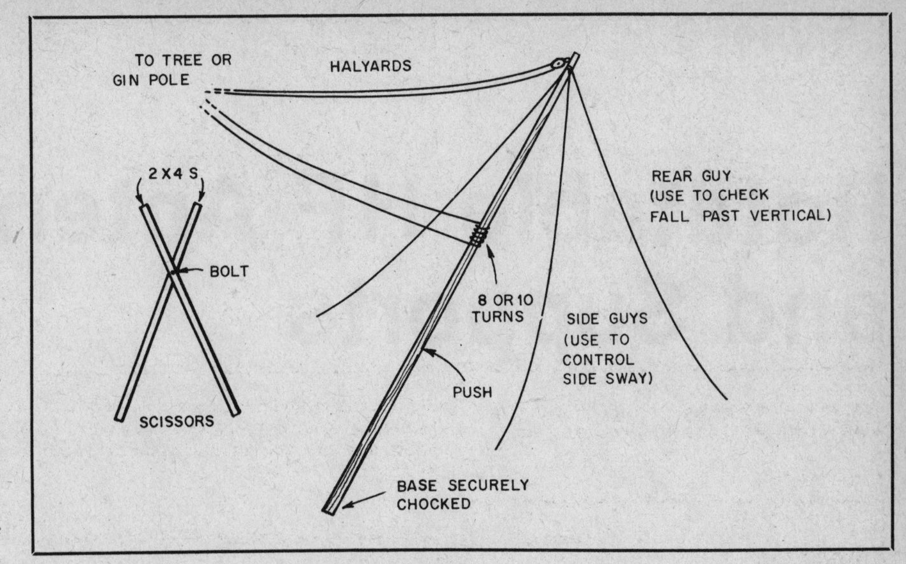


Fig. 74 — Pulling on a gin line fastened slightly above the center point of the mast and on the halyards will assist materially in erecting a tall mast. The tensions should be such as to keep the mast in as straight a line as possible. The scissors may be used to push on the under side and to serve as a rest in case a pause in raising becomes necessary.

## BIBLIOGRAPHY

Source material and more extended discussion of topics covered in this chapter can be found in the references given below.


Elengo, "Predicting Sag in Long Wire Antennas," *QST*, Jan. 1966.

Gann, "A Center-Fed 'Zepp' for 80 and 40," *QST*, May 1966.

Gordon, "Invisible Antennas," *QST*, Nov. 1965.

Gue, "An 80-Meter Inverted Vee," *QST*, June 1968.

McCoy, "An Easy-to-Make Coax-Fed Multiband Trap Dipole," *QST*, Dec. 1964.

Brown, "The Phase and Magnitude of Earth Currents Near Radio Transmitting Antennas," *Proceedings of the I.R.E.* Feb. 1935.

Brown, Lewis and Epstein, "Ground Systems as a Factor in Antenna Efficiency," *Proceedings of the I.R.E.* June 1937.

"Some Notes on Ground Systems for 160 Meters," *QST*, April 1965.

Elliott, "Phased Verticals for 40," *QST*, April 1972.

Hubbell, "Feeding Grounded Towers as Radiators," *QST*, June 1960.

Hopps, "A 75-Meter DX Antenna," *QST*, March 1979, p. 44.

Atchley, "Putting the Quarter-Wave Sloper to Work on 160," *QST*, July 1979, p. 19.

DeMaw, "Additional Notes on the Half Sloper," *QST*, July 1979, p. 20.

Mathison, "Inexpensive Traps for Wire Antennas," *QST*, Feb. 1977.

Bell, "Trap Collinear Antenna," *QST*, Aug. 1963.

Berg, "Multiband Operation with Parallelled Dipoles," *QST*, July 1956.

Buchanan, "The Multimatch Antenna System," *QST*, March 1955.

Greenberg, "Simple Trap Construction for the Multiband Antenna," *QST*, Oct. 1956.

Lattin, "Multiband Antennas Using Decoupling Stubs," *QST*, Dec. 1960.

Lattin, "Antenna Traps of Spiral Delay Line," *QST*, Nov. 1972.

Richard, "Parallel Dipoles of 300-Ohm Ribbon," *QST*, March 1957.

Sevick, "The W2FMI Ground-Mounted Short Vertical," *QST*, March 1973.

Shafer, "Four-Band Dipole with Traps," *QST*, Oct. 1958.

Wrigley, "Impedance Characteristics of Harmonic Antennas," *QST*, Feb. 1954.

# Chapter 9

# Rotatable HF Antennas and Supports

Constructing an antenna that is to be rotated requires materials that are strong, lightweight and easy to obtain. Procurement is often the most difficult portion of the project, but that can usually be overcome with some careful searching of the Yellow Pages section of the telephone book for the nearest large metropolitan area. (Such telephone books may be available in the reference section of your local library.)

The materials required to build a suitable rotatable antenna will vary, depending on many factors. Perhaps the most important factor that determines the type of hardware needed is the weather conditions normally encountered. High winds usually don't cause as much damage to an antenna as does ice — or even heavy ice along with high winds. Aluminum sizes should be selected so that the various sections of tubing will telescope to provide the necessary total length. Table 1 gives the specifications for aluminum that will meet the needs for most amateur installations.

The boom size for a rotatable Yagi or quad should be selected to provide stability to the entire system. The best diameter for the boom depends on several factors, but mostly the element weight, number of elements and overall length. Tubing diameters of 1-1/4 inches can easily support three-element 10-meter arrays and perhaps a two-element 15-meter system. For larger 10-meter antennas or for harsh weather conditions, and for antennas up to three elements on 20 meters or four elements on 15 meters, a 2-inch diameter boom will be adequate. It is not recommended that 2-inch diameter booms be made any longer than 24 feet unless additional support is given to reduce both vertical and horizontal bending forces. Suitable reinforcement for a long 2-inch boom can consist of a truss or a truss and lateral support, as shown in Fig. 1.

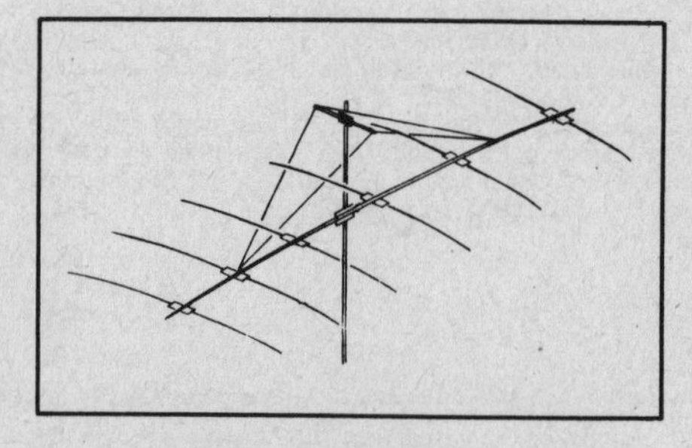

Fig. 1 — A long boom needs both vertical and horizontal support. The cross bar mounted above the boom can support a double truss which will help keep the antenna in position

A boom length of 24 feet is about the point where a 3-inch diameter begins to be very worthwhile. This dimension provides a considerable amount of improvement in overall mechanical stability as well as increased clamping surface area for element hardware. The latter is extremely important if heavy icing is commonplace and rotation of elements around the boom is to be avoided. Pinning an element to the boom with a large bolt helps in this regard. However, on the smaller diameter booms, the elements sometimes work loose and tend to elongate the pinning holes in both the element and the boom. After some time the elements shift their positions slightly (sometimes from day to day!) and give a rather ragged appearance to the system, even though this doesn't generally harm the electrical performance.

A 3-inch diameter boom with a wall thickness of 0.065 inch is very satisfactory for antennas up in size to about a five-element, 20-meter array that is spaced on a 40-foot long boom. A truss is recommended for any boom longer than 24 feet.

### YAGI ELEMENTS THAT LAST

While the length of an antenna element depends to some extent on its diameter, the only laws that specify the minimum diameter of an element are the laws of nature. That is, the element must be rugged enough to survive whatever weather conditions it will be subjected to.

Fig. 2 shows designs for Yagi elements that should survive all but the most extreme weather conditions. No guarantees are made as to their ability to handle a heavy load of ice in high winds, but if you lose an antenna made with elements like these you'll have plenty of company among your neighbors with commercially made antennas!

In each case where a smaller-diameter length of tubing is telescoped inside a larger-diameter one, cut a slit about 1/8-inch wide at the end of the larger diameter piece and secure the connection with a hose clamp. It's a good idea to coat the inside of the joint with Penetrox or a similar substance to ensure a good electrical bond. Antenna elements have a tendency to vibrate when they are mounted on a tower, so it's essential to dampen the vibrations by running a piece of clothesline through the length of the element. Cap or tape the end of the element to secure the clothesline. If there is a

**Table 1**
**Standard Sizes of Aluminum Tubing**

6061-T6 (61S-T6) round aluminum tube in 12-foot lengths

| *OD (in.)* | *Wall Thickness in.* | *stubs ga.* | *ID (in.)* | *Approx. Weight (lb) per ft* | *per length* |
|---|---|---|---|---|---|
| 3/16 | .035 | no. 20 | .117 | .019 | .228 |
| | .049 | no. 18 | .089 | .025 | .330 |
| 1/4 | .035 | no. 20 | .180 | .027 | .324 |
| | .049 | no. 18 | .152 | .036 | .432 |
| | .058 | no. 17 | .134 | .041 | .492 |
| 5/16 | .035 | no. 20 | .242 | .036 | .432 |
| | .049 | no. 18 | .214 | .047 | .564 |
| | .058 | no. 17 | .196 | .055 | .660 |
| 3/8 | .035 | no. 20 | .305 | .043 | .516 |
| | .049 | no. 18 | .277 | .060 | .720 |
| | .058 | no. 17 | .259 | .068 | .816 |
| | .065 | no. 16 | .245 | .074 | .888 |
| 7/16 | .035 | no. 20 | .367 | .051 | .612 |
| | .049 | no. 18 | .339 | .070 | .840 |
| | .065 | no. 16 | .307 | .089 | 1.068 |
| 1/2 | .028 | no. 22 | .444 | .049 | .588 |
| | .035 | no. 20 | .430 | .059 | .708 |
| | .049 | no. 18 | .402 | .082 | .984 |
| | .058 | no. 17 | .384 | .095 | 1.040 |
| | .065 | no. 16 | .370 | .107 | 1.284 |
| 5/8 | .028 | no. 22 | .569 | .061 | .732 |
| | .035 | no. 20 | .555 | .075 | .900 |
| | .049 | no. 18 | .527 | .106 | 1.272 |
| | .058 | no. 17 | .509 | .121 | 1.452 |
| | .065 | no. 16 | .495 | .137 | 1.644 |
| 3/4 | .035 | no. 20 | .680 | .091 | 1.092 |
| | .049 | no. 18 | .652 | .125 | 1.500 |
| | .058 | no. 17 | .634 | .148 | 1.776 |
| | .065 | no. 16 | .620 | .160 | 1.920 |
| | .083 | no. 14 | .584 | .204 | 2.448 |
| 7/8 | .035 | no. 20 | .805 | .108 | 1.308 |
| | .049 | no. 18 | .777 | .151 | 1.810 |
| | .058 | no. 17 | .759 | .175 | 2.100 |
| | .065 | no. 16 | .745 | .199 | 2.399 |
| 1 | .035 | no. 20 | .930 | .123 | 1.476 |
| | .049 | no. 18 | .902 | .170 | 2.040 |
| | .058 | no. 17 | .884 | .202 | 2.424 |
| | .065 | no. 16 | .870 | .220 | 2.640 |

| *OD (in.)* | *Wall Thickness in.* | *stubs ga.* | *ID (in.)* | *Approx. Weight per ft* | *per length* |
|---|---|---|---|---|---|
| 1 | .083 | no. 14 | .834 | .281 | 3.372 |
| 1 1/8 | .035 | no. 20 | 1.055 | .139 | 1.668 |
| | .058 | no. 17 | 1.009 | .228 | 2.736 |
| 1 1/4 | .035 | no. 20 | 1.180 | .155 | 1.860 |
| | .049 | no. 18 | 1.152 | .210 | 2.520 |
| | .058 | no. 17 | 1.134 | .256 | 3.072 |
| | .065 | no. 16 | 1.120 | .284 | 3.408 |
| | .083 | no. 14 | 1.084 | .357 | 4.284 |
| 1 3/8 | .035 | no. 20 | 1.305 | .173 | 2.076 |
| | .058 | no. 17 | 1.259 | .282 | 3.384 |
| 1 1/2 | .035 | no. 20 | 1.430 | .180 | 2.160 |
| | .049 | no. 18 | 1.402 | .260 | 3.120 |
| | .058 | no. 17 | 1.384 | .309 | 3.708 |
| | .065 | no. 16 | 1.370 | .344 | 4.128 |
| | .083 | no. 14 | 1.334 | .434 | 5.208 |
| | *.125 | 1/8" | 1.250 | .630 | 7.416 |
| | *.250 | 1/4" | 1.000 | 1.150 | 14.823 |
| 1 5/8 | .035 | no. 20 | 1.555 | .206 | 2.472 |
| | .058 | no. 17 | 1.509 | .336 | 4.032 |
| 1 3/4 | .058 | no. 17 | 1.634 | .363 | 4.356 |
| | .083 | no. 14 | 1.584 | .510 | 6.120 |
| 1 7/8 | .058 | no. 17 | 1.759 | .389 | 4.668 |
| 2 | .049 | no. 18 | 1.902 | .350 | 4.200 |
| | .065 | no. 16 | 1.870 | .450 | 5.400 |
| | .083 | no. 14 | 1.834 | .590 | 7.080 |
| | *.125 | 1/8" | 1.750 | .870 | 9.960 |
| | *.250 | 1/4" | 1.500 | 1.620 | 19.920 |
| 2 1/4 | .049 | no. 18 | 2.152 | .398 | 4.776 |
| | .065 | no. 16 | 2.120 | .520 | 6.240 |
| | .083 | no. 14 | 2.084 | .660 | 7.920 |
| 2 1/2 | .065 | no. 16 | 2.370 | .587 | 7.044 |
| | .083 | no. 14 | 2.334 | .740 | 8.880 |
| | *.125 | 1/8" | 2.250 | 1.100 | 12.720 |
| | *.250 | 1/4" | 2.000 | 2.080 | 25.440 |
| 3 | .065 | no. 16 | 2.870 | .710 | 8.520 |
| | *.125 | 1/8 | 2.700 | 1.330 | 15.600 |
| | *.250 | 1/4 | 2.500 | 2.540 | 31.200 |

*These sizes are extruded; all other sizes are drawn tubes. Shown here are standard sizes of aluminum tubing that are stocked by most aluminum suppliers or distributors in the United States and Canada. Note that all tubing comes in 12-foot lengths and also that any diameter tubing will fit snugly into the next larger size, if the larger size has a 0.58-inch wall thickness. For example, 5/8-inch tubing has an outside diameter of 0.625 inches and will fit into 3/4-inch tubing with a .058-inch wall which has an inside diameter of 0.634 inch. The .009-inch clearance is just right for a slip fit or for slotting the tubing and then using hose clamps. The 6061-T6 type of aluminum is of relatively high strength and has good workability, plus being highly resistant to corrosion, and will bend without taking a "set." Check the Yellow Pages for aluminum dealers.

U bolt going through the center of the element, the clothesline may be cut into two pieces.

## A DL-STYLE YAGI FOR 10 METERS

Guenter Schwarzbeck, DL1BU, has made extensive comparisons between various Yagi and quad antenna designs on an antenna test range and has documented his work in the pages of *CQ-DL*. Gun has been especially interested in the parameters of bandwidth, front-to-back ratio and gain. He has designed 7-element Yagis for the 28- and 50-MHz bands that perform well in all three respects. The design has been copied using sizes of aluminum tubing commonly available here. The 28-MHz version requires a 40-foot boom; the 50-MHz version is described in Chapter 11. A gamma match is used, but the constructor may substitute a favorite matching system. It is worth noting that the dimensions of the driven element are not especially critical on any Yagi design, and resonable adjustments to the length of the driven element can be made in the interests of impedance matching without affecting the other characteristics of the antenna. The same does not hold true for the parasitic elements, as departures from the design

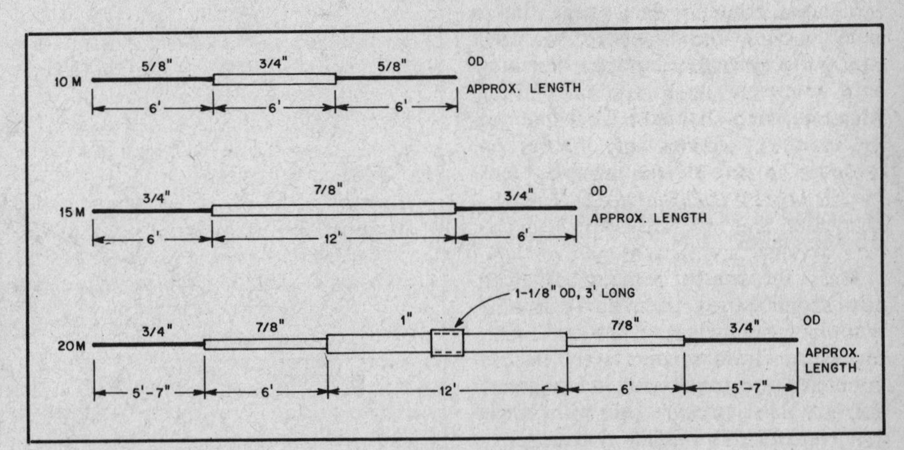


Fig. 2 — Element designs for Yagi antennas. (Use 0.058 in.-wall aluminum tubing except for end pieces where thinner-wall tubing may be used.)

Fig. 3 — A 7-element 6-meter Yagi of the "DL design" is mounted 4 feet above a 7-element 10-meter Yagi of similar design.

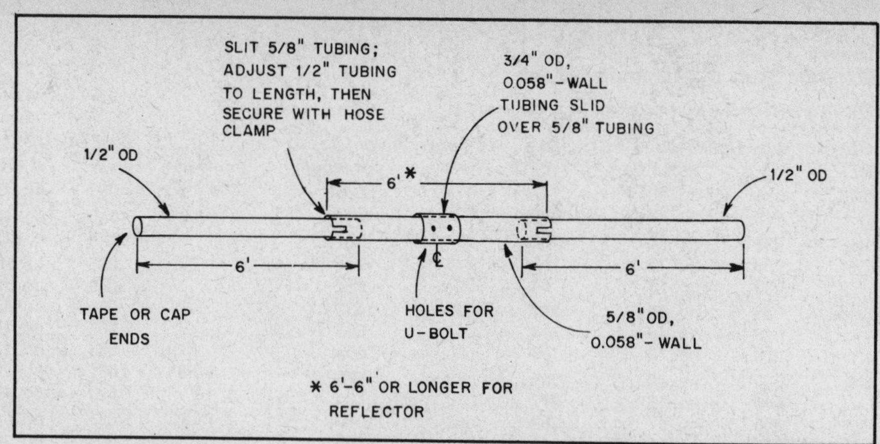


Fig. 4 — Construction of a typical element for the DL-style 10-meter Yagi.

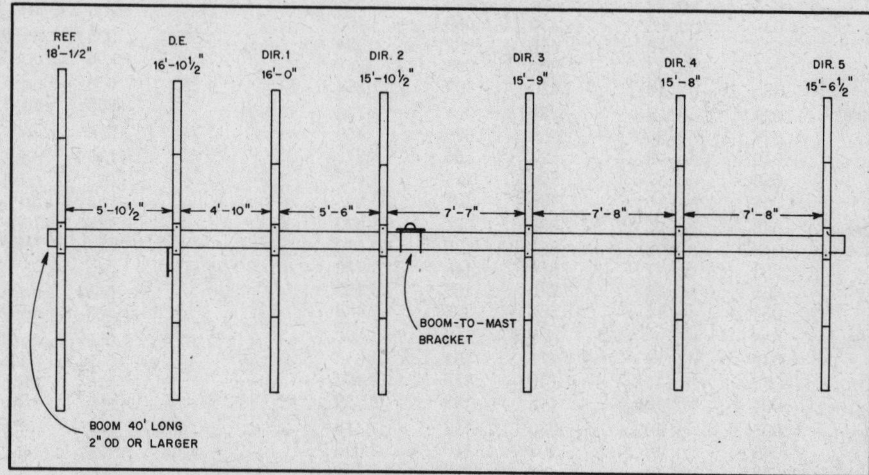


Fig. 5 — Dimensions for the DL-style 10-meter Yagi.

can result in significant differences in performance.

Building the 28-MHz version of this 7-element antenna is an ambitious project. The minimum requirement for boom material is 2 inches OD, and additional bracing is essential (see Fig. 3). If the elements are secured to the boom by a single U bolt, the suggested double thickness of tubing should be used at the center of the element. See Figs. 4 and 5 for details. While the dimensions given are to cover the bottom half of the 28.0-29.7 MHz band, the antenna has good gain, but a reduced front-to-back ratio, in the top half.

The 3-dB beamwidth of this 7-element antenna is on the order of 40° in the horizontal plane, which means that a pointing error of as little as 20° will result in a 3-dB loss in signal strength. Operators used to antennas with shorter booms and broader patterns will find that their antenna rotators are getting more of a workout when they switch to an antenna like this.

## A BEAM ANTENNA FOR 15 METERS

When the amateur is interested in constructing an array that can be rotated, aluminum tubing is generally used for the elements. The mechanical problems encountered are usually not much greater for several elements than with single-element rotatable antenna systems.

This four-element Yagi antenna provides appreciable power gain and exhibits significant back-rejection characteristics. Overall dimensions are given in Fig. 6. Construction is straightforward using commonly available tubing material which normally is 12 feet long. The center of each element is made from a 12-foot

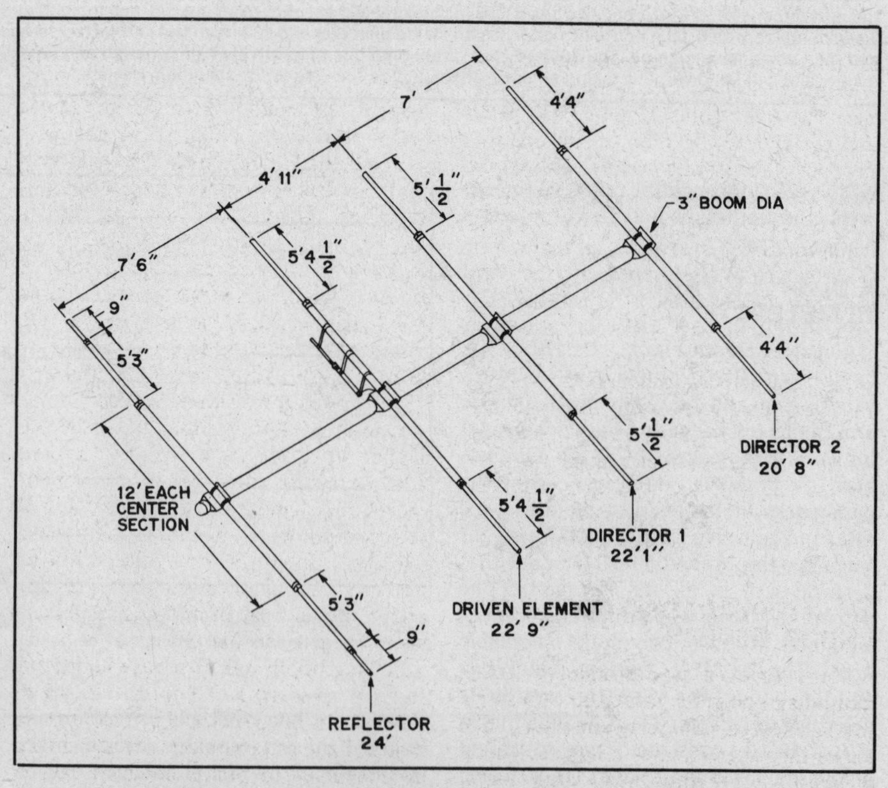


Fig. 6 — Overall dimensions for the 4-element 15-meter array.

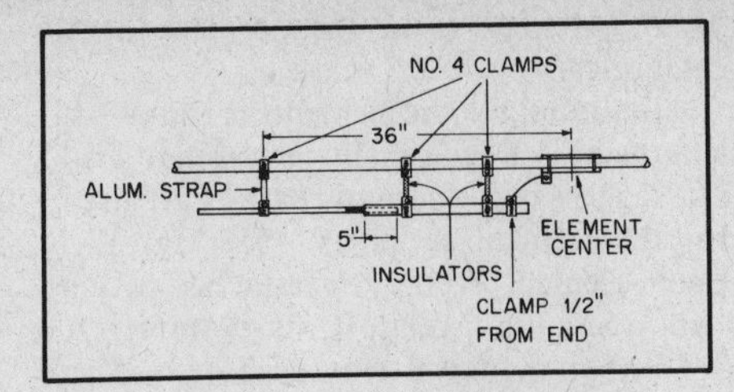


Fig. 7 — Gamma-matching-section dimensions for the 15-meter 4-element array (also see Fig. 11).

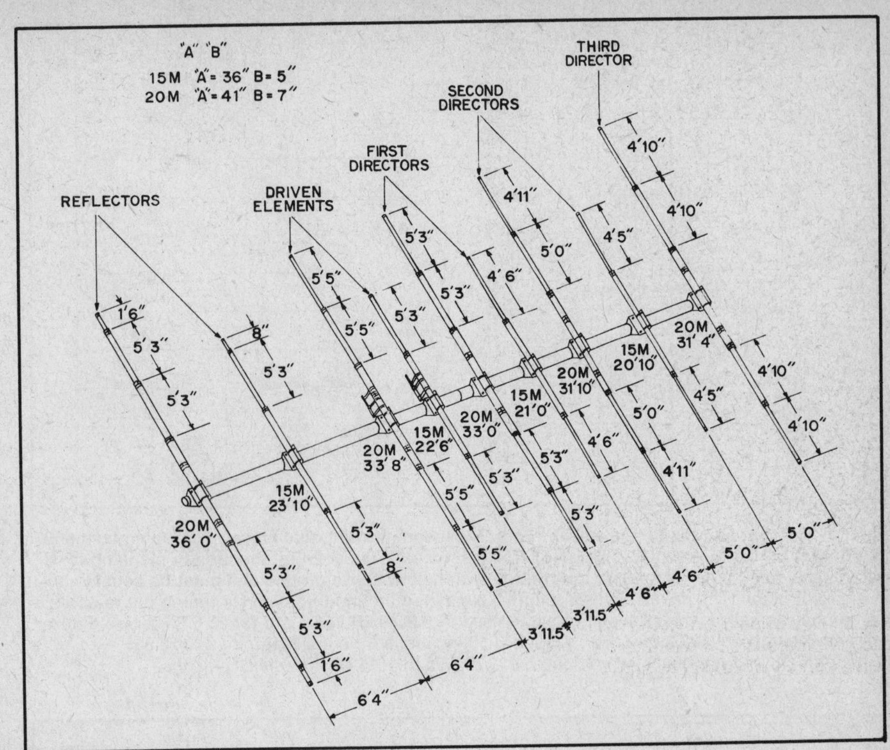


Fig. 8 — Interlaced 20- and 15-meter Yagi antenna. Design dimensions courtesy of Wilson Electronics, Pittman, Nevada.

length of 6061-T6 aluminum alloy that is 1-1/8 inches in diameter. The overall length of the element is determined by the distance the telescoping section is extended beyond the end of the center piece. Each of these telescoping sections is one inch OD and 6 feet long to provide the proper fit. Table 1 provides a guide for determining proper sizes for element material diameters. When telescoping sections are needed, the difference between joining pieces (in terms of diameters) should be about 0.009 inch. An additional section of 7/8-inch OD material is used at the tips of the reflector element to meet the dimensions specified. The two 7/8-inch pieces extend about 9 inches beyond the 1-inch diameter stock.

Each element is held in place with two U bolts that clamp it to a 6-inch long piece of aluminum stock. These pieces of angle material are then fastened to the boom with automotive muffler clamps. The size of the muffler clamp depends on the size of the boom. For this model, a 2-inch-diameter boom size should be satisfactory for all but the roughest climate conditions. A 3-inch-diameter boom does have advantages, however, as explained earlier.

Matching a feed line to the driven element can be accomplished by using the dimensions given in Fig. 7. Final adjustment of the gamma system should be made after the antenna is mounted in place atop the mast by placing a power meter (or SWR indicator) in series with the feed line at the input connector and adjusting the capacitor along with the tap point for minimum reflected power as described in Chapter 5.

The mechanical dimensions for the Yagi described here were developed by Wilson Electronics, Pittman, Nevada. While gain measurements are impossible without a test facility, the estimated power gain of this system is on the order of 9 dB. The front-to-back ratio is typically 20 or 25 dB.

## AN INTERLACED YAGI FOR 20 AND 15 METERS

Many times it is desirable to install more than one antenna on top of a single tower or mast. Stacking antennas, one above the other, creates a large stress on the mast and the rotor. With large arrays, it is desirable to reduce the weight and wind loading characteristics in every possible way to lower damage possibilities from ice and wind. One simple solution to the problem is to mount two complete antennas on one boom.

Installing elements for two different antennas on one boom has been popular for many years. Most commercially manufactured triband antennas use this technique. The question which develops however, is whether or not interaction between elements for different bands causes detrimental effects. It is generally accepted that interaction, if any, is very minimal between bands which are not harmonically related. The example shown here is a Wilson Electronics Model DB-54 designed to operate on both 15 and 20 meters. Two driven elements are required and each is fed independently with separate transmission lines. The boom is 40 feet long and is 3 inches OD. Smaller boom sizes are not recommended.

Construction details of this system are similar to those given for the 15-meter antenna described earlier in this chapter except the element center sections for 20 meters begin with 1-1/4 inch material and telescope down in size. The 15-meter elements are identical to the ones described earlier. All of the critical dimensions are given in Fig. 8.

A long boom needs to have additional support given to it if appreciable sag or droop is to be avoided. The truss can be made of any suitable steel wire and should be connected to points about 10 feet in each direction from the boom-to-mast plate. Turnbuckles should be used at the mast to create suitable tension for the wires. Each of the interlaced arrays can be treated as separate antennas for the purposes of tune-up. Since there is little (if any) interaction between elements, the 15-meter section could be removed from the boom if only a 20-meter monoband system is needed.

## THE THREE-ELEMENT MONOBAND YAGI

Perhaps the most popular type of antenna used on 20, 15 and 10 meters is the three-element array. Three elements offer the best compromise between gain, size, weight, wind loading and front-to-back ratio. Construction techniques should be the same as specified with the two antennas described earlier in this chapter.

Fig. 9 gives suitable dimensions for a three-element antenna. Hardware sizes should approximate the values given in Fig. 10. A gamma-matching system is easy to construct and adjust correctly. The dimensions shown in Fig. 11 are typical of the requirements for a working system. Should the builder desire to use a hairpin match arrangement, Chapter 5 should be consulted.

## A LINEAR LOADED 20-METER BEAM

There are two reasons for considering a shortened antenna. The first has to do with considerations of space and appearance. The second is the cost factor.

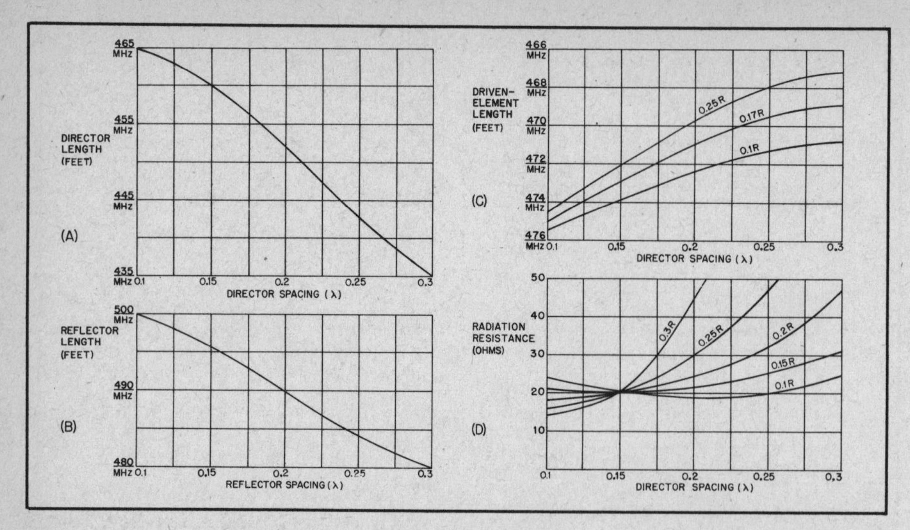


Fig. 9 — Element lengths for 3-element Yagis. These lengths will hold closely for tubing elements suported at or near the center. The radiation resistance (D) is useful information in planning for a matching system, but it is subject to variation with height above ground and must be considered an approximation. The driven-element length (C) may require modification for tuning out reactance if a gamma- or hairpin-match feed system is used. A 0.2D-0.2R beam cut for 28.6 MHz would have a director length of 452/28.6 = 15.8 = 15 feet 10 inches, a reflector length of 490/28.6 = 17.1 = 17 feet 1 inch, and a driven-element length of 470.5/28.6 = 16.45 = 16 feet 5 inches.

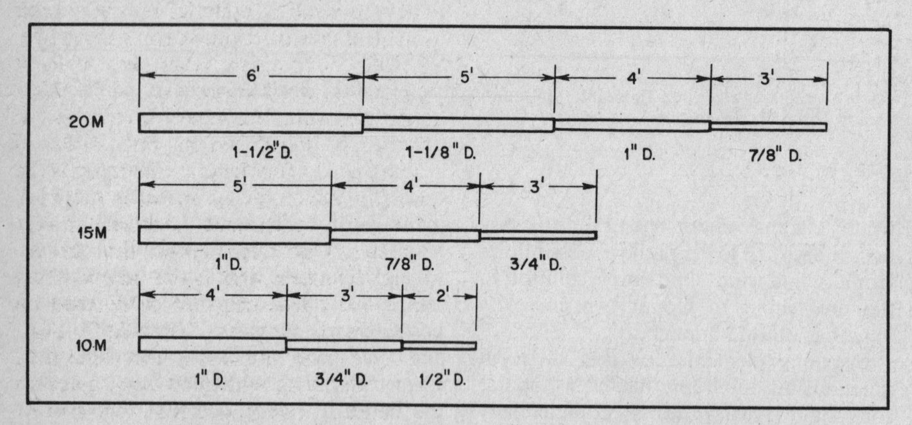


Fig. 10 — Suggested proportions for one side of tapered Yagi elements. The other side is identical, of course, and the center section of the element can be a single piece twice as long as the length shown here for the first (largest diameter) section. Appropriate overall element lengths may be determined from the graphs of Fig. 9. See Table 1 for aluminum tubing details.

This antenna is a winner in both categories.

Aluminum for the elements is four 12-ft lengths of 7/8-inch diameter tubing. Two 2-ft lengths of aluminum channel are used for the element supports. (See Fig. 12.) Each element section is attached using 1-in. ceramic standoff insulators. A 3/4-in. birch dowel was used to stiffen and prevent crushing of the tubing where the attachment is made. Element supports are grooved and attached to a 10-ft TV mast section using a single 5/16-in. U bolt.

The loading wire is no. 12 copper wire. It is supported by two bakelite blocks on each element section. The blocks are anchored to the tubing by small sheet-metal screws. Ends of the wire sections are formed into a loop to hold their position.

Dimensions for the antenna are given in Fig. 13. Details of the driven element can be seen in Fig. 14.

## CONSTRUCTION OF QUADS

The sturdiness of a quad is directly proportional to the *quality* of the material used and the care with which it is constructed. The size and type of wire selected for use with a quad antenna is important because it will determine the capability of the spreaders to withstand high winds and ice. One of the more common problems confronting the quad owner is that of broken wires. A solid conductor is more apt to break than stranded wire under constant flexing conditions. For this reason, copper stranded wire is recommended. For 20-, 15- or 10-meter operation, wire size no. 14 or 12 is a good choice. Soldering of the stranded wire at points where flexing is likely to occur should be avoided.

Connecting the wires to the spreader arms may be accomplished in many ways.

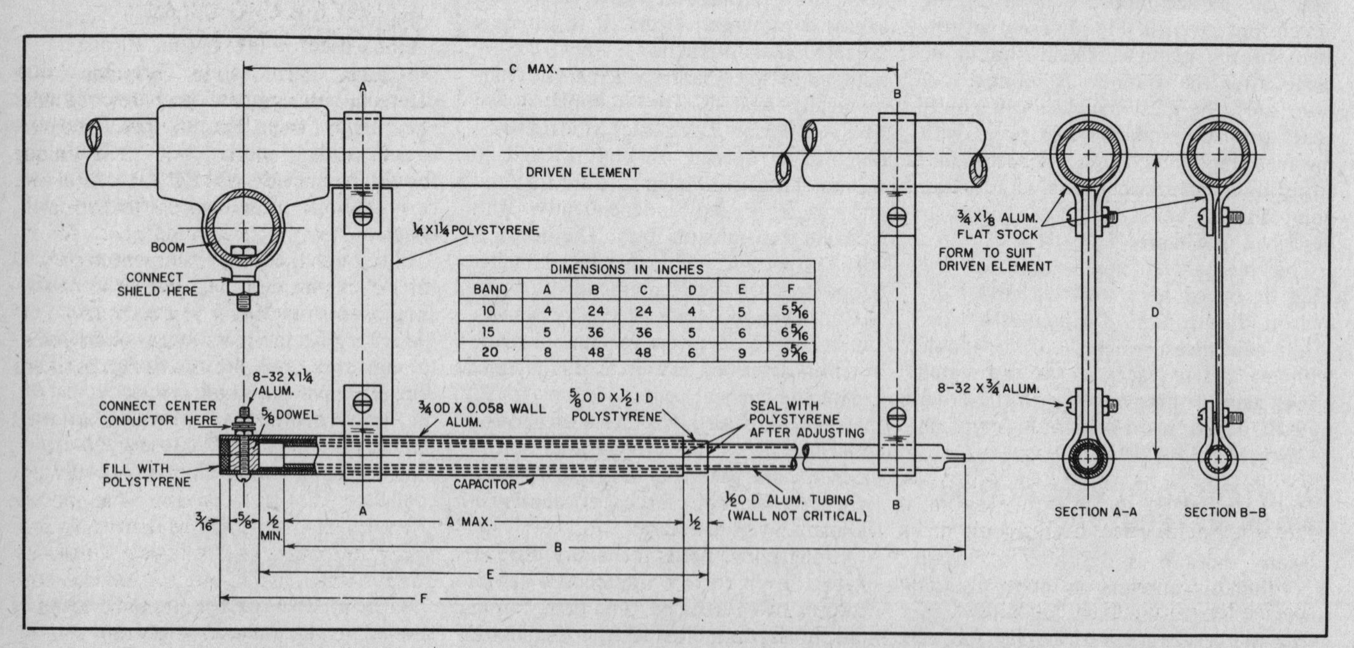


DIMENSIONS IN INCHES

| BAND | A | B | C | D | E | F |
|---|---|---|---|---|---|---|
| 10 | 4 | 24 | 24 | 4 | 5 | 5 5/16 |
| 15 | 5 | 36 | 36 | 5 | 6 | 6 5/16 |
| 20 | 8 | 48 | 48 | 6 | 9 | 9 5/16 |

Fig. 11 — Constructional details of a gamma-matching section for 52-ohm coax line. The reactance-compensating capacitor is in tubular form. It is made by dividing the gamma rod or bar into two telescoping sections separated by a length of polystyrene tubing, which serves as the dielectric.

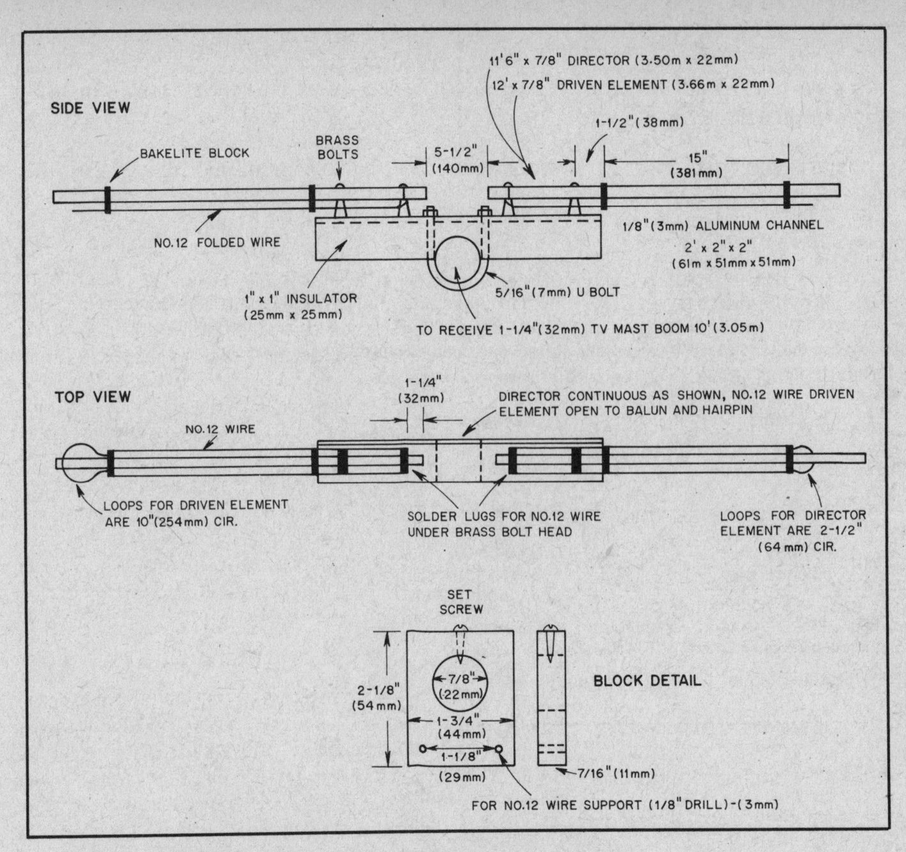


Fig. 12 — Construction details for the linear-loaded 20-meter beam.

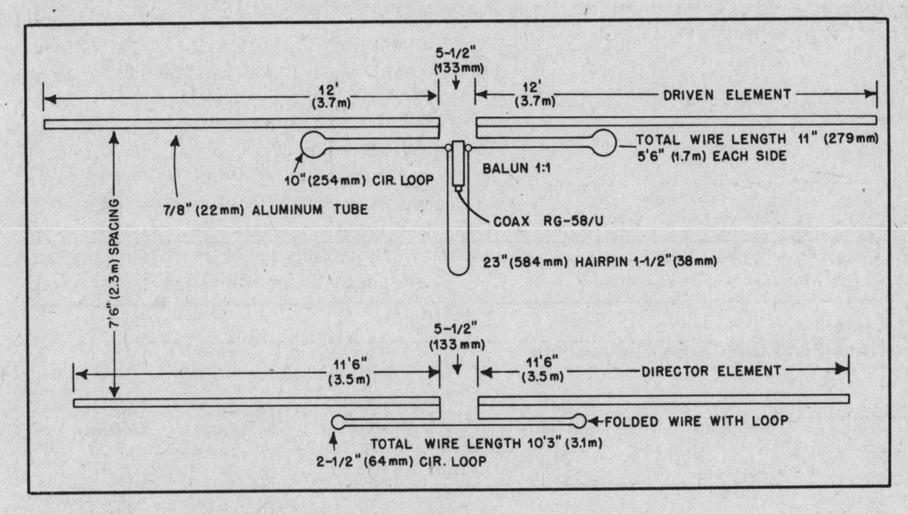


Fig. 13 — Dimensions for the two-element 20-meter beam.

Fig. 14 — The driven element of the linear-loaded beam.

The simplest method is to drill holes through the fiberglass at the approximate points on the arms and route the wires through the holes. Soldering a wire loop across the spreader, as shown later, is recommended. However, care should be taken to prevent solder from flowing to the corner point where flexing could break it.

Dimensions for quad elements and spacing have been given in texts and *QST* over the years. It is generally felt that quads are not very critical in their tuning, nor is element spacing very critical. Table 2 is a collection of dimensions that will suit almost every amateur need for a quad system.

A boom diameter of 2 inches is recommended for systems having two or three elements for 20, 15 and 10 meters. When the boom length becomes 20 feet or longer, as encountered with four- and five-element antennas, a 3-inch diameter boom is highly recommended. Wind creates two forces on the boom, vertical and horizontal. The vertical load on the boom can be reduced with a guy-wire truss cable. The horizontal forces on the boom are more difficult to relieve, and the larger size of 3-inch diameter tubing is desirable.

There are, generally speaking, three grades of materials which can be used for quad spreaders. The least expensive material is bamboo. Bamboo, however, is also the weakest material normally used for quad construction. It has a short life, typically only a few years, and will not withstand a harsh climate very well. Additionally, bamboo is heavy in contrast to fiberglass, which weighs only about a pound per 13-foot length. Fiberglass is the most popular type of spreader material, and will withstand normal winter climates. One step beyond the conventional fiberglass arm is the pole-vaulting arm. For quads designed to be used on 40-meters, surplus "rejected" pole-vaulting poles are highly recommended. Their ability to withstand large amounts of bending is very desirable. The cost of these poles is high, and they are difficult to obtain. Those interested should check with sporting goods manufacturers.

## A THREE-BAND QUAD ANTENNA SYSTEM

Quads have been popular with amateurs during the past few decades because of their light weight, relatively small turning radius, and their unique ability to provide good DX performance even though mounted close to the earth. A two-element, three band quad, for instance, with the elements mounted only 35 feet above ground, will give good performance in situations where a triband Yagi will not. Fig. 15 shows a large quad antenna which can be used as a design basis for either smaller or larger arrays.

Five sets of element spreaders are used to support the three-element 20-meter, four-element 15-meter, and five-element 10-meter wire-loop system. The spacing between elements has been chosen to provide optimum performance consistent with boom length and mechanical construction. Each of the parasitic loops is closed (ends soldered together) and requires no tuning. All of the loop sizes are listed in Table 3, and are designed for a

**Table 2**
**Quad Dimensions**

Two-element quad (WØHTH). Spacing given below; boom length given below.

| | *40 Meters* | *20 Meters* | *15 Meters* | *10 Meters* |
|---|---|---|---|---|
| Reflector | 144' 11-1/2" | 72' 4" | 48' 8" | 35' 7" |
| Driven Element | 140' 11-1/2" | 70' 2" | 47' 4" | 34' 7" |
| Spacing | 30' | 13' | 10' | 6' 6" |
| Boom length | 30' | 13' | 10' | 6' 6" |
| Feed method | Directly with 23' RG-11/U, then any length of RG-8/U coax. | Directly with 11' 7" RG-11/U, then any length RG-8/U coax. | Directly with 7' 8-1/2" RG-11/U, then any length RG-8/U coax. | Directly with 5' 8" RG-11/U, then any length RG-8/U coax. |

(Note that a spider or boomless quad arrangement could be used for the 10/15/20 meter parts of the above dimensions, yielding a triband antenna)

Four-element quad* (WØAIW (20 M) /WØHTH** KØKKU/KØEZH/W6FXB). Spacing: equal; 10 ft. Boom length: 30 ft.

| | *20 Meters* Phone | *20 Meters* CW | *15 Meters* | *10 Meters* |
|---|---|---|---|---|
| Reflector | 72' 1-1/2" | 72' 5" | 48' 8" | 35' 8-1/2" |
| Driven Element | 70' 1-1/2" | 70' 5" | 47' 4" | 34' 8-1/2" |
| Director 1 | 69' 1" | 69' 1" | 46' 4" | 33' 7-1/4" |
| Director 2 | 69' 1" | 69' 1" | 46' 4" | 33' 7-1/4" |
| Feed Method | Directly with 50-ohm coax. | | Directly with 50-ohm coax. | Directly with 5' 9" RG-11/U, then any length RG-8/U coax. |

*Common boom used to form a triband array.
**The 2-element 40-meter quad given above is added to form a four-band quad array.

Four-element quad (WØHTH/K8DYZ*/K8Y1B*/W7EPA*). Spacing: equal; 13' 4". Boom length: 40 ft.

| | *20 Meters* | *15 Meters* | *10 Meters* |
|---|---|---|---|
| Reflector | 72' 5" | 48' 4" | 35' 8-1/2" |
| Driven Element | 70' 5" | 47' 0" | 34' 8-1/2"* |
| Director 1 | 69' 1" | 46' 1" | (Directors 1-3 all |
| Director 2 | 69' 1" | 46' 1" | 33' 7")* |
| Feed method | Directly with 50-ohm coax. | Directly with 7' 9" RG-11/U, then any length 50-ohm coax. | Directly with 50-ohm coax. |

*For the 10-meter band the driven element is placed between the 20/15 reflector and 20/15 driven element. The 10-meter reflector is placed on the same frame as the 20/15-meter reflectors and the remaining 10-meter directors are placed on the remaining 20/15-meter frames. The 10-meter portion is then a 5-element quad.

Six-element quad (WØYDM, W7UMJ). Spacing: equal; 12 ft. Boom length: 60 ft.

| | *20 Meters* |
|---|---|
| Reflector | 72' 1-1/2" |
| Driven Element | 70' 1-1/2" |
| Directors 1, 2 and 3 | 69' 1" |
| Director 4 | 69' 4" |
| Feed Method | Directly with 50-ohm coax. |

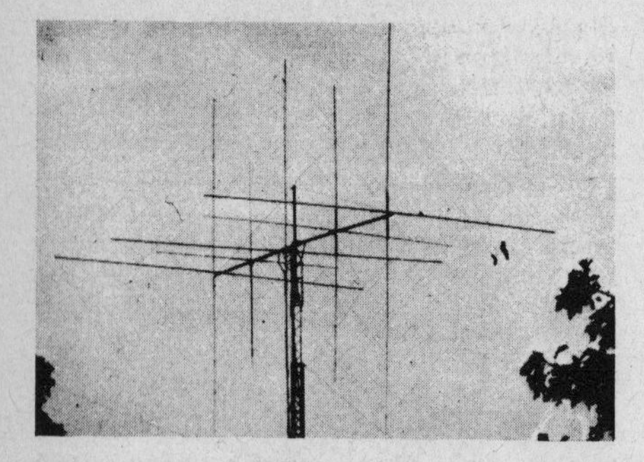

Fig. 15 — The three-band quad antenna.

**Table 3**
**Three-Band Quad Loop Dimensions**

| *Band* | *Reflector* | *Driven Element* | *First Director* | *Second Director* | *Third Director* |
|---|---|---|---|---|---|
| 20 Meters | (A) 72' 8" | (B) 71' 3" | (C) 69' 6" | — | — |
| 15 Meters | (D) 48' 6-1/2" | (E) 47' 7-1/2" | (F) 46' 5" | (G) 46' 5" | — |
| 10 Meters | (H) 36' 2-1/2" | (I) 35' 6" | (J) 34'7" | (K) 34' 7" | (L) 34' 7" |

Letters indicate loops identified in Fig. 16

center frequency of 14.1, 21.1 and 28.3 MHz. Since quad antennas are rather broadtuning devices, excellent performance is obtained in both the cw and ssb band segments of each band (with the possible exception of the very high end of 10 meters). Changing the dimensions to favor a frequency 200 kHz higher in each band to create a "phone" antenna is not necessary.

The most obvious problem related to quad antennas is the ability to build a structurally sound system. If high winds or heavy ice are a normal part of the environment, then special precautions are necessary if the antenna is to survive a winter season. Another stumbling block for would-be quad builders is the installation of a three-dimensional system (assuming a Yagi has only two important dimensions) on top of a tower — especially if the tower needs guy wires for support. With proper planning, however, many of these obstacles can be overcome, i.e., a tram system may be used.

One question which comes up quite often is whether to mount the loops in a diamond or a square configuration. In other words, should one spreader be horizontal to the earth, or should the wire be horizontal to the ground (spreaders mounted in the fashion of an x)? From the electrical point of view, it is probably a

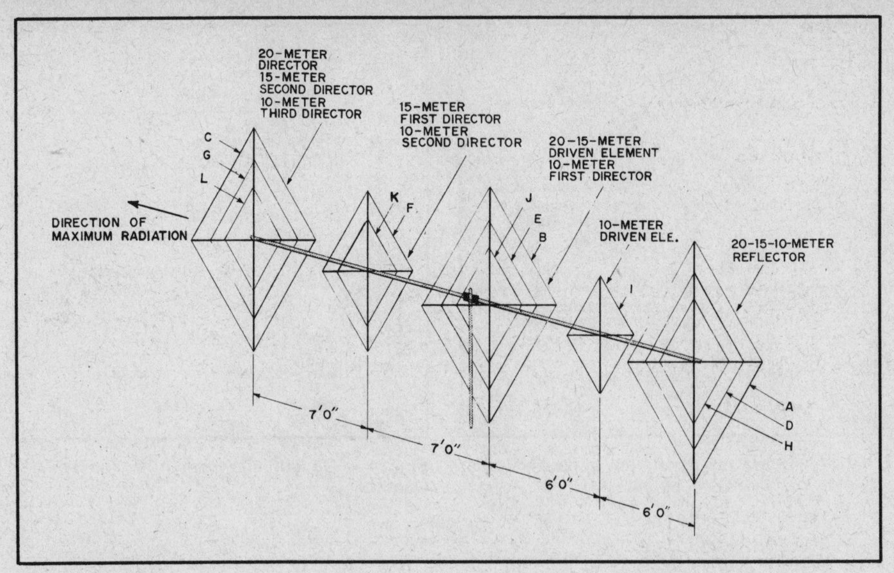


Fig. 16 — Dimensions of the three-band quad, not drawn to scale. See Table 3 for dimensions of lettered wires.

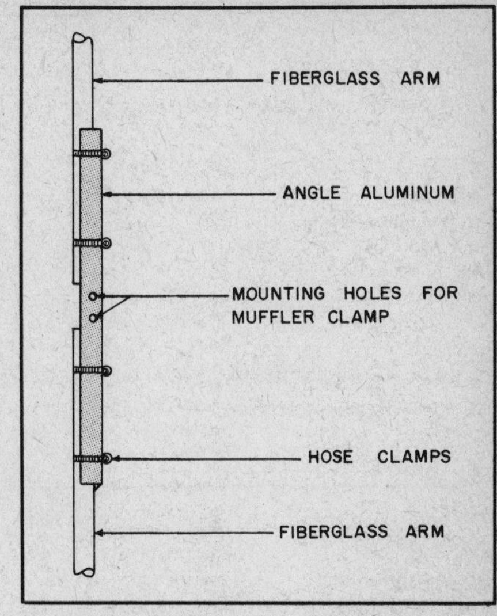


Fig. 17 — Details of one of two assemblies for a spreader frame. The two assemblies are jointed to form an × with a muffler clamp mounted at the position shown.

trade-off. Some authorities indicate that separation of the current points in the diamond system gives slightly more gain than is possible with a square layout. It should be pointed out, however, that there never has been any substantial proof in favor of one or the other, electrically.

From the mechanical point of view there is no question which version is better. The diamond quad, with the associated horizontal and vertical spreader arms, is capable of holding an ice load much better than a system where no vertical support exists to hold the wire loops upright. Stated differently, the vertical poles of a diamond array, if sufficiently strong, will hold the rest of the system erect. When water droplets are accumulating and forming into ice, it is very reassuring to see water running down the wires to a corner and dripping off, rather than just sitting there on the wires and freezing. The wires of the loop (or several loops, in the case of a multiband antenna) help support the horizontal spreaders under a load of ice. A square quad will droop severly under heavy ice conditions because there is nothing to hold it up straight.

Another consideration enters into the selection of a design for a quad. The support itself, if guyed, will require a diamond quad to be mounted a short distance higher on the mast or tower than an equivalent square array if the guy wires are not to interfere with rotation.

The quad array shown in Figs. 15 and 16 uses fiberglass spreaders available from Kirk Electronics, Chester, Connecticut. Bamboo is a suitable substitute (if economy is of great importance). However, the additional weight of the bamboo spreaders over fiberglass is an important consideration. A typical 12-foot bamboo pole weighs about 2 pounds; the fiberglass type weighs less than a pound. By multiplying the difference times eight for a two-element array, 12 times for a three-element antenna, and so on, it quickly becomes apparent that fiberglass is worth the investment if weight is an important factor. Properly treated, bamboo has a useful life of three or four years, while fiberglass life is probably 10 times longer.

Spreader supports (sometimes called spiders) are available from many different manufacturers. If the builder is keeping the cost at a minimum, he should consider building his own. The expense is about half that of a commercially manufactured equivalent and, according to some authorities, the homemade arm supports described below are less likely to rotate on the boom as a result of wind pressure.

A 3-foot long section of 1-inch-per-side steel angle stock is used to interconnect the pairs of spreader arms. The steel is drilled at the center to accept a muffler clamp of sufficient size to clamp the assembly to the boom. The fiberglass is attached to the steel angle stock with automotive hose clamps, two per pole. Each quad-loop spreader frame consists of two assemblies of the type shown in Fig. 17.

Connecting the wires to the fiberglass can be done in a number of different ways. Holes can be drilled at the proper places on the spreader arms and the wires run through them. A separate wrap wire should be included at the entry/exit point to prevent the loop from slipping. Details are presented in Fig. 18. Some amateurs have experienced cracking of the fiberglass, which might be a result of drilling holes through the material. However, this seems to be the exception rather than the rule. The model described here has no holes in the spreader arms; the wires are

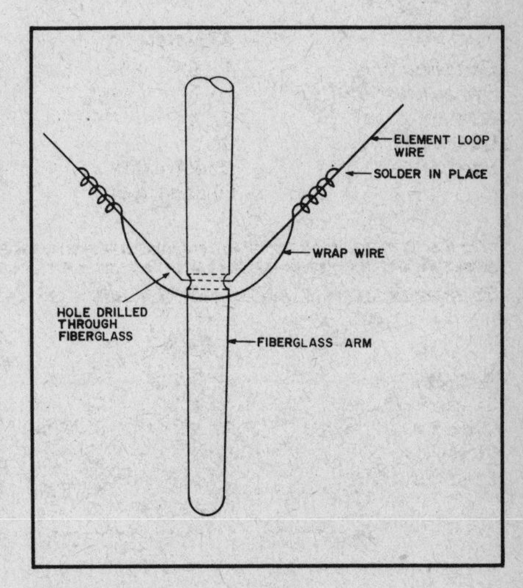


Fig. 18 — A method of assembling a corner of the wire loop of a quad element to the spreader arm.

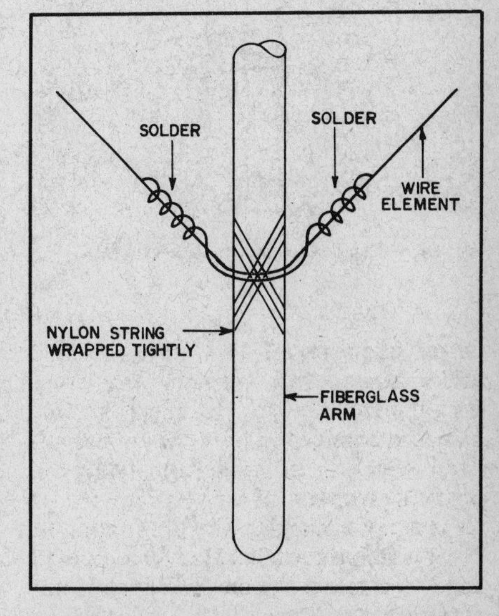


Fig. 19 — An alternative method of assembling the wire of a quad loop to the spreader arm.

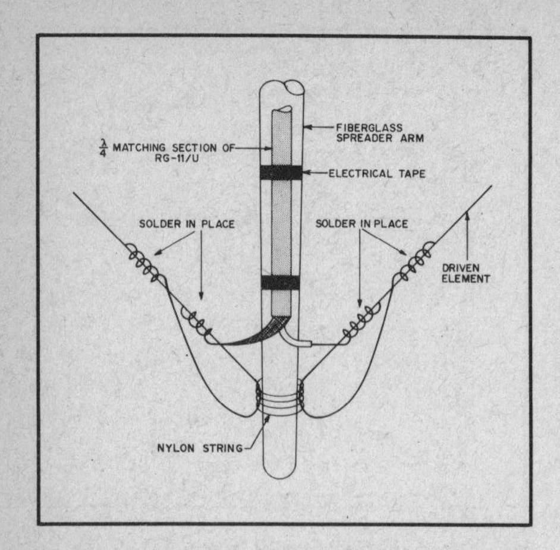


Fig. 20 — Assembly details of the fed element of a quad loop.

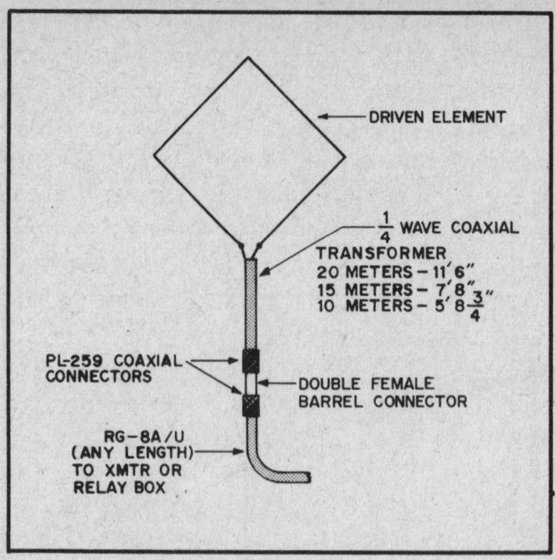


Fig. 23 — Showing installation of quarter-wave 75-ohm transformer section.

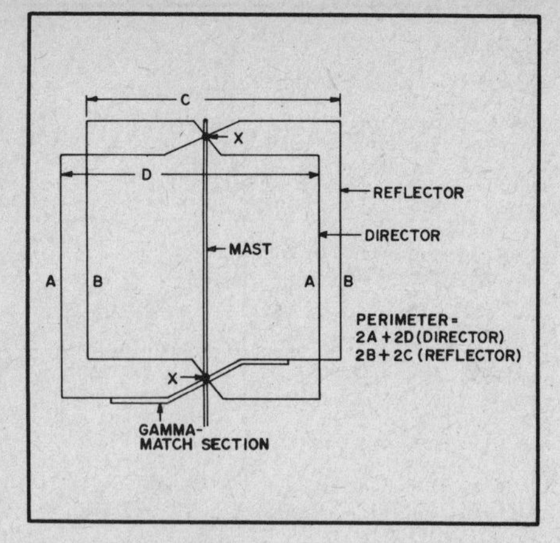


Fig. 24 — General arrangement for the Swiss Quad.

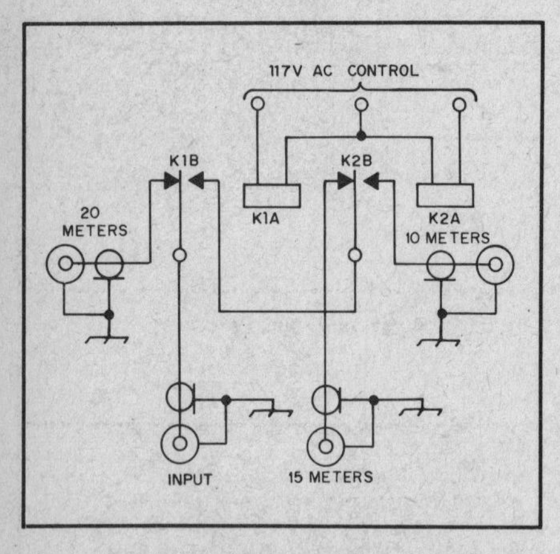


Fig. 21 — Suitable circuit for relay switching of bands for the three-band quad. A 3-wire control cable is required. K1, K2 — any type of relay suitable for rf switching, coaxial type not required (Potter and Brumfeld MR11A acceptable; although this type has double-pole contacts, mechanical arrangements of most single-pole relays make them unacceptable for switching of rf).

Fig. 22 — The relay box is mounted on the boom near the center. Each of the spreader-arm fiberglass poles is attached to steel angle stock with hose clamps.

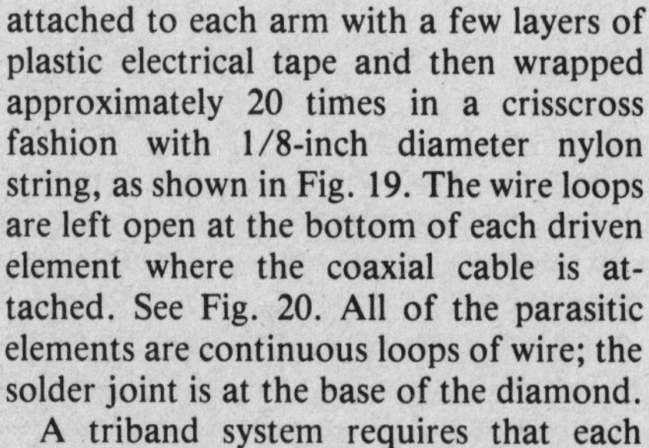

attached to each arm with a few layers of plastic electrical tape and then wrapped approximately 20 times in a crisscross fashion with 1/8-inch diameter nylon string, as shown in Fig. 19. The wire loops are left open at the bottom of each driven element where the coaxial cable is attached. See Fig. 20. All of the parasitic elements are continuous loops of wire; the solder joint is at the base of the diamond.

A triband system requires that each driven element be fed separately. Two methods are possible. First, three individual sections of coaxial cable may be used. Quarter-wave transformers of 75-ohm line are recommended for this service. Second, a relay box may be installed at the center of the boom. A three-wire control system may be used to apply power to the proper relay for the purpose of changing bands. The circuit diagram of a typical configuration is presented in Fig. 21 and its installation is shown in Fig. 22.

The quarter-wave transformers mentioned above are necessary to provide a match between the wire loop and a 50-ohm transmission line. It is simply a section of 75-ohm coax cable placed in series between the 50-ohm line and the antenna feed point, as shown in Fig. 23. A pair of PL-259 connectors and a barrel interconnector may be used to splice the cables together. The connectors and the barrel should be wrapped well with plastic tape and then sprayed with acrylic for protection against the weather.

Every effort must be placed upon proper construction if freedom from mechanical problems is to be expected. Hardware must be secure or vibrations created by the wind may cause unjoining of assemblies. Solder joints should be clamped in place to keep them from flexing, which might fracture a connection point.

## A 10-METER SWISS QUAD

The Swiss Quad is a two-element array with both elements driven. One element is longer than the other and is called the "reflector," while the shorter one is called the "director." Spacing between elements is usually 0.1 wavelength. The impedance of the antenna, using the 0.1-wavelength dimensions, is approximately 50 ohms.

Fig. 24 is a drawing of the components of the beam. In its usual form, lengths of aluminum or copper tubing are bent to form the horizontal members. The element perimeters are completed with vertical wires. At the crossover points (X, Fig. 24), which are connected together, voltage nodes occur.

The formulas for the element sizes are based on the square (perimeter) and *not* the lengths of the wires. For the reflector the perimeter is equal to 1.148 × wavelength, and for the director 1.092 × wavelength, or: Perimeter (inches) =

$$\frac{984}{\text{f MHz}} \times 12 \times 1.148 \text{ (reflector)}$$

As an example: Perimeter for 28.1 MHz

$$= \frac{984}{28.1} = 35;\ 35 \times 12 = 420.2$$

$$420.2 \times 1.148 = 482 \text{ in. (reflector)}$$

These formulas apply only to the use of horizontal members of aluminum or copper tubing. Using PVC tubing and wire elements, the overall lengths of the perimeters are different and the correct lengths given later were determined experimentally.

One of the advantages of this antenna over the more conventional quad type is that plumber's delight type construction can be used. This means that both elements, at the top and bottom of the beam, can be grounded to the supporting mast. The structure is lightweight but strong, and an inexpensive TV rotator carries it nicely. Another feature is the small turning radius, which is less than half that of a 3-element Yagi.

**Table 4**
**Materials List, Swiss Quad**

Four 10-ft lengths 1/2-inch rigid PVC pipe.
Two 10-ft lengths 3/4-inch rigid PVC pipe.
One 10-ft length 1-inch rigid PVC pipe.
Twelve feet 1-1/8-inch or larger steel or aluminum tubing.
Epoxy cement (equal parts of resin and hardener).
100 ft annealed copper wire, 14 or 15 gauge.

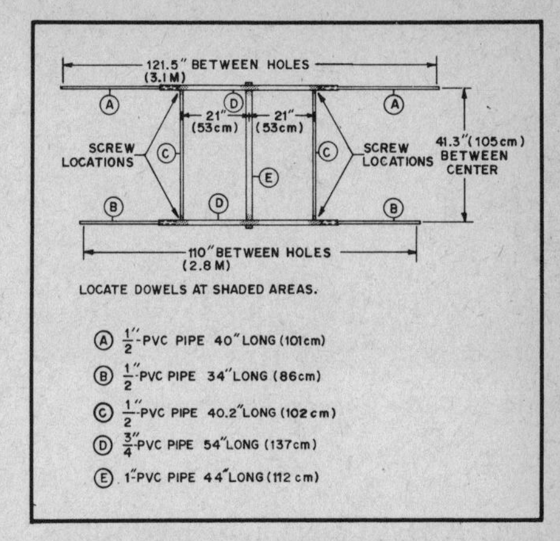


Fig. 25 — Dimensions and layout of the insulating frame.

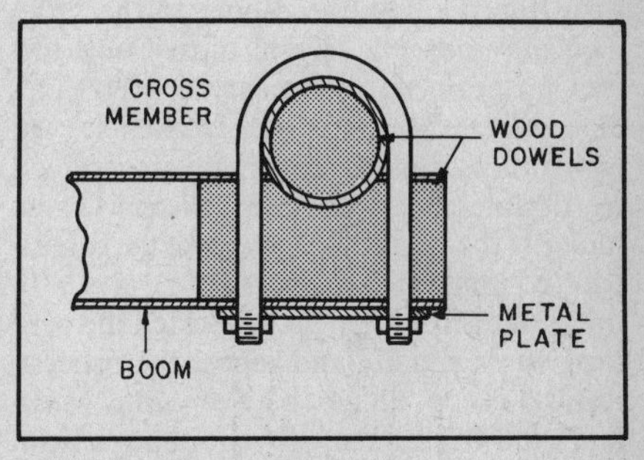


Fig. 26 — Boom end-joint detail.

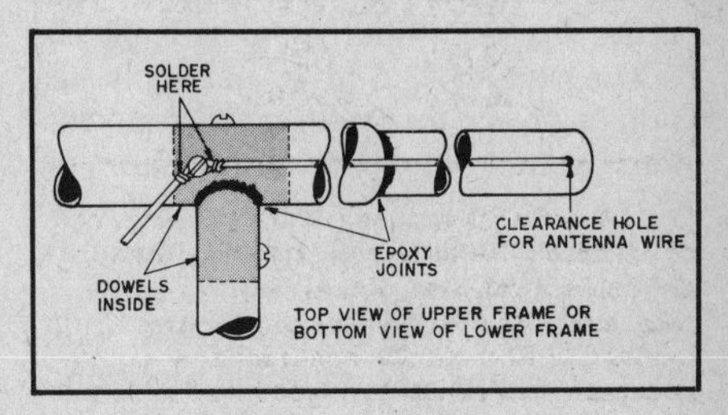


Fig. 27 — Details of the frame and wire assembly.

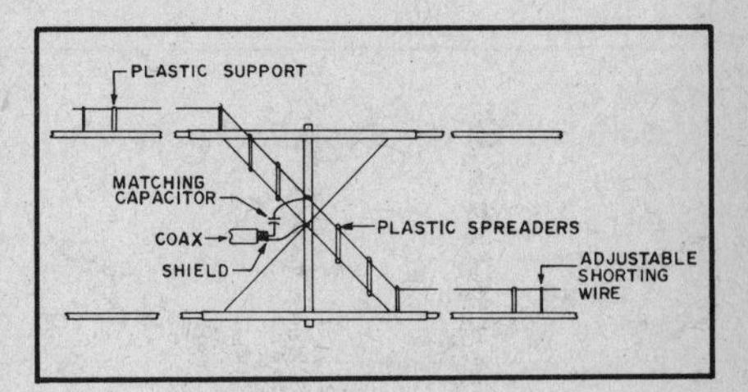


Fig. 28 — Details of the double gamma match.

The antenna described below is made entirely of wire that is supported by two insulating frames constructed from rigid plastic water pipe. Rigid polyvinylchloride (PVC) water pipe is readily available from plumbing supply houses and from the large mail-order firms. The standard 10-foot lengths are just right for building the 10-meter Swiss Quad. You can cut and drill PVC pipe with woodworking tools. PVC plastic sheds water, an advantage where winter icing is a problem. Heat from the intense summer sun has not softened or deformed the original quad structure.

To build the wire version of the Swiss Quad you will need the materials listed in Table 4 plus some wood screws and U bolts. Also required are a few scraps of wood dowel rod and some old toothbrushes.

Cut the PVC pipe to the lengths shown in Fig. 25. Also cut several short lengths of dowel rod for reinforcement at the points indicated. These are held in place by means of epoxy cement. The bond is improved if the PVC surface is roughened with sandpaper and wiped clean before the cement is applied. A tack inserted through a tiny hole in the pipe will hold each dowel in place while the epoxy cures.

Reasonable care is required in forming the boom end joints so that the two sections of 3/4-inch pipe are parallel. The joining method used at WØERZ is illustrated in Fig. 26. Parallel depressions were filed near each end of each boom with a half-round rasp. These cradles are about 0.4 inch deep and their centers are 41.3 inches apart. Holes are drilled for the U bolts and the joints are completed with the U bolts and the epoxy cement. Draw the bolts snug, but not so tight as to damage the PVC pipe. Final assembly of the insulating frames should be done on a level surface. Chalk an outline of the frame on the work surface so that any misalignment will be easy to detect and correct. If the 1/2-inch pipe sections fit too loosely into the lateral members, shim them with two bands of masking tape before applying the epoxy cement.

Supports for the gamma-matching section can be made from old toothbrush handles or other scraps of plastic. Space the supports about 10 inches apart so that they support the gamma wire 2.5 inches on top of the lower PVC pipe. Attach the spacers with epoxy cement. Strips of masking tape can be used to hold the spacers in place while the epoxy is curing.

There are several ways to attach the frames to the vertical mast. The mounting hardware designed for the larger TV antennas should be quite satisfactory. Metal plates about 5 inches square can be drilled to accept four U bolts. Two U bolts should be used around the boom and two around the mast. A piece of wooden dowel inside the center of the boom prevents crushing the PVC pipe when the U bolts are tightened. The plates should not interfere with the element wires that must cross at the exact center of the frame. A 12-foot length of metal tubing serves as the vertical support. The galvanized steel tubing used as a top rail in chain link fences would be satisfactory.

When the epoxy resin has fully cured, you are ready to add the wire elements to produce the configuration shown in Fig. 24. Start on the top side of the upper frame. Cut two pieces of copper wire (no. 14 or larger) at least 30.5 feet long and mark their centers. Thread the ends downward through holes spaced as shown in Fig. 25 so that the wires cross at the top of the upper frame. Following the detail in Fig. 27, drill pilot holes through the PVC pipe and drive four screws into the dowels. The screws must be 41.3 inches apart and equidistant from the center of the frame. With the centers of the two wires together, bend the wires 45 degrees around each screw and anchor with a short wrap of wire. Now pull the wires through the holes at the ends of the pipes until taut. A soldered wire wrap just below each hole prevents the element wires from sliding back through the holes.

Attach the wired upper frame about two feet below the top of the vertical mast. Make a bridle from stout nylon cord (or fiberglass-reinforced plastic clothesline), tying it from the top of the mast to each of four points on the upper frame to reduce sagging.

Now cut two 11.5 foot lengths of wire and attach them to the bottom of the lower frame. Also cut a 9-foot length for the gamma-matching section. If insulated wire is used, bare six inches at each end of the gamma wire. Details of the double gamma match are shown in Fig. 28. Attach the wired lower frame to the mast about nine feet below the upper frame and parallel to it. The ultimate spacing between the upper and lower frames, determined during the tuning process, will result in moderate tension in the vertical wires. Join the vertical wires to complete the elements of your Swiss Quad. All vertical wires must be of equal length. Do not solder the wire joints until you have tuned the elements.

For tuning and impedance matching you will need a dip meter, a VSWR indicator, and the station receiver and exciter. Stand the Swiss Quad vertically in a clear space with the lower frame at least two feet above ground. Using the dipper as a resonance indicator, prune a piece of 50-ohm coaxial cable to an integral multiple of a half-wavelength at the desired frequency. RG-8/U and RG-58/U with polyethylene insulation have a velocity propagation factor of 0.66. At 28.6 MHz, a half-wavelength section (made from the above cables ) is approximately 11.35 feet

Fig. 29 — The log-periodic dipole array.

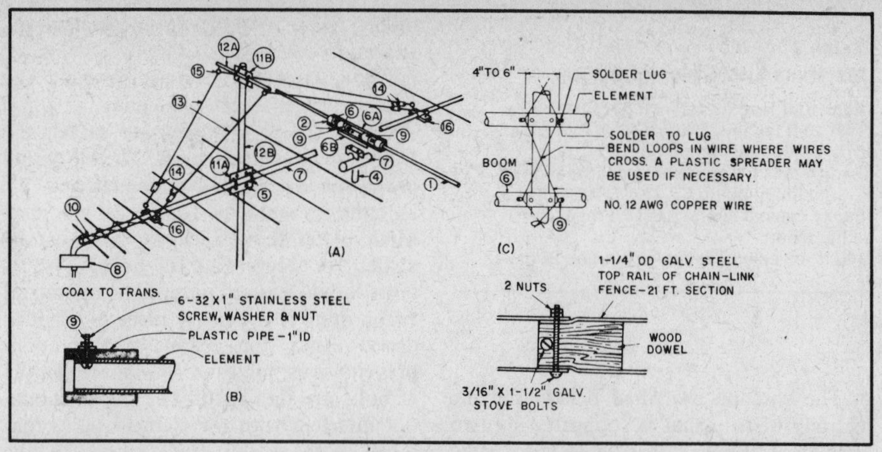


Fig. 30 — Construction diagram of the log-periodic array. At B and C are shown the method of making electrical connection to each half element, and at D is shown how the boom sections are joined.

long. (Coaxial cable using polyfoam insulation has a velocity factor of 0.80.) Connect one end to the midpoint of the gamma section and the other to a two-turn link. Couple the dipper to the link. You may observe several dips. Look for two pronounced dips near 26 MHz and 31.4 MHz. Measure the frequencies at which these dips occur using your receiver to double-check the dip meter. Then multiply the frequencies and take the square root of this product; that is $\sqrt{f1 \times f2}$. If the result is less than 28.6, shorten the vertical wires equally and repeat the process until $\sqrt{f1 \times f2}$ lies between 28.6 MHz and 28.8 MHz. Your Swiss Quad is now tuned for the 10-meter band.

Remove the link and connect the VSWR bridge in its place. Connect your exciter to the input terminals of the bridge, tune to 28.6 MHz and apply just enough power to obtain a full-scale forward voltage indication. Measure the VSWR. Now slide the two shorting wires of the matching section to new positions, equidistant from the center of the wire elements, and measure the VSWR. Continue adjusting the shorting wires until minimum VSWR is obtained. Insert a 100-pF variable capacitor between the center conductor of the coaxial cable feeder and the midpoint of the gamma wire. Adjust the capacitor for minimum VSWR indication. It may be necessary to readjust both the shorting wires and the capacitor to obtain a satisfactory impedance match. With patience, a perfect match (VSWR = 1:1) can be achieved. Solder the shorting wires.

The variable capacitor may be replaced with a short length of RG-59/U coaxial cable. Each foot of this cable has a capacitance of approximately 20 pF. Measure or estimate the value to which the variable capacitor was finally set, add ten percent, and cut a corresponding length of RG-59/U. Solder the shield braid to the midpoint of the gamma wire and the center wire to the center conductor of the 50-ohm transmission line, leaving the other end of the coaxial-cable capacitor open. You will probably observe that the VSWR has increased. Snip short lengths from the open end of the capacitor until the original low VSWR is obtained. When the antenna is raised to 40 feet the VSWR should be less than 1.5:1 over the entire 10-meter band.

**Table 5**
**Array Dimensions, Feet**

| *El. no.* | $l_n$ | *h* | $d_{n-1,n}$ *(spacing)* | *Nearest Resonant* |
|---|---|---|---|---|
| 1 | 38.0 | 19 | 0 | |
| 2 | 34.2 | 17.1 | 3.862 = $d_{12}$ | 14 MHz |
| 3 | 30.78 | 15.39 | 3.475 = $d_{23}$ | |
| 4 | 27.7 | 13.85 | 3.13 = $d_{23}$ | |
| 5 | 24.93 | 12.465 | 2.815 = $d_{23}$ | |
| 6 | 22.44 | 11.22 | 2.533 = $d_{23}$ | 21 MHz |
| 7 | 20.195 | 10.098 | 2.28 = $d_{23}$ | |
| 8 | 18.175 | 9.088 | 2.05 = $d_{23}$ | |
| 9 | 16.357 | 8.179 | 1.85 = $d_{23}$ | 28 MHz |
| 10 | 14.72 | 7.36 | 1.663 = $d_{23}$ | |
| 11 | 13.25 | 6.625 | 1.496 = $d_{23}$ | |
| 12 | 11.924 | 5.962 | 1.347 = $d_{11,12}$ | |

**Table 6**
**Materials List, Log-Periodic Dipole Array**

| *Material Description* | *Quantity* |
|---|---|
| 1) Aluminum tubing — .047" wall thickness | |
| 1" — 12' or 6' lengths | 126 lineal feet |
| 7/8" — 12' lengths | 96 lineal feet |
| 7/8" — 6' or 12' lengths | 66 lineal feet |
| 3/4" — 8' lengths | 16 lineal feet |
| 2) Stainless-steel hose clamps — 2" max. | 48 ea. |
| 3) Stainless-steel hose clamps — 1-1/4" max. | 26 ea. |
| 4) TV-type U bolts | 14 ea. |
| 5) U-bolts, galv. type | |
| 5/16" × 1-1/2" | 4 ea. |
| 1/4" × 1" | 2 ea. |
| 6) 1" ID polyethelene water-service pipe — 160 psi test, approx. 1-1/4" OD | 20 lineal feet |
| A) 1-1/4" × 1-1/4" × 1/8" aluminum angle — 6' lengths | 30 lineal feet |
| B) 1" × 1/4" aluminum bar — 6' lengths | 12 lineal feet |
| 7) 1-1/4" top rail of chain-link fence | 26.5 lineal feet |
| 8) 1:1 toroid balun | 1 ea. |
| 9) 6 — 32 × 1" stainless-steel screws | 24 ea. |
| 6 — 32 stainless-steel nuts | 48 ea. |
| No. 6 solder lugs | 24 ea. |
| 10) No. 12 copper feeder wire | 60 lineal feet |
| 11) | |
| A) 12" × 8" × 1/4" alum. plate | 1 ea. |
| B) 6" × 4" × 1/4" alum. plate | 1 ea. |
| 12) | |
| A) 3/4" galv. pipe | 3 lineal feet |
| B) 1" galv. pipe — mast | 5 lineal feet |
| 13) Galv. guy wire | 50 lineal feet |
| 14) 1/4" × 2" turnbuckles | 4 ea. |
| 15) 1/4" × 1-1/2" eye bolts | 2 ea. |
| 16) TV guy clamps and eye bolts | 2 ea. |

**Table 7**
**Element Material Requirements, Log Periodic Dipole Array**

| El. No. | 1" tubing Lth. | Qty. | 7/8" tubing Lth. | Qty. | 3/4" tubing Lth. | Qty. | 1-1/4" angle Lth. | 1" bar Lth. |
|---|---|---|---|---|---|---|---|---|
| 1 | 6' | 2 | 6' | 2 | 8' | 2 | 3' | 1' |
| 2 | 6' | 2 | 12' | 2 | — | — | 3' | 1' |
| 3 | 6' | 2 | 12' | 2 | — | — | 3' | 1' |
| 4 | 6' | 2 | 8.5' | 2 | — | — | 3' | 1' |
| 5 | 6' | 2 | 7' | 2 | — | — | 3' | 1' |
| 6 | 6' | 2 | 6' | 2 | — | — | 3' | 1' |
| 7 | 6' | 2 | 5' | 2 | — | — | 2' | 1' |
| 8 | 6' | 2 | 3.5' | 2 | — | — | 2' | 1' |
| 9 | 6' | 2 | 2.5' | 2 | — | — | 2' | 1' |
| 10 | 3' | 2 | 5' | 2 | — | — | 2' | 1' |
| 11 | 3' | 2 | 4' | 2 | — | — | 2' | 1' |
| 12 | 3' | 2 | 4' | 2 | — | — | 2' | 1' |

Tape the capacitor to the PVC pipe boom, then wrap a few bands of tape around the sections where the wires run along the sides of the pipes. Check the soldered joints and mechanical connections. Coat the soldered joints and the cable ends with a weatherproof sealing compound (e.g., silicone bathtub caulk) and hoist your new Swiss Quad up the tower.

## A LOG-PERIODIC DIPOLE ARRAY

The antenna system shown in Figs. 29 and 30 was originally described in November 1973 *QST*. Additional information on the design of a log-periodic dipole array (LPDA) is given in Chapter 6.

The characteristics of the triband antenna are:

Frequency range, 13-30 MHz
Half-power beamwidth, 43° (14 MHz)
Operating bandwidth, B = 30/13 = 2.3
Design parameter $\tau$ = 0.9
Relative element spacing constant $\sigma$ = .05
Apex half-angle $\propto$ = 25°, cot $\propto$ = 2.0325
Bandwidth of active group, $B_{ar}$ = 1.4
Bandwidth of structure, $B_s$ = 3.22
Boom length, L = 26.5 ft
Longest element $\ell_1$ = 38 ft (a tabulation of element lengths and spacings is given in Table 5)
Total weight, 116 pounds
Wind-load area, 10.7 sq. ft
Required input impedance (mean resistance), $R_o$ = 67 ohms, Zt = 6-inch jumper no. 18 wire
Average characteristic dipole impedance: $Za_{14\ MHz}$ = 450 ohms; $Za_{21\ MHz}$ = 420 ohms; $Za_{28\ MHz}$ = 360 ohms
Mean spacing factor $\sigma'$ = .0527
Impedance of the feeder: $Zo_{14\ MHz}$ = 95 ohms; $Zo_{21\ MHz}$ = 97 ohms; $Zo_{28\ MHz}$ = 103 ohms
Using a toroid balun at the input terminals and a 72-ohm coax feeder the SWR is 1.4 to 1 (maximum).

The mechanical assembly uses materials readily available from most local hardware stores or aluminum supply houses. The materials needed are given in Table 6. In the construction diagram, Fig. 30, the materials are referenced by their respective material list number. The photograph shows the overall construction picture, and the drawings show the details. Table 7 gives the required tubing lengths to construct the elements.

## THE TELERANA

The Telerana (Spanish for "spiderweb") is a rotatable log-periodic antenna that is lightweight, easy to construct and relatively inexpensive to build. The array consists of 13 dipole elements properly spaced and transposed along an open-wire, interelement feeder having an impedance of approximately 400 ohms. See Figs. 31 and 32. The array is fed at the forward (smallest) end with a 4:1 balun and RG-8/U cable placed inside the front arm and leading to the transmitter. An alternative feed method is to use open wire or ordinary TV cable and a tuner, eliminating the balun. The direction of gain or forward lobe is away from the small end.

The frame (Fig. 33) used to support the array consists of four 15-ft fiberglass vaulting poles slipped over short nipples at the hub, appearing like wheel spokes (Fig. 34). Instead of being mounted directly into the fiberglass, short metal tubing sleeves are inserted into the outer ends of the arms and the necessary holes drilled to receive the wires and nylon.

A shopping list is provided in Table 8. The center hub is made from a 1-1/4-inch

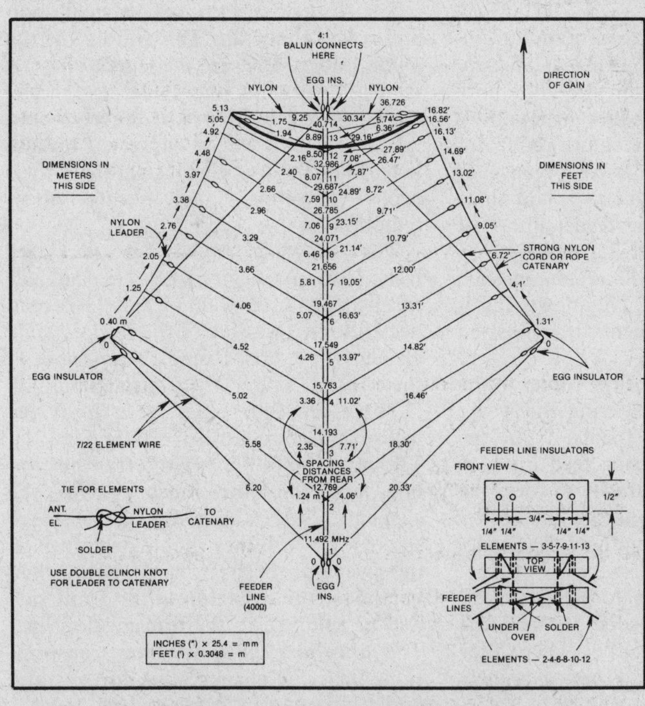


Fig. 31 — Configuration of the YV5DLT spiderweb antenna. Nylon monofilament line is used from the ends of the elements to the nylon cords. Solder all metal-to-metal connections. Use nylon line to tie every point where the lines cross. The forward fiberglass feeder lies on the feeder line and is tied to it. Note that both metric and English measurements are shown except for the illustration of the feed-line insulator. Use soft-drawn copper wire for elements 2 through 12. Element 1 should have no. 7/22 flexible wire or no. 14 Copperweld.

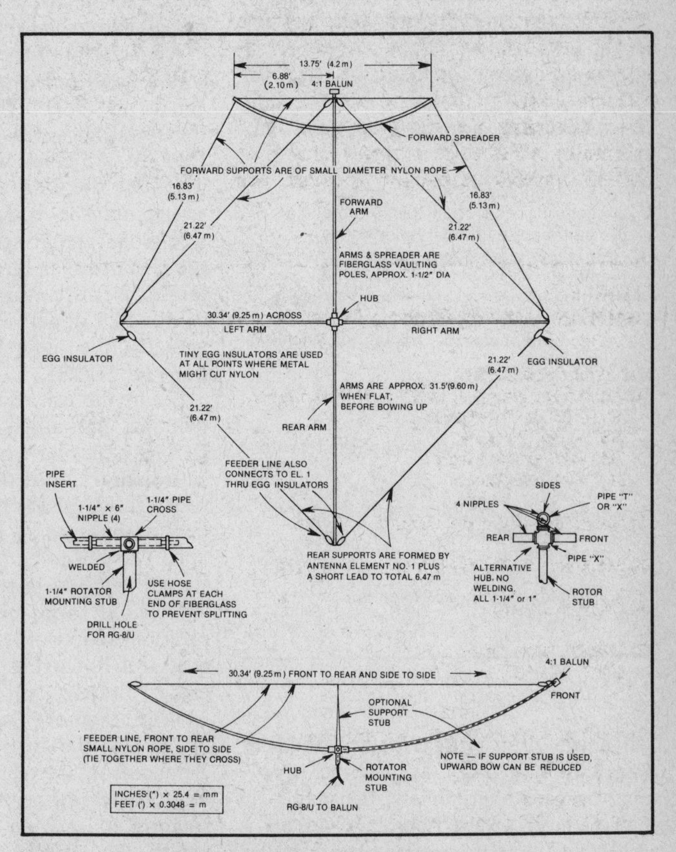


Fig. 32 — The frame construction for the YV5DLT spiderweb antenna. Two different hub arrangements are illustrated.

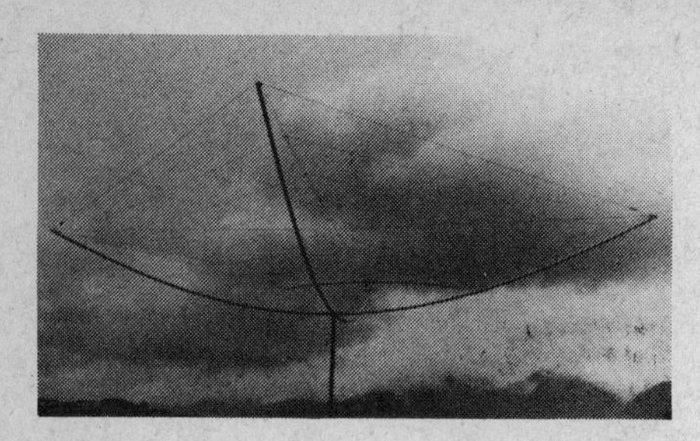

Fig. 33 — The spiderweb antenna, as shown in this somewhat deceptive photo, might bring to mind a rotatable clothesline. Of course it is much larger than the clothesline is, as indicated by Figs. 31 and 32. It can be lifted by hand.

Fig. 34 — The simple arrangement of the hub of the YV5DLT spiderweb. See Fig. 31 and the text for details.

**Table 8**
**Shopping List for the Telerana**

1 — 1-1/4-inch galvanized, 4-outlet cross or X.
4 — 8-inch nipples.
4 — 15-ft long arms. Vaulting poles suggested. These must be strong and all of the same strength (150 lb) or better.
1 — spreader, 14.8 ft long (must not be metal).
1 — 4:1 balun unless open-wire or TV cable is used.
12 — feed-line insulators made from Plexiglas or fiberglass.
36 — small egg insulators.
328 ft copper wire for elements, flexible 7-22 is suggested.
65.6 ft (20 m) no. 14 Copperweld wire for interelement feed line.
164 ft (50 m) strong 1/8-inch dia cord.
1 — roll of nylon monofilament fishing line, 50 lb test or better.
4 — metal tubing inserts to go into the ends of the fiberglass arms.
2 — fiberglass fishing-rod blanks.
4 — hose clamps.

galvanized four-outlet cross or X and four 8-inch nipples (Fig. 34). A 1-inch dia X may be used alternatively, depending on the diameter of the fiberglass. A hole is drilled in the bottom of the hub to allow the cable to be passed through after welding the hub to the rotator mounting stub.

All four arms of the array must be 15 feet long. They should be strong and springy for maintaining the tautness of the array. If vaulting poles are used, try to obtain all of them with identical strength ratings.

The front spreader should be approximately 14.8 feet long. It can be much lighter than the four main arms, but must be strong enough to keep the lines rigid. If tapered, the spreader should have the same measurements from the center to each end. *Do not use metal for this spreader.*

Building the frame for the array is the first construction step. Once that is prepared, then everything else can be built onto it. Assemble the hub and the four arms, letting them lie flat on the ground with the rotator stub inserted into a hole in the ground. The tip-to-tip length should be about 31.5 feet each way. A hose clamp is used at each end of the arms to prevent splitting. Insert the metal inserts at the outer ends of the arms, with 1 inch protruding. The mounting holes should have been drilled at this point. If the egg insulators and nylon cords are mounted to these tube inserts, the whole antenna can be disassembled simply by bending up the arms and pulling out the inserts with everything still attached.

Choose the arm to be at the front end. Mount two egg insulators at the front and rear to accommodate the interelement feeder. These insulators should be as close as possible to the ends.

At each end of the crossarm on top, install a small pulley and string nylon cord across and back. Tighten the cord until the upward bow reaches 39.4 inches above the hub. All cords will require retightening after the first few days because of stretching. The crossarm can be laid on its side while preparing the feeder line. For the front-to-rear bow-string, it is important to use a wire that will not stretch such as no. 14 Copperweld. This bowstring is actually the interelement transmission line. See Fig. 35.

Secure the rear ends of the feeder to the two rear insulators, soldering the wrap. Before securing the fronts, slip the 12-insulators onto the two feed lines. A rope can be used temporarily to form the bow and to aid in mounting the feeder line. The end-to-end length of the feeder should be 30.24 feet.

Now, lift both bows to their upright position and tie the feeder line and the crossarm bowstring together where they cross, directly over and approximately 39.4 inches above the hub.

The next step is to install the no. 1 rear element from the rear egg insulators to the right and left crossarms using other egg insulators to provide the proper element length. Be sure to solder the element halves to the transmission line. Complete

Fig. 35 — The elements, balun, transmission line and main bow of the spiderweb antenna.

this portion of the construction by installing the nylon cord catenaries from the front arm to the crossarm tips. Use egg insulators where needed to prevent cutting the nylon cords.

In preparing the fiberglass front spreader, keep in mind that it should be 14.75 feet long before bowing and is approximately 13.75 feet when bowed. Secure the center of the bowstring to the end of the front arm. Lay the spreader on top of the feed line, then tie the feeder to the spreader with nylon fish line. String the catenary from the spreader tips to the crossarm tips.

At this point of assembly antenna elements 2 through 13 should be prepared. There will be two segments for each element. At the outer tip make a small loop and solder the wrap. This will be for the nylon leader. Measure the length plus 0.4 inch for wrapping and soldering the element segment to the feeder. Seven-strand no. 22 antenna wire is suggested for use here. Slide the feed-line insulators to their proper position and secure them temporarily.

The drawings show the necessary transposition scheme. Each element half of elements 3, 5, 7, 9, 11 and 13 is connected to its own side of the feeder, while elements 2, 4, 6, 8, 10 and 12 cross over to the opposite side of the transmission line.

There are four holes in each of the transmission-line insulators (see Fig. 31). The inner holes are for the transmission line, and the outer ones are for the elements. Since the array elements are slanted forward, they should pass through the insulator from front to back, then back over the insulator to the front side and be soldered to the transmission line. The drawings show how the transpositions have the element end go over and under the opposite line.

Everywhere lines cross, they are tied together with nylon line, whether copper/nylon or nylon/nylon. This makes the array much more rigid. All elements should be mounted loosely before you try to align the whole thing. Tightening any line or element affects all others. There will be plenty of walking back and forth before the array is aligned properly. Do not expect it to be very taut.

## THE LOG-YAG ARRAY

The Log-Yag array utilizes an LPDA-driven group of elements, designed to cover a desired bandwidth, in conjunction with parasitic elements to achieve higher gains and greater directivity than would be realized with either the LPDA or Yagi array alone. The Yagi array requires a long boom and wide element spacing for wide bandwidth and high gain. This is because the Q of the Yagi system increases as the number of elements is increased and/or as the spacing between adjacent elements is decreased. An increase in the Q of the Yagi array means that the total bandwidth of that array is decreased, and optimum gain, front-to-back ratio and side lobe rejection are obtainable only over small portions of the band.

The Log-Yag system overcomes this difficulty by using a multiple driven element "cell" designed in accordance with the principles of the log-periodic dipole array. Since this log cell exhibits both gain and directivity by itself, it is a more effective radiator than a simple dipole driven element. The front-to-back ratio and gain of the log cell can be improved with the addition of a parasitic reflector and director. It is not necessary for the parasitic element spacings to be large with respect to wavelength, as in the Yagi array, since the log cell is the determining factor in the array bandwidth. In fact, the element spacings within the log cell may be small with respect to a wavelength without appreciable deterioration of the cell gain. For example, decreasing the relative spacing constant ($\sigma$) from 0.1 to 0.5 $\lambda$ will decrease the gain by less than 1 dB.

The array design takes the form of a 4-element log cell, parasitic reflector spaced at 0.085 $\lambda_{max}$ and parasitic director spaced at 0.15 $\lambda_{max}$ where $\lambda_{max}$ is the longest free-space wavelength within the array passband. It has been found that array gain is almost unaffected with reflector spacings from 0.08 $\lambda$ to 0.25 $\lambda$ and the increase in boom length is not justified. The function of the reflector is to improve the front-to-back ratio of the log cell while the director sharpens the forward lobe and decreases the half-power beamwidth. As the spacing between the parasitic elements and the log cell decreases, the parasitic elements must increase in length.

The log cell is designed to meet upper and lower band limits with $\sigma = 0.05\ \lambda$. The design parameter $\tau$ is dependent on the structure bandwidth, $B_s$. When the log-periodic design parameters have been found, the element length and spacings can be determined. Further discussion of log-periodic antennas can be found in Chapter 6.

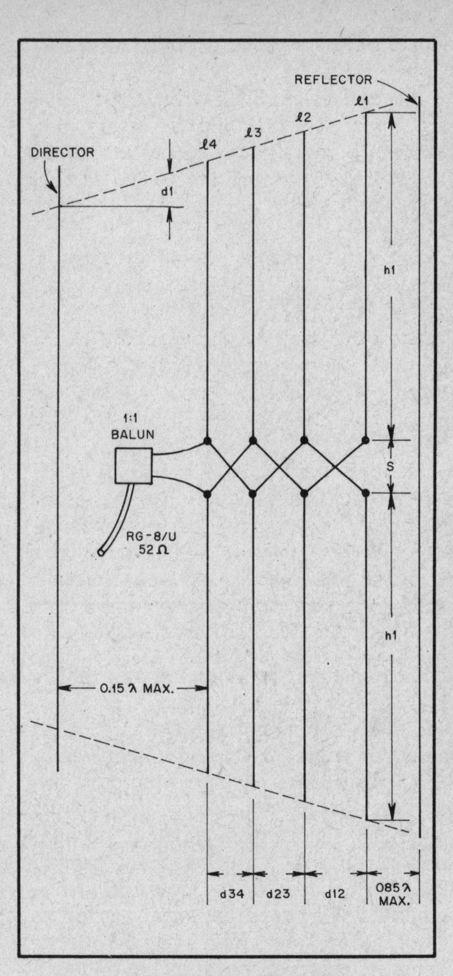


Fig. 36 — Layout of the Log-Yag array.

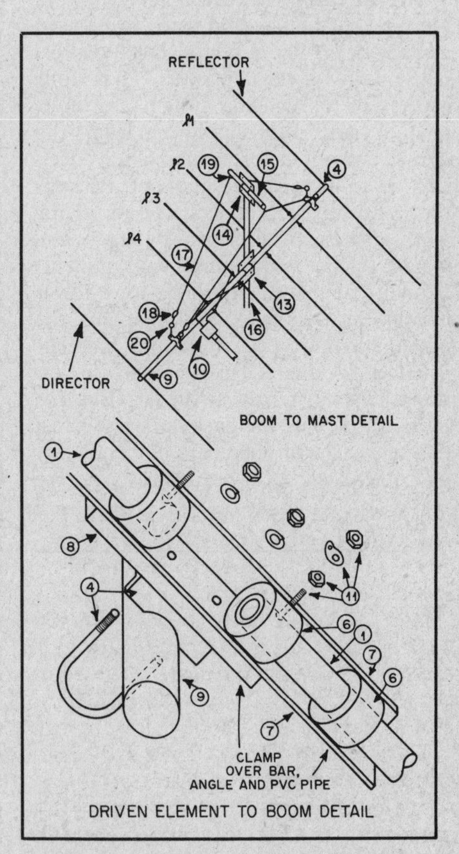


Fig. 37 — Assembly details. The numbered components refer to Table 12.

Fig. 38 — The attachment of the elements to the boom.

Fig. 39 — From the front to the back of the Log-Yag array. Note the truss provides lateral and vertical support.

Array layout and construction details can be seen in Figs. 36 through 39. Characteristics of the array are given in Table 9.

The method of feeding the antenna is identical to that of feeding the log-periodic dipole array without the parasitic elements. As shown in Fig. 36, a balanced feeder is required for each log-cell element, and all adjacent elements are fed with a 180° phase shift by alternating connections. Since the Log-Yag array will be covering a relatively small bandwidth, the radiation resistance of the narrow-band log cell will vary from 80 to 90 ohms (tubing elements) depending on the operating bandwidth. The addition of parasitic elements lowers the log-cell radiation resistance. Hence, it is recommended that a 1-to-1 balun be connected at the log-cell input terminals and 52-ohm coaxial cable be used for the feed line. The measured radiation resistance of the 14-MHz Log-Yag is 37 ohms, 14.0 to 14.35 MHz. It is assumed that tubing elements will be used. However, if a wire array is used then the radiation resistance $R_o$ and antenna-feeder input impedance $Z_o$ must be calculated so that the proper balun and coax may be used. The procedure is outlined in detail in Chapter 6.

Table 10 has array dimensions. Tables 11 and 12 contain lists of the materials

**Table 9**
**Log-Yag Array Characteristics**

| | |
|---|---|
| 1) Frequency range | 14-14.35 MHz |
| 2) Operating bandwidth | B = 1.025 |
| 3) Design parameter | $\tau$ = .946657 |
| 4) Apex half angle | $a$ = 14.92°, cot = 3.753 |
| 5) Half-power beam width | 42° (14-14.35 MHz) |
| 6) Bandwidth of structure | $B_s$ = 1.17875 |
| 7) Free-space wavelength | $\lambda_{max}$ = 70.28 |
| 8) Log cell boom length | L = 10.0 ft |
| 9) Longest log element | $\ell 1$ = 35.14 ft (a tabulation of element lengths and spacings given in Table 10) |
| 10) Forward gain over dipole | 11.5 dB (theoretical) |
| 11) Front-to-back ratio | 32 dB (theoretical) |
| 12) Front-to-side ratio | 45 dB (theoretical) |
| 13) Input impedance | $Z_o$ = 37 ohms |
| 14) SWR | 1.3 to 1 (14 — 14.35 MHz) |
| 15) Total weight | 96 pounds |
| 16) Wind-load area | 8.5 sq. ft |
| 17) Feed-point impedance | $Z_o$ = 37 ohms |
| 18) Reflector length | 36.4 ft @ 6.0 ft spacing |
| 19) Director length | 32.2 ft @ 10.5 ft spacing |
| 20) Total boom length | 26.5 ft |

**Table 10**
**Log-Yag Array Dimensions**

| *Element* | *Length Feet* | *Spacing Feet* |
|---|---|---|
| Reflector | 36.4 | 6.0 (Ref. to $\ell 1$) |
| $\ell 1$ | 35.14 | 3.51 (d12) |
| $\ell 2$ | 33.27 | 3.32 (d23) |
| $\ell 3$ | 31.49 | 3.14 (d34) |
| $\ell 4$ | 29.81 | 10.57 ($\ell 4$ to dir.) |
| Director | 32.2 | |

necessary to build the Log-Yag array.

## TOWER SELECTION AND INSTALLATION

Probably the most important part of any Amateur Radio installation is the antenna system. It determines the effectiveness of the signal transmitted at a particular power level. In terms of dollar investment, the antenna provides double duty; while it can provide gain during transmitting periods (which can also be accomplished by increasing transmitter power), it has the same beneficial effect on received signals. Therefore a tall support for a gain antenna that can be rotated is very desirable. This is especially true if DX contacts are of prime interest on the 20-, 15- and 10-meter bands.

Of the two important features of an antenna system, height and antenna gain, height is usually considered the most important, if the antenna is horizontally polarized. The typical amateur installation consists of a three-element triband beam (tribander) for 20, 15 and 10 meters mounted on a tower that may be as low as 25 or 30 feet, or as high as 65 or 70 feet. Some systems use large antennas on much taller towers.

The selection of a tower, its height and the type of antenna and rotor to be used all may seem like a complicated matter, particularly for the newcomer. These four aspects of an antenna system are interrelated, and one should consider the overall system before making any decisions as to a specific component. Perhaps the most important consideration for many amateurs is the effect of the antenna system on the surrounding environment. If plenty of space is available for a tower installation and there is little chance of the antenna irritating neighbors, the amateur is indeed fortunate. This amateur's limitations will be mostly financial. But for most, the size of the backyard, the effect of the system on members of the family and neighbors, local ordinances, and the proximity of power lines and poles influence the overall selection of antenna components considerably.

The amateur must consider the practical limitations for installation. Some points for consideration are given below:

1) A tower should never be installed in a position whereby it could conceivably fall onto a neighbor's property.

2) The antenna must be located in such a position that *it cannot possibly tangle with power lines during normal operation*, or if a disastrous windstorm comes along.

3) Sufficient yard space must be available to position a guyed tower *properly*. The guy anchors should be between 60 and 80 percent of the tower height in distance from the base of the tower.

4) Provisions must be made to keep neighborhood children from climbing the support. Chicken wire around the tower base will serve this need nicely.

5) Local ordinances should be checked to determine if any legal restrictions are on record.

Other important considerations are:

6) The total dollar value to be invested.

7) The size and weight of the antenna desired.

8) The overall yearly climate.

9) Ability of the owner to climb a fixed tower.

The selection of a tower support is usually dictated more by circumstances

**Table 11**
**Element Material Requirements, Log-Yag Array**

| *Element* | *1-In. Tubing Lth. (Ft)* | *Qty.* | *7/8-In. Tubing Lth. (Ft)* | *Qty.* | *3/4-In. Tubing Lth. (Ft)* | *Qty.* | *1-1/4-In. Angle Lth. (Ft)* | *1 × 1/4-In. Bar Lth. (Ft)* |
|---|---|---|---|---|---|---|---|---|
| Reflector | 12 | 1 | 6 | 2 | 8 | 2 | None | None |
| $\ell 1$ | 6 | 2 | 6 | 2 | 8 | 2 | 3 | 1 |
| $\ell 2$ | 6 | 2 | 6 | 2 | 8 | 2 | 3 | 1 |
| $\ell 3$ | 6 | 2 | 6 | 2 | 6 | 2 | 3 | 1 |
| $\ell 4$ | 6 | 2 | 6 | 2 | 6 | 2 | 3 | 1 |
| Director | 12 | 1 | 6 | 2 | 6 | 2 | None | None |

**Table 12**
**Materials List, Log-Yag Array**

1) Aluminum tubing — 0.047 in. wall thickness
 1 in. — 12 ft lengths, 24 lin. ft
 1 in. — 12 ft or 6 ft lengths, 48 lin. ft
 7/8 in. — 12 ft or 6 ft lengths, 72 lin. ft
 3/4 in. — 8 ft lengths, 72 lin. ft
 3/4 in. — 6 ft lengths, 36 lin. ft
2) Stainless steel hose clamps — 2 in. max., 8 ea.
3) Stainless steel hose clamps — 1-1/4 in. max., 24 ea.
4) TV-type U bolts — 1-1/2 in., 6 ea.
5) U bolts, galv. type: 5/16 in. × 1-1/2 in., 4 ea.
6) U bolts, galv. type: 1/4 in. × 1 in., 2 ea.
7) 1 in. ID water-service polyethylene pipe 160 lb/in.$^2$test, approx. 1-3/8 in. OD, 7
8) 1-1/4 in. × 1-1/4 in. × 1/8 in. aluminum angle — 6 ft lengths, 12 lin. ft
9) 1 in. × 1/4 in. aluminum bar — 6 ft lengths, 6 lin. ft
10) 1-1/4 in. top rail of chain-link fence, 26.5 lin. ft
11) 1:1 toroid balun, 1 ea.
12) No. 6-32 × 1 in. stainless steel screws, 8 ea.
 No. 6-32 stainless steel nuts, 16 ea.
 No. 6 solder lugs, 8 ea.
13) No. 12 copper feed wire, 22 lin. ft
14) 12 in. × 6 in. × 1/4 in. aluminum plate, 1 ea.
15) 6 in. × 4 in. × 1/4 in. aluminum plate, 1 ea.
16) 3/4 in. galv. pipe, 3 lin. ft
17) 1 in. galv. pipe — mast, 5 lin ft
18) Galv. guy wire, 50 lin. ft
19) 1/4 in. × 2 in. turnbuckles, 4 ea.
20) 1/4 in. × 1-1/2 in. eye bolts, 2 ea.
21) TV guy clamps and eyebolts, 2 ea.

than by desire. The most economical system, in terms of feet-per-dollar investment, is a guyed tower.

Once a decision has been tentatively made, the next step is to write to the manufacturer (several are listed in Table 13) and request a specification sheet. Meanwhile, lay out any guy anchor points needed to ensure that they will fit on the assigned property. The specification sheet for the tower should give a wind-load capability; an antenna can then be selected that will not overload it. If a tentative decision on the antenna type is made, a note to the antenna manufacturer giving the complete set of details for installation is not a bad idea. Be sure to give complete details of your plans, including all specifications of the antenna system planned. Remember, *the manufacturer will not custom design a system directly for your needs,* but may offer comments.

It is often very helpful to the novice tower installer to visit other local amateurs who have installed towers. Look over their hardware and ask questions. If possible, have a few local experienced amateurs look over your plans — before you commit yourself. They may be able to offer a great deal of help. If someone in your area is planning to install a tower and antenna system, be sure to offer your assistance. There is no substitute for experience when it comes to tower work, and your experience there may prove invaluable to you later.

## THE TOWER

The most common variety of tower is the guyed tower made of stacked identical sections. The information in Fig. 40 is based on data taken from the Unarco-Rohn catalog. A list of tower manufacturers can be found in Table 13. Rohn calls for a maximum vertical separation between sets of guy wires of 35 feet. At A, the tower is 70 feet high, and there are two sets of evenly spaced guy wires. At B, the tower is 80 feet high, and there are three sets of evenly spaced guy wires. Exceeding the vertical spacing requirements could result in the tower buckling.

This may not seem like a reasonable possibility unless you understand the functioning of the guy wires. The guy wires restrain the tower against the force of the wind and translate the lateral force of the wind into a downward compression that forces the tower down onto the base. Normally, the manufacturers specify the initial tension in the guy wires. This is another force that is translated into the downward compression on the tower. If there are not enough guys and if they are not properly spaced, a heavy gust of wind may turn out to be the "straw that breaks the camel's back." Fig. 40C is an overhead view of a guyed tower. Manufacturers usually call for equal angular spacing between radials. If it is necessary to deviate from this spacing, you would be well advised to contact the engineering staff of the manufacturer or a civil engineer.

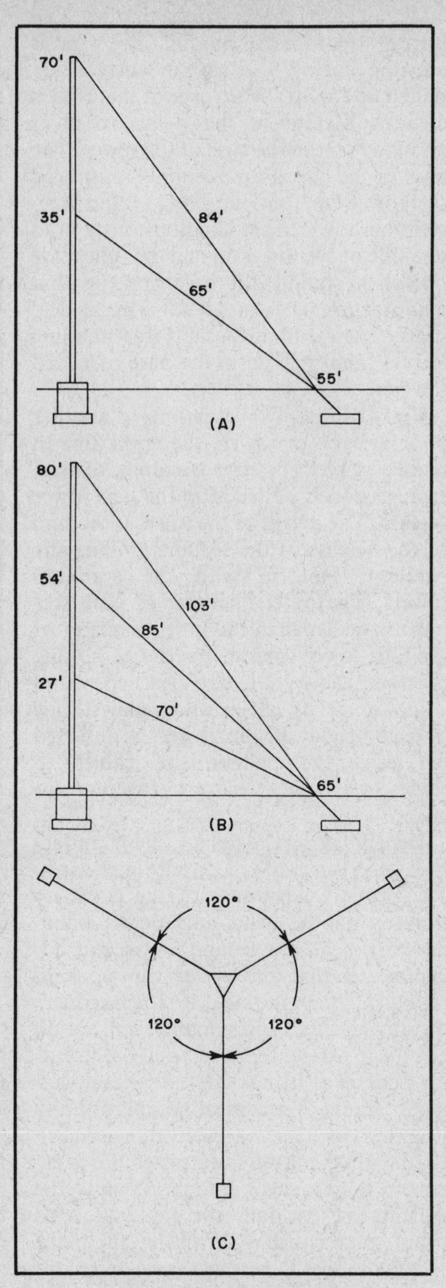


Fig. 40 — Diagram depicting proper method of installation of a typical guyed tower.

Some types of towers are not normally guyed — these are usually referred to as free-standing or self-supporting towers. The principles involved are the same regardless of the manufacturer's choice of names. The wind blowing against the side of the tower creates an overturning moment that would topple the tower if it were not for the anchoring at the base. Fig. 41 details the action and reaction involved. The tower is restrained by the base. As the wind blows against one side of the tower, the opposite side is compressed downward much as in the guyed-type setup. Because there are no guys to

**Table 13**
**Tower Manufacturers**

Aluma Tower
Box 2806
Vero Beach, FL 32960

Hy-Gain Division
Telex Communications, Inc.
9600 Adrich Ave., S.
Minneapolis, MN 55420

Tristao Tower Company
P. O. Box 3715
Visalia, CA 93278

Wilson Systems, Inc.
4286 S. Polaris Ave.
Las Vegas, NV 89103

Unarco-Rohn
P. O. Box 2000
Peoria, IL 61601

Tri-Ex Tower Corp.
7182 Rasmussen Ave.
Visalia, CA 93277

Universal Manufacturing Company
12357 E. 8 Mile Rd.
Warren, MI 48089

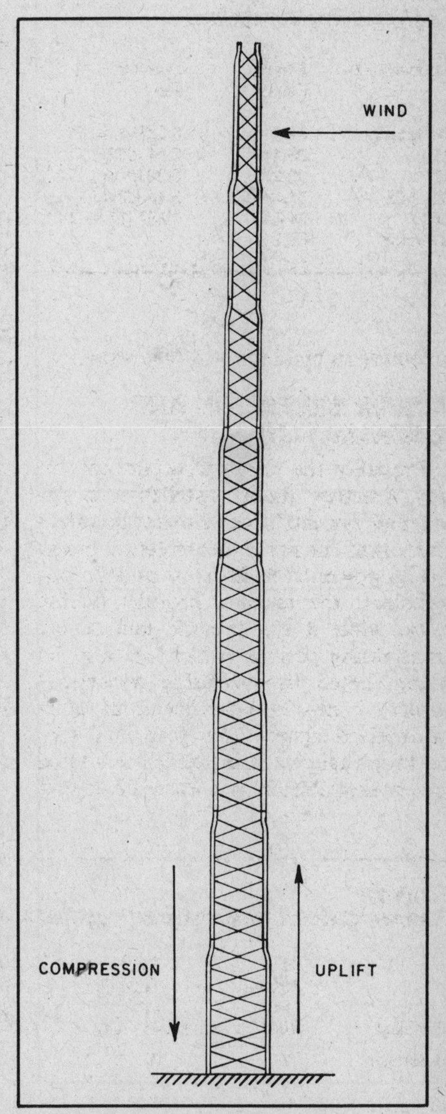


Fig. 41 — Diagram of typical free-standing (unguyed) tower. Arrows indicate the directions of the forces acting upon the structure. See text for discussion.

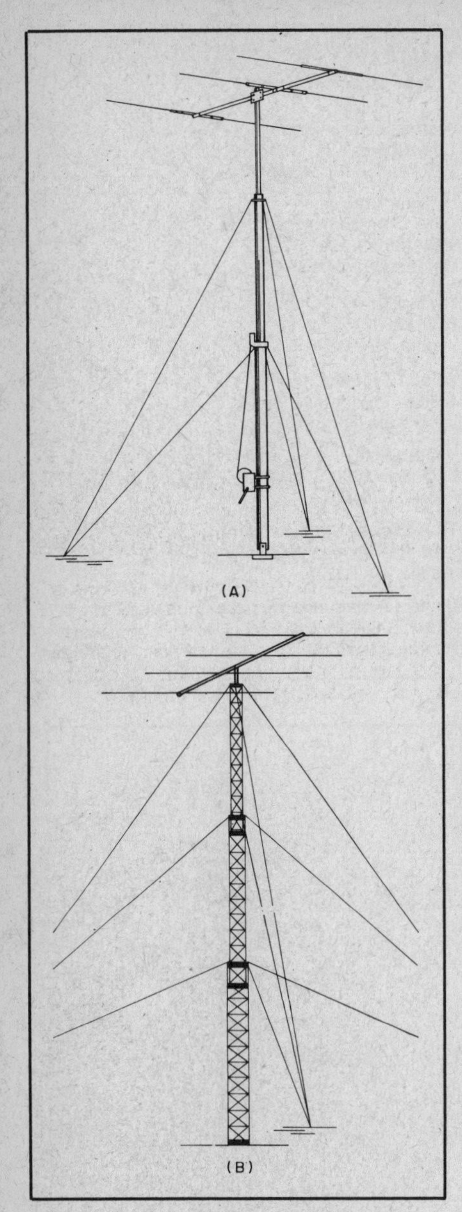


Fig. 42 — Two examples of "crank-up" towers.

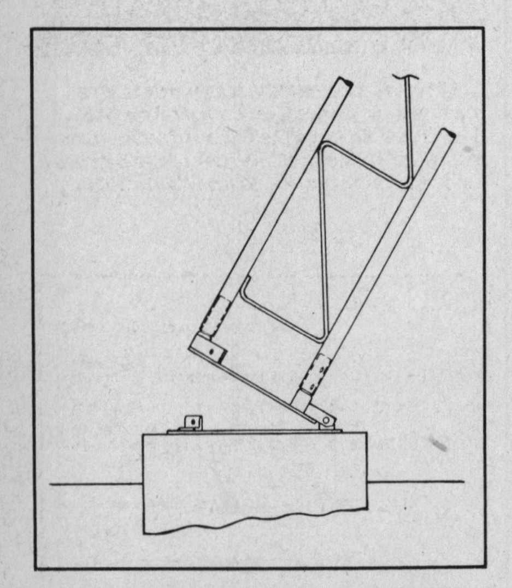

Fig. 43 — Fold-over or tilting base. There are several different variations of hinged sections permitting widely different types of installation. Great care should be exercised when raising or lowering a tilting tower.

restrain the top, the side that the wind is blowing against is simultaneously being pulled up (uplift). The force of the wind is creating a moment that tends to pivot about a point in the base of the tower. The base of the guyed tower simply must hold the tower up, but the base of the free-standing tower must simultaneously hold one side of the tower up and the other side *down*! It should not be surprising that manufacturers often call for a great deal more concrete in the base of free-standing towers than they do in the base of guyed towers.

Fig. 42 shows two variations of another popular type of tower, the crank-up. In regular guyed or free-standing towers, each section is bolted atop the next lower section. The height of the tower is the sum of the heights of the sections (minus any overlap). Not so with the crank-up towers. The outer diameter of each section is smaller than the inner diameter of the next lower section. Instead of bolting together, the sections are attached with a complex set of cables and pulleys. The overall height of the tower is adjusted by using the pulleys and cables to "telescope" the sections together or apart.

Depending on the design, the manufacturer may or may not require guy wires. The primary advantage of the crank-up tower is that the owner must do the antenna work near the ground. A second advantage is that the tower can be kept retracted except during use, which reduces the guying needs (presumably, you would not try to extend the tower and use it during periods of high winds). The disadvantages include mechanical complexity and (usually) cost. It is extremely dangerous to climb on a crank-up tower, even if it is extended only a small amount. Should the hoisting system fail, the inner sections could come crashing down like the blade of a guillotine! (There are cases on record where amateurs have lost their lives by climbing extended crank-up towers on which the hoisting system failed.)

Some towers have another convenience feature — a hinged section that permits the owner to fold over all or a portion of the tower. The primary benefit is in allowing antenna work to be done closer to ground level without the necessity of removing the antenna and lowering it. Fig. 43 shows a hinged base; of course, the hinged section can be designed for portions of the tower other than the base. Also, a hinged feature can be added to a crank-up tower.

Misuse of hinged sections during tower erections is a common problem among radio amateurs. Unfortunately, these episodes often end in accidents. If you do not have a good grasp of the fundamentals of physics, it might be wise to avoid hinged towers (or to consult an expert). It is often far easier (and safer) to erect a regular guyed tower or self-supporting

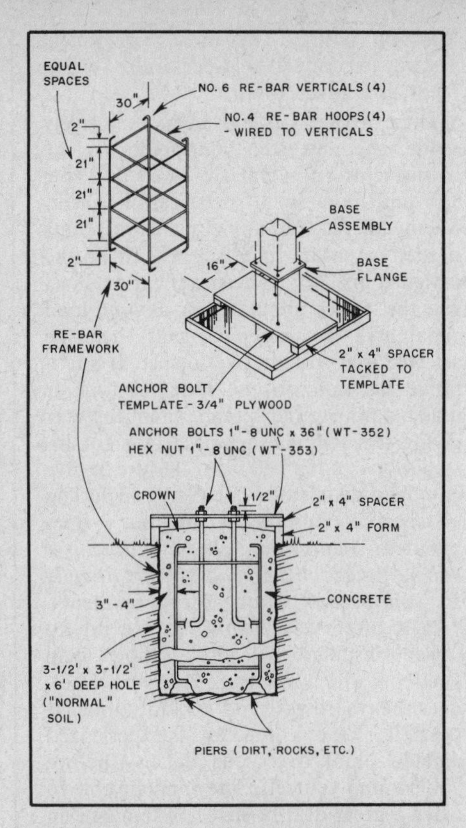


Fig. 44 — Plans for installing concrete base for Wilson ST-77B. Although the instructions and dimensions will vary from one tower to the next, this is representative of the type of concrete base specified by most manufacturers.

tower with gin pole and climbing belt than it is to try to "walk up" an unwieldy hinged tower.

## THE BASE

Each manufacturer will provide customers with detailed plans for properly constructing the base. Fig. 44 is an example of one such plan. This plan calls for a hole that is 3.5 × 3.5 × 6.0 feet. Steel reinforcement bars are lashed together and placed in the hole. The bars are positioned such that they will be completely embedded in the concrete, yet will not contact any metallic object in the base itself. This is done to minimize the possibility of a direct discharge path for lightning through the base. Should such a discharge occur, the concrete base would likely explode and bring about the collapse of the tower.

A strong wooden form is constructed around the top of the hole. The hole and the wooden form are filled with concrete so that the resultant block will be 4 inches above grade. The anchor bolts are embedded in the concrete before it hardens. Usually it's easier to ensure that the base is level and properly aligned by attaching the mounting base and the first section of the tower to the concrete anchor bolts. Each manufacturer will provide specific detailed instructions for the

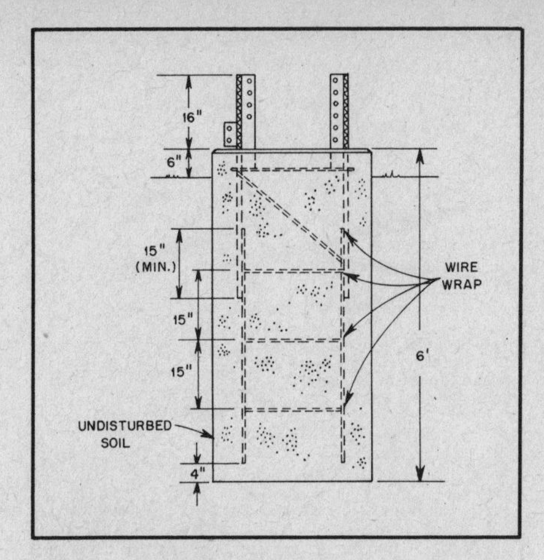


Fig. 45 — Another example of a concrete base (Tri-Ex LM-470).

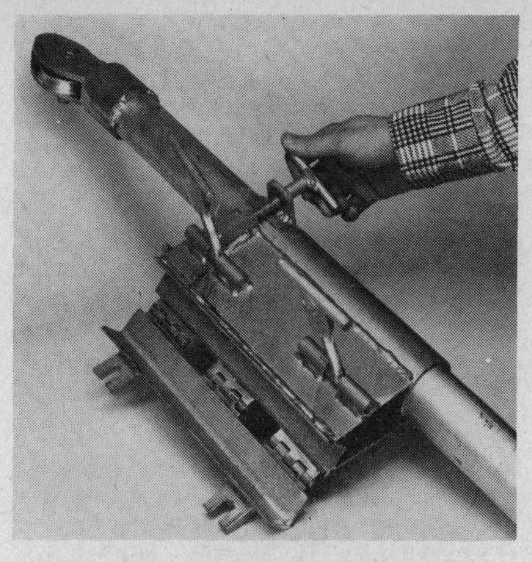

Fig. 46 — A gin pole is a mechanical device which can be clamped to a tower leg to aid in the assembly of sections as well as the installation of antenna. The aluminum tubing extends through the clamp and may be slipped into position before the tubing clamp is tightened. A rope should be routed through the tubing and over the pulley mounted at the top.

Fig. 47 — The assembly of tower sections is made simple when a gin pole is used to lift each one into position. Note that the safety belts of both climbers are fastened below the pole thereby preventing the strap from slipping over the top section. *(photo by K1THQ)*

proper mounting procedure. Fig. 45 provides a slightly different design for a tower base.

The one assumption so far is that you have normal soil. "Normal soil" is a mixture of clay, loam, sand and small rocks. A technical discussion is beyond the scope of this article, but you may want to adopt more conservative design parameters for your base (usually, more concrete) if your soil is sandy, swampy or extremely rocky. If you have any doubts about your soil, contact your local agricultural extension office and ask for a more technical description of your soil. Once you have that information, contact the engineering department of your tower manufacturer or a civil engineer.

## TOWER INSTALLATION

The installation of a tower is not difficult when the proper techniques are known. A guyed tower, in particular, is not hard to erect since each of the individual sections are relatively light in weight and can be handled with only a few helpers and some good-quality rope. A gin pole is a handy device for working with tower sections. The gin pole shown in Fig. 46 is designed to fit around the leg of Rohn No. 25 tower and clamp in place. The tubing, which is about 12 feet long, has a pulley on one end. A rope is routed through the tubing and over the pulley. When the gin pole is attached to the tower and the tubing is extended into place and locked, the rope may be used to haul tower sections and the antenna into place. Fig. 47 shows the basic process.

One of the most important aspects of any tower-installation project is the safety of all persons involved. The use of hard hats is highly recommended for all assistants helping from the ground. Helpers should always stand clear of the tower base to prevent being hit by a dropped tool or hardware. A good grade of climber's safety belt, such as is shown in Fig. 48, should be used by each person working on the tower. When climbing the tower, if more than one person is involved, one should climb into position before the other begins climbing. The same procedure is required for climbing down a tower after the job is completed. The purpose is to have the nonclimbing person stand relatively still so as not to drop any tools or objects on the climbing person, or unintentionally obstruct his movements. When two persons are working on top of a tower, only one should change position (unbelt and move) at a time.

For most installations, a good-quality 1/2-inch-diameter manila hemp rope will be able to handle adequately the work load for the hoisting tasks. The rope must be periodically inspected to assure that no tearing or chafing has developed, and if the rope should get wet from rain, it should be hung out to dry at the first opportunity. Safety knots should be used to assure that none come loose during the hoisting of a tower section or antenna.

### Attaching Guy Wires

In typical Amateur Radio installations a guy wire may experience "pulls" in excess of 1000 pounds. Under such circumstances, you do not merely twist the wires together and expect them to hold. Fig. 49 depicts the traditional method for fixing the end of a piece of guy wire. A thimble is used to prevent the wire from breaking because of a sharp bend at the point of intersection. Three cable clamps follow to hold the wire securely. As a final backup measure, the individual strands of the free end are unraveled and wrapped around the guy wire. It is a lot of work, but it is necessary to ensure a firm connection.

Fig. 48 — A good-quality safety belt is a requirement for working on a tower. The belt should contain large steel loops for the strap snap. Leather loops at the rear of the belt are handy for holding tools. *(photo by K1THQ)*

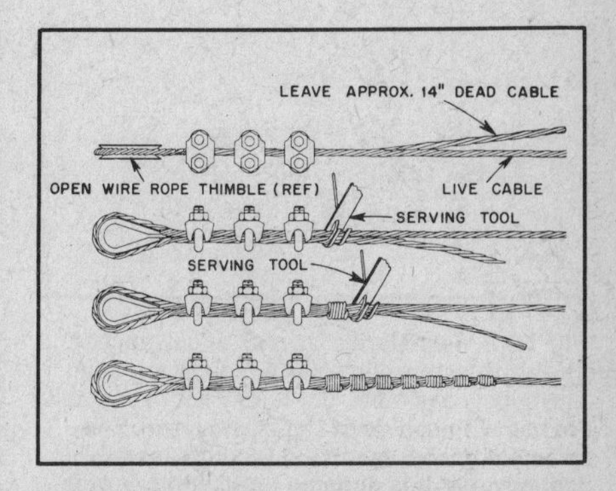


Fig. 49 — Traditional method for securing the end of a guy wire.

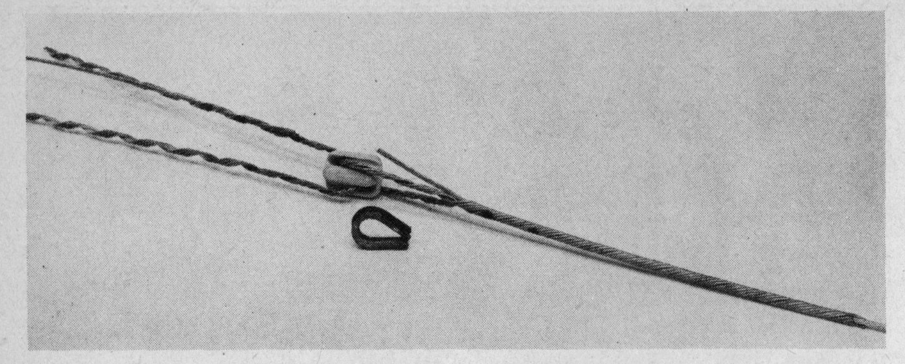

Fig. 50 — Alternative method for attaching guy wires using dead ends. The dead end on the right is completely assembled (the end of the guy wire was left extending from the grip for illustrative purposes). On the left, one side of the dead end has been partially attached to the guy wire. In front, a thimble for use where a sharp bend might cause the guy wire or dead end to break.

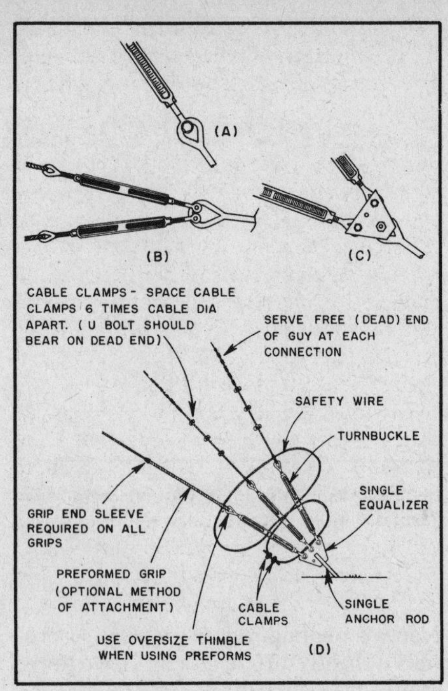


Fig. 53 — Variety of means available for attaching guy wires and turnbuckles to anchors.

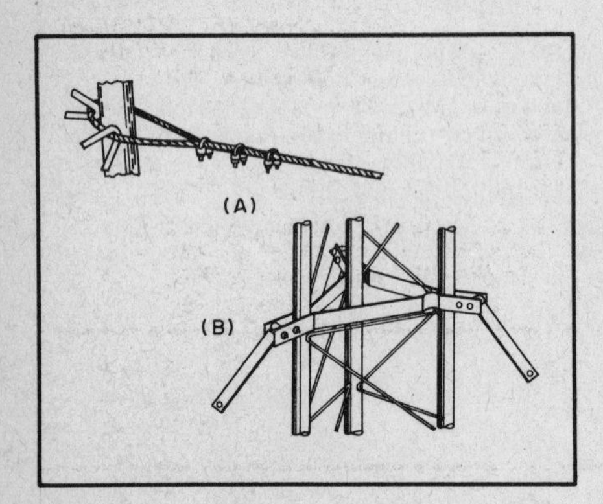


Fig. 51 — Two methods of attaching guy wires to tower. See text for discussion.

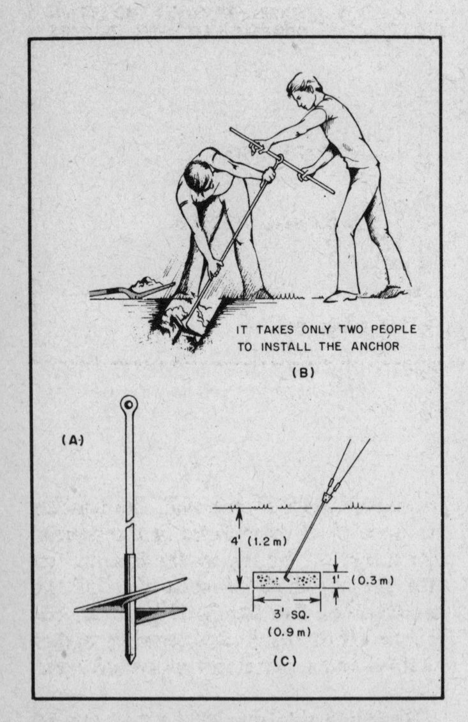


Fig. 52 — Two standard types of guy anchors. The earth screw shown at A is easy to install and widely available, but it may not be suitable for use with soil deviating from normal. The concrete anchor is more difficult to install properly, but it is suitable for use with a wide variety of soil conditions and will satisfy most building code requirements.

Fig. 50 shows the use of a device that replaces the clamps and twisted strands of wire. These devices are known as dead ends. They are far more convenient to use than are clamps. You must cut the guy wire to the proper length. The dead end is installed into whatever the guy wire is being attached to (use a thimble, if needed). One side of the dead end is then wrapped around the guy wire. The other side of the dead end follows. The savings in time and trouble more than make up for the slightly higher cost.

Guy wire comes in different sizes, strengths and types. Typically, 3/16-inch EHS guy wire will be adequate for the moderate tower installation found at most Amateur Radio stations. Some amateurs prefer to use 5/32-inch "aircraft cable." Although this cable is somewhat more flexible than 3/16-inch EHS, it is only about 70% as strong. It is recommended that you stay with standard guy wire and that you use nothing smaller than 3/16-inch EHS.

Fig. 51 shows two different methods for attaching guy wires to towers. At A, the guy wire is simply looped around the tower leg and terminated in the usual manner. At B, a "torque bracket" has been added. There probably isn't much difference in performance for wind forces that are tending to "push the tower over." If you happen to have more projected area (antennas, feed lines, etc.) on one side of the tower than the other, then the force of the wind will cause the tower to tend to twist into the ground. The torque bracket will be far more effective in resisting this twisting motion than will the simpler installation. The trade-off, of course, is in terms of initial cost.

There are two main types of anchors used for guy wires. Fig. 52A depicts an earth screw. It usually measures 4 to 6 feet long. The screw blade at the bottom typically measures 6 to 8 inches in diameter. Fig. 52B illustrates two people installing the anchor. The shaft is tilted such that it will be in line with the guy wires. Earth screws are suitable for use in "normal" soil where permitted by local building codes.

The alternative to earth screws is the concrete block anchor. Fig. 52C shows the installation of this type of anchor; it is suitable for any soil condition, with the possible exception of a bed of lava rock or coral. Consult the instructions from the manufacturer for the precise method of installation.

Turnbuckles and associated hardware are used to attach guy wires to anchors and to provide a convenient method of adjusting tension on the guy wires. Fig. 53A shows a turnbuckle of a single guy wire attached to the eye of the anchor. Turnbuckles are usually fitted with either two eyes or one eye, and one jaw. The eyes are the oval ends, while the jaws are U-shaped with a bolt through the tips. Fig. 53B depicts two turnbuckles attached to the eye of an anchor. The procedure for installation is to remove the bolt from the jaw, pass the jaw over the eye of the anchor and reinstall the bolt through the jaw, through the eye of the anchor, and through the other side of the jaw. For two or more guys attached to one anchor, it is recommended that you install an equalizer plate (Fig. 53C). In addition to providing a convenient point to attach the turnbuckles, the plate will pivot slightly and tend to equalize the tension on the guy wires. Once the installation is complete, a safety wire should be passed through the turnbuckles in a "figure-eight" fashion to prevent the turnbuckle from working loose.

### The Tower Shield

A tower can be legally classified as an

"attractive nuisance" that could cause injuries unless some precautions are taken. The tower shield should eliminate the worry.

Generally, the "attractive nuisance" doctrine is based on the theory that one who maintains upon his premises an agency or condition that is dangerous to children of tender years by reason of their inability to appreciate danger and which may reasonably be expected to attract children to the premises, is under a duty to exercise reasonable care to protect them against dangers of the attraction.

The tower shield is simply composed of panels that enclose the tower and make climbing practically impossible. These panels are 5 feet in height and are wide enough to fit snugly between the tower legs and flat against the rungs. A height of 5 feet is sufficient in most every case. The panels are constructed from 18-gauge galvanized sheet metal obtained and cut to proper dimensions from a local sheet-metal shop. A lighter gauge could probably be used, but the extra physical weight of the heavier gauge is an advantage if no additional means of securing the panels to the tower rungs are used. The three types of metals used for the components of the shield are supposedly rust proof and nonreactive. The panels are galvanized sheet steel, the brackets aluminum, and the screws and nuts are brass. The tower shield consists of three panels, one for each of the three sides, supported by two brackets. These brackets are constructed from 6-inch pieces of thin aluminum angle stock. Two of these pieces are bolted together to form a Z bracket (see Figs. 54, 55 and 56). The Z brackets are bolted together with flat-head (binding-head) brass machine screws.

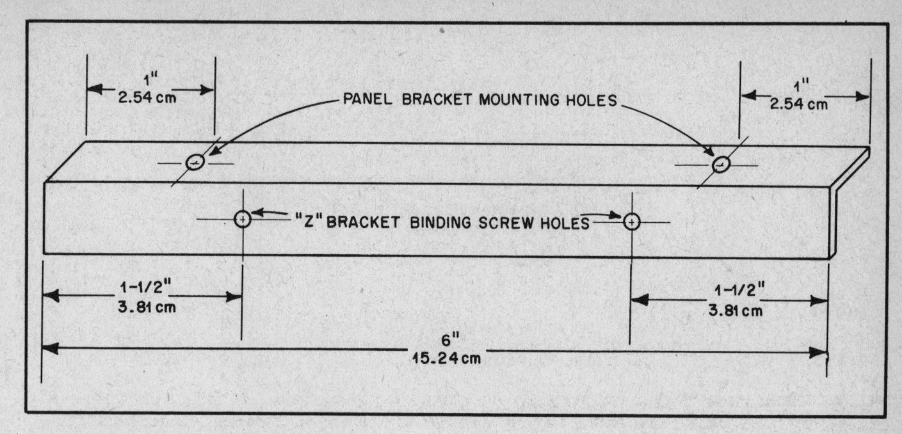


Fig. 54 — Z-bracket component pieces.

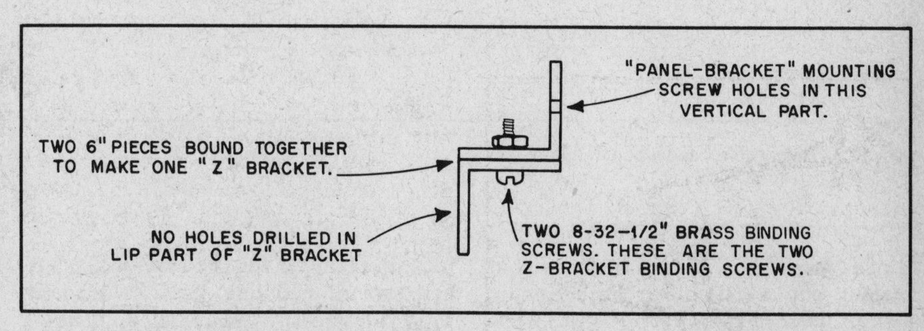


Fig. 55 — Assembly of the Z bracket.

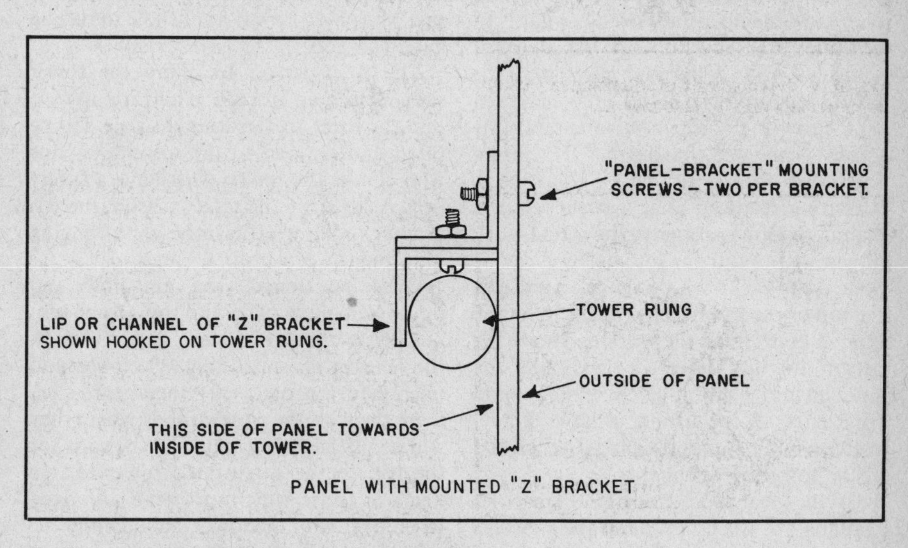


Fig. 56 — Installation of the shield on a tower rung.

The panels are laid flat for measuring, marking and drilling. The first measurement is from the top of the upper mounting rung on the tower to the top of the bottom rung. These mounting rungs are selected to position the panel on the tower. This distance from rung to rung is then marked on the panel. Using the same size brass screws and nuts, the top vertical portion of each Z bracket is bolted to the panel. The mounting-screw holes are drilled about one inch from the end of the Z brackets so that an offset clearance occurs between the Z-bracket binding-screw holes and the panel-bracket mounting-screw holes. The panel holes are drilled to match the Z-bracket holes.

The panels are held on the tower by their own weight. They are not easy to grasp because they fit snugly between the tower legs. If the need exists for added safety against deliberate removal of the panels, this can be accomplished by means of tie wires. A small hole can be drilled in the panel just above, just below, and in the center of each Z bracket. Run a piece of heavy galvanized wire through the top hole, around the Z bracket, and then back through the hole just below the Z bracket. Twist together the two ends of the wire. One tie wire should be sufficient for each panel, but use two if desired.

The completed panels are rather bulky and difficult to handle. A feature that is useful if the panels have to be removed often for tower climbing or accessibility is a pair of removable handles. The removable handles can be constructed from one threaded rod and eight nuts. (See Fig. 57.) The two pairs of handle holes were drilled in the panels a few inches below the top Z bracket and several inches above the bottom Z bracket. For panel placement or removal, the handles are hooked in these holes in the panels. The hook, on the top of the handle, fits into the top hole of each pair of the handle holes. The handle is optional, but for the effort required it certainly makes removal and replacement much safer and easier.

The installed tower shield can be seen in Fig. 58. This relatively simple device could prevent an accident.

## ANTENNA INSTALLATION

All antenna installations are different in some respects. Therefore, thorough plan-

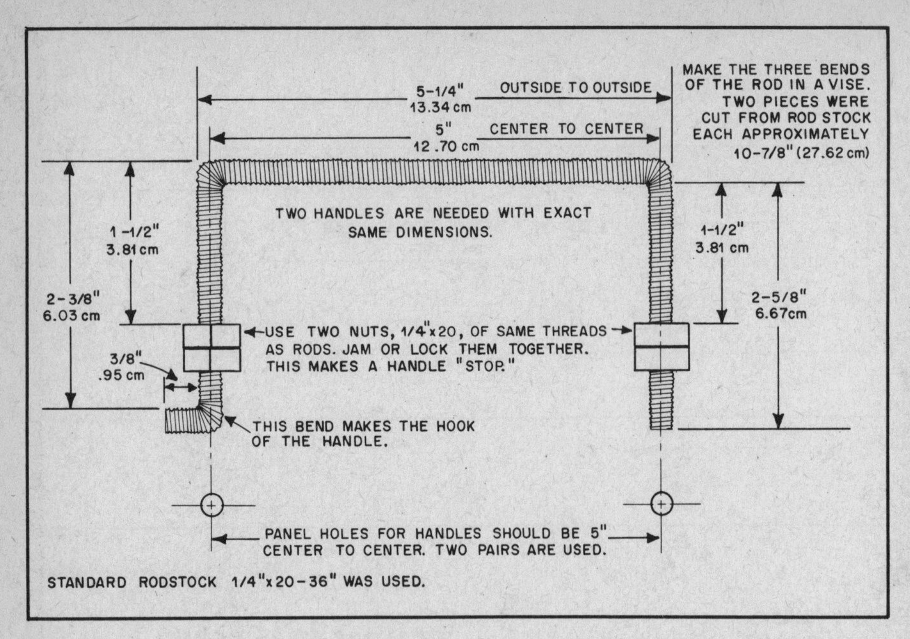


Fig. 57 — Removable handle construction.

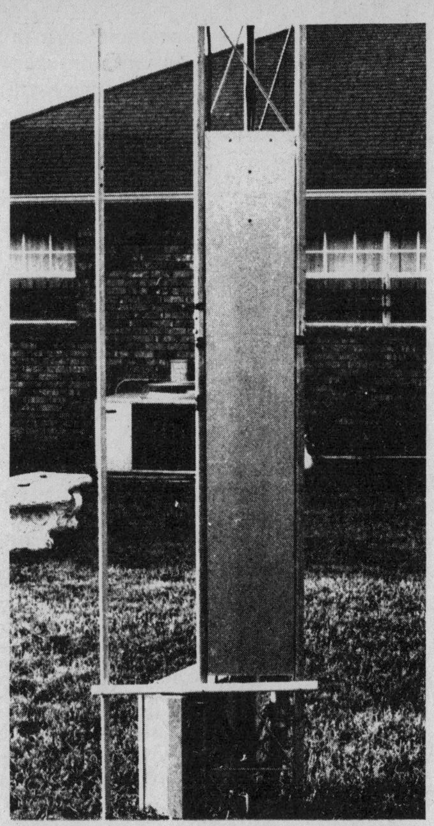

Fig. 58 — Installed tower shield. Note the holes for using the handles.

ning is the most important first step in installing any antenna. At the beginning, before anyone climbs the tower, the whole process should be thought through. The procedure should be discussed to be sure each crew member understands what is to be done. Plan how to work out all bugs. Consider what tools and parts must be assembled and what items must be taken up the tower. Extra trips up and down the tower can be avoided by using forethought.

Getting ready to raise a beam requires planning. Done properly, the actual work of getting the antenna into position can be done quite easily with only one person at the top of the tower. The trick is to let the ground crew do all the work and leave the person on the tower free to guide the antenna into position. Because the ground crew does all the lifting, a large pulley, preferably on a gin pole placed at the top of the tower, is essential.

Often, local radio clubs have gin poles available for use by their members. Stores that sell tower sections to amateurs and commercial customers frequently will rent them.

A gin pole should be placed along the side of the tower so the pulley is no more than 2 feet above the top of the tower or the point at which the beam is to be placed. Normally this height is sufficient to allow the antenna to be positioned easily. An important reason that the pulley is placed at this level, however, is that there can be considerable strain on the pole when the antenna is maneuvered past the guy wires.

The working rope (halyard) through the pulley must be a little longer than twice the tower height so the ground crew can raise the antenna from ground level. The rope should be 1/2 inch, or better yet, 5/8 inch, in diameter for the sake of strength and ease of handling. Smaller diameter rope is less easily manipulated. It has a tendency to jump out of the pulley track and foul up the operation. Needless to say, Murphy's Law is applicable to raising antennas, too!

The first person to climb the tower should carry an end of the halyard up with him (or her!) so that the gin pole can be lifted and secured to the tower. Anyone who climbs the tower *must* have a safety belt. Aside from the safety reason, there is no way to work efficiently while hanging onto the tower with one hand.

Once positioned, the gin pole and pulley will allow parts and tools to be sent quickly to the tower man. A useful trick for sending up small items like bolts and pliers is for a ground crew member to stick them through the rope strands where they will be sufficiently gripped by the rope for the trip to the top of the tower. Large items or those which might be dislodged by contact with the tower should either be taped or tied to the halyard.

Take heed of this caution: *Remember, once someone is on the tower, no one should be allowed to stand near the base of the tower!* Anyone who has seen a falling soldering gun explode on impact will testify as to the foolishness of standing near a tower when someone is working above.

## Raising the Antenna

A little technique can save much effort when the work of raising the antenna gets underway. First, the halyard is passed through the gin-pole pulley and the leading end is returned to the ground crew where it is to be tied to the antenna. The assembled antenna should be placed in a clear area of the yard (or on the roof) so it points toward the tower. The halyard is then passed *under* the front elements of the beam to a position past the midpoint of the antenna where it is securely tied to the boom (Fig. 59A). Note that once the antenna is installed, the tower man must be able to reach and untie the halyard from the boom. Therefore don't tie the rope more than an arm's length along the boom from the center of the antenna. If necessary, a large loop may be placed around the first element located beyond the midpoint of the boom, with the knot tied near the center of the antenna. The rope may then be untied easily after completion of the installation. The halyard should be tied to the boom at the front of the antenna by means of a short piece of light rope or twine.

While the antenna is being raised, the ground crew does all the pulling. As soon as the front of the antenna reaches the top of the mast, the tower man unties the light rope and prevents the front of the antenna from falling, as the ground crew continues to lift the antenna (Fig. 59B). When the center of the antenna is even with the top of the tower, the tower man puts one bolt through the mast and the antenna mounting bracket on the boom. The single bolt acts as a pivot point and the ground crew continues to lift the back of the antenna with the halyard (Fig. 59C). After the antenna is horizontally position-

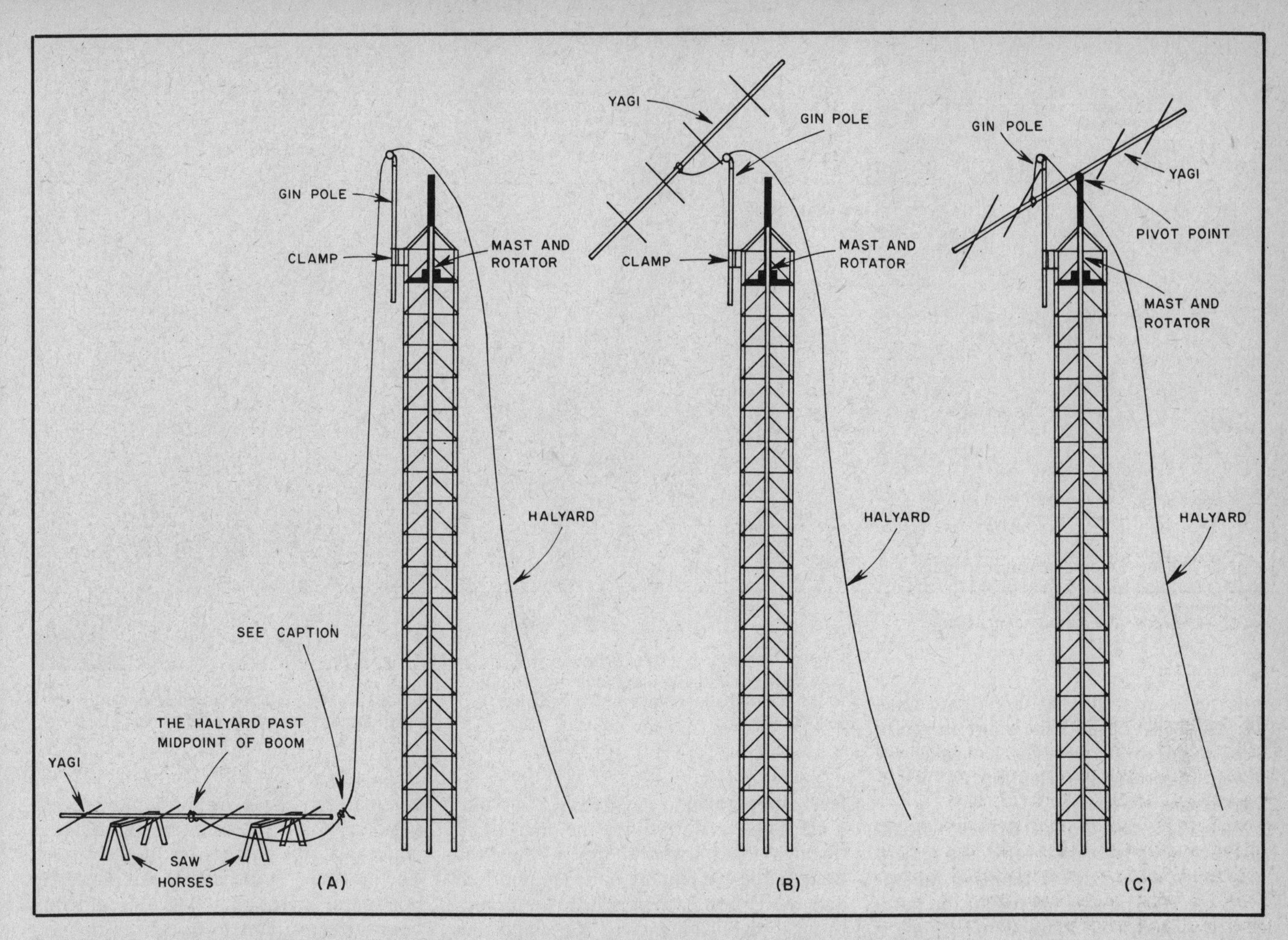


Fig. 59 — Raising a Yagi antenna to the top of a tower. In drawing A the Yagi is placed in a clear area and pointing toward the tower. The halyard is passed under the elements, then secured to the boom beyond the midpoint. B shows the antenna approaching the top of the mast. The person on the tower guides it after the lifting rope has been untied from the front of the antenna. In drawing C the antenna is pulled into a horizontal position by the ground crew. The tower person inserts the pivot bolt and secures it. Note: A short piece of rope is tied around the halyard and the boom at the front of the antenna. It is removed by the tower person when the antenna reaches the top.

ed, the tower man secures the rest of the mounting bolts and unties the halyard. By using this technique, the tower man performs no lifting.

### Avoiding Guy Wires

Although the same basic methods of installing a Yagi apply to any tower, guyed towers pose a special problem. Steps must be taken to avoid snagging the antenna on the guy wires. Let's consider, therefore, some ways in which this difficulty can be circumvented. With proper precautions, even large antennas can be pulled to the top of a tower, even if the mast is guyed at several levels.

Sometimes one of the top guys can provide a track to support the antenna as it is pulled upward. Insulators in the guys, however, may offer obstructions that could cause the beam to get hung up. A better track made with rope is an alternative, provided there is sufficient rope available. One end of the rope is secured outside the guy anchors. The other end is passed over the top of the tower and back down to an anchor near the first anchor. So arranged, the rope forms a narrow V track strung outside the guy wires. Once the V track is secured, the antenna may simply be pulled up the track.

Another method is to tie a rope to the back of the antenna (but within reach of the center). The ground crews then pull the antenna out away from the guys as the antenna is raised. With this method, some crew members are pulling up the antenna to raise it while others are pulling down and out to have the beam clear the guys. Obviously, the opposing crews must be coordinated or they can literally tear the antenna apart. The beam is especially vulnerable when it begins to tip into the horizontal position. If the crew pulling out and down continues to pull against the antenna, the boom can be broken. Another problem with this approach is that the antenna may rotate on the axis of the boom as it is raised. To prevent such rotation, long lengths of twine may be tied to an element, one piece on each side of the boom. Ground personnel may then use these to stabilize the antenna. Where this is done, provision should be made for untying the twine once the antenna is in place.

A third method is to tie the halyard to the center of the antenna. A crew member, wearing a safety belt, walks the antenna up the tower as the crew on the ground raises it. Because the halyard is tied at the center balance point, the man on the tower can rotate the elements around the guys. A tag line can be tied to the bottom end of the boom so that a man on the ground can help move the antenna around the guys. The tag line must be removed while the antenna is still vertical and you can reach it.

### THE PVRC MOUNT

The above methods will probably not be satisfactory for large arrays. The best way to handle large Yagis is to assemble them on top of the tower. The way to do that is by using the "PVRC mount." Many members of the Potomac Valley Radio Club have successfully used this method to install their large antennas. Simple and ingenious, the idea involves offsetting the boom from the mast to per-

Fig. 60 — The PVRC mount, boom-to-mast plate, mast and rotator ready to go. The mast and rotator are installed first.

Fig. 62 — Working at the 70-foot level. A gin pole makes pulling up and mounting the boom to the boom-to-mast plate a safe and easy procedure.

Fig. 63 — Mounting the last element prior to positioning the boom in a horizontal plane.

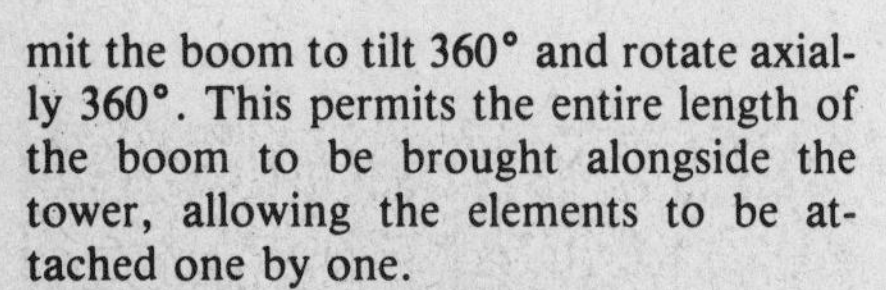

mit the boom to tilt 360° and rotate axially 360°. This permits the entire length of the boom to be brought alongside the tower, allowing the elements to be attached one by one.

See Figs. 60 through 64. The mount itself consists of a short length of pipe of the same (or greater) diameter as the rotating mast, a steel plate and the hardware to hold it all together. The plate is drilled for eight U bolts: four to attach the plate to the pipe, and four to attach the pipe to the antenna boom-to-mast plate. Additionally, four pinning bolts are used to ensure that the antenna would end up level and parallel to the ground. Two pinning bolts pass through the mast, two through the horizontal pipe. When the horizontal pipe pinning bolts are removed and the U bolts are loosened, the boom-to-mast plate can be tilted 360°, allowing either half of the boom to come alongside the tower.

After carefully marking all critical dimensions, the antenna elements are removed from the boom. Once the rotator and mast have been installed on the tower, a gin pole is used to bring the adapter plate and pipe to the top of the tower. There, the "top crew" unpins the horizontal pipe and tilts the antenna boom-to-mast plate to place it in the vertical plane. The boom is attached to the boom-to-mast plate at the balance point *of the assembled antenna.* It is important that the boom be rotated axially so that the bottom side of the boom is closest to the tower. This will ensure that the antenna elements will be parallel to the side of the tower during installation, allowing the boom to be tilted without the elements striking the tower.

During installation it may be necessary to remove one guy wire temporarily to allow for tilting of the boom. As a safety precaution, a temporary guy should be mounted to the same leg of the tower just low enough so that the antenna will not hit it.

The elements are assembled on the boom starting with those closest to the center of the boom, working out alternately to the farthest director and reflector. This procedure *must* be followed. If all the elements are put first on one half of the boom, it will be dangerous (if not impossible) to put on the remaining elements. By starting at the middle, and working outward the antenna weight will never be so far removed from the balance point that tilting of the boom becomes impossible.

When the last element is attached, the boom is brought parallel to the ground, the horizontal pipe is pinned and the U bolts tightened. Now, all the antenna elements are positioned vertically. Next,

Fig. 61 — Close-up of the PVRC mount. Two of the four locking pins (bolts) may be seen at the midline of the left hand vertical plate. The other two pins are located along the axis of the short pipe section; the head of the right hand bolt blends in with the U-bolt lock nut to the rear.

Fig. 64 — The mast-to-pipe U bolts are loosened and the boom is turned to a horizontal position. This puts the elements in a vertical plane. Then, the pipe U bolts are tightened and pinning bolts secured. The boom U bolts are then loosened and the boom turned axially 90°.

Fig. 65 — This tripod tower supports a rotary beam antenna. Besides saving yard space, a roof-mounted tower can be more economical than a ground-mounted tower. A ground lead fastened to the lower part of the frame is for lightning protection. The rotator control cable and the coaxial line are dressed along two of the legs. *(photo courtesy Jane Wolfert)*

Fig. 67 — Three lengths of 2 × 6 wood mounted on the outside of the roof and reinforced under the roof by three identical lengths provide a durable means for anchoring the tripod. Liberal coatings of tar guard against weathering and leaks.

loosen the U bolts that hold the boom and rotate the boom axially 90°, bringing the elements parallel to the ground. Then tighten the boom U bolts and double-check all the hardware.

Many long-boom Yagis employ a truss to prevent boom sag. With the type of mount just described the truss must be attached to a pipe that is independent of the rotating mast. A short length of pipe is attached to the boom as close as possible to the balance point. The truss will now move along with the boom whenever the boom is tilted or twisted.

### The Tower Alternative

A cost-saving alternative to the ground mounted tower is the roof mounted tripod. Units suitable for small hf or vhf antennas are readily available. Perhaps the biggest problem with a tripod is determining how to fasten it securely to the roof.

One method of mounting a tripod on a roof is to nail 2 × 6s to the underside of the rafters. Bolts can be extended from the leg mounts through the roof and the 2 × 6s. To avoid exerting too much pressure on the area of the roof between rafters, place another set of 2 × 6s on top of the roof (a mirror image of the ones within the attic). Installation details are shown in Figs. 65 through 68.

The 2 × 6s are cut 4 inches longer than the outside distance between two rafters. Bolts are cut from a length of 1/4-inch threaded rod. Nails are used to hold the boards in place during installation. Roofing tar is used to seal against leaks.

Find a location on the roof that will allow the antenna to turn without obstruction from such things as trees, TV antennas and chimneys. Determine the rafter locations. (Chimneys and vent pipes make good reference points.) Now the tower is set in place atop three 2 × 6s. A plumb line run from the top center of the tower can be used to center it on the peak of the roof. Holes for the mounting bolts can now be drilled through the roof.

Before proceeding, the bottom of the 2 × 6s and the area of the roof under them should be given a coat of roofing tar. Leave about 1/8 inch of clear area around the holes to ensure easy passage of the bolts.

Now put the tower back in place and insert the bolts and tighten them. Apply tar to the bottom of the legs and the wooden supports, including the bolts.

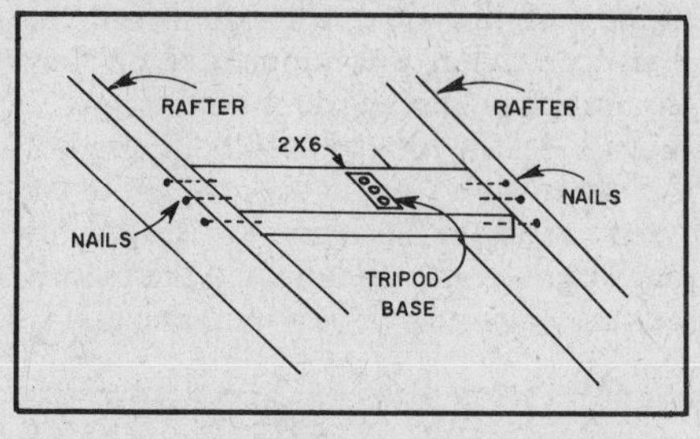


Fig. 66 — This cutaway view illustrates how the tripod tower is secured to the roof rafters. The leg to be secured to the cross piece is placed on the outside of the roof. Another crossmember is fastened to the underside of the rafters. Bolts, inserted through the roof and the two cross pieces, hold the inner crossmember in place because of pressure applied. The author nailed the inner cross piece to the rafter for added strength.

Fig. 68 — The strengthened anchoring for the tripod. Bolts are placed through a 2 × 6 on the underside of the roof and through the 2 × 6 on the top of the roof, as shown in Fig. 67.

For added security the tripod can be guyed. Guys should be anchored to the frame of the house.

If a rotator is to be mounted above the tripod, pressure will be applied to the bearings. Wind load on the antenna will be translated into a "pinching" of one side of the bearings. Make sure that your rotator is capable of handling this additional stress.

## ROTATOR SYSTEMS

There are not many choices when it comes to antenna rotators for the amateur antenna system. However, making the correct decision as to how big the rotator needs to be is very important if trouble-free operation is desired. There are basically four grades of rotators available to the amateur. The lightest duty rotator is the TV type, typically used to turn TV antennas. These rotators will handle, without much difficulty, a small three-element tribander (20-, 15- and 10-meter) array or a single 15- or 10-meter monoband three-element antenna. The important consideration with a TV rotator is that it lacks braking or holding capability. High winds will turn the rotator motor via the gear train in a reverse fashion. Sometimes broken gears result. The next grade up from the TV class of rotator usually includes a braking arrangement whereby the antenna is held in place when power is not applied to the rotator. Generally speaking, the brake will prevent gear damage on windy days. If adequate precautions are taken, this group of rotators is capable of holding and turning stacked monobands arrays, or up to a five-element 20-meter system. The next step up in strength is more expensive. This class of rotator will turn just about anything the most demanding amateur might want to install.

A description of antenna rotators would not be complete without the mention of the prop-pitch class. The prop-pitch rotator system consists of a surplus aircraft propeller-blade pitch motor coupled to an indicator system as well as a power supply. There are mechanical problems of installation, however. It has been said that a prop-pitch rotator system, properly installed, is capable of turning a house. This is no doubt true! Perhaps in the same class as the prop-pitch motor but with somewhat less capability is the electric motor of the type used for opening garage doors. These have been used successfully in turning quite large arrays.

Proper installation of the antenna rotator can provide many years of trouble-free service; sloppy installation can cause problems such as a burned-out motor, slippage, binding and casting breakage. Most rotators are capable of accepting mast sizes of different diameters, and suitable precautions must be taken to shim an undersized mast to assure dead-center rotation. It is very desirable to mount the rotator inside and as far below the top of the tower as possible. The mast will absorb the torsion developed by the antenna during high winds, as well as during starting and stopping. A mast length of 10 feet or more between the rotator and the antenna will add greatly to the longevity of the entire system. Some amateurs have used a long mast from the top to the base of the tower. Installation and service can be accomplished at ground level. Another benefit of mounting the rotator 10 or more feet below the antenna is that any misalignment among the rotator, mast and the top of the tower is less significant. A tube at the top of the tower through which the mast protrudes will almost completely eliminate any lateral forces on the rotator casting. All the rotator need do is support the downward weight of the antenna system and turn the array. While the normal weight of the antenna and the mast is usually not more than a couple of hundred pounds, even with a large system, one can ease this strain on the rotator by installing a thrust bearing at the top of the tower. The bearing is then the component that holds all of the weight, and the rotator need perform only the rotating task.

A problem often encountered in the amateur installation is that of misalignment between the direction indication of the rotator control box and the heading of the antenna. This is caused by mechanical slippage in the system caused by loose bolts or antenna boom-to-mast hardware. Many texts suggest that the boom be pinned to the mast with a heavy-duty bolt and, likewise, the rotator be pinned to the mast. There is a trade-off here. The amateur might not like to climb the tower and straighten out the assembly after each heavy windstorm. However, if there is sufficient wind to cause slippage in the couplings, the wind could break a casting. The slippage will act as a clutch release, which *may* prevent serious damage.

## DELAYED-ACTION BRAKING FOR THE HAM-M ROTATOR

On most rotators equipped with braking capabilities, the brake is applied almost instantly after power is removed from the rotator motor to stop the array from rotating and hold it at a chosen bearing. Because of inertia, however, the array itself does not stop rotating instantly. The larger and heavier the antenna, the more it will tend to continue its travel, in which case the mast may absorb the torsion, the entire tower may twist back and forth, or the brake of the rotator may shear. A more suitable system is to remove power from the rotator motor during rotation before the desired bearing is reached, allowing the beam to coast to a slower speed or to a complete stop before the brake is applied. Delayed-action braking may be added to the Ham-M rotator system by adding a couple of components inside the control head case. Fig. 69 is a partial schematic diagram showing the necessary changes.

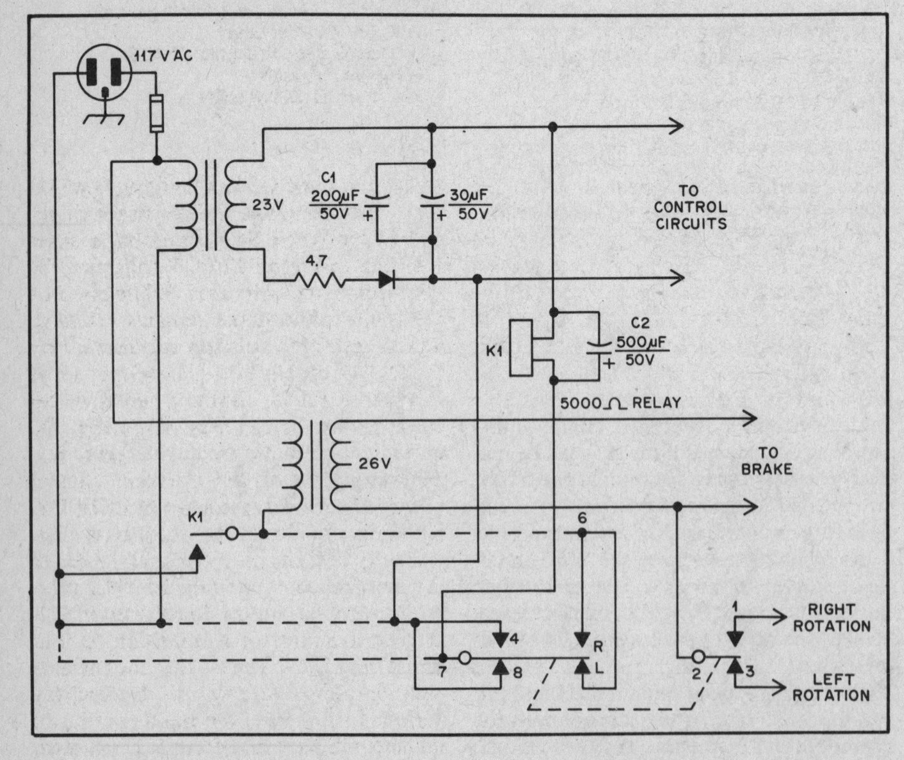

Fig. 69 — Partial schematic diagram for delayed braking in a Ham-M rotator.

### Circuit Operation

The 5000-ohm relay is energized by the operating switch and held closed after release for approximately 1-3/4 seconds by means of the 500-μF capacitor. The relay contacts supply primary 120 V to the main transformer, which continues to hold the brake off after rotation power is removed.

Note the addition of the 200-μF capacitor in parallel with the original 50-μF filter. This is required because the 500-μF capacitor across the relay coil increases the control voltage, thereby causing approximately a 15° error between readings. The additional 200-μF capacitor increases the control voltage such that identical readings are obtained

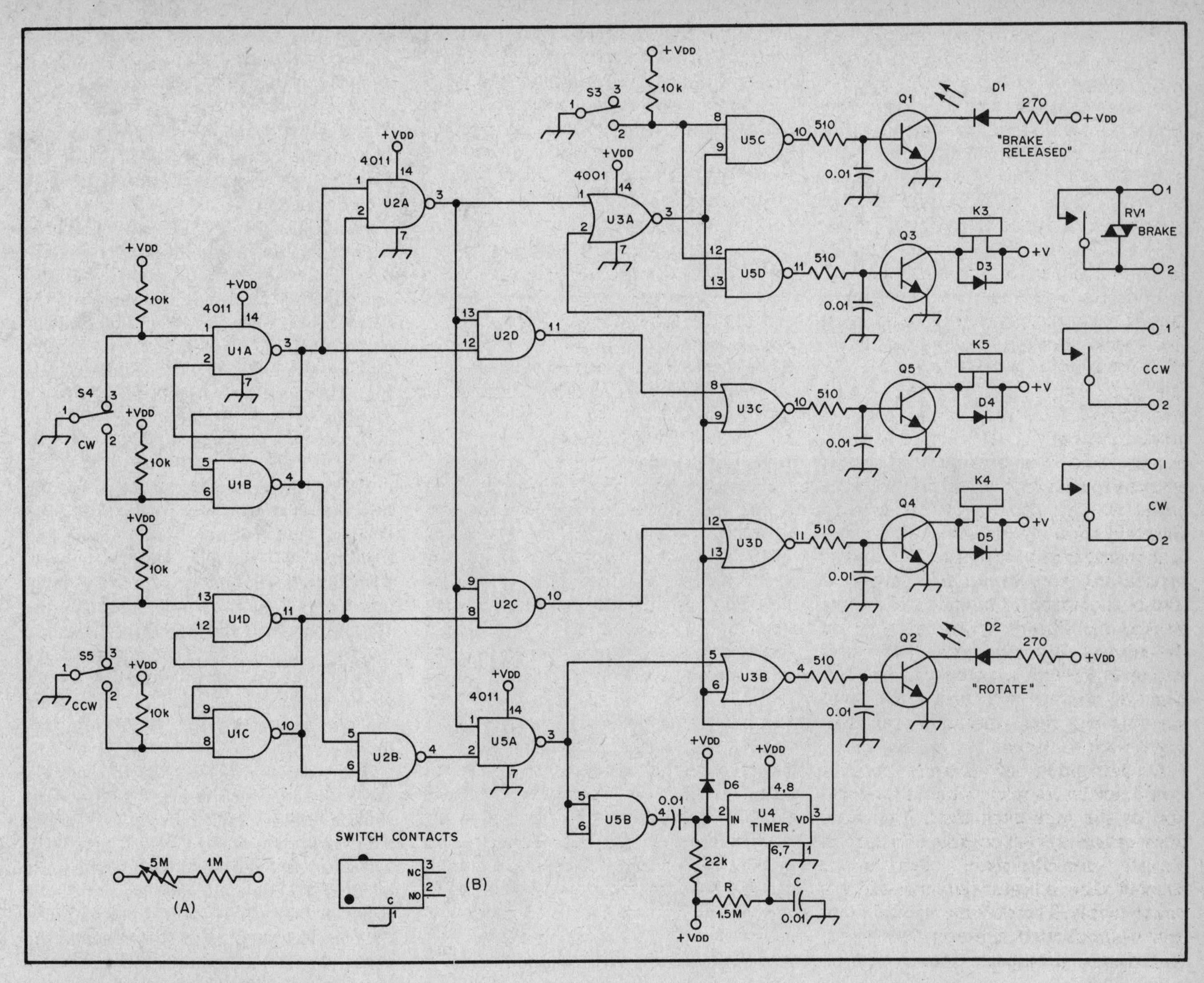


Fig. 70 — This schematic diagram shows the circuit for a brake-delay system for protecting the Ham-II rotator and antenna.

D1, D2 — Light-emitting diode, Motorola type MLED600 or equiv.
D3-D6, incl. — Silicon signal diode, 1N914 or equiv.
K3-K5, incl. — Switching relay, 12 V dc, 1200 ohms, 10 mA; contact rating 1 A; 125 V ac; Radio Shack 275-003 or equiv.
Q1-Q5, incl. — Silicon npn transistor, 2N3904 or equiv.
RV1 — Varistor, GE 750 or equiv.
U1, U2, U5 — CMOS quad NAND-gate IC RCA CD-4011A or equiv.
U3 — CMOS quad NOR-gate IC, RCA CD-4001A, or equiv.
U4 — Timer IC, 555 or equiv.

during rotation or at rest. This modification also causes the unit to read position whenever it is plugged in. An ON/OFF switch is easily added if you do not have a master switch for your station.

To increase the indicator lamp life, change the lamps to 28 V types. The relay is approximately 25 × 35 mm and fits nicely near the left front just above the screwdriver-adjust calibration control. The capacitors are fitted easily near the rear of the meter.

## A DELAYED BRAKE RELEASE FOR THE HAM-II

Not only is it wise to delay braking in a rotator system, it is even more important that rotation in the opposite direction is not initiated until the system is at rest. The circuit presented here offers the protection of delayed braking, and it also disables the direction-selector switches. In this manner, the antenna system coasts to a stop before rotation may begin in the opposite direction. The automatic delay prevents damage to the antenna system and rotator, even during a contest when the operator's attention is not on the rotator control.

Fig. 70 presents the brake-delay circuit schematic diagram. S3, S4 and S5 are the existing Ham-II control unit brake release and direction switches. S4 selects clockwise (cw) rotation and S5 selects counter-clockwise (ccw) rotation. These switches are replaced by K3, K4 and K5, respectively, in the modified control unit.

A pair of NAND gates in U1 form a debouncing circuit for each direction switch to prevent false triggering of the brake from contact bounce. Pressing S4 causes pin 3 of U2 to go high ($+V_{DD}$), or to a logic 1, which forces pin 3 of U3 low (0), pin 11 of U5 high, and energizes both the brake relay K3 and the BRAKE RELEASED LED, D1. In addition, pressing only S4 forces pin 10 of U2 low and pin 11 of U3 high, energizing K4, the cw rotation control relay. When S4 is released, a short pulse appears at pin 2 of U4, triggering the monostable multivibrator. While pin 3 of U4 is high, the brake remains released, and the selection switches are disabled by the logic 1 on pin 9 of U3 and pin 13 of U3. In a similar fashion, pressing S5 energizes the brake relay K3, LED, D1 and the cw rotation control relay, K5. Whenever one of the direction control relays is energized, the ROTATE LED, D2, illuminates to indicate the rotor is turning.

The circuit has been designed to detect the simultaneous selection of both rotation directions using a NAND gate in U2. If both are pressed, a transition to 0 at pin 4 of U2 triggers the monostable multivibrator, forcing a brake-delay period. In this way, the rapid rocking of the antenna back and forth is eliminated. After the end of the delay cycle, if both

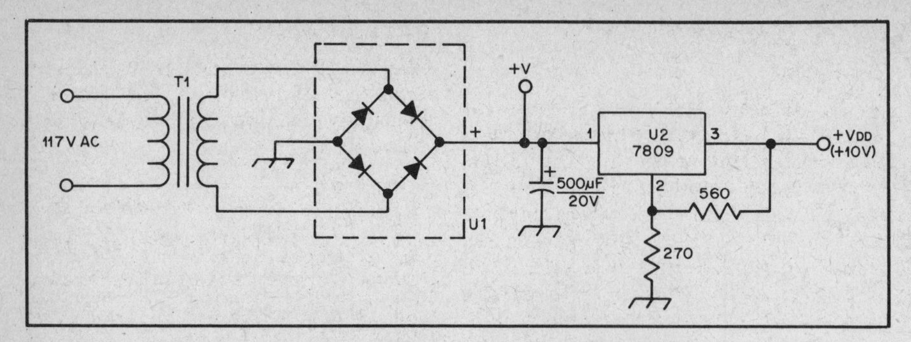


Fig. 71 — Regulated power supply for the delayed brake release system.
T1 — Power transformer; pri. 117 V; sec. 12 V, 300 mA; Radio Shack 273-1385 or equiv.
U1 — Bridge rectifier, 50 PIV, 1.5 A; Radio Shack 267-1151 or equiv.
U2 — Monolithic three-terminal positive-voltage regulator, 9 V, 500 mA; Fairchild 7809 or equiv.

Fig. 72 — Modification of the Ham-II control unit showing the Vector circuit board and components.

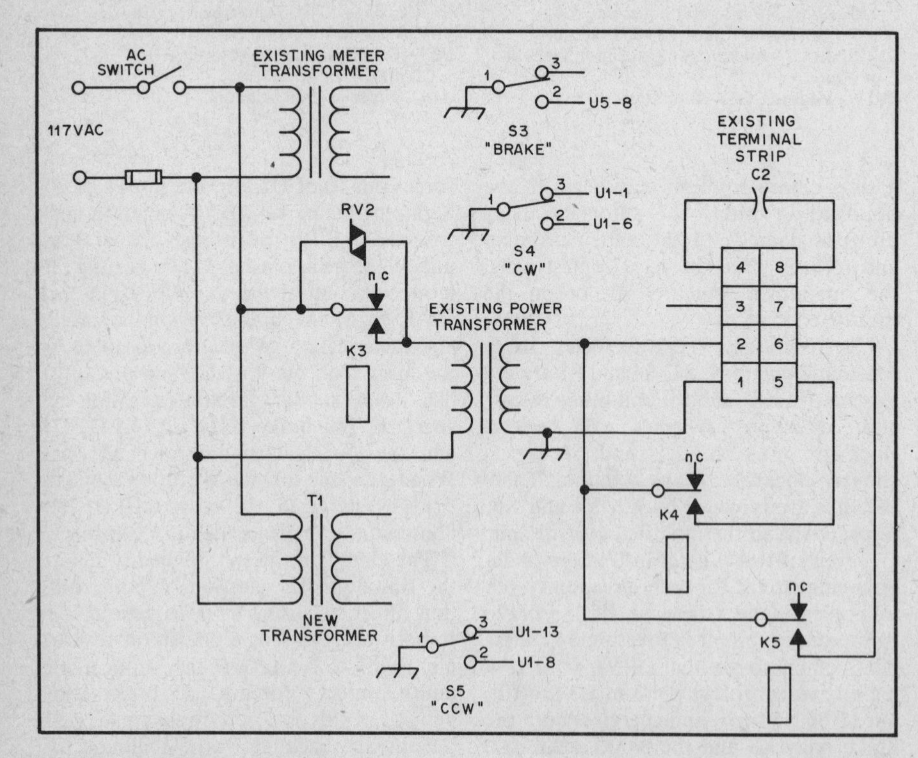


Fig. 73 — The Ham-II circuit modifications.

direction switches are still pressed, neither control relay is energized since both pins 8 and 12 of U3 are high, keeping Q4 and Q5 off.

If a longer delay is desired the brake can be released manually with S3. D1 signals when the brake is energized, but no delay cycle is initiated.

The delay timer (NE555) is connected in a monostable multivibrator configuration. The components R and C at pins 6 and 7 of U4 determine the length of the delay. The values shown provide a delay period of about 3 seconds. An alternative is to use a potentiometer for R as shown in Fig. 70A to yield a variable delay of 2 to 8 seconds.

**Construction**

CMOS integrated circuits were used in this design because of their high-noise margin, low power dissipation, and tolerance of varying supply voltage. CMOS units will operate with a $V_{DD}$ ranging from 3 to 15 volts, although the 10-volt regulator shown in Fig. 71 is used in this unit. TTL circuits may be substituted but some immunity to rf would be sacrificed, and, of course, the pin connections of the devices are different.

The transistor drivers Q1 through Q5 are necessary since these CMOS devices cannot draw enough current to energize either the relays or the LEDs. The 0.01-µF capacitor on the base of each transistor is included to eliminate false keying of the relays by stray rf. An added precaution is the transient suppressor shown across the contacts of K3. The brake relay connects the line voltage to the primary of the brake and rotation power transformer. Without the suppressor, the contacts of K3 would pit badly because of arcing when the relay contacts open.

The circuit as shown in Fig. 72 is constructed on a Vector IC breadboard circuit card using IC sockets and standard wire-wrap techniques. One could just as easily use a homemade printed-circuit board or any other fabrication technique since the layout is not critical.

Fig. 73 illustrates the Ham-II circuit modifications. Relays K3, K4 and K5 replace S3, S4 and S5 in the original diagram, and the primary of a small 12-V ac power transformer is connected to the control-unit ac power switch.

There is more than enough room beneath the Ham-II chassis to mount the delay-circuit card. It may be necessary to relocate the phasing capacitor, C2, above the chassis. The wires that were originally connected to S3, S4 and S5 are relocated, connecting them to the corresponding relay contacts. The switches are connected to the delay-circuit inputs. These switches are single-pole double-throw microswitches with the contact configuration shown in Fig. 70B. In our unit the LEDs are mounted below the switches in the front

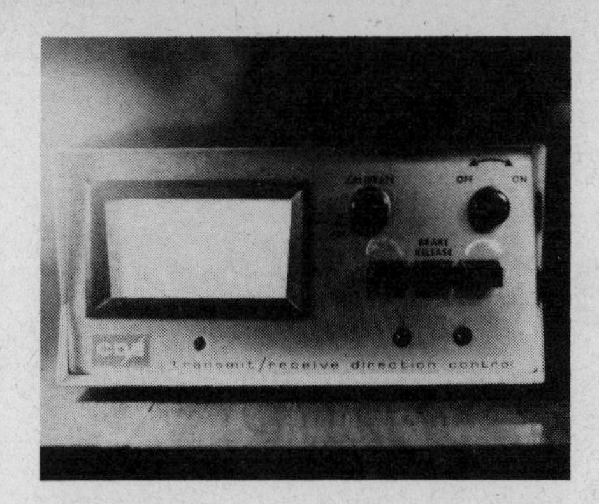

Fig. 74 — A view of the control panel of the Ham-II rotator.

panel as pictured in Fig. 74.

### Operation

The modified rotator control unit is used in the same manner as always except that the operation of S3, the brake release, is now automatic. Both LEDs, D1 and D2, are illuminated during rotation and D1 (BRAKE RELEASED) remains on through the brake-delay cycle after rotation. Because the antenna will coast approximately 10 degrees, the operator must release the rotation switch about 10 degrees before the antenna reaches the desired direction. With practice, the early release becomes natural.

## BIBLIOGRAPHY

Source material and more extended discussion of topics covered in this chapter can be found in the references given below.


Bergren, A. L., "The Multielement Quad," *QST,* May 1963.

Erhorn, P. C., "Element Spacing in 3-Element Beams," *QST,* October 1947.

Reynolds, F., "Simple Gamma Match Construction" *QST,* July 1957

Rhodes, P. D., "The Log-Periodic Dipole Array," *QST,* November 1973.

*Structural Standards for Steel Antenna Towers and Antenna Supporting Structures,* EIA Standard RS-222-C, Electronic Industries Association, March 1976, available from EIA, 2001 Eye St., N.W., Washington, DC 20006, price $7.40.

Chapter 10

# HF Antennas for Restricted Space

It is not always practical to erect full-size antennas for the hf bands. Those who live in apartment buildings may be restricted to the use of miniscule radiators because of house rules, or simply because the required space for full-size antennas does not exist. Other amateurs may desire small antennas for aesthetic reasons, perhaps to prevent the neighbors in their residential areas from becoming annoyed at the sight of a high tower and full-size beam antenna. There are many reasons why some amateurs prefer to use physically shortened antennas, and this chapter discusses various ways of building and using them effectively.

Few compromise antennas are capable of delivering the performance one can expect from the full-size variety. But the patient and skillful operator can often do as well as some who are equipped with high power and full-size antennas. The former may not be able to "bore a hole" in the band as often, and with the commanding dispatch enjoyed by those who are better equipped, but DX can be worked successfully when band conditions are suitable.

## "INVISIBLE" ANTENNAS

We amateurs don't regard our antennas as eyesores; in fact, we almost always regard them as works of art! But, there are occasions when having an outdoor or visible antenna can present problems.

When we are confronted with restrictions — self-imposed or otherwise — we can take advantage of a number of options toward getting on the air and radiating at least a moderately effective signal. In this context, a poor antenna is certainly better than no antenna at all!

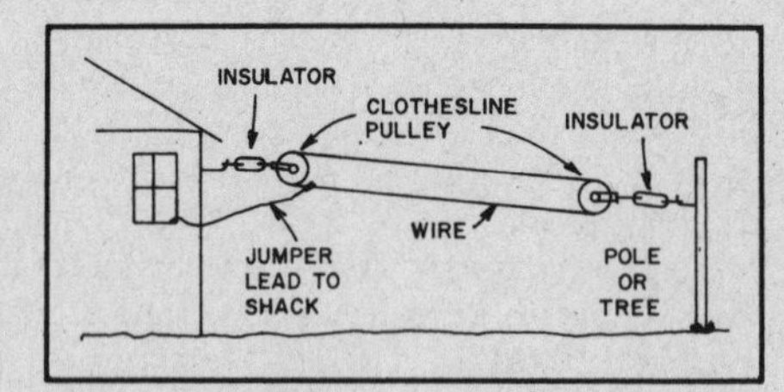


Fig. 1 — The clothesline antenna is more than what it appears to be.

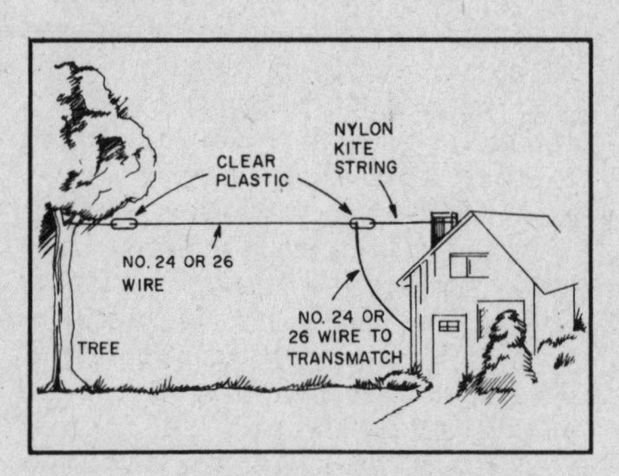


Fig. 2 — The "invisible" end-fed antenna.

This section describes a number of techniques that enable us to use indoor antennas or "invisible" antennas out of doors. Many of these systems will yield good to excellent results for local and DX contacts, depending on band conditions at any given time. *The important consideration is that of not erecting any antenna that can present a hazard (physical or electrical) to humans, animals and buildings. Safety first!*

## CLOTHESLINE ANTENNA

There are some areas of the world where clotheslines are attached to pulleys (Fig. 1) so that the user can load the line and retrieve the laundry from a back porch. Laundry lines of this variety are accepted parts of the neighborhood "scenery," and can be used handily as amateur antennas by simply insulating the pulleys from their support points. This calls for the use of a conducting type of clothesline, such as heavy-gauge stranded electrical wire with Teflon or vinyl insulation. A high-quality, flexible steel cable (stranded) is suitable as a substitute if one doesn't mind cleaning it each time clothing is hung on it.

A jumper wire can be brought from one end of the line to the ham shack when the station is being operated. If a good electrical connection exists between the wire clothesline and the pulley, a permanent connection can be made by connecting the lead-in wire between the pulley and its insulator. A Transmatch can be used to match the "invisible" random-length wire to the transmitter and receiver.

## INVISIBLE "LONG WIRE"

In reality, an antenna is not a classic "long wire" unless it is one wavelength or greater in length. Yet many amateurs refer to (relatively) long physical spans of conductor as "long wires." For the purpose of this discussion we will assume we have a fairly long span of wire, and refer to it as an "end-fed wire."

If we use small-diameter enameled wire for our end-fed antenna, chances are that it will be very difficult to see against the sky and neighborhood scenery. The lighter the wire gauge the more "invisible" the antenna will be. The limiting factor is fragility. A good compromise is no. 24 or no. 26 magnet wire for spans up to 130 feet. Lighter gauge wire can be used for shorter spans, such as 30 or 60 feet. The major threat to the longevity of fine wire is icing. Also, birds may fly into the wire and break it. Therefore, this style of antenna may require frequent service or replacement.

Fig. 2 illustrates how we might install an invisible end-fed wire. It is important that the insulators also be lacking in prominence. Tiny Plexiglas blocks work well in this regard. Small-diameter clear plastic medical vials are suitable also. Some amateurs simply use rubber bands for end insulators, but they will deteriorate rapidly from sun and air pollutants. They are entirely adequate for short-term operation with an invisible antenna, however.

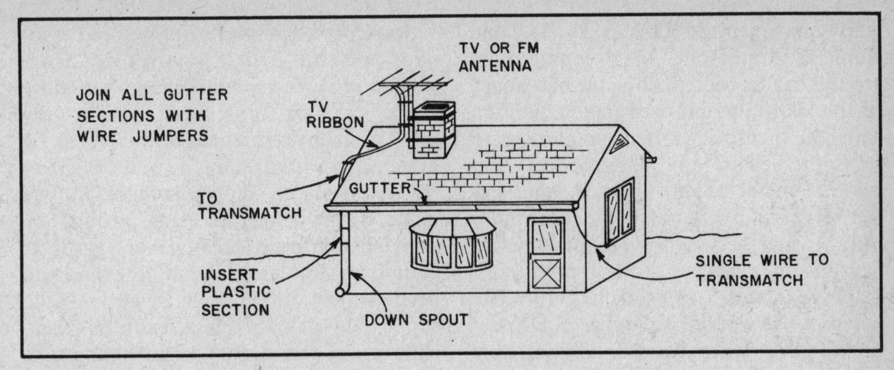


Fig. 3 — Rain gutters and TV antenna installations can be used as inconspicuous Amateur Radio antennas.

## RAIN GUTTER AND TV ANTENNAS

A great number of amateurs have taken advantage of standard house fixtures when contriving inconspicuous antennas. A very old technique is the use of the gutter and downspout system on the building. This can be seen in Fig. 3, where a lead wire is routed to the operating room from one end of the gutter trough. We must assume that the wood on which the gutter is affixed is dry and of good quality to provide a reasonable insulation factor. The rain-gutter antenna may perform quite poorly during wet weather or when there is ice and snow on it and the house roof.

We need to ensure that all joints between gutter and downspout sections are electrically bonded with straps of braid or flashing copper to provide good continuity in the system. Poor joints can cause rectification and subsequent TVI and other harmonic interference. Also, it is prudent to insert a section of plastic downspout about 8 feet above ground. This will prevent humans from receiving rf shocks or burns while the antenna is being used. Improved performance may result if the front and back gutters of the house are joined by a jumper wire to increase the area of the antenna.

Fig. 3 also shows a TV or fm antenna that can be employed as an invisible amateur antenna. Many of these antennas can be modified easily to accommodate the 144- or 220-MHz bands, thereby permitting the use of the 300-ohm line as a feeder system. Some fm antennas can be used on 6 meters by adding no. 10 buswire extensions to the ends of the elements, and adjusting the match for a VSWR of 1:1. If 300-ohm line is used it will require a balun or Transmatch to interface the line with the station equipment.

For operation in the hf bands we can tie the TV or fm-antenna feeders together at the transmitter end of the span and treat the overall system as a random-length wire. If this is done, the 300-ohm line will have to be on TV standoff insulators and spaced well away from phone and power company service-entrance lines. The TV or fm radio must, naturally, be disconnected from the system when it is used for amateur work! Similarly, masthead amplifiers and splitters must be removed from the line if the system is to be used for amateur operation.

## FLAGPOLE ANTENNAS

We can exhibit our patriotism and have an invisible amateur antenna at the same time by disguising our antenna as shown in Fig. 4. The vertical antenna is a wire that has been placed inside a plastic or fiberglass pole.

As shown, the flagpole antenna is structured for a single amateur band, and it is assumed that the height of the pole corresponds to a quarter wavelength for the chosen band. The radials and feed line can be buried in the ground as shown. In a practical installation, the sealed end of the coax cable would protrude slightly into the lower end of the plastic pole.

If a large-diameter fiberglass pole were available, we might be able to conceal a four-band trap vertical inside it. Or, we might use a metal pole and bury at its base a water-tight box, which contained fixed-tuned matching networks for the bands of interest. The networks could be selected remotely by means of a stepping relay inside the box. A 30-footflag pole would provide good results in this kind of system, provided it was used in conjunction with a buried radial system. At least one commercial antenna (Delta Corp.) is used in this manner, but with an elaborate continuously adjustable matching network (and VSWR indicator) that is operated remotely.

Still another technique is one that employs a wooden flagpole. A small-diameter wire can be stapled to the pole and routed underground to the coax feeder or matching box. The halyard could by itself constitute the antenna wire if it were made from heavy-duty insulated hookup wire. There are countless variations for this type of antenna, and they are limited only by the imagination of the amateur. Detailed plans for a flagpole antenna can be found at the end of this chapter.

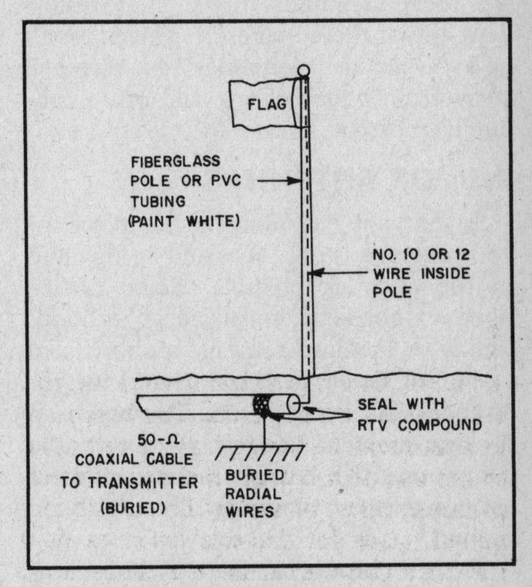


Fig. 4 — A flagpole antenna.

## OTHER INVISIBLE ANTENNAS

Some amateurs have used the metal fence on apartment verandas as antennas, and have had good results on the higher hf bands (20, 15 and 10 meters). We must presume that the fences were not connected to the steel framework of the building, but rather were insulated by the concrete floor to which they were affixed. These veranda fences have also been used effectively as ground systems (counterpoises) for hf-band vertical antennas that were put in place temporarily after dark.

One New York City amateur used the fire escape on his apartment building as a 40-meter antenna, and reported high success in working DX stations with it. Another apartment dweller made use of the aluminum frame on his living-room picture window as an antenna for 10 and 15 meters. He worked it against the metal conductors of the baseboard heater in the same room.

There have been many jokes told over the past decades about "bedspring antennas." The idea is by no means absurd. Bedsprings and metal end boards have been used to advantage as antennas by many apartment dwellers as 20-, 15-, and 10-meter radiators. A counterpoise ground can be routed along the baseboard of the bedroom and used in combination with the bedspring. It is important to remember that any independent (insulated) metal object of reasonable size can serve as an antenna if the transmitter can be matched to it. An amateur in Detroit once used his Shopsmith craft machine (about 5 feet tall) as a 10-meter antenna. He worked a number of DX stations with it when band conditions were good.

A number of operators have used metal curtain rods and window screens for vhf work, and found them to be acceptable for local communications. Best results with any of these makeshift antennas will be had when the "antennas" are kept well away from house wiring and other conductive objects.

## INDOOR ANTENNAS

Beyond any question, the best place for your antenna is outdoors and as high and in the clear as possible. Some of us, however, for legal, social, neighborhood, family or landlord reasons, are restricted to indoor antennas. Having to settle for an indoor antenna is certainly a handicap for the amateur seeking effective radio communication, but that is not significant enough reason to abandon all operation in despair.

First, we should be aware of the reasons why indoor antennas *do not* work well. Principal faults are (1) low height above ground — the antenna cannot be placed higher than the highest peak of the roof, a point usually low in terms of wavelength at hf; (2) the antenna must function in a lossy rf environment that involves close coupling to electrical wiring, guttering, plumbing and other parasitic conductors, besides dielectric losses in such nonconductors as wood, plaster and masonry; (3) sometimes the antenna must be made small in terms of a wavelength and (4) usually it cannot be rotated. These are appreciable handicaps. Nevertheless, global communication with an indoor antenna is still possible.

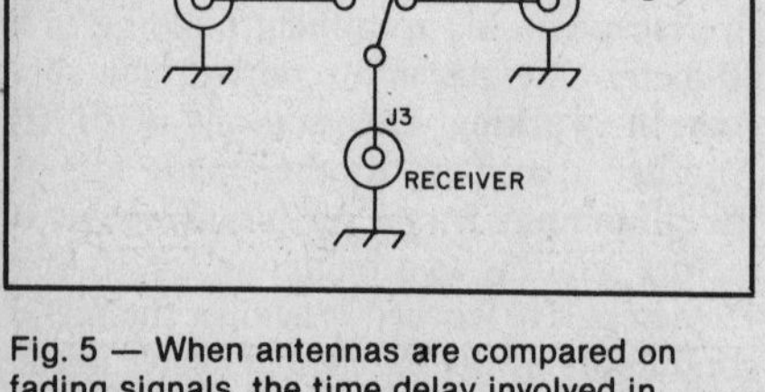

Fig. 5 — When antennas are compared on fading signals, the time delay involved in disconnecting and reconnecting coaxial cables is too long for accurate measurements. A simple slide switch will do well for switching coaxial lines at hf. The four components can be mounted in a tin can or any small metal box. Leads should be short and direct. J1 through J3 are coaxial connectors.

Some practical points *in favor* of the indoor antenna include (1) freedom from weathering effects and damage caused by wind, ice, rain, dust and sunlight (the SWR of an attic antenna, however, can be affected somewhat by a wet or snow-covered roof); (2) indoor antennas can be made from materials that would be altogether impractical outdoors, such as aluminum foil and thread (the antenna need support only its own weight); (3) the supporting structure is already in place, eliminating the need for antenna masts and (4) the antenna is readily accessible in all weather conditions, simplifying pruning or tuning, which can be accomplished without climbing or tilting over a tower.

### Empiricism

A typical house or apartment involves such a complex electromagnetic environment that it is impossible to predict theoretically which location or orientation of the indoor antenna will work best. This is where good old-fashioned cut-and-try, use-what-works-best empiricism pays off. But to properly determine what really is most suitable requires an understanding of some antenna-measuring fundamentals.

Unfortunately, many amateurs do not know how to evaluate performance scientifically or compare one antenna with another. Typically, they will put up one antenna and try it out on the air to see how it "gets out" in comparison with a previous antenna. This is obviously a very poor evaluation method because there is no way to know if the better or worse reports are caused by changing band conditions, different S-meter characteristics or any of several other factors that could influence the reports received.

Many times the difference between two antennas or between two different locations for identical antennas amounts to only a few decibels, a difference that is hard to discern unless instantaneous switching between the two is possible. Those few decibels are not important under strong-signal conditions, of course, but when the going gets rough, as is often the case with an indoor antenna, a few dB can make the difference between copy and no copy.

Very little in the way of test equipment is needed for antenna evaluation other than a communications receiver. You can even do a qualitative comparison by ear, *if* you can switch antennas instantaneously. Differences of less than 2 dB, however, are still hard to discern. The same is true of S meters. Signal-strength differences of less than a decibel are not usually visible. If you want that last fraction of a decibel, you should use an ac VTVM at the receiver audio output (with the agc turned off, of course).

In order to compare two antennas, switching the coaxial transmission line from one to the other becomes necessary. No elaborate coaxial switch is needed; even a simple double-throw toggle or slide switch will provide more than 40 dB of isolation at hf. See Fig. 5. Switching by means of manually connecting and disconnecting coaxial lines is not recommended because that takes too long. Fading can cause strength changes during the changeover interval.

Whatever difference shows up in the strength of the received signal will be the difference in performance between the two antennas in the direction of that signal. For this test to be valid, both antennas must have nearly the same feed-point impedance, a condition that is reasonably well met if the SWR is below 2.0 on both.

On ionospheric-propagated signals (sky wave) there will be constant fading and for a valid comparison it will be necessary to make an average of the difference between the two antennas. Occasionally, the inferior antenna will deliver a stronger signal to the receiver, but in the long run the law of averages will put the better antenna ahead.

Of course with a ground-wave signal, such as that from a station across town, there will be no fading problems. A ground-wave signal will enable the operator to evaluate properly the antenna under test in the direction of the source and will be valid for ionospheric propagated signals at low elevation angles in that direction. On 10 meters, all sky-wave signals arrive and leave at low angles. But on the lower bands, particularly 80 and 40 meters, we often use signals propagated at high elevation angles, almost up to the zenith. For these angles a ground-wave test will not provide a proper evaluation of the antenna and use of sky-wave signals becomes necessary.

## DIPOLES

At hf the most practical indoor antenna is usually the dipole. Any attempt to get more gain with parasitic elements will usually fail because of close proximity of the ground or coupling to house wiring. Beam-antenna dimensions determined outdoors will not usually be valid for an attic antenna because the roof structure will cause dielectric loading of the parasitic elements. It is usually more worthwhile to spend time optimizing the location and performance of a dipole than to try to improve results with parasitic elements.

Most attics are not long enough to accommodate a doublet for 75 meters. Many are not even large enough for a 40-meter aerial. This means some folding of the dipole will be necessary. The final shape of the antenna will depend on the dimensions and configuration of the attic. Remember that the center of the dipole carries the most current and therefore does most of the radiating. This part should be as high and unfolded as possible. Because the dipole ends radiate less energy than the center, their orientation is not very important. They do carry a maximum voltage, nevertheless, so care should be taken to position the ends far enough from other conductors to avoid arcing. The dipole may end up being L shaped, Z shaped, U shaped or some indescribable corkscrew shape, depending on what space is available. Fig. 6 shows some possible configurations. Multiband operation may be had with open wire feeders and a Transmatch.

If coaxial feed is used, some pruning to establish minimum SWR at the band center will be required. Tuning the antenna outdoors and then installing it inside is usually not feasible. The behavior of the antenna will not be the same when placed in the attic. Resonance will be somewhat affected if the antenna is bent. Even if it is placed in a straight line, parasitic conductors and dielectric loading by nearby wood structures will affect the impedance.

Trap and loaded dipoles are shorter than the full sized versions, but are comparable performers. Trap dipoles are discussed in Chapter 8; loaded dipoles later in this chapter.

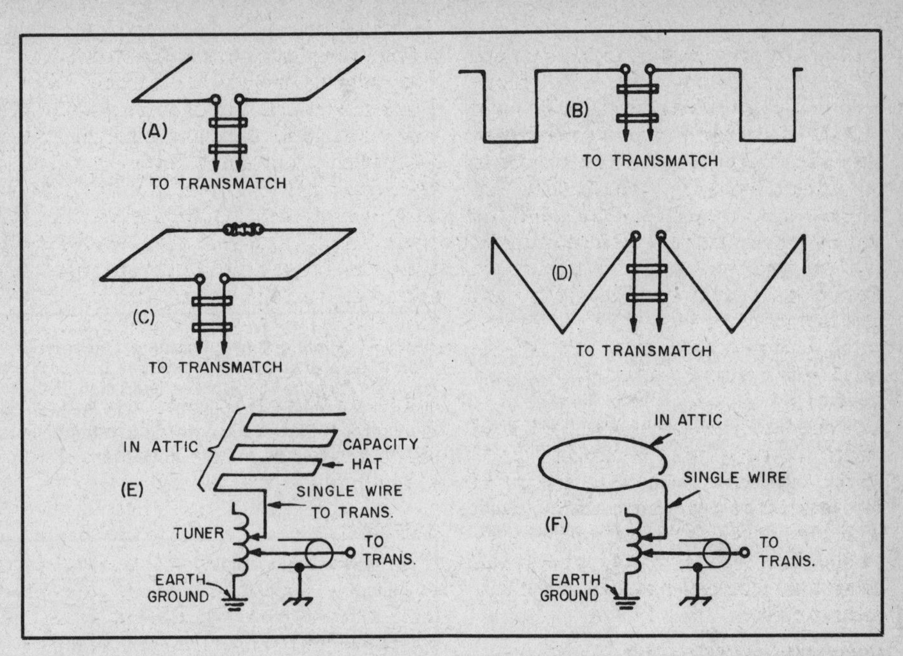


Fig. 6 — Various configurations for small indoor antennas. A discussion of installation is contained in the text.

### Orientation

Theoretically a vertical dipole is more effective at low radiation angles, but practical experience shows that the horizontal dipole is usually a better indoor antenna. A horizontal dipole, however, exhibits directional effects. Theoretically there are nulls off the ends, especially at low radiation angles. Your dipole, therefore, should be broadside to directions in which you are interested in working. If you want 360-degree coverage, place two dipoles at right angles to each other with provision at the operating position for switching between the two. In fact, this is a good idea even if you are not interested in 360-degree coverage, because the radiation patterns will inevitably be distorted in an unpredictable manner by parasitic condutors. There will be little coupling between the dipoles if they are oriented at right angles to each other as shown in Figs. 7A and 7B. There will be some coupling with the arrangement shown in Fig. 7C but even this orientation is preferable to a single dipole.

You may find that one dipole is consistently better in nearly all directions, in which case you will want to remove the inferior dipole, perhaps placing it someplace else. In this manner the best spots in the house or attic can be determined experimentally.

## PARASITIC CONDUCTORS

Inevitably, any conductor in your house near a quarter wave in length or longer will be parasitically coupled to your antenna. The word *parasitic* is particularly appropriate in this case because these conductors will absorb energy and leave less for radiation into space. Unlike the parasitic elements in a beam antenna, conductors such as house wiring and plumbing are usually connected to lossy objects such as earth, electrical appliances, masonry or other objects that can dissipate energy. Even where this energy is reradiated, it is not likely to be in the right phase in the desired direction.

There are, however, some things that can be done about parasitic conductors. The most obvious is to reroute them at right angles to the antenna or close to the ground, or even underground — procedures that are usually not feasible in a finished home. Where these conductors cannot be rerouted, other measures can be taken. Electrical wiring can be broken up with rf chokes to prevent the flow of radio-frequency currents while permitting 60-Hz current (or audio, in the case of telephone wires) to flow unimpeded. A typical rf choke for a power line can be 100 turns of no. 10 insulated wire closewound on a length of 1-inch dia plastic pipe. Of course one choke will be needed for each conductor. A three-wire line calls for three chokes.

## THE RESONANT BREAKER

Obviously, rf chokes cannot be used on conductors such as metal conduit or water pipes. But it is still possible, surprising as it may seem, to obstruct rf currents on such conductors without breaking the metal.

Fig. 8 discloses how this is done. A figure-eight loop is inductively coupled to the parasitic conductor and is resonated to the desired frequency with a variable capacitor. The result is a very high impedance induced in series with the pipe, conduit or wire. This impedance will

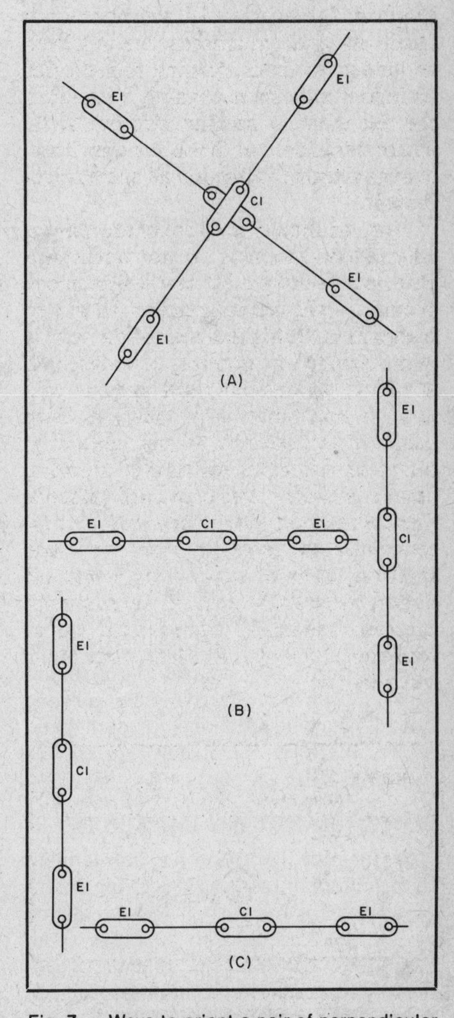


Fig. 7 — Ways to orient a pair of perpendicular doublets for 360-degree coverage. The orientations of A and B will result in no mutual coupling between the two dipoles, but there will be some coupling with the configuration shown at C. End (E1) and center (C1) insulators are shown.

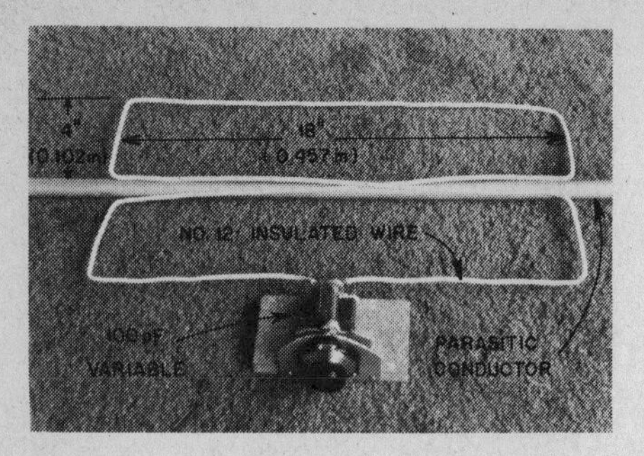


Fig. 8 — A "resonant breaker" such as shown above can be used to obstruct radio-frequency currents in a conductor without the need to break the conductor physically. A vernier dial is recommended for use with the variable capacitor because tuning is quite sharp. The 100-pF capacitor is in series with the loop. This resonant breaker tunes through 10, 15 and 20 meters. Larger models may be constructed for 40 or 80 meters.

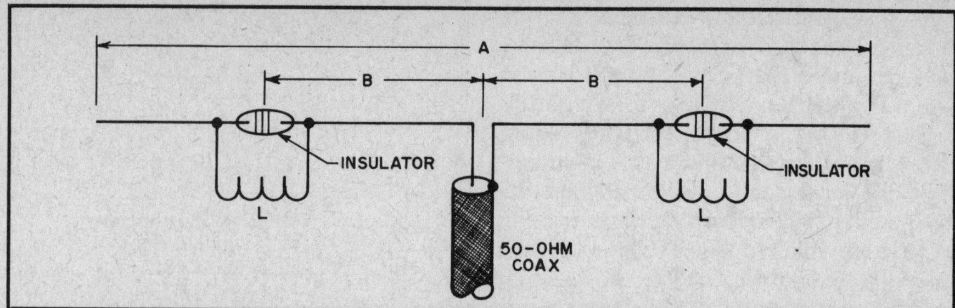


Fig. 9 — Dipole antenna lengthened electrically with off-center loading coils. For a fixed dimension A, greater efficiency will be realized with greater distance B, but as B is increased, L must be larger in value to maintain resonance. If the two coils are placed at the ends of the antenna, in theory they must be infinite in size to maintain resonance. Capacitive loading of the ends, either through proximity of the antenna to other objects or through the addition of capacitance hats, will reduce the required value of the coils.

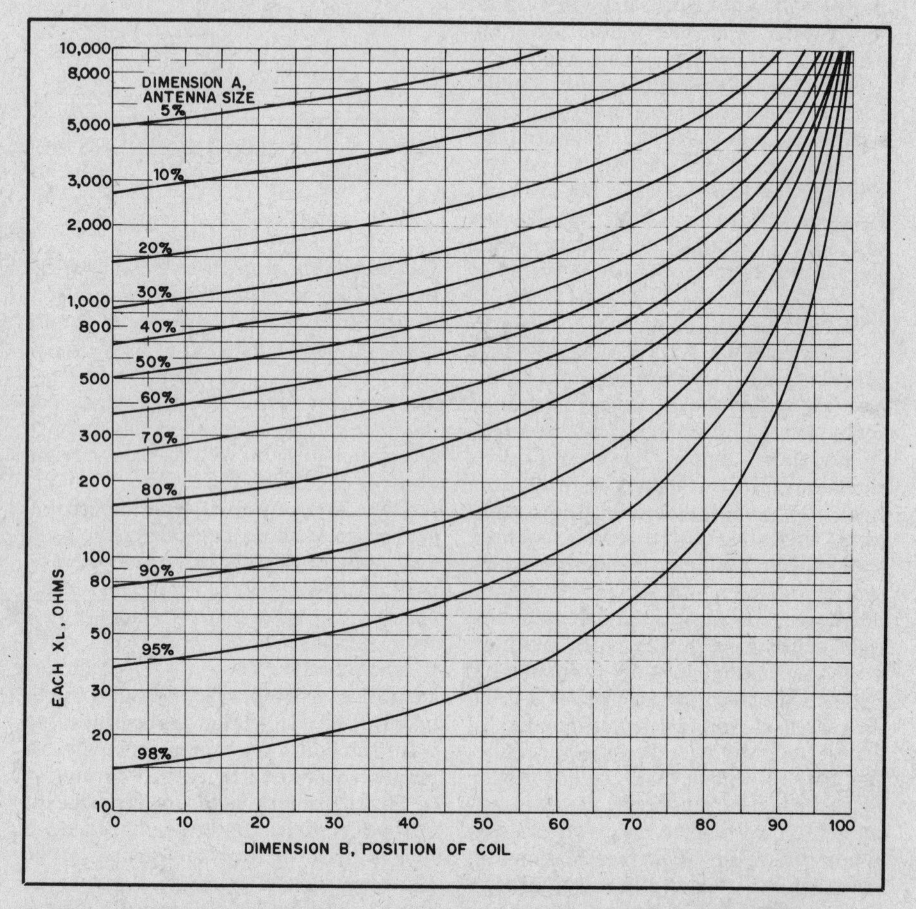


Fig. 10 — Chart for determining approximate inductance values for off-center-loaded dipoles. At the intersection of the appropriate curve from the body of the chart for dimension A and the proper value for the coil position from the horizontal scale at the bottom of the chart, read the required inductive reactance for resonance from the scale at the left. See Fig. 9. Dimension A is expressed as percent length of the shortened antenna with respect to the length of a half-wave dipole of the same conductor material. Dimension B is expressed as the percentage of coil distance from the feed point to the end of the antenna. For example, a shortened antenna which is 50% or half the size of a half-wave dipole (one-quarter wavelength overall) with loading coils positioned midway between the feed point and each end (50% out) would require coils having an inductive reactance of approximately 950 ohms at the operating frequency for antenna resonance.

block the flow of radio-frequency currents. The figure-eight coil can be thought of as two turns of an air-core toroid and since the parasitic conductor threads through the hole of this core, there will be tight coupling between the two. Inasmuch as the figure-eight coil is parallel resonated, transformer action will reflect a high impedance in series with the linear conductor.

Before you bother with a "resonant breaker" of this type, first be sure that there is a significant amount of rf current flowing in the parasitic conductor. The relative magnitude of this current can be determined with an rf current probe of the type described in Chapter 15. According to the rule of thumb regarding parasitic conductor current, if it measures less than 1/10 of that measured near the center of the dipole, the parasitic current is generally not large enough to be of concern.

The current probe is also needed for resonating the breaker after it is installed. Normally, the resonant breaker will be placed on the parasitic conductor near the point of maximum current. When it is tuned through resonance, there will be a sharp dip in rf current, as indicated by the current probe. Of course, the resonant breaker will be effective only on one band. You will need one for each band where there is significant current as measured by the probe.

## POWER-HANDLING CAPABILITY

So far, our discussion has been limited to the indoor antenna as a receiving antenna, except for the current measurements, where it is necessary to supply a small amount of power to the antenna. These measurements will not indicate the full power-handling capability of the antenna. Any tendency to flash over must be determined by running full power or, preferably, somewhat more than the peak power you intend to use. The antenna should be carefully checked for arcing or rf heating before you do any operating. Bear in mind that attics are indeed vulnerable to fire hazards. A potential of several hundred volts exists at the ends of a dipole fed by the typical Amateur Radio transmitter. If an external amplifier is used, there could be a few thousand volts at the ends of the dipole. Keep your antenna elements well away from other objects. *Safety first!*

## OFF-CENTER-LOADED DIPOLE ANTENNAS

Physically shortened dipoles are practical and should be of interest to the indoor-antenna user. When there is insufficient area to mount a full-size dipole, one can install a loading inductor in each leg of the doublet and tune the system to resonance by adjusting the number of coil turns. Fig. 9 is a drawing of such an antenna. The longer the overall length, dimension A, and the farther the loading coils are positioned from the center of the antenna, dimension B, the greater the efficiency of the antenna. The greater is distance B (for a fixed overall antenna size), however, the larger the inductors must be to maintain resonance. Approximate inductance values for single-band resonance may be determined with the aid of Fig. 10, but the final values will depend on the proximity of surrounding objects in individual installations and must be determined experimentally. The use of high-Q low-loss coils is suggested. A dip meter, Macromatcher, or SWR indicator is recommended for use during adjustment of the system.

**Outdoor Antennas**

It is possible to reduce the physical size of an antenna by 50% or more and still obtain good results. Use of an outdoor off-center-loaded dipole, as described in the previous section, will permit a city-size lot to accommodate a doublet antenna on the lower frequency hf bands, 40, 80, or even 160 meters. Short vertical antennas can also be made effective by using lumped inductance to obtain resonance, and by using a capacitance hat to increase the feed-point impedance of the system. As is the case with full-size vertical quarter-wave antennas, the ground-radial network should be as effective as possible. Ground-mounted vertical radiators should be used in combination with several buried radials. Above-ground vertical antennas should be worked against at least four quarter-wavelength radials.

Fig. 11 — Jerry Sevick, W2FMI, adjusts the 6-foot, 40-meter vertical.

Fig. 12 — Construction details for the top hat. For a diameter of 7 feet, half-inch aluminum tubing is used. The hose clamp is of stainless steel and available at Sears. The rest of the hardware is aluminum.

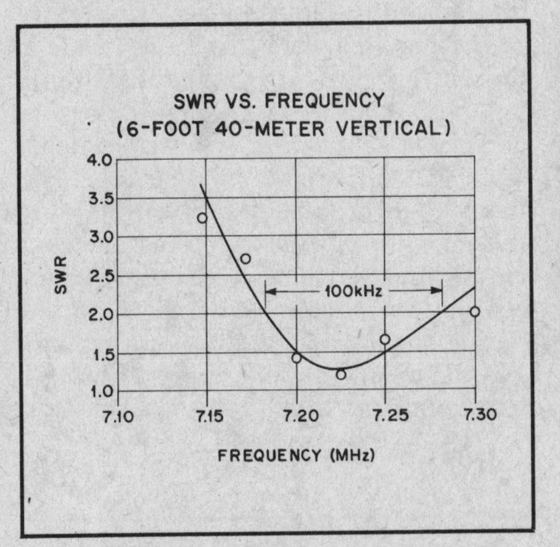


Fig. 13 — Standing-wave ratio of the 6-foot vertical using a 7-foot top hat and 14 turns of loading 6 inches below the top hat.

Fig. 14 — Base of the vertical antenna showing the 60 radials. The aluminum disc is 15 inches in diameter and 1/4-inch thick. Sixty tapped holes for 1/4-20 aluminum hex-head bolts form the outer ring and 20 form the inner ring. The insulator is polystyrene material (phenolic or Plexiglas suitable) with a 1-inch diameter. Also shown is the impedance bridge used for measuring input resistance.

## A 6-FOOT-HIGH 40-METER VERTICAL ANTENNA

Figs. 11 through 14 give details for building short, effective vertical quarter-wavelength radiators. The information gathered and presented here was provided by Sevick, W2FMI. (See the bibliography for reference to his *QST* articles on the subject of shortened antennas.)

A short vertical antenna, properly designed and installed, approaches the efficiency of a full-size resonant quarter-wave antenna. Even a 6-foot vertical on 40 meters can produce an exceptional signal. Theory tells us that this should be possible, but the practical achievement of such a result requires an understanding of the problems of ground losses, loading, and impedance matching, treated in the theory chapters of this book.

The key to success with shortened vertical antennas lies in the efficiency of the ground system with which the antenna is used. A system of 60 wire radials is recommended for best results, though the builder may want to reduce the number at some expense to performance. The radials can be tensioned and pinned at the far ends to permit on-the-ground installation, which will enable the amateur to mow the lawn without the wires becoming entangled in the mower blades. Alternatively, the wires can be buried in the ground where they will not be visible. There is nothing critical about the wire size for the radials. No. 28, 22, or even 16-gauge, will provide the same results. The radials should be at least 0.2 wavelength long (27 feet or greater).

A top hat is formed as illustrated in Fig. 12. The diameter is 7 feet, and a continuous length of wire is connected to the spokes around the outer circumference of the wheel. A loading coil consisting of 14 turns of B&W 3029 Miniconductor stock (2-1/2 inch dia, 6 TPI, no. 12 wire) is installed 6 inches below the top hat (see Fig. 11). This antenna exhibits a feed-point impedance of 3.5 ohms at 7.21 MHz. For operation above or below this frequency, the number of coil turns must be decreased or increased, respectively. Matching is accomplished by increasing the feed-point impedance to 14 ohms through addition of a 4:1 transformer, then matching 14 ohms to 50 ohms (feeder impedance) by means of a pi network. Bandwidth for this antenna is approximately 100 kHz (range of frequency where SWR is less than 2:1).

More than 200 contacts with the 6-foot antenna have strongly indicated the efficiency and capability of a short vertical. Invariably at distances greater than 500 or 600 miles, the short vertical yielded excellent signals. Similar antennas can be scaled and constructed for bands other than 40 meters. The 7-ft-dia top hat was tried on an 80-m vertical, with an antenna height of 22 ft. The loading coil had 24 turns and was placed 2 feet below the top hat. On-the-air results duplicated those on 40 meters. The bandwidth was 65 kHz.

Short verticals such as these have the ability to radiate and receive almost as well as a full-size quarter-wave antenna. The differences are practically negligible. Trade-offs are in lowered input im-

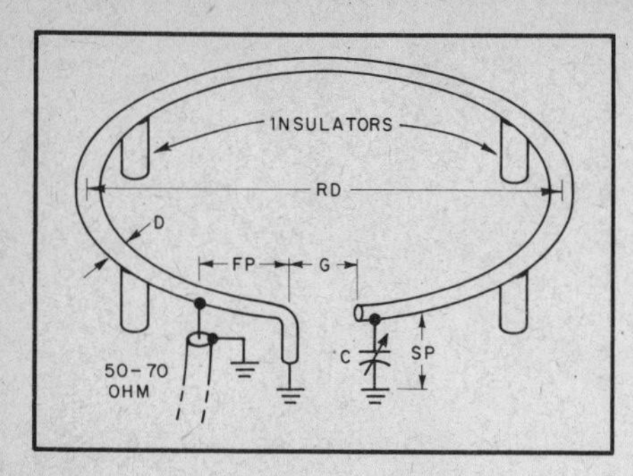


Fig. 15 — RD = 0.078 λ (28°); SP = 0.11D (2.5°); FP = 0.25SP (see Note 1); C = (see Note 2); D = (see Note 3); G = (see Table 1). *Notes:* (1) Actual dimension must be found experimentally. (2) Value to resonate the antenna to the operating freq. (3) D ranges upward from 1/2". The larger D is, the higher the efficiency is. Use largest practical size, e.g., 1/2" for 10 meters, 5" or 6" for 80 or 160 meters.

pedances and bandwidths. However, with a good image plane and a proper design, these trade-offs can be entirely acceptable.

## THE DDRR ANTENNA

Another physically small but effective antenna is the DDRR (directional discontinuity ring radiator), described in *Electronics,* January 1963. An in-depth mathematical analysis of this low-profile antenna was given by Dome in July 1972 *QST*. Fig. 15 shows details for constructing a DDRR antenna.

In this example the radiating element of the antenna is made from 2-inch diameter automobile exhaust pipe. Most muffler shops can supply the materials as well as bend the pipe to specifications. Table 1 lists the dimensions required for operation from 160 to 2 meters, inclusive. The following technique illustrates how a 40-meter model can be assembled (from English, W6WYQ, Dec. 1971 *QST*).

In forming the ring to these dimensions, four 10-foot lengths of tubing are used. A 10-degree bend is made at 9-inch intervals in three of the lengths. The fourth length is similarly treated, except for the last 18 inches, which is bent at right angles to form the upright leg of the ring. One end of each section is flared so that the sections can be coupled together by slipping the end of one into the flare of its mate.

The required flares are easily made at the muffler shop with the aid of the forming tools. Another task which can best be completed at the shop is to weld a flange onto the end of the upright leg. This flange is to facilitate attaching the leg to the mounting plate which provides a chassis for the tuning mechanism and the coaxial-feed coupler. After bending and flaring is complete, the ring is assembled and minor adjustments made to bring it into round and to the proper dimensions. This can best be done by drawing a circle on the floor with chalk and fitting the ring inside the circle. The circle must be slightly larger than the center-to-center diameter so that the reference line can be seen easily. For example, with two-inch tubing the diameter of the reference circle must be 9 feet, 2 inches. When a satisfactory fit is obtained between the tubing ring and the chalk ring, drill a 1/4-inch hole through each of the joints to accept a 1/4-inch bolt. These bolts will clamp the sections together. Also, they can be used to attach the insulators which support the ring at a fixed height above the ground plane.

Insulators for the antenna are made from 11-inch lengths of 2-inch PVC pipe inserted into a standard cap of the same material. The PVC caps are first drilled through the center to accept the 1/4-inch bolt previously installed at the joints. The caps are then slipped onto the bolts and nuts are installed and tightened to secure the caps in place. The 11-inch length of pipe, when inserted into the cap and pressed firmly until it touches bottom, results in a total insulator length of 12 inches. Four insulators are required, one at each of the joints and one near the open end of the ring for support. It is wise to locate this insulator as far back from the end of the ring as possible because of the increasing high rf voltage that develops as the end of the ring is approached. (Because of the danger of rf burns in the event of accidental contact with the antenna, precautions should be taken to prevent random access to the completed installation.) As a final measure, the bottom ends of the insulators are sealed to prevent moisture from forming on the inside surfaces. Standard PVC caps may be used here, but plastic caps from 15-ounce aerosol cans fit well.

A mounting plate is required to provide good mechanical and electrical connections for the grounded leg of the radiator, the coaxial feed-line connection, and the tuning mechanism. If you are using aluminum tubing, you should use an aluminum plate, and for steel tubing, a steel plate to lessen corrosion from the contacting of dissimilar metals. Dimensions for the plate are shown in Fig. 16. The important consideration here is that good, solid mechanical and electrical connections are made between the ground side at the coaxial connector, the ring base and the tuning capacitor.

In the installation shown in Fig. 17 the 9-foot ring resonated easily with approximately 20 pF of capacitance between the high-impedance end of the ring and the base plate or ground. Any variable capacitor which will tune the system to resonance and which will not arc under full power should be satisfactory. Remember, the rf voltage at the high impedance end of this antenna can reach 20 to 30 kV with high power, so if you are us-

**Table 1**
**Dimensions for 1/4-Wavelength DDRR Elements**

| *Band (Meters)* | *160* | *80* | *40* | *20* | *15* | *10* | *6* | *2* |
|---|---|---|---|---|---|---|---|---|
| Feed Point (FP) | 12" | 6" | 6" | 2" | 1.5" | 3" | 1" | 1/2" |
| Gap (G) | 16" | 7" | 5" | 3" | 2.5" | 2" | 1.5" | 1" |
| Capacitor, pF (C) | 150 | 100 | 70 | 35 | 15 | 15 | 10 | 5 |
| Spacing (Height) (SP) | 48" | 24" | 11" | 6" | 4 3/4" | 3" | 1 1/2" | 1" |
| Tubing Diameter (D) | 5" | 4" | 2" | 1" | 3/4" | 3/4" | 1/2" | 1/4" |
| Ring Diameter (RD) | 36' | 18' | 9' | 4.5' | 3'4" | 2'4" | 16 1/4" | 6" |

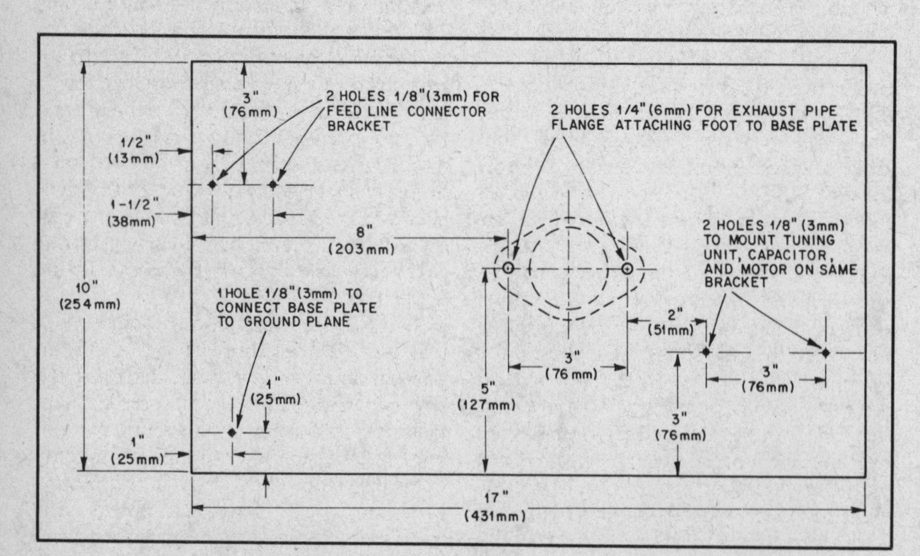


Fig. 16 — Drawing of the base plate, which can be made from either steel or aluminum, as described in the text. The lower right portion of the plate may be used for the mounting of the tuning capacitor (and motor, if used).

Fig. 17 — The chicken-wire ground plane is evident in the background. The base plate can be seen at lower right. Note the relative positions of the 52-ohm coaxial feed at the left end of the plate, the flange on the foot of the post, and the tuning unit at the right hand end of the plate.

ing the maximum legal limit, you would do well to consider using a vacuum variable capacitor. To provide for full band coverage, a 35-pF variable capacitor was coupled to a reversible, slow-speed motor which enabled the antenna to be tuned remotely from the operating position in the antenna pictured. An indicated SWR of 1.1 to 1 was obtained easily over the entire 40-meter band. The motor used was a surplus item made by Globe Industries of Dayton, Ohio. At 20 volts dc the shaft of this motor turns at about 1 rpm, which is ideal for DDRR tuning. The gears used were surplus items. If you cannot obtain gears, string and pulley drive will do almost as well, or you can mount both the motor and the capacitor in line and use direct coupling. Of course, if you operate on a fixed frequency, or within a 40- to 50-kHz segment of the band, you can dispense with the motor entirely and simply tune the capacitor manually. In any case, the tuning unit must be protected from the weather. A plastic refrigerator box may be used to house the tuning capacitor and its drive motor.

**Electrical Connections and the Ground Plane**

The connection between the open end of the ring and the tuning capacitor is made with no. 12 wire or larger. On the end of the base plate opposite the tuning unit, and directly under the ring about 8 inches from the grounded post, install a bracket for a coaxial connector. The connector should be oriented so that the feed line will lead away from the ring at close to 90 degrees. Install a clamp on the ring directly above the coaxial connector. Connect a lead of no. 12 or larger wire from the coaxial connector to the clamp. This wire must have a certain amount of flexibility to accommodate the movement necessary when adjusting the match. The matching point must be found by experimentation. It will be affected by the nature and quality of the ground plane over which the antenna is operating. The antenna will function over earth ground, but a ground-plane surface of chicken wire (laid under the antenna and bonded to the base plate) will provide a constant ground reference and improved performance. In a roof-top location, sheet-metal roofing should provide an excellent ground plane. A poor ground usually results in a matching point for the feed line far out along the circumference of the circle. In the installation shown a near-perfect match was obtained with the feed line connected to the ring about 12 inches from the grounded post. During testing, when the antenna was set up on a concrete surface without the ground plane, a match was found when the feed line was connected nearly 7 feet from the post!

As shown in the photo, the compactness of the antenna is readily apparent. The ground plane is made up of three 12-foot lengths of chicken wire, each 4 feet wide, which are bonded along the edges at about 6-inch intervals. In this installation the antenna, with the ground plane, could be dismantled in about 30 minutes. If portability is not important, it is best to bond all of the joints in the tubing so that good electrical continuity is assured.

After all construction is completed, the antenna should be given a coat of primer paint to minimize rust. If it suits you, there is no reason why a final coat of enamel could not be applied.

**Tuning Procedure**

Once the mechanical construction is completed, the antenna should be erected in its intended operating location. Coupling to the station may be accomplished with either 52- or 72-ohm coaxial cable. Tune and load the transmitter as with any antenna. While observing an SWR meter in the line, operate the tuning motor. Indication of resonance is the noticeable decrease in indicated reflected power. At this point, note the loading of the transmitter; it will probably increase markedly as antenna resonance is approached. Retune the transmitter and move the feed-point tap on the antenna for a further reduction in indicated reflected power. There is interaction between the movement at the feed tap and the resonance point; therefore, it will be necessary to operate the tuning motor each time the tap is adjusted until the lowest SWR is achieved. Don't settle for anything less than 1.1 to 1. With a good ground and proper tuning and matching, this ratio can be achieved and maintained over the entire band. Once the proper feed point has been located, the only adjustment necessary when changing frequency is retuning the antenna to resonance by means of the motor. If the antenna is to be fixed tuned, provide an insulated shaft extension of 18 inches or so to the tuning-capacitor shaft for manual adjustment. This not only provides insulation from the high rf voltage but also minimizes body-capacitance effects during the tuning process.

Fig. 18 — This short two-band Yagi can be turned by a light-duty rotator.

## SHORTENED YAGI BEAM ANTENNA WITH LOADING COILS

At some sacrifice in bandwidth it is practical to shrink the element dimensions of Yagi antennas. Resonance can be established through the use of loading inductors in the elements, or by using inductors in combination with capacitance hats. Though not as effective in terms of gain, the short Yagi beam offers the advantage of small size, less weight, and lower wind loading when compared to a full-size array. When tower-mounted, the two-band Yagi of Figs. 18 through 22 requires only 22.5 feet maximum turning diameter for its 16-foot elements and boom. The antenna consists of interlaced elements for 15 and 20 meters. A driven element and reflector are used for 15-meter operation. The 20-meter section is comprised of a driven element and a director. Both driven elements are gamma matched. A low-cost TV antenna rotator has sufficient torque to handle this lightweight array. (From September 1973 *QST*.)

A misconception among amateurs is that any element short of full size is no good in an antenna system. Reducing the size of an antenna by 50 percent does lower the efficiency by a decibel or two, but the gain capability of a parasitic array outweighs this small loss in efficiency. Mounting the antenna above the interference-generating neighborhood can greatly reduce susceptibility to man-made noise and certainly aids in the reduction of rf heating to trees, telephone poles and

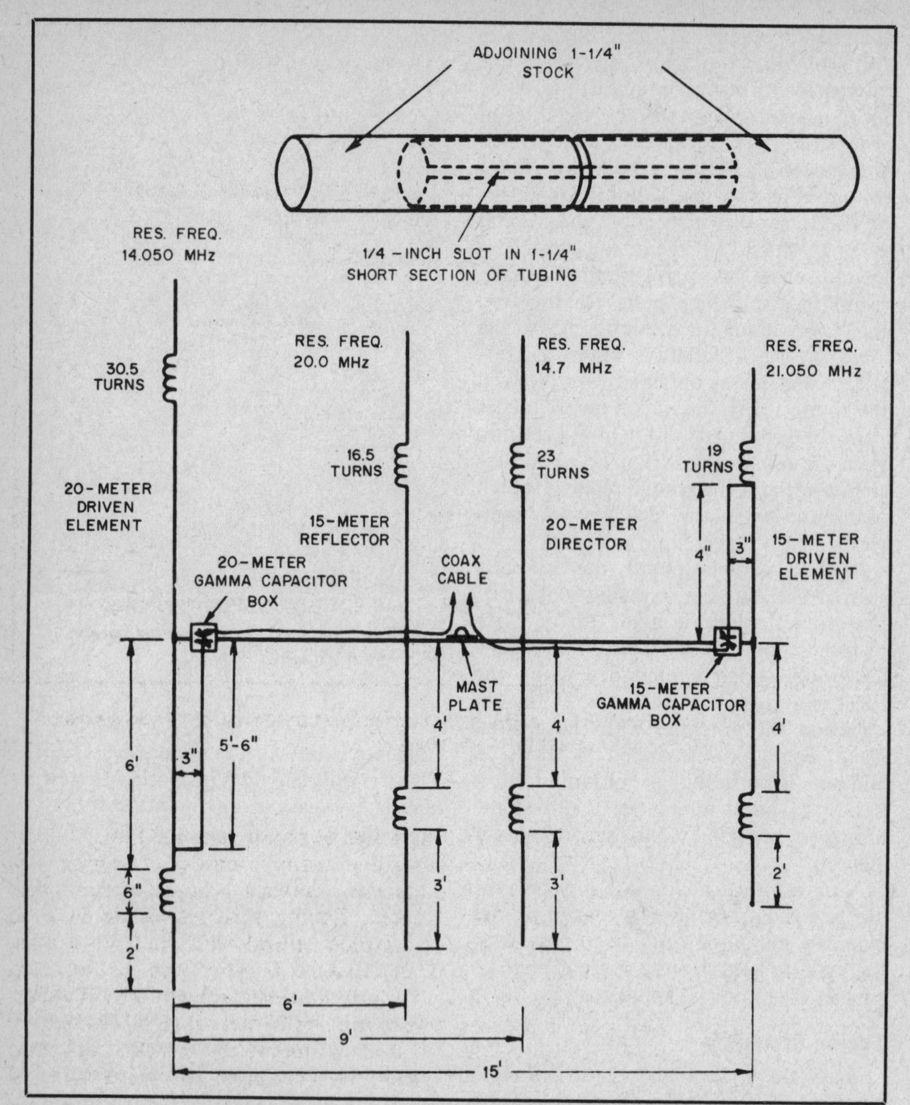


Fig. 19 — Constructional details for the 20- and 15-meter beam. The coils for each side of the element are identical. The gamma capacitors are each 140-pF variable units. The capacitors are insulated from ground within the container. Since the antenna is one-half size for each band, the tuning is somewhat critical. The builder is encouraged to follow carefully the dimensions given here.

Fig. 20 — The gamma assembly is held in place by means of a small U bolt. The capacitors are mounted on etched circuit board.

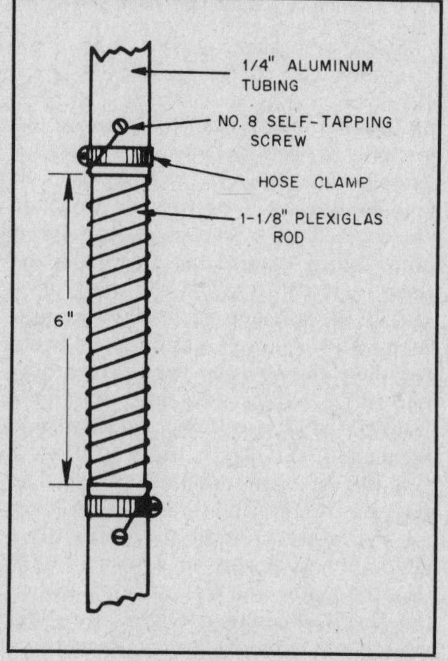


Fig. 21 — Construction details for the loading coils.

buildings. Placing the antenna above these energy-absorbing objects is very desirable.

The dual-band beam has four elements, the longest of which is 16 feet. All of the elements and the boom are made from 1-1/4-inch diameter aluminum tubing available at most hardware stores. A complete parts list is given in Table 2. Element sections and boom pieces are joined together by slotting a 10-inch length of 1-1/4-inch tubing with a nibbling tool and compressing it for a snug fit inside the element and boom tubing. Coupling details are shown in Fig. 19.

The loading coils are wound on 1-1/8-inch diameter Plexiglas rod. The rod slips into the element tubing and is held in place with compression clamps. Be sure to slit the end of the aluminum where the compression clamps are placed. See Fig. 21. The model shown in the photographs has coils made of surplus Teflon-insulated miniature audio coaxial cable with the shield braid and inner conductor shorted together. A suitable substitute would be no. 14 enameled copper wire wound to the same dimensions as those given in Fig. 19.

All of the elements are secured to the boom with common TV U-bolt hardware. Plated bolts are desirable to prevent rust from forming. A 1/4-inch thick boom-to-mast plate is constructed from a few pieces of sheet aluminum cut into 10-inch square sheets and held together with no. 8 hardware. Several cookie tins could be used if sheet aluminum is not available. A plate from a large electrical box might even be used as a boom-to-mast bracket. Since it is galvanized, it is quite resistant to harsh weather.

A boom strut (sometimes called a truss) is recommended because the weight of the elements is sufficient to cause the boom to sag a bit. A 1/8-inch diameter nylon line is

Fig. 22 — The boom-to-mast plate.

plenty strong. A U-bolt clamp is placed on the mast several feet above the antenna and provides the attachment point for the center of the truss line. To reduce the possibility of water accumulating in the

**Table 2**
**Complete Parts List for the Short Beam**

| *Qty* | *Material* |
|---|---|
| 9 | Eight-foot lengths of aluminum tubing, 1-1/4" dia |
| 11 | U bolts |
| 2 | Variable capacitors, 140 pF (E. F. Johnson) |
| 4' | Plexiglas cast rod, 1-1/8" dia |
| 16 | Stainless steel hose clamps, 1-1/2 dia |
| 1 | Aluminum plate, 8-inches square |
| 10' | Aluminum solid rod, 1/4" dia |
| 2 | Refrigerator boxes, 4 × 4 × 4 inches |
| 25' | Nylon rope, 1/8" dia |
| 16 | No. 8 sheet metal screws |
| 16 | No. 8 solder lugs |
| 8 | Plastic (or rubber) end caps, 1-1/4" dia |

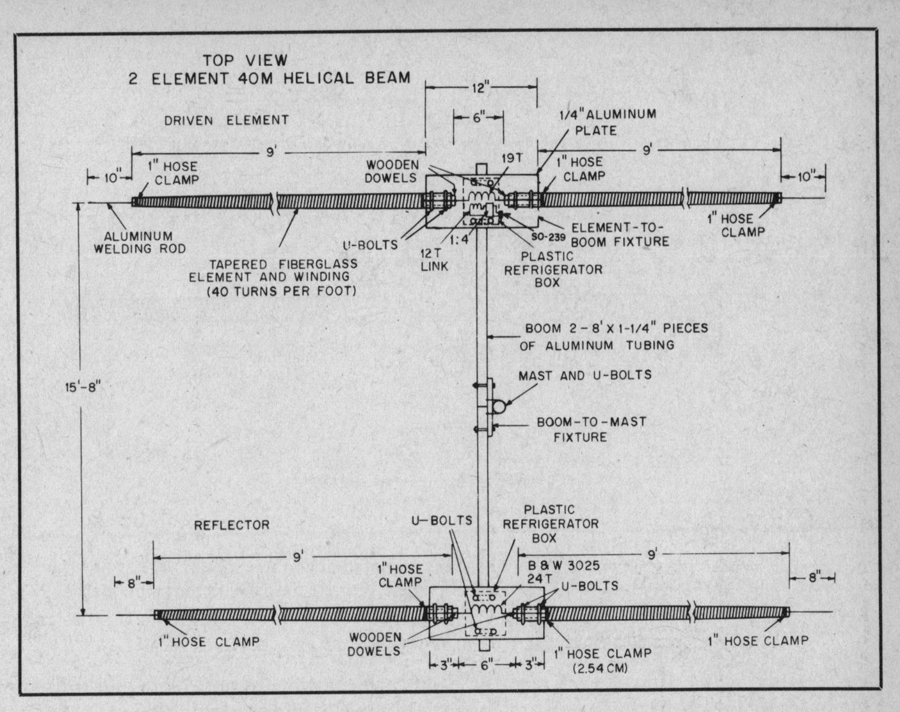


Fig. 24 — Overall dimensions for the 40-meter short beam. The boom consists of two pieces of standard 1-1/4 inch dia Do-It-Yourself aluminum tubing.

element tubing and subsequently freezing (rupture may be the end result), crutch caps are placed over the element ends. Rubber tips suitable for keeping steel-tubing furniture from scratching hardwood floors would serve the same purpose.

A heavy-duty steel mast should be used, such as a 1-inch-diameter galvanized water pipe. Steel TV mast is also acceptable. Any conventional TV type antenna rotator should hold up under load conditions presented by this antenna. Nevertheless, certain precautions should be taken to assure continued trouble-free service. For instance, whenever possible, mount the rotator inside the tower and extend the mast through the tower top sleeve. This procedure relieves the rotator from having to handle lateral pressures during windy weather conditions. A thrust bearing is desirable to reduce downward forces on the rotator bearing.

The monoband nature of the beam requires the use of two coaxial feed lines. The coaxial cable is attached to the 15-meter element (at the front of the beam) at the gamma-capacitor box. The other end of the cable is connected to a surplus 28-V dc single-pole coaxial switch. The cable for the 20-meter element is connected in a similar fashion. The switch allows the use of a single feed line from the shack to a point just below the antenna where the switch is mounted. It is a simple matter to provide voltage to the switch for operation on one of the two bands. At the price of coaxial cable today, a double run of feed line represents a substantial investment and should be avoided if possible.

Fig. 23 — The short beam with helically wound elements for 40 meters is shown here mounted on top of a 40-foot tower. A nylon-rope cross strut was not used with this installation and a slight amount of boom sag is noticeable.

An etched circuit board was mounted inside an aluminum Minibox to provide support and insulation for each of the gamma tuning capacitors. Plastic refrigerator boxes available from most department stores would serve just as well. The capacitor housing is mounted to the boom by means of U bolts.

The builder is encouraged to follow the dimensions given in Fig. 19 as a starting point for the position of the gamma rods and shorting bar. Placing the antenna near the top of the tower and then tilting it to allow the capacitors to be reached makes it possible to adjust the capacitors for minimum SWR as indicated by an SWR meter (or power meter) connected in the feed line at the relay. If the SWR cannot be reduced below some nominal figure of approximately 1.1:1, a slight repositioning of the gamma short might be required. The dimensions given are for operation at 14.050 and 21.050 MHz. The SWR climbs above 2:1 about 50 kHz in either direction from the center frequency. Although tests were not conducted at more than 150 watts input to the transmitter there is no reason why the system would not operate correctly with a kilowatt of power supplied to it.

After many months of testing this antenna, several characteristics were noted. During this period the antenna withstood several wind and ice storms. Performance is what can be expected from a two-element Yagi. The front-to-back ratio on 20 meters is a bit less than 10 dB. On 15 meters the front-to-back ratio is considerably better — on the order of 15 dB. Gain measurements were not made.

## A YAGI ANTENNA WITH HELICALLY WOUND ELEMENTS

Another practical approach in building shortened Yagi antennas is the use of helically wound elements. Bamboo poles or fiberglass quad antenna spreaders are utilized as forms for the spirally wound elements. The 40-meter beam illustrated in Figs. 23 through 25 is only 28 percent of full size. The elements measure 18 feet, tip to tip, and the boom is 16 feet in length. The feed-point impedance is approximately 12 ohms, thereby permitting the use of a 4:1 broadband balun transformer to match the antenna to a 50-ohm coaxial feed line.

This antenna can be built for any 50-kHz segment of the 40-meter band and will operate with an SWR of less than 2.5:1 across that range. An SWR of 1 can be obtained at the center of the 50-kHz range to which the beam is adjusted, and a gradual rise in SWR will occur as the frequency of operation is changed toward the plus or minus 25-kHz points from center frequency.

Ten-inch-long stubs of aluminum welding rod or no. 8 aluminum clothesline

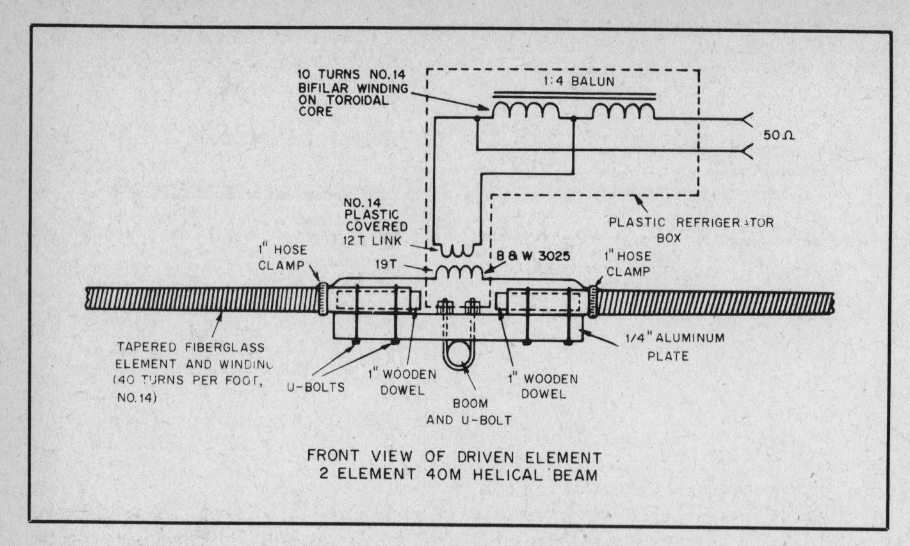


Fig. 25 — Schematic diagram of the balun assembly mounted inside the plastic utility box. The core is a single T-200-2 Amidon. The 12-turn link is wound directly over the 19-turn Miniductor.

wire are used at the tips of each element to help lower the Q (in the interest of increased bandwidth). The stubs are useful in trimming the elements to resonance after the beam is elevated to its final height above ground. Coarse adjustment of the elements is effected by means of tapped inductors located at the center of each element. Plastic refrigerator boxes are mounted at the center of each element to protect the loading inductors and balun transformer from the effects of weather. Two coats of exterior spar varnish should be applied to the helically wound elements after they are adjusted to resonance. This will keep the turns in place and offer protection against moisture. Details for a 40-meter version of this style antenna are given here, but the same approach can be used in fabricating short beams for the other hf bands. Performance with the test model was excellent. The antenna was mounted 36 feet above ground (rotatable) on a steel tower. Many European stations were worked nightly on 40-meter cw. While using 100 watts rf output power with this antenna, reports from Europe ranged between RST 559 and RST 589. Similar good results were obtained when working South American stations and U.S. amateurs on the West Coast. All tests were conducted from Newington.

### Construction Details

The construction of the 40-meter beam is very simple and requires no special tools or hardware. Two fiberglass 15-meter quad arm spreaders are mounted on an aluminum plate with U bolts, as shown in Fig. 24. A wooden dowel is inserted approximately 6 inches into the end of each fiberglass arm to prevent the U bolts from crushing the poles. The aluminum mounting plate is equipped with U-bolt hardware for attachment to the 1-1/4-inch diameter boom.

A plastic refrigerator box is mounted on each element support plate and is used to house a Miniductor coil. No. 14 copper wire is used for the elements. The wire is wound directly onto the fiberglass poles at a density of *40 turns per foot* (not turns per inch) for a total of 360 evenly spaced turns. The wire is attached at each end with an automotive hose clamp of the proper size to fit the fiberglass spreader. Since the fiberglass is tapered, care must be taken to keep the turns from sliding in the direction of the end tips. Several pieces of plastic electrical tape were wrapped around the pole and wire at intervals of about every foot. All of the element half sections are identical in terms of wire and pitch. Coil dimensions and type are given in Figs. 24 and 25.

The driven-element matching system consists of a 4:1 balun transformer and a tightly coupled link to the main-element Miniductor. Complete details are given in Fig. 25.

Mounted at the end of each element held in place by the hose clamp is a short section of stiff wire material used for final tuning of the system. Since the overall antenna is very small in relation to a full-sized array, the SWR points of 2:1 are rather close to each other. The antenna shown in the photograph provides an SWR of less than 2:1 within about 30 kHz either side of resonance. This particular antenna was tuned for 7.040 MHz and can be used throughout the cw portion of the band. Tuning the antenna for phone-band operation should not be difficult and the procedure outlined below should be suitable.

### Tuning

The parasitic element was adjusted to be about four percent lower in frequency than the driven element. A dip oscillator was coupled to the center loading coil and the stiff-wire element tips were trimmed (a quarter of an inch at a time) until resonance was indicated at 6.61 MHz. For phone-band use, the ends could be snipped for 6.91 MHz. Adjusting the driven element is simple. Place an SWR meter or power meter at the input connector and cut the end wires (or add some if necessary) to obtain the best match between the line and the antenna.

## SHORT HELICALLY WOUND VERTICAL ANTENNAS

The concept of size reduction can be applied to vertical antennas as well as to Yagi beams. One has the option of using lumped L and C to achieve resonance in a shortened system, or the antenna can be helically wound to provide a linear distribution of the required inductance, as shown in Fig. 26. No capacitance other than the amount existing between the radiator and ground is used in establishing resonance at the operating frequency. Shortened quarter-wavelength vertical antennas can be made by forming a helix on a long cylindrical form of reasonable dielectric constant. The diameter of the helix must be very small in terms of wavelength in order to prevent the antenna from radiating in the axial mode. Acceptable form diameters for hf-band operation are from 1 inch to 10 inches when considering the practical aspects of antenna construction. Insulating poles of fiberglass, PVC tubing, treated bamboo or wood, or phenolic are suitable for use in building helically wound radiators. If wood or bamboo is used the builder should treat the material with at least two coats of exterior spar varnish prior to winding the antenna element. The completed structure should be given two more coats of varnish, regardless of the material used for the coil form. Application of the varnish will weatherproof the antenna and prevent the coil turns from changing position.

No strict rule has been established concerning how short a helically wound vertical can be before a significant drop in

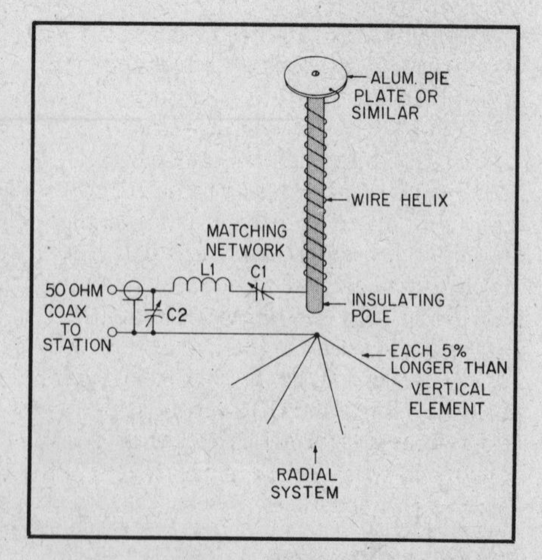


Fig. 26 — Details on how to build and hook up a helically wound short vertical antenna.

performance is experienced. As a general recommendation, one should use the greatest amount of length consistent with available space. A guideline might be to maintain an element length of 0.05 wavelength or more for antennas which are electrically a quarter wavelength long. Thus, use 13 feet or more of stock for an 80-meter antenna, 7 feet for 40 meters, and so on.

A quarter-wavelength helically wound vertical can be used in the same manner as a full-size vertical. That is to say, it can be worked against an above-ground wire-radial system (four or more radials), or it can be used in a ground-mounted manner where the radials are buried or lying on the ground. Some operators have reported good results when using antennas of this kind with four helically wound radials cut for resonance at or slightly lower than the operating frequency. The latter technique should capture the attention of those persons who must use indoor antennas.

### Winding Information

There is no hard-and-fast formula for determining the amount of wire needed to establish resonance in a helical antenna. Experience has indicated that a section of wire approximately one half wavelength long, wound on the insulating form with a linear pitch (equal spacing between turns) will come close to yielding a resonant quarter wavelength. Therefore, an antenna for use on 160 meters would require approximately 260 feet of wire, spirally wound on the support. No specific rule exists concerning the size or type of wire one should use in making a helix. It is reasonable to assume that the larger wire sizes are preferable in the interest of minimizing $I^2R$ losses in the system. For power levels up to 1000 watts it is wise to use a wire size of no. 16 or greater. Aluminum clothesline wire is suitable for use in systems where the spacing between turns is greater than one wire diameter. Antennas requiring close-spaced turns can be made from enameled magnet wire or no. 14 vinyl-jacketed, single-conductor house wiring stock.

A short rod or metal disc should be made for the top or high-impedance end of the vertical. This is a necessary part of the installation to assure reduction in antenna Q. This broadens the bandwidth of the system and helps prevent extremely high amounts of rf voltage from appearing at the far end of the radiator. (Some helical antennas have acted like Tesla coils when used with high-power transmitters, and have actually caught fire at the high-impedance end when a stub or disc was not used.) Since the Q-lowering device exhibits some additional capacitance in the system, it must be in place before the antenna is tuned.

### Tuning and Matching

Once the element is wound it should be mounted where it will be used, with the ground system installed. The feed end of the radiator can be connected temporarily to the ground system. Use a dip meter and check the antenna for resonance by coupling the dipper to the last few turns near the ground end of the radiator. Add or remove turns until the vertical is resonant at the desired operating frequency.

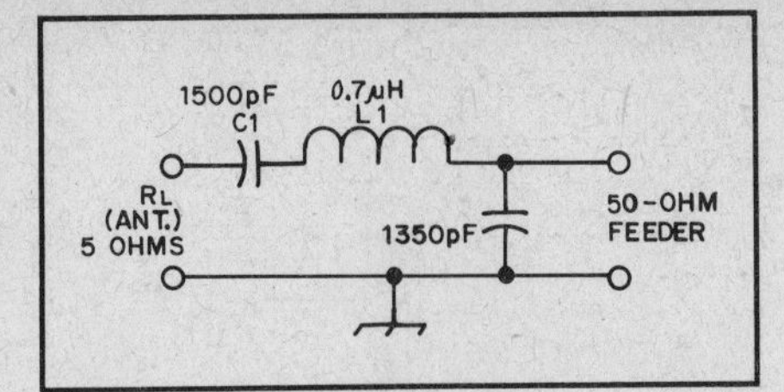


Fig. 27 — An L network suitable for matching the low feed impedance of helically wound verticals to a 50-ohm coaxial cable at 7 MHz. The loaded Q of this network is 3.

It is impossible to predict the absolute value of feed impedance for a helically wound vertical. The value will depend upon the length and diameter of the element, the ground system used with the antenna, and the size of the disc or stub atop the radiator. Generally speaking, the radiation resistance will be very low — approximately 3 to 10 ohms. An L network of the kind shown in Fig. 27 can be used to increase the impedance to 50 ohms. Constants are given for 40-meter operation at 7.0 MHz. The $Q_L$ (loaded Q) of the network inductors is low to provide reasonable bandwidth, consistent with the bandwidth of the antenna. Network values for other operating bands and frequencies can be determined by using the reactance values listed below. The design center for the network is based on a radiation resistance of 5 ohms. If the exact feed impedance is known, the following equations can be used to determine precise component values for the matching network. (See Chapter 4 for additional information on L-network matching.)

$$X_{C1} = QR_L$$

$$X_{C2} = 50\sqrt{\frac{R_L}{50 - R_L}}$$

$$X_{L1} = X_{C1} + \left(\frac{R_L 50}{X_{C2}}\right)$$

where
$X_{C1}$ = Capacitive reactance of C1
$X_{C2}$ = Capacitive reactance of C2
$X_{L1}$ = Inductive reactance of L1
Q = Loaded Q of network
$R_L$ = Radiation resistance of antenna

Example: Find the network constants for a helical antenna whose feed impedance is 5 ohms at 7 MHz, Q = 3:

$$X_{C1} = 3 \times 5 = 15 \text{ and}$$

$$X_{C2} = 50\sqrt{\frac{5}{50 - 5}} = 50\ \sqrt{0.111}$$

$$= 50 \times 0.333 = 16.666 \text{ and}$$

$$X_{L1} = 15 + \left(\frac{250}{16.66}\right)$$

$$= 15 + 15 = 30$$

Therefore, C1 = 1500 pF, C2 = 1350 pF, and L1 = 0.7 μH. The capacitors can be made from parallel or series combinations of transmitting micas. L1 can be a few turns of large Miniductor stock. At rf power levels of 100 W or less, large compression trimmers can be used at C1 and C2 because the maximum rms voltage at 100 W (across 50 ohms) will be 50. At, say, 800 W there will be approximately 220 volts rms developed across 50 ohms. This suggests the use of small transmitting variables at C1 and C2, possibly paralleled with fixed values of capacitance to constitute the required amount of capacitance for the network. By making some part of the network variable it will be possible to adjust the circuit for an SWR of 1 without knowing precisely what the antenna feed impedance is. Actually, C1 is not required as part of the matching network. It is included here to bring the necessary value for L1 into a practical range.

Fig. 26 illustrates the practical form a typical helically wound ground-plane vertical might take. Performance from this type antenna is comparable to that of many full-size quarter-wavelength vertical antennas. The major design trade-off is in usable bandwidth. All shortened antennas of this variety are narrow-band devices. At 7 MHz, in the example illustrated here, the bandwidth between the 2:1 SWR points will be on the order of 50 kHz, half that amount on 80 meters, and twice that amount on 20 meters. Therefore, the antenna should be adjusted for operation in the center of the frequency spread of interest.

## THE FLAGPOLE DELUXE

The flagpole deluxe is a trap vertical that covers 40 through 10 meters from which you can fly a 4 × 6-foot flag. The antenna is built from aluminum tubing with slim, home-built traps, covered by standard PVC pipe, and topped with a toilet-tank ball. The PVC pipe has nothing to do with the antenna itself; it is the flagpole. The base is small and simple, thus easily concealed by brickwork, rocks or flowers.

The system described by Schnell in

**Table 3**

**Parts List for the Flagpole Deluxe**

1 — Insulator block, 1 × 1-1/2 × 3/8" (Formica, Plexiglas, plastic, etc.).
1 — Length aluminum tubing, 1-1/4 × 0.058 × 144" 6061-T5.
1 — Length aluminum tubing 1-1/8 × 0.058 × 144" 6061-T5.
1 — Length aluminum tubing 1-1/8 × 0.125 × 36" 6061-T5.
1 — Piece of aluminum rod 1 × 1" (round).
1 — Fiberglass round rod 1 × 24".
4 — 2" U bolts.
5 — 25-pF Centralab transmitting capacitors 850S series.
1 — Piece of construction aluminum channel stock, 3/4 × 1-3/4 × 14", 1/8" wall thickness
1 — Piece of construction aluminum T-bar stock, 3 × 1-1/2 × 14", 1/8" wall thickness.
8 — 1-1/4" hose clamps.
1 — 2-foot length 1-1/4" EMT conduit.
1 — 3-foot length 1-1/4" EMT conduit.
1 — 12-foot length of 2" PVC schedule water pipe.
1 — 2"-to-2" PVC coupling.
1 — 1-1/2" PVC cap.
1 — 12-foot length of 1-1/2" PVC schedule 40 water pipe.
1 — PVC reducer, 2" to 1-1/2".
1 — Tank ball 5".
1 — 10-foot length copper water pipe (for ground, if required).
1 — Piece of copper or brass foil, 0.005 × 1 × 12"; this may be salvaged from an old transformer or shim stock, or may be purchased from auto supply or metal suppliers.
2 — Insulating sleeves made from plastic tubing 1-1/8" inside diameter, 1/2" wall thickness. These can be made from two PVC 1-1/4-to-3/4" bushings. These are obtainable wherever PVC pipe is sold. The bushings will have to be reamed very slightly to have a press fit over the 1-1/8 × 0.125" aluminum tubing. Either file or turn the flange off on a lathe.

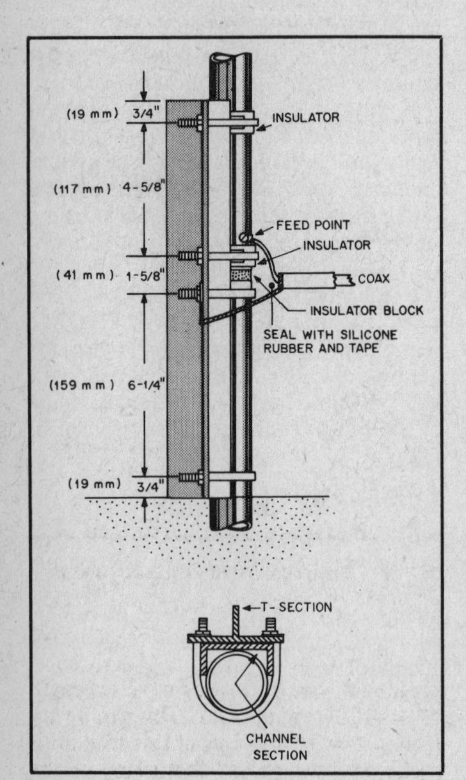


Fig. 28 — U- and T-bar stock details for the base assembly.

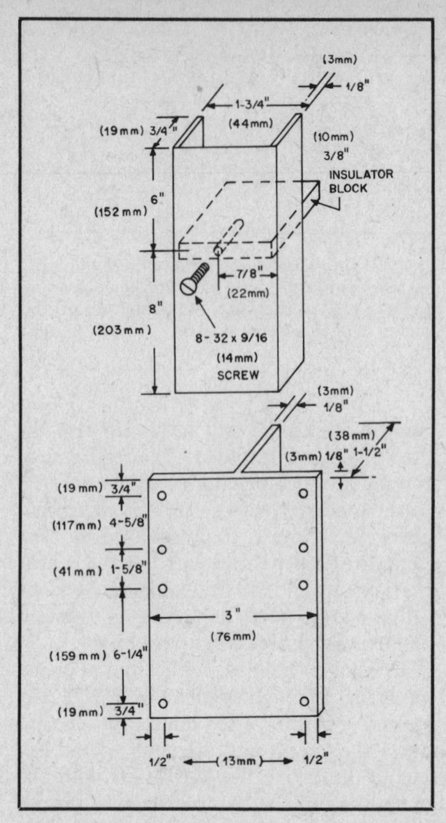


Fig. 29 — Construction details of the T bar and U channel for the base mounting assembly.

March 1978 *QST* used two ground rods and the "skin" of a mobile home for a ground system. You should try to put down a system of ground radials if at all possible. A ground rod does not make a very good rf ground. Alternatively, a large metal structure, such as a mobile home, can serve as a counterpoise. Table 3 is a complete list of the necessary parts.

### Preliminary Construction

Fig. 28 shows details of the base mounting assembly. Drill and countersink a hole to accommodate a no. 8-32 × 9/16-in. flat-head screw. Drill and tap the insulator block and mount it inside the U channel. Drill eight 3/8-in. holes in the T bar as shown; the 2-in. U bolts will straddle the U channel. Assemble the U and T bars per Fig. 29, leaving the U bolts loose.

Fit the insulator sleeves on the 36-in. length of 1-1/8-in., 0.125-in. wall tubing (Fig. 29). These insulators should fit snugly and the U bolts which clamp them to the base assembly will hold them tight. Drill and tap the tubing for a no. 10 screw, 1/4 in. above the top insulator; this will be the feed point.

### Traps

Refer to Fig. 30. Cut one 7-1/2-in. and two 8-in. lengths of 1-in. fiberglass rod, and through all three pieces drill a 1/8-in. hole lengthwise. In one end of the 7-1/2 in. rod, drill a 3/4-in. hole 3/4 in. deep. In the other two pieces, drill a hole in one end 1-1/2 in. deep. Use masonry drill bits for these holes.

In the 7-1/2-in. rod, drill two 1/8-in. holes (3/16 in deep) for pegs, as indicated in Fig. 30. Insert 1/8-in. pegs in these holes. These pegs will be used to hold the windings in place.

Referring to Fig. 31, wind the 10-meter trap. Begin by soldering one end of a suitable length of no. 16 wire (Formvar or Thermaleze) to one end of a 25-pF capacitor. Push the wire through the rod from the large hole end and push the capacitor all the way in, then pull the wire snugly over the end of the rod to the nearest peg and wind five turns, close spaced. Coming off the next peg, dress the wire to the capacitor and solder to it by means of a no. 6 solder lug. This completes assembly of the 10-meter trap.

Following the same procedure, wind the 15- and 20-meter traps. Note that these traps are space wound and that the last turn is spaced a little more than the others; Figs. 31 and 32 illustrate. The 15-meter trap has 13 turns spaced to a length of 1-3/8 in. and the 20-meter trap has 23 turns spaced to a length of 2-3/8 in., with the last turn of each spaced slightly more than the rest. This aids in later tuning of the traps.

Referring to Fig. 33, scrape the enamel off both ends of each winding from the pegs to the ends of the fiberglass rods, and carefully tin each one with a hot soldering iron. Cut six foil tabs, each 3/4 × 1-1/2 in. and run a solder-path lengthwise along each tab, on one side only. Now solder these tabs onto the previously tinned wires (Fig. 33).

Cut six lengths of the 1-1/8-in. tubing, each six in. long. In one end only of two of the tubes cut a slot 3-1/2 in. long, using two hacksaw blades together in the hacksaw (Fig. 34). This will make a slot about 1/8 in. wide. On the other four tubes, make this slot 3 in. long. Remove one hacksaw blade and cut three more slots in the same end as the wide slot, spacing them 90 degrees apart. Clean and deburr all cuts. These slots permit tight compression when the hose clamps are applied.

Carefully push these bushings (the slotted tubes) into place on the trap forms, up against the pegs; the longest slotted bushings go onto the 10-meter trap. Then gently press the foil tabs down against the tubing. The bushings must be in place for the following operation (Fig. 35).

The traps should be checked with a dip meter and adjusted for resonance at the following frequencies: 28.0 MHz, 20.5 MHz and 14.0 MHz. Do not couple too tightly to the traps as the dipper can be "pulled," resulting in erroneous readings. Adjust the traps by carefully spreading or compressing the last turn on each trap.

### Construction of Adjustable Section

Refer to Fig. 36 and cut the 12-foot

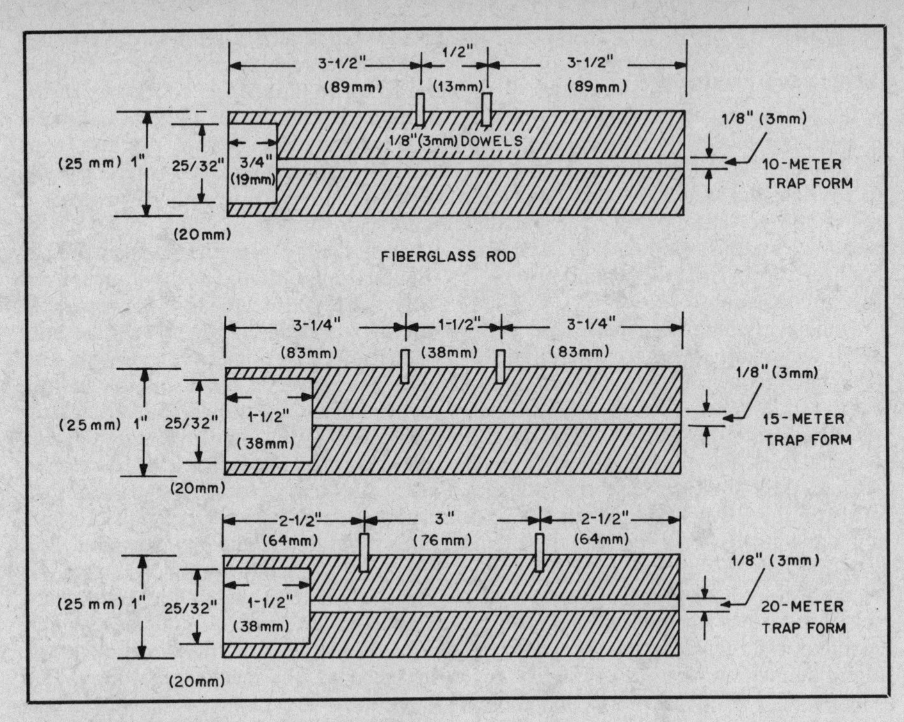


Fig. 30 — Cross-sectional views of the three traps.

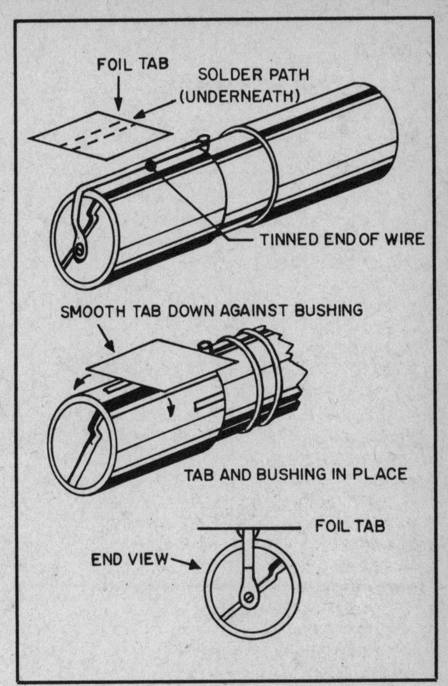


Fig. 33 — Further details of trap construction showing the foil tab and bushings.

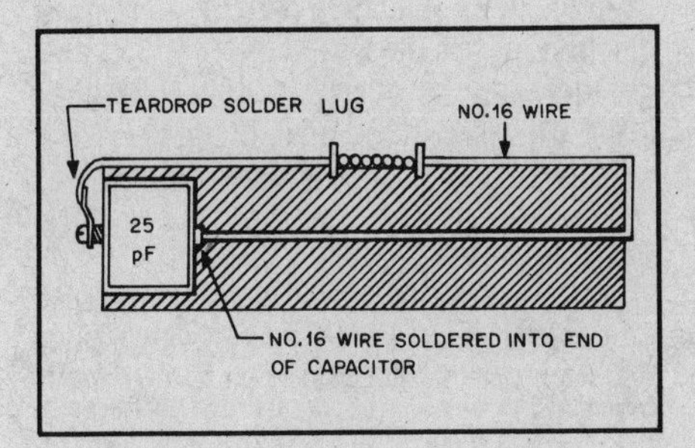


Fig. 31 — Ten-meter trap winding detail.

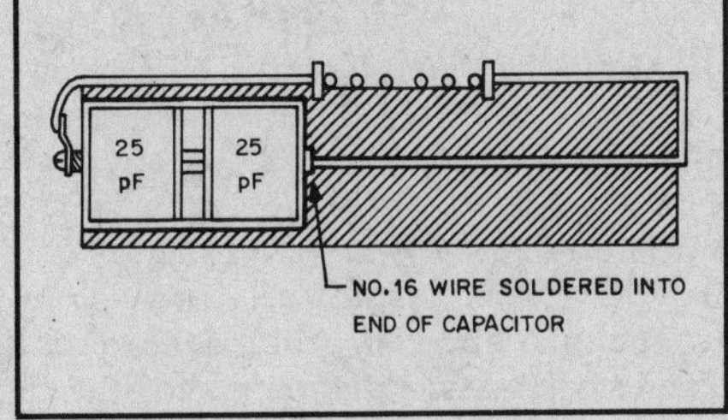


Fig. 32 — Fifteen- and 20-meter trap winding details.

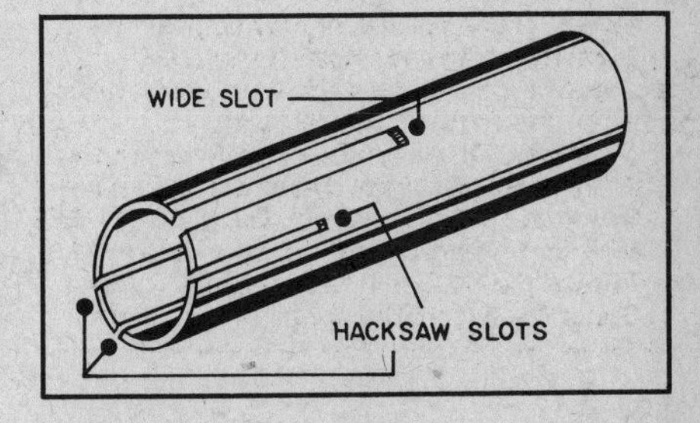


Fig. 34 — Method of slotting the tubing for use as bushings.

length of 1-1/4-in. tubing into four lengths, as follows: 5 ft, 6 in.; 1 ft. 9 in.;* 2 ft. 10-1/2 in.;* 1 ft, 10-1/2 in. The starred (*) items may be cut in half and a bushing inserted to permit adjustment of the 15- and 20-meter bands (Fig. 35). If you choose this method, four more hose clamps will be required.

The tubes just cut are all slotted and deburred on both ends, inside and out. If this deburring is not done the aluminum may seize or gall and it then becomes difficult (if not impossible) to separate or adjust. A stainless-steel hose clamp is placed over each slotted end of each tube.

Cut and deburr a 6-foot length of the 1-1/8-inch tubing and insert the 1-inch piece of round aluminum rod, which should be drilled and tapped for 1/4-20 thread all the way through. Secure this in the end of the tubing by drilling two no. 36 holes and tapping for 6-32 screws.

**Assembly**

Fig. 36 provides an overall picture of assembly of the vertical. Nothing is critical except that care must be exercised when attaching the traps to the 1-1/4-inch tubing sections. Start by spreading *one* of the slots in the tubing ends to expand the diameter slightly. The foil on each trap must be started smoothly between the trap bushing and the tubing sections. This is made easier by carefully pushing the foil tightly against the brushing before sliding the brushing onto the trap itself.

Start by sliding the 21-inch, 15-meter tubing section onto the capacitor end of the 10-meter trap, up against the peg, and tighten the hose clamp securely. In the same manner, clamp the 34-1/2-inch, 20-meter section onto the capacitor end of the 20-meter trap. Tighten all clamps securely, then slide the 15- and 20-meter sections onto the bottoms of their respective traps. See Fig. 36. Next, clamp the 6-foot length of 1-1/4-inch tubing to the bottom of the 10-meter trap, the 36-inch length of 1-1/8-inch tubing into the last tubing just installed and adjust the total length to 87-3/4 inches from the 10-meter trap to the end of the tubing.

Install the ball on the end of the 6-foot length of 1-1/8-inch tubing, which has the aluminum plug in it, using a 2-inch length of 1/4-20 threaded stock. This can be cut from a 1/4-20 bolt. Install this section of tubing into the end of the tubing on the 20-meter trap and adjust to the dimension shown in Fig. 36. Recheck all clamps for

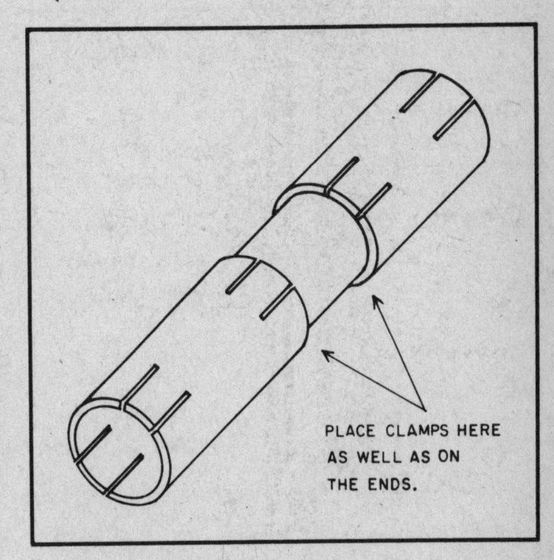


Fig. 35 — Bushings mounted in place on the 20-meter trap.

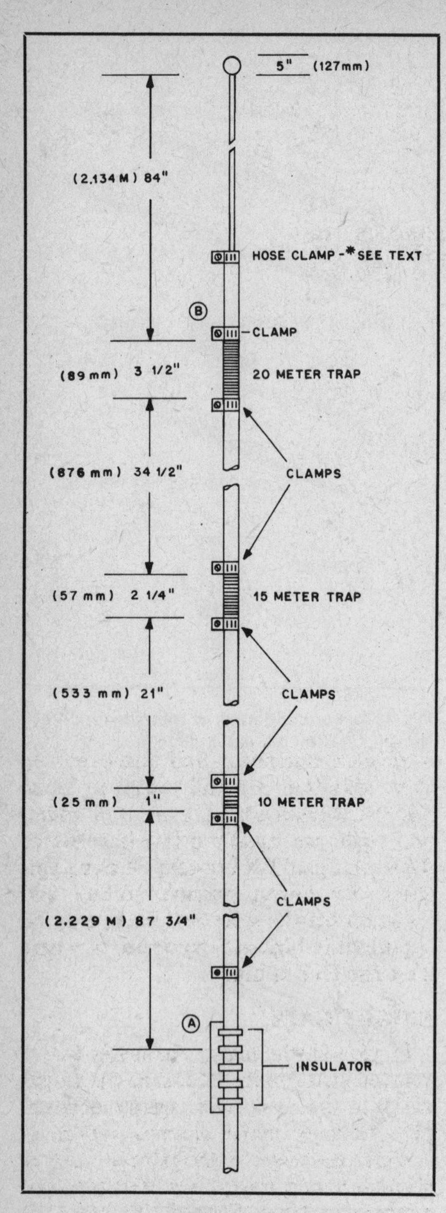


Fig. 36 — The complete Flagpole Deluxe vertical antenna.

tightness.

### Testing and Adjustment

Do your tuning in the final position if it is at all possible. Set the 3-foot length of 1-1/4-inch EMT conduit in concrete next to your ground system, leaving 7-3/4 inch above the concrete. Mount the antenna assembly on this conduit (Fig. 28), and connect a strap from the ground system to the base assembly.

Once everything is in place and tightened use a dip meter to determine the resonant frequencies of the vertical on each band. When making adjustments, remember that any change made on one band affects all *lower* frequency bands. Adjust the 10-meter section first, and measure the SWR by attaching the coax and applying very low power to the antenna through an appropriate SWR indicator. Continue with the other bands; very little trimming or adding should be required. If adding length is necessary, the alternate method described earlier of cutting a section of tubing in half and adding a sleeve is recommended.

After the 40-meter section is adjusted, remove the hose clamp and secure the tubing with four no. 6 sheet-metal screws. This is necessary because the 1-1/2-inch PVC pipe will not clear the hose clamp. After all adjustments have been made, SWR should not exceed 2:1 at any band edge.

### Camouflaging

The time has come to turn the vertical into a flagpole, or vice versa. Lay the entire antenna (including the base mounting assembly) flat on the ground, with the 2-inch PVC alongside it. Carefully measure the distance from the top of the base assembly (Point A, Fig. 36) to the top of the 20-meter trap (Point B) and cut the PVC to length. Install the 2-inch to 1-1/2-inch reducer assembly with PVC cement. Next measure the distance from the ridge inside the reducer to the top of the tubing of the 40-meter section; make sure you have the end of the 2-inch PVC even with the top of the base assembly. Cut the 1-1/2-inch PVC to length and cement it into the reducer.

Drill a hole in the PVC cap to clear the 1/4-20 stud in the end of the 40-meter section. Remove the base assembly and slide the antenna through the PVC pipe until the stud comes out the top; slip the cap onto the stud and screw the top ball on tight. Then slide the antenna back until the cap slides all the way on in place on the end of the 1-1/2-inch PVC. Do *not* cement the cap in place. One word of caution — make certain all clamps are tight before sliding the antenna back and forth in the PVC, to prevent any change in dimensions. The end of the 2-inch PVC should now be even with the base mounting assembly when the antenna is reattached. A small pulley and rope should be added for flying a flag from the vertical.

## BIBLIOGRAPHY

Source material and more extended discussion of topics covered in this chapter can be found in the references given below.

Dome, R. B., "A Study of the DDRR Antenna," *QST*, July 1972.
English, W. E., "A 40-Meter DDRR Antenna," *QST*, December 1971.
Hall, J. "Off-Center-Loaded Dipole Antennas," *QST*, Sept. 1974.
Myers, R. M. and DeMaw, D., "The HW-40 Micro Beam," *QST*, February 1974.
Myers, R. M. and Greene, C., "A Bite Size Beam," *QST*, September 1973.
Sevick, J., "The Ground-Image Vertical Antenna," *QST*, July 1971.
Sevick, J., "The W2FMI Ground-Mounted Short Vertical," *QST*, March 1973.
Sevick, J., "The Constant-Impedance Trap Vertical," *QST*, March 1974.
Sevick, J., "Short Ground-Radial Systems for Short Verticals," *QST*, April 1978.

## Chapter 11

# VHF and UHF Antenna Systems

Improving an antenna system is one of the most productive moves open to the vhf enthusiast. It can increase transmitting range, improve reception, reduce interference problems, and bring other practical benefits. The work itself is by no means the least attractive part of the job. Even with high-gain antennas, experimentation is greatly simplified at vhf and uhf because an array is a workable size, and much can be learned about the nature and adjustment of antennas. No large investment in test equipment is necessary.

## OBJECTIVES

Whether we buy or build our antennas, we soon find that there is no one "best" design for all purposes. Selecting the antenna best suited to our needs involves much more than scanning gain figures and prices in a manufacturer's catalog. The first step should be to establish priorities.

### GAIN

Shaping the pattern of an antenna to concentrate radiated energy, or received-signal pickup, in some directions at the expense of others is the only way to develop gain. This is best explained by starting with the hypothetical *isotropic antenna,* which would radiate equally in all directions. A point source of light illuminating the inside of a globe uniformly, from its center, is a visual analogy. No practical antenna can do this, so all antennas have "gain over isotropic" (*dBi*). A half-wave dipole in free space has 2.14 dBi. If we can plot the radiation pattern of an antenna in all planes, we can compute its gain, so quoting it with respect to isotropic is a logical base for agreement and understanding. It is rarely possible to erect a half-wave antenna that has anything approaching a free-space pattern, and this fact is responsible for much of the confusion about true antenna gain.

Radiation patterns can be controlled in various ways. One is to use two or more driven elements, fed in phase. Such *collinear* arrays provide gain without markedly sharpening the frequency response, compared to that of a single element. More gain per element, but with a sacrifice in frequency coverage, is obtained by placing *parasitic* elements, longer and shorter than the driven one, in the plane of the first element, but not driven from the feed line. The reflector and directors of a *Yagi* array are highly frequency sensitive, and such an antenna is at its best over frequency changes of less than one percent of the operating frequency.

### FREQUENCY RESPONSE

Ability to work over an entire vhf band may be important in some types of work. The response of an antenna element can be broadened somewhat by increasing the conductor diameter, and by tapering it to something approximating a cigar shape, but this is done mainly with simple antennas. More practically, wide frequency coverage may be a reason to select a collinear array, rather than a Yagi. On the other hand, the growing tendency to channelize operations in small segments of our bands tends to place broad frequency coverage low on the priority list of most vhf stations.

### RADIATION PATTERN

Antenna radiation can be made omnidirectional, bidirectional, practically undirectional, or anything between these conditions. A vhf net operator may find an omnidirectional system almost a necessity, but it may be a poor choice otherwise. Noise pickup and other interference problems tend to be greater with such antennas, and those having some gain are especially bad in these respects. Maximum gain and low radiation angle are usually prime interests of the weak-signal DX aspirant. A clean pattern, with lowest possible pickup and radiation off the sides and back, may be important in high-activity areas, or where the noise level is high.

### HEIGHT GAIN

In general, the higher the better in vhf antenna installations. If raising the antenna clears its view over nearby obstructions, it may make dramatic improvements in coverage. Within reason, greater height is almost always worth its cost, but height gain (see Chapter 1) must be balanced against increased transmission-line loss. The latter is considerable, and it increases with frequency. The best available line may be none too good, if the run is long in terms of wavelength. Give line-loss information, shown in table form in Chapter 3, close scrutiny in any antenna planning.

### PHYSICAL SIZE

A given antenna design for 432 MHz will have the same gain as one for 144 MHz, but being only one-third the size it will intercept only one-third as much energy in receiving. The 432-MHz antenna has less pickup efficiency (see Chapter 2). Thus, to be equal in communication effectiveness, the 432-MHz array should be at least equal in *size* to the 144-MHz one, which will require roughly three times as many elements. With all the extra difficulties involved in going higher in frequency, it is well to be on the big side in building an antenna for the higher bands.

## DESIGN FACTORS

Having sorted out objectives in a

general way, we face decisions on specifics, such as polarization, type of transmission line, matching methods and mechanical design.

## POLARIZATION

Whether to position the antenna elements vertically or horizontally has been a moot point since early vhf pioneering. Tests show little evidence on which to set up a uniform polarization policy. On long paths there is no consistent advantage, either way. Shorter paths tend to yield higher signal levels with horizontal in some kinds of terrain. Man-made noise, especially ignition interference, tends to be lower with horizontal. Verticals are markedly simpler to use in omnidirectional systems, and in mobile work.

Early vhf communication was largely vertical, but horizontal gained favor when directional arrays became widely used. The major trend to fm and repeaters, particularly in the 144-MHz band, has tipped the balance in favor of verticals in mobile work and for repeaters. Horizontal predominates in other communication on 50 MHz and higher frequencies. A circuit loss of 20-dB or more can be expected with cross-polarization.

## TRANSMISSION LINES

There are two main categories of transmission lines: balanced and unbalanced. The former include open-wire lines separated by insulating spreaders, and twin-lead, in which the wires are embedded in solid or foam insulation. Line losses result from ohmic resistance, radiation from the line and deficiencies in the insulation. Large conductors, closely spaced in terms of wavelength, and using a minimum of insulation, make the best balanced lines. Impedances are mainly 300 to 500 ohms. Balanced lines are best in straight runs. If bends are unavoidable, the angles should be as obtuse as possible. Care should be taken to prevent one wire from coming closer to metal objects than the other. Wire spacing should be less than 1/20 wavelength.

Properly built, open-wire line can operate with very low loss in vhf and even uhf installations. A total line loss under 2 dB per hundred feet at 432 MHz is readily obtained. A line made of no. 12 wire, spaced 3/4 inch or less with Teflon spreaders, and running essentially straight from antenna to station, can be better than anything but the most expensive coax, at a fraction of the cost. This assumes the use of baluns to match into and out of the line, with a short length of quality coax for the moving section from the top of the tower to the antenna. A similar 144-MHz setup could have a line loss under 1 dB.

Small coax such as RG-58/U or -59/U should never be used in vhf work if the run is more than a few feet. Half-inch lines (RG-8/U or -11/U) work fairly well at 50 MHz, and are acceptable for 144-MHz runs of 50 feet or less. If these lines have foam rather than solid insulation they are about 30 percent better. Aluminum-jacket lines with large inner conductors and foam insulation are well worth their cost. They are readily waterproofed, and can last almost indefinitely. Beware of any "bargains" in coax for vhf or uhf uses. Lost transmitter power can be made up to some extent by increasing power, but once lost, a weak signal can never be recovered in the receiver.

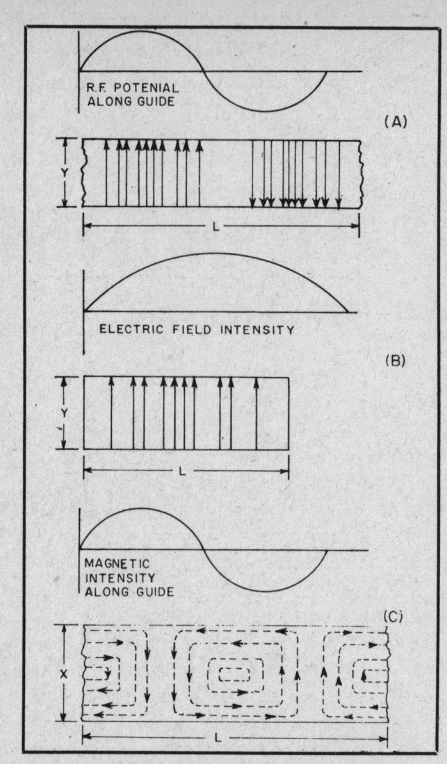

Fig. 1 — Field distribution in a rectangular waveguide. The $TE_{1,0}$ mode of propagation is depicted.

Effects of weather should not be ignored. A well-constructed open-wire line works well in nearly any weather, and it stands up well. Twin-lead is almost useless in heavy rain, wet snow or icing. The best grades of coax are impervious to weather. They can be run underground, fastened to metal towers without insulation, or bent into any convenient position, with no adverse effects on performance.

## G-LINE

Conventional two-conductor transmission lines become quite lossy in the upper uhf and microwave ranges. If the station and antenna are separated by more than a hundred feet, the RG-8/U family of coax is almost useless for serious work. Unless one can obtain the very best rigid coax with proper fittings, it may be worthwhile to explore alternative methods for conveying rf energy between the station and antenna.

Most uhf amateurs are aware that there is a single-conductor transmission line, invented by Goubau, and called "G-line" in his honor. Papers by the inventor appeared some years ago, in which seemingly fantastic claims for line loss were made; under 1 dB per hundred feet in the microwave region, for example. Especially attractive was the statement that the matching device was broad-band in nature, making it appear that a single G-Line installation might be made to serve on both 432 and 1296 MHz.

The basic idea is that a single conductor can be an almost lossless transmission line at ultra-high frequencies, if a suitable launching device is used. A similar launcher is placed at the other end. Basically, the launcher is a cone-shaped device that is a flared extension of the coaxial cable shield. In effect, the cone gets the energy accustomed to traveling on the inner conductor, as the outer conductor is gradually removed. These launch cones should be at least three wavelengths long. The line should be large and heavily insulated, such as no. 14 vinyl. Propagation along a G-Line is similar to "ground wave," or "surface wave" propagation over perfectly conducting earth. The dielectric material confines the energy to the vicinity of the wire, preventing radiation. The major drawback of G-Line is that it is very sensitive to bends. If any must be made, they should be in the form of an arc of large radius, this being preferable to even an obtuse-angle change in the direction of run. The line must be kept several inches away from any metal and should be supported with as few insulators as possible.

## WAVEGUIDES

Above 2 GHz, coaxial cable is a losing proposition for communications work. Fortunately, at this frequency the wavelength is short enough to allow practical, efficient energy transfer by an entirely different means. A waveguide is a conducting tube through which energy is transmitted in the form of electromagnetic waves. The tube is not considered as carrying a current in the same sense that the wires of a two-conductor line do, but rather as a *boundary* that confines the waves to the enclosed space. Skin effect prevents any electromagnetic effects from being evident outside the guide. The energy is injected at one end, either through capacitive or inductive coupling or by radiation, and is received at the other end. The waveguide then merely confines the energy of the fields, which are propagated through it to the receiving end by means of reflections against its inner walls.

Analysis of waveguide operation is based on the assumption that the guide material is a perfect conductor of electricity. Typical distributions of electric and magnetic fields in a rectangular guide are shown in Fig. 1. It will be observed that the intensity of the electric field is greatest (as indicated by closer spacing of the lines of force) at the center along the X dimension, Fig. 1(C), diminishing to zero at the end walls. The latter is a necessary condition, since the existence of any electric field parallel to the walls at the surface would cause an infinite current to flow in a perfect conductor. This represents an impossible situation.

### Modes of Propagation

Fig. 1 represents a relatively simple distribution of the electric and magnetic fields. There is in general an infinite number of ways in which the fields can arrange themselves in a guide so long as there is no upper limit to the frequency to be transmitted. Each field configuration is called a mode. All modes may be separated into two general groups. One group, designated TM (transverse magnetic), has the magnetic field entirely transverse to the direction of propagation, but has a component of electric field in that direction. The other type, designated TE (transverse electric) has the electric field entirely transverse, but has a component of magnetic field in the direction of propagation. TM waves are sometimes called E waves, and TE waves are sometimes called H waves, but the TM and TE designations are preferred.

The particular mode of transmission is identified by the group letters followed by two subscript numerals; for example, $TE_{1,0}$, $TM_{1,1}$, etc. The number of possible modes increases with frequency for a given size of guide. There is only one possible mode (called the dominant mode) for the lowest frequency that can be transmitted. The dominant mode is the one generally used in practical work.

### Waveguide Dimensions

In the rectangular guide the critical dimension is X in Fig. 1, this dimension must be more than one-half wavelength at the lowest frequency to be transmitted. In practice, the Y dimension usually is made about equal to 1/2 X to avoid the possibility of operation at other than the dominant mode.

Other cross-sectional shapes than the rectangle can be used, the most important being the circular pipe. Much the same considerations apply as in the rectangular case.

Wavelength formulas for rectangular and circular guides are given in the following table, where X is the width of a rectangular guide and r is the radius of a circular guide. All figures are in terms of the dominant mode.

| | *Rectangular* | *Circular* |
|---|---|---|
| Cutoff wavelength | 2X | 3.41r |
| Longest wavelength transmitted with little attenuation | 1.6X | 3.2r |
| Shortest wavelength before next mode becomes possible | 1.1X | 2.8r |

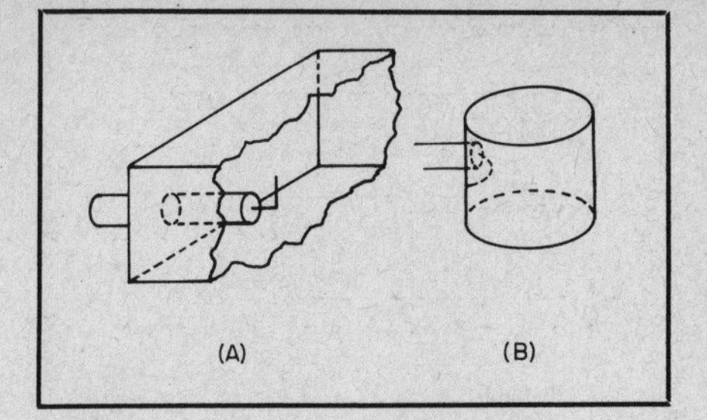


Fig. 2 — Coupling to waveguide and resonators.

### Coupling to Waveguides

Energy may be introduced into or extracted from a waveguide or resonator by means of either the electric or magnetic field. The energy transfer frequently is through a coaxial line, two methods for coupling to which are shown in Fig. 2. The probe shown at A is simply a short extension of the inner conductor of the coaxial line, so oriented that it is parallel to the electric lines of force. The loop shown at B is arranged so that it encloses some of the magnetic lines of force. The point at which maximum coupling will be secured depends upon the particular mode of propagation in the guide or cavity; the coupling will be maximum when the coupling device is in the most intense field.

Coupling can be varied by turning the probe or loop through a 90-degree angle. When the probe is perpendicular to the electric lines the coupling will be minimum; similarly, when the plane of the loop is parallel to the magnetic lines the coupling will have its minimum value.

If a waveguide is left open at one end it will radiate energy. This radiation can be greatly enhanced by flaring the waveguide to form a pyramidal horn antenna. The horn acts as a transition between the confines of the waveguide and free space. To effect the proper impedance transformation the horn must be at least one half wavelength on a side. A horn of this dimension (cutoff) has a unidirectional radiation pattern with a null toward the waveguide transition. The gain at the cutoff frequency is 3 dB, increasing 6 dB with each doubling of frequency. Horns are used extensively in microwave work, both as primary radiators and as feed elements for elaborate focusing systems.

### Evolution of a Waveguide

Suppose an open-wire line is used to convey rf energy from a generator to a load. If the line has any appreciable length it must be supported mechanically. The line must be well insulated from the supports if high losses are to be avoided. Since high-quality insulators are difficult to realize at microwave frequencies, the logical alternative is to support the transmission line with quarter-wavelength stubs, shorted at the far end. The open end of such a stub presents an infinite impedance to the transmission line, provided the shorted stub is nonreactive. However, the shorting link has finite length and, therefore, some inductance. This inductance can be nullified by making the rf current flow on the surface of a plate rather than a thin wire. If the plate is large enough, it will prevent the magnetic lines of force from encircling the rf current.

Infinitely many of these quarter-wave stubs may be connected in parallel without affecting the standing waves of voltage and current. The transmission line may be supported from the top as well as the bottom, and when infinitely many supports are added, they form the walls of a waveguide at its cutoff frequency. Fig. 3 illustrates how a rectangular waveguide evolves from a two-wire parallel transmission line. This simplified analysis also shows why the cutoff dimension is a half wavelength.

While the operation of waveguides is

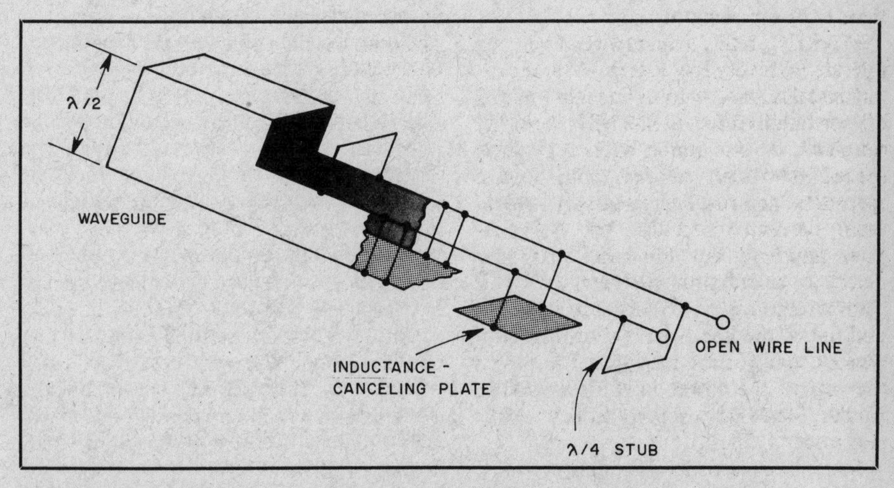


Fig. 3 — At its cutoff frequency a rectangular waveguide can be analyzed as a parallel two-conductor transmission line supported from top and bottom by infinitely many quarter-wavelength stubs.

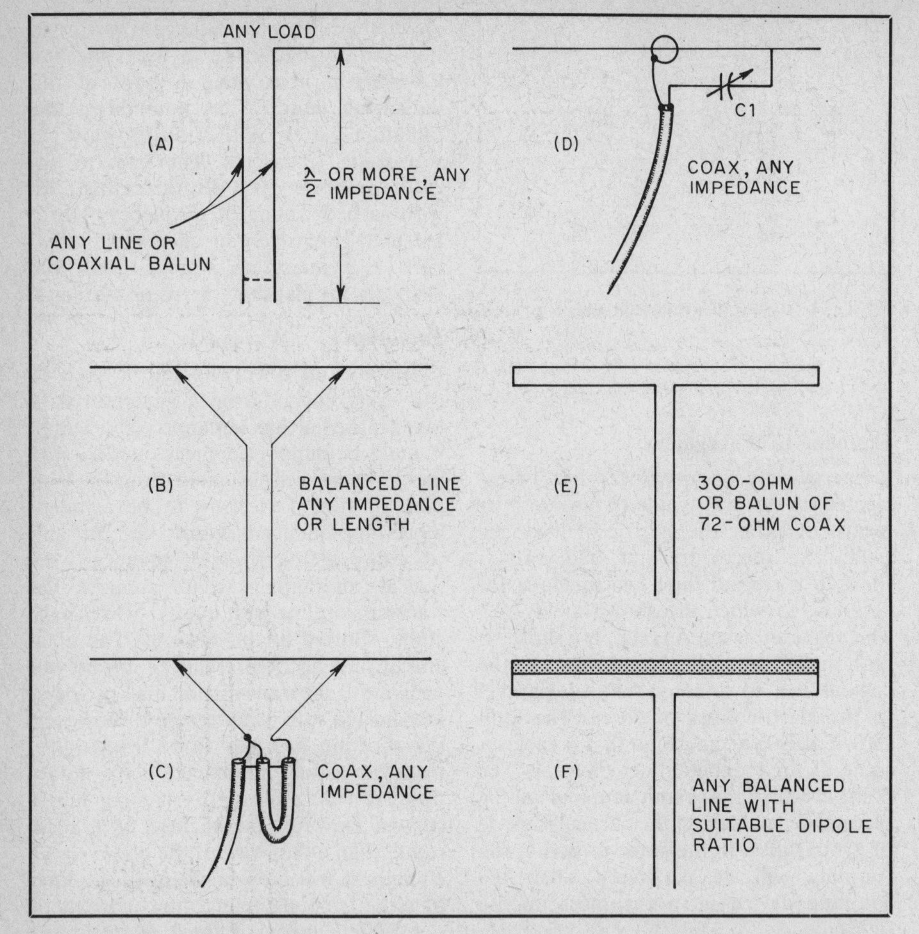


Fig. 4 — Matching methods commonly used in vhf antennas. The universal stub, A, combines tuning and matching. The adjustable short on the stub, and the points of connection of the transmission line, are adjusted for minimum reflected power in the line. In the delta match, B and C, the line is fanned out to tap on the dipole at the point of best impedance match. Impedances need not be known in A, B and C. The gamma-match, D, is for direct connection of coax. C1 tunes out inductance in the arm. A folded dipole of uniform conductor size, E, steps up antenna impedance by a factor of four. Using a larger conductor in the unbroken portion of the folded dipole, E, gives higher orders of impedance transformation.

usually described in terms of fields, current flows on the inside walls, just as fields exist between the conductors of a two-wire transmission line. At the waveguide cutoff frequency, the current is concentrated in the center of the walls, and disperses toward the floor and ceiling as the frequency increases.

## IMPEDANCE MATCHING

Theory and practice in impedance matching are given in detail in earlier chapters, and theory, at least, is the same for frequencies above 50 MHz. Practice may be similar, but physical size can be a major modifying factor in choice of methods. Only the matching devices used in practical construction examples later in this chapter will be discussed in detail here. This should not rule out consideration of other methods, however, and a reading of relevant portions of Chapter 5 is recommended.

### UNIVERSAL STUB

As its name implies, the double-adjustment stub of Fig. 4A is useful for many matching purposes. The stub length is varied to resonate the system, and the transmission line attachment point is varied until the transmission line and stub impedances are equal. In practice this involves moving both the sliding short and the point of line connection for zero reflected power, as indicated on an SWR bridge connected in the line.

The universal stub allows for tuning out any small reactance present in the driven part of the system. It permits matching the antenna to the line without knowledge of the actual impedances involved. The position of the short yielding the best match gives some indication of amount of reactance present. With little or no reactive component to be tuned out, the stub will be approximately a half-wavelength from load to short.

The stub should be of stiff bare wire or rod, spaced no more than 1/20 wavelength. Preferably it should be mounted rigidly, on insulators. Once the position of the short is determined, the center of the short can be grounded, if desired, and the portion of the stub no longer needed can be removed.

It is not necessary that the stub be connected directly to the driven element. It can be made part of an open-wire line, as a device to match into or out of the line with coax. It can be connected to the lower end of a delta match, or placed at the feed point of a phased array. Examples of these uses are given later.

### DELTA MATCH

Probably the first impedance match was made when the ends of an open line were fanned out and tapped onto a half-wave antenna at the point of most efficient power transfer, as in Fig. 4B. Both the side length and the points of connection either side of the center of the element must be adjusted for minimum reflected power in the line, but as with the universal stub, the impedances need not be known. The delta makes no provision for tuning out reactance, so the universal stub is often used as a termination for it, to this end.

Once thought to be inferior for vhf applications because of its tendency to radiate if improperly adjusted, the delta has come back to favor, now that we have good methods for measuring the effects of matching. It is very handy for phasing multiple-bay arrays with open lines, and its dimensions in this use are not particularly critical. It should be checked out carefully in applications like that of Fig. 4C, having no tuning device.

### GAMMA MATCH

An application of the same principle to direct connection of coax is the gamma match, Fig. 4D. There being no rf voltage at the center of a half-wave dipole, the outer conductor of the coax is connected to the element at this point, which may also be the junction with a metallic or wooden boom. The inner conductor, carrying the rf current, is tapped out on the element at the matching point. Inductance of the arm is tuned out by means of C1, resulting in electrical balance. Both the point of contact with the element and the setting of the capacitor are adjusted for zero reflected power, with a bridge connected in the coaxial line.

The capacitor can be made variable temporarily, then replaced with a suitable fixed unit when the required capacitance value is found, or C1 can be mounted in a waterproof box. Maximum should be about 100 pF for 50 MHz and 35 to 50 pF for 144. The capacitor and arm can be combined in one coaxial assembly, with the arm connecting to the driven element by means of a sliding clamp, and the inner end of the arm sliding inside a sleeve connected to the inner conductor of the coax. An assembly of this type can be constructed from concentric pieces of tubing, insulated by plastic sleeving. Rf voltage across the capacitor is low, once the match is adjusted properly, so with a good dielectric, insulation presents no great problem, if the initial adjustment is made with low power. A clean, permanent high-conductivity bond between arm and element is important, as the rf current is high at this point.

### FOLDED DIPOLE

The impedance of a half-wave antenna broken at its center is 72 ohms. If a single conductor of uniform size is folded to make a half-wave dipole as shown in Fig. 4E, the impedance is stepped up four times. Such a folded dipole can thus be fed directly with 300-ohm line with no appreciable mismatch. Coaxial line of 70 to 75 ohms impedance may also be used if a 4:1 balun is added. (See balun information presented below.) Higher impedance step-up can be obtained if the unbroken portion is made larger in cross-section than the fed portion, as in Fig. 4F.

### HAIRPIN MATCH

The feed-point resistance of most multi-element Yagi arrays is less than 50 ohms. If the driven element is split and fed at the center, it may be shortened from its resonant length to add capacitive reactance at the feed point. Then, shunting the feed point with a wire loop resembling a hairpin causes a step-up of the feed-point resistance. The hairpin match does not appear in Fig. 4, but is used in two of the 50-MHz arrays described in a later section of this chapter.

### BALUNS AND TRANSMATCHES

Conversion from balanced loads to unbalanced lines, or vice versa, can be performed with electrical circuits, or their equivalents made of coaxial line. A balun made from flexible coax is shown in Fig. 5A. The looped portion is an electrical half-wavelength. The physical length depends on the propagation factor of the line used, so it is well to check its resonant frequency, as shown at B. The two ends are shorted, and the loop at one end is coupled to a dip-meter coil. This type of balun gives an impedance step-up of 4:1 in impedance, 50 to 200 ohms, or 75 to 300 ohms, typically.

Coaxial baluns giving 1:1 impedance transfer are shown in Fig. 6. The coaxial sleeve, open at the top and connected to the outer conductor of the line at the lower end (A) is the preferred type. In B, a conductor of approximately the same size as the line is used with the outer conductor to form a quarter-wave stub. Another piece of coax, using only the outer conductor, will serve this purpose. Both baluns are intended to present an infinite impedance to any rf current that might otherwise tend to flow on the outer conductor of the coax.

The functions of the balun and the impedance transformer can be handled by various tuned circuits. Such a device, commonly called an antenna coupler or Transmatch, can provide a wide range of impedance transformations. Additional selectivity inherent to the Transmatch can reduce RFI problems.

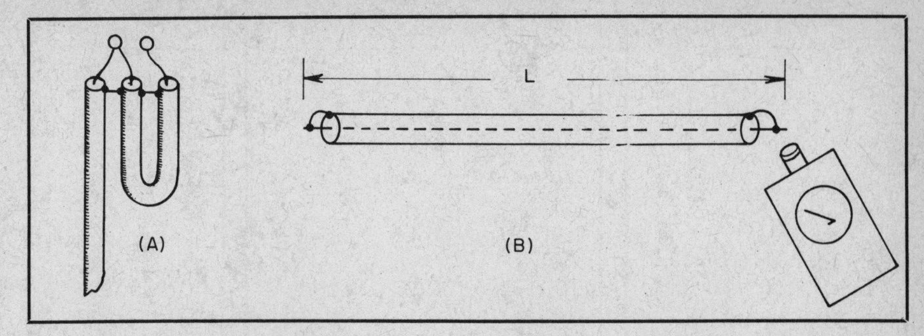


Fig. 5 — Conversion from unbalanced coax to a balanced load can be done with a half-wave coaxial balun, A. Electrical length of the looped section should be checked with a dip meter, with ends shorted, B. The half-wave balun gives a 4:1 impedance step-up.

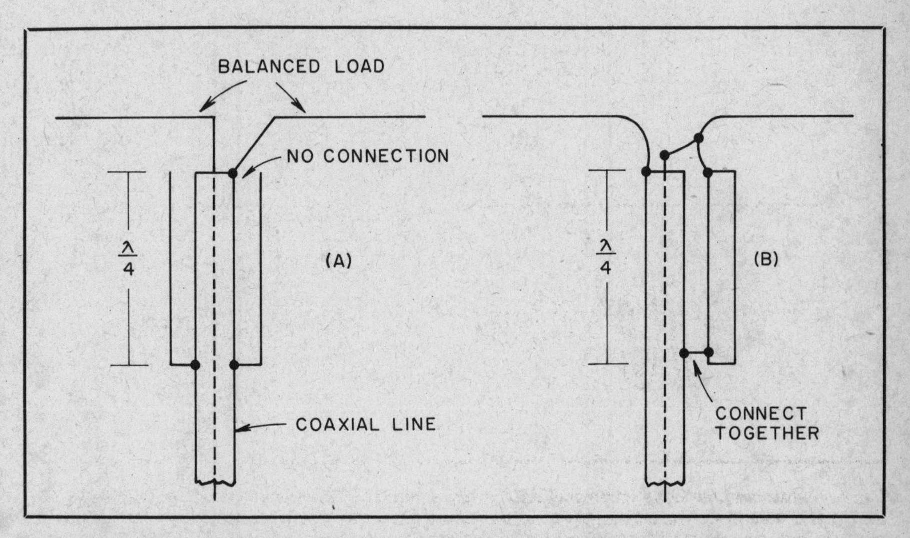


Fig. 6 — The balun conversion function, with no impedance change, is accomplished with quarter-wave lines, open at the top and connected to the coax outer conductor at the bottom. The coaxial sleeve, A, is preferred.

## THE VHF, UHF YAGI

The small size of vhf and, especially, uhf arrays opens up a wide range of construction possibilities. Finding components is becoming difficult for home constructors of ham gear, but it should not hold back antenna work. Radio and TV distributors have many useful antenna parts and materials. Hardware stores, metals suppliers, lumber yards, welding-supply and plumbing-supply houses and even junkyards should not be overlooked. With a little imagination, the possibilities are endless.

#### Boom Materials

Wood is very useful in antenna work. It is available in a great variety of shapes and sizes. Rug poles of wood or bamboo make fine booms. Round wood stock (doweling) is found in many hardware stores in sizes suitable for small arrays. Square or rectangular boom and frame materials can be cut to order in most lumber yards if they are not available from the racks in suitable sizes.

There is no rf voltage at the center of a half-wave dipole or parasitic element, so no insulation is required in mounting elements that are centered in the support, whether the latter is wood or metal. Wood is good for the framework of multibay arrays for the higher bands, as it keeps down the amount of metal in the active area of the array.

Wood used for antenna construction should be well-seasoned and free of knots or damage. Available materials vary, depending on local sources. Your lumber dealer can help you better than anyone else in choosing suitable materials. Joining wood members at right angles is often done advantageously with gusset plates. These can be of thin outdoor-grade plywood or Masonite. Round materials can be handled in ways similar to those used with metal components, with U clamps and with other hardware.

Metal booms have a small "shortening effect" on elements that run through them. With materials sizes commonly employed, this is not more than one percent of the element length, and may not be noticeable in many applications. It is just perceptible with 1/2-inch tubing booms

**Table 1**
**NBS 50.1 MHz Yagi Dimensions**

| Boom Length | Boom Diameter | Element Diameter | Insulated Elements | Ref. | Driven | Dir. 1 | Dir. 2 | Dir. 3 | Dir. 4 | Dir. 5 | Dir. 6 | Dir. 7 | Dir. 8 | Dir. 9 | Dir. 10 |
|---|---|---|---|---|---|---|---|---|---|---|---|---|---|---|---|
| 7' 10"(0.4 λ) | 1-1/4" | 1/2" | YES | 9' 7" | 9' 1-3/4" | 9' 5/8" | | | | | | | | | |
| | | | NO | 9' 7-3/4" | 9' 1-3/4" | 9' 1-3/8" | | | | | | | | | |
| 15' 8-1/2"(0.8 λ) | 2" | 3/4" | YES | 9' 6-1/2" | 9' 1-3/4" | 8' 9-1/8" | 8'8-3/8" | 8'9-1/8" | | | | | | | |
| | | | NO | 9'7-3/4" | 9' 1-3/4" | 8'10-1/4" | 8'9-5/8" | 8'10-1/4" | | | | | | | |
| 23' 6-7/8"(1.2λ) | 2" | 3/4" | YES | 9'6-1/2" | 9' 1-3/4" | 8'9-1/8" | 8'7-3/4" | 8'7-3/4" | 8' 9-1/8" | | | | | | |
| | | | NO | 9'7-3/4" | 9' 1-3/4" | 8'10-1/4" | 8'8-7/8" | 8'8-7/8" | 8' 10-1/4" | | | | | | |
| 39' 3-3/8"(2.2λ) | 2" | 3/4" | YES | 9'6-1/2" | 9' 1-3/4" | 8'9-7/8' | 8'7" | 8'5-3/8" | 8' 3-1/2" | 8' 1-3/4" | 8' 1-3/4" | 8' 1-3/4" | 8' 1-3/4" | 8' 3-1/2" | 8' 5-3/8" |
| | | | NO | 9'7-3/4" | 9' 1-3/4" | 8'11" | 8'11" | 8'6-1/2" | 8' 4-5/8" | 8' 3" | 8' 3" | 8' 3" | 8' 3" | 8' 4-5/8" | 8' 6-1/2" |

**Table 2**
**NBS 144.1-MHz Yagi Dimensions**

| Boom Length | Boom Diameter | Element Diameter | Insulated Elements | Ref | Driven • | Dir. 1 | Dir. 2 | Dir. 3 | Dir. 4 | Dir. 5 | Dir. 6 | Dir. 7 | Dir. 8 | Dir. 9 | Dir. 10 | Dir. 11 | Dir. 12 | Dir. 13 | Dir. 14 | Dir. 15 |
|---|---|---|---|---|---|---|---|---|---|---|---|---|---|---|---|---|---|---|---|---|
| 5'5-9/16"(0.8λ) | 1" | 3/16" | YES | 3'4" | 3'2-3/16" | 3'7/8" | 3'11/16" | 3'7/8" | | | | | | | | | | | | |
| | | | NO | 3'4-5/8" | | 3'1-1/2" | 3'1-3/8" | 3'1-1/2" | | | | | | | | | | | | |
| 8'2-5/16"(1.2λ) | 1" | 3/16" | YES | 3'4" | | 3'7/8" | 3'7/16" | 3'7/16" | 3'7/8" | | | | | | | | | | | |
| | | | NO | 3'4-5/8" | | 3'1-1/2" | 3'1-1/8" | 3'1-1/8" | 3'1-1/2" | | | | | | | | | | | |
| 15'1/4 (2.2λ) | 1 1/4" | 3/16" | YES | 3'4" | | 3'1-1/8" | 3'5/16" | 2'11-13/16" | 2'11-1/4" | 2'10-9/16" | 2'10-9/16" | 2'10-9/16" | 2'10-9/16" | 2'11-1/4" | 2'11-13/16" | | | | | |
| | | | NO | 3'4-13/16" | | 3'1-15/16" | 3'1-1/8" | 3'5/8" | 3' | 2'11-3/8" | 2'11-3/8" | 2'11-3/8" | 2'11-3/8" | 3' | 3'5/8" | | | | | |
| 21'10-1/16"(3.2λ) | 1 1/2" | 3/16" | YES | 3'4" | | 3'7/8" | 3'9/16" | 2'11-3/4" | 2'11-1/8" | 2'10-7/8" | 2'10-9/16" | 2'10-5/16" | 2'10-5/16" | 2'10-5/16" | 2'10-5/16" | 2'10-5/16" | 2'10-5/16" | 2'10-5/16" | 2'10-5/16" | 2'10-5/16" |
| | | | NO | 3'5-1/16" | | 3'1-15/16" | 3'1-3/8" | 3'13/16" | 3' 3/16" | 3' | 2'11-5/8" | 2'11-3/8" | 2'11-3/8" | 2'11-3/8" | 2'11-3/8" | 2'11-3/8" | 2'11-3/8" | 2'11-3/8" | 2'11-3/8" | 2'11-3/8" |
| 28'8-1/8"(4.2λ) | 1 1/2" | 3/16" | YES | 3'3-3/8" | | 3'9/16" | 3'9/16" | 3'3/8" | 2'11-5/8" | 2'11-1/2" | 2'11-1/8" | 2'10-13/16" | 2'10-9/16" | 2'10-9/16" | 2'10-9/16" | 2'10-9/16" | 2'10-9/16" | 2'10-9/16" | | |
| | | | NO | 3'4-1/2" | | 3'1-5/8" | 3'1-5/8" | 3'1-7/16" | 3'11/16" | 3'9/16" | 3'3/16" | 2'11-7/8" | 2'11-5/8" | 2'11-5/8" | 2'11-5/8" | 2'11-5/8" | 2'11-5/8" | 2'11-5/8" | | |

**Table 3**
**NBS 220.1-MHz Yagi Dimensions**

| Boom Length | Boom Diameter | Element Diameter | Insulated Elements | Ref | Driven • | Dir. 1 | Dir. 2 | Dir. 3 | Dir. 4 | Dir. 5 | Dir. 6 | Dir. 7 | Dir. 8 | Dir. 9 | Dir. 10 | Dir. 11 | Dir. 12 | Dir. 13 | Dir. 14 | Dir. 15 |
|---|---|---|---|---|---|---|---|---|---|---|---|---|---|---|---|---|---|---|---|---|
| 3'6-15/16"(0.8λ) | 1" | 3/16" | YES | 2'2-1/16" | 2'1" | 1'11-13/16" | 1'11-11/16" | 1'11-13/16" | | | | | | | | | | | | |
| | | | NO | 2'2-3/4" | | 2'1/2" | 2'3/8" | 2'1/2" | | | | | | | | | | | | |
| 5'4-3/8"(1.2λ) | 1" | 3/16" | YES | 2'2-1/16" | | 1'11-13/16" | 1'11-9/16" | 1'11-9/16" | 1'11-3/16" | | | | | | | | | | | |
| | | | NO | 2'2-3/4" | | 2'1/2" | 2'1/4" | 2'1/4" | 2'1/2" | | | | | | | | | | | |
| 9'10"(2.2λ) | 1" | 3/16" | YES | 2'2-1/16" | | 2'1/16" | 1'11-5/16" | 1'10-15/16" | 1'10-1/2" | 1'10-1/8" | 1'10-1/8" | 1'10-1/8" | 1'10-1/8" | 1'10-1/2" | 1'10-15/16" | | | | | |
| | | | NO | 2'2-3/4" | | 2'3/4" | 2'1/16" | 1'11-5/8" | 1'11-1/4" | 1'10-7/8" | 1'10-7/-8" | 1'10-7/8" | 1'10-7/8" | 1'11-1/4" | 1'11-5/8" | | | | | |
| 14'3-11/16"(3.2λ) | 1 1/4" | 3/16" | YES | 2'2-1/16" | | 1'11-13/16" | 1'11-9/16" | 1'10-15/16" | 1'10-1/2" | 1'10-5/16" | 1'10-1/8" | 1'9-7/8" | 1'9-7/8" | 1'9-7/8" | 1'9-7/8" | 1'9-7/8" | 1'9-7/8" | 1'9-7/8" | 1'9-7/8" | 1'9-7/8" |
| | | | NO | 2'3" | | 2'3/4" | 2'7/16" | 1'11-7/8" | 1'11-7/16" | 1'11-1/4" | 1'11" | 1'10-13/16" | 1'10-13/16" | 1'10-13/16" | 1'10-13/16" | 1'10-13/16" | 1'10-13/16" | 1'10-13/16" | 1'10-13/16" | 1'10-13/16" |
| 18'9-5/16"(4.2λ) | 1 1/2" | 3/16" | YES | 2'1-11/16" | | 1'11-5/8" | 1'11-5/8" | 1'11-7/16" | 1'10-7/8" | 1'10-3/4" | 1'10-1/2" | 1'10-5/16" | 1'10-1/8" | 1'10-1/8" | 1'10-1/8" | 1'10-1/8" | 1'10-1/8" | 1'10-1/8" | | |
| | | | NO | 2'2-3/4" | | 2'11/16" | 2'11/16" | 2'1/2" | 2' | 1'11-13/16" | 1'11-9/16" | 1'11-3/8" | 1'11-3/16" | 1'11-3/16" | 1'11-3/16" | 1'11-3/16" | 1'11-3/16" | 1'11-3/16" | | |

**Table 4**
**NBS 432.1-MHz Yagi Dimensions**

| Boom Length | Boom Diameter | Element Diameter | Insulated Elements | Ref | Driven • | Dir. 1 | Dir. 2 | Dir. 3 | Dir. 4 | Dir. 5 | Dir. 6 | Dir. 7 | Dir. 8 | Dir. 9 | Dir. 10 | Dir. 11 | Dir. 12 | Dir. 13 | Dir. 14 | Dir. 15 |
|---|---|---|---|---|---|---|---|---|---|---|---|---|---|---|---|---|---|---|---|---|
| 2'8-13/16"(1.2λ) | 1" | 3/16" | YES | 1'1-3/16" | 1'23/32" | 11-13/16" | 11-5/8" | 11-5/8" | 11-13/16" | | | | | | | | | | | |
| | | | NO | 1'1-15/16" | | 1'17/32" | 1'11/32" | 1'11/32" | 1"17/32" | | | | | | | | | | | |
| 5'1/8"(2.2λ) | 1" | 3/16" | YES | 1'1-3/16" | | 11-29/32" | 11-7/16" | 11-1/4" | 11" | 10-13/16" | 10-13/16" | 10-13/16" | 10-13/16" | 11" | 11-1/4" | | | | | |
| | | | NO | 1'1-15/16" | | 1'21/32" | 1'3/16" | 1' | 11-3/4" | 11-17/32" | 11-17/32" | 11-17/32" | 11-17/32" | 11-3/4" | 1' | | | | | |
| 7'3-15/32"(3.2λ) | 1" | 3/16" | YES | 1'1-3/16" | | 11-27/32" | 11-5/8" | 11-1/4" | 11" | 10-29/32" | 10-13/16" | 10-11/16" | 10-11/16" | 10-11/16" | 10-11/16" | 10-11/16" | 10-11/16" | 10-11/16" | 10-11/16" | 10-11/16" |
| | | | NO | 1'1-15/16" | | 1'9/16" | 1'11/32" | 1' | 11-3/4" | 11-5/8" | 11-17/32 | 11-13/32" | 11-13/32" | 11-13/32" | 11-13/32" | 11-13/32" | 11-13/32" | 11-13/32" | 11-13/32" | 11-13/32" |
| 9'6-25/32"(4.2λ) | 1" | 3/16" | YES | 1'1" | | 11-22/32" | 11-22/32" | 11-19/32" | 11-1/4" | 11-5/32" | 11" | 10-29/32" | 10-13/16" | 10-13/16" | 10-13/16" | 10-13/16" | 10-13/16" | 10-13/16" | | |
| | | | NO | 1'1-3/4" | | 1'7/16" | 1'7/16" | 1'11/32" | 1' | 11-7/8" | 11-3/4" | 11-5/8" | 11-17/32" | 11-17/32" | 11-17/32" | 11-17/32" | 11-17/32" | 11-17/32" | | |

used on 432 MHz, for example. Formula lengths can be used as given, if the matching is adjusted in the frequency range one expects to use. The center frequency of an all-metal array will tend to be 0.5 to 1 percent higher than a similar system built of wooden supporting members.

## Element Materials

Antennas for 50 MHz need not have elements larger than 1/2-inch diameter, though up to 1 inch is used occasionally. At 144 and 220 MHz the elements are usually 1/8 to 1/4 inch in diameter. For 420 MHz, elements as small as 1/16 inch in diameter work well, if made of stiff rod. Aluminum welding rod, 3/32 to 1/8 inch in diameter is fine for 420-MHz arrays, and 1/8 inch or larger is good for the 220-MHz band. Aluminum rod or hard-drawn wire works well at 144 MHz. Very strong elements can be made with stiff-rod inserts in hollow tubing. If the latter is slotted, and tightened down with a small clamp, the element lengths can be adjusted experimentally with ease.

Sizes recommended above are usable with formula dimensions given in Table 1. Larger diameters broaden frequency response; smaller ones sharpen it. Much smaller diameters than those recommended will require longer elements, especially in 50-MHz arrays.

## Element and Boom Dimensions

Tables 1 through 4 list element and boom dimensions for several Yagi configurations for operation on 50, 144, 220 and 432 MHz. These figures are based on information contained in the *NBS Technical Note 688,* which offers element dimensions for maximum-gain Yagi arrays as well as other types of antennas. The original information provides various element and boom diameters. The information shown in the tables represents a highly condensed set of antenna designs, however, making use of standard and readily available material. Element and boom diameters have been chosen so as to produce lightweight, yet very rugged, antennas.

Since these antennas are designed for maximum forward gain, the front-to-back pattern ratios may be a bit lower than those for some other designs. Ratios on the order of 15 to 25 dB are common for these antennas and should be more than adequate for most installations. Additionally, the patterns are quite clean, with the side lobes well suppressed. The driven-element lengths for the antennas represent good starting-point dimensions. The type of feed system used on the array may require longer or shorter lengths, as appropriate. Generally speaking, a balanced feed system is preferred in order to prevent pattern skewing and the possibility of unwanted side lobes, which can occur with an unbalanced feed system

Element spacing for the various arrays is presented in Fig. 7 in terms of the wavelength of the boom, as noted in the first column of Tables 1 through 4. The 0.4-, 0.8-, 2.2- and 3.2-wavelength boom antennas have equally spaced elements for both reflector and directors; 1.2- and 4.2-wavelength boom antennas have different reflector and director spacings. As all of the antenna parameters are interrelated, changes in element diameter, boom diameter and element spacing will require that a new design be worked out. As the information presented in the *NBS Technical Note 688* is straightforward, the serious antenna experimenter should have no difficulty designing antennas with different dimensions from those presented in the tables. For antennas with the same element and boom diameters, but for different frequencies within the band, standard scaling techniques may be applied.

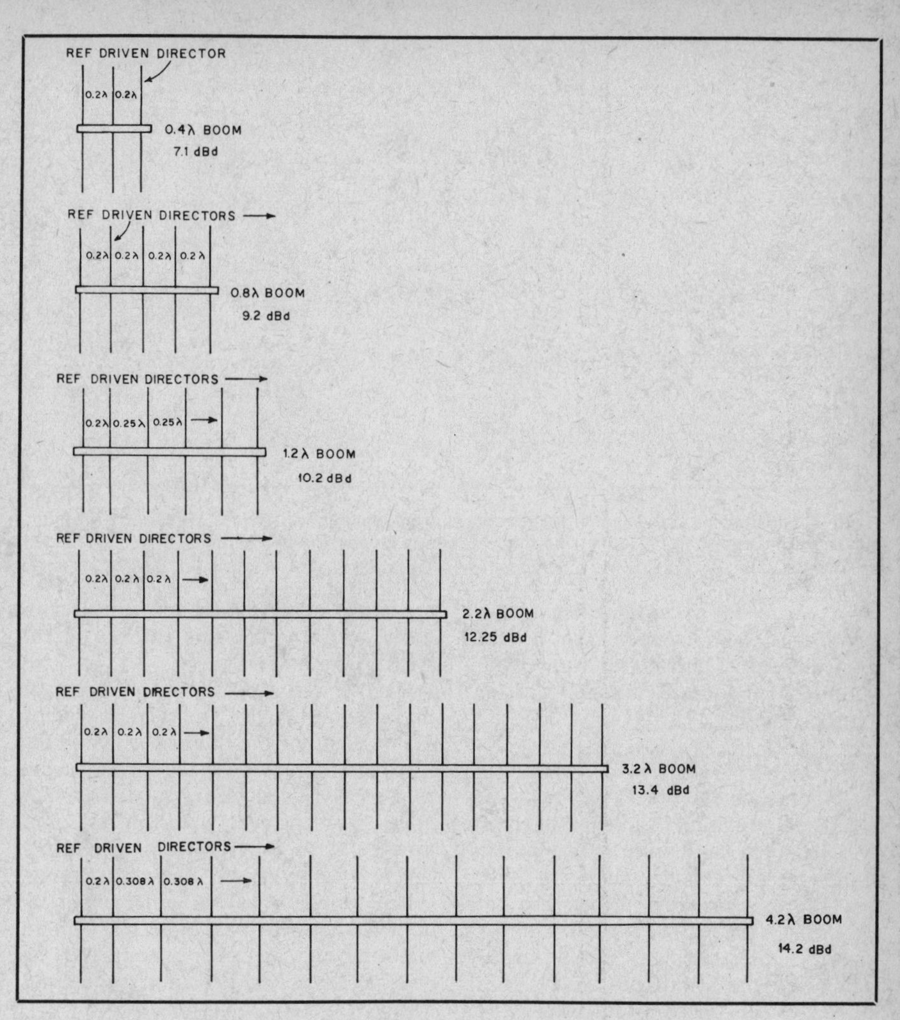

Fig. 7 — Element spacing for the various arrays, in terms of boom wavelength.

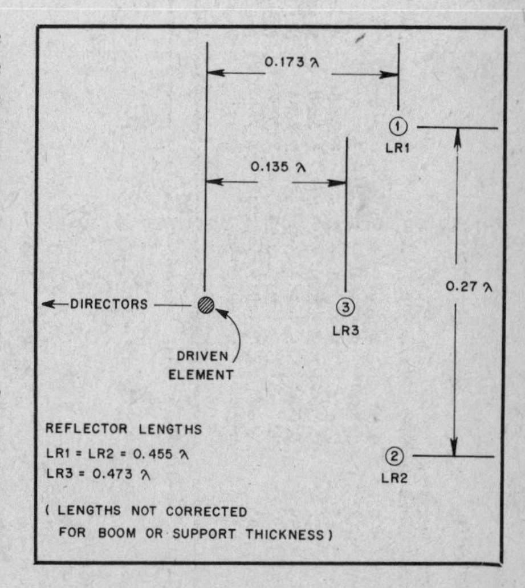

Fig. 8 — A trigonal reflector, shown here in the edge view, can add up to 0.75 dB to the gain of an NBS Yagi.

## Trigonal Reflectors

One of the experiments documented in *NBS Technical Note 688* concerned reflector arrangements other than single cylindrical elements. The report claims that a 4.2-wavelength Yagi will gain 0.75 dB when the reflector configuration of Fig. 8 is used in place of the conventional system. This modification was not applied to the shorter arrays, but the report suggests a similar advantage would be had. In

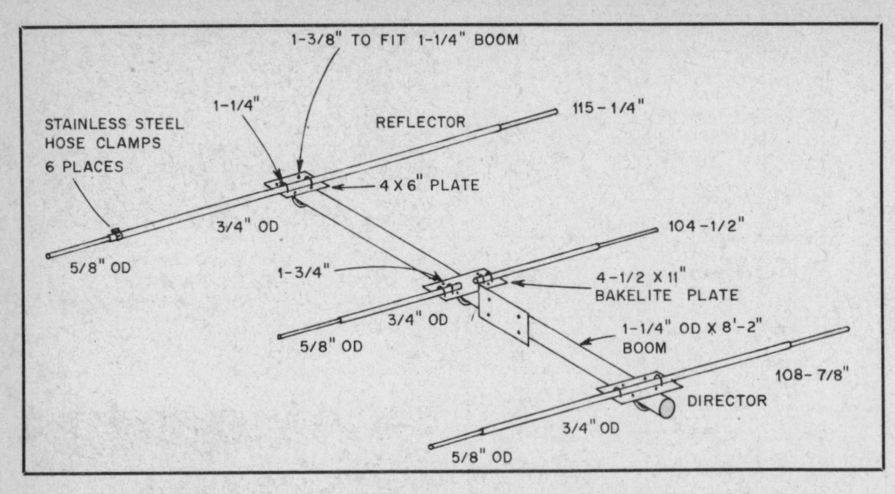


Fig. 9 — Dimensional drawing of a 3-element maximum-gain Yagi for 50 MHz. The feed system is similar to that shown in Fig. 13, except that the hairpin loop is 1 inch shorter.

cases where one is restricted to a short boom, a trigonal reflector might produce a significant improvement — possibly equivalent to adding a director to a conventional design.

## A LOW-COST YAGI FOR 50 MHz

The antenna described here can be considered a "beginner" antenna in that it will permit a newcomer to explore the 6-meter band with a minimum of labor and expense. However, it can also be a permanent fixture for serious efforts if space is limited, because it is optimized for gain and is the best that can be done with three elements on an 8-foot boom. In its first week on the air (November 1981) this model produced contacts with American Samoa and The Gambia from Connecticut with only modest power. Dennis Lusis, W1LJ, built this antenna in the ARRL laboratory.

Figs. 9, 10 and 11 show the important construction details. The element lengths depart slightly from those specified in the NBS chart because of the larger tubing used. A balanced feed system to a split driven element was used on the unit shown, but an all-metal (plumber's delight) construction using a gamma match would be a justifiable simplification. A 3-element Yagi has a fairly broad major lobe, so any slight pattern skewing resulting from feed imbalance should not be objectionable, if noticeable.

This antenna can be adjusted near the ground if it is pointed straight up at the sky away from utility lines or other antennas. The reflector can be supported by two wooden sawhorses or chairs, and the boom can be steadied with a pair of broomsticks. Adjust the driven element length and the matching system for minimum reflected power, which should be near zero. The dimensions given in Fig. 9 are optimized for 50.1 MHz, but the antenna will perform well up to 51 MHz. If this antenna is stacked above an hf beam, the separation should be at least 4 feet for a good impedance match, or 10 feet for low-angle radiation. See the later section on stacking.

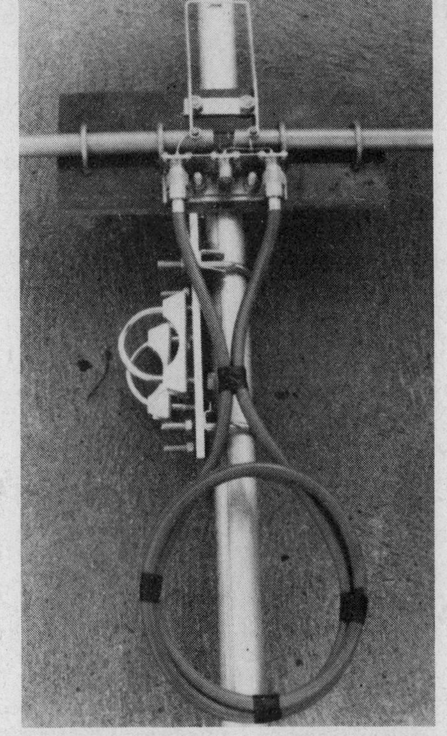

Fig. 11 — This photo shows how the driven element and feed system are attached to the boom. The center of the hairpin loop may be connected to the boom electrically and mechanically if desired.

Fig. 10 — The 3-element beam, ready for installation. One person can easily handle this array.

## A 5-ELEMENT YAGI FOR 50 MHz

The antenna described here was designed from information contained in Table 1 and Fig. 7. This antenna has a theoretical gain of 9.2 dB over a half-wavelength dipole and should exhibit a front-to-back ratio of roughly 18 dB. The pattern is quite clean, with side lobes well suppressed. A hairpin matching system is used, and if the dimensions are followed closely no adjustment should be necessary. The completed antenna is rugged, yet lightweight, and should be easy to install on any tower or mast.

### Mechanical Details

Constructional details of the antenna are given in Figs. 12 and 13. The boom of the antenna is 17 feet long and is made from a single piece of 2-inch aluminum irrigation tubing that has a wall thickness of

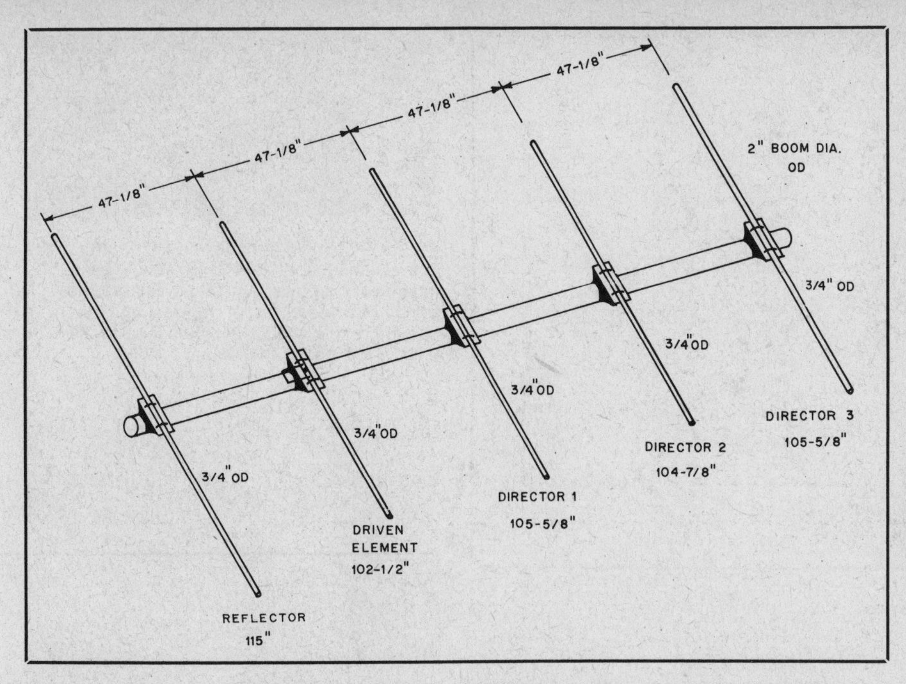


Fig. 12 — Dimensional drawing of the 5-element, 50-MHz Yagi antenna described in the text.

Fig. 14 — Close-up of the driven element and feed system for the 5-element 6-meter Yagi. The phasing line is coiled and taped to the boom.

Fig. 15 — Showing the element-to-boom clamp. U bolts are used to hold the element to the plate, and 2-inch plated muffler clamps hold the plates to the boom.

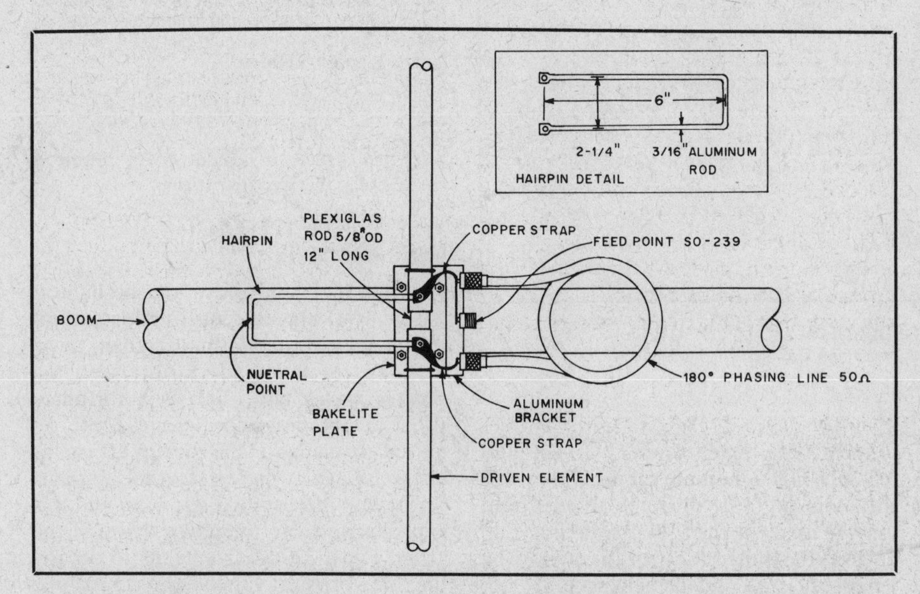


Fig. 13 — Detailed drawing of the feed system used with the 50-MHz Yagi. Phasing-line lengths are: for cable with 0.80 velocity factor — 7′ 10-3/8″; for cable with 0.66 velocity factor — 6′ 5-3/4″.

0.047 inch. Irrigation tubing is normally supplied in 20 foot sections so several feet may be removed from the length.

Elements are constructed from 3/4-inch OD aluminum tubing of the 6061-T6 variety, with a wall thickness of 0.058 inch. Each element, with the exception of the driven element, is made from a single length of tubing. The driven element is split in the center and insulated from the boom to provide a balanced feed system. The reflector and directors have short lengths of 7/8-inch aluminum tubing telescoped over the center of the elements for reinforcement purposes. Boom-to-element clamps were fashioned from 3/16-inch thick aluminum-plate stock, as shown in Figs. 14 and 15. Two muffler clamps hold each plate to the boom, and two U bolts affix each element to the plate. Exact dimensions of the plate are not critical, but should be great enough to accommodate the two muffler clamps and two U bolts. The element and clamp structure may seem to be a bit over-engineered. However, the antenna is designed to withstand the severe weather conditions common to New England. This antenna has withstood many windstorms and several ice storms, and no maintenance has been required.

The driven element is mounted to the boom on a Bakelite plate of similar dimension to the reflector and director element-to-boom plates. A piece of 5/8-inch Plexiglas rod, 12 inches in length, is inserted into each half of the driven element. The Plexiglas piece allows the use of a single clamp on each side of the element and also seals the center of the elements against moisture. Self-tapping screws are used for connection to the driven element. A length of 1/4-inch polypropylene rope is inserted into each element, and end caps are placed on the elements. The rope dampens element vibrations, which could lead to element or hardware fatigue.

## Feed System

Details of the feed system are shown in Figs. 13 and 14. A bracket fashioned from a piece of scrap aluminum is used to mount the three SO-239 connectors to the driven element plate. A half-wavelength phasing line connects the two element halves, providing the necessary 180-degree phase difference between them. The "hairpin" is connected directly across the element halves. It should be noted that the exact center of the hairpin is electrically neutral and may be fastened to the boom or allowed to hang. Phasing-line lengths are: for cable with 0.80 velocity factor — 7′ 10-3/8″; for cable with 0.66 velocity factor — 6′ 5-3/4″. It will be noted that the driven element is the shortest element in this array. While this may seem a bit

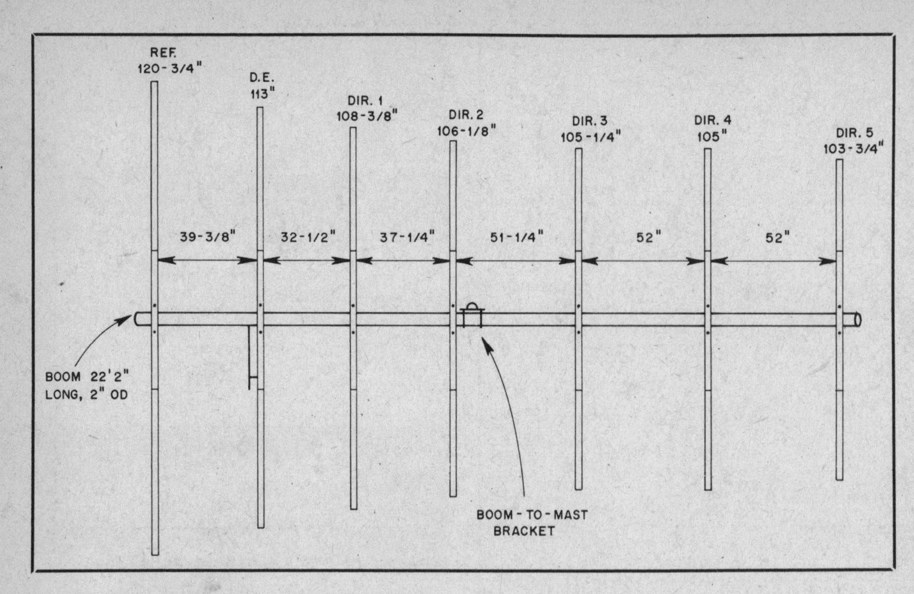


Fig. 16 — Dimensions of DL-style 6-meter beam.

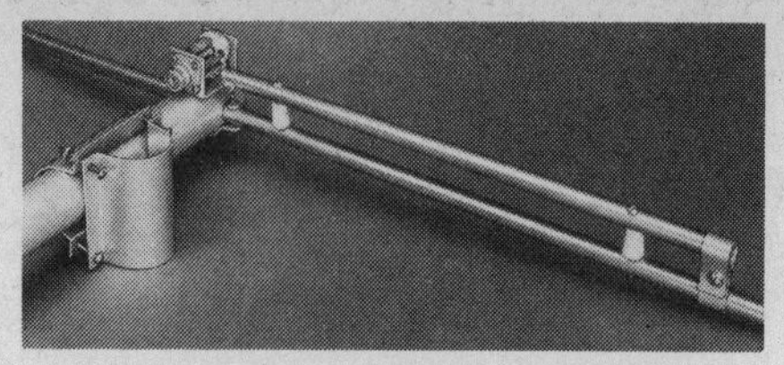

Fig. 18 — Typical gamma-match construction. The variable capacitor, 50 pF, should be mounted in an inverted plastic cup or other device to protect it from the weather. The gamma arm is about 12 inches long for 50 MHz, 5 inches for 144 MHz.

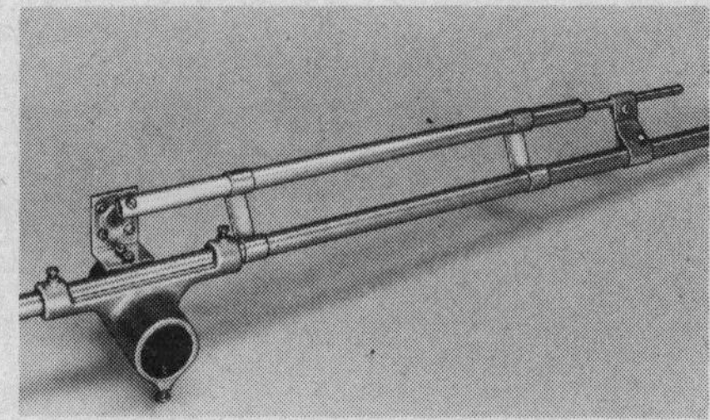

Fig. 19 — Gamma-matching section using tubular capacitor. The sheet-aluminum clip at the right is moved along the driven element for matching. The small rod can be slid in and out of the 15-inch tube for adjustment of series capacitance. The rod should be about 14 inches long for 50 MHz.

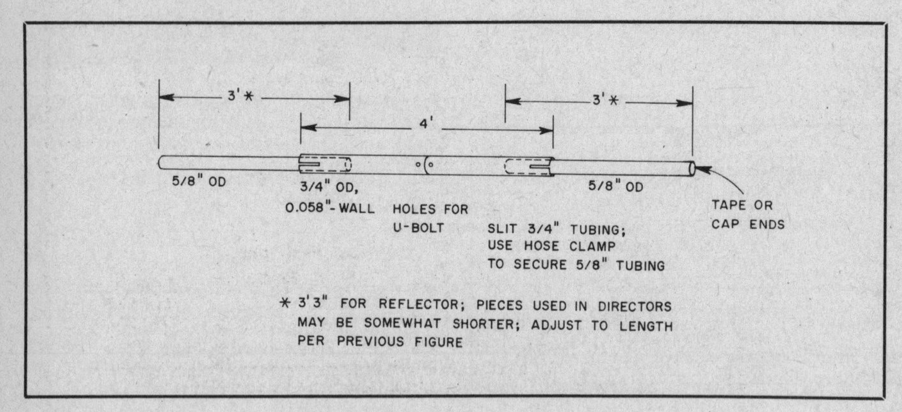


Fig. 17 — Construction of a typical element for the DL-style beam.

unusual, it is necessary with the hairpin matching system.

## DL-STYLE LONG YAGI FOR 50 MHz

The 6-meter antennas described previously are based on the NBS designs for maximum forward gain. Optimization of front-to-back ratio or bandwidth rather than gain requires adjustment of the element lengths, diameters and spacing. One cannot simultaneously optimize all three characteristics. However, a large array can be assembled to produce the maximum values of gain, front-to-back ratio and bandwidth of a smaller array. Guenter Schwarzbeck, DL1BU, applied that philosophy to the 7-element Yagi described here. The gain is roughly equal to that of the 5-element NBS design, but the front-to-back ratio and bandwidth are somewhat greater.

An aluminum boom just over 22 feet long is required, and it should have an OD of 2 inches. Depending on the wall thickness of the material selected, some additional bracing may be required. The elements are made from 3/4-inch and 5/8-inch OD aluminum tubing with a wall thickness of 0.058 inch. A thinner wall may be used for the 5/8-inch tubing. Fig. 16 gives the dimensions for the array. The length of the driven element is not especially critical, and may be varied over a considerable range to effect an impedance match. A standard gamma match, such as pictured in Fig. 18 or 19, is used, or one may substitute one of the more elaborate balanced feed systems. Each element is fastened to the boom by a single U bolt extending through holes drilled at the center of the element. This is detailed in Fig. 17. It may be desirable to reinforce the element where the holes are drilled; one way to do this is to put a short piece of 7/8-inch OD tubing around the center of the 3/4-inch tubing so the U bolt goes through a double thickness of tubing.

The antenna is dimensioned for 50.1 MHz, but will work well up to 52 MHz. The 3-dB beamwidth is on the order of 40 degrees in the azimuthal plane. This is a sharp pattern, so a good rotator and an accurate direction indicator are required for most effective operation.

## A 15-ELEMENT YAGI FOR 432 MHz

At 144 MHz and above, the mechanical problems of antenna construction and installation are much simpler than those associated with 50 MHz, owing to the shorter wavelength. This is a fortunate situation, because more elements are required to maintain the pickup efficiency as the frequency increases. Another name for pickup efficiency is *aperture,* and this is proportional to the physical size of the array. For a 220-MHz antenna to occupy the same physical volume as a similar one designed for 50 MHz, it must have four to five times as many elements. The array described in this section illustrates the construction of Yagis for the uhf range.

This 432-MHz Yagi antenna was designed using the information shown in Table 4 and Fig. 7. The theoretical gain for this antenna is 14.2 dB over a dipole, with a front-to-back pattern ratio of approximately 22 dB. The pattern is very sharp and quite clean, as would be expected from a well-tuned array of this size. Four of these antennas in a "box" or "H" array would serve well for terrestrial work, while eight would make a respectable EME system

### Mechanical Details

Dimensions for the antenna are given in Figs. 20 and 21. The boom of the antenna is made from a length of 1-inch aluminum

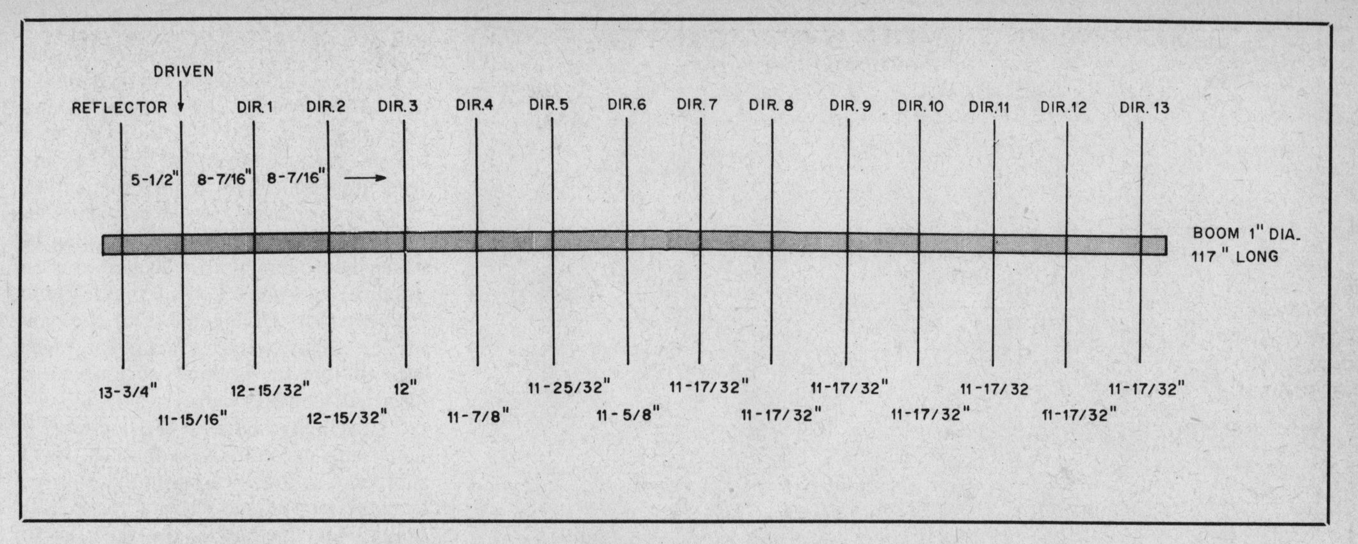


Fig. 20 — Dimensional drawing of the 15-element Yagi for 432 MHz. The balance point of the antenna is between the fifth and sixth director. The antenna was designed from the information presented in Table 4 and Fig. 7 of this chapter.

tubing. Each of the elements is mounted through the boom, and only the driven element is insulated. Auveco 8715 external retaining rings secure each of the parasitic elements in place. These rings are available at most hardware supply houses.

The driven element is insulated from the boom by a length of 1/2-inch Teflon rod. The length of rod is drilled to accept the 1/4 inch thick aluminum tubing driven element. A press fit was used to secure the Teflon piece in the boom of the antenna. An exact fit can be obtained by drilling the hole slightly undersized and enlarging the hole with a hand reamer, a small amount at a time. Should the hole turn out to be oversized, a small amount of RTV (silicone seal) can be used to secure the Teflon in place. Although it wasn't tried with this antenna, it should be possible to use a driven element that is not insulated from the boom. Small changes in the position of the matching rods and/or clamps might be necessary.

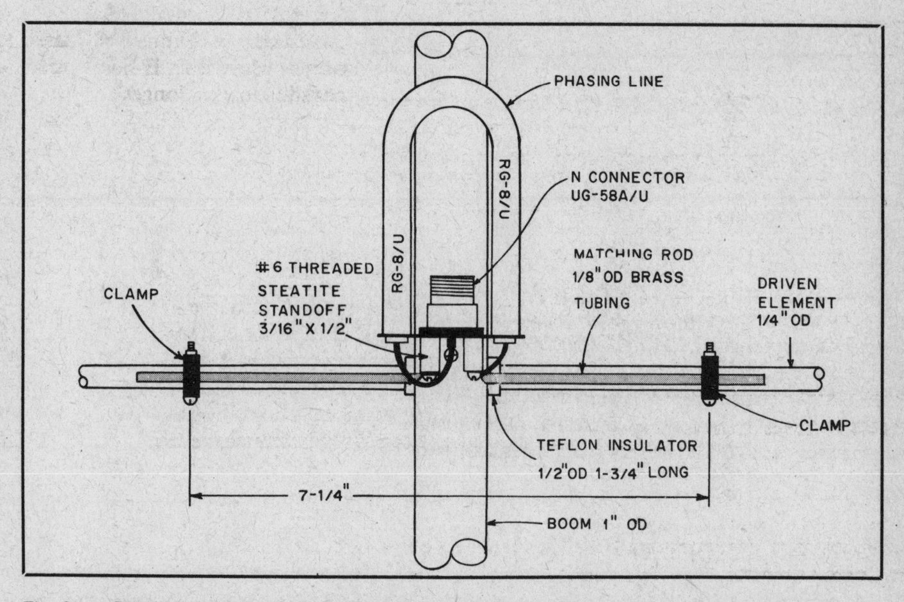


Fig. 21 — Detailed drawing of the feed system used with the Yagi. A small copper plate is attached to the coaxial-connector plate assembly for connection of the phasing line braid. As indicated, the braid is soldered to the plate. A coat of clear lacquer or enamel is recommended for waterproofing the feed system.

Details of the feed system are shown in Fig. 21. This is a form of the T match, where the driven element is shortened from its resonant length to provide the necessary capacitance to tune out the reactance of the matching rods. With this system no variable capacitors are required, as in the more conventional T-match systems used at hf. This is a definite plus in terms of antenna endurance in harsh-weather environments.

The center pin of the UG-58A/U N-type connector is attached to one of the matching rods. A half wavelength of 50-ohm, foam-dielectric cable is used to provide the 180-degree phase shift from one half of the element to the other. An alternative to the large and cumbersome cable used here would be the miniature copper Hardline with Teflon dielectric material, such as RG-401/U.

Each of the matching rods is secured to two threaded steatite standoffs at the center of the antenna. These standoffs provide tie points for the ends of the phasing lines, the center pin of the coaxial connector and the ends of the matching rods. Solder lugs are used for each of the connections for easy assembly or disassembly. The clamps that connect the matching rods to the driven element are constructed from pieces of aluminum measuring 1/4 × 1/2 × 1-5/16 inches. These pieces are drilled and slotted so that when the screws are tightened the pieces will compress slightly to provide a snug fit. Alternatively, simple clamps could be fashioned from strips of aluminum.

**Adjustment**

If the dimensions given in the drawings are followed closely, little adjustment should be necessary. If adjustment is necessary, as indicated by an SWR greater than 1.5 to 1, move the clamps a short distance along the driven element. Keep in mind that the clamps should be located equidistant from the center of the boom.

## STACKING YAGIS

Where suitable provision can be made for supporting them, two Yagis mounted one above the other and fed in phase may be preferable to one long Yagi having the same theoretical or measured gain. The pair will require a much smaller turning space for the same gain, and their lower radiation angle can provide interesting results. On long ionospheric paths a stacked pair occasionally may show an

Fig. 22 — The complete 15-element Yagi for 432 MHz, ready for installation atop the tower.

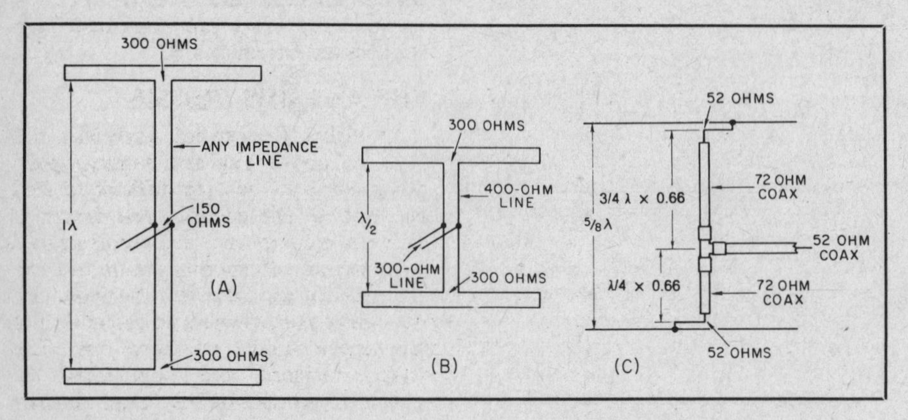


Fig. 23 — Three methods of feeding stacked vhf arrays. A and B are for bays having balanced driven elements, where a balanced phasing line is desired. Array C has an all-coaxial matching and phasing system. If the lower section is also 3/4 wavelength, no transposition of bays is needed.

*apparent* gain much greater than the 2 to 3 dB that can be measured locally as the gain from stacking.

Optimum spacing for Yagis of five elements or more is one wavelength, but this may be too much for many builders of 50-MHz antennas to handle. Worthwhile results can be obtained with as little as one-half wavelength (10 feet), and 5/8 wavelength (12 feet) is markedly better. The difference between 12 and 20 feet may not be worth the added structural problems involved in the wider spacing, at 50 MHz, at least. The closer spacings give lower measured gain, but the antenna patterns are cleaner than will be obtained with one-wavelength spacing. The extra gain with wider spacings is usually the objective on 144 MHz and higher bands, where the structural problems are not severe.

Yagis can also be stacked in the same plane for sharp azimuthal directivity. A spacing of 5/8 wavelength between the ends of the inner elements yields the maximum gain before the pattern breaks up into minor lobes.

If individual bays of a stacked array are properly designed they will look like noninductive resistors to the phasing system that connects them. The impedances involved can thus be treated the same as resistances in parallel, if the phasing lines are a half wavelength or a multiple thereof. The latter point is important because the impedance at the end of a transmission line is repeated at every half wavelength along it.

In Fig. 23 we have three sets of stacked dipoles. Whether these are merely dipoles or the driven elements of Yagi bays makes little difference for the purpose of these examples. Two 300-ohm antennas at A are one wavelength apart, resulting in a feed impedance of approximately 150 ohms at the center. (It will be slightly less than 150 ohms, because of coupling between bays, but we can neglect this for practical purposes.) This value holds regardless of the impedance of the phasing line. Thus, we can use any convenient type of line for phasing, so long as the *electrical* length is right.

The velocity factor of the line must be taken into account. As with coax, this is subject to so much variation that it is well to make a resonance check if there is any doubt. The method is the same as for coax, Fig. 5B. A half-wavelength of line is resonant both open and shorted, but the shorted condition (both ends) is usually the more readily checked.

The impedance-transforming quality of a quarter-wavelength of line can be employed in combination matching and phasing lines, as shown in B and C of Fig. 23. In B, two bays spaced a half-wavelength are phased and matched by a 400-ohm line, acting as a double Q section, so that a 300-ohm main transmission line is matched to two 300-ohm bays. The two halves of this phasing line could each be 3 or 5 quarter-wavelengths long equally well, if these lengths serve any useful purpose. An example would be the stacking of two Yagis, where the desirable spacing is more than one-half wavelength.

A double Q section of coaxial line is illustrated in Fig. 23C. This is useful for feeding stacked bays which were originally set up for 52-ohm feed. A spacing of 5/8 wavelength is optimum for small Yagis, and this is the equivalent of an electrical full wavelength of solid-dielectric coax, such as RG-11/U. If our phasing line is made electrically one quarter-wavelength on one side of the feed and three quarters on the other, one driven element should be reversed with respect to the other to keep the rf currents in phase. If the number of quarter-wavelengths is the same on either side of the feed point, the two elements should be in the same position, not reversed as shown in C.

One marked advantage of coaxial phasing lines is that they can be wrapped around the vertical support, taped or grounded to it, or arranged in any way that is convenient mechanically. The spacing between bays can be set at the most desirable value, and the phasing line placed anywhere necessary to use up the required electrical length.

In stacking horizontal Yagis one above the other on a single support, certain considerations apply whether the bays are for different bands or for the same band. As a rule of thumb, the minimum desirable spacing is one-half the boom length for

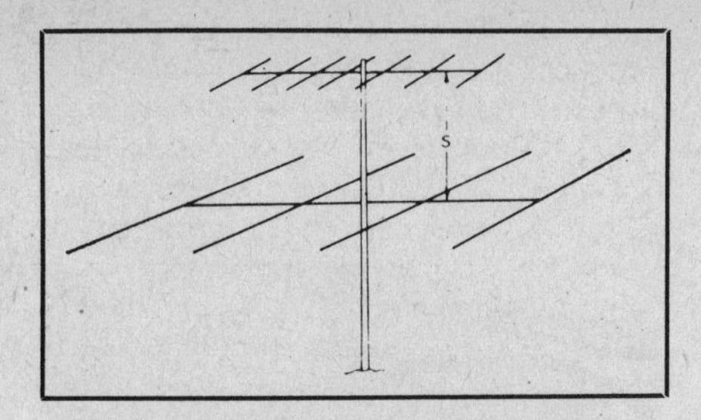


Fig. 24 — In stacking Yagi arrays one above the other the minimum spacing between bays, S, should be about half the boom length of the smaller array. Wider spacing is desirable, in which case it should be a half wavelength, or some multiple thereof, at the frequency of the smaller array. If the beams shown are for 50 and 144 MHz, S should be 40 inches minimum, with 80 inches preferred. Similar conditions apply for stacking bays for a single band.

two bays on the same band, or half the boom length of the higher frequency array where two bands are involved.

In the stacked two-band array of Fig. 24, the 50-MHz, 4-element Yagi is going to "look like ground" to the 7-element 144-MHz Yagi above it, if it has any effect at all. It is well known that the impedance of an antenna varies with height above ground, passing through the free-space value at a quarter wavelength and multiples thereof. At one-quarter wavelength and at the *odd* multiples thereof, ground also acts like a reflector, causing considerable radiation straight up. This effect is least at the half-wave points, where the impedance also passes through the free-space value. Preferably, then, the spacing S should be a *half* wavelength, or multiple thereof, at the frequency of the smaller antenna. The half-the-boom-length rule gives about the same answer in this example. For this length of 2-meter antenna, 40 inches would be the minimum desirable spacing, but 80 inches would be better.

The effect of spacing on the larger array is usually negligible. If spacing closer than half the boom length or a half wavelength must be used, the principal thing to watch for is variation in feed impedance of the smaller antenna. If the smaller antenna has an adjustable matching device, closer spacings can be used in a pinch, if the matching is adjusted for minimum SWR. Very close spacing and interlacing of elements should be avoided unless the builder is prepared to go through an extensive program of adjustments of both element lengths and matching.

## QUADS FOR VHF

Though it has not been used to any great extent in vhf work, the quad antenna has interesting possibilities. It can be built of very inexpensive materials, yet its performance should be at least equal to other arrays of its size. Adjustment for resonance and impedance matching can be accomplished readily. Quads can be stacked horizontally and vertically, to provide high gain, without sharply limiting the frequency response.

### THE 2-ELEMENT QUAD

The basic 2-element quad array for 144 MHz is shown in Fig. 25. The supporting frame is 1 by 1-inch wood, of any kind suitable for outdoor use. Elements are no. 8 aluminum wire. The driven element is one wavelength (83 inches) long, and the reflector 5 percent longer, or 87 inches. Dimensions are not critical, as the quad is relatively broad in frequency response.

The driven element is open at the bottom, its ends fastened to a plastic block, which is mounted at the bottom of the forward vertical support. The top portion of the element runs through the support and is held firmly by a screw running into the wood and then bearing on the aluminum wire. Feed is by means of 52-ohm coax, connected to the driven element loop.

The reflector is a closed loop, its top and bottom portions running through the rear vertical support. It is held in position with screws, top and bottom. The loop can be closed by fitting a length of tubing over the element ends, or by hammering them flat and bolting them together, as shown in the sketch.

The elements in this model are not adjustable, though this can easily be done by the use of stubs. It would then be desirable to make the loops slightly smaller, to compensate for the wire in the adjusting stubs. The driven-element stub would be trimmed for length and the point of connection for the coax would be adjustable for best match. The reflector stub could be adjusted for maximum gain or front-to-back ratio, whichever quality the builder wishes to optimize.

In the model shown only the spacing is adjusted, and this is not particularly critical. If the wooden supports are made as shown, the spacing between the elements can be adjusted for best match, as indicated in an SWR meter connected in the coaxial line. The spacing has little effect on the gain, from 0.15 to 0.25 wavelength, so the variation in impedance with spacing can be used for matching. This also permits use of either 52- or 72-ohm coax for the transmission line.

#### Stacking

Quads can be mounted side by side or one above the other, or both, in the same general way as described for other antennas. Sets of driven elements can also be mounted in front of a screen reflector. The recommended spacing between adjacent element sides is a half wavelength. Phasing and feed methods can be similar to those employed with other antennas described in this chapter.

#### Adding Directors

Parasitic elements ahead of the driven element work in a manner similar to those

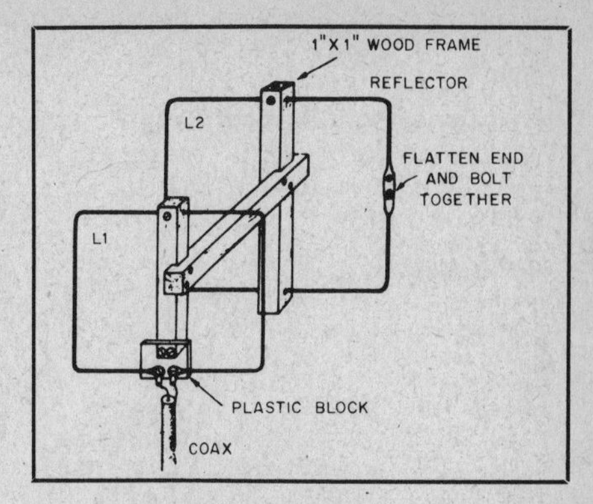


Fig. 25 — Mechanical details of a 2-element quad for 144 MHz. Driven element, L1, is one wavelength long; reflector, L2, 5 percent longer. Sets of elements of this type can be stacked horizontally and vertically for high gain with broad frequency response. Bay spacing recommended is 1/2 wavelength between adjacent element sides. Example shown may be fed directly with 52-ohm coax.

in a Yagi array. Closed loops can be used for directors, by making them 5 percent shorter than the driven element, or about 79 inches. Spacings can be similar to those for conventional Yagis. In an experimental model the reflector was spaced 0.25 wavelength and the director 0.15. A square array using four 3-element bays worked out extremely well.

## VHF AND UHF QUAGIS

At higher frequencies, especially 420 MHz and above, Yagi arrays using dipole-driven elements become difficult to feed and match. The cubical quad described earlier overcomes the feed problems to a large extent. When many parasitic elements are used, however, the loops are not nearly as convenient to assemble and tune as are straight cylindrical ones. The quagi, designed and popularized by Wayne Overbeck, N6NB, is an antenna having a full-wave loop-driven element and reflector, and straight rod directors.

### HOW TO BUILD A QUAGI

There are few tricks to quagi building. The designer mass produced as many as 16 in one day. Tables 5 and 6 give the dimensions for various frequencies up to 446 MHz.

The boom is *wood* or any other non-conductor (e.g., fiberglass). If a metal boom is used, a new design and new element lengths will be required. Many vhf antenna builders go wrong by failing to follow this rule: If the original uses a metal boom, use the same size and shape metal boom when you duplicate it. If it calls for a wood boom, use a non-conductor. Many amateurs dislike wood booms, but in a salt-air environment they outlast aluminum (and surely cost less). Varnish the boom for added protection.

The 2-meter version is usually built on a 14-foot, 1 × 3 inch boom, with the boom cut down to taper it to one inch at both

**Table 5**

**Dimensions, Eight-Element Quagi**

| *Element Lengths* | *144.5 MHz* | *147 MHz* | *222 MHz* | *432 MHz* | *446 MHz* |
|---|---|---|---|---|---|
| Reflector (all no. 12 TW wire, closed) | 86-5/8" (loop) | 85" | 56-3/8" | 28" | 27-1/8" |
| Driven element (no. 12 TW, fed at bottom) | 82" (loop) | 80" | 53.5" | 26-5/8" | 25-7/8" |
| Directors | 35-15/16" to 35" in 3/16" steps | 35-5/16" to 34-3/8" in 3/16" steps | 23-3/8" to 22-3/4" in 1/8" steps | 11-3/4" to 11-7/16" in 1/16" steps | 11-3/8" to 11" in 1/16" steps |
| *Spacing* | | | | | |
| R-DE | 21" | 20-1/2" | 13-5/8" | 7" | 6.8" |
| DE-D1 | 15-3/4" | 15-3/8" | 10-1/4" | 5-1/4" | 5.1" |
| D1-D2 | 33" | 32-1/2" | 21-1/2" | 11" | 10.7" |
| D2-D3 | 17-1/2" | 17-1/8" | 11-3/8" | 5.85" | 5.68" |
| D3-D4 | 26.1" | 25-5/8" | 17" | 8.73" | 8.46" |
| D4-D5 | 26.1" | 25-5/8" | 17" | 8.73" | 8.46" |
| D5-D6 | 26.1" | 25-5/8" | 17" | 8.73" | 8.46" |
| *Stacking Distance Between Bays* | 11' | 10'10" | 7'1-1/2" | 3'7" | 3'5-5/8" |

**Table 6**

**432-MHz, 15-Element, Long-Boom Quagi Construction Data**

| *Element Lengths — Inches* | | *Interelement Spacing — Inches* | |
|---|---|---|---|
| R — 28" loop | D7 — 11-3/8 | R-DE — 7 | D6-D7 — 12 |
| DE — 26-5/8" loop | D8 — 11-5/16 | DE-D1 — 5-1/4 | D7-D8 — 12 |
| D1 — 11-3/4 | D9 — 11-5/16 | D1-D2 — 11 | D8-D9 — 11-1/4 |
| D2 — 11-11/16 | D10 — 11-1/4 | D2-D3 — 5-7/8 | D9-D10 — 11-1/2 |
| D3 — 11-5/8 | D11 — 11-3/16 | D3-D4 — 8-3/4 | D10-D11 — 9-3/16 |
| D4 — 11-9/16 | D12 — 11-1/8 | D4-D5 — 8-3/4 | D11-D12 — 12-3/8 |
| D5 — 11-1/2 | D13 — 11-1/16 | D5-D6 — 8-3/4 | D12-D13 — 13-3/4 |
| D6 — 11-7/16 | | | |

Boom — 1 × 2-inch × 12-ft Douglas fir, tapered to 5/8 inch at both ends.
Driven element — No. 12 TW copper-wire loop in square configuration, fed at center bottom with type N connector and 52-ohm coax.
Reflector — No. 12 TW copper-wire loop, closed at bottom.
Directors — 1/8-inch rod passing through boom.

ends. Clear pine is best because of its light weight, but construction-grade Douglas fir works well. At 220 MHz, the boom is under 10 feet long and most builders use 1 × 2 or (preferably) 3/4 by 1-1/4-inch pine molding stock. On 432 MHz the boom must be 1/2-inch thick or less. Most builders use strips of 1/2-inch exterior plywood for 432 MHz.

The quad elements are supported at the current maxima (the top and bottom, the latter beside the feed point) with Plexiglas or small strips of wood. See Fig. 26. The quad elements are made of no. 12 copper wire, commonly used in house wiring. Some builders may elect to use no. 10 wire on 144 MHz and no. 14 on 432 MHz, although this will change the resonant frequency slightly. Solder a type-N connector (an SO-239 is often used at 2 meters) at the midpoint of the driven element bottom side, and close the reflector loop.

The directors are mounted through the boom. They can be made of almost any metal rod or wire of about 1/8-inch diameter. Welding rod or aluminum clothesline wire will work well if straight. (The designer used 1/8-inch stainless-steel rod obtained from an aircraft surplus store.)

A TV-type U bolt mounts the antenna on a mast. The designer uses a single machine screw, washers and nut to secure the spreaders to the boom so the antenna can be quickly "flattened" for travel. In permanent installations two screws are recommended.

**Construction Reminders**

Here are a couple of hints based on the experiences of some who have built the quagi. First, remember that at 432 MHz even a 1/8-inch measuring error will deteriorate performance. Cut the loops and elements as carefully as possible. No precision tools are needed, but be careful about accuracy. Also, make sure to get the elements in the right order. The longest director goes closest to the driven element.

Finally, remember that a balanced

Fig. 26 — A close-up view of the feed method used on a 432-MHz quagi. This arrangement produces a low SWR and an actual gain in excess of 13 dB over an isotropic antenna with a 4-foot 10-inch boom! The same basic arrangement is used on lower frequencies, but wood may be substituted for the Plexiglas spreaders. The boom is 1/2-inch exterior plywood.

antenna is being fed with an unbalanced line. Every balun the designer tried introduced more losses than the feed imbalance problem. Some builders have tightly coiled several turns of the feed line near the feed point to limit radiation further down the line. In any case, the feed line should be kept at right angles to the antenna. Run it from the driven element directly to the supporting mast and then up or down perpendicularly for best results.

**QUAGIS FOR 1296 MHz**

The quagi principle has recently been extended to the 1296-MHz band, where conventional Yagis are extremely difficult to get working. Fig. 27 and Table 7 give the design information for antennas having 10, 15 and 25 elements.

At 1296 MHz even slight variations in design or building materials may cause substantial changes in performance. The 1296-MHz antennas described here will work every time — but only if you use the same materials and build them exactly as described here. This is not to discourage experimentation. Innovation and experimentation are part of Amateur Radio. But if you want to modify these 1296-MHz antenna designs, you might consider building one antenna *exactly* as described here, so that you have a reference against which to compare your variations.

The quagis (and the cubical quad) are built on 1/4-in. thick Plexiglas booms. The driven element and reflector (and also the directors in the case of the cubical quad) are made of insulated no. 18 AWG solid-copper bell wire, available at hardware and electrical supply stores. Other types and sizes of wire will work equally well, but the dimensions will vary with the wire diameter. Even removing the insulation usually necessitates changing the loop lengths.

Quad loops are approximately square (Fig. 28) although the shape is relatively noncritical. However, the element lengths *are* critical. At 1296 MHz, variations of

Fig. 27 — A view of the 10-element version of the 1296-MHz quagi. It is mounted on a 30-in. Plexiglas boom with a 3- × 3-in. square of Plexiglas to support the driven element and reflector. Note how the driven element is attached to a standard UG-290 BNC connector. The elements are held in place with silicone sealing compound.

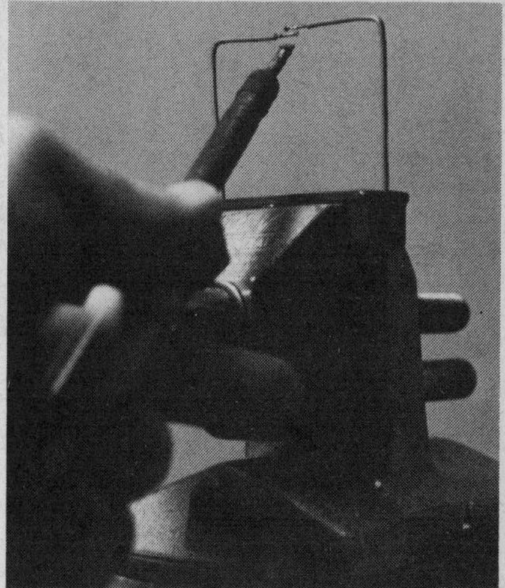

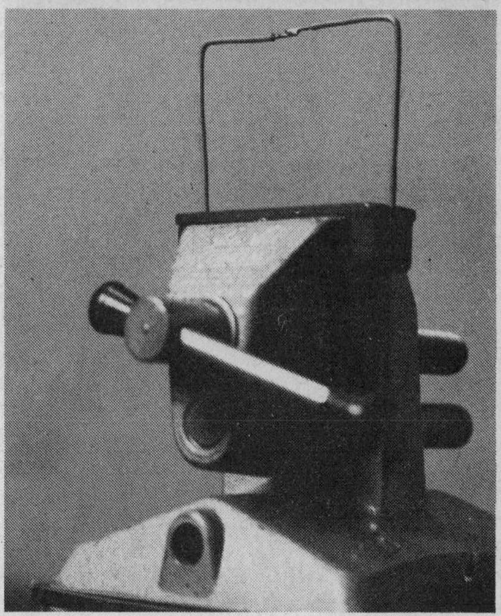

Fig. 28 — These photos show the construction method used for the 1296-MHz quad type parasitic elements. The two ends of the no. 18 AWG copper wire are brought together with an overlap of 1/8 in. and soldered.

**Table 7**

**Dimensions, 1296-MHz Quagi Antennas**

Note: All lengths are gross lengths. See text and photos for construction technique and recommended overlap at loop junctions. All loops are made of no. 18 AWG solid-covered copper bell wire. The Yagi-type directors are 1/16-in. brass brazing rod. See text for a discussion of director taper.

*Feed:* Direct with 52-ohm coaxial cable to UG-290 connector at driven element; run coax symmetrically to mast at rear of antenna.

*Boom:* 1/4-in.-thick Plexiglas, 30 in. long for 10-element quad or quagi and 48 in. long for 15-element quagi; 84 in. for 25-element quagi.

**10-Element Quagi for 1296 MHz**

| *Element* | *Length (in.)* | *Construction* | *Element* | *Interelement Spacing (in.)* |
|---|---|---|---|---|
| Reflector | 9.5625 | (loop) | R-DE | 2.375 |
| Driven El. | 9.25 | (loop) | DE-D1 | 2.0 |
| Director 1 | 3.91 | (brass rod) | D1-D2 | 3.67 |
| Director 2 | 3.88 | (brass rod) | D2-D3 | 1.96 |
| Director 3 | 3.86 | (brass rod) | D3-D4 | 2.92 |
| Director 4 | 3.83 | (brass rod) | D4-D5 | 2.92 |
| Director 5 | 3.80 | (brass rod) | D5-D6 | 2.92 |
| Director 6 | 3.78 | (brass rod) | D6-D7 | 4.75 |
| Director 7 | 3.75 | (brass rod) | D7-D8 | 3.94 |
| Director 8 | 3.72 | (brass rod) | | |

**15-Element Quagi for 1296 MHz**

The first 10 elements are the same lengths (inches) as above, but the spacing from D6 to D7 is 4.0 in. here; D7 to D8 is also 4.0 in.

| Element | Length (in.) | Element | Spacing (in.) |
|---|---|---|---|
| Director 9 | 3.70 | D8-D9 | 3.75 |
| Director 10 | 3.67 | D9-D10 | 3.83 |
| Director 11 | 3.64 | D10-D11 | 3.06 |
| Director 12 | 3.62 | D11-D12 | 4.125 |
| Director 13 | 3.59 | D12-D13 | 4.58 |

**25-Element Quagi for 1296 MHz**

The first 15 elements use the same element lengths and spacings as the 15-element model. The additional directors are evenly spaced at 3.0-in. intervals and taper successively by 0.02 in. per element. Thus, D23 is 3.39 in.

1/16 in. may alter the performance measurably, and a 1/8-in. departure can cost several decibels of gain. The loop lengths given are *gross* lengths. Cut the wire to these lengths and then solder the two ends together. There is a 1/8-in. overlap where the two ends of the reflector (and director) loops are joined, as shown in Fig. 28.

The driven element is the most important of all. The no. 18 wire loop is soldered to a standard UG-290 chassis-mount BNC connector as shown in the photographs. This exact type of connector must be used to ensure uniformity in construction. Any substitution may alter the driven-element electrical length. One end of the 9.25-in. driven loop is pushed as far as it will go into the center pin and soldered. Then the loop is shaped and threaded through small holes drilled in the Plexiglas support. Finally, the other end is fed into one of the four mounting holes on the BNC connector and soldered. In most cases, the best VSWR is obtained if the end of the wire just passes through the

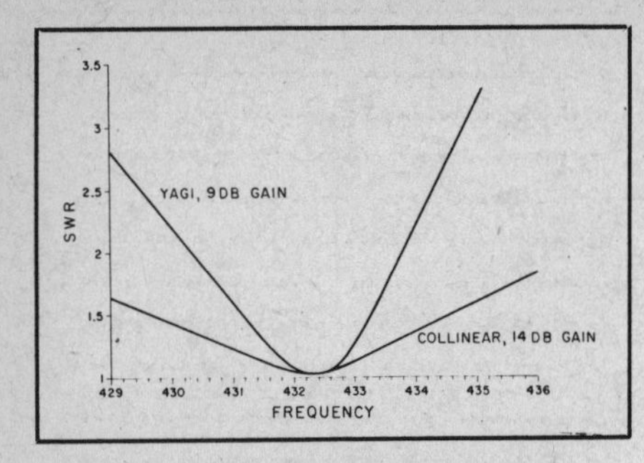


Fig. 29 — Comparison of the frequency responses of a small Yagi antenna and a large collinear array. A Yagi of comparable gain would have a still sharper frequency response.

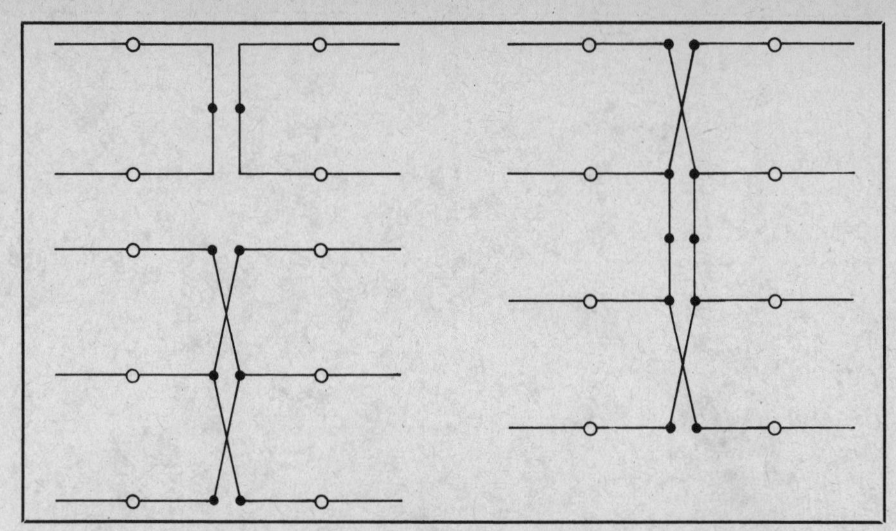

Fig. 30 — Element arrangements for 8-, 12- and 16-element collinear arrays. Parasitic reflectors, omitted here for clarity, are 5 percent longer and 0.2 wavelength in back of the driven elements. Feed points are indicated by black dots. Open circles are recommended support points. The elements can run through wood or metal booms, without insulation, if supported at their centers in this way. Insulators at the element ends (points of high rf voltage) tend to detune and unbalance the system.

hole so that it is flush with the opposite side of the connector.

## COLLINEAR ANTENNAS

Information given thus far is mainly on parasitic arrays, but the collinear antenna has much to recommend it. Inherently broad in frequency response, it is a logical choice where coverage of an entire band is wanted (see Fig. 29). This tolerance also makes a collinear easy to build and adjust for any vhf application, and the use of many driven elements is popular in very large phased arrays, such as may be required for moonbounce (EME) communication.

### Large Collinear Arrays

Bidirectional curtain arrays of four, six and eight half-waves in phase are shown in Fig. 30. Usually reflector elements are added, normally at about 0.2 wavelength in back of each driven element, for more gain and a unidirectional pattern. Such parasitic elements are omitted from the sketch in the interest of clarity.

When parasitic elements are added, the feed impedance is low enough for direct connection to open line or twin-lead, connected at the points indicated by black dots. With coaxial line and a balun, it is suggested that the universal stub match, Fig. 4A, be used at the feed point. All elements should be mounted at their electrical centers, as indicated by open circles in Fig. 30. The framework can be metal or insulating material, with equally good results. The metal supporting structure is entirely in back of the plane of the reflector elements. Sheet-metal clamps can be cut from scraps of aluminum to make this kind of assembly, which is very light in weight and rugged as well. Collinear elements should always be mounted at their centers, where rf voltage is zero — never at their ends, where the voltage is high and insulation losses and detuning can be very harmful.

Collinear arrays of 32, 48, 64 and even 128 elements can be made to give outstanding performance. Any collinear should be fed at the center of the system, for balanced current distribution. This is very important in large arrays, which are treated as sets of six or eight driven elements each, and fed through a balanced harness, each section of which is a resonant length, usually of open-wire line. A 48-element collinear array for 432 MHz, Fig. 31, illustrates this principle.

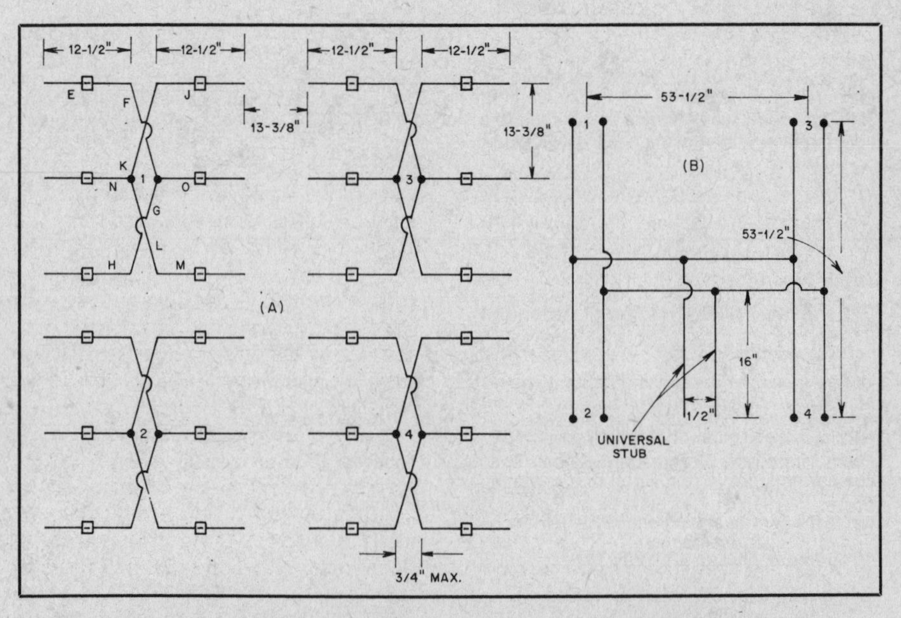


Fig. 31 — Large collinear arrays should be fed as sets of no more than eight driven elements each, interconnected by phasing lines. This 48-element array for 432 MHz (A) is treated as if it were four 12-element collinears. Reflector elements are omitted for clarity. The phasing harness is shown at B.

A reflecting plane, which may be sheet metal, wire mesh, or even closely spaced elements of tubing or wire, can be used in place of parasitic reflectors. To be effective, the plane reflector must extend on all sides to at least a quarter-wavelength beyond the area occupied by the driven elements. The plane reflector provides high front-to-back ratio, a clean pattern and somewhat more gain than parasitic elements, but large physical size rules it out for amateur use below 420 MHz. An interesting space-saving possibility lies in using a single plane reflector with elements for two different bands mounted on opposite sides. Reflector spacing from the driven element is not critical. About 0.2 wavelength is common.

## THE CORNER REFLECTOR

When a single driven element is used, the reflector may be bent to form an angle, giving an improvement in the radia-

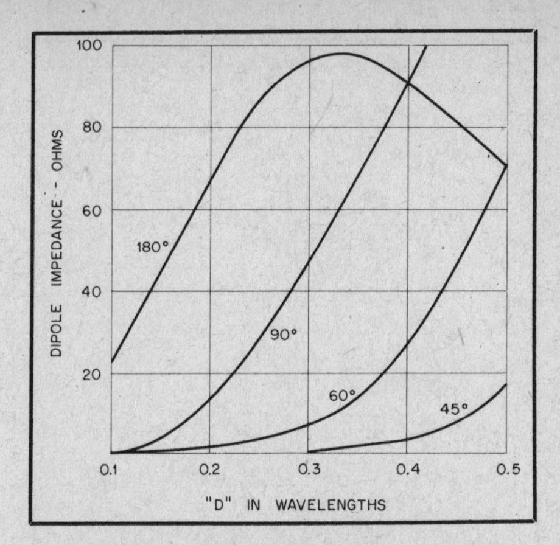


Fig. 32 — Feed impedance of the driven element in a corner-reflector array, for corner angles of 180 (flat sheet), 90, 60 and 45 degrees.

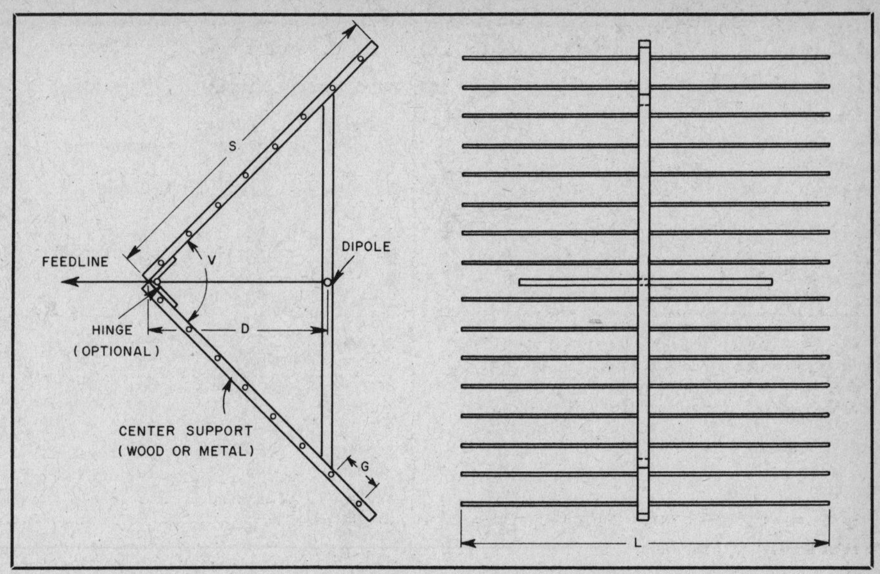


Fig. 33 — Construction of a corner-reflector array. Frame can be wood or metal. Reflector elements are stiff wire or tubing. Dimensions for several bands are given in Table 8. Reflector element spacing, G, is the maximum that should be used for the frequency; closer spacings are optional. Hinge permits folding for portable use.

tion pattern and gain. At 220 and 420 MHz its size assumes practical proportions, and at 902 MHz and higher, reflectors approaching semi-infinite dimensions may be used, resulting in more gain and sharper patterns. The corner reflector can be used at 144 MHz, though usually at less than optimum size. For a given aperture it does not equal a parabola in gain, but it is simple to construct, broadband, and offers gains from about 10 dB to 15 dB, depending on the angle and size.

The corner angle can be 90, 60 or even 45 degrees, but the side length must be increased as the angle is narrowed. For a 90-degree corner, the driven element spacing can be anything from 0.25 to 0.7 wavelength, 0.35 to 0.75 wavelength for 60 degrees, and 0.5 to 0.8 wavelength for 45 degrees. In each case the gain variation is about 1.5 dB. Since the spacing is not critical as to gain, it may be varied to obtain impedance matching. Close spacing means lowered radiation resistance, but a folded dipole radiator could be used to raise it to a usable level.

Radiation resistance vs. spacing is shown in Fig. 32. The maximum gain obtained with minimum spacing is the primary mode, the one generally used at 144, 220 and 420 MHz to prevent side length from being excessive. A 90-degree corner, for example, should have a minimum side length (S, Fig. 33) equal to twice the dipole spacing, or 1-wavelength long for 0.5-wavelength spacing. A side length greater than 2 wavelengths is ideal. Gain with a 60- or 90-degree corner with 1-wavelength sides runs about 10 dB. A 60-degree corner with 2-wavelength sides has about 12 dB gain, and a 45-degree corner with 3-wavelength sides has about 13 dB gain.

Reflector length (L) should be a minimum of 0.6 wavelength. Less than 0.6 wavelength causes radiation to increase to the sides and rear, and gain decreases.

Reflector spacing (G) should not exceed 0.06 wavelength for best results. A spacing of 0.06 wavelength results in a back lobe that is about 6% of the forward lobe (down 12 dB). A small mesh screen or solid sheet is preferable at the higher frequencies to obtain maximum efficiency and highest front-to-back ratio, and to simplify construction. A spacing of 0.06 wavelength at 1296 MHz, for example, would require mounting reflector rods about every half-inch along the sides. Rods or spines may be used to reduce wind loading. The support used for mounting the reflector rods may be of insulating or conductive material. Rods or mesh weave should be parallel to the radiator.

A suggested arrangement for a corner reflector is shown in Fig. 33. The frame may be made of wood or metal, with a hinge at the corner to facilitate portable work or assembly atop a tower. Hinged sides would also be useful in experimenting with different angles. Table 8 gives the principal dimensions for corner-reflector arrays for 144 to 2300 MHz. The arrays for 144, 220 and 420 MHz have side lengths of twice to four times the driven element spacing. The 902-MHz corners use side lengths of three times the element spacing, the 1296-MHz corners use side lengths of four times the spacing, and the 2304-MHz corners employ side lengths of six times the spacing. Reflector lengths of 2.0, 3.0 and 4 wavelengths are used on the 902, 1296 and 2304 MHz reflectors, respectively. A 4 × 6-wavelength reflector approximates an infinite sheet.

A corner may be used for several bands, or perhaps for uhf television reception, as well as amateur uhf work. For operation on more than one frequency, side length and reflector length should be selected for the lowest frequency and reflector spacing for the highest frequency. The type of driven element plays a part in determining bandwidth, as does the spacing to the corner. A fat cylindrical element (small wavelength-to-diameter ratio) or triangular dipole ("bow tie") gives more bandwidth than a thin driven element. Wider spacings of driven element to corner give greater bandwidth. A small increase in gain can be obtained for any corner by mounting collinear elements in a corner of sufficient size, but the simple feed of a dipole is lost if more than two elements are used.

A dipole radiator is usually employed with a corner reflector. This requires a balun to convert from coaxial line to balanced antenna. This is easily constructed of coaxial line on the lower vhf bands, but becomes more difficult at the higher frequencies. This problem may be overcome by using a ground-plane corner reflector, a natural for vertical polarization. The corner is mounted on the ground plane and a 1/4 wavelength radiator mounted on the ground plane is used, permitting direct connection to a coax line if the proper spacing is used. The capture area is reduced, but at the higher frequencies second or third mode radiator spacing and larger reflectors could be employed to obtain more gain and offset the loss in effective aperture. A "J" antenna could be used to maintain the capture area and provide matching to a coaxial line. A ground-plane corner with monopole radiator is shown in Fig. 34.

For vertical polarization work, four 90-degree corners built back-to-back (with common reflectors) could be used for scanning 360 degrees of horizon with modest gain by feed-line switching to select the sector desired.

## TROUGH REFLECTORS

To reduce the overall dimensions of a

**Table 8**

**Dimensions of Corner-Reflector Arrays for 144, 220, 420, 902, 1240(1) and 2300 MHz**

| Band (MHz) | Side Length S (In.) | Dipole to Vertex D (In.) | Reflector Length L (In.) | Reflector Spacing G (In.) | Corner Angle V (Deg.) | Feed Impedance (Ohms) |
|---|---|---|---|---|---|---|
| 144* | 65 | 27.5 | 48 | 7-3/4 | 90 | 70 |
| 144 | 80 | 40 | 48 | 4 | 90 | 150 |
| 220* | 42 | 18 | 30 | 5 | 90 | 70 |
| 220 | 52 | 25 | 30 | 3 | 90 | 150 |
| 220 | 100 | 25 | 30 | screen | 60 | 70 |
| 420 | 27 | 8.75 | 16.25 | 2-5/8 | 90 | 70 |
| 420 | 54 | 13.5 | 16.25 | screen | 60 | 70 |
| 902** | 20 | 6.5 | 25.75 | 0.65 | 90 | 70 |
| 902** | 51 | 16.75 | 25.75 | screen | 60 | 65 |
| 902** | 78 | 25.75 | 25.75 | screen | 45 | 70 |
| 1215(1) | 18 | 4.5 | 27.5 | 0.5 | 90 | 70 |
| 1215(1) | 48 | 11.75 | 27.5 | screen | 60 | 65 |
| 1215(1) | 72 | 18.25 | 27.5 | screen | 45 | 70 |
| 2300(2) | 15.5 | 2.5 | 20.5 | 0.25 | 90 | 70 |
| 2300(2) | 40 | 6.75 | 20.5 | screen | 60 | 65 |
| 2300(2) | 61 | 10.25 | 20.5 | screen | 45 | 70 |

*Side length and number of reflector elements somewhat below optimum — slight reduction in gain
**Dimensions given for 915 MHz.
(1) Presently 1215 MHz, but 1240 after WARC. Dimensions given for 1296 MHz.
(2) Dimensions given for 2304 MHz.

Notes
902 MHz: side length S is 3 × D, dipole-to-vertex distance
reflector length L is 2.0 λ
reflector spacing G is 0.05 λ
wavelength is 12.9 inches for 915 MHz

1215 MHz: side length S is 4 × D, dipole to vertex distance
reflector length L is 3.0 λ
reflector spacing G is 0.05 λ
wavelength is 9.11 inches for 1296 MHz

2300 MHz: side length S is 6 × D, dipole to vertex distance
reflector length L is 4.0 λ
reflector spacing G is 0.05 λ
wavelength is 5.12 inches for 2304 MHz

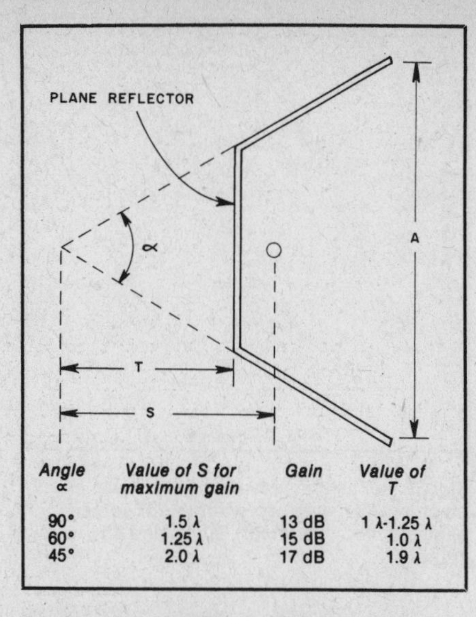


| Angle ∝ | Value of S for maximum gain | Gain | Value of T |
|---|---|---|---|
| 90° | 1.5 λ | 13 dB | 1 λ-1.25 λ |
| 60° | 1.25 λ | 15 dB | 1.0 λ |
| 45° | 2.0 λ | 17 dB | 1.9 λ |

Fig. 35 — Trough reflector. This is a useful modification of the corner reflector shown in Fig. 33. The vertex has been cut off and replaced by a simple plane section. The tabulated data shows the gain obtainable for greater values of S than those covered by Table 8, assuming that the reflector is of adequate size.

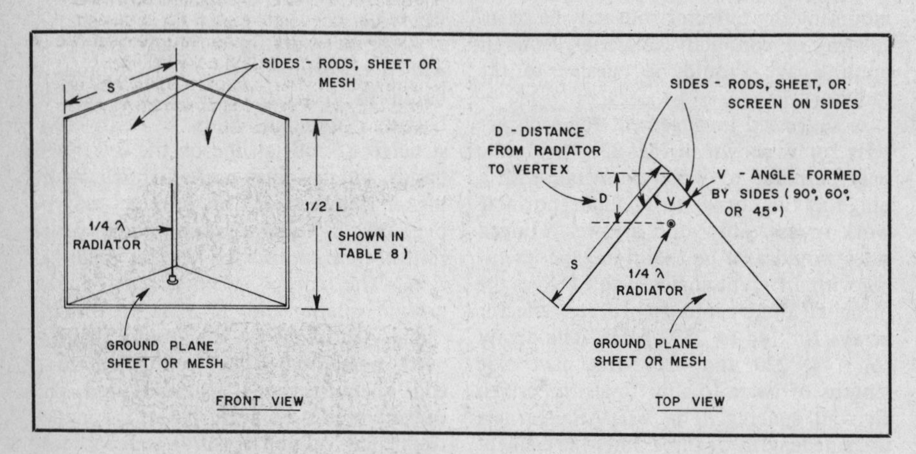


Fig. 34 — A ground-plane corner-reflector antenna for vertical polarization, such as in fm communications.

large corner reflector the vertex can be cut off and replaced with a plane reflector. Such an arrangement is known as a trough reflector. See Fig. 35. Performance similar to that of the large corner reflector can thereby be had, provided that the dimensions of the trough do not exceed the limits indicated in the figure. The resulting antenna has a performance very little different from the corner-reflector type and presents fewer mechanical problems since the plane center portion is relatively easy to mount on the mast and the sides are considerably shorter.

The gain of both corner reflectors and trough reflectors may be increased still further by stacking two or more and arranging them to radiate in phase, or alternatively by adding further collinear dipoles within a wider reflector similarly fed in phase. Not more than two or three radiating units should be used since the great virtue of the simple feeder arrangement would then be lost.

## TROUGH REFLECTORS FOR 432 AND 1296 MHz

Dimensions are given in Fig. 36 for 432 and 1296 MHz trough reflectors. The gain to be expected is 15 dB and 17 dB, respectively. A very convenient arrangement, especially for portable work, is to use a metal hinge at each angle of the reflector. This permits the reflector to be folded flat for transit. It also permits experiments to be carried out with different apex angles.

A housing will be required at the dipole center to prevent the entry of moisture and, in the case of the 432-MHz antenna, to support the dipole elements. The dipole may be moved in and out of the reflector to get either minimum standing wave ratio or, if this cannot be measured, maximum gain. If a two-stub tuner or other matching device is used, the dipole may be placed to give optimum gain and the matching device adjusted to give optimum match. In the case of the 1296-MHz antenna, the dipole length can be adjusted by means of the brass screws at the ends of the elements. Locking nuts are essential.

The reflector should be made of sheet aluminum for 1296 MHz but can be constructed of wire mesh (with twists parallel to the dipole) for 432 MHz. To increase the gain by 3 dB a pair can be stacked so that the reflectors are just not touching (to avoid a slot radiator being formed by the edges). The radiating dipoles must then be fed in phase and suitable feeding and matching must be arranged. A two-stub tuner can be used for matching either a single- or double-reflector system.

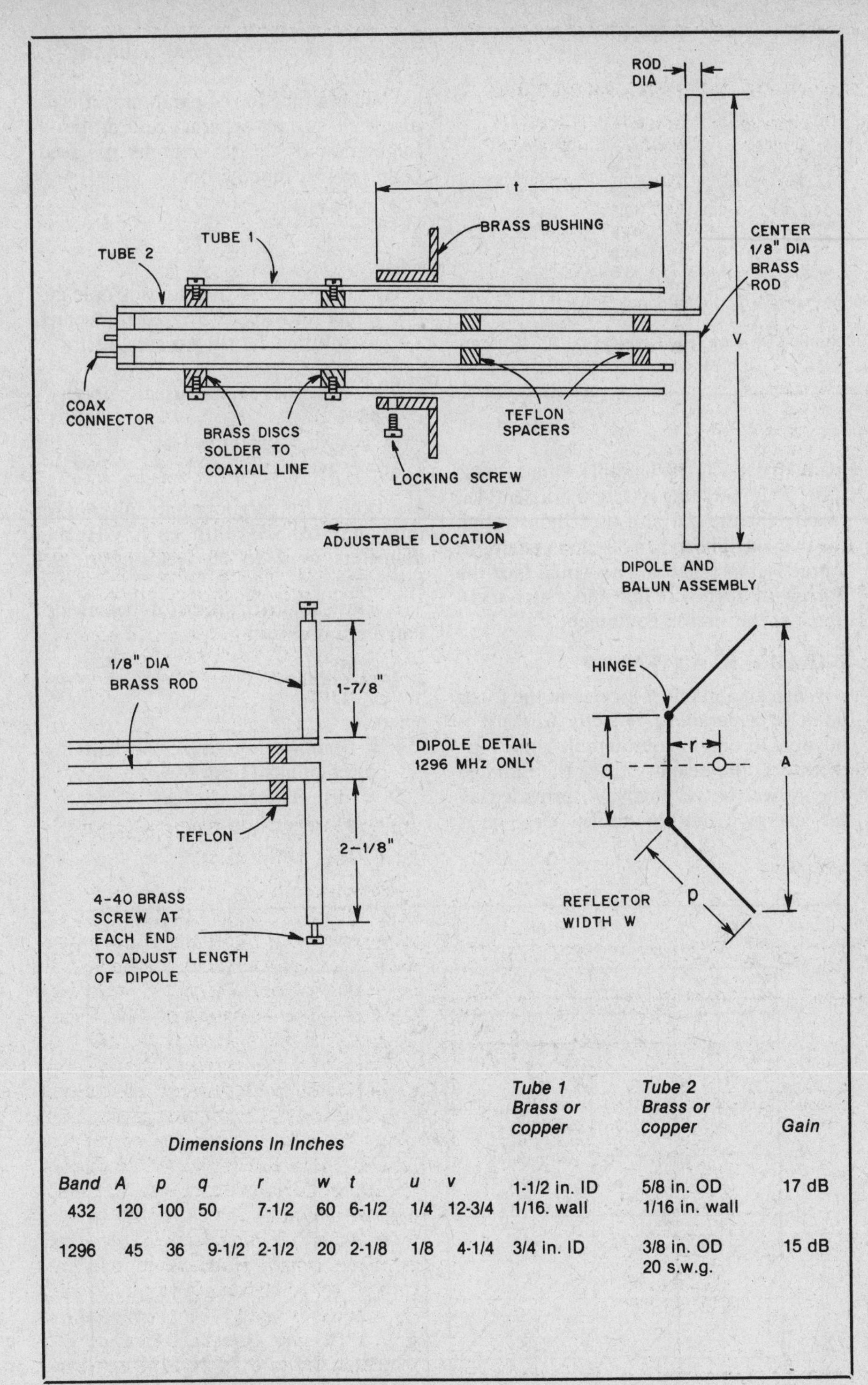


| Band | A | p | q | r | w | t | u | v | Tube 1 Brass or copper | Tube 2 Brass or copper | Gain |
|---|---|---|---|---|---|---|---|---|---|---|---|
| 432 | 120 | 100 | 50 | 7-1/2 | 60 | 6-1/2 | 1/4 | 12-3/4 | 1-1/2 in. ID 1/16. wall | 5/8 in. OD 1/16 in. wall | 17 dB |
| 1296 | 45 | 36 | 9-1/2 | 2-1/2 | 20 | 2-1/8 | 1/8 | 4-1/4 | 3/4 in. ID | 3/8 in. OD 20 s.w.g. | 15 dB |

Fig. 36 — Practical construction information for trough-reflector antennas for 432 and 1296 MHz.

Fig. 37 — A horn antenna is standard equipment with the Microwave Associates Gunnplexer 10-GHz transceiver. The antenna is fabricated from metal-plated plastic and mounts to UG-39/U waveguide flange.

Fig. 38 — An experimental two-sided pyramidal horn constructed in the ARRL laboratory. A pair of muffler clamps allows mounting the antenna on a mast. This model has sheet-aluminum sides, although window screen would work as well. Temporary elements could be made from cardboard covered with aluminum foil. The horizontal spreaders are Plexiglas rod. This antenna radiates horizontally polarized waves.

## SIMPLE HORN ANTENNAS FOR THE MICROWAVE BANDS

Horn antennas were introduced briefly in the section on coupling energy in and out of waveguides. A commercially manufactured horn is pictured in Fig. 37. For amateur purposes, horns begin to show usable gain with practical dimensions in the 902- to 928-MHz band.

It isn't necessary to feed a horn with waveguide. If only two sides of a pyramidal horn are constructed, the antenna may be fed at the apex with a two-conductor transmission line. The impedance of this arrangement is on the order of 300 to 400 ohms. A 60-degree two-sided pyramidal horn having 18-inch sides is shown in Fig. 38. This antenna has a theoretical gain of 15 dBi at 1296 MHz, although the feed system detailed in Fig. 39 probably degrades this value somewhat. A quarter-wave 150-ohm matching section made from two parallel lengths of twin-lead connects to a bazooka balun

Fig. 39 — Matching system used to test the horn. Better performance would be realized with open-wire line, RG-8/U coax and a cable-mount type N connector. See text.

**Table 9**

**Gain, Parabolic Antennas***

| *Frequency* | *24"* | *48"* | *72"* | *10'* | *15'* | *20'* | *30'* |
|---|---|---|---|---|---|---|---|
| 420 MHz | 6.0 | 12.0 | 15.5 | 20.0 | 23.5 | 26.0 | 29.5 |
| 902 | 12.5 | 18.5 | 22.0 | 26.5 | 30.0 | 32.5 | 36.0 |
| 1215 | 15.0 | 21.0 | 24.5 | 29.0 | 32.5 | 35.0 | 38.5 |
| 2300 | 20.5 | 26.5 | 30.0 | 34.5 | 38.0 | 40.5 | 44.0 |
| 3300 | 24.0 | 30.0 | 33.5 | 37.5 | 41.5 | 43.5 | 47.5 |
| 5650 | 28.5 | 34.5 | 38.0 | 42.5 | 46.0 | 48.5 | 52.0 |
| 10 GHz | 33.5 | 39.5 | 43.0 | 47.5 | 51.0 | 53.5 | 57.0 |

*Gain over an isotropic antenna (subtract 2.1 dB for gain over a dipole antenna). Reflector efficiency of 55% assumed.

Reference: Gabriel Antenna Calculator, Gabriel Electronics Division, The Gabriel Co., 125 Crescent Rd., Needham Heights, Massachusetts.

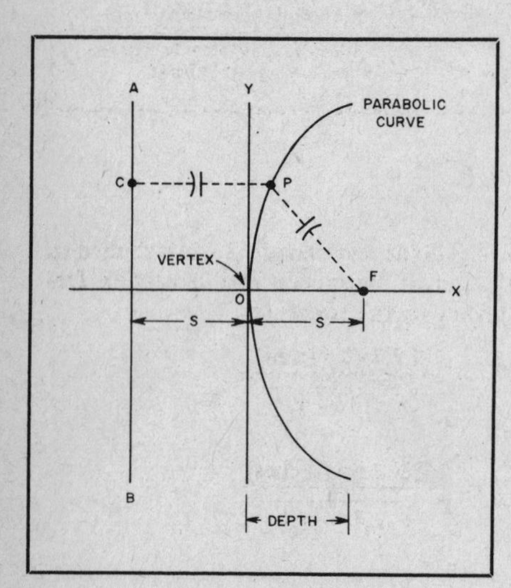


Fig. 40 — Details of the parabolic curve, $Y^2 = 4SX$. This curve is the locus of points which are equidistant from a fixed point, the focus (F), and a fixed line (AB) which is called the "directrix". Hence, FP = PC. The focus (F) contains the coordinates S and O.

made from RG-58/U cable and a brass tube. This matching system was assembled strictly for the purpose of demonstrating the two-sided horn in a 50-ohm system. In a practical installation one would feed the horn with open-wire line and match to 50 ohms at the station equipment.

## PARABOLIC ANTENNAS

When an antenna is located at the focus point of a parabolic reflector (dish) it is possible to obtain considerable gain. Furthermore, the beamwidth of the radiated energy will be very narrow, provided all the energy from the driven element is directed toward the focal point of the reflector.

Gain is a function of parabolic reflector diameter, surface accuracy and proper illumination of the reflector by the feed. Gain may be found from

$$G = \frac{k (\pi D)^2}{\lambda}$$

where

G = power ratio over an isotropic antenna (must be converted to decibels — subtract 2.1 dB for gain over a dipole)

k = efficiency factor, usually about 55%

D = dish diameter in feet

$\lambda$ = wavelength in feet

See Table 9 for parabolic antenna gain for the bands 420 MHz through 10 GHz and diameters of 2 to 30 feet (0.6 to 9.1 meters).

A close approximation of beamwidth may be found from

$$\phi = \frac{70° \lambda}{D}$$

where

$\phi$ = beamwidth in degrees at half-power points (3 dB down)

D = dish diameter in feet

$\lambda$ = wavelength in feet

**Table 10**

**f/D vs Subtended Angle at Focus of a Parabolic Reflector Antenna**

| *f/D* | *Subtended Angle (Deg.)* |
|---|---|
| 0.2 | 203 |
| 0.25 | 181 |
| 0.3 | 161 |
| 0.35 | 145 |
| 0.4 | 130 |
| 0.45 | 117 |
| 0.5 | 106 |
| 0.55 | 97 |
| 0.6 | 88 |
| 0.65 | 80 |
| 0.7 | 75 |
| 0.75 | 69 |
| 0.8 | 64 |
| 0.85 | 60 |
| 0.9 | 57 |
| 0.95 | 55 |
| 1.00 | 52 |

Taken from graph "f/D vs Subtended Angle at Focus" on page 170 of the 1966 *Microwave Engineers' Handbook and Buyers Guide.* Graph courtesy of K. S. Kelleher, Aero Geo Astro Corp., Alexandria, Virginia.

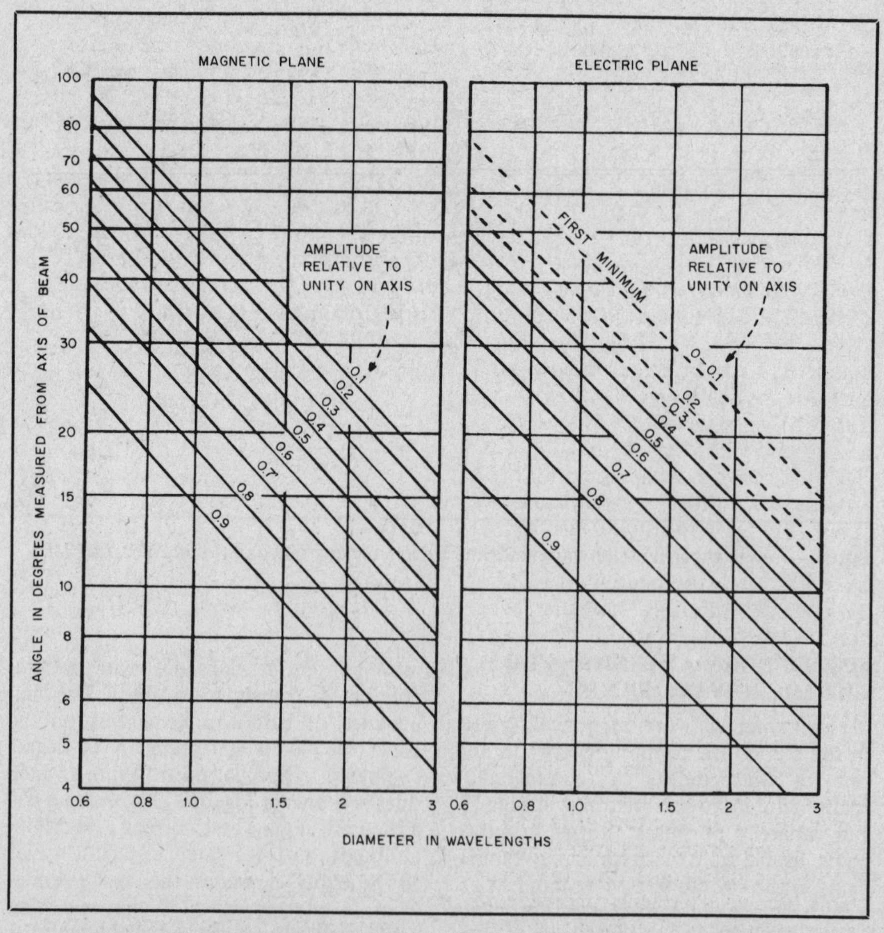


Fig. 41 — This graph can be used in conjunction with Table 10 to select the proper diameter waveguide to illuminate a parabolic reflector.

At 420 MHz and higher, the parabolic antenna becomes a practical gain antenna. A simple, single feed point eliminates phasing harnesses and balun requirements. Gain is dependent on good surface accuracy, which is more difficult to achieve with increasing frequency. Surface errors should not exceed 1/8 wavelength in amateur work. At 430 MHz 1/8 wavelength is 3.4 inches, but at 10 GHz becomes 0.1476 inch! Mesh can be used for the reflector surface to reduce weight and wind loading, but hole size should be less than 1/12 wavelength. At 430 MHz the use of 2-inch (hole diameter) chicken wire is acceptable. Fine mesh aluminum screening works well as high as 10 GHz. A support form may be fashioned to provide the proper parabolic shape by plotting a curve, Fig. 40, from

$$y^2 = 4SX$$

as shown in the figure.

Optimum illumination occurs when power at the reflector edge is 10 dB less than that at the center. A circular waveguide feed of correct diameter and length for the frequency and correct beamwidth for the dish focal length/diameter (f/D) ratio will provide optimum illumination at 902 MHz and higher, but is impractical at 432 MHz, where a dipole and plane reflector are often used. An f/D ratio between 0.4 and 0.6 is considered ideal for maximum gain and simple feeds.

The focal length of a dish may be found by

$$f = \frac{D^2}{16d}$$

where
 f = focal length
 D = diameter
 d = depth distance from plane at mouth of dish to vertex (see Fig. 40)

The units of f are the same as those used to measure the depth and diameter.

Table 10 gives the subtended angle at focus for dish f/D ratios from 0.2 to 1.0. A dish, for example, with a typical f/D of 0.4 requires a 10-dB beamwidth of 130 degrees. A circular waveguide feed with a diameter of approximately 0.7 wavelength will provide nearly optimum illumination, but will not uniformly illuminate the reflector in both the magnetic (TM) and electric (TE) planes. Fig. 41 shows data for plotting radiation patterns from circular guides. The waveguide feed aperture can be modified to change the beamwidth. One approach used successfully by some experimenters is the use of a disc at a short distance behind the aperture as shown in Fig. 42. As the distance between the aperture and disc is changed, the TM plane patterns become alternately broader and narrower than from an unmodified aperture. A disc about two wavelengths in diameter apparently is as effective as one much larger. Some experimenters have noted a 1 to 2 dB increase in dish gain with this modified feed. Rectangular waveguide feeds can also be used, but dish illumination is not as uniform as with round guide feeds.

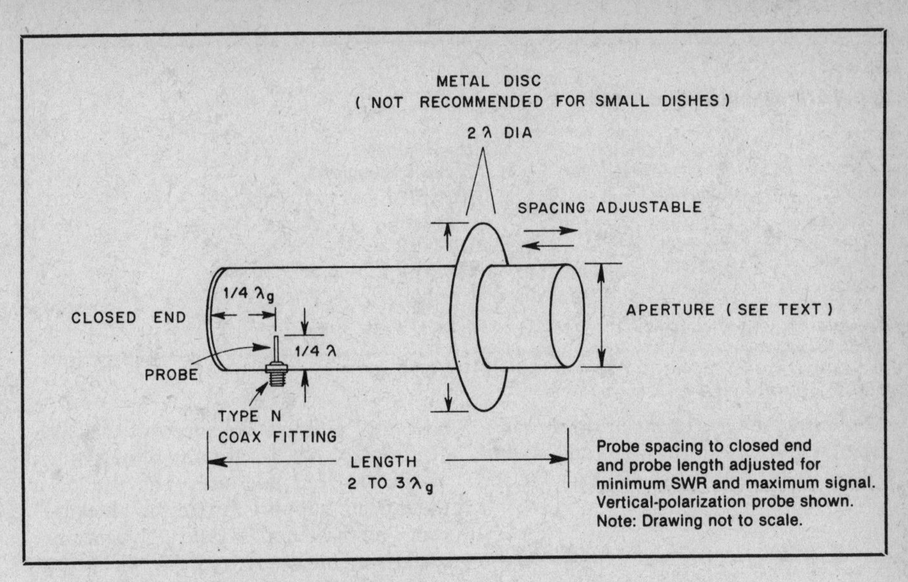


Fig. 42 — Details of a circular waveguide feed.

The circular feed can be made of copper, brass, aluminum or even tin in the familiar form of a coffee or juice can, but the latter must be painted on the outside to prevent corrosion. The circular feed must be within a proper size (diameter) range for the frequency being used. This feed operates in the dominant circular waveguide mode known as the $TE_{11}$ mode. The guide must be large enough to pass the $TE_{11}$ mode with no attenuation but smaller than the diameter that will permit the next higher $TM_{01}$ mode to propagate. To support the desirable $TE_{11}$ mode in circular waveguide, the cutoff frequency, $F_c$, is defined by

$$f_c\ (TE_{11}) = \frac{175{,}698.51}{d\ (mm)}$$

$$= \frac{6917.26}{d\ (inches)}$$

where
 $f_c$ = cutoff frequency in MHz for $TE_{11}$ mode
 d = waveguide inside diameter

Circular waveguide will support the $TM_{01}$ mode whose cutoff frequency is defined by

$$f_c\ (TM_{01}) = \frac{229{,}485.13}{d\ (mm)}$$

$$= \frac{9034.85}{d\ (inches)}$$

The wavelength in a waveguide always exceeds the free-space wavelength and is called guide wavelength, $\lambda_g$. It is related to the cutoff frequency and operating frequency by the equation

$$\lambda_g = \frac{299{,}792.5\ (mm)}{\sqrt{f_o^2 - f_c^2}}$$

$$= \frac{11{,}802.85\ (inches)}{\sqrt{f_o^2 - f_c^2}}$$

where
 $\lambda_g$ = guide wavelength
 $f_o$ = operating frequency
 $f_c$ = $TE_{11}$ waveguide cutoff frequency

An inside diameter range of about 0.66 to 0.76 wavelength is suggested. The lower frequency limit (longer dimension) is dictated by proximity to cutoff frequency. The higher frequency limit (shorter dimension) is dictated by higher order waves. See Table 11 for recommended inside diameter dimensions for 902- to 10,000-MHz amateur bands.

The probe that excites the waveguide and makes the transition from coaxial cable to waveguide is 1/4 wavelength long and spaced from the closed end of the guide by 1/4 guide wavelength. The length of the feed should be two to three guide wavelengths. The latter is preferred if a second probe is to be mounted for polarization change or for "polaplexer" work where duplex communications (simultaneous transmission and reception) is possible because of the isolation between two properly located and oriented probes. The second probe for polarization switching or "polaplexer" work should be spaced 3/4 guide wavelength from the closed end and mounted at right angles to the first probe.

The feed aperture is located at the focal point of the dish and aimed at the center of the reflector. The feed mountings

**Table 11**
**Circular Waveguide Dish Feeds**

| Freq. (MHz) | Inside Diameter Circular Waveguide Range (in.) | Inside Diameter Circular Waveguide Range (mm) |
|---|---|---|
| 915 MHz | 8.52-9.84 | 216.4-250 |
| 1296 | 6.02-6.94 | 152.9-176.3 |
| 2304 | 3.39-3.91 | 86.1- 99.3 |
| 3400 | 2.29-2.65 | 58.2- 67.3 |
| 5800 | 1.34-1.55 | 34.0- 39.4 |
| 10,250 | 0.76-0.88 | 19.3- 22.4 |

should permit adjustment of the aperture either side of the focal point and should present a minimum of blockage to the reflector. Correct distance to the dish center places the focal point about 1 inch inside the feed aperture. The use of a nonmetalic support minimizes blockage. PVC pipe, fiberglass and Plexiglas are materials commonly used. A simple test by placing a material in the kitchen microwave oven will reveal if it is satisfactory up to 2450 MHz. PVC pipe has tested satisfactorily and appears to work well at 2300 MHz. A simple, clean looking mounting for a 4-foot dish with 18-inch focal length, for example, can be made by mounting a length of 4-inch PVC pipe using a PVC flange at the center of the dish. At 2304 MHz the circular feed is approximately 4 inches ID, making a snug fit with the PVC pipe. Precautions should be taken to keep rain and small birds from entering the feed.

Fig. 43 — Coffee-can 2304-MHz feed described in text and Fig. 42 mounted on a 4-foot dish.

*As a safety measure, one should never look into the open end of a waveguide when power is applied, or stand directly in front of a dish while transmitting.* Tests and adjustments in these areas should be done while receiving or at extremely low levels of transmitter power (less than 0.1 watt). The U.S. Government has set a limit of 10 mW/cm$^2$ averaged over a 6-minute period as the safe maximum. Other authorities believe a lower level should be used. More than this causes thermal heating of the skin tissue. Heating effect is especially dangerous to the eyes. The accepted safe level of 10 mW/cm$^2$ is reached in the near field of a parabolic antenna if the level at $2D^2/\lambda$ is 0.242 mW/cm$^2$. The equation for power density is

Power density =

$$\frac{3\lambda P}{64D^2} = \frac{158.4\ P}{D^2}\ \text{mW/cm}^2$$

where
P = average power in kilowatts
D = antenna diameter in feet
$\lambda$ = wavelength in feet

New commercial dishes are expensive, but surplus ones can sometimes be purchased at low cost. Some amateurs build theirs, while others modify uhf TV dishes or circular metal snow sleds for the amateur bands. Fig. 43 shows a dish using the homemade feed just described. Photos showing a highly ambitious dish project under construction by ZL1BJQ appear in Figs. 44 and 45. Practical details for constructing this type of antenna are given in Chapter 12. R. Knadle, K2RIW, described modern uhf-antenna test procedures in "UHF Antenna Ratiometry," *QST,* February 1976. Dimensions for a 1296-MHz symmetrical excitation system for a parabolic reflector may be found in the ARRL 1982 *Radio Amateur's Handbook* or the RSGB *Radio Communication Handbook,* 5th Edition, Vol. 2.

Fig. 44 — Aluminum framework for a 23-foot (7 m) dish under construction by ZL1BJQ.

Fig. 45 — Detailed look at the hub assembly for the ZL1BJQ dish. Most of the structural members are made from 3/4-inch T section.

# Omnidirectional Antennas for VHF and UHF

Local work with mobile stations requires an antenna with wide coverage. Most mobile work is on fm, and the polarization used with this mode is vertical. Some simple vertical systems are described in this section. Additional material on antennas of this type is presented in the Mobile and Portable Antenna chapter.

## GROUND-PLANE ANTENNAS FOR 144, 220 AND 420 MHz

For the fm operator living in the primary coverage area of a repeater the ease of construction and low cost of a quarter-wavelength ground-plane antenna make it an ideal choice. Three different types of construction are detailed in Figs. 46 through 49; the choice of construction method will depend upon the materials at hand and the desired style of antenna mounting.

The 2-meter model shown in Fig. 46 uses a flat piece of sheet aluminum, to which the radials are connected with machine screws. A 45-degree bend is made in each of the radials. This bend can be made with the aid of an ordinary bench vise. An SO-239 chassis connector is mounted at the center of the aluminum plate with the threaded part of the connector facing down. The vertical portion of the antenna is made of no. 12 copper wire soldered directly to the center pin of the SO-239 connector.

The 220-MHz version, Fig. 47, uses a slightly different technique for mounting and sloping the radials. In this case the corners of the aluminum plate are bent down at a 45-degree angle with respect to the remainder of the plate. The four radials are held to the plate with machine screws, lockwashers and nuts. A mounting tab has been included in the design of this antenna as part of the aluminum base. A compression-type hose clamp could be used to secure the antenna to a mast. As with the 2-meter version the vertical portion of the antenna is soldered directly to the SO-239 connector.

A very simple method of construction, shown in Figs. 48 and 49, requires nothing more than an SO-239 connector and some no. 4-40 hardware. A small loop, formed at one end of the radial, is used to attach the radial directly to the mounting holes of the coaxial connector. After the radial is fastened to the SO-239 with 4-40 hardware, a large soldering iron or propane torch is used to solder the radial and the mounting hardware to the coaxial connector. The radials are bent to a 45-degree angle and the vertical portion is soldered to the center pin to complete the antenna. The antenna can be mounted by passing the feed line through a mast of 3/4 inch ID plastic or aluminum tubing. A compression hose clamp can be used to secure the PL-259 connector, attached to the feed line, in the end of the mast. Dimensions for the 144-, 220- and 440-MHz bands are give in Fig. 48.

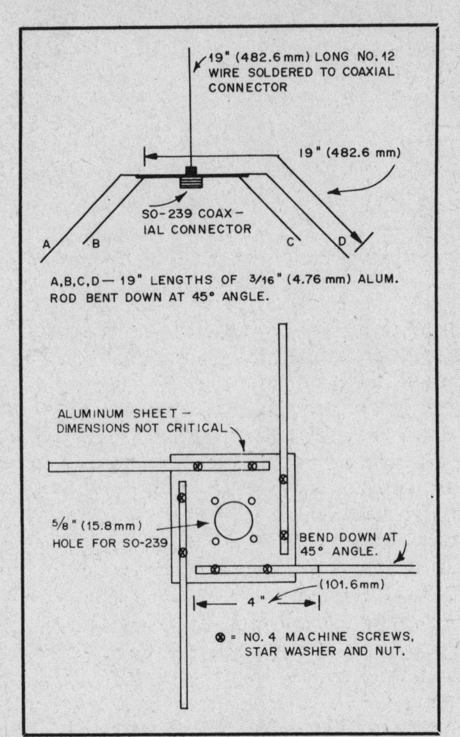


Fig. 46 — These drawings illustrate the dimensions for the 144 MHz ground-plane antenna. The radials are bent down at a 45-degree angle.

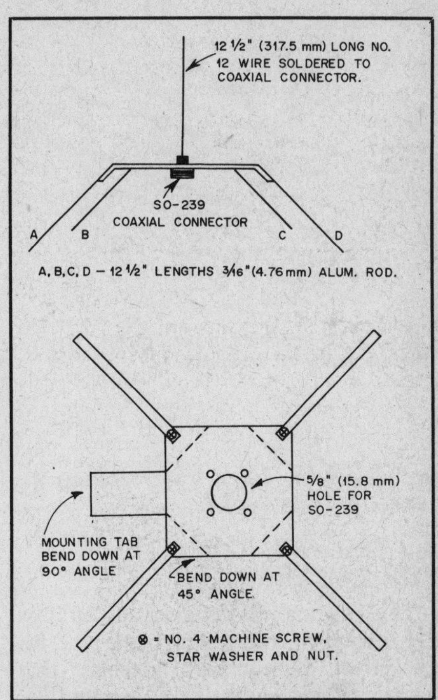


Fig. 47 — Dimensional information for the 220-MHz ground-plane antenna. The corners of the aluminum plate are bent down at a 45-degree angle rather than bending the aluminum rod as in the 144-MHz model. Either method is suitable for these antennas.

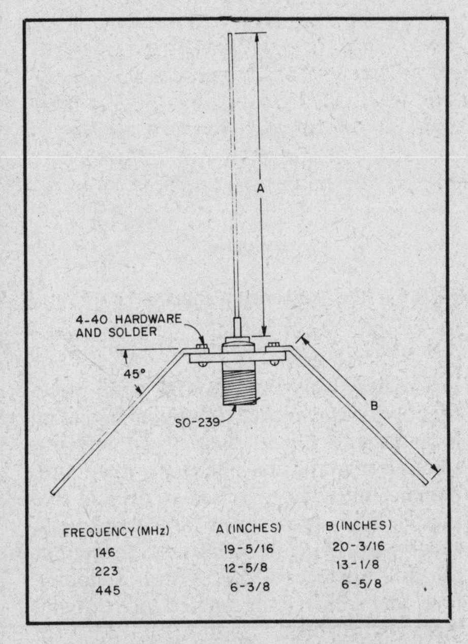

| FREQUENCY (MHz) | A (INCHES) | B (INCHES) |
| --- | --- | --- |
| 146 | 19-5/16 | 20-3/16 |
| 223 | 12-5/8 | 13-1/8 |
| 445 | 6-3/8 | 6-5/8 |

Fig. 48 — Simple ground-plane antenna for the 144-, 220-, and 440-MHz bands. The vertical element and radials are 3/32- or 1/16-inch brass welding rod. Although 3/32-inch rod is preferred for the 2-meter antenna, 10 or 12 ga. copper wire can also be used.

Fig. 49 — A 440-MHz ground plane constructed using only an SO-239 connector, no. 4-40 hardware and 1/16-inch brass welding rod.

If these antennas are to be mounted outside it is wise to apply a small amount of RTV sealant or similar material around the areas of the center pin of the connec-

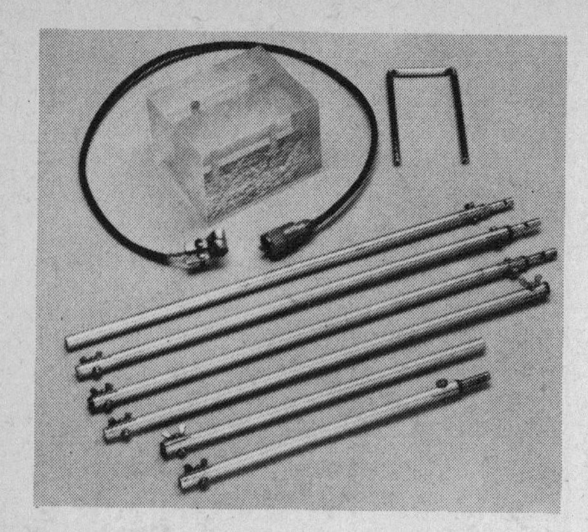

Fig. 50 — Components of the 2-meter J antenna, ready for assembly.

tor to prevent water from entering the connector and coax line.

## A J-FED VERTICAL FOR 2 METERS

This antenna was designed and built by Lee Aurick, W1SE, to satisfy the need for an effective, compact antenna that operates independently of ground or radials. It comprises a half-wave radiator and a quarter-wave matching section. Common materials are used.

### Construction

Figs. 50 and 51 show the simple assembly. The radiating part of the antenna is 38 inches long, the matching section is 19 inches long. Both are made from 3/8-inch OD tubing. One side of the matching section must be added to the antenna section for a total of 57 inches. From this length is subtracted the length of one side of the "U" mounted within the plastic. In this instance, the length of the matching section within the plastic is 2-1/4 inches long. On the other side of the matching section this length is, similarly, subtracted from the 19-inch length. Therefore, each side is shortened by the length within the plastic and is, respectively, 54-3/4 inches, and 16-3/4 inches long. The U-shaped base of the antenna within the plastic is made from 1/4-inch OD aluminum tubing, bent to form a U, and spaced 2-1/8 inches, center-to-center. The U projects 1-7/8 inches above the plastic. Total length is 4-1/8 inches.

The length of the larger size tubing mounted above the plastic base is divided into four pieces on the longer side, and into two pieces on the shorter side. The result is four pieces approximately 13-11/16 inches long, and two pieces approximately 8-3/8 inches long. Four lengths of 1/4-inch tubing are cut to 2 inches. These are inserted 1 inch into the larger tubing and are held with no. 6 screws, 1/2-inch long. One end is fastened in place with hex nuts, but the other end is secured with wing nuts for "no-tools" assembly and disassembly.

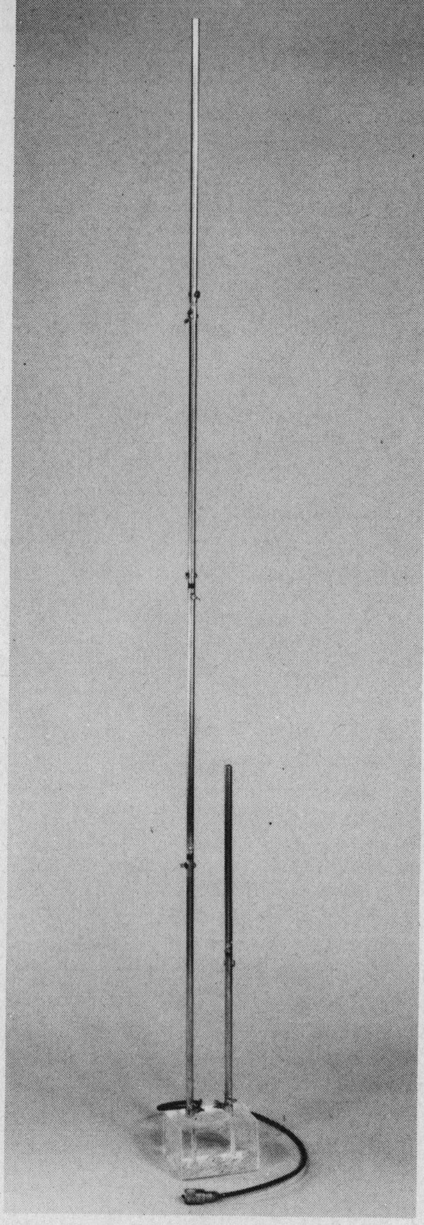

Fig. 51 — The completed J antenna.

The base was made from a scrap piece of polystyrene picked out of a bin at a local plastics house. Just about any insulating material may be used. Care and patience must be exercised in drilling the plastic. The plastic must be worked very slowly, and with frequent rests to permit the drill bit to cool off. If not, a sloppy hole and charred or melted plastic will meet your efforts.

The plastic base used in this project had a groove that ran the width of the block. This enabled the bottom of the U to clear whatever the block might be placed upon. Small rubber feet cemented to the base will also accomplish this.

### Matching

The "J" is easy to match. Two narrow strips of aluminum were cut from scraps to make the connectors. These were bent

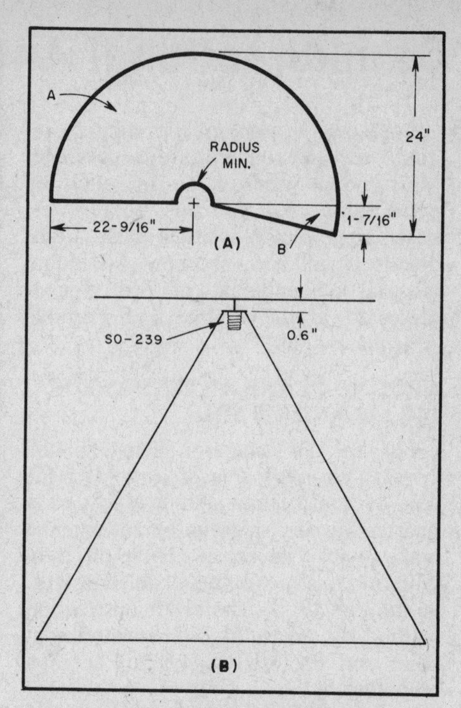


Fig. 52 — At A, cutting dimensions for the 140-to 450-MHz discone. The size of "radius min." will depend on the type of connector used at the apex of the cone. Not shown at A is the 15-in. diameter disc to be cut. At B, placement of the disc above the cone. Dimensions given may be scaled for other frequencies.

to fit tightly around the aluminum sections and were fastened with no. 6 screws. The center conductor is connected to one, and the shield to the other. It doesn't seem to make much difference which is connected to which; both ways seemed to work equally well.

An SWR bridge is connected in the line to indicate the point of match, and the connectors are moved up and down until the match point is located. Matched to 1:1 at 146 MHz, the SWR remained below 1.2:1 throughout the band. Pivoting the connectors toward or away from each other has a slight effect upon the SWR, and may be used for "fine tuning" the match. The antenna shown yielded unity SWR at the point where the large tubing met the plastic base.

## THE DISCONE — A VHF-UHF TRIBANDER

This broadband, vertically polarized antenna can be assembled easily and inexpensively. The model described can be used on 144, 220 and 420 MHz. It was presented by Dave Geiser, WA2ANU, in December 1978 *QST*.

The discone antenna functions as a wide-bandwidth, impedance-matching transformer, coupling a low-impedance transmission line to the higher impedance of free space. In the process, it radiates with a pattern similar to that of a quarter-wavelength vertical antenna above a ground plane. Waves form at the feed point (cone apex) and travel on the antenna surface to the edges of the cone and disc. The dimensions and geometry of the antenna are chosen so as to make the impedance at the edges similar to that of free space. We know that maximum energy transfer occurs when impedances are matched, so the antenna radiates. A discone antenna acts like a high-pass filter. Below some cutoff frequency, the SWR will increase rapidly. Above this frequency, the antenna SWR remains low up to a maximum of 10 times the cutoff value, depending on the design proportions. The unit described here shows less than 2:1 SWR from 140 to 450 MHz. At 1300 MHz, the SWR measured 5:1. Fig. 52 gives dimensional information for the antenna. The slant height and diameter of the cone are the same, about 110 percent of a quarter wavelength at the lowest operating frequency. Diameter of the disc is about 66 percent of a quarter wavelength.

### Construction

The radiating surfaces are cut from hardware cloth. It cost less than $5 for a 5-foot-long piece of 2-foot-wide material. The galvanized-steel wire that makes up the hardware cloth is spaced 1/4 inch.

Cutting information for the discone may be ascertained from Fig. 52. A felt-tip marking pen is useful for drawing a pattern on the hardware cloth. Forming the cone may require some help. Leather-palmed gloves will protect your hands from the sharp ends of the wire. Use bread-wrapper ties to hold the edges together. To make sure it stays in place, solder the seam in a few locations. This is only for mechanical reasons — current flows down the cone, not around it, so electrical continuity isn't required. After the seam is soldered, the ties should be removed.

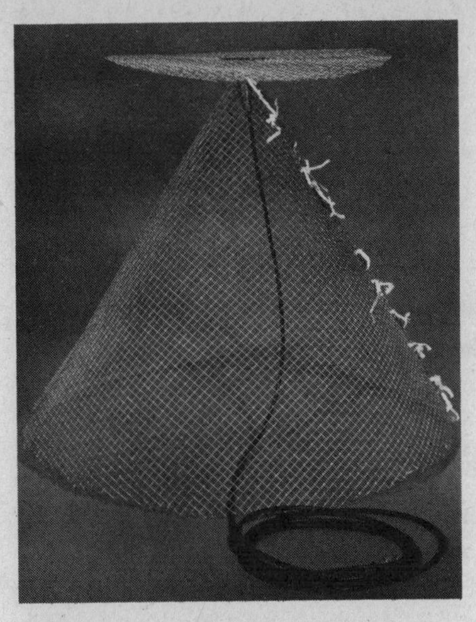

Fig. 53 — A completed discone antenna suitable for use on the 144-, 220- and 420-MHz bands. This antenna wouldn't last long in the outdoors, but is fine for indoor use. For outside installation a more robust construction is required.

A 1 × 3-inch piece of sheet copper is supported by the SO-239 connector, and is soldered to the disc. The connector is soldered to the cone with its threaded end pointed down. The disc is supported about 1/2 inch above the cone.

The lower edge of the cone is at the same potential all the way around, allowing the antenna to be mounted on a metal surface, although this will change the radiation pattern somewhat. A support mast that extends into the cone will have little effect on the antenna performance. Lower frequency discones may be built using a number of individual wires to make the cone and disc. The disc may be approximated with metal rods if desired. If necessary, thin, nonconductive insulators may be used to support the disc. A photograph of the antenna appears in Fig. 53.

## BIBLIOGRAPHY

Source material and more extended discussion of the topics covered in this section can be found in the references below.


Evans, D. and Jessop G., *VHF-UHF Manual* (3rd ed.). London: The Radio Society of Great Britain, 1976.

Foot, N., "Cylindrical Feed Horn for Parabolic Reflectors," *Ham Radio*, May 1976.

Goubau, *Proceedings of the IRE*, 39, 619-624 (1951); *Journal of Applied Physics*, 21, 1119-1128 (1950).

Hatherell, G. A., "Putting the G Line to Work," *QST*, June 1974.

Jasik, H., ed. *Antenna Engineering Handbook* (1st ed.). New York: McGraw-Hill, Inc., 1961.

Moreno, T., *Microwave Transmission Design Data.* New York: McGraw-Hill, Inc., 1948.

*Radio Communciations Handbook* (5th ed.). London: The Radio Society of Great Britain, 1976.

Southworth, G., *Principles and Applications of Waveguide Transmission.* New York: D. Van Nostrand Co., Inc., 1950.

Tillitson, G., "The Polarization Diplexer — A Polaplexer," *Ham Radio*, March 1977.

Viezbicke, P. P., "Yagi Antenna Design," *NBS Technical Note 688*, National Bureau of Standards, Boulder, Colo., December 1976.

Vilardi, D., "Simple and Efficient Feed for Parabolic Antennas," *QST*, March 1973.

Vilardi, D., "Easily Constructed Antennas for 1296 MHz," *QST*, June 1969.

## Chapter 12

# Antennas for Space Communications

There are two basic modes of space communications: satellite and earth-moon-earth (EME), also called moonbounce. Both require consideration of the effects of polarization and angle of elevation, along with the azimuth of either a transmitted or received signal. Normally, provisions for polarization are unnecessary on the hf bands, since the original polarization direction is lost after the signal passes through the ionosphere. A vertical antenna will receive a signal emanating from a horizontal one, and the converse is true when transmitting and receiving antennas are interchanged. Neither is it worth the effort to make provisions for tilting the antenna, since the elevation angle is so unpredictable. However, with satellite communications the polarization changes, and a signal that would disappear into the noise on a normal antenna might be S9 on one that is insensitive to polarization direction. Angle of elevation is also important from the standpoint of tracking and avoiding indiscriminate ground reflections that might cause nulls in signal strength.

Fig. 1 — This vhf crossed Yagi antenna design by KH6IJ was presented in January 1973 *QST*. Placement of the phasing harness and T connector is shown in the lower half of the photograph. Note that the gamma match is mounted somewhat off element center for better balance of rf voltages on elements.

Those are the characteristics common to satellite and EME communications. There are also characteristics unique to each mode, and these cause the antenna requirements to differ in several ways — some subtle, others profound. Each mode will be dealt with separately after some basic information pertaining to all space communications is presented.

## CIRCULAR POLARIZATION

The ideal antenna for random polarization would be one with a circularly polarized radiation pattern. Two commonly used methods for obtaining circular polarization are the crossed Yagi, as shown in Fig. 1 and the helical antenna, as described later in this chapter.

Polarization sense may also be a factor, especially if the satellite uses a circularly polarized antenna. In physics, clockwise rotation of an *approaching* wave is called "right circular polarization," but the IEEE standard uses the term "clockwise circular polarization" for a *receding* wave. Either clockwise or a counterclockwise sense can be selected by reversing the phasing harness of a crossed Yagi antenna. The sense of a helical antenna is fixed, being determined by its physical construction.

Mathematically, linear and circular polarization are special cases of elliptical polarization. Consider two electric-field vectors at right angles to each other. The frequencies are the same, but the magnitudes and phase angles can vary. If either one or the other of the magnitudes is zero, linear polarization results. If the magnitudes are the same and the phase angle between the two vectors (in time) is 90 degrees, then the polarization is circular. Any combination between these two limits gives elliptical polarization.

### Crossed Linear Antennas

A dipole radiates a linearly polarized signal, and the polarization direction depends upon the orientation of the antenna. Fig. 2 shows the electric field or E-plane patterns of horizontal and vertical dipoles at A and B. If the two out-

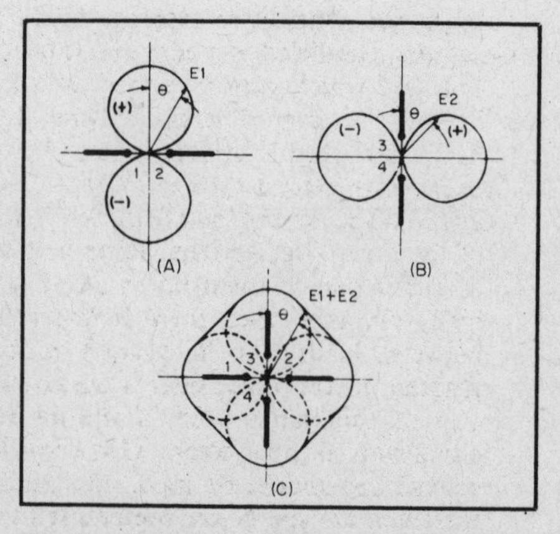


Fig. 2 — Radiation patterns looking head-on at dipoles.

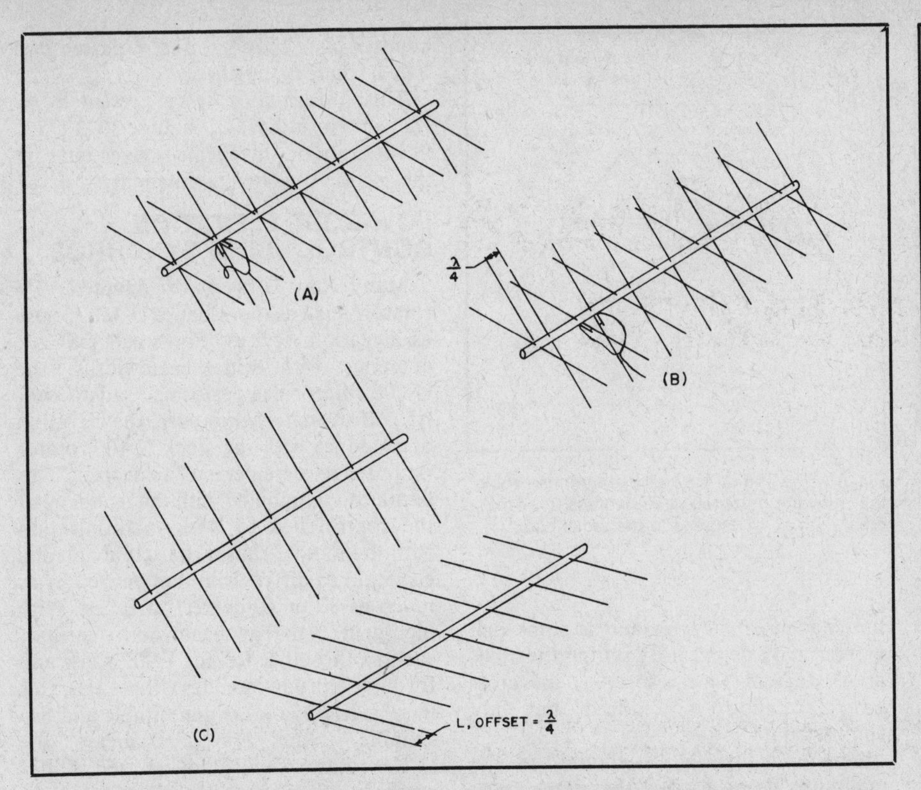


Fig. 3 — Evolution of the circularly polarized Yagi. The simplest form of crossed Yagi, A, is made to radiate circularly by feeding the two driven elements 90 degrees out of phase. Antenna B uses the same line length for both feeds, but has the elements of one bay a quarter-wavelength forward of those in the other. Antenna C has separate booms. The elements in one set are perpendicular to those in the other, and the set on the right has its elements a quarter-wavelength forward of those on the left.

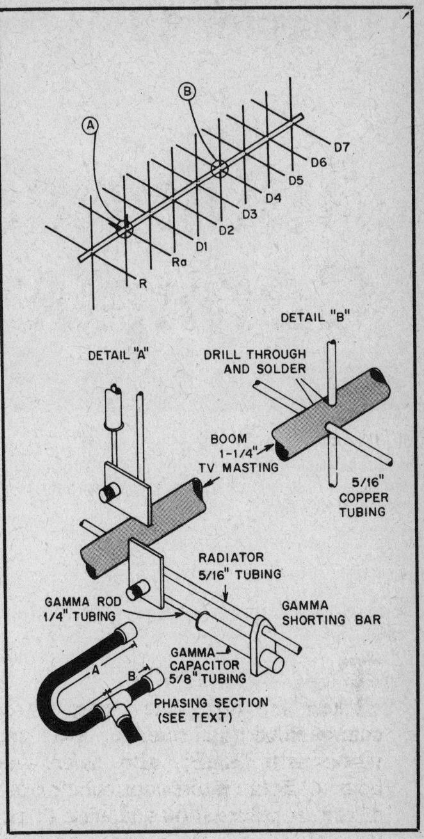


Fig. 4 — Construction details of a crossed Yagi antenna.

puts are combined with the correct phasing (90 degrees), a circularly polarized wave results, and the electric field pattern is shown in Fig. 2C. Notice that since the electric fields must be identical in magnitude, the power from the transmitter must be equally divided between the two antennas; hence the gain of each one is decreased by 3 dB when taken alone in the plane of its orientation.

As previously mentioned, a 90-degree phase shift must exist between the two antennas. The simplest way to obtain the shift is to use two feed lines with one section a quarter-wavelength longer than the other one. These two separate feed lines are then paralleled to a common transmission line which goes to either the transmitter or receiver. Therein lies one of the headaches of this system, since the impedance presented to the common transmission line by the parallel combination of the other two sections is one half that of either one of them taken alone (normally not true when there is interaction between loads, as in phased arrays). Another factor to consider is the attenuation of the cables used in the harness, along with the connectors. Good low-loss coaxial line should be used, and connectors such as type N are preferable to the UHF variety.

Another method for obtaining circular polarization is to use equal-length feed lines and place one antenna a quarter-wavelength ahead of the other. The advantage of equal-length feed lines is that identical load impedances will be presented to the common feeder. With the phasing-line method, any slight mismatch at one antenna will be magnified by the extra quarter wavelength of transmission line. This will upset the current balance between the two antennas, resulting in a loss of circularity.

Three arrangements for obtaining circular polarization from two Yagi antennas are shown in Fig. 3. A practical construction method for implementing the system of Fig. 3A is given in Fig. 4. A variation of the system at C of Fig. 3 is presented in a later section.

## ANTENNA POSITIONING

Where high-gain antennas are required in space communications, precise and accurate azimuth and elevation control and indication are usually needed also. High gain implies narrow beamwidth in at least one plane. Low-orbit satellites such as OSCAR 8 move across the sky very quickly, so azimuth and elevation tracking are essential if high-gain antennas are used. However, these satellites are fairly easy to access with moderate power and broad-coverage antennas. The low-power, high-gain approach is more sophisticated, but the high-power, low-gain solution may be more practical and economical. Some EME arrays are fixed, but these are limited to narrow time windows for communication. The az-el positioning systems described in the following systems are adaptable to either satellite or modest EME arrays. Figs. 5 and 6 illustrate one of the more ambitious attacks on the prob-

Fig. 5 — An aggressive approach to steering a giant EME antenna — a 5-inch gun turret from a destroyer.

Fig. 6 — The gun mount of Fig. 5 with its warhead attached — a homemade 42-foot parabolic dish. This is part of the arsenal of Ken Kucera, KAØY.

lem of positioning a large moonbounce antenna.

## AN AZ-EL MOUNT FOR CROSSED YAGIS

The basic criteria in the design of this system were low cost and ease of assembly. In the matter of choice between a crossed Yagi system and a helical antenna, the main factor was that Yagi antennas can be bought off the dealer's shelf, but most helical antennas cannot.

Fig. 7 shows the overall assembly of the array. The antennas used are Hy-Gain Model 341 eight-element Yagis. Fig. 8 is a head-on view of the array, showing the antennas mounted at 90 degrees with respect to each other and 45 degrees with respect to the cross arm.

Coupling between the two Yagis is minimal at 90 degrees and is somewhat greater at 45 degrees. By setting the angle at 45 degrees with respect to the cross arm, coupling is reduced but not eliminated.

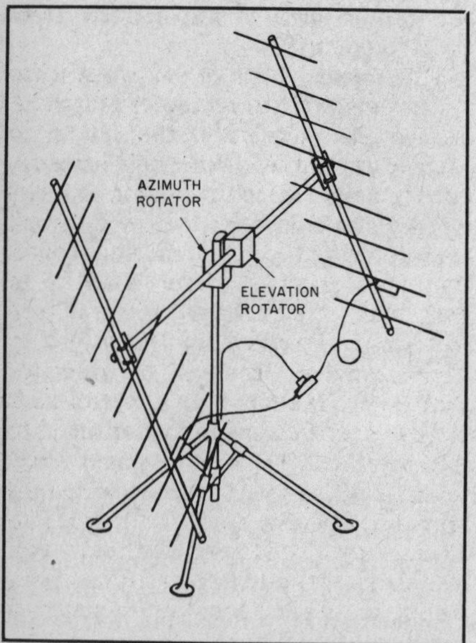

Fig. 7 — The antenna system can be assembled using off-the-shelf components such as Hy-Gain Yagis, Cornell-Dubilier or Blonder-Tongue rotators and a commercially made tripod.

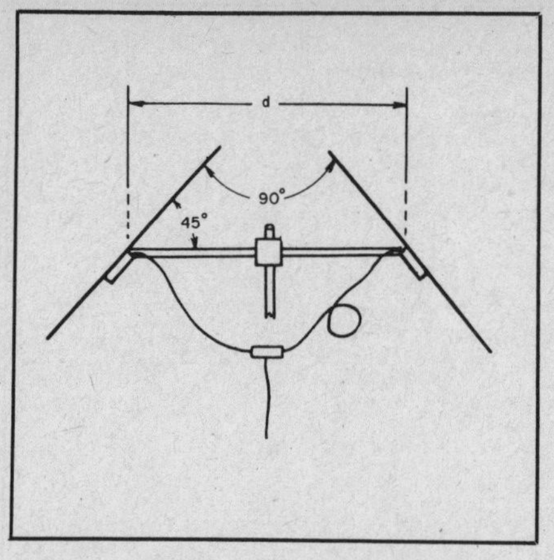


Fig. 8 — An end-on view of the antennas show that they are mounted at 90 degrees to each other, and at 45 degrees to the cross boom.

Length d in Fig. 8 should be the minimum necessary for the elements to clear the tripod base when the array is pointed straight up and rotated. In the array shown in Fig. 7, a 5-foot section of TV mast serves the purpose.

### The Mounting Tripod

A mounting tripod could be made by using aluminum railing, called "NuRail," which comes with all manner of swivels, crosses and T fittings. However, the cheapest method is to purchase a TV tripod such as the HDT-5 made by South River Metal Products Co., South River, NJ 08820. This tower sells for such a low price that there is little point in constructing your own. Spread the legs of the tripod more than usual to assure greater support, but be sure that the elements of the antenna will clear the base in the straight-up position.

### Elevation-Azimuth Rotators

The azimuth rotator is a Cornell-Dubilier AR-20. The elevation rotator is a Blonder-Tongue Prism-matic PM-2. The latter is one of the few on the market that allow the cross arm of the array to be rotated on its axis when supported at the center.

Fig. 9 shows the detail of the method of mounting the two rotators together. Notice that the flat portion of the AR-20 makes an ideal mounting surface for the PM-2. If you want to utilize commercially fabricated components throughout, a mounting plate similar to that shown in Fig. 9B can be purchased. Blonder-Tongue makes an adapter plate for their heavy-duty CATV antennas. It is called a YSB Stacking Block. The PM-2 rotator fits horizontally on this plate even though this was not the intended application. The adapter plate may be used to fasten two PM-2 rotators together.

This information was presented by Katashi Nose, KH6IJ, in June 1973 *QST*. A variation on this theme is given later, in the section on satellite antennas.

## LOW-COST ELEVATION CONTROL USING SYNCHROS

Many amateurs have adapted TV rotators such as the Alliance U-100 for use as elevation devices. For small OSCAR antennas with wide beamwidths, the U-100 rotator has performed satisfactorily. Unfortunately, however, the elevation of antennas with the stock U-100 rotator is limited to increments of 10 degrees. This limitation, combined with the tendency of the control box to lose synchronization with the motor, causes the actual antenna elevation to differ from that desired by as much as 30 or 40 degrees at times. With high-gain, narrow-beamwidth arrays, such as those needed for EME work and for high-altitude satellites (Phase III), this large a discrepancy is undesirable and unsuitable. (Note: For EME arrays, the U-100 rotator should not be used. Other rotators designed specifically for use in the horizontal position should be used instead. The elevation readout system described here will still provide superior accuracy when used with most rotators.)

This system uses a pair of *synchro transformers* to provide an accurate, continuous readout of the elevation angle of the antenna array. A U-100 rotator control unit is modified so that the motor can be operated to provide a continuously variable angle of antenna elevation. Jim Bartlett, K1TX described this system in June 1979 *QST*.

The synchro is actually a specialized form of transformer. See Fig. 10A. It

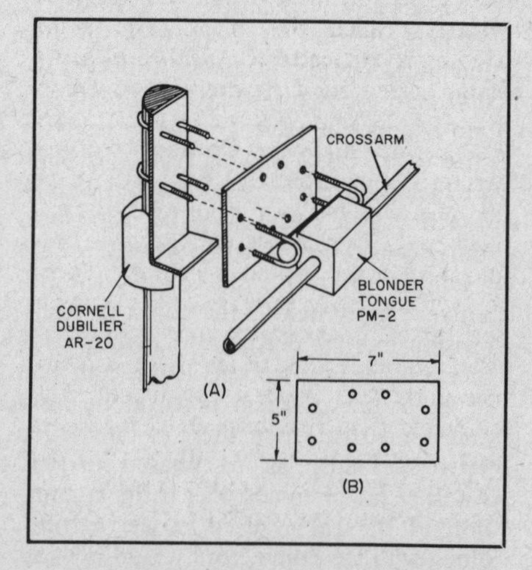


Fig. 9 — The method of mounting two rotators together. A pair of PM-2 rotators may also be used. The adapter plate (B) may be fabricated from 1/4-inch-thick aluminum stock, or a ready-made plate is available from Blonder-Tongue.

might be best described as a transformer having three secondary windings and a single *rotating* primary winding. Synchros are sometimes called "one-by-threes" for this reason. When two synchros are connected together as in Fig. 10B and power is applied, the shaft attached to the rotating primary in one synchro will track the position of the shaft and winding in the other. When two synchros are used together in such an arrangement, the system is called a *synchro repeater loop*.

Usually in repeater loops, one synchro transformer is designated as the one where motion is initiated, while the other simply *repeats* this motion. When we have two synchro transformers in such a repeater loop, the individual units can be thought of as "transmitter" and "receiver," or *synchro generator* and *synchro motor*, respectively. In our application, where one unit is located at the antenna array and another is used as an indicator in the radio room, we will call the antenna unit the generator and the indicator unit the motor.

The synchro generator is so named because it "generates" and transmits, by electrical energy, a rotational force to the synchro motor. The motor, also sometimes called the *receiver, follower* or *repeater,* receives this energy from the generator, and turns its shaft accordingly.

**Physical Characteristics**

Synchro transformers, both generator and motor types, resemble small electric motors, with only minor differences. Generator and motor synchros are identical in design for all practical purposes. The only difference between them is the presence of an *inertia damper* — a special flywheel — on units specifically designated as synchro motors. For antenna use, the inertia damper is not a necessity.

Fig. 10A shows the synchro transformer schematically. In each synchro, there are two elements: the fixed secondary windings, called the *stator*, and the rotatable primary, called the *rotor*. The rotor winding is connected to a source of alternating current, and the shaft is coupled to a controlling shaft or load — in our case, the antenna array or elevation readout pointer. An alternating field is set up by the rotor winding as a result of the ac applied to it. This causes voltages to be induced into the stator windings. These voltages are representative of the angular position of the rotor.

The stator consists of many coils of wire placed in slots around the inside of a laminated field structure, much like that in an electric motor. The stator coils are divided into three groups spaced 120 degrees around the inside of the field with some overlap to provide a uniform magnitude of attractive force on the rotor. The leads from the rotor and stator windings are attached to insulated terminal strips, usually located at the rear of the motor or generator housing. The rotor connections are labeled R1 and R2; and the stator connections, S1, S2 and S3. These are shown in Fig. 10A. These rotor and stator designations are standard identifications.

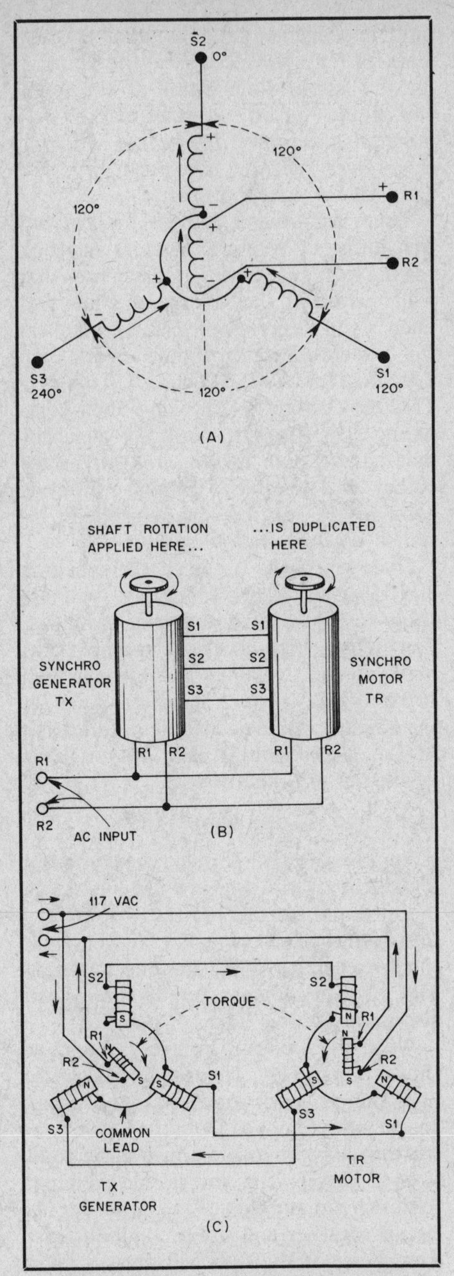


Fig. 10 — At A, a schematic diagram of the synchro transformer. Connection of two synchros in a repeater loop is shown at B. The drawing at C shows the instantaneous forces in the repeater loop with the rotor shafts at different positions. The "TX" and "TR" notations stand for *torque transmitter* and *torque receiver*, respectively. Synchros are sometimes listed in catalogs by these "type" symbols.

**Synchro Transformer Action**

Synchros operate much like transformers. The main difference between them is that in a synchro, the primary winding (rotor) can be rotated through 360 degrees. The ac that should be applied to the synchro rotor coil varies, but the most common ratings are 115 V/60 Hz, 115 V/400 Hz, and 26 V/400 Hz. The 400-Hz varieties are easier to find on the surplus market but more difficult to use, since a 400-Hz supply must be built. K1TX used 60-Hz synchros for this project, and the 90 volts required was obtained by using two surplus transformers back to back (one 6.3 volt and one 5 volt). Regardless of the voltage or line frequency used, synchros should be fused, and *isolated from the ac mains by a transformer*. This is important to ensure a safe installation.

The voltages induced into the stator windings are determined by the position of the rotor. As the rotor changes position and different values are induced, the direction of the resultant fields changes. We could also say that the magnetic field rotates in synchronization with the rotor.

When a second synchro transformer is connected to the first, forming a generator/motor pair or repeater loop, the voltages induced in the three generator stator coils are also induced into the respective motor stator coils. As long as the two rotor shafts are in the same position, the voltages induced in the stator windings of the generator and motor units are equal. However, these voltages are of opposite polarity because of the connection scheme used between the two units. Since this results in a zero potential difference between the stators in the two synchro units, no current flows in either set of stator coils. With the absence of current flow, there is no magnetic field set up by the stator windings and the system is static. This means that there is no force applied to either rotor. This situation exists whenever the two rotors are aligned in identical angular positions regardless of the specific angle of displacement from the zero point (S2).

The repeater action of the two-synchro system occurs when one rotor is moved, causing the voltages in the system to become unbalanced. When this happens, current flows through the stator coils setting up magnetic fields that tend to pull the rotors together so that the static condition again exists. A torque is set up in both units, causing the two rotors to turn in opposite directions until they tend to align themselves. However, the generator shaft is usually attached to a control shaft or large load (relative to that attached to the motor shaft) so that it cannot freely rotate. Thus, as long as the motor rotor is fairly free to move, it will be brought into alignment with the generator rotor. Fig. 10C shows the instantaneous forces present in a repeater loop with unaligned rotors.

**Selecting the Synchros**

The operating voltage on synchros is not critical. Most units will function with

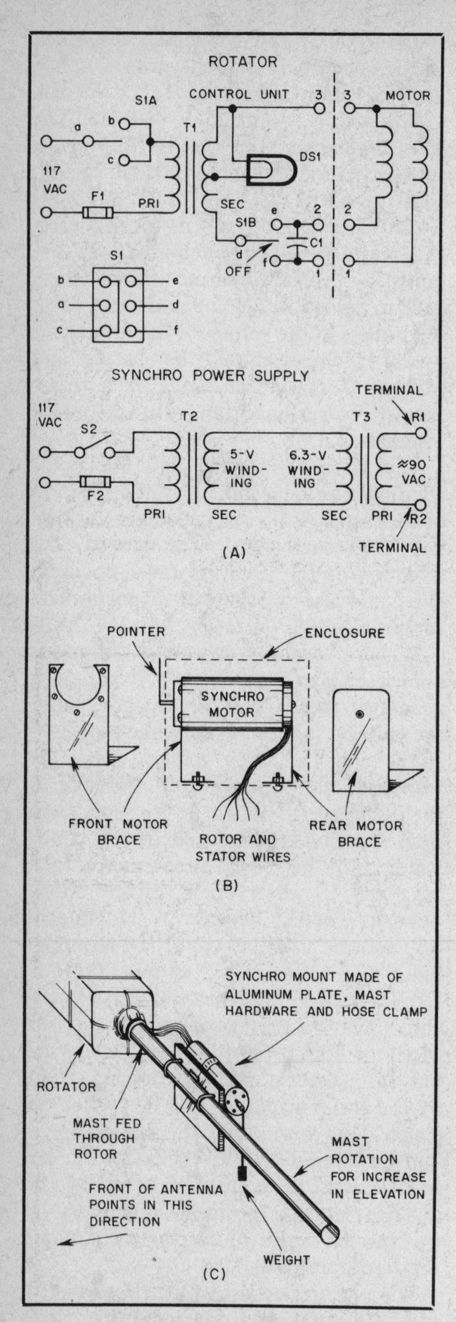


Fig. 11 — Shown at A are the circuits for the modified control unit and the synchro power supply. T1, DS1 and C1 are from the stock U-100 control box. See text. At B, the mounting method used to secure the synchro motor is shown. Details of the synchro generator mounting are shown at C. See text for description of materials.
F1, F2 — 1-A, 250-V fuse.
S1 — Dpdt momentary-contact, center-off toggle switch.
S2 — Spst toggle switch.
T2, T3 — Transformers selected for proper voltage to synchro rotor.

voltages as much as 20 percent above or 30 percent below their nominal ratings. Make sure the transformer(s) you use will handle the necessary current. Fig. 11A shows how to connect two filament transformers to obtain 90 V for the units used in this project.

Synchro transformers you are likely to find in surplus catalogs and at flea markets may not necessarily be suitable for this application. Some of the types you should *not* buy are ones marked *differential generator, differential synchro,* or *resolver synchro.* These synchros are designed for different uses.

There are several sources for synchro transformers.[1] Write for catalogs and look for the least expensive set of synchros that will operate at the voltage and line frequency you desire. Some catalog numbers for synchros that should work are TM20K228, TM17K743, TM20K531, TM20K523 and TM20K524. When comparing specifications, look for synchros that have a high *torque gradient* or accuracy. It is possible to obtain accuracy as good as ±1 degree with a properly installed synchro readout system.

Once you have obtained the synchros and designed a power supply, you are ready to begin construction of the elevation system. First check the synchros out; hook up the two units as shown in Fig. 10B and verify proper operation. Then set the synchros aside and begin modification of the U-100 control unit if you have decided to use this rotator.

## The U-100 Control Unit

Remove the transformer, capacitor and pilot light (if you want to use one) from the unit and discard the rest. Mount the transformer and capacitor in a small, shallow enclosure, like the one shown in Fig. 12. The synchro power supply will also go inside this box.

Wire the rotator control circuit as shown in Fig. 11A. The transformer, pilot light and capacitor shown in that drawing are the ones removed from the U-100 control unit. Be sure to add a fuse at the point shown. If you wish, you can tie the input to this circuit and that of the power supply circuit together and use a common fuse. The rating of the fuse will depend upon the current drain of the synchros used, but a 1-A fuse should be ample to handle the whole thing. The fourth wire in the U-100 system is not used.

Test the complete control unit before putting the U-100 motor up on the mast. Connect the motor to the modified control unit and check to see that it rotates properly in both directions when S1 is activated. This switch should be a dpdt, momentary-on, center-off toggle switch. When you are satisfied that the rotator is operating properly, install the synchro power supply inside the rotator-control enclosure. Some type of multiconnector plug/jack combination should be used at the rear of the cabinet so that the rotator and synchro control wires can be easily disconnected from the control box. A total of eight wires is used between the control unit and the synchro and rotor mounted at the antennas. An 8-pin, octal-type connector set and standard 8-wire rotator cable was used in this project. A suitable alternative connector set is Calectro F3-248 (male cord) and F3-268 (female chassis).

[1]Herbach & Rademan, Inc. (See catalog, volume 44, no. 6.) 401 East Erie Ave., Philadelphia, Pa. 19134, tel. 215-426-1700. Also see American Design Components, *First Source for Electromechanical Equipment and Components, 39 Lispenard St., New York, NY 10013, 212-966-5650.*

## Mechanical Details

The synchro motor used to provide the elevation readout was mounted inside an old cube-shaped chassis that had been gutted of parts. Two aluminum brackets support the motor inside the box, as shown in Fig. 10B. The motor is positioned to allow the shaft to protrude through the front panel of the enclosure. The pointer is fashioned from a scrap of copper sheet, and soldered to the edge of a washer. This is secured to the shaft between two nuts. A large protractor that fits the front of the enclosure serves as the dial.

## Mounting and Calibration

The synchro generator mounting is next. An aluminum plate is drilled and fitted with standard mast hardware, as shown in Fig. 11C. Two slots are made between the clamps, and a large stainless-steel hose clamp is fed through the slots and tightened around the generator casing. The generator is mounted close to the U-100, and so that it is directly behind the elevation mast when the antennas are pointed at the horizon. The U-100 and azimuth rotator are mounted in the normal fashion, as shown in Fig. 13. Elevation of the antennas causes generator-shaft rotation through the use of a weighted rod fastened to the synchro shaft, as shown in Fig. 11C.

As the antenna array is elevated, the synchro generator is rotated through an arc from behind the mast through a position directly below it to one in front of it. During the swing through this arc, gravity keeps the weighted rod perpendicular and the synchro shaft turns in proportion to the elevation angle. (If high winds are common in your area, you may wish to keep the "plumb-line" swing arm short so that gusts won't cause fluctuations in the elevation readout.)

The easiest way to calibrate the system is to attach the antennas and synchro to the mast when the U-100 is at the end of rotation (at a stop). Do this so that any movement must be in the direction that will elevate the array with respect to the horizon. Finally, with the antennas pointing at the horizon, set the synchro motor pointer to zero degrees at one end of the protractor scale. The proper "zero" end depends upon the specific mounting scheme used at the antenna. However, if the generator is mounted as shown in Fig. 11C, and all connections are properly

Fig. 12 — The completed control/readout unit for antenna elevation. The dial face was made from a plastic protractor.

made, the elevation needle should swing from right to left as the antennas move from zero through 90 to 180 degrees.

## A RADIO-COMPASS ELEVATION READOUT SYSTEM

As described by Bartlett in September 1979 *QST*, an MN-98 Canadian radio compass combined with a Sperry R5663642 synchro transmitter will make a highly precise elevation indicator. These components, displayed in Fig. 14, are available from Fair Radio Sales Co., P.O. Box 1105, Lima, OH 45802. The AY-201 transmitter is *not* suitable for this project.

Place the MN-98 indicator face down on a soft cloth draped over your workbench, and remove the rear cover of the indicator unit. Disconnect the four wires that go to the glass-metal feedthrough located on the back panel. This should free the rear cover, which should be put aside for now. Next, drill a small hole in the rear of the case, next to the edge of the feedthrough. (See Fig. 15A.) Do this *carefully,* making sure that the drill bit doesn't push through into the inside of the indicator shell and get tangled in the wiring. When the bit breaks through the metal casing, the pressurized seal will be broken. Now, using a small screwdriver and a hammer, tap each of the individual glass feedthrough inserts, cracking them. Try to keep the screwdriver from pushing the broken pieces of glass down into the enclosure where they could get lodged in the dial mechanism. Instead, attempt to shake all the pieces of glass out onto the workbench where they can be swept up. The remaining part of the feedthrough can be removed now by heating with a soldering iron and prying with a screwdriver or needle-nosed pliers.

After the feedthrough has been removed, you should gently pull the ends of the wires out through the hole left by the feedthrough. Clip off the feedthrough terminal pins. There should be five wires — a group of three and two others. The group of three will most likely be blue, yellow and black. The other two wires twisted together should be red and black. Fig. 15B shows how these should be connected to the terminals on the synchro transmitter in a five-wire system.

### Construction of the System

Fig. 15C shows the schematic diagram of a simple 6.3-V ac power supply for the indicator system. Since the synchro and indicator were originally designed to

Fig. 14 — The MN-98 Canadian radio compass and Sperry R5663642 synchro transmitter. Note the small knob at the upper right-hand corner of the indicator face. This can be used to calibrate the indicator/transmitter system without making any changes at the antenna end. By turning this knob, you can rotate the degree markings around the outside of the dial face so that any desired heading can be placed in line with the pointer.

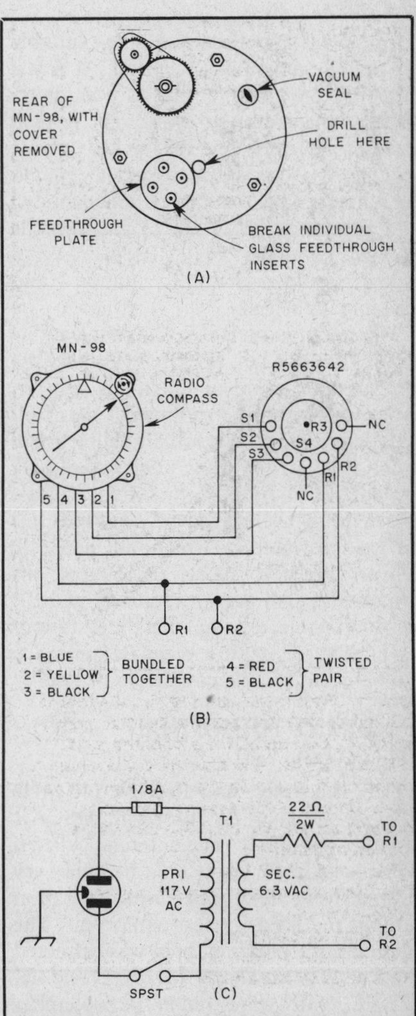


Fig. 15 — The rear of the MN-98 Canadian radio compass is shown at A. The drawing at B shows the interconnecting method used between the MN-98 and the Sperry synchro transmitter. The schematic diagram at C shows the power supply used with this indicator system. T1 can be Radio Shack 273-1384 or any junk-box 6.3-V filament transformer. At D, the drawing shows a method that can be used to prevent wind or ice from disturbing elevation readings.

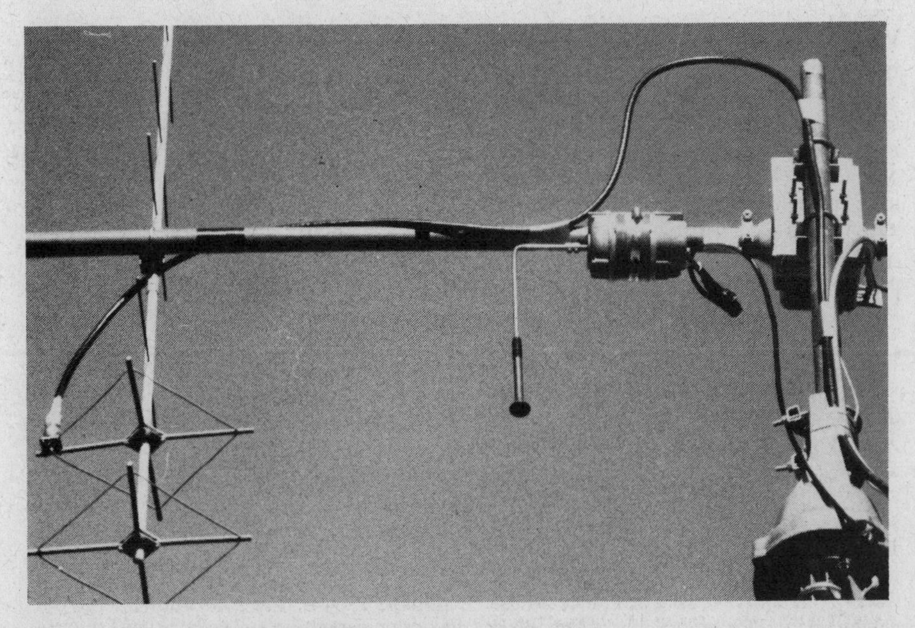

Fig. 13 — Close-up photo of the synchro transformer mounting method. The weighted arm is kept short to minimize wind effects on elevation readout.

operate at 26 V and 400 Hz, a 6.3-V transformer is acceptable for use at 60 Hz, A 22-Ω resistor was placed in series with the synchros to limit current and thus eliminate a slightly irritating buzzing sound in the indicator unit at certain pointer positions.

The indicator, along with the power supply, can be mounted in a small metal enclosure. Be sure to include a fuse, on/off switch, and three-wire line cord. At the synchro transmitter end (at the antenna), provide some kind of shield to keep wind, rain, ice and snow from affecting the system. Fig. 15D shows a possible solution. A small weight, cut in the shape of a large pie section and drilled to fit the synchro transmitter shaft, could be mounted on the shaft and shielded with a small margarine tub which is taped or glued to the outside of the synchro casing. This arrangement should allow free movement of the weight, yet keep high winds or heavy icing from affecting the indicator. The synchro transmitter should be mounted to the mast in such a way that it will rotate with the antennas, causing the weight to turn the shaft.

# Antennas for Satellite Work

This section contains a number of antenna systems that have proved practical for satellite communications. Some of the simpler ones bring space communications into the range of any amateur's budget.

### RECEIVING ANTENNAS FOR 29.4 MHz

Fig. 16 shows three antennas suitable for satellite downlink reception. At A is a turnstile, an antenna that is omnidirectional in the azimuthal plane. The vertical pattern depends on the height above ground. (This subject is treated in detail in the chapter on antenna fundamentals.) The circular polarization of the turnstile at high elevation angles reduces signal fading from satellite rotation and ionospheric effects.

The antenna at B is a simple rotatable dipole for use when the satellite is near the horizon and some directivity is helpful. When horizontally mounted, the full-wave loop at C gives good omnidirectional reception for elevation angles above 30 degrees. It should be mounted at least one-eighth-wavelength above ground. It is difficult to predict which antenna will deliver the best signal under any circumstances. All are inexpensive, and the most effective amateur satellite stations have all three, with a means of selecting the best one for the conditions at hand.

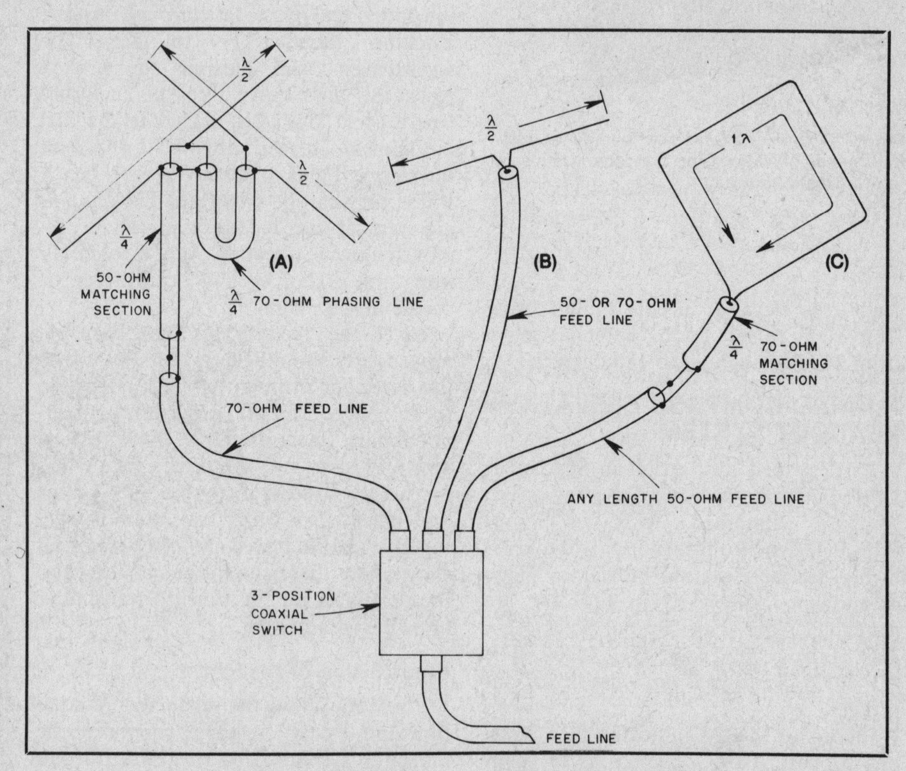


Fig. 16 — Any one of three 10-meter antennas — a turnstile (A), rotary dipole (B), or horizontal-loop (C) — may be selected for OSCAR downlink reception.

## A 146-MHz TURNSTILE ANTENNA

Here is a simple and effective 146-MHz antenna suitable for use with OSCAR Modes A, B and J. The antenna, called a turnstile-reflector array, can be built very inexpensively and put into operation without the need for test equipment. The information contained here was extracted from a *QST* article by Martin Davidoff, K2UBC, which appeared in the September 1974 issue.

Experience with several OSCAR satellites has shown that rapid fading can be a severe problem for satellite communicators. Fortunately, the ground station has control over two important parameters affecting fading — cross polarization between the ground-station antenna and OSCAR antenna, and nulls in the ground-station antenna pattern. Fading because of cross polarization can be reduced by using a circularly polarized ground-station antenna. Fading because of radiation pattern nulls can be overcome by (1) using a rotatable, tiltable array and continuously tracking the satellite or (2) using an antenna with a broad null-free pattern. The turnstile-reflector array solves this problem since it is circularly polarized at high angles and produces a balloon-like pattern.

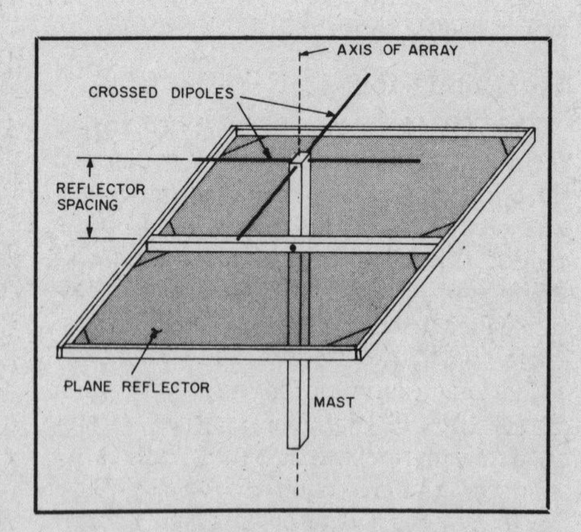


Fig. 17 — The turnstile-reflector (TR) array consists of crossed dipoles above a screen reflector.

### Construction

The mast used to support the two dipoles is constructed from wood and is 2 inches square and 8 feet long. Dipoles are formed from no. 12 copper wire, aluminum rod, or tubing. The reflecting screen is 20-gauge hexagonal chicken wire, 1-inch mesh, stapled to a four-foot-square frame made from furring strips. Hardware cloth can be used in place of the chicken wire. Corner bracing of the reflector screen will help provide mechanical stability. Spar varnish applied to the wooden members will help extend the life expectancy. See the drawing in Fig. 17.

Dimensions for the two dipole antennas and the phasing network are shown in Fig. 18. Spacing between the dipole antennas and the reflecting screen affect the antenna pattern. Choose the spacing for the pattern that best suits your needs from data in Chapter 2, and construct the antenna accordingly.

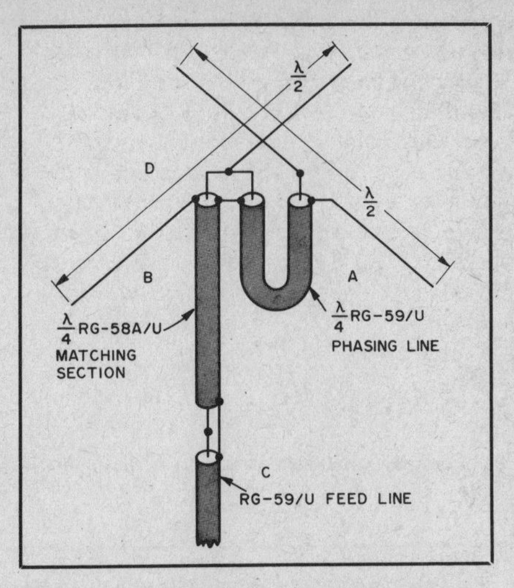


Fig. 18 — Dimensions and connections for the turnstile antenna. The phasing line is 13.3 inches of RG-59/U coax. A similar length of RG-58/U cable is used as a matching section between the turnstile and the feed line.

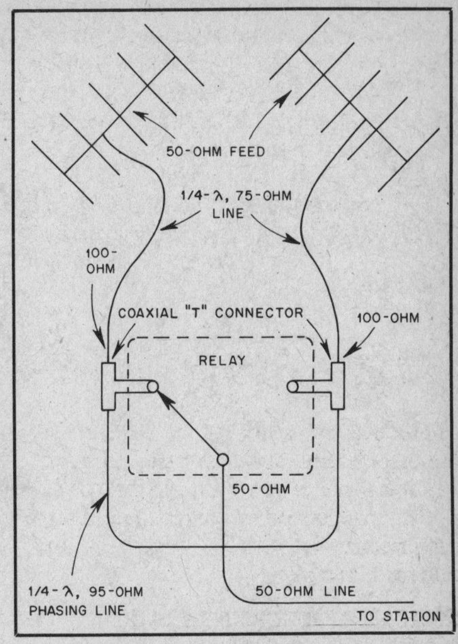


Fig. 21 — A drawing of the switchable polarization antenna system, complete with cable specifications. When calculating the length of individual cable be sure to include the velocity factor of the cables.

## CIRCULARLY POLARIZED ANTENNAS FOR 2-METER AND 70-CM SATELLITE WORK

The antenna described here provides switchable right-hand or left-hand circular polarization and positioning for both azimuth and elevation. This system makes use of commercially available antennas (KLM 144-148-9 and KLM 420-450-14), rotators (Alliance U-100 and CDE TR-44) and relays (Inline Instruments type 101) which are combined in a system that offers total flexibility. This setup is suited for operation through OSCAR 8 and the Phase III satellites. As shown in the accompanying photographs, the whole assembly is built on a heavy-duty, TV type of tripod so that it may be roof-mounted. The idea for this system came from Clarke Greene, K1JX.

Fig. 19 — A circularly polarized antenna system for satellite communications on 146 and 435 MHz. The array is assembled from KLM log-periodic Yagis.

### System Outline

The antennas displayed in Fig. 19 are actually two totally separate systems sharing the same azimuth and elevation positioning systems. Each system is identical in the way it performs — one system for 2 meters and one for 70 cm. This arrangement is quite handy for Mode B and J work since both antenna systems track automatically. Individual control lines allow independent control of the polarization sense for each system. This is mandatory, as often a different polarization sense is required for the uplink and downlink. Also, throughout any given "pass" of the satellite the sense is apt to switch several times.

Fig. 20 — The polarization sense of the antenna is controlled by the coaxial relays and phasing lines. The 146- and 435-MHz systems are controlled independently.

### Mechanical Details

The TR-44 rotator is mounted inside the tripod by means of a rotator plate of the type commonly used with a top section of Rohn 25 tower. See Fig. 20. U bolts around the tripod legs secure the plate to the tripod. A length of 1-inch galvanized water pipe (used as the mast) extends from the top of the rotator out through a homemade aluminum bearing at the peak of the tripod. Since a relatively small diameter mast is used, several pieces of shim material are required between it and the body of the rotator to assure that it will be aligned in the bearing through 360 degrees of rotation. This is covered in detail in any TR-44, CD-45, Ham-M or Ham-IV rotator instruction sheet.

The Alliance U-100 elevation rotator is mounted to the 1-inch water-pipe mast by means of a 1/8-inch aluminum plate. TV U-bolt hardware provides a perfect fit for this mast material. The cross arm that supports the two 2-meter and 70-cm antennas is a piece of 1-1/4-inch thick fiberglass rod, 6 feet in length. Other materials can be used. However, most cannot match the strength of fiberglass. This should be a consideration if you live in an area that is frequented by ice storms. Although it is relatively expensive (about $3 per foot), one piece should last a lifetime.

### Electrical Details

Since the antenna systems are identical, this description will apply to either. A simple way of obtaining a circularly polarized pattern is to use two Yagi antennas with the elements mounted at right angles to each other and feed the antennas 90 degrees out of phase. In many cases this is accomplished by mounting the horizontal and vertical elements on the same boom. It is also possible to use two separate antennas mounted apart from each other as shown in the photographs. One advantage of this system is that the weight distribution on each side of the elevation rotator is equal. As long as the separation between antennas is small, performance should be as good as when having both sets of elements on a single boom.

In order to obtain circular polarization, one antenna must be fed 90 degrees out of phase with respect to the other. For switchable right-hand and left-hand polarization some means must be included to shift a 90-degree phasing line in series with either antenna. Such a scheme is shown in Fig. 21. Since two antennas are essentially connected in parallel, the feed impedance will be half that of either antenna alone. The antennas used in this system have a 50-ohm feed impedance.

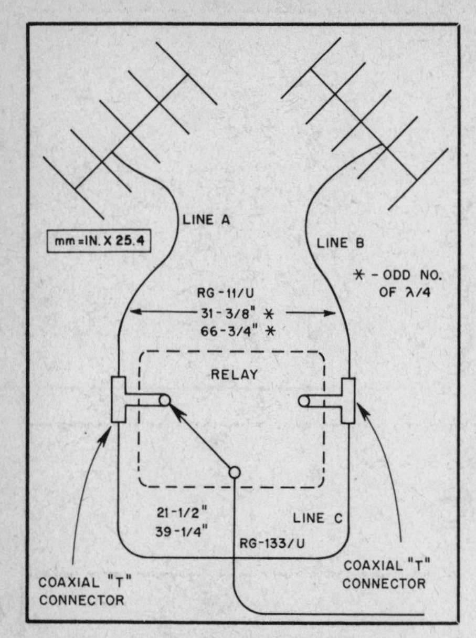


Fig. 22 — This is a drawing of the basic antenna system for switchable right- or left-hand circular polarization. The quarter-wavelength lines between the antennas and the relay step the antenna 50-ohm impedance up to 100 ohms. The phasing line is made from 95-ohm coaxial cable so as to provide a good match to the 100-ohm system. See text for a detailed description of the system. The shorter lengths are for 435.15 MHz and the longer lengths are for 145.925 MHz.

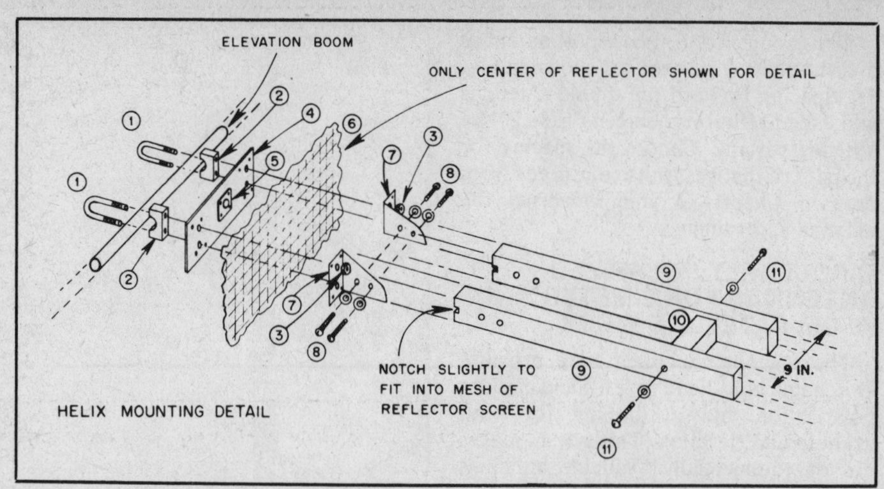


Fig. 24 — The details of the helix mounting arrangement. See Table 1 for a number-keyed parts list.

**Table 1**

**Parts List for the Helix Mounting Detail Shown in Fig. 24**

| *Piece No.* | *Description* | *Comments* |
|---|---|---|
| 1 | U bolt, TV type | Use to bolt antenna to elevation boom |
| 2 | U bolt spacer | As above |
| 3 | U bolt nut with lock washer | As above |
| 4 | Reflector mounting plate (see Fig. 25) | Rivet through reflector to boom brackets |
| 5 | Coaxial receptacle, N type | Rivet to mounting plate |
| 6 | 1- × 2-in. heavy gauge wire mesh | Reflector, cut approx. 22 in. square |
| 7 | Helix boom-to-reflector brackets | Rivet through reflector to mounting plate |
| 8 | No. 8-32 bolts with nuts and washers | Bolt boom brackets to boom |
| 9 | Boom, approx. 1- × 1-in. tomato stake, 6 feet long | |
| 10 | Boom spacer, 1- × 1-in. tomato stake | Boom to bolt; cut to give 9-in. spacing |
| 11 | No. 8 wood screws with washers | Attach spacers to boom (three places) |

Notes:
1) Mount reflector mounting plate to boom brackets, leaving 9-inch clearance for boom.
2) Wire mesh may be bent to provide clearance for U bolts.
3) When positioning the reflector mounting plate, try to center the coaxial receptacle in the wire-mesh screen.

For this reason both antennas make use of a quarter-wavelength transformer between the antenna and the relay. This quarter wavelength of 75-ohm line steps up the 50-ohm impedance of each antenna to roughly 100 ohms. As shown in the drawing, each fixed contact of the relay is also connected to the quarter-wavelength (90-degree) section of cable that acts as the phasing line. The phasing line was constructed from RG-133/U cable, which has a characteristic impedance of 95 ohms. This provides a very close match to the 100-ohm impedance of the system. If RG-133/U proves difficult to locate, RG-63/U (125-ohm impedance) may be used with a slightly higher mismatch. As can be seen in the drawing, the phasing line is always in series with the system feed point and one of the antennas. As shown, the antenna on the left receives energy 90 degrees ahead of the one on the right. If the relay were switched, just the opposite would be true.

In reality it is not necessary to use single quarter wavelengths of line. For example, the 75-ohm impedance transforming lines between each antenna and the relay could be any odd multiple of one quarter wavelength, such as 3/4, 5/4, 7/4 wavelength, etc. The same is true for the 95- or 125-ohm phasing line. One must keep track while using different lengths for the phasing line. This is especially true when figuring out which position of the relay will yield right- or left-hand polarization. The builder is apt to find that it will be necessary to use one of the odd multiples of a quarter wavelength since a single quarter wavelength of line, when the velocity factor is taken into consideration, will be extremely short. The lengths used in this particular system are shown in Fig. 22. The builder should try to use the shortest lengths practical, since the higher the multiple of quarter wavelengths of line the narrower the SWR bandwidth will become.

## A SWITCHABLE-SENSE HELICAL ANTENNA

A helical antenna is another effective means to generate circular polarization. Constructing a set of helix antennas for the 70-cm band is very easy. One antenna wound for RHCP, the other LHCP, a uhf spdt antenna switch or relay, and some good hardline is all that is needed to complete the system. Only readily available, inexpensive materials are used for con-

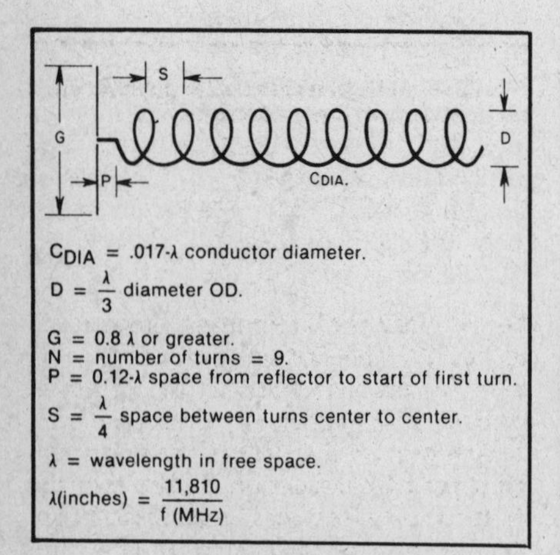


Fig. 23 — A 70-cm, 9-turn helix, with formulas for determining the critical dimensions. For a 430 MHz design frequency, $C_{DIA}$ = 0.47 inch, D = 9.16 inches, G = 22.0 inches, P = 3.3 inches and S = 6.9 inches.

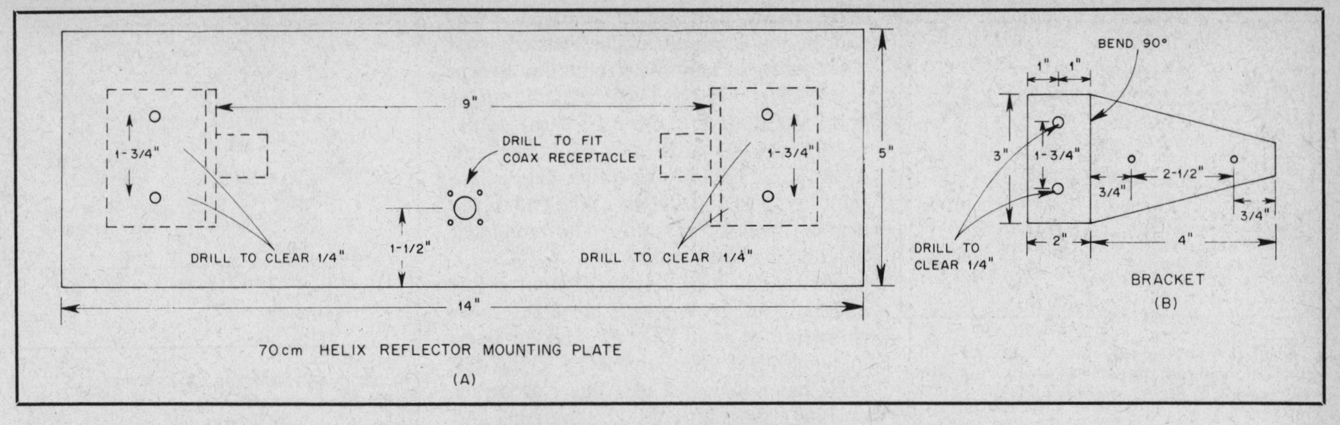


Fig. 25 — (A) The helix reflector mounting plate (part no. 4 in Table 1). (B) The boom brackets (part no. 7 in Table 1).

struction, and the most comforting part is that the dimensions are not critical. Fig. 23 shows the helix formulas and dimensions. This broadband beauty, with a 70-percent bandwidth, is ideal for a high-gain, broad-beamwidth satellite-tracking antenna. With this switchable antenna system and 50 to 100 watts of rf output, you should have a respectable signal on the new Phase III satellite.

A close-up detail of the complicated portion of the helix is shown in Fig. 24. A good starting point for construction is the reflector, made of heavy wire mesh. This type of wire mesh is used in most uhf TV "bow tie" antennas. Hardware or wire companies can supply this material in four-foot widths. It is 14-gauge galvanized steel and sells at approximately $1.60 per foot. To build two antennas you will need a piece of mesh two by four feet. Trim the mesh so that no sharp ends stick out, or you may end up with a four-sided porcupine.

The next step is to make the reflector mounting plates and boom brackets. Follow the dimensions shown in Fig. 25. Heavy aluminum material is recommended; 0.060 inch is the minimum recommended thickness. Thicker material will be more difficult to bend, but two bends of 45 degrees spaced about 1/4 inch apart will work fine for the brackets in this case. The measurements shown are for TV-type 1-3/4 inch U-bolt clamps. If you use another size, change the dimensions to suit. Drill the four holes in the reflector mounting plate and mount the coax receptacle, using pop rivets or nuts and bolts. It is advisable to check clearance between the coax receptacle and the elevation boom before final assembly. The thickness of the U-bolt spacers will affect this clearance. Mount a short piece of pipe, the same size as the elevation boom you will be using, to the U bolts, wire mesh reflector, reflector mounting plate and boom brackets, as shown in Fig. 24. Position the plate in the center of the wire-mesh reflector. It will be necessary to bend some of the mesh to clear the U bolts. Loosely tighten the U bolts so the plate can be adjusted to fit the mesh.

The wood boom assembly shown in Fig. 24 is two 6-foot lengths of tomato stake joined together in three places. Mount one spacer in the center and the other spacers one foot in from each end. Position the notched ends of the boom to fit into the mesh. When the correct alignment is obtained, clamp the assembly together and drill holes for rivets or bolts through the reflector mounting plate, brackets and wood boom assembly. When drilling the boom holes, place the reflector flat on the floor and use a square so the boom is perpendicular to the reflector. Mark the boom through the holes in the boom bracket. When the assembly is complete give the wood boom a coat of marine varnish.

The most unusual aspect of this antenna is its use of coaxial cable for the helix conductor. Coax is readily available, inexpensive, light in weight, and easy to shape into the coil required for the helix. Nine turns will require about 22 feet, but allow 25 feet and trim off the excess. The antenna of Fig. 28 uses an FM-8 type of coaxial cable, but any type may be used that is near the 1/2-inch diameter required, and has a center conductor and shield that can be soldered together. Strip about 4 inches at one end of the cable down to the center conductor, but leave enough braid to solder to the center conductor. This will become an electrical short. The exposed center conductor should be measured 3.3 inches from the short and the excess cut off. This is the dimension P in Fig. 23. Wind the 25-foot length of coax in a coil about 10 inches in diameter. Check Fig. 26 to determine which way to wind the coil for RHCP or LHCP. Slip the coil over the boom and move the stripped end of the cable toward the coax receptacle, which will become the starting point of the nine turns. Solder the center conductor to the coax receptacle, and start the first turn 3.3 inches from the point of connection at the coax receptacle. Tie-wraps are used to fasten the coax to the wood

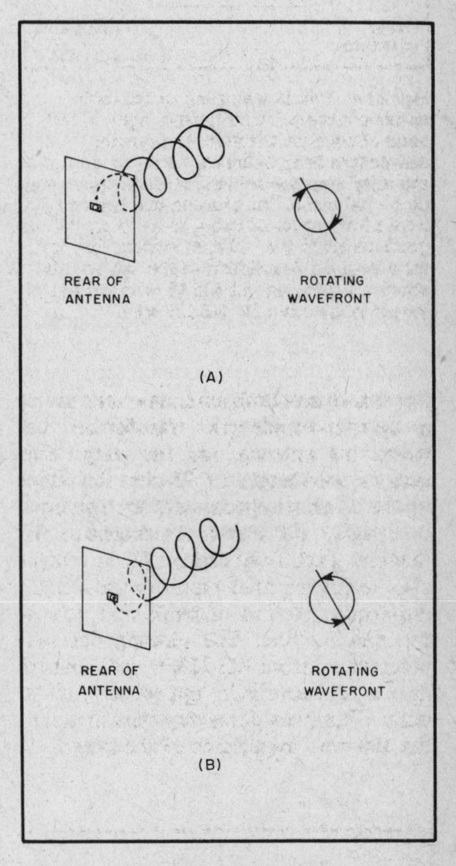


Fig. 26 — (A) Right-hand circular polarization. (B) Left-hand circular polarization.

boom. Mark the boom using dimension S in Fig. 23. The first tie-wrap will be only half this distance when it first comes in contact with the boom, then each successive turn on that side of the boom will be spaced by dimension S. Use two tie-wraps so they form an X around the boom and coax. Once the first wrap is secure, wind each turn and fasten the cable one point at a time. Before each turn is tightened, make sure the dimensions are correct. When you reach nine turns, check

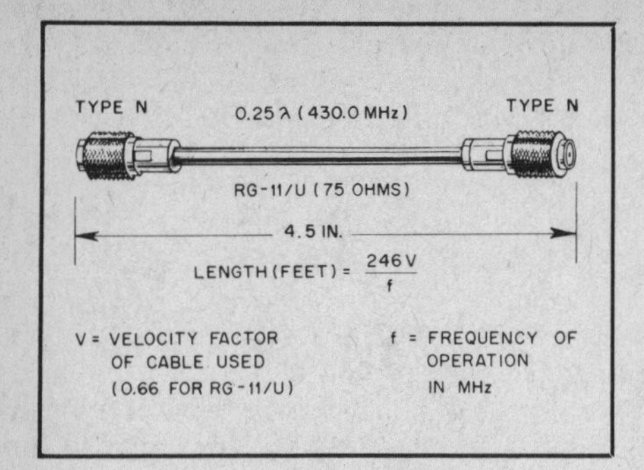


Fig. 27 — A coaxial matching section of 75-ohm cable matches the 140-ohm impedance of the helix to the 52-ohm feed line.

all dimensions again. Cut the coax at the ninth turn, strip the end, and solder a short. The exposed solder connections at each end of the coax conductor may be taped and sprayed or covered with an RTV-type covering to weatherproof them.

The coaxial 75-ohm quarter-wavelength matching section shown in Fig. 27 is connected in series with the feed line at the antenna feed point. The impedance of the helix is approximately 140 ohms. To match the 52-ohm transmission line, a transformer of 85.3 ohms is required. The 75-ohm cable used here is close enough to this value for a good match. The transformer should be connected directly to the female connector mounted on the reflector mounting plate. Use a double-female adapter to connect the feed line to the matching transformer. Wrap the connectors with plastic electrical tape, then spray with an acrylic resin for waterproofing.

Fig. 28 — A close-up view of the 70-cm helical antenna, designed and built by Bernie Glassmeyer, W9KDR.

To mount these antennas on an elevation boom, a counterbalance is required. The best way to do this is to mount an arm about 2 feet long to the elevation boom, at some point that is clear of the rotator, mast and other antennas. Point the arm away from the direction the helices are pointing, and add weight to the end of the arm until balance is found. The completed antenna is shown in Fig. 28.

Do not run long lengths of coax to this antenna, unless you use hardline. Even short runs of good RG-8/U coax are quite lossy; 50 feet of foam RG-8/U at 430 MHz has a loss of 2 dB. There are other options if you must make long runs and can't use hardline. Some amateurs mount the converters, transverters, amplifiers and filters at the antenna. This could be done with the helix antenna very easily; the units could be mounted behind the reflector, which will also add counterbalance. If this approach is used, check local electrical codes before running any power lines to the antenna.

## 50-OHM FEED FOR A HELIX

Joe Cadwallader, K6ZMW, disclosed this feed method in June 1981 *QST*. Terminate the helix in an N connector mounted on the ground screen at the periphery of the helix (Fig. 29). Simply connect the helix conductor to the N connector as close to the ground screen as possible (Fig. 30). Then adjust the first turn of the helix to maintain uniform spacing of the turns.

This modification goes a long way toward curing a deficiency of the helix — the 140 Ω nominal feed-point impedance. The traditional quarter-wavelength matching section has proved difficult to fabricate and maintain. But if the helix is fed at the periphery, the first half turn of the helix conductor (leaving the N connector) acts much like a transmission line — a single conductor over a perfectly conducting ground plane. The impedance of such a transmission line is

$$Z_o = 138 \log \frac{4h}{d}$$

where $Z_o$ is the impedance of the line, h is the height of the center of the conductor above the ground plane and d is the conductor diameter (both h and d must be in the same units of measure). Clearly, the impedance of the helix is 140 Ω a turn or two away from the feed point. But as the helix conductor swoops down toward the feed connector (and the ground plane), h is getting smaller; therefore, the impedance is dropping. The 140 Ω nominal impedance of the helix is being transformed to a lower value. For any particular conductor diameter, an optimum height can be found that will produce a feed-point impedance equal to 50 Ω. The height should be kept very small, and the diameter should be large. Apply power to the helix and measure the VSWR at the operating frequency; adjust the height for an optimum match.

Typically, the conductor diameter may not be large enough to result in a 50-Ω match at practical (small) values of h. In this case a strip of thin brass shim stock can be soldered to the first quarter turn of the helix conductor (Fig. 31). This effectively produces a larger diameter conductor which causes the impedance to drop further. The edges of this strip can be slit every 1/2 in. or so, and bent up or down (toward or away from the ground plane)

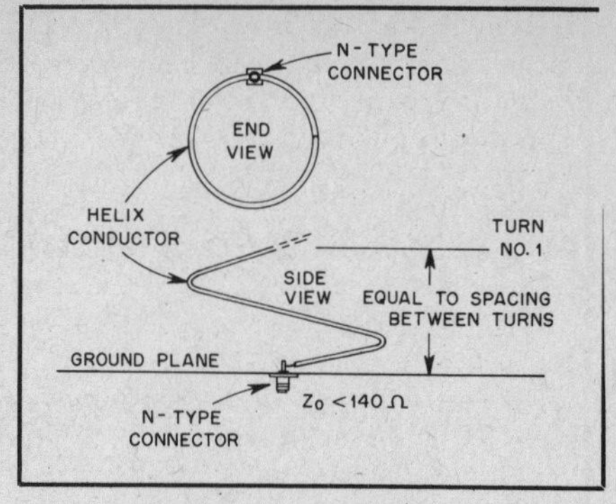

Fig. 29 — End view and side view of peripherally fed helix.

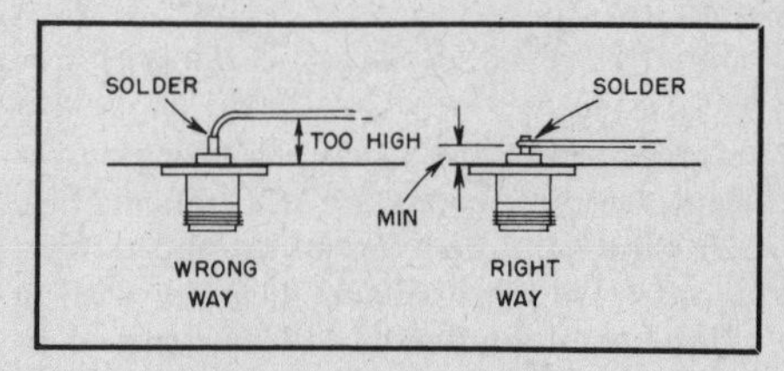


Fig. 30 — Wrong and right ways to attach helix to N connector for 50-Ω feed.

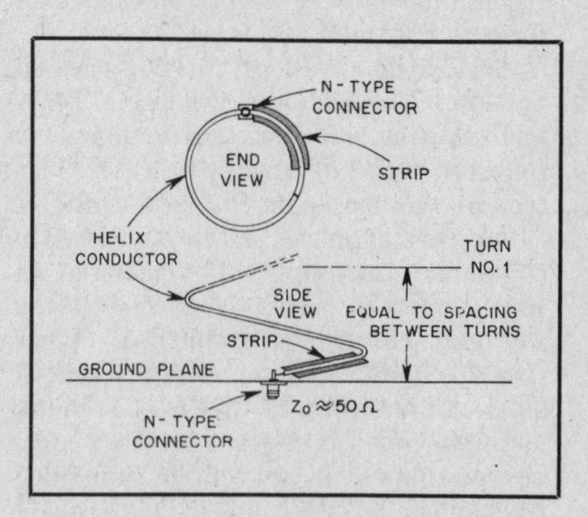


Fig. 31 — End view and side view of peripherally fed helix with metal strip added to improve transformer action.

to tune the line for an optimum match.

This approach will yield a perfect match to nearly any coax. The usually wide bandwidth of the helix (70% for VSWR less than 2 to 1) will be reduced slightly to about 40% for the same conditions. This is not enough to be of any consequence for most amateur work. The improvements in assembly, adjustment and performance are well worth the effort to make the helix more practical to build and tune.

# Antenna Systems for EME

The tremendous path loss incurred over an EME circuit places stringent requirements on the station performance. Low-noise receiving equipment, maximum legal power and large antenna arrays are required for successful EME operation. Although it may be possible to copy some of the better-equipped stations while using a single high-gain Yagi antenna, it is doubtful whether such an antenna could provide reliable two-way communication. Antenna gain of at least 20 dB is required for reasonable success. Generally speaking, more antenna gain will yield the most noticeable improvement in station performance, as the increased gain will aid both the received and transmitted signals.

Several types of antennas have become popular among EME enthusiasts. Perhaps the most popular antenna for 144-MHz work is an array of either four or eight long-boom (14- to 15-dB gain) Yagis. The four-Yagi array would provide approximately 20 dB gain, and the eight-antenna system would show an approximate 3-dB increase over the four-antenna array. At 432 MHz, eight or 16 long-boom Yagis are used. Yagi antennas are available commercially or can be constructed from readily available materials. Information on maximum-gain Yagi antennas is presented in Chapter 11.

A moderately sized Yagi array has the advantage that it is relatively easy to construct and can be positioned in azimuth and elevation with commercially available equipment. Matching and phasing lines present no particular problems. The main disadvantage of a Yagi array is that the polarization plane of the antenna cannot be conveniently changed. One way around this would be to use cross-polarized Yagis and a relay-switching system to select the desired polarization. This represents a considerable increase in system cost and complexity. Polarization shift at 144 MHz is fairly slow and the added complexity of the cross-polarized antenna system may not be worth the effort. At 432 MHz, where the shift is at a somewhat faster rate, an adjustable polarization system offers a definite advantage over a fixed one.

A photograph of the Yagi antenna system used at K1ZZ is shown in Fig. 32. The system consists of four 2-meter Cushcraft Boomer antennas mounted on a 70-foot, Rohn 25 tower. A CDE Ham-III rotator is used for positioning the antenna in azimuth and a TET KR-500 rotator is used for elevation control. The gain of this array is approximately 20 dB, taking into account phasing line losses. The more elaborate Yagi arrays of OH6NU/OH6NM and K1WHS are shown in Figs. 33 and 34.

Fig. 32 — The EME antenna system used at K1ZZ — four Cushcraft 2-meter Boomers with associated stacking and wiring harness. This system is mounted atop a 70-foot Rohn 25 tower.

Fig. 33 — The Yagi array used for EME by OH6NU/OH6NM.

Fig. 34 — K1WHS uses this system for serious moonbounce work.

Quagi antennas (made from both quad and Yagi elements) are also popular for EME work. Slightly more gain per unit boom length is possible as compared to the conventional Yagi. Additional information on the quagi is presented in Chapter 11.

The collinear is another popular type of antenna for EME work. A 40-element collinear array has approximately the same frontal areas as an array of four Yagis. The collinear array would produce approximately 1 to 2 dB less gain. Of course the depth dimension of the collinear array is considerably less than for the long-boom Yagis. An 80-element collinear would be marginal for EME communications, providing approximately 19-dB gain. Many operators choosing this type of antenna use 160-element or larger systems. As with Yagi and quagi antennas, the collinear cannot be easily adjusted for polarity changes. From a constructional standpoint there may be little difference in complexity and material costs between the collinear and Yagi arrays.

The parabolic dish is another antenna that is used extensively for EME work. Unlike the other antennas described, the major problems associated with dish antennas are mechanical ones. Dishes 20 feet in diameter are required for successful EME operation on 432 MHz. Structures of this size and wind/ice loading place a severe strain on the mounting/positioning systems. Extremely rugged mounts are required for large dish antennas, especially when used in windy locations. However, several aspects of the parabolic dish antennas make the extra mechanical problems worth the effort. For example, the dish antenna is inherently broadband and may be used on several different bands by simply changing the feed. An antenna that is suitable for 432 MHz work is also usable for each of the higher amateur bands. Additional gain is available as the frequency of operation is increased. Another advantage of this antenna is in the feed system. The polarization of the feed, and therefore the polarization of the antenna, can be adjusted with little difficulty. It should be a relatively easy matter to devise a system whereby the feed could be rotated remotely from the shack. Changes in polarization of the signal could be compensated for at the operating position! As polarization changes can account for as much as 30 dB of signal attenuation, the rotatable feed could make

Fig. 35 — A newcomer to EME stands in awe of the K2UYH 28-foot dish.

the difference between working a station and not. A photograph of the parabolic dish antenna used at K2UYH is shown in Fig. 35. More information on parabolic dish antennas is available in Chapter 11.

Antennas suitable for EME work are by no means limited to the types described thus far. Rhombic, quad arrays, helices and others have been used. These types have not gained the popularity of the Yagi, quagi, collinear and parabolic dish, however.

## A 12-FOOT STRESSED PARABOLIC DISH

Very few antennas evoke as much interest among uhf amateurs as the parabolic dish, and for good reason. First, the parabola and its cousins — Cassegrain, hog horn and Gregorian — are probably the ultimate in high-gain antennas. The highest gain antenna in the world (148 dB) is a parabola. This is the 200-inch Mt. Palomar telescope. (The very short wavelength of light rays causes such a high gain to be realizable.) Second, the efficiency of the parabola does not change as it gets larger. With collinear arrays, the loss of the phasing harness increases as the size increases. The corresponding component of the parabola is lossless air between the feed horn and the reflecting surface. If there are few surface errors, the efficiency of the system stays constant regardless of antenna size.

Some amateurs reject parabolic antennas because of the belief that these are all heavy, hard to construct, have large wind-loading surfaces, and require precise surface accuracies. However, with modern construction techniques, a prudent choice of materials, and an understanding of accuracy requirements, these disadvantages can be largely overcome. A parabola may be constructed with a 0.6 f/d (focal length/diameter) ratio, producing a rather flat dish, which makes it easy to surface and allows the use of recent advances in high-efficiency feed horns. This results in greater gain for a given size of dish over conventional designs. Such an antenna is shown in Fig. 36. This parabolic dish is lightweight, portable, easy to build, and can be used for 432- and 1296-MHz mountaintopping, as well as on 2300, 3300 and 5600 MHz. Disassembled, it fits into the trunk of a car, and it can be reassembled in 45 minutes.

Fig. 36 — A stressed parabolic dish designed by K2RIW (shown at the right) set up for reception of Apollo or Skylab signals near 2280 MHz. A preamplifier is shown taped below the feed horn. From *QST,* August 1972.

The usually heavy structure which supports the surface of most parabolic dish antennas has been replaced in this design by aluminum spokes bent into a near-parabolic shape by string. These strings serve the triple function of guying the focal point, bending the spokes, and reducing the error at the dish perimeter (as well as at the center) to zero. By contrast, in conventional designs, the dish perimeter, which has a greater surface area than the center, is furthest from the supporting center hub so it often has the greatest error. This error is pronounced when the wind blows. Here, each of the spokes is basically a cantilevered beam with end loading. The equations of beam bending predict a near-perfect parabolic curve for extremely small deflections. Unfortunately the deflections in this dish are not that small, and the loading is not perpendicular. For these reasons, mathematical prediction of the resultant curve is quite difficult. A much better solution is to measure the surface error with a template and make the necessary correction by bending each of the spokes to fit. The uncorrected surface is accurate enough for 432- and 1296-MHz use. Trophies taken by this parabola in antenna-gain contests were won using a completely natural surface with no error correction.

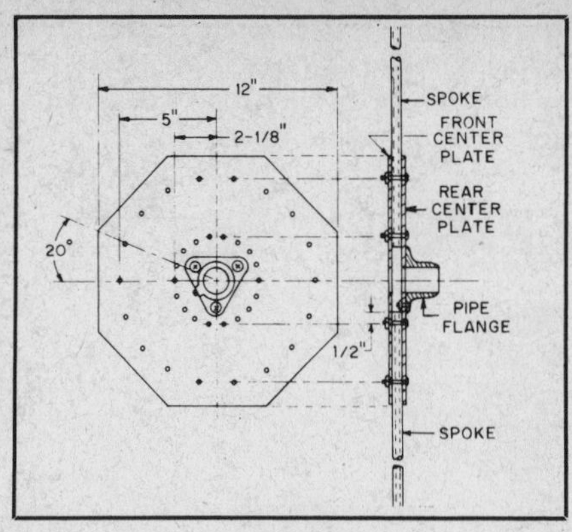


Fig. 37 — Center plate details. Two center plates are bolted together to hold the spokes in place.

**Table 2**

**Materials List for the 12-Foot Dish**

1) Aluminum tubing, 12-feet × 1/2-inch OD × .049-inch wall, 6061-T6 alloy, 9 required to make 18 spokes.
2) Octagonal mounting plates 12 × 12 × 1/8 inches, 2024-T3 alloy, 2 required.
3) 1-1/4-inch ID pipe flange with setscrews.
4) 1-1/4 inches × 8 feet, TV mast tubing, 2 required.
5) Aluminum window screening, 4 × 50 feet.
6) 130-pound-test Dacron trolling line (available from Finney Sports, 2910 Glansman Rd., Toledo, OH 43614.)
7) 38 feet of no. 9 galvanized fence wire (perimeter), Mongomery Ward Farm and Garden Catalog.
8) Two hose clamps, 1-1/2 inch; two U bolts; 1/2 × 14-inch bakelite rod or dowel; water-pipe grounding clamp; 18 eye bolts; 18 S hooks.

By placing the transmission line inside the central pipe which supports the feed horn, the area of the shadows or blockages on the reflector surface is much smaller than in other feeding and supporting systems, thus increasing gain. For 1296 MHz a backfire feed horn may be constructed to take full advantage of this feature. On 432 MHz a dipole and reflector assembly produces 1.5 dB additional gain over a corner-reflector type feed horn. Since the preamplifier is located right at the horn on 2300 MHz, a conventional feed horn may be used.

### Construction

Table 2 is a list of materials required for construction. Care must be exercised when drilling holes in the connecting center plates so that assembly difficulty will not be experienced later. See Fig. 37. A notch in each plate will allow them to be assembled in the same relative position. The two plates should be clamped together and drilled at the same time. Each of the 18 half-inch-diameter aluminum spokes has two no. 28 holes drilled at its root to accept 6-32 machine screws which go through the center plates. The

6-foot long spokes are created by cutting standard 12-foot lengths of tubing in half. A fixture built from a block of aluminum assures that the holes are drilled in exactly the same position in each spoke. The front and back center plates constitute an I-beam-like structure, which gives the dish center considerable rigidity. Fig. 38 shows a side view of the complete antenna. Aluminum alloy (6061-T6) is used for the spokes, while 2024-T3 aluminum alloy sheet, 1/8-inch thick, serves for the center plates. Aluminum has approximately three times the strength-to-weight ratio of wood used in other designs. Additionally, aluminum does not become water-logged or warped. The end of each of the 18 spokes has an eyebolt facing the dish focus point which serves a double purpose: to accept the no. 9 galvanized fence wire which is routed through the screw eyes to define the dish perimeter, and to facilitate rapid assembly by accepting the S hooks which are tied to the end of each of the lengths of 130-pound-test Dacron fishing string. The string bends the spokes into a parabolic curve; the dish may be adapted for many focal lengths by tightening the strings. Dacron was chosen because it has the same chemical formula as Mylar. This is a low-stretch material which keeps the dish from changing shape. The galvanized perimeter wire has a five-inch overlap area which is bound together with baling wire after the spokes have been hooked to the strings.

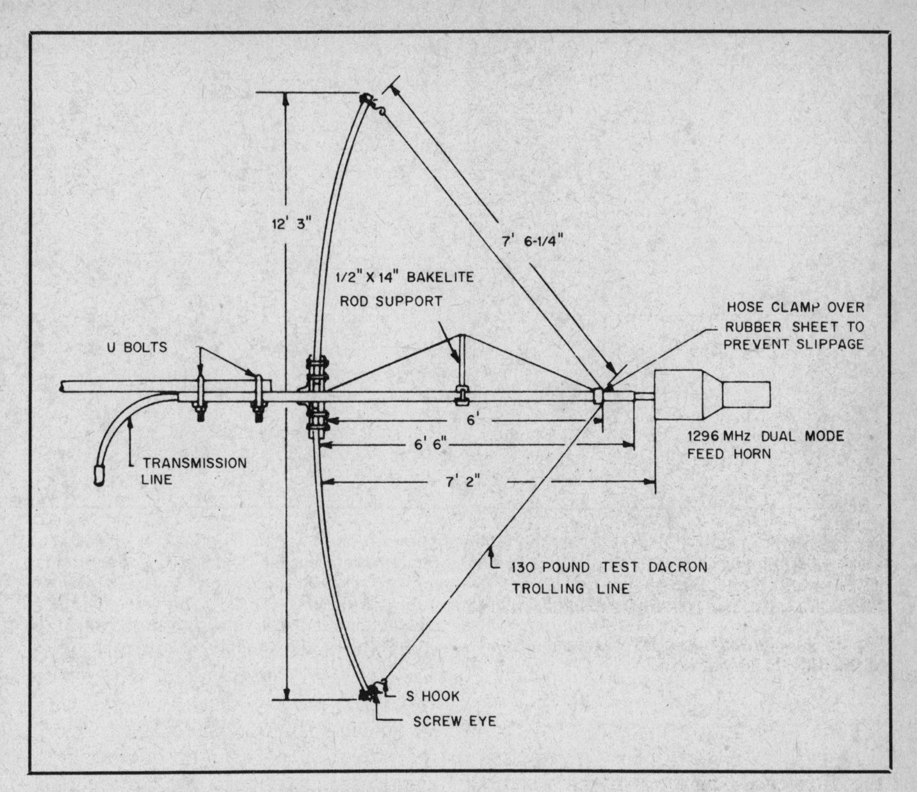


Fig. 38 — Side view of the stressed parabolic dish.

The aluminum window screening is bent over the perimeter wire to hold it in place on the back of the spokes. It was thought originally that the spokes in front of the screening might cause surface perturbations and decrease the gain. However, the total spoke area is small. Placing the aluminum screening in front of the spokes requires the use of 200 pieces of baling wire to hold the screening in place. This procedure increases the assembly time by at least an hour. For contest and mountaintop operating (when the screening is on the back of the spokes) no other fastening technique is required than bending the screen overlap around the wire perimeter.

**Surface**

A 4-foot-wide roll of aluminum screening 50 feet long is cut into appropriate lengths and laid parallel with a 3-inch overlap between the top of the unbent spokes and hub assembly. The overlap seams are sewn together on one half of the dish using heavy Dacron thread and a sailmaker's curved needle. Every seam is sewn twice, once on each edge of the overlapped area. The seams on the other half are left open to accommodate the increased overlap which occurs when the spokes are bent into a parabola. The perimeter of the screening then is trimmed. Notches are cut in the three-inch overlap to accept the screw eyes and S hooks. The first time this dish was assembled, the screening strips were anchored to the inside surface of the dish and the seams sewn in this position. It is easier to fabricate the surface by placing the screen on the back of the dish frame with the structure inverted. The spokes are sufficiently strong to support the complete weight of the dish when the perimeter is resting on the ground.

The 4-foot wide strips of aluminum screening conform to the compound bend of the parabolic shape very easily. If the seams are placed parallel to the E-field polarization of the feed horn, minimum feedthrough will occur. This feedthrough, even if the seams are placed perpendicular to the E-field, is so small that it is negligible. Some constructors may be tempted to cut the screening into pie-shaped sections. This procedure will increase the seam area and construction time considerably. The dish surface appears most pleasing from the front when the screening perimeter is slipped between the spokes and the perimeter wire, and is then folded back over the perimeter wire. When disassembly is desired, the screening is removed in one piece, folded in half, and rolled.

**The Horn and Support Structure**

The feed horn is supported by the 1-1/4-inch aluminum television mast. The television line which is inserted into this tubing is connected first to the front of the feed horn which then slides back into the tubing for support. A setscrew assures that no further movement of the feed horn occurs. During antenna-gain competition the setscrew is omitted, allowing the 1/2-inch semi-rigid (CATV cable) transmission line to move in or out while adjusting the focal length for maximum gain. The TV mast is held firmly at the center plates by two setscrews attached to the pipe flange which is mounted on the rear plate. On 2300 MHz the dish is focused for best gain by loosening these setscrews on the pipe flange and sliding the dish along the TV mast tubing (the dish is moved instead of the feed horn).

All of the fishing strings are held in position by attaching them to a hose clamp which is permanently connected to the TV tubing. A piece of rubber sheet under the hose clamp prevents slippage and keeps the hose clamp from cutting the fishing string. A second hose clamp is mounted below the first as double protection against slippage.

The high-efficiency 1296-MHz dual-mode feed horn, detailed in Fig. 39, weights 5-3/4 pounds. This weight causes some bending of the mast tubing; however this is corrected by a 1/2-inch-diameter Bakelite support. It is mounted to a pipe grounding clamp with a no. 8-32 screw inserted in the end of the rod. The Bakelite rod and grounding clamp are mounted midway between the hose clamp and the center plates on the mast. A double run of fishing string slipped over the notched upper end of the Bakelite rod counteracts bending.

The success of high-efficiency parabolic antennas is determined primarily by the

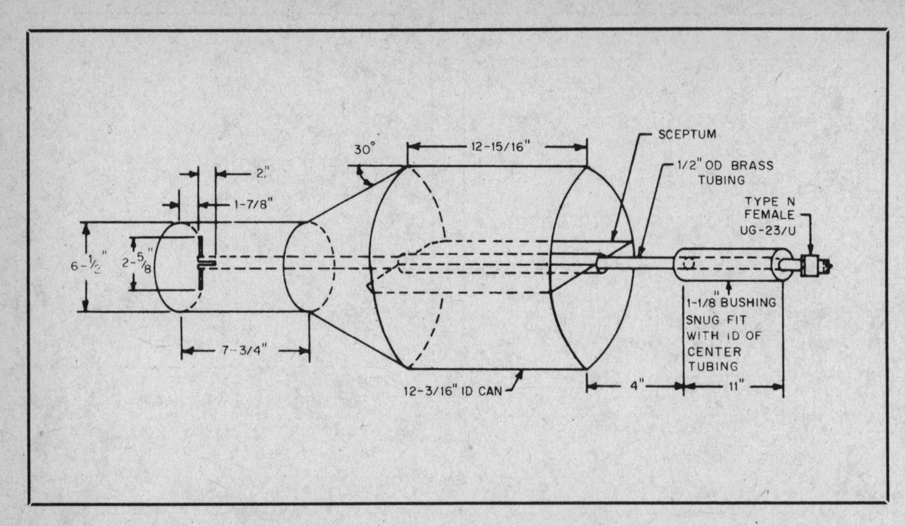


Fig. 39 — Backfire type 1296-MHz feed horn, linear polarization only. The small can is a Quaker State oil container; the large can is a 50-pound shortening container (obtained from a restaurant, "Gold Crisp" brand). Brass tubing, 1/2-inch OD, extends from UG-23/U connector to dipole. Center conductor and dielectric are obtained from 3/8-inch Alumafoam coaxial cable. The dipole is made from 3/32-inch copper rod. The sceptum and 30-degree section are made from galvanized sheet metal. Styrofoam is used to hold the sceptum in position. The primary gain is 12.2 dB over isotropic.

feed-horn effectiveness. The multiple diameter of this feed horn may seem unusual. This newly designed and patented dual-mode feed, by Dick Turrin, W2IMU, achieves efficiency by launching two different kinds of waveguide modes simultaneously, which causes the dish illumination to be more constant than conventional designs. The illumination drops off rapidly at the perimeter, reducing spillover. The feedback lobes are reduced by at least 35 dB because the current at the feed perimeter is almost zero; the phase center of the feed system stays constant across the angles of the dish reflector. The larger diameter section is a phase corrector and should not be changed in length. Theory predicts that almost no increase in dish efficiency can be achieved without increasing the feed size in a way that would increase complexity, as well as blockage. The feed is optimized for a 0.6 f/d dish. The dimensions of the feeds are slightly modified from the original design in order to accommodate the cans. Either feed type can be constructed for other frequencies by changing the scale of all dimensions.

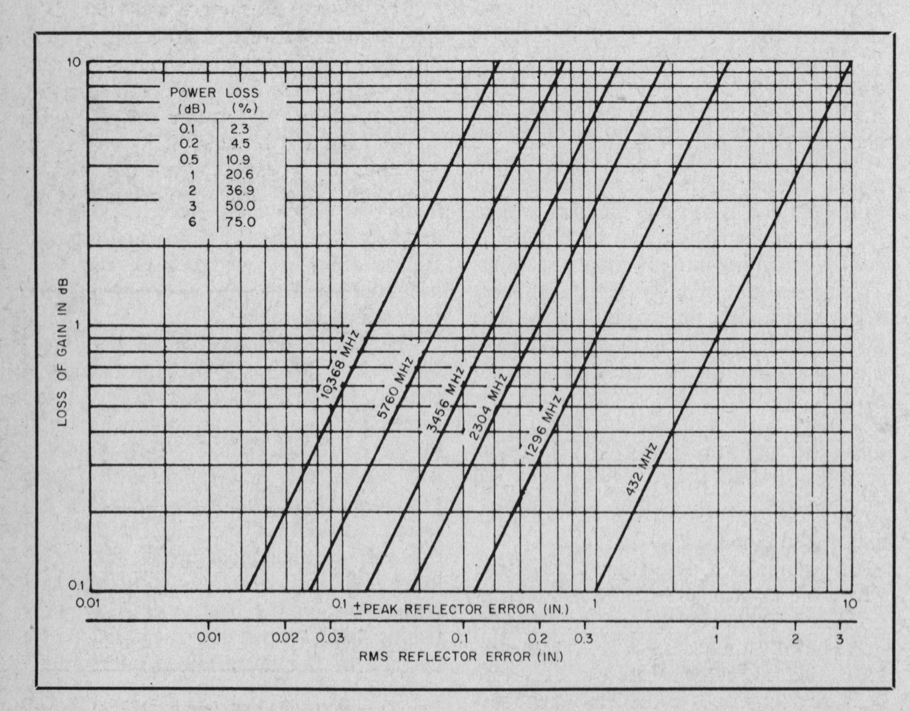


Fig. 40 — Gain deterioration vs. reflector error. Basic information obtained from J. Ruze, British *IEE*.

## Multiband Use

Many amateurs construct multiple-band antennas by putting two dishes back to back on the same tower. This is inefficient. The parabolic reflector is a completely frequency-independent surface and studies have shown that a 0.6 f/d surface can be steered seven beamwidths by moving only the feed horn from side to side before the gain diminishes one dB. Therefore, the best dual-band antenna can be built by mounting separate horns side by side. At worst the antenna may have to be moved a few degrees (usually less than a beamwidth) when switching between horns, and the unused horn increases the shadow area slightly. In fact, the same surface can function simultaneously on two frequencies, making cross-band operation possible with the same dish.

## Order of Assembly

1) A single spoke is held upright behind the rear center plate with the screw eye facing forward. Two 6-32 machine screws are pushed through the holes in the rear center plate, through the two holes of the spoke, and into the corresponding holes of the front center plate. Lock washers and nuts are placed on the machine screws and hand tightened.

2) The remaining spokes are placed between the machine screw holes. Make sure that each screw eye faces forward. Machine screws, lock washers, and nuts are used to mount all 18 spokes.

3) The no. 6-32 nuts are tightened using a nut driver.

4) The mast tubing is attached to the spoke assembly, positioned properly, and locked down with the setscrews on the pipe flange at the rear center plate. The S hooks of the 18 Dacron strings are attached to the screw eyes of the spokes.

5) The ends of two pieces of fishing string (which go over the Bakelite rod support) are tied to a screw eye at the forward center plate.

6) The dish is laid on the ground in an upright position and no. 9 galvanized wire is threaded through the eyebolts. The overlapping ends are lashed together with bailing wire.

7) The dish is placed on the ground in an inverted position with the focus downward. The screening is placed on the back of the dish and the screening perimeter is fastened as previously described.

8) The extension mast tubing (with counterweight) is connected to the center plate with U bolts.

9) The dish is mounted on a support (if one is used) and the transmission line is

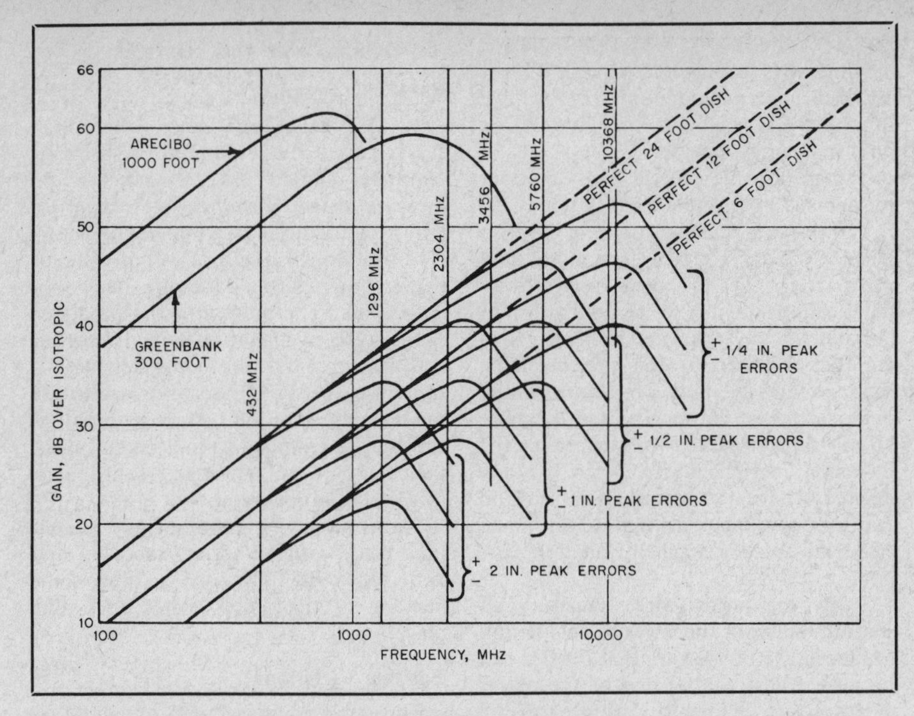


Fig. 41 — Parabolic-antenna gain versus size, frequency, and surface errors. All curves assume 60-percent aperture efficiency and 10-dB power taper. Reference: J. Ruze, British *IEE*.

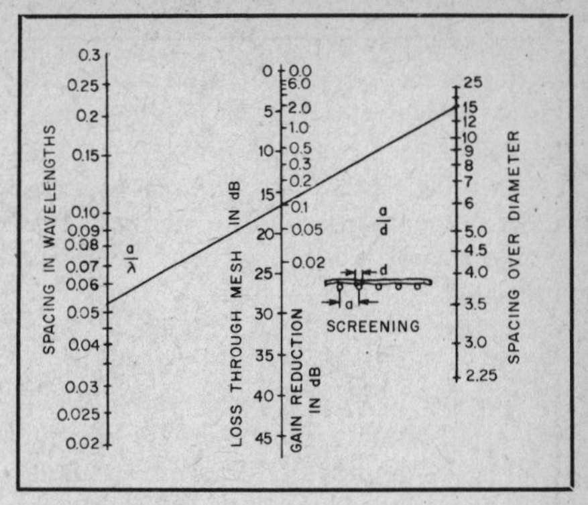


Fig. 42 — Surfacing material quality.

routed through the tubing and attached to the horn.

## PARABOLA GAIN VERSUS ERRORS

"How accurate must a parabolic surface be?" is a frequently asked question. According to the Rayleigh limit for telescopes, little gain increase is realized by making the mirror accuracy greater than ± 1/8 wavelength peak error. John Ruze of the MIT Lincoln Laboratory, among others, has derived an equation for parabolic antennas and built models to prove it. The tests show that the tolerance loss can be predicted within a fraction of a decibel, and less than 1 dB of gain is sacrificed with a surface error of ± 1/8 wavelength. An eighth of a wavelength is 3.4 inches at 432 MHz, 1.1 inches at 1296 MHz and 0.64 inch at 2300 MHz.

Some confusion about requirements of greater than 1/8-wavelength accuracy may be the result of technical literature describing highly accurate surfaces for reasons of low side-lobe levels. We are concerned more with forward gain than with low side-lobe levels; therefore, these stringent requirements do not apply. When a template is held up against a surface, positive and negative (±) peak errors can be measured. The graphs of dish-accuracy requirements are frequently plotted in terms of rms error, which is a mathematically derived function much smaller than ± peak error (typically 1/3). These small rms accuracy requirements have discouraged many constructors who confuse them with ± peak errors.

Fig. 40 may be used to predict the resultant gain of various dish sizes with typical errors. There are a couple of surprises, as shown in Fig. 41. As the frequency is increased for a given dish, the gain increases 6 dB per octave until the tolerance errors become significant. Then gain deterioration occurs rapidly. Maximum gain is realized at the frequency where the tolerance loss is 4.3 dB. Notice that at 2304 MHz, a 24-foot dish with ± 2-inch peak errors has the same gain as a 6-foot dish with ± 1-inch peak errors. Quite startling, when it is realized that a 24-foot dish has 16 times the area of a 6-foot dish. Each time the diameter or frequency is doubled or halved, the gain changes 6 dB. Each time all the errors are halved, the frequency of maximum gain is doubled. With this information, the gain of other dish sizes with other tolerances may be predicted.

These curves are adequate to predict gain, assuming a high-efficiency feed horn is used (as described earlier) which realizes 60-percent aperture efficiency. At frequencies below 1296 MHz where the horn is large and causes considerable blockage, the curves are a little optimistic. A properly built dipole and splasher feed will have about 1.5 dB less gain when used with a 0.6 f/d dish than will the dual-mode feed system described.

The worst kind of surface distortion is where the surface curve in the radial direction is not parabolic but gradually departs in a smooth manner from an exact parabola. The decrease in gain can be severe because a large area is involved. If the surface is checked with a template, and if reasonable construction techniques are employed, deviations will be under control and the curves will represent an upper limit to the gain that can be realized. If a 24-foot dish with ± 2-inch peak errors is being used with 432-MHz and 1296-MHz multiple feed horns, the constructor might be discouraged from trying a 2300-MHz feed because there is 15 dB of gain degradation. However, the dish will have 29 dB of gain remaining on 2300 MHz, making it worthy of consideration.

The near-field range of the 12-foot 3-inch antenna is 703 feet at 2300 MHz. By using the sun as a transmitter and observing receiver noise powei, it was discovered that the antenna had two main lobes about 4 degrees apart. The template showed a surface error (insufficient spoke bending at 3/4 radius), and a correction was made. A recheck showed one main lobe, and the sun noise was almost 3 dB stronger.

### Other Surfacing Materials

The choice of surface materials is a compromise between rf reflecting properties and wind loading. Aluminum screening, with its very fine mesh (and weighing 4.3 pounds per 100 square feet) is useful beyond X-band because of its very close spacing. It is easy to roll up and is therefore ideal for a portable dish.

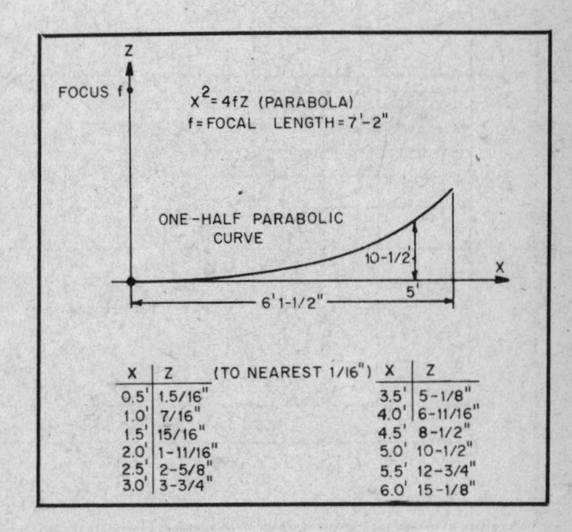


| X | Z | X | Z |
|---|---|---|---|
| 0.5' | 1.5/16" | 3.5' | 5-1/8" |
| 1.0' | 7/16" | 4.0' | 6-11/16" |
| 1.5' | 15/16" | 4.5' | 8-1/2" |
| 2.0' | 1-11/16" | 5.0' | 10-1/2" |
| 2.5' | 2-5/8" | 5.5' | 12-3/4" |
| 3.0' | 3-3/4" | 6.0' | 15-1/8" |

Fig. 43 — Parabolic template for 12-foot, 3-inch dish.

However, this close spacing causes it to be a 34-percent filled aperture, which will cause the wind force at 60 miles per hour to be more than 400 pounds on this 12-foot dish. Those amateurs considering a permanent installation of this dish should look into other surfacing materials.

One-inch hexagonal chicken wire, which is an 8-percent filled aperture, is very desirable for 432-MHz operation. It weighs 10 pounds per 100 square feet and exhibits 81 pounds of force with 60-mile-per-hour winds. However, measurement on a large piece reveals 6 dB of feed-through at 1296 MHz. Therefore, on 1296 MHz, one fourth of the power will feed through the surface material, but this will cause a loss of only 1.3 dB forward gain. Since the low-wind-loading material will provide a 30-dB gain potential, it is a very good trade-off. Chicken wire is very poor material for 2300 MHz and higher, since the hole dimensions become comparable to a half wavelength. As with all surfacing materials, minimum feedthrough will occur when the E-field polarization is parallel to the longest dimension of the surfacing holes. Half-inch hardware cloth weighs 20 pounds per 100 square feet. It has a wind-loading characteristic of 162 pounds with 60-mile-per-hour winds. The filled aperature is 16 percent and this material is useful to 2300 MHz.

A rather interesting material worthy of investigation is 1/4-inch reinforced plastic (described in Montgomery Ward Farm and Garden Catalog). It weighs only 4 pounds per 100 square feet. The plastic melts with many universal solvents such as lacquer thinner. If a careful plastic-melting job is done, what will be left is the 1/4-inch spaced aluminum wires with a small blob of plastic at each junction to hold the matrix together.

There are some general considerations to be made in selecting surface materials:

1) Joints of screening do not have to make electrical contact. The horizontal wires reflect the horizontal wave. Skew polarizations are merely a combination of horizontal and vertical components which are thus reflected by the corresponding wires of the screening. To a horizontally polarized wave, the spacing and diameter of only the horizontal wires determine the reflection coefficient (see Fig. 42). Many amateurs have the mistaken impression that screening materials that do not make electrical contact at their junctions are poor reflectors.

2) By measuring wire diameter and spacings between the wires, a calculation of percentage of aperture that is filled can be made. This will be one of the major determining factors of wind pressure when the surfacing material is dry. Under ice and snow conditions, smaller aperture materials may become clogged, which could make the surfacing material act as one solid sail. The ice and snow will have a rather minor effect on the reflecting properties of the surface, however.

3) Amateurs who live in areas where ice and snow are prevalent should consider a de-icing scheme such as weaving enameled wire through the screening and passing a current through it, fastening water-pipe heating tape behind the screening, or soldering heavy leads to the screening perimeter and passing current through the screening itself.

### Parabolic Template

At 2300 MHz and higher, where high surface accuracy is required, a parabolic template should be constructed to measure surface errors. A simple template may be constructed (see Fig. 43) by taking a 12-foot 3-inch length of 4-foot wide tar paper and drawing a parabolic shape on it with chalk. The points for the parabolic shape were calculated at 6-inch intervals and these points were connected with a smooth curve. For those who wish to use the template with the surface material installed, the template should be cut along the chalk line and stiffened by cardboard or a wood lattice frame.

Surface-error measurements should take place with all spokes installed and deflected by the fishing strings, since some bending of the center plates does take place.

### Possible Variations

The stressed parabolic antenna, as described, is a new construction technique for which a patent application has been filed. Because of its newness, all of its possibilities have not been explored. For instance, a set of fishing strings or guy wires could be set up behind the dish for error correction as long as it does not permanently bend the aluminum spokes. This technique would also protect the dish against wind loading from the rear. An extended piece of TV mast would be an ideal place to hang a counterweight and attach the back guys. It would strengthen the structure.

Chapter 13

# Portable and Mobile Antennas

Few amateurs construct their own antennas for hf mobile use since safety reasons dictate a very sound mechanical construction. However, most installations using commercially made components still have to be optimized for the particular installation and type of operation desired.

## HF-MOBILE FUNDAMENTALS

The drawing of Fig. 1 shows a typical bumper-mounted center-loaded whip suitable for operation in the hf range. The antenna could also be mounted on the car body proper (such as a fender), and mounts are available for this purpose. The base spring acts as a shock absorber for the base of the whip since the continual flexing while in motion would otherwise weaken the antenna. A short heavy mast section is mounted between the base spring and loading coil. Some models have a mechanism which allows the antenna to be tipped over for adjustment or for fastening to the roof of the car when not in use. It is also advisable to extend a couple of guy lines from the base of the loading coil to clips or hooks fastened to the rain trough on the roof of the car. Nylon fishing line of about 40-pound test strength is suitable for this purpose. The guy lines act as safety cords and also reduce the swaying motion of the antenna considerably. The feed line to the transmitter is connected to the bumper and base of the antenna. Good low-resistance connections are important here.

Tune-up of the antenna is usually accomplished by changing the height of the adjustable whip section above the precut loading coil. First, tune the receiver and try to determine where the signals seem to peak up. Once this frequency is found, check the SWR with the transmitter on, seeking the frequency of lowest SWR. Shortening the adjustable section will increase the resonant frequency and making it longer will lower the frequency. It is important that the antenna be away from surrounding objects such as overhead wires, 10 feet or more, since considerable detuning can occur. Once the setting is found where the SWR is lowest at the center of the desired operating frequency range, the length of the adjustable section should be recorded.

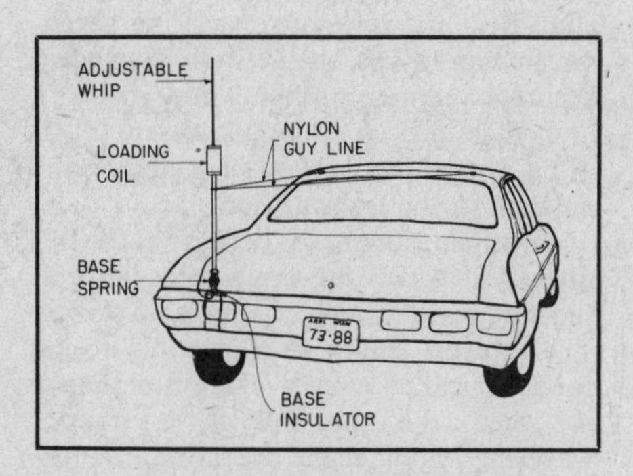


Fig. 1 — A typical bumper-mounted hf-mobile antenna. Note the nylon guy lines.

Propagation conditions and ignition noise are usually the limiting factors for mobile operation on 20, 15 and 10 meters, while antenna-size restrictions greatly affect operation on 160, 80 and 40 meters. From this standpoint, perhaps the optimum band for hf-mobile operation is 40 meters. The popularity of the regional mobile nets on 40 meters is perhaps the best indication of its suitability. For local work, 10 meters is also useful since antenna efficiency is high and relatively simple antennas without loading coils are possible.

As the frequency of operation is lowered, an antenna of fixed length looks at its base feed point like a decreasing resistance in series with an increasing capacitive reactance. The capacitive reactance must be tuned out, which necessitates the use of a series inductive reactance or loading coil. The amount of inductance required will be determined by the placement of the coil in the antenna system. Base loading requires the lowest value of inductance for a fixed-length antenna, and as the coil is placed farther up the whip, the necessary value increases. This is because the capacitance of the shorter antenna section (above the coil) to the car body is now lower (higher capacitive reactance), requiring a bigger inductance in order to tune the antenna to resonance. The advantage is that the current distribution in the whip is improved, which increases the radiation resistance. The disadvantage is that requirement of a larger coil also means the coil losses go up. Center loading has been generally accepted as a good compromise with minimal construction problems.

The difficulty in constructing suitable loading coils increases as the frequency of operation is lowered for typical antenna lengths used in mobile work. Since the required resonating inductance gets larger and the radiation resistance decreases at lower frequencies, most of the power may be dissipated in the coil resistance and in other ohmic losses. This is one reason why it is advisable to buy a commercially made loading coil with the highest power rating possible, even though only low-power

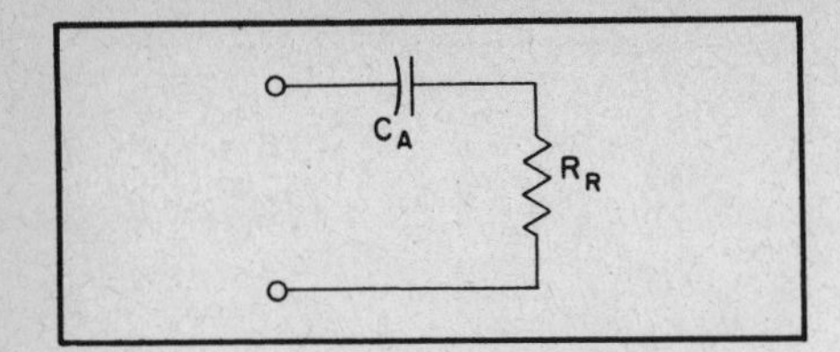


Fig. 2 — At frequencies below the resonant frequency, the whip antenna will show the capacitive reactance as well as resistance. $R_R$ is the radiation resistance, and $C_A$ represents the capacitive reactance.

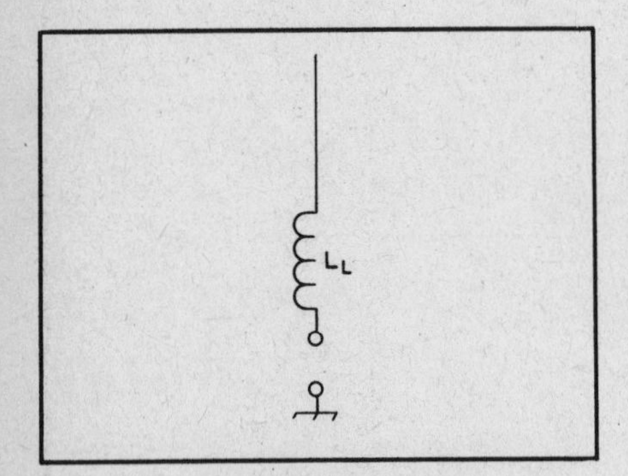


Fig. 3 — The capacitive reactance at frequencies lower than the resonant frequency of the whip can be canceled out by adding an equivalent inductive reactance in the form of a loading coil in series with the antenna.

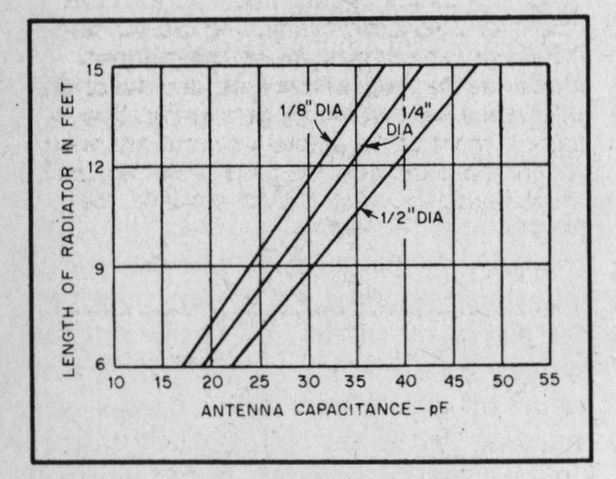


Fig. 4 — Graph showing the approximate capacitance of short vertical antennas for various diameters and lengths. These values should be approximately halved for a center-loaded antenna.

operation is contemplated. Percent-wise, the coil losses in the higher power loading coils are usually less, with subsequent improvement in radiating efficiency, regardless of the power level used. Of course, the above philosophy also applies to homemade loading coils, and design considerations will be considered in a later section.

Once the antenna is tuned to resonance, the input impedance at the antenna terminals will look like a pure resistance. Neglecting losses, this value drops from nearly 15 ohms on 15 m to 0.1 ohm on 160 m for an 8-foot whip. When coil and other losses are included, the input resistance increases to approximately 20 ohms on 160 m and 16 ohms on 15 m. These values are for relatively high-efficiency systems. From this it can be seen that the radiating efficiency is much poorer on 160 m than on 15 m under typical conditions.

Since most modern gear is designed to operate with a 50-ohm transmission line, a matching network may be necessary with the high-efficiency antennas previously mentioned. This can take the form of either a broad-band transformer, tapped coil, or an LC-matching network. With homemade or modified designs, the tapped-coil arrangement is perhaps the easiest one to build, while the broad-band transformer requires no adjustment. As the losses go up, so does the input resistance, and in less efficient systems the matching network may be eliminated.

## THE EQUIVALENT CIRCUIT OF A TYPICAL MOBILE ANTENNA

In the previous section, some of the general considerations were discussed, and these will now be taken up in more detail. It is customary in solving problems involving electric and magnetic fields (such as antenna systems) to try to find an equivalent network to replace the antenna. In many cases, the network may be an accurate representation only over a limited frequency range. However, this is often a valuable method in matching the antenna to the transmission line.

Antenna resonance is defined as the frequency at which the input impedance at the antenna terminals is a pure resistance. The shortest length at which this occurs for a vertical antenna over a ground plane is when the antenna is a quarter wavelength long at the operating frequency; the impedance value for this length (neglecting losses) is about 36 ohms. The idea of resonance can be extended to antennas shorter (or longer) than a quarter wave, and only means that the input impedance is purely resistive. As pointed out previously, when the frequency is lowered, the antenna looks like a series RC circuit, as shown in Fig. 2. For the average 8-ft whip, the reactance of $C_A$ may range from about 150 ohms at 21 MHz to as high as 8000 ohms at 1.8 MHz, while the radiation resistance $R_R$ varies from about 15 ohms at 21 MHz to as low as 0.1 ohm at 1.8 MHz.

For an antenna less than 0.1 wavelength long, the approximate radiation resistance may be determined from the following:

$$R_R = 273\,(\ell f)^2 \times 10^{-8}$$

where $\ell$ is the length of the whip in inches, and f is the frequency in megahertz.

Since the resistance is low, considerable current must flow in the circuit if any appreciable power is to be dissipated as radiation in the resistance $R_R$. Yet it is apparent that little current can be made to flow in the circuit so long as the comparatively high series reactance remains.

**Antenna Capacitance**

The capacitive reactance can be canceled out by connecting an equivalent inductive reactance, $L_L$, in series, as shown in Fig. 3, thus tuning the system to resonance.

The capacitance of a vertical antenna shorter than a quarter wavelength is given by

$$C_A = \frac{17\ell}{\left[\left(\ln \frac{24\ell}{D}\right) - 1\right]\left[1 - \left(\frac{f\ell}{234}\right)^2\right]}$$

where

$C_A$ = capacitance of antenna in pF
$\ell$ = antenna height in feet
D = diameter of radiator in inches
f = operating frequency in MHz

$$\ln \frac{24\ell}{D} = 2.3 \log_{10} \frac{24\ell}{D}$$

The graph of Fig. 4 shows the approximate capacitance of whip antennas of various average diameters and lengths. For 1.8, 4 and 7 MHz, the loading-coil inductance required (when the loading coil is at the base) will be approximately the inductance required to resonate in the desired band with the whip capacitance taken from the graph. For 14 and 21 MHz, this rough calculation will give more than the required inductance, but it will serve as a starting point for the final experimental adjustment that must always be made.

## LOADING COIL DESIGN

To minimize loading-coil loss, the coil should have a high ratio of reactance to resistance, i.e., high Q. A 4-MHz loading coil wound with small wire on a small-diameter solid form of poor quality, and enclosed in a metal protector, may have a Q as low as 50, with a resistance of 50 ohms or more. High-Q coils require a large conductor, "air-wound" construction, large spacing between turns, the best insulating material available, a diameter not less than half the length of the coil (not always mechanically feasible), and a minimum of metal in the field. Such a coil for 4 MHz may show a Q of 300 or more, with a resistance of 12 ohms or less.

The coil could then be placed in series with the feed line at the base of the antenna to tune out the unwanted capacitive reactance, as shown in Fig. 3. Such a method is often referred to as base loading, and many practical mobile antenna systems have been built in this way.

Over the years, the question has come up as to whether or not more efficient designs are possible compared with simple base loading. While many ideas have been tried with varying degrees of success, only

**Table 1**
**Approximate Values for 8-foot Mobile Whip**

**Base Loading**

| f(kHz) | Loading L(μH) | RC(Q50) Ohms | RC(Q300) Ohms | RR Ohms | Feed R* Ohms | Matching L(μH) |
|---|---|---|---|---|---|---|
| 1800 | 345 | 77 | 13 | 0.1 | 23 | 3 |
| 3800 | 77 | 37 | 6.1 | 0.35 | 16 | 1.2 |
| 7200 | 20 | 18 | 3 | 1.35 | 15 | 0.6 |
| 14,200 | 4.5 | 7.7 | 1.3 | 5.7 | 12 | 0.28 |
| 21,250 | 1.25 | 3.4 | 0.5 | 14.8 | 16 | 0.28 |
| 29,000 | — | — | — | — | 36 | 0.23 |
| **Center Loading** | | | | | | |
| 1800 | 700 | 158 | 23 | 0.2 | 34 | 3.7 |
| 3800 | 150 | 72 | 12 | 0.8 | 22 | 1.4 |
| 7200 | 40 | 36 | 6 | 3 | 19 | 0.7 |
| 14,200 | 8.6 | 15 | 2.5 | 11 | 19 | 0.35 |
| 21,250 | 2.5 | 6.6 | 1.1 | 27 | 29 | 0.29 |

RC = Loading-coil resistance; RR = Radiation resistance.
*Assuming loading coil Q = 300, and including estimated ground-loss resistance.
Suggested coil dimensions for the required loading inductance are shown in a following table.

**Table 2**
**Suggested Loading-Coil Dimensions**

| Req'd L(μH) | No. Turns | Wire Size | Dia. In. | Length In. |
|---|---|---|---|---|
| 700 | 190 | 22 | 3 | 10 |
| 345 | 135 | 18 | 3 | 10 |
| 150 | 100 | 16 | 2-1/2 | 10 |
| 77 | 75 | 14 | 2-1/2 | 10 |
| 77 | 29 | 12 | 5 | 4-1/4 |
| 40 | 28 | 16 | 2-1/2 | 2 |
| 40 | 34 | 12 | 2-1/2 | 4-1/4 |
| 20 | 17 | 16 | 2-1/2 | 1-1/4 |
| 20 | 22 | 12 | 2-1/2 | 2-3/4 |
| 8.6 | 16 | 14 | 2 | 2 |
| 8.6 | 15 | 12 | 2-1/2 | 3 |
| 4.5 | 10 | 14 | 2 | 1-1/4 |
| 4.5 | 12 | 12 | 2-1/2 | 4 |
| 2.5 | 8 | 12 | 2 | 2 |
| 2.5 | 8 | 6 | 2-3/8 | 4-1/2 |
| 1.25 | 6 | 12 | 1-3/4 | 2 |
| 1.25 | 6 | 6 | 2-3/8 | 4-1/2 |

To obtain dimensions in millimeters, multiply inches by 25.4.

a few have been generally accepted and incorporated into actual systems. These are center loading, continuous loading, and combinations of the latter with more conventional antennas.

## BASE LOADING AND CENTER LOADING

If a whip antenna is short compared to a wavelength *and the current is uniform along the length,* $\ell$, the electric field strength, E, at a distance, d, away from the antenna is given approximately by

$$E = \frac{120\pi I \ell}{d\lambda}$$

where I is the antenna current in amperes, and λ is the wavelength in the same units as d and $\ell$. A uniform current flowing along the length of the whip is an idealized situation, however, since the current is greatest at the base of the antenna and goes to a minimum at the top. In practice, the field strength will be less than that given by the above equation, being a function of the current distribution in the whip.

The reason that the current is not uniform in a whip antenna can be seen from the circuit approximation shown in Fig. 5. A whip antenna over a ground plane is similar in many respects to a tapered coaxial cable where the center conductor remains the same diameter as the length increases, but with an increasing diameter of the outer conductor. The inductance per unit length of such a cable would increase farther along the line while the capacitance per unit length would decrease. In Fig. 5 the antenna is represented by a series of L-C circuits in which C1 is greater than C2, which is greater than C3, and so on. L1 is less than L2, which is less than succeeding inductances.

The net result is that most of the antenna current returns to ground near the base of the antenna, and very little near the top. Two things can be done to improve this distribution and make the current more uniform. One would be to increase the capacitance of the top of the antenna to ground through the use of top loading or a "capacitive hat," as discussed in Chapter 2. Unfortunately, the wind resistance of the hat makes it somewhat unwieldly for mobile use. The other method is to place the loading coil farther up the whip, as shown in Fig. 6, rather than at the base. If the coil is resonant (or nearly so) with the capacitance to ground of the section above the coil, the current distribution is improved as also shown in Fig. 6. The result with both top loading and center loading is that the radiation resistance is increased, offsetting the effect of losses and making matching easier.

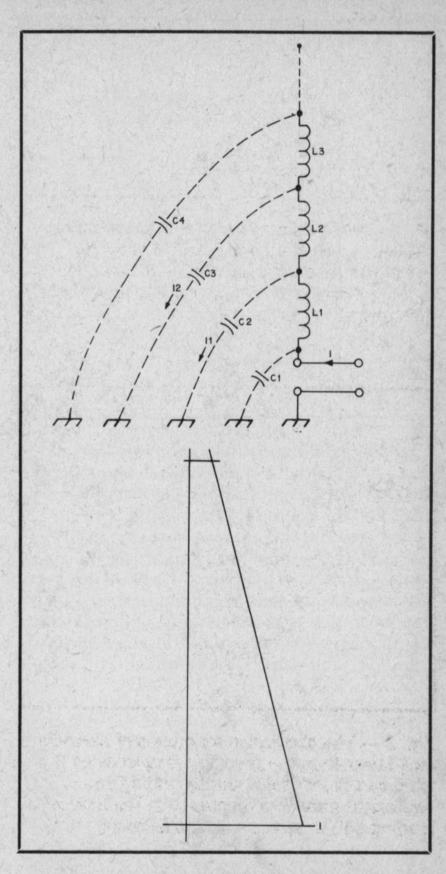


Fig. 5 — A circuit approximation of a simple whip over a perfectly conducting ground plane. The shunt capacitance per unit length gets smaller as the height increases, and the series inductance per unit length gets larger. Consequently, most of the antenna current returns to the ground plane near the base of the antenna, giving the current distribution shown at the bottom.

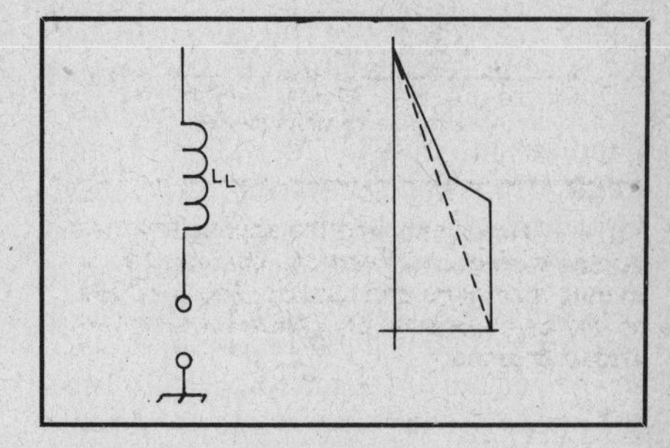


Fig. 6 — Improved current distribution resulting from center loading.

Table 1 shows the approximate loading-coil inductance for the various amateur bands. Also shown in the table are approximate values of radiation resistance to be expected with an 8-ft whip, and the resistances of loading coils — one group having a Q of 50, the other a Q of 300. A comparison of radiation and coil resistances will show the importance of reducing the coil resistance to a minimum, especially on the three lower frequency bands. Table 2 shows suggested loading-coil dimensions for the inductance values given in Table 1.

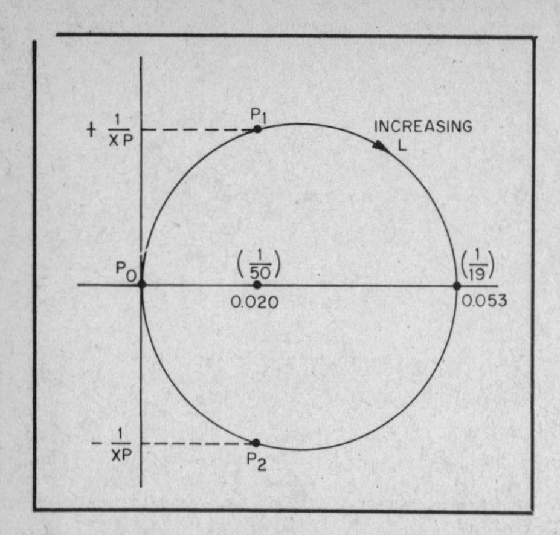


Fig. 7 — Admittance diagram of the RLC circuit consisting of the whip capacitance, radiation resistance and loading coil discussed in text. The horizontal axis represents conductance, and the vertical axis susceptance. The point Po is the input admittance with no whip loading inductance. Points P1 and P2 are described in the text. The conductance equals the reciprocal of the resistance, if no reactive components are present. For a series RX circuit, the conductance is given by:

$$G = \frac{R}{R^2 + X^2}$$

and the susceptance is given by:

$$B = \frac{-X}{R^2 + X^2}$$

Consequently, a parallel equivalent GB circuit of the series RX one can be found which makes computations easier. This is because conductances and susceptances add in parallel the same way resistances and reactances add in series.

## CONTINUOUSLY LOADED ANTENNAS

The design of high-Q air-core inductors for rf work is complicated by the number of parameters which must be optimized simultaneously. One of these factors which affects coil Q adversely is radiation. Therefore, the possibility of cutting down the other losses while incorporating the coil radiation into that from the rest of the antenna system is an attractive one.

The general approach has been to use a coil made from heavy wire (no. 14 and larger), with length-to-diameter ratios as high as 21. British experimenters have reported good results with 8-foot overall lengths on the 1.8- and 3.5-MHz bands. The idea of making the entire antenna out of one section of coil has also been tried with some success. This technique is referred to as linear loading. Further information on linear loaded antennas is contained in Chapter 10.

While going to extremes in trying to find a perfect loading arrangement may not improve antenna performance very much, a poor system with lossy coils and high-resistance connections is also to be avoided.

## MATCHING TO THE TRANSMITTER

Most modern transmitters require a 50-ohm output load and since the feed-point impedance of a mobile whip is quite low, a matching network may be necessary. While calculations are helpful in the initial design, considerable experimenting is often necessary in final tune-up. This is particularly true for the lower bands, where the antenna is electrically short compared with a quarter-wave whip. The reason is that the loading coil is required to tune out a very large capacitive reactance, and even small changes in component values result in large reactance variations. Since the feed-point resistance is low to begin with, the problem is even more aggravated. This is one reason why it is advisable to guy the antenna, and make sure that no conductors such as overhead wires are near the whip during tune-up.

Transforming the low resistance of the whip up to a value suitable for a 50-ohm system can be accomplished with an rf transformer or with a shunt-feed arrangement, such as an L network. The latter may only require one extra component at the base of the whip, since the circuit of the antenna itself may be used as part of the network. The following example illustrates the calculations involved.

Assume that a center-loaded whip antenna, 8.5 feet in overall length, is to be used on 7.2 MHz. From Table 1, we see that the feed-point resistance of the antenna will be approximately 19 ohms, and from Fig. 4 that the capacitance of the whip, as seen at its base, is approximately 24 pF. Since the antenna is to be center loaded, the capacitance value of the section above the coil will be cut approximately in half, to 12 pF. From this, it may be calculated that a center-loading inductor of 40.7 μH is required to resonate the antenna, that is, cancel out the capacitive reactance. (This figure agrees with the approximate value of 40 μH shown in Table 1. The resulting feed-point impedance would then be 19 + j0 ohms — a good match, *if* one happens to have a supply of 19-ohm coax.

Solution: The antenna can be matched to a 50-ohm line either by tuning it above or below resonance and then canceling out the undesired component with an appropriate shunt element, inductive or capacitive. The way in which the impedance is transformed up can be seen by constructing an admittance plot of the series RLC circuit consisting of the loading coil, antenna capacitance, and feed-point resistance. Such a plot is shown in Fig. 7 for a constant feed-point resistance of 19 ohms. There are two points of interest, P1 and P2, where the input conductance is 20 millisiemens, which corresponds to 50 ohms. The undesired susceptance is shown as 1/Xp and − 1/Xp, which must be canceled with a shunt element of the opposite sign but with the same numerical value. The value of the canceling shunt reactance, Xp, may be found from the formula:

$$Xp = \frac{Rf \cdot Z_o}{\sqrt{Rf(Z_o - Rf)}}$$

where Xp is the reactance in ohms, Rf is the feed-point resistance, and $Z_o$ is the feed-line impedance. For $Z_o$ = 50 ohms and Rf = 19 ohms, Xp = ± 39.1 ohms. A coil or good-quality mica capacitor may be used as the shunt element. With the tune-up procedure described later, the value is not critical, and a fixed-value component may be used.

To arrive at point P1, the value of the center loading-coil inductance would be less than that required for resonance. The feed-point impedance would then appear capacitive, and an inductive shunt matching element would then be required. To arrive at point P2, the center loading coil would be more inductive than required for resonance, and the shunt element would need to be capacitive. The value of the center loading coil required for the shunt-matched and resonated condition may be determined in henrys from the equation:

$$L = \frac{1}{4\pi^2 f^2 C} \pm \frac{Xs}{2\pi f}$$

where addition is performed if a capacitive shunt is to be used, or subtraction performed if the shunt is inductive, and where f is the frequency in *hertz,* C, the capacitance of the antenna section being matched in *farads,* and

$$Xs = \sqrt{Rf(Z_o - Rf)}.$$

For the example given, Xs is found to be 24.3 ohms, and the required loading inductance is either 40.2 μH or 41.3 μH, depending on the type of shunt. The various matching configurations for this example are shown in Fig. 8. At A is shown the antenna as tuned to resonance with $L_L$, a 40.7-μH coil, but with no provisions included for matching the resulting 19-ohm impedance to the 50-ohm line. At B, $L_L$ has been reduced to 40.2 μH to make the antenna appear capacitive, and $L_M$, having a reactance of 39.1 ohms, added in shunt to cancel the capacitive reactance and transform the feed-point impedance to 50 ohms. The arrangement at C is similar to that at B except that $L_L$ has been increased to 41.3 μH, and $C_M$, a shunt capacitor having a reactance of 39.1 ohms added, which also results in a 50-ohm nonreactive termination for the feed line.

The values determined for the loading coil in the above example point out an important consideration concerning the matching of short antennas — that relatively small changes in values of the loading components will have a greatly magnified effect on the matching requirements. A change of less than 3 percent in loading-coil inductance value necessitates a completely different matching network! Likewise, calculations show that a 3 percent change in antenna capacitance will give similar results, and the value of the precautions mentioned earlier becomes clear. The sensitivity of the circuit with regard to frequency variations is also quite critical and an excursion around practically the entire circle in Fig. 7 may represent only 600 kHz, centered around 7.2 MHz for the above example. This is why tuning up a mobile antenna can be very frustrating unless a systematic procedure is followed.

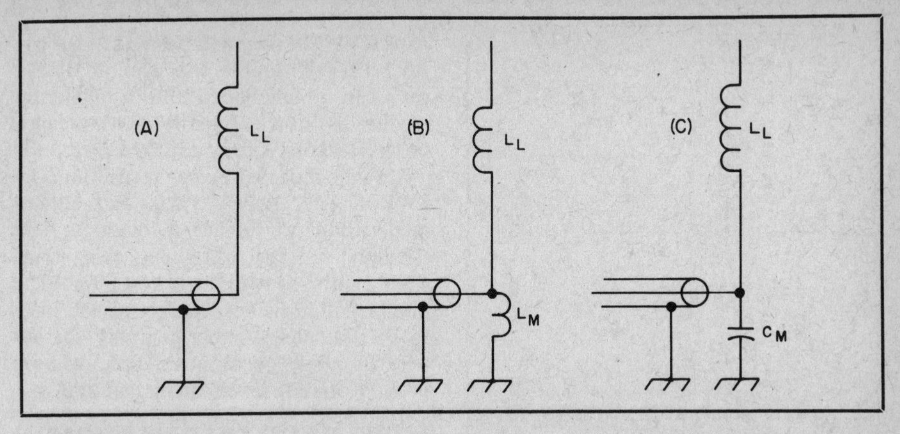


Fig. 8 — At A, a whip antenna which is resonated with a center loading coil. At B and C, the value of the loading coil has been altered slightly to make the feed-point impedance appear reactive, and a matching component is added in shunt to cancel the reactance. This provides an impedance transformation to match the $Z_0$ of the feed line. An equally acceptable procedure, rather than altering the loading-coil inductance, is to adjust the length of the top section above the loading coil for the best match, as described in the tune-up section of the text.

### Tune-Up

Assume that inductive shunt matching is to be used with the antenna in the previous example, which means that a coil of 0.86 μH will be placed across the whip terminals to ground. With a 40-μH loading coil in place, the adjustable section above the loading coil should be adjusted for minimum height. Signals in the receiver will sound weak and the whip should be lengthened a bit at a time until they start to peak up. Turn the transmitter on and check the SWR at a few frequencies to find where a minimum occurs. If it is below the desired frequency of operation, shorten the whip slightly and check again. It should be moved approximately 1/4 inch at a time until the SWR is minimum at the center of the desired range. If the frequency where the minimum SWR exists is above the desired frequency, repeat the above only lengthen the whip slightly.

If a shunt capacitance is to be used, a value of 565 pF would correspond to the needed 39.1 ohms of reactance for a frequency of 7.2 MHz. With a capacitive shunt, start with the whip in its longest position and shorten it until signals peak up.

## TOP-LOADING CAPACITANCE

Because the coil resistance varies with the inductance of the loading coil, the resistance can be reduced, beneficially, by reducing the number of turns on the coil. This can be done by adding capacitance to that portion of the mobile antenna that is *above* the loading coil (Fig. 9). To achieve resonance, the inductance of the coil is reduced proportionally. "Capacitance hats," as they are often called, can consist of a single stiff wire, two wires or more, or a disc made up from several wires like the spokes of a wheel. A solid metal disc can also be used. The larger the capacitance hat, in terms of volume, the greater the capacitance. The greater the capacitance, the smaller the amount of inductance needed in the loading coil for a given resonant frequency.

There are two schools of thought concerning the attributes of center-loading and base-loading. It has not been established that one system is superior to the other, especially in the lower part of the hf spectrum. For this reason both the base- and center-loading schemes are popular. Capacitance-hat loading is applicable to either system. Since more inductance is required for center-loaded whips to make them resonant at a given frequency, capacitance hats should be particularly useful in improving their efficiency.

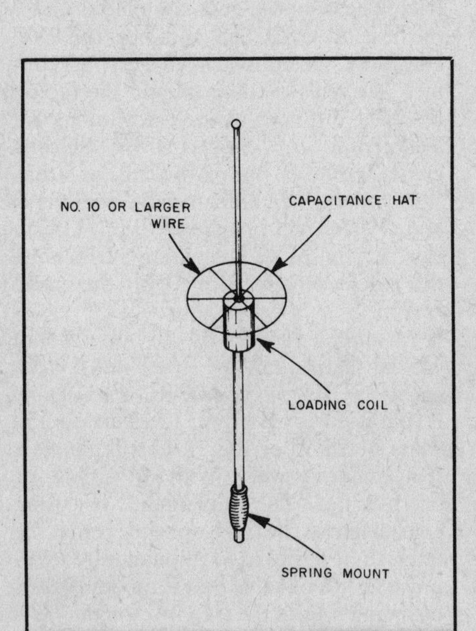


Fig. 9 — A capacitance "hat" can be used to improve the performance of base- or center-loaded whips. A solid metal disc can be used in place of the skeleton disc shown here.

## TAPPED-COIL MATCHING NETWORK

Some of the drawbacks of the previous circuits can be eliminated by the use of the tapped-coil arrangement shown in Fig. 10. While tune-up is still critical, a smaller loading coil that cuts down losses is required. For operation on 75 meters with a center-loaded whip, L2 will have approximately 18 turns of no. 14 wire, spaced one wire thickness between turns, and wound on a 1-inch diameter form. Initially, the tap will be approximately 5 turns above the ground end of L2. Coil L2 can be inside the car body, at the base of the antenna, or it can be located at the base of the whip, outside the car body. The latter method is preferred. Since L2 helps determine the resonance of the overall antenna, L1 should be tuned to resonance in the

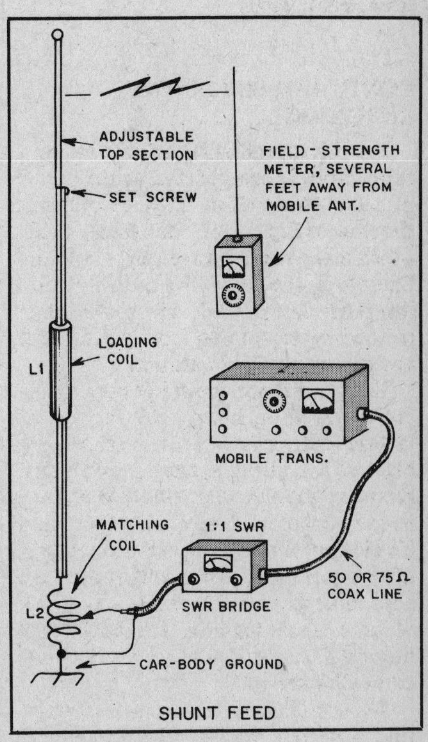


Fig. 10 — A mobile antenna using shunt-feed matching. Overall antenna resonance is determined by the combination of L1 and L2. Antenna resonance is set by pruning the turns of L1, or adjusting the top section of the whip, while observing the field-strength meter or SWR indicator. Then, adjust the tap on L2 for lowest SWR.

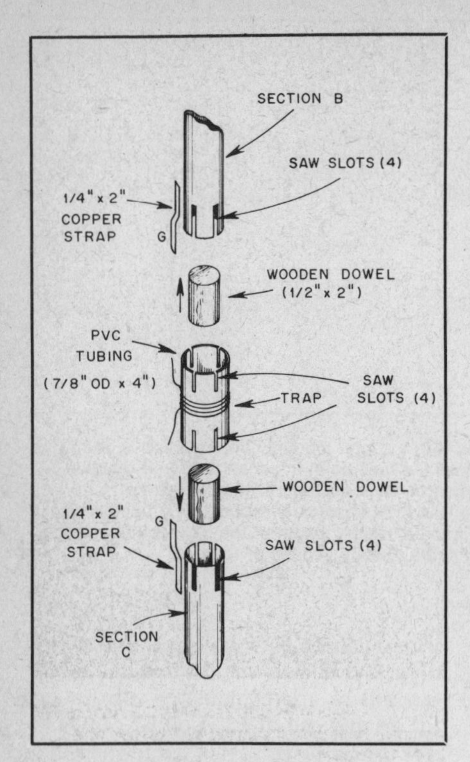


Fig. 11 — Breakdown view of the PVC trap. The hose clamps that go on the ends of the PVC coil form are not shown.

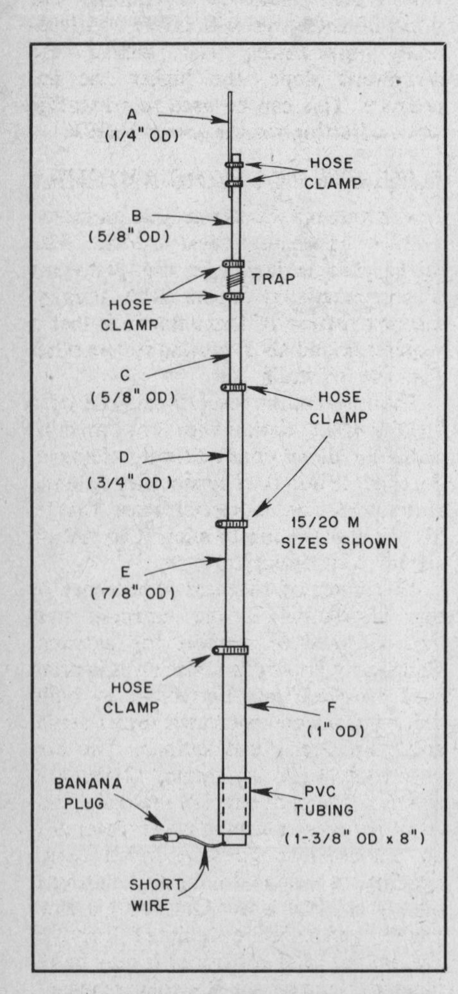


Fig. 12 — Assembly details for the two-band trap vertical. The coaxial-cable trap capacitor is taped to the lower end of section "B."

desired part of the band with L2 in the circuit. The adjustable top section of the whip can be telescoped until a maximum reading is noted on the field-strength meter. The tap is then adjusted on L2 for the lowest reflected-power reading on the SWR bridge. Repeat these two adjustments until no further increase in field strength can be obtained, this point should coincide with the lowest SWR. The number of turns needed for L2 will have to be determined experimentally for 40- and 20-meter operation. There will be proportionately fewer turns required.

## A TWO-BAND TRAP VERTICAL ANTENNA FOR THE TRAVELING HAM

This antenna was originally built for use in the 20- and 15-meter bands while operating from an RV camper or on DXpeditions to the Caribbean. Short lengths of aluminum tubing that telescope into one another were used to fabricate the antenna; the longest tube is 39 inches. A 2-inch ID piece of aluminum tubing or a heavy-duty cardboard mailing tube will serve nicely as a container for shipping or carrying. Iron-pipe thread protectors can be used as plugs for the ends of the carrying tube. The trap, mounting plate and coaxial feed line should fit easily into a suitcase with the operator's personal effects.

The ends of the sections are cut with a hacksaw to permit securing the joints by means of stainless-steel hose clamps. The trap is held in place by two hose clamps that compress the PVC coil form and the 1/2-inch tubing sections onto 1/2-inch dowel-rod plugs (Fig. 11). Strips of flashing copper (parts "G" of Fig. 11) slide inside sections B and C of the vertical. The opposite ends are placed under the hose clamps, which compress the PVC coil form. This provides an electrical contact between the trap coil and the tubing sections. The ends of the coil winding are soldered to the copper strips. Silicone grease should be put on the ends of straps "G" where they enter tubing sections B and C. This will retard corrosion. Grease can be applied to all mating surfaces of the telescoping sections for the same reason.

A suitable length of 50- or 75-ohm coaxial cable can be used as a trap capacitor. RG-58/U and RG-59/U cable is suggested for rf power levels below 150 watts. RG-8/U or RG-11/U will handle a few hundred watts without arcing or overheating. The advantage in using coaxial line as the trap capacitor is that the trap can be adjusted to resonance by selecting a length of cable that is too long, then trimming it until the trap is resonant. This is possible because each type of coax exhibits a specific amount of capacitance between the conductors. See Chapter 3 for a table that lists coaxial cable characteristics.

The trap (after final adjustment) should be protected against weather conditions. A plastic drinking glass can be inverted and mounted above the trap, or several coats of high-dielectric glue (Polystyrene Q-Dope) can be applied to the coil winding. If a coaxial-cable trap capacitor is used, it should be sealed at each end by applying noncorrosive RTV compound.

The trap is tuned to resonance prior to installing it in the antenna. It should be resonant in the center of the desired operating range, i.e., at 21,050 kHz if you prefer to operate from 21,000 to 21,100 kHz. Tuning can be done while using an accurately calibrated dip meter. If the dial isn't accurate, locate the dipper signal using a calibrated receiver *while the dipper is coupled to the coil and is set for the dip.*

A word of caution is in order here: Once the trap is installed in the antenna, it will not yield a dip at the same frequency as before. This is because it becomes absorbed in the overall antenna system and will appear to have shifted much lower in frequency. For the 20/15-meter vertical, the apparent resonance will drop some 5 MHz. Ignore this condition and proceed with the installation.

### The Tubing Sections

An illustration of the assembled two-band vertical is shown in Fig. 12. The tubing diameters indicated are suitable for 15- and 20-meter use. The longer the overall antenna, the larger should be the tubing diameter to ensure adequate strength.

A short length of test-lead wire is used at the base of the antenna to join it to the coaxial connector on the mounting plate. A banana plug is attached to the end of the wire to permit connection to a uhf style of bulkhead connector. This method aids in easy breakdown of the antenna. A piece of PVC tubing slips over the bottom of section "F" to serve as an insulator between the antenna and the mounting plate.

If portable operation isn't planned, fewer tubing sections will be required. Only two sections need be used below the trap, and two sections will be sufficient above the trap. Two telescoping sections are necessary in each half of the antenna to permit resonating the system during final adjustment.

### Other Bands

Those who are interested in other frequency pairings will find pertinent data in Table 3. Information is included on building a trap vertical for the 10-MHz WARC-79 band, plus 20 meters. One additional band can be accommodated for any of the combinations shown by using a top resonator.

Assuming that a trap vertical has been built for two bands, what might be done to obtain capability for one additional band without using the trap concept? The

**Table 3**
**Dimensions for Various Frequency Pairings, Two-Band Trap Vertical**

| *Tubing (Fig. 1)* | *Band (MHz)* | *A* | *B* | *C* | *D* | *E* | *F* | *C1 (pF)* | *L1 (Approx. µH)* |
|---|---|---|---|---|---|---|---|---|---|
| | 21/28 | 25 | 16 | 25 | 25 | 25 | 33 | 18 | 1.70 |
| Tubing Length | 14/21 | 38 | 33 | 37 | 37 | 37 | 33 | 25 | 2.25 |
| (inches) | 10/14 | 42 | 42 | 54 | 54 | 54 | 49 | 39 | 3.25 |

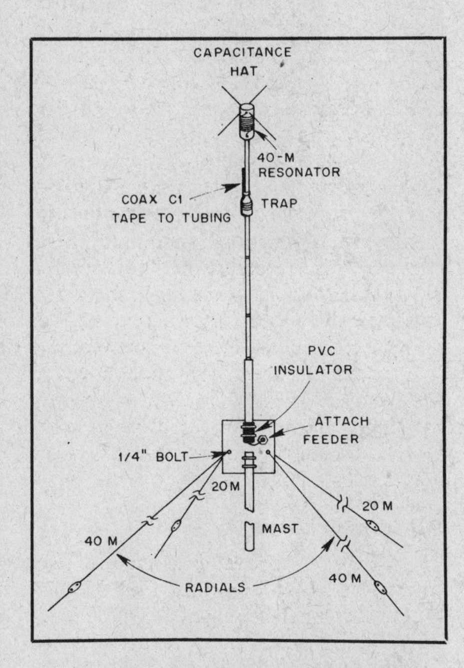


Fig. 13 — The assembled trap vertical, showing how a resonator can be placed at the top of the radiator to provide operation on an additional band (see text).

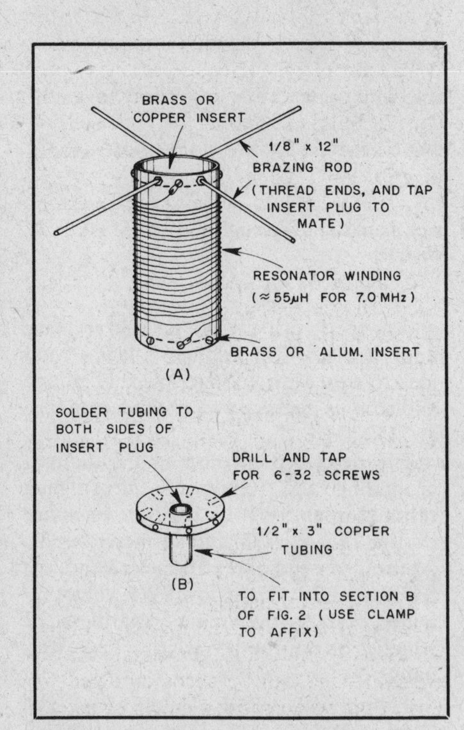


Fig. 14 — Details for building a homemade top-loading coil and capacitance hat. The completed resonator should be protected against the weather to prevent detuning and deterioration.

simple way to achieve this is to place a coil and capacitance hat on the top of the trap vertical. This is equivalent to "top loading" any vertical antenna. Suppose we had a trap vertical for 20 and 15 meters, but also wanted to use the system on 40 meters. The popular way to do this is to construct a 40-meter loading coil which can be installed as shown in Fig. 13. A number of commercial trap verticals use this technique. The loading coil is called a "resonator" because it makes the complete antenna resonant at the lowest chosen operating frequency (40 meters in our example). The coil turns must be adjusted while the antenna is assembled and installed in its final location. The remainder of the antenna has to be adjusted for proper operation on all of the bands *before* the resonator is trimmed for 40-meter resonance. If the capacitance-hat wires are short (approximately 12 inches), we can assume a capacitance of roughly 10 pF, which gives us an $X_C$ of 2275 ohms. Therefore, the resonator will also have an $X_L$ of 2275 ohms. This becomes 51 µH for operation at 7.1 MHz, since $L_{\mu H} = X_L/2\pi f$. The resonator coil should be wound for roughly 10% more inductance than needed to allow some leeway for trimming it to resonance. Alternatively, the resonator can be wound for 51 µH and the capacitance-hat wires shortened or lengthened until resonance in the selected part of the 40-meter band is achieved.

As was true of the traps, the resonator should be wound on a low-loss form. The largest conductor size practical should be used to minimize losses and elevate the power-handling capability of the coil. Details of how a homemade resonator might be built are provided in Fig. 14. The drawing in Fig. 13 shows how the antenna would look with the resonator in place.

### Ground System

There's nothing as rewarding as a *big* ground system. That is, the more radials the better, up to the point of diminishing returns. Some manufacturers of multiband trap verticals specify two radial wires for each band of operation. Admittedly, an impedance match can be had that way, and performance will be reasonably good. So during temporary operations where space for radial wires is at a premium, use two wires for each band. Four wires for each band will provide greatly improved operation. The slope of the wires will affect the feed-point impedance. The greater the downward slope, the higher the impedance. This can be used to advantage when adjusting for the lowest VSWR.

Fig. 15 — Ready to go! The W8UCG deluxe RV antenna is shown mounted at the rear of a 31-ft Airstream.

## A DELUXE RV 5-BAND ANTENNA

This antenna was designed to mount on a 31-ft Airstream travel trailer. With minor changes it can be used with any other recreational vehicle (RV). Perhaps the best feature of this antenna is that it requires no radials or ground system other than the RV itself.

The installation involves the use of a Hustler 4BTV vertical with the normal installation dimensions radically changed. See Figs. 15 and 16. The modified antenna is mounted on a special mast that is hinged near the top to allow it to rest on the RV roof during travel.

The secret of the neat appearance of this installation is the unusual mast material used to support the antenna. Commonly known as Unistrut, it is often used by electrical contractors to build switching and control panels for industrial and commercial installations. The size selected is 1-5/8 in. square 12-gauge U channel. The open edges of the U are folded in for greater strength; the material is an extremely tough steel which resists bending (as well as drilling and cutting!). The U channel is available with a zinc-plated, galvanized or painted finish to prevent rust and corrosion; it may be repainted to match any RV color scheme.

The supporting mast is secured to the rear frame or bumper of the RV by means

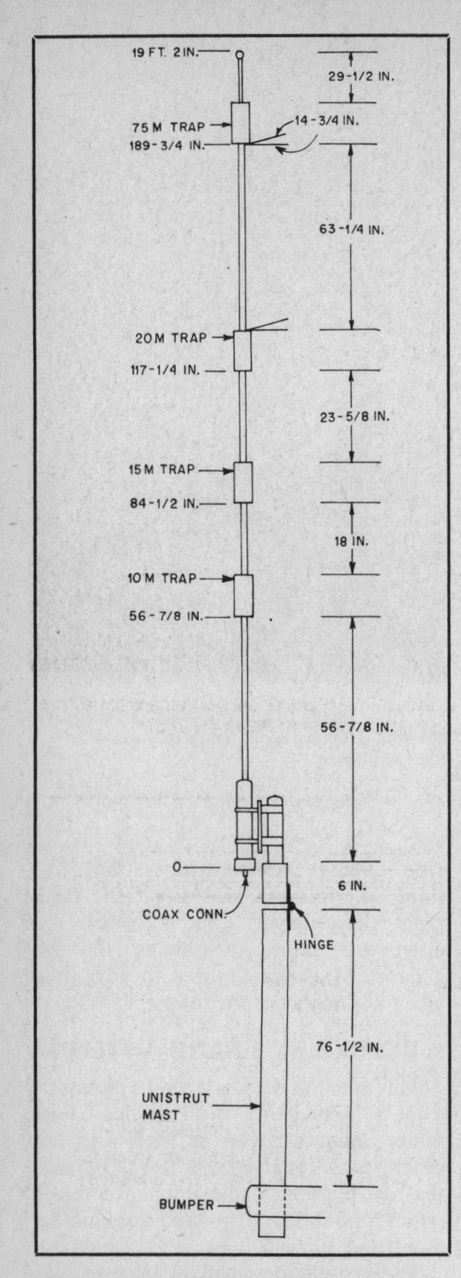


Fig. 16 — The dimensions of the modified Hustler 4BTV antenna. Refer to the text concerning the SWR bandwidth of the antenna on 75 meters.

of 3/8-in. diameter bolts. A 3/4-in. wrap-around strap was attached at the RV center trim line. Any brackets mounted higher detract from the neat appearance and will necessitate a complete change of dimensions for proper tuning. Install the antenna on the curb side of the vehicle to hide the lowered antenna behind the awning and provide greater safety to the person raising the antenna. This precaution is primarily for safety when stopping alongside a highway to meet a schedule. (Caution — beware of overhead power lines!)

### Mounting the Hustler

The 4BTV base is U bolted to a 19-in. piece of 1-5/16 in. OD galvanized steel pipe (1-in. water pipe). This is inserted into and welded along the edges of a short piece of Unistrut, which is attached to the lower portion of the mast by means of a heavy-duty, welded-on hinge. It is not feasible to bolt these pieces together, as the inside of the pipe must be completely clear to accept the end of a 54-in. piece of 1/2-in. water pipe (7/8 in. OD) that is used as a removable raising fixture and handle. This handle is wrapped with vinyl tape at the top end and also about 12 inches back to form a loose-fitting shim that provides a better fit inside the 1-in. water pipe. At 15 inches from the top end, a thicker wrap of tape acts as a stop to allow the handle to be inserted the same distance into the base support pipe in every instance. A short projecting bolt near the bottom end of the handle provides a means of lifting it off the lock pin (a bolt) which is mounted inside the Unistrut on an L-shaped bracket. See Figs. 17 and 18.

The top-hat spider rods should be installed only on one side of the antenna so as not to poke holes in the top of the trailer. No effect on antenna performance will be noted. Bring the coaxial cable into the trailer at a point close to the antenna. This is preferable to running it beneath the trailer, where it can be more easily damaged and where ground-loop paths for rf current may be created. Be sure to use drip loops at both the antenna and at the point of entry into the RV. Silicone rubber sealant should be used at the outside connector end and at the RV entry hole. Clear acrylic spray will provide corrosion protection for any hardware used. Lock washers and locknuts should be used on all bolts; good workmanship will result in first-class appearance and long, trouble-free service.

To ensure optimum results, great care should be taken to obtain a good ground return from the antenna all the way back to the transceiver. Clean, tight connections, together with heavy-duty tinned copper braid, should be used across the mast hinge, the mast-to-RV frame and to bond the frame to the equipment chassis. This is a *must* if the vertical quarter-wave antenna is to work properly.

### Antenna Pruning and Tuning

The antenna must be carefully tuned to resonance on each band starting with 10 meters. The most radical departure from the manufacturer's antenna dimensions (for home use) takes place with the 10-meter section. There, 30 in. of tubing is cut off. Only 3 in. need be removed from the 15-meter section. The 40-meter (top) section has to be lengthened, however, because of the radical shortening of the 10-meter section. The easiest way to do this, short of buying a longer piece of 1-1/4-in. OD aluminum tubing, is merely to lengthen one of the top-hat spider rods. A 15-1/2 in. length of 1/2-in. diameter aluminum tubing (with one end flattened

Fig. 17 — The handle and raising fixture, seated on the lock pin.

Fig. 18 — In this photo the open edges of the Unistrut channel are facing the viewer. The lock pin may be seen in the middle of the channel.

and properly drilled) can be held in place under the RM-75S resonator. It is a good idea to start with a longer piece of tubing and trim as necessary to obtain resonance at 7150 kHz.

The described method of installation, grounding and tuning of the antenna resulted in an SWR of 1.0:1 at resonance on the 75-, 40-, 20- and 15-meter bands. At the lower end of the 10-meter band, the SWR is 1.05:1. These low SWR values remain exactly the same regardless of whether or not the RV is grounded externally.

This system design also provides full band coverage on the 40- through 10-meter bands with an SWR of less than 2:1. Band coverage on 75 meters is limited to approximately 100 kHz because of the short overall length of the resonator coil/whip. The tip rod is adjustable to enable you to select your favorite 100-kHz band segment.

## A LADDER MAST

A temporary support is sometimes needed for an antenna system such as for antenna testing, site selection, emergency exercises or Field Day. Ordinary aluminum extension ladders are ideal candidates for this service. They are strong, light, extendable, weatherproof and easily transported. Additionally, they are readily available and can be returned to normal use once the project is concluded. A ladder tower will support a lightweight tri-band beam and rotator.

With patience and ingenuity one person can erect this assembly. One of the biggest problems is holding the base down while "walking" the ladder to a vertical position. Use 1/4 inch polypropylene rope to guy the tower; the rope guys are arranged in the standard fashion with three at each level. If you have help available, the ladder can be walked up in its retracted position and extended once the antenna and rotator are attached. The lightweight pulley system found on most extension ladders will not be strong enough to lift the ladder extension and survive. You will need to replace this with a heavy-duty pulley and rope. Make sure when attaching the guy ropes that they do not foul the operation of the sliding upper section of the ladder.

There is one hazard to avoid. Do not climb or stand on the ladder once it has been extended — even so much as one rung. If the locking mechanism should fail, you would likely lose a limb or two, if not your life. The risk simply isn't worth it. *Never* stand on the ladder and attempt to raise or lower the upper section. Do all the extending and retracting with the heavy-duty rope and pulley!

If you are going to raise the ladder by yourself, here are some pointers that may help. First of all, make sure that the rung latching mechanism is operating properly before you begin. The base must be hinged such that it does not slip along the ground during the erection activity. The guy ropes should be tied and positioned in such a way that they serve as safety constraints in the event that you lose control. You should have available a device (such as another ladder) for supporting the ladder during the rest periods. (See Fig. 19.) After the ladder is erect and the lower section guys tied and tightened, raise the upper portion one rung at a time. Do not raise the upper section higher than it is designed to go (you may pay a high price for the extra height).

Finding suitable guy anchors can be an exercise in creativity. Fence posts, trees and heavy pipes are all possibilities. If nothing of sufficient strength is available, you can drive anchor posts or pipes into

Fig. 19 — Walking the ladder up to its vertical position. Keith, VE2AQU, supports the mast with a second ladder while Chris, VE2FRJ, checks the ropes. (photo by Keith Baker, VE2XL)

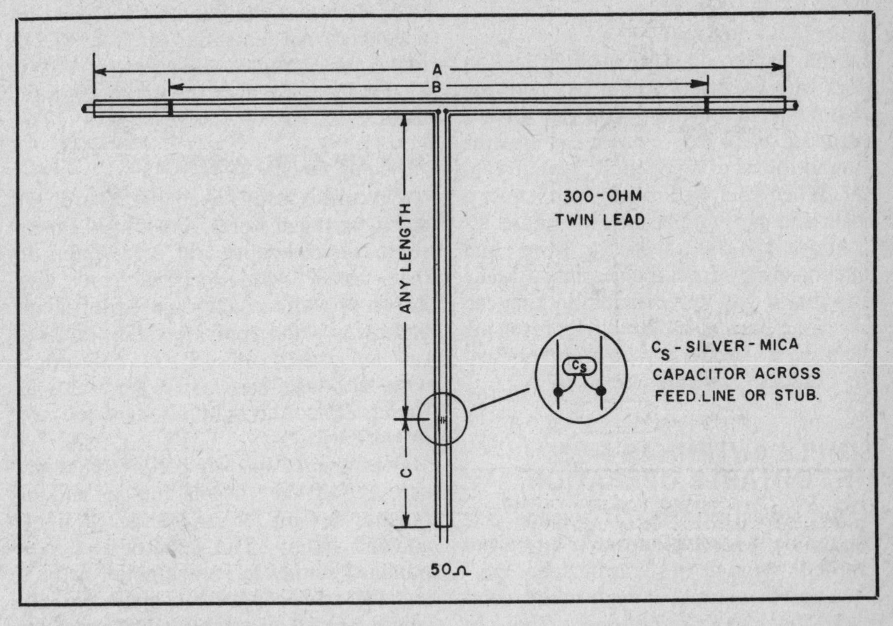


Fig. 20 — A twin-lead folded dipole makes an excellent portable antenna that is easily matched to 50-ohm stations. See text and Table 4 for details.

**Table 4**
**Twin-Lead Dipole Dimensions and Capacitor Values**

| *Frequency* | *Length A* | *Length B* | *Length C* | $C_s$ | *Stub Length* |
|---|---|---|---|---|---|
| 3.75 MHz | 124'9-1/2" | 104'11-1/2" | 13'0" | 289 pF | 37'4" |
| 7.15 | 65'5-1/2" | 55'1/2" | 6'10" | 151 pF | 19'7" |
| 10.125 MHz | 46'2-1/2" | 38'10-1/2" | 4'10" | 107 pF | 13'10" |
| 14.175 MHz | 33'0" | 27'9-1/4" | 3'5-1/2" | 76 pF | 9'10-1/2" |
| 18.118 MHz | 25'10" | 21'8-3/4" | 2'8-1/2" | 60 pF | 7'9" |
| 21.225 MHz | 22'-1/2" | 18'6-1/2" | 2'3-1/2" | 51 pF | 6'7" |
| 24.94 MHz | 18'9" | 15'9-1/2" | 1'11"-1/2" | 43 pF | 5'7-1/2" |
| 28.5 MHz | 16'5" | 13'9-3/4" | 1'8-1/2" | 38 pF | 4'11" |

meters = ft × 0.3048
mm = in. × 25.4

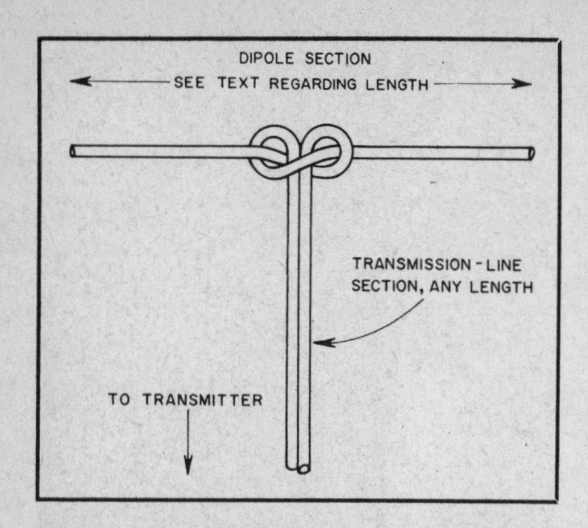


Fig. 21 — This electrician's knot, often used inside lamp bases and appliances in lieu of a plastic grip, can also serve to prevent the transmission-line section of a zip-cord antenna from unzipping itself under the tension of dipole suspension. To tie the knot, first use the right-hand conductor to form a loop, passing the wire behind the unseparated zip cord and off to the left. Then pass the left-hand wire of the pair behind the wire extending off to the left, in front of the unseparated pair, and thread it through the loop already formed. Adjust the knot for symmetry while pulling on the two dipole wires.

the soil. Sandy soil is the most difficult to work with because it does a very poor job of holding the anchor. You can drive a discarded back axle from a car into the ground to use as an anchor; it is substantial. When you are finished with it, use a chain and car bumper jack to remove it.

Above all else, keep the tower and antenna away from power lines. Make sure that if you lose control, nothing can touch the lines. Disassemble by reversing the process. Ladder towers are a good bet for "quickie" antenna supports.

## SIMPLE ANTENNAS FOR HF PORTABLE OPERATION

The typical portable hf antenna is a random-length wire flung over a tree and end-fed through a Transmatch. QRP Transmatches can be made quite compact, but each additional piece of necessary equipment makes portable operation less attractive. The station can be simplified by using resonant impedance-matched antennas for the bands of interest. Perhaps the simplest antenna of this type is the half-wave dipole, center-fed with 50- or 75-ohm coax. Unfortunately, RG-58, RG-59 or RG-8 cable is quite heavy and bulky for backpacking, and the miniature cables such as RG-174 are too lossy. A practical solution to the coax problem is to use folded dipoles made from lightweight TV twin lead. The characteristic impedance of this type of dipole is near 300 ohms, but it can be transformed to a 50-ohm source or load by means of a simple matching stub. Fig. 20 illustrates the construction method and important dimensions for the twin-lead dipole. Shorting connections must be made some distance inside the ends of the radiator, as shown in the drawing. (This subject is covered more fully in Chapter 2.) The ends may also be shorted.

A silver-mica capacitor is shown for the reactive element, but an open-end stub of twin lead can serve as well, provided it is dressed at right angles to the transmission line for some distance. The stub method has the advantage of easy adjustment of the system resonant frequency.

The dimensions and capacitor values for twin-lead dipoles for the hf bands are given in Table 4. To preserve the balance of the feeder, a 1:1 balun must be used at the end of the feed line. In most applications the balance is not critical, and the twin lead can be connected directly to a coaxial output jack, one lead to the center contact, and one lead to the shell. Because of the transmission-line effect of the shorted radiator sections, a folded dipole exhibits a wider bandwidth than a single-conductor type. The antennas described here are not as broad as a standard folded dipole because the impedance-transformation mechanism is frequency selective. However, the bandwidth should be adequate. An antenna cut for 14.175 MHz, for example, will present a VSWR of less than 2:1 over the entire 20-meter band.

### THE ZIP-CORD ANTENNA

Zip cord is readily available at hardware and department stores, and it's not expensive. The nickname, zip cord, refers to that parallel-wire electrical cord with brown or white insulation used for lamps and many small appliances. The conductors are usually no. 18 stranded copper wire, although larger sizes may also be found. Zip cord is light in weight and easy to work with.

For these reasons, zip cord is sometimes used as both the transmission line and the radiator section for an emergency dipole antenna system. The radiator section is obtained simply by "unzipping" or pulling the two conductors apart for the length needed to establish resonance for the operating frequency band. The initial dipole length can be determined from the equation $\ell = 468/f$, where $\ell$ is the length in feet and f is the frequency in megahertz. (It would be necessary to unzip only half the length found from the formula, since each of the two wires becomes half of the dipole.) The insulation left on the wire will have some loading effect, so a bit of length trimming may be needed for exact resonance at the desired frequency.

For installation, many amateurs like to use the electrician's knot shown in Fig. 21 at the dipole feed point to keep the transmission-line part of the system from unzipping itself under the tension of dipole suspension. This way, if zip cord of sufficient length for both the radiator and the feed line is obtained, a solder-free installation can be made right down to the input end of the line. Granny knots (or any other variety) can be used at the dipole ends with cotton cord to suspend the system. You end up with a lightweight, low-cost antenna system that can serve for portable or emergency use.

But just how efficient is a zip-cord antenna system? Since it is easy to locate the materials and simple to install, how about using such for a more permanent installation? Upon casual examination, zip cord looks about like 72-ohm balanced feed line. Does it work as well? Ask several amateurs these questions and you're likely to get answers ranging all the way from, "Yes, it's a very good antenna system!" to "Don't waste your time and money — it's not worth it!" Myths and hearsay seem to prevail, with little factual data.

#### Zip Cord as a Transmission Line

In order to determine the electrical characteristics of zip cord as a radio-frequency transmission line, a 100-foot roll was subjected to tests in the ARRL laboratory with an rf impedance bridge. Zip cord is properly called parallel power cord. The variety tested was manufactured for GC Electronics, Rockford, IL, being 18-gauge, brown, plastic-insulated type SPT-1, GC cat. no. 14-118-2G42. Undoubtedly, minor variations in the electrical characteristics will occur among similar cords from different manufacturers, but the results presented here are probably typical.

The characteristic impedance was determined to be 107 ohms at 10 MHz, dropping in value to 105 ohms at 15 MHz and to a slightly lower value at 29 MHz. The nominal value is 105 ohms at hf. The velocity factor of the line was determined to be 69.5 percent.

Who needs a 105-ohm line, especially to feed a dipole? A dipole in free space exhibits a feed-point resistance of 73 ohms, and at heights above ground of less than 1/4 wavelength the resistance is even lower. An 80-meter dipole at 35 feet, for example, will exhibit a feed-point resistance of about 40 ohms. Thus, for a resonant antenna, the SWR in the zip-cord transmission line can be 105/40 or 2.6:1, and maybe even higher in some installations. Depending on the type of transmitter in use, the rig may not like working into the load presented by the zip-cord antenna system.

But the really bad news is still to come — line loss! Fig. 22 is a plot of line attenuation in decibels per hundred feet of line versus frequency. Chart values are based on the assumption that the line is perfectly matched (sees a 105-ohm load as its terminating impedance).

In a feed line, losses up to about 1 decibel or so can be tolerated, because at the receiver a 1-dB difference in signal

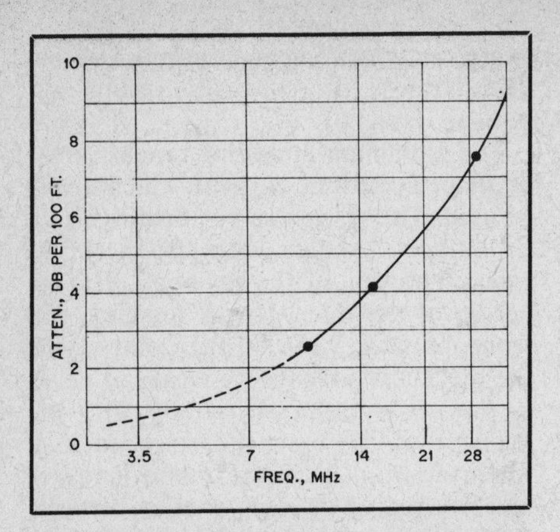


Fig. 22 — Attenuation of zip cord in decibels per hundred feet when used as a transmission line at radio frequencies. Measurements were made only at the three frequencies where plot points are shown, but the curve has been extrapolated to cover all high-frequency amateur bands.

strength is just barely detectable. But for losses above about 1 dB, beware. Remember that if the total losses are 3 dB, half of your power will be used just to heat the transmission line. Additional losses over those charted in Fig. 2 will occur when standing waves are present. See Chapter 3. The trouble is, you can't use a 50- or 75-ohm SWR instrument to measure the SWR in zip-cord line accurately.

Based on this information, we can see that a hundred feet or so of zip-cord transmission line on 80 meters might be acceptable, as might 50 feet on 40 meters. But for longer lengths and higher frequencies, the losses become appreciable.

### Zip Cord Wire as the Radiator

For years, amateurs have been using ordinary copper house wire as the radiator section of an antenna, erecting it without bothering to strip the plastic insulation. Other than the loading effects of the insulation mentioned earlier, no noticeable change in performance has been noted with the insulation present. And the insulation does offer a measure of protection against the weather. These same statements can be applied to single conductors of zip cord.

The situation in a radiating wire covered with insulation is not quite the same as in two parallel conductors, where there may be a leaky dielectric path between the two conductors. In the parallel line, it is this current leakage that contributes to line losses. The current flowing through the insulation on a single radiating wire is quite small by comparison, and so as a radiator the efficiency is high.

In short, communications can certainly be established with a zip-cord antenna in a pinch on 160, 80, 40, 30 and perhaps 20 meters. For higher frequencies, especially with long line lengths for the feeder, the efficiency of the system is so low that its value is questionable.

## VHF QUARTER-WAVELENGTH VERTICAL

Ideally, a vhf vertical antenna should be installed over a perfectly flat plane reflector to assure uniform omnidirectional radiation. This suggests that the center of the automobile roof is the best place to mount it for mobile use. Alternatively, the flat portion of the auto rear-trunk deck can be used, but will result in a directional pattern because of car-body obstruction. Fig. 23 illustrates at A and B how a Millen high-voltage connector can be used as a roof mount for a 144-MHz whip. The hole in the roof can be made over the dome light, thus providing accessibility through the upholstery. RG-59/U and the 1/4-wave matching section L, Fig. 23C, can be routed between the car roof and the ceiling upholstery and brought into the trunk compartment, or down to the dashboard of the car. Some operators install an SO-239-type coax connector on the roof for mounting the whip. The method is similar to that of Fig. 23.

It has been established that quarter-wavelength vertical antennas for mobile work through repeaters are not as effective as 5/8-wavelength verticals are. The 1/4-wavelength types cause considerably more "picket fencing" (rapid flutter) of the signal than is the case with the 5/8-wavelength type of antenna. The flutter that takes place when vertical polarization is used can be caused by vertical conductive objects being between the mobile antenna (near field) and the station being worked (or the repeater). As the vehicle moves past these objects there is a momentary blockage or partial blockage of the signal path. Another cause is alternate reinforcement and cancellation of signals arriving over two or more paths at various points along the route being traveled.

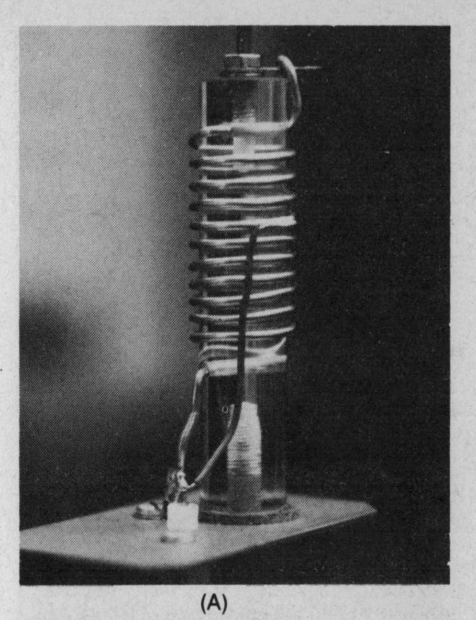

(A)

(B)

Fig. 24 — (A) A photograph of the 5/8-wavelength vertical base section. The matching coil is affixed to an aluminum bracket which screws onto the inner lip of the car trunk. (B) The completed assembly. The coil has been wrapped with vinyl electrical tape to prevent dirt and moisture from degrading performance.

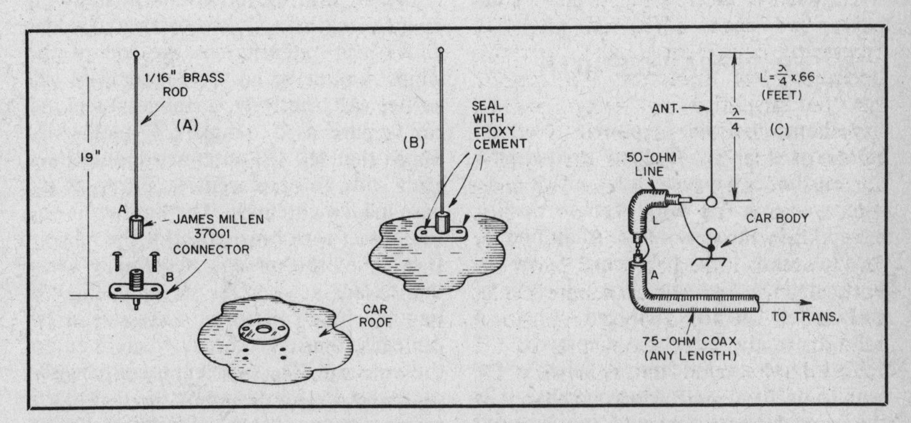


Fig. 23 — At A and B, an illustration of how a quarter-wavelength vertical antenna can be mounted on a car roof. The whip section should be soldered into the cap portion of the Millen connector, then screwed to the base socket. This handy arrangement permits removing the antenna when desired. Epoxy cement should be used at the two mounting screws to prevent moisture from entering the car. Diagram C is discussed in the text.

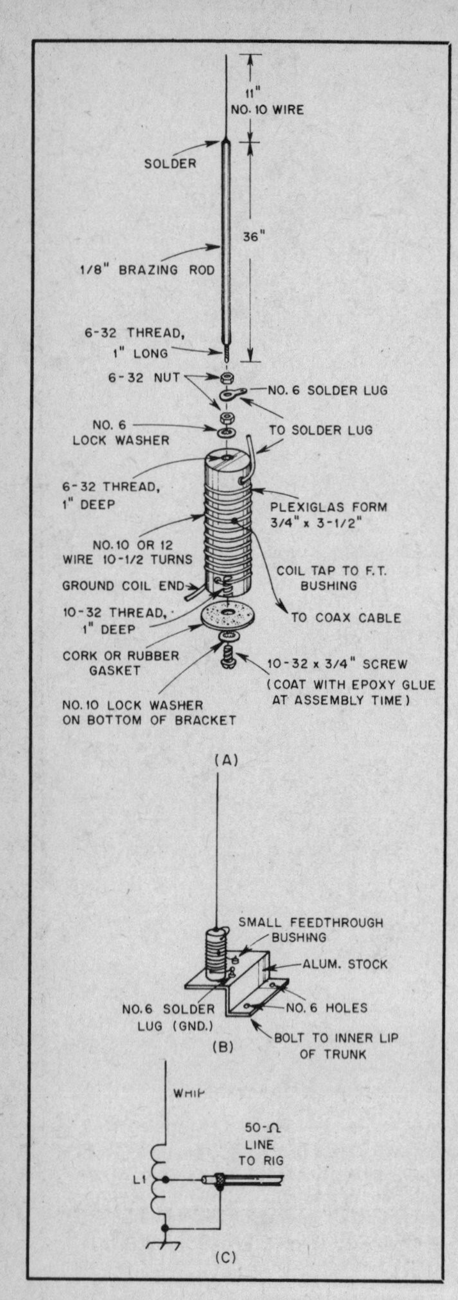


Fig. 25 — Structural details for the 2-meter 5/8-λ antenna are provided at A. The mounting bracket is shown at B and the equivalent circuit is given at C.

## 2-METER 5/8-WAVELENGTH VERTICAL

Perhaps the most popular vertical antenna for fm mobile and fixed-station use is the 5/8-wavelength vertical. As compared to a 1/4-wavelength vertical, it has some gain over a dipole. Additionally, the so-called "picket-fencing" type of flutter that results when the vehicle is in motion is greatly reduced when a 5/8-wavelength radiator is employed.

This style of antenna is suitable for mobile or fixed-station use because it is small, omnidirectional and can be used with radials or a solid-plane ground (such as is afforded by the car body). If radials are used, they need be only 1/4 wavelength or slightly longer.

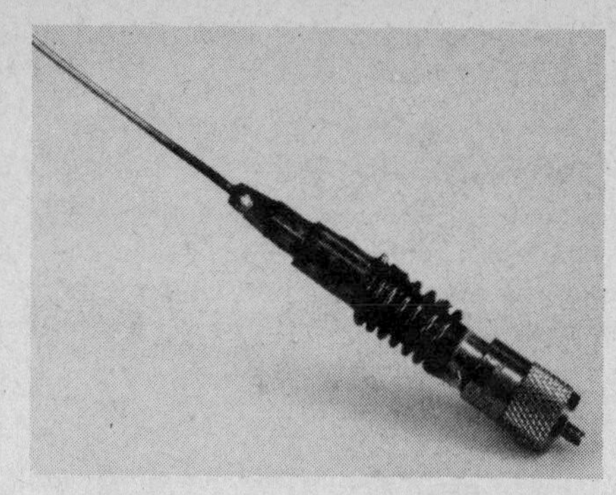

Fig. 26 — The 220-MHz 5/8-wavelength mobile antenna. The bottom end of the coil is soldered to the coaxial connector.

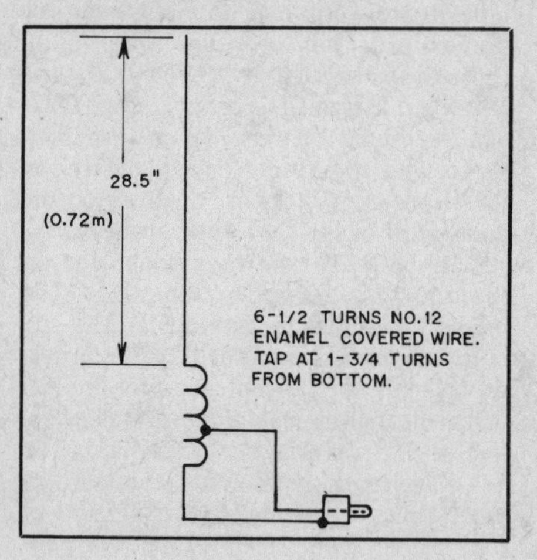


Fig. 27 — Diagram of the 220-MHz mobile antenna.

### Construction

The antenna shown here is made from low-cost materials. Fig. 24A and B shows the base coil and aluminum mounting plate. The coil form is a piece of low-loss solid rod, such as Plexiglas or phenolic. The dimensions for this and other parts of the antenna are given in Fig. 25. A length of brazing rod is used as the whip section.

The whip should be 47 inches long. However, brazing rod comes in standard 36-inch lengths, so it is necessary to solder an 11-inch extension to the top of the whip. A piece of no. 10 copper wire will suffice. Alternatively, a stainless-steel rod can be purchased to make a 47-inch whip. Shops that sell CB antennas should have such rods for replacement purposes on base-loaded antennas. The limitation one can expect with brazing rod is the relative fragility of the material, especially when the threads are cut for screwing the rod into the base-coil form. Excessive stress can cause the rod to break where it enters the coil form. The problem is complicated somewhat in this design by the fact that a spring is not used at the antenna mounting point. Innovators can find all manner of solutions to the problems just outlined by changing the physical design and using different materials when constructing the overall antenna. The main purpose of this description is to provide dimensions and tune-up data.

The aluminum mounting bracket must be shaped to fit the car with which it will be used. The bracket can be used to effect a "no-holes" mount with respect to the exterior portion of the car body. The inner lip of the vehicle trunk (or hood for front mounting) can be the point where the bracket is attached by means of no. 6 or no. 8 sheet-metal screws. The remainder of the bracket is bent so that when the trunk lid or car hood is raised and lowered, there is no contact between the bracket and the moving part. Details of the mounting unit are seen in Fig. 25 at B. A 14-gauge metal thickness (or greater) is recommended for best rigidity.

There are 10-1/2 turns of no. 10 or no. 12 copper wire wound on the 3/4-inch diameter coil form. The tap on L1 is placed approximately four turns below the whip end. A secure solder joint is imperative.

### Tune-Up

After the antenna has been affixed to the vehicle, insert an SWR indicator in the 50-Ω transmission line. Turn on the 2-meter transmitter and experiment with the coil-tap placement. If the whip section is 47 inches long, an SWR of 1:1 can be obtained when the tap is at the right location. As an alternative to the foregoing method of adjustment, place the tap at four turns from the top of L1, make the whip 50 inches long and trim the whip length until an SWR of 1:1 is secured. Keep the antenna well separated from other objects during tune-up, as they may detune the antenna and yield false adjustments for a match.

## A 5/8-WAVELENGTH 220-MHz MOBILE ANTENNA

This antenna, shown in Figs. 26 and 27, was developed to fill the gap between a homemade 1/4-wavelength mobile antenna and a commercially made 5/8-wavelength model. There have been other antennas made using modified CB models. This still presents the problem of cost in acquiring the original antenna. The major cost in this setup is the whip portion. This can be any tempered rod that will spring easily.

### Construction

The base insulator portion is constructed of 1/2-inch Plexiglas rod. A few minutes' work on a lathe was sufficient to shape and drill the rod. The bottom 1/2-inch of the rod is turned down to a diameter of 3/8 inch. This portion will now fit into a PL-259 UHF connector. A hole, 1/8-inch diameter, is drilled through the center of the rod. This hole will contain the wires that make the connections between the center conductor of the connector and the coil tap. The connection between the whip and the top of the coil is also run through this opening. A stud is force-fitted into the top of the Plexiglas rod. This allows the whip to be detached from the insulator portion.

The coil should be initially wound on a form slightly smaller than the base insulator. When the coil is transferred to the Plexiglas rod it will keep its shape and will not readily move. After the tap point has been determined, a longitudinal hole is drilled into the center of the rod. A no. 22 wire can then be inserted through the center of the insulator into the connector. This method is also used to attach the whip to the top of the coil. After the whip has been fully assembled, a coating of epoxy cement is applied. It seals the entire assembly and provides some additional strength. During a full winter's use there was not any sign of cracking or mechanical failure. The adjustment procedure is the same as for the 2-meter antenna just described.

## A PORTABLE QUAD FOR 2 METERS

Figs. 28 and 29 show a portable quad for home construction. Both driven and reflector elements fold back on top of each other, resulting in a package about 17 inches long. The wire loop elements may be held in place around the boom with an elastic band. To support the antenna once it has been erected, the container is used as a stand. To provide more stability, four small removable struts slip into holes in the base of the container. Both the support rods and struts fit inside the container when the antenna is disassembled. Figs. 30 and 31 show a method of attaching the spreaders to the boom. A mechanical stop is machined into the hub, and elastic bands are used to hold the spacers erect. The bands are attached to an additional strut to hold the spacers open. When not in use, the strut pulls out and sits across the hub, and the spacers can be folded back. Details are shown in Fig. 31.

### Building Materials

The portable quad antenna may be fabricated from any one of several plastic or wood materials. The most inexpensive method is to use wood doweling, available at most hardware stores. Wood is inexpensive and easily worked with hand tools; 1/4-inch doweling may be used for the quad spacers, and 3/8- or 1/2-inch doweling may be used for the boom and support elements. A hardwood is recommended for the hub assembly, since a softwood may tend to crack along its grain if the hub is impacted or dropped. Plastics will also work well, but the cost will rise sharply if the material is purchased from a supplier. Plexiglas is an excellent choice for the hub. Fiberglass or phenolic rods are also excellent for the quad elements and support.

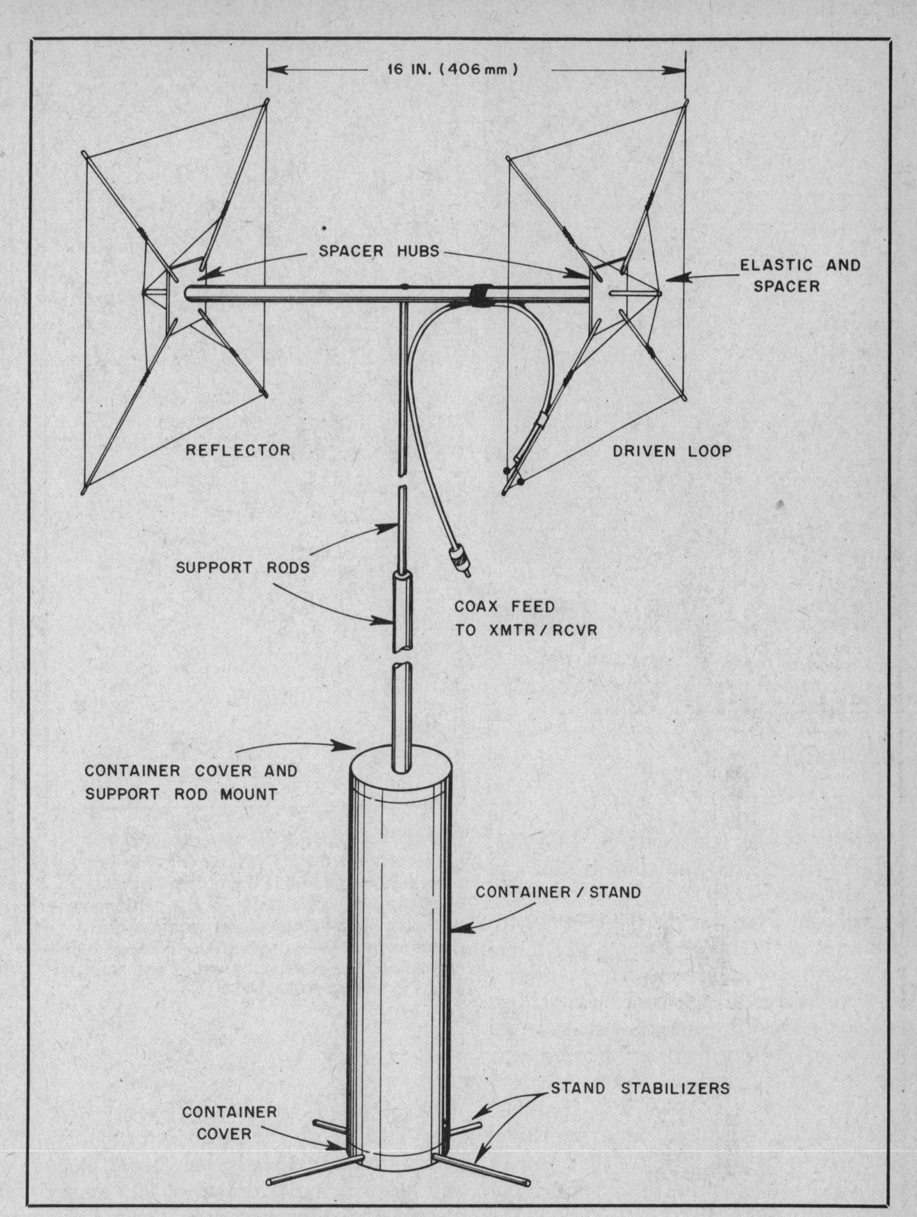


Fig. 28 — The basic portable quad assembly. An element spacing of 16 in. is used so that the quad spacers will fold neatly between the hubs.

Fig. 29 — The portable quad in stow configuration. Two long dowels are used as support rods. Four smaller dowels are used to stabilize the container.

Fig. 30 — This version of the portable quad uses mechanical stops machined into the hub; elastic bands hold the spacers open.

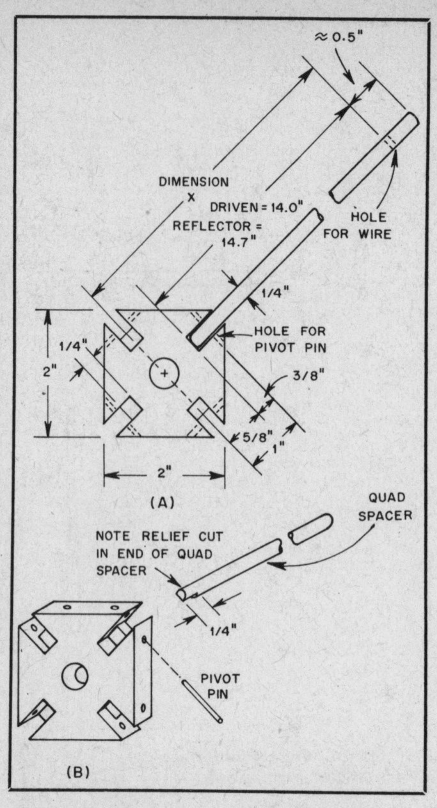


Fig. 31 — A detail of the spacer hub with spacer lengths for the director and reflector is shown at A. The hub is made from 1/4-inch plastic or hardwood material. The center-hole diameter can be whatever is necessary to match the diameter of your boom. The version of the hub with mechanical stops and elastic bands is shown at B.

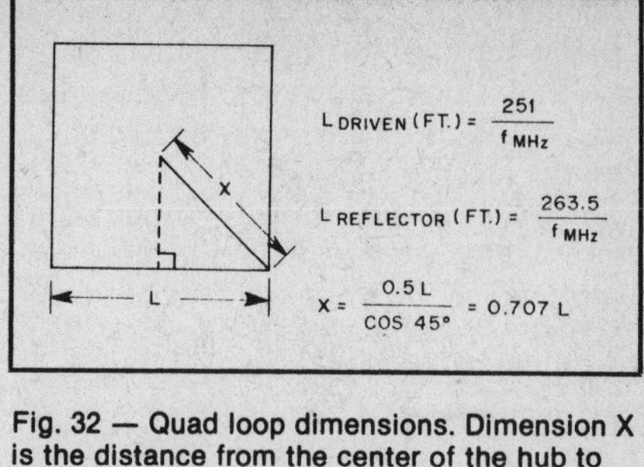


Fig. 32 — Quad loop dimensions. Dimension X is the distance from the center of the hub to the hole drilled in each spacer for the loop wire. At 146 MHz, dimension X for the driven element is 1.216 feet (14.6 inches), and dimension X for the reflector is 1.276 feet (15.3 inches).

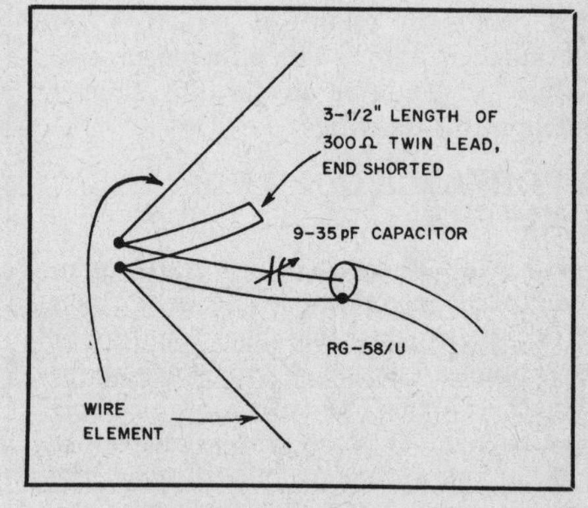


Fig. 33 — Matching system for the portable quad. The stub may be taped to the element.

The loops were made with no. 18 AWG copper wire. If no insulation is used on the wire and wood doweling is used for the spacers, a coat of spar varnish in and around the spacer hole through which the wire runs is recommended. The loop wire is terminated at one element by attaching to heavy-gauge copper-wire posts inserted into tightly fitting holes in the element. For the driven element, three posts are used to allow the RG-58/U feed-line braid, center conductor and matching capacitor to be attached. A single post is used on the reflector to complete the loop circuitry.

Fig. 32 shows how to calculate quad loop dimensions. The boom is 16 inches long. The feed-point matching system is detailed in Fig. 33. The matching system uses a 3.5-inch length of 300-Ω twin-lead as a shorted stub. Adjustment of the match is made at the 9- to 35-pF variable capacitor that is connected in series with the coaxial feed line.

The storage container was made from a heavy cardboard tube originally used to store roll paper. Any rigid cylindrical housing of the proper dimensions may be used. Two wood end pieces were fabricated to cap the cardboard cylinder. The bottom end piece is cemented in place and has four holes drilled at 90° angles around the circumference. These holes hold 4-inch struts, which provide additional support when the antenna is erected. The top end piece is snug fitting and removable. It is of sufficient thickness (about 5/8 inch) to provide sufficient support for the antenna-supporting elements. A mounting hole for the supporting elements is drilled in the center of the top end piece. This hole is drilled only about three-quarters of the way through the end piece and should provide a snug fit for the antenna support. One or more antenna support elements may be used, depending on the height the builder wishes to have. Keep in mind, however, that the structure will be more prone to blow over the higher above the ground it gets! Doweling and snug-fitting holes are used to mate the support elements and the antenna boom.

## BIBLIOGRAPHY

Source material and more extended discussion of topics covered in this chapter can be found in the references given below.

Belcher, D. K., "RF Matching Techniques, Design and Example," *QST*, October 1972.

Belrose, J. S. "Short Antennas for Mobile Operation," *QST*, September 1953.

Stephens, R. E., "Admittance Matching the Ground-Plane Antenna to Coaxial Transmission Line," Technical Correspondence, *QST*, April 1973.

# Chapter 14

# Direction-Finding Antennas

The use of radio for direction-finding purposes (RDF) is almost as old as its application for communications. Radio amateurs have learned RDF techniques and found much satisfaction by participating in hidden transmitter (fox) hunts. Other hams have discovered RDF through an interest in boating or aviation where radio direction finding is used for navigation and emergency location systems. There are less pleasant applications as well, such as tracking down noise sources or illegal operation from unidentified stations. The ability of certain RDF antennas to reject signals from selected directions has also been used to advantage in reducing noise and interference.

When transmitter hunting is done for fun or sport, there are many variations to the rules. In the USA the hunt normally takes place from an automobile. The pursuit takes place away from main roads and population centers. European amateurs emphasize the sport aspect of foxhunting. There the chase finds the participant on foot. A portable RDF receiver, map and compass are used to locate a number of foxes located in the hunt area. Competition is keen, with regular national and European regional events scheduled each year.

## THE TECHNIQUE OF TRIANGULATION

It is impossible, using amateur techniques, to pinpoint the whereabouts of a transmitter from a single receiving location. With a directional antenna you can determine the direction of a signal source, but not how far away it is. To find the distance you can then travel in the direction determined above until you discover the transmitter location. That technique does not normally work very well. It is far

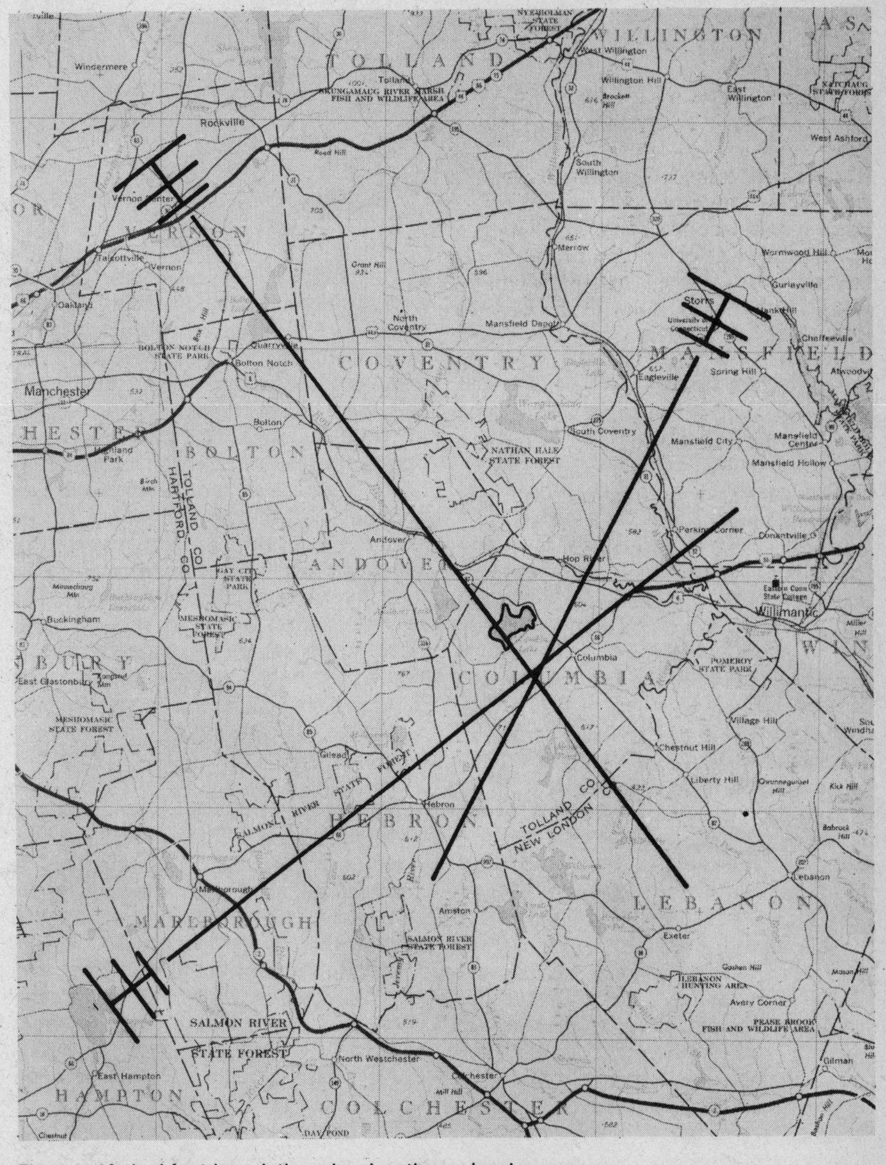


Fig. 1 — Method for triangulation when locating a signal source.

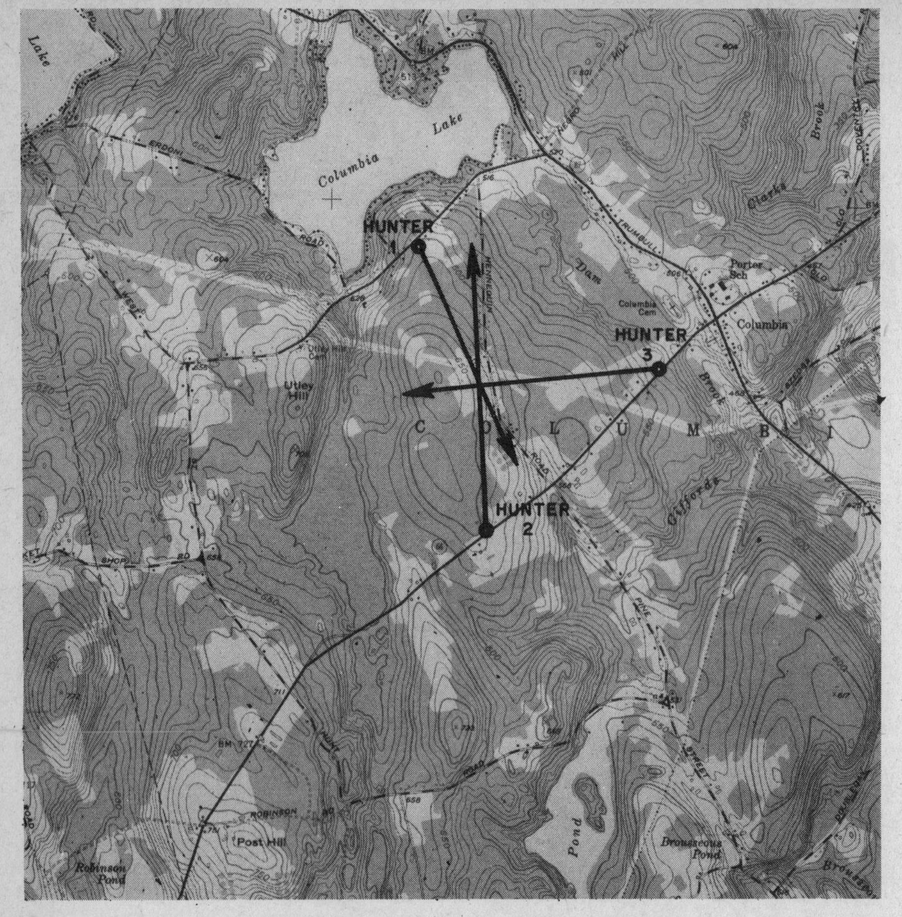


Fig. 2 — Portable or mobile equipment is used to pinpoint the exact location.

better to take at least one additional direction measurement from another receiving location.

Consider a hypothetical case of malicious interference. Let's assume that an interference committee has been formed and that the members are equipped with receivers and directional, rotatable antennas for the hf band in question — 10, 15 or 20 meters. (We'll deal with the lower frequencies later.)

If the source of the malicious interference is within the ground-wave contour of the interference-hunters' stations, and if the "hunters" are a few miles apart, triangulation can be used to determine the general location of the offender. This is illustrated pictorially in Fig. 1. Stations 1, 2 and 3 rotate their beam antennas to obtain maximum response from the interfering signal. The beam headings are noted, compared, and plotted on an area map. At the point where the three headings converge is the approximate location of the interfering transmitter.

This kind of exercise is rather impractical when dealing with sky-wave signals. Fading complicates things by preventing accuracy in obtaining a peak signal reading on the receiver S meter. Furthermore, the beamwidth of an hf-band beam is too broad to yield good resolution at great distances. So, what is our next step in pinpointing the exact geographical point from which the offending signal is being sent? We can comb the area at close range with hand-held loop antennas. This will enable us to find the exact neighborhood of the offender and finally arrive at his exact location. This can be done with relative ease by placing an hf-band transceiver or receiver in an automobile, bringing the vehicle to the general area of the interfering signal, then standing outside the car and rotating the loop for maximum signal response. The scheme requires two or three mobile stations, which can be situated as shown in Fig. 2, but only a mile or less from the target. A pocket compass is necessary to determine the headings for correct triangulation.

To avoid ambiguity from the bidirectional response of loop antennas we can build our direction finders to yield a cardioid (heart-shaped) response. This is done by adding a sense antenna to the loop, along with a phasing network. We'll see how that's done later.

The problem of locating transmitters on 40, 75 and 80 meters is a bit more complicated. This is because beam antennas are somewhat impractical that low in frequency, even though some of the super stations have antennas of that type.

The practical answer to direction finding at these lower frequencies lies in the use of an Adcock antenna for effective searching from one's home. Alternatively, a small loop with sense antenna can be built for the frequency of interest. It should be used out of doors so that house wiring and plumbing will not distort the directive pattern. This may require placing a receiver on a table in the backyard and running an extension cord to it for power. Once a triangulation has been effected, it will be time for the committee to move afield with the mobile units for close-in hunting.

## Loop Circuits and Criteria

No single word describes a loop of high performance better than "symmetry." For us to obtain an undistorted response pattern from this type of antenna, we must build it in the most symmetrical manner possible. The next key word is "balance." The better the electrical balance, the deeper the loop null and the sharper the maxima.

The physical size of the loop for 7 MHz and below is not of major consequence: A 4-foot loop will exhibit the same electrical characteristics as one which is only 1 inch in diameter. The smaller the loop, however, the lower its efficiency. This is because its aperture samples a smaller section of the wave front. Thus, if we use very small loops, we will need to employ preamplifiers to compensate for the reduced efficiency.

## Loop Choices

The earliest loop antennas were of the "frame antenna" variety. These were unshielded antennas which were built on a wooden frame in a rectangular format. The loop conductor could be a single turn of wire (on the larger units) or several turns if the frame was small. Later, shielded versions of the frame antenna were popularized to provide electrostatic shielding — an aid to noise reduction from such sources as precipitation static.

Some years after wire loops gained widespread application, we began using magnetic-core loop antennas. The advantage was reduced size, and this appealed to the designers of aircraft and portable radios. Most of these antennas contain ferrite bars or cylinders, which provide high inductance and Q with a small number of coil turns. A 7-inch rod, 0.5-inch OD, of Q2 ferrite ($\mu_1 = 125$) is suitable for a loop core from the bc band through 10 MHz. Maximum response is off the *broadside* of a rod loop, whereas the maximum occurs in the *plane* of a frame type of loop. The performance of the two antennas is otherwise similar.

## Frame Loops

Fig. 3 illustrates the details of a practical frame type of loop antenna. The circuit at A is a 5-turn system which is tuned to resonance by C1. If the layout is symmetrical, we should be able to obtain good balance. L2 helps to achieve our objective by eliminating the need for direct coupling to the feed terminals of L1. If the loop feed was attached in parallel with C1, which is common practice, the chance for imbalance would be considerable.

L2 can be situated just inside or slightly outside of L1; a 1-inch separation works nicely. The receiver or preamplifier can be connected to terminals A and B of L2 as shown at B of Fig. 3. C2 controls the amount of coupling between the loop and the preamplifier. The lighter the coupling, the higher the loop Q, the narrower the frequency response, and the greater the gain requirement from the preamplifier. It should be noted that we are making no attempt to match the loop impedance to the preamplifier: The characteristic impedance of small loops is very low — on the order of 1 ohm or less.

A supporting frame for the loop of Fig. 3 can be structured as shown in Fig. 4. The dimensions given are for a 1.8-MHz frame antenna. For use on 75 or 40 meters, L1 of Fig. 3A will require fewer turns, or the size of the wooden frame will have to be made somewhat smaller than that of Fig. 4.

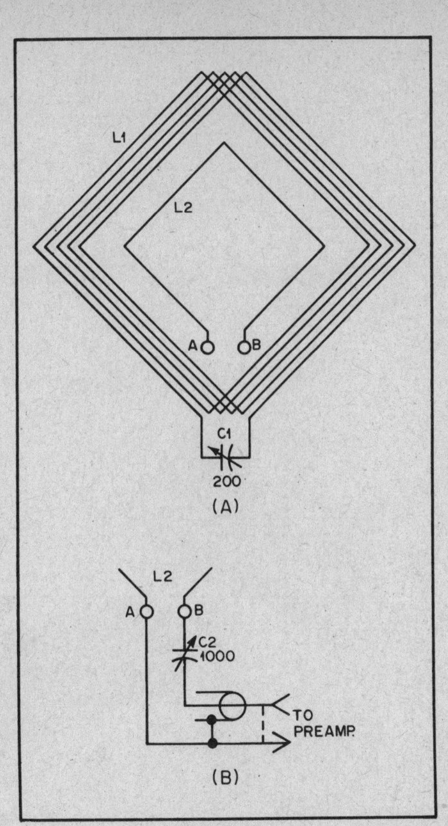


Fig. 3 — A multiturn frame antenna is shown at A. L2 is the coupling loop. The drawing at B shows how L2 is connected to a preamplifier.

Fig. 5 — An assembled table-top model of the W1FB 4T-ES electrostatically shielded loop which is made from a length of RG-58/U cable.

### Shielded Frame Loops

If electrostatic shielding is desired we can adopt the format shown in Fig. 5. In this example, the loop conductor is made from RG-58/U coaxial cable with ample turns to permit resonance at the operating frequency. A 3-turn link connected to the loop feed terminals can be probed with a dip-meter to check the resonant frequency (tuning capacitor C1 of Fig. 3A must be connected and set at midrange for this test).

Larger single-turn loops of this kind can be fashioned from aluminum-jacketed Hardline, if that style of coax is available. In either case, the shield conductor must be opened at the electrical center of the loop, as shown in Fig. 6 at A and B. The design example is based on 1.8-MHz operation.

In order to realize the best performance from an electrostatically shielded loop antenna, we must operate it near to and directly above an effective ground plane. An automobile roof (metal) qualifies nicely for small shielded loops. For fixed-station use, a chicken-wire ground screen can be placed below the antenna at a suggested distance of 1 to 6 feet.

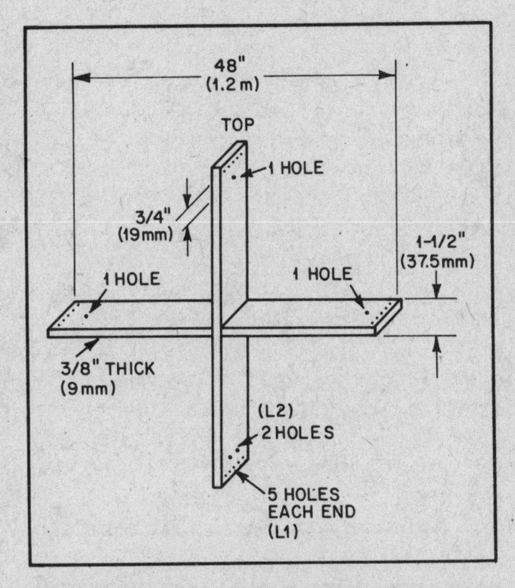


Fig. 4 — A wooden frame of this type can be used to contain the wire of the loop shown in Fig. 3.

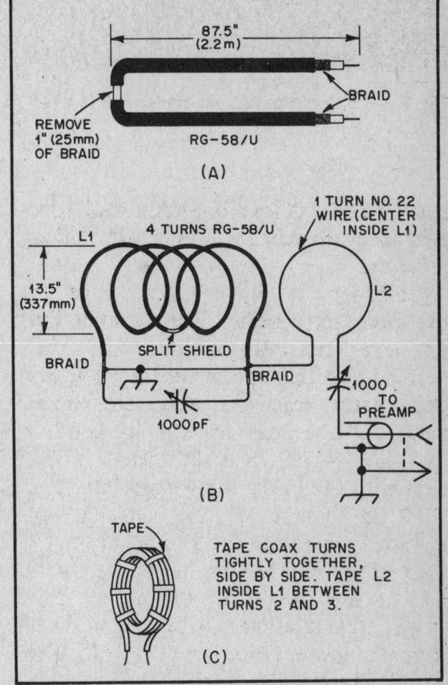


Fig. 6 — Components and assembly details of the 4T-ES shielded loop.

### Ferrite-Core Loops

Fig. 7 contains a diagram for a rod loop. The winding (L1) has the appropriate number of turns to permit resonance with C1 at the operating frequency. L1 should be spread over approximately 1/3 of the core center. Litz wire will yield the best Q, but Formvar magnet wire can be used if desired. A layer of 3M Company glass tape (or Mylar tape) is recommended as a covering for the core before adding the wire. Masking tape can be used if nothing else is available.

L2 functions as a coupling link over the exact center of L1. C1 is a dual-section variable, although a differential capacitor might be better toward obtaining optimum balance (not tried). The loop Q is controlled by means of C2, which is a mica compression trimmer.

Electrostatic shielding of rod loops can be effected by centering the rod in a U-shaped aluminum, brass or copper channel which extends slightly beyond the ends of the rod loop (1 inch is suitable). The open side (top) of the channel can't be closed, as that would constitute a shorted-turn condition and render the antenna useless. This can be proved by shorting across the center of the channel with a screwdriver blade when the loop is

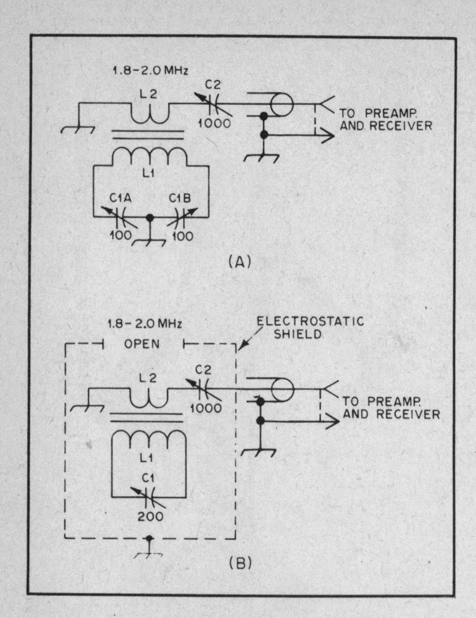


Fig. 7 — Diagram of a ferrite loop (A). C1 is a dual-section air variable. The circuit at B shows a rod loop contained in an electrostatic shield channel (see text). A low-noise preamplifier is shown in Fig. 12.

Fig. 8 — The assembly at the top of the picture is a shielded ferrite-rod loop for 160 meters. Two rods have been glued end to end (see text). The other units in the picture are a low-pass filter, broadband preamplifier (lower center) and a Tektronix step attenuator (lower right). These were part of the test setup used when this type of antenna was evaluated.

tuned to an incoming signal. The shield-braid gap in the coaxial loop of Fig. 6 is maintained for the same reason.

A photograph of a shielded rod loop is offered in Fig. 8. It was developed experimentally for 160 meters and uses two 7-inch ferrite rods which were glued end-to-end with epoxy cement. The longer core resulted in improved sensitivity during weak-signal reception. The other items in the photograph were used during the evaluation tests and are not pertinent to this discussion. All of the loops discussed thus far have bidirectional responses, as shown in Fig. 9.

**Obtaining a Cardioid Pattern**

Although the bidirectional pattern of loop antennas can be used effectively in tracking down signal sources by means of triangulation, an essentially unidirectional loop response will help to reduce the time spent when on a "hunting" trip. Adding a sense antenna to our loop is simple to do, and it will provide the desired cardioid response we need. The theoretical pattern for this combination is shown in Fig. 10.

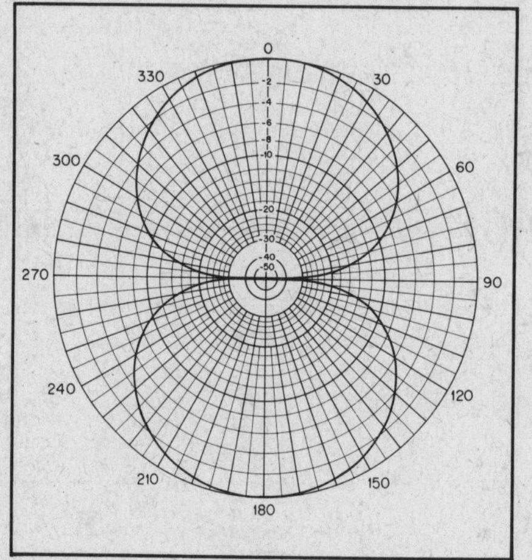


Fig. 9 — Directive pattern for a small loop antenna. The sharp nulls are used for direction finding.

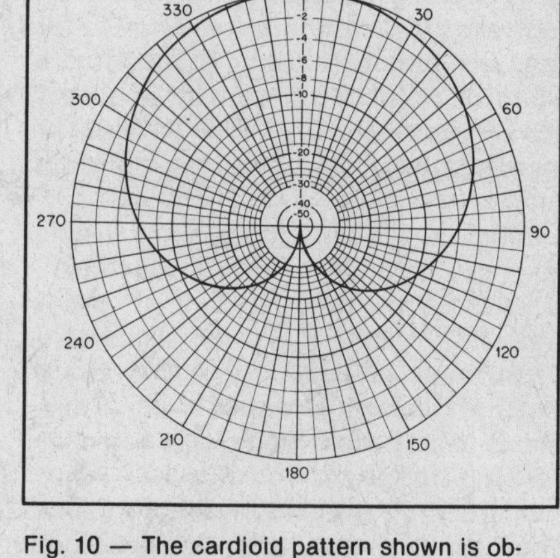


Fig. 10 — The cardioid pattern shown is obtained by combining the loop pattern of Fig. 9 with the omnidirectional pattern of the sense antenna. See text and Fig. 11.

Fig. 11 shows how this can be accomplished. The link from the rod loop or frame loop is connected via coaxial cable to the primary of T1, which is a tuned toroidal transformer with a split secondary winding. C3 is adjusted for peak signal response at the frequency of interest (as is C4), then R1 is adjusted for minimum back response of the loop. It will be necessary to readjust C3 and R1 several times to compensate for the interaction of these controls. The adjustments are repeated until no further null depth can be obtained. Tests at ARRL Hq. showed that null depths as great as 40 dB could be obtained with the circuit of Fig. 11 on 75 meters. A near-field weak-signal source was used during the tests. The greater the null depth the lower the signal output from the system, so plan to include a preamplifier with 25 to 40 dB of gain. Q1 of Fig. 11 will deliver approximately 15 dB of gain. The circuit of Fig. 12 can be used following T2 to obtain an additional 24 dB of gain. In the interest of maintaining a good noise figure, even at 1.8 MHz, Q1 should be a low-noise device. A Siliconix U310 JFET would be ideal in this circuit, but a 2N4416, an MPF102, or a 40673 MOSFET would also be satisfactory. The sense antenna can be mounted 6 to 15 inches from the loop. The vertical whip need not be more than 12 to 20 inches long. Some experimenting may be necessary in order to secure the best results. It will depend also on the operating frequency of the antenna.

## THE ADCOCK ANTENNA

Loops are adequate in applications where only the ground wave is present.

The performance of an RDF system for sky-wave reception can be improved by the use of an Adcock antenna. A basic version is shown in Fig. 13.

The response of the Adcock array to vertically polarized waves is similar to a conventional loop, and the directive pattern is essentially the same (see Fig. 9). Response of the array to a horizontally polarized wave is considerably different from that of a loop, however. The currents induced in the horizontal members tend to balance out regardless of the orientation of the antenna. This effect was verified in practice when good nulls were obtained with an experimental model under sky-wave conditions. The same circumstances produced poor nulls with small loops (both conventional and ferrite-loop models). Generally speaking, the Adcock antenna seems to have attractive properties for amateur RDF applications. Unfortunately, its portability leaves something to be desired, making it more suitable to fixed or semi-portable applications. While a metal support for the mast and boom could be used, wood, PCV or fiberglass are preferable because they are nonconductors and would hence cause less pattern distortion.

Since the array is balanced, a coupler is required to match the unbalanced input of a typical receiver. Fig. 14 shows how to modify a link-coupled network such as is found in Chapter 4. C2 and C3 are null-clearing capacitors. A low-power signal source is placed some distance from the Adcock antenna and broadside to it. C2 and C3 are then adjusted until the deepest null is obtained. The coupler can be placed below the wiring-harness junction on the boom. Connection can be made by means of a short length of 300-ohm twin-lead.

## A 2-METER ANTENNA FOR DIRECTION FINDING

A simple antenna design that yields a cardioid pattern is depicted in Fig. 15. Two quarter-wavelength vertical elements are spaced one quarter-wavelength apart and are fed 90 degrees out of phase. Each radiator is shown with two radials approximately 5 percent shorter than the radiators.

Construction is simple and straightforward. Fig. 15B shows a female BNC connector (Radio Shack 278-105) that has been mounted to a small piece of pc-board material. The BNC connector is held "upside down" and the vertical radiator is soldered to the center solder lug. A 12 in. piece of brass tubing provides a snug fit over the solder lug. A second piece of tubing, slightly smaller in diameter, is telescoped inside the first. The outer tubing is crimped slightly at the top after the inner tubing is installed. This provides positive contact between the two tubes. For 146 MHz the length of the radiators calculates to about 19 in. You should be

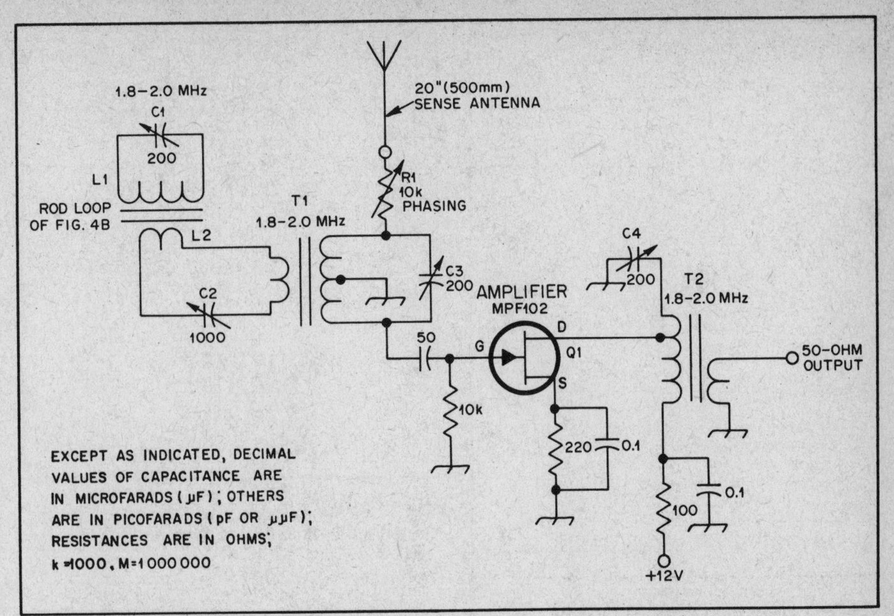


Fig. 11 — Schematic diagram of a rod-loop antenna with a cardioid response. The sense antenna, phasing network and a preamplifier are shown also. The secondary of T1 and the primary of T2 are tuned to resonance at the operating frequency of the loop. T68-2 or T68-6 Amidon toroid cores are suitable for both transformers. Amidon also sells the ferrite rods for this type of antenna.

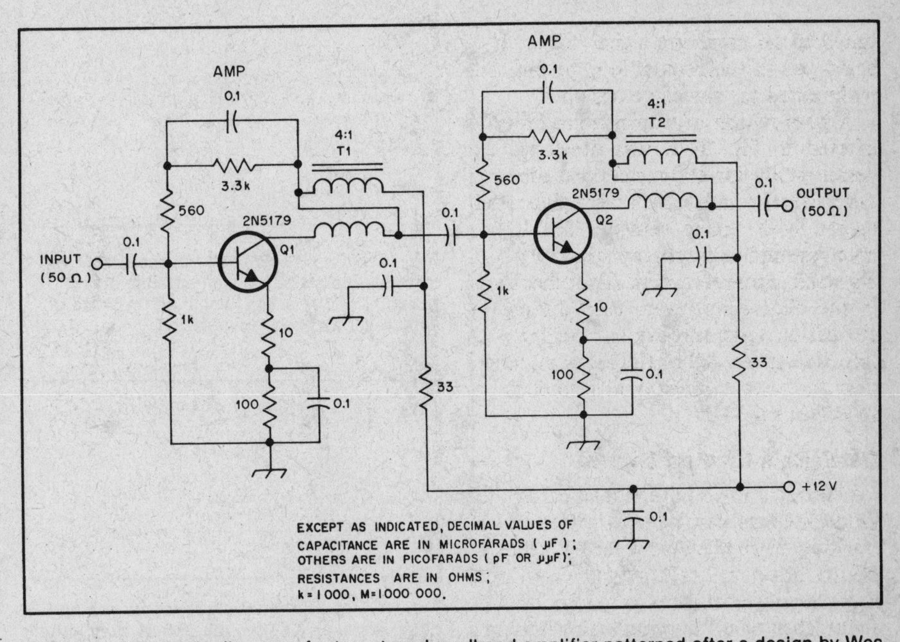


Fig. 12 — Schematic diagram of a two-stage broadband amplifier patterned after a design by Wes Hayward, W7ZOI. T1 and T2 have a 4:1 impedance ratio and are wound on FT-50-61 toroid cores (Amidon) which have a $\mu_i$ of 125. They contain 12 turns of no. 24 enam., bifilar wound. The capacitors are disc ceramic. This amplifier should be built on double-sided circuit board for best stability.

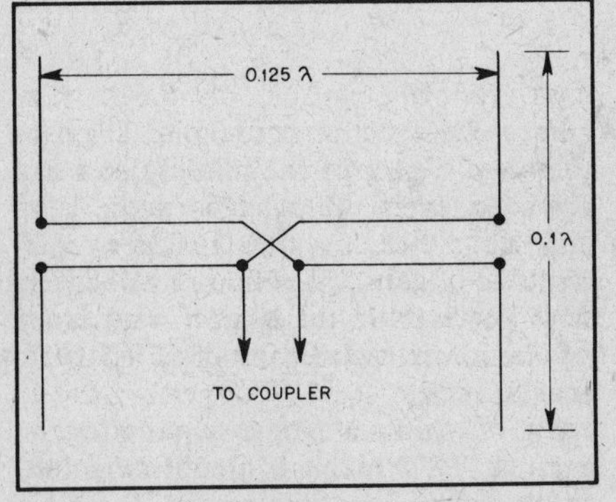


Fig. 13 — A simple Adcock antenna.

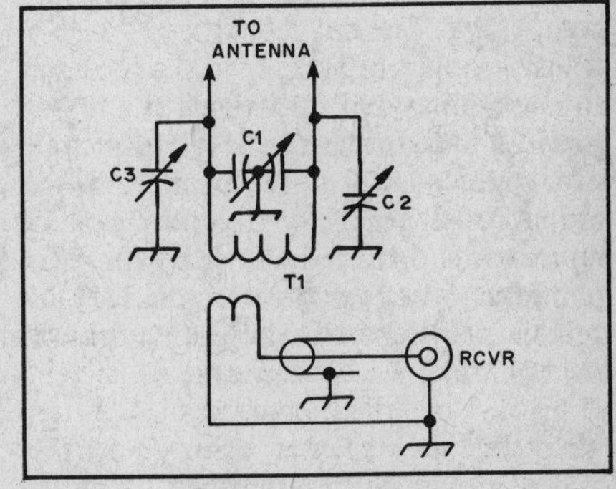


Fig. 14 — A coupler for use with the Adcock antenna.

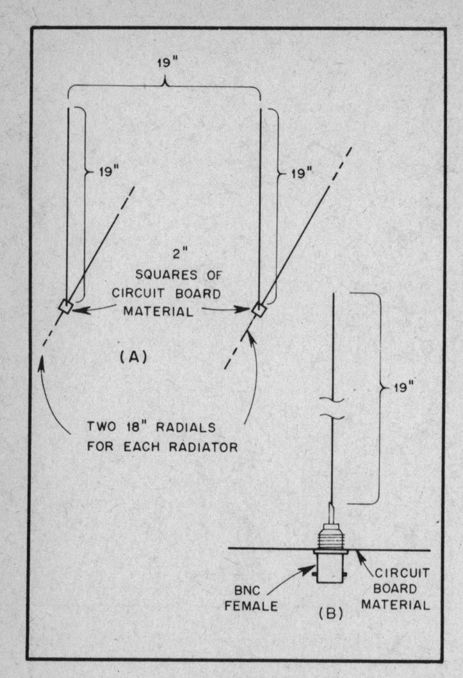


Fig. 15 — At A is a simple configuration that can produce a cardioid pattern. At B is a convenient way of fabricating a sturdy mount for the radiator using BNC connectors.

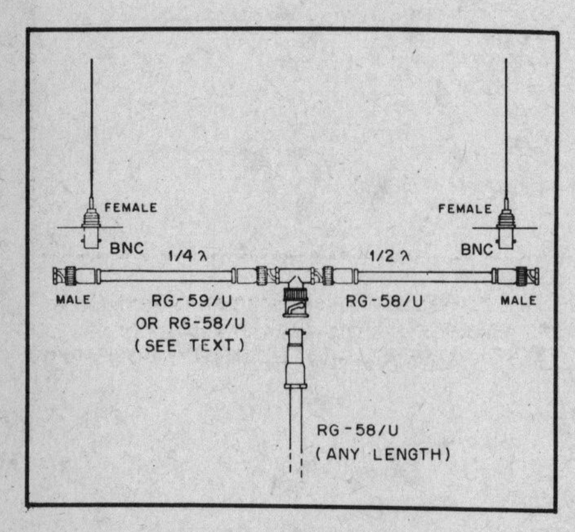


Fig. 16 — The phasing harness for the two verticals that produce a cardioid pattern. The phasing sections must be measured from the center of the T connector to the point that the vertical radiator emerges from shield portion of the upside-down BNC female; i.e., don't forget to take the length of the connectors into account when constructing the harness. With this phasing system, the null will be in a direction that runs along the boom toward the quarter-wavelength section.

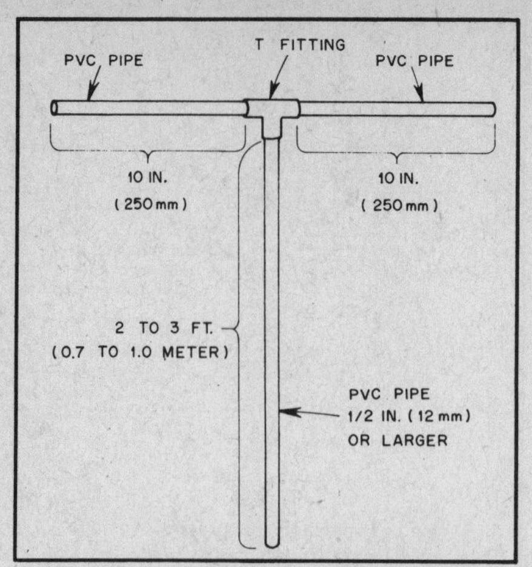


Fig. 17 — A simple mechanical support for the DF antenna made of PVC pipe and fittings.

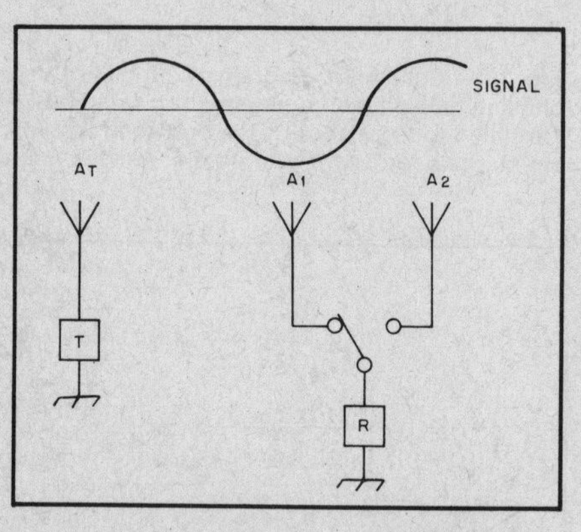


Fig. 18 — Rapid switching between antennas samples the phase at each antenna creating a pseudo-Doppler effect, which an fm detector will detect as phase modulation.

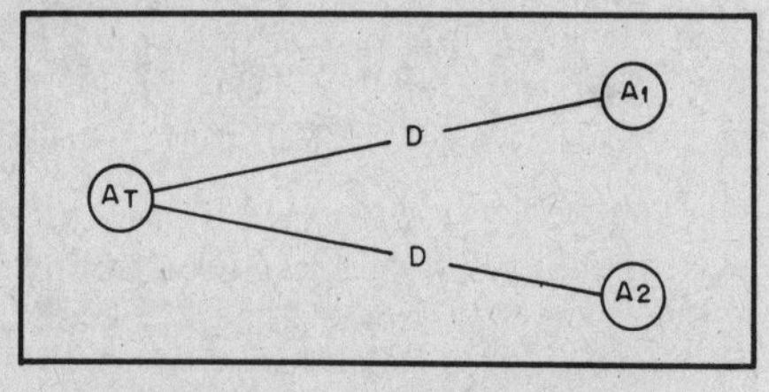


Fig. 19 — If both receiving antennas are an equal distance (D) from the transmitting antenna, there will be no difference in the phase angles of the signals in the receiving antennas; therefore, the detector will not detect any phase modulation, and the audio tone will disappear from the output of the detector.

able to find small brass tubing at a hobby store. If none is available in your area, you might consider brazing rods. It will probably be necessary to solder a short piece to the top of these since they come in 18 in. sections. Two 18 inch radials are added to each element by soldering them to the board.

### The Phasing Harness

The phasing harness is shown in Fig. 16. The half-wave section is 26.7 in. long if polyethelene dielectric coaxial cable is used (velocity factor of 0.66). The quarter-wave section is 13.35 in. The connectors are part of the phasing line and are included in the overall length. Tests done at ARRL show a deeper null resulting under some conditions when the 50-Ω RG-58/U in the quarter-wave phasing section is replaced by 75-Ω RG-59/U coaxial cable.

Fig. 17 shows a simple support for the antenna. PVC tubing is used throughout. Additionally, you will need a T fitting, two end caps and possibly some cement. (Don't cement the parts together if you want the option of disassembly for transportation.) Cut the PVC for the dimensions shown. Drill a small hole through the pc board near the female BNC of each element assembly. Measure 19 in. along the boom (horizontal) and mark the two end points. Drill a small hole vertically through the boom at each mark. Use a small nut and bolt to attach each element assembly to the boom.

### Tuning

Tuning of the array is accomplished by adjusting the length and spacing of the vertical elements. The objective is to obtain the deepest null possible. During the tuning process the antenna should be set up in the open, far from any metal objects.

## THE DOUBLE-DUCKY DIRECTION FINDER

Most amateurs use antennas having pronounced directional effects, either a null or a peak in signal strength, for direction finding. Fm receivers are designed to try to eliminate the effects of amplitude variations and are difficult to use for direction finding without looking at an S meter. Most modern portable transceivers do not have S meters.

This "Double-Ducky" direction finder (DDDF) was designed by David Geiser, WA2ANU, and described in *QST* for July 1981. It is different in that it switches between two nondirectional antennas, creating phase modulation on the incoming signal that is heard easily on the fm receiver (Fig. 18). When the two antennas are exactly the same distance (phase) from the transmitter (Fig. 19), the tone disappears.

In theory the antennas may be very close to each other, but in practice the amount of phase modulation increases directly with the spacing, up to spacings of a half wavelength. While a half-wavelength separation on 2 meters (40 inches) is pretty large for a mobile array, a quarter wavelength gives entirely satisfactory results, and even a one-eighth wavelength (10 inches) is acceptable.

Think in terms of a fixed spacing between the antennas, mount them on a ground plane and rotate that ground plane. The ground plane held above the hiker's head or car roof reduces the needed height of the array and the directional-distorting effects of the searcher's body or other conducting objects.

The DDDF is bidirectional and, as described, its tone null points both to and away from the signal origin. An L-shaped search path would be needed to resolve the ambiguity. Use the technique of triangulation described earlier.

### Specific Design

It is not possible to find a long-life mechanical switch operable at a fairly high audio rate, such as 1000 Hz. Yet we want an audible tone, and the 400- to 1000-Hz range is perhaps most suitable considering audio amplifiers and average hearing. Also, if we wish to use the transmit function of a transceiver, we need a switch that will carry perhaps 10 watts without much problem.

A solid-state switch, the PIN (positive-intrinsic-negative) diode, has been developed within the last few years. The intrinsic region of this type of diode is ordinarily bare of current carriers and, with a bit of reverse bias, looks like a low-capacitance open space. A bit of forward bias (20 to 50 mA) will load the intrinsic region with current carriers that are happy to dance back and forth at a 148-MHz rate, looking like a resistance of an ohm or so. In a 10-watt circuit, little enough power is dissipated in the diode for it to survive.

Because only two antennas are used, the obvious approach (Fig. 20) is to connect one diode "forward" to one antenna, to connect the other "reverse" to the second antenna and to drive the pair with square-wave audio frequency ac. Rf chokes (Ohmite Z144, J. W. Miller RFC-144 or similar vhf units) are used to let the audio through to bias the diodes while blocking rf. Of course, the reverse bias on one diode is only equal to the forward bias on the other, but in practice this seems sufficient.

A number of PIN diodes were tried in the particular setup built. These were the Hewlett-Packard HP5082-3077, the Alpha LE-5407-4, the KSW KS-3542 and the Microwave Associates M/A-COM 47120. All worked well, but the HP diodes were used because they provided a slightly lower SWR (about 3:1).

A type 567 IC is used as the square-wave generator. The output does have a dc bias that is removed with a nonpolarized coupling capacitor. This minor inconvenience is more than rewarded by the ability of the IC to work well with between 7 and 15 volts (a nominal 9-V minimum is recommended).

The nonpolarized capacitor is also used for dc blocking when the function switch is set to XMIT. D3, a light-emitting diode (LED), is wired in series with the transmit bias to indicate selection of the XMIT mode. In that mode there is a high battery current drain (20 mA or so).

S1 should be a center-off locking type toggle switch. An ordinary center-off switch may be used but beware. If the switch is left on XMIT you will soon have dead batteries.

Cables going from the antenna to the coaxial T connector were cut to an electrical 1/2 wavelength to help the open circuit, represented by the reverse-biased diode, look open at the coaxial T. (The length of the line within the T was included in the calculation.)

The length of the line from the T to the control unit is not particularly critical. If possible, keep the total of the cable length from the T to the control unit to the transceiver under 8 feet, because the capacitance of the cable does shunt the square-wave generator output.

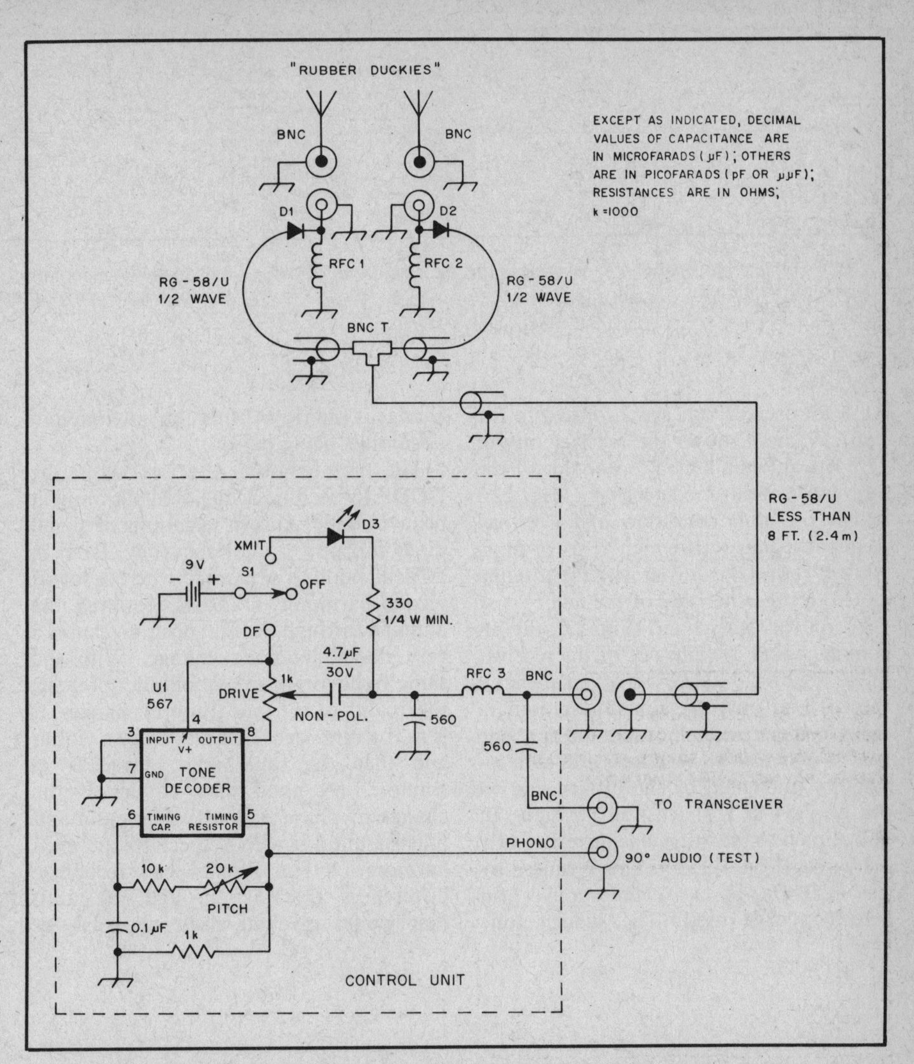


Fig. 20 — Schematic diagram of the DDDF circuit. Construction and layout are not critical. Components inside broken lines should be housed inside a shielded enclosure. Most of the components are available from Radio Shack, except D1, D2, the antennas and RFC1-RFC3, which are discussed in the text. S1 — See text.

Ground-plane dimensions are not critical. See Fig. 21. Slightly better results may be obtained with a larger ground plane. Increasing the spacing between the pickup antennas will give the greatest improvement. Every doubling (up to a half wavelength) will cut the width of the null in half. A 1° wide null can be obtained with 20-inch spacing.

## Usage Instructions

Switch the control unit to DF and advance the drive potentiometer until a tone is heard on the desired signal. Do not advance the drive high enough to distort or "hash up" the voice. Rotate the antenna for a null in the fundamental tone. Note that a tone an octave higher may appear. The cause of the effect is shown in Fig. 22.

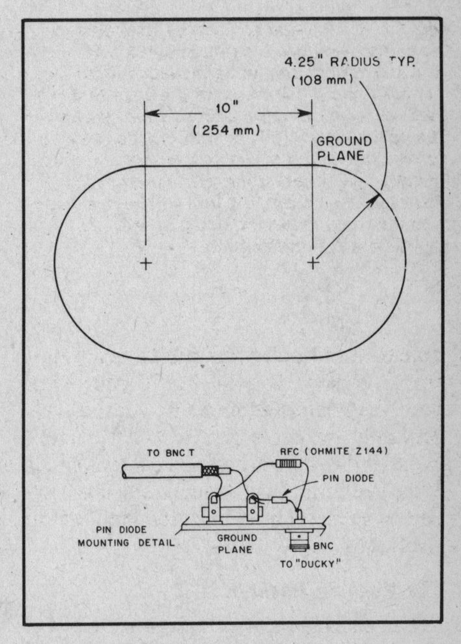


Fig. 21 — Ground-plane layout and detail of parts at the antenna connectors.

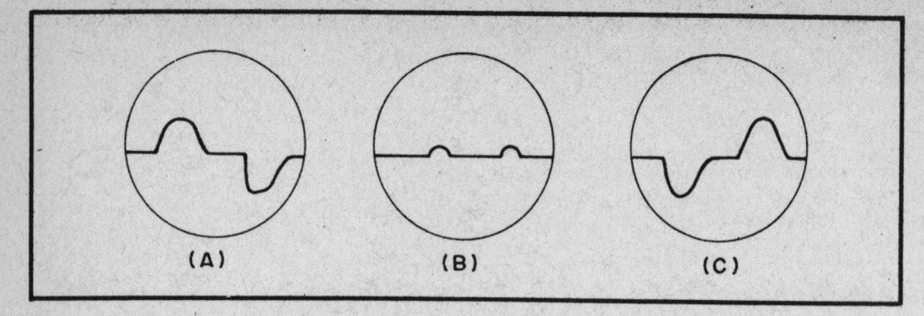


Fig. 22 — Typical on-channel responses. See text for discussion of the meaning of the patterns.

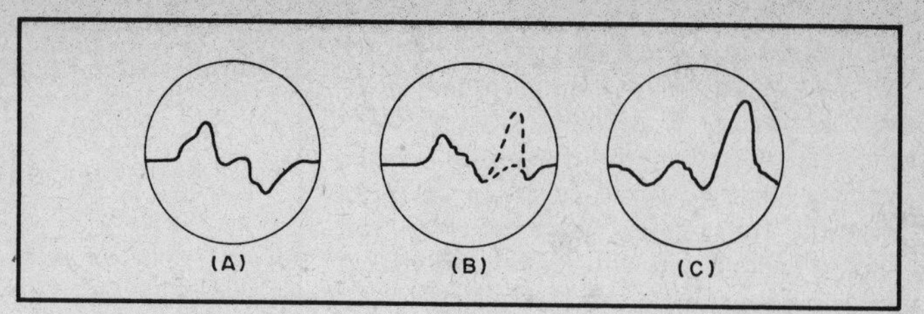


Fig. 23 — Representative off-channel responses. See text for discussion of the meaning of the patterns.

At A an oscilloscope synchronized to the "90° Audio" shows the receiver output with the antenna aimed to one side of the null (on a well-tuned receiver). Fig. 22B shows the null condition and a twice-frequency (one octave higher) set of pips, while C shows the output with the antenna aimed to the other side of the null.

If, on the other hand (Fig. 23), the incoming signal is quite out of the receiver linear region (10 kHz or so off frequency), the off-null antenna aim may present a fairly symmetrical af output to one side (A). It may also show a near null with instability (indicated by the broken line on the display) at a sharp null position (B) and, aimed to the other side, give a greatly increased af output (C). This is caused by the different parts of the receiver fm detector curve used. The sudden tone change is the tip-off that the antenna null position is being passed.

The user should practice with the DDDF to become acquainted with how it behaves under known situations of signal direction, power and frequency. Even in difficult nulling situations where a lot of second-harmonic af exists, rotating the antenna through the null position causes a very distinctive tone change. With the same frequencies and amplitudes present, the quality of the tone (timbre) changes. It is as if a note were first played by a violin, and then the same note played by a trumpet. (A good part of this is the change of phase of the fundamental and odd harmonics with respect to the even harmonics.) The listener can recognize differences (passing through the null) that would give an electronic analyzer indigestion.

## BIBLIOGRAPHY

Source material and more extended discussion of topics covered in this chapter can be found in the references given below and in the textbooks listed at the end of Chapter 2.

*Radio Direction Finding*, published by the Happy Flyers, 1811 Hillman Ave., Belmont, CA 94002.

DeMaw, D., "Beat the Noise with a Scoop Loop," *QST*, July 1977.

Dorbuck, T., "Radio Direction-Finding Techniques," *QST*, August 1975.

Geiser, D. T., "Double-Ducky Direction Finder," *QST*, July 1981.

Holter, N. K., "Radio Foxhunting in Europe," parts 1 and 2, *QST*, August and November 1976.

O'Dell, P., "Simple Antenna and S-Meter Modification for 2-Meter FM Direction Finding," Basic Amateur Radio, *QST*, March 1981.

## Chapter 15

# Antenna and Transmission-Line Measurements

The principal quantities to be measured on transmission lines are line current or voltage, and standing-wave ratio. Measurements of current or voltage are made for the purpose of determining the power input to the line. SWR measurements are useful in connection with the design of coupling circuits and the adjustment of the match between the antenna and transmission line, as well as in the adjustment of matching circuits.

For most practical purposes a *relative* measurement is sufficient. An uncalibrated indicator that shows when the largest possible amount of power is being put into the line is just as useful, in most cases, as an instrument that measures the power accurately. It is seldom necessary to know the actual number of watts going into the line unless the overall efficiency of the system is being investigated. An instrument that shows when the SWR is close to 1 to 1 is all that is needed for most impedance-matching adjustments. Accurate measurement of SWR is necessary only in studies of antenna characteristics such as bandwidth, or for the design of some types of matching systems, such as a stub match.

Quantitative measurements of reasonable accuracy demand good design and careful construction in the measuring instruments. They also require intelligent use of the equipment, including a knowledge not only of its limitations but also of stray effects that often lead to false results. A certain amount of skepticism regarding numerical data resulting from amateur measurements with simple equipment is justified until the complete conditions of the measurements are known. On the other hand, purely qualitative or relative measurements are easy to make and are reliable for the purposes mentioned above.

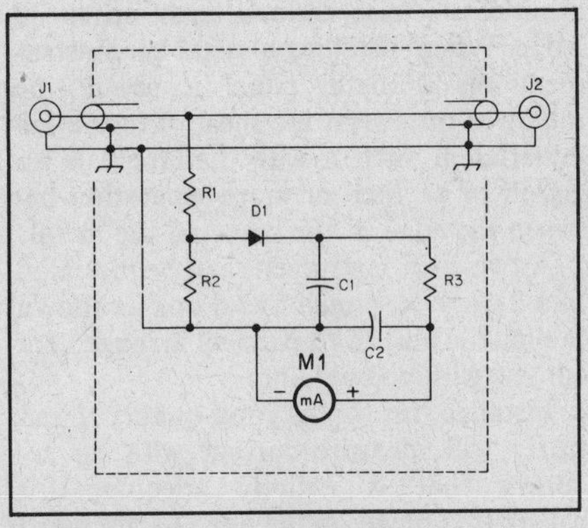

Fig. 1 — Rf voltmeter for coaxial line.
C1, C2 — 0.005- or 0.01-μF ceramic.
D1 — Germanium diode, 1N34A.
J1, J2 — Coaxial fittings, chassis-mounting type.
M1 — 0-1 milliammeter (more sensitive meter may be used if desired; see text).
R1 — 6.8 kΩ, composition, 1 watt for each 100 watts of rf power.
R2 — 680 Ω, 1/2 or 1 watt composition.
R3 — 10 kΩ, 1/2 watt (see text).

### LINE CURRENT AND VOLTAGE

A current or voltage indicator that can be used with coaxial line is a useful piece of equipment. It need not be elaborate or expensive. Its principal function is to show when the maximum power is being taken from the transmitter; for any given set of line conditions (length, SWR, etc.) this will occur when the transmitter coupling is adjusted for maximum current or voltage at the input end of the line. Although the final-amplifier plate or collector current meter is frequently used for this purpose, it is not always a reliable indicator. In many cases, particularly with a screen-grid tube in the final stage, minimum loaded plate current does not occur simultaneously with maximum power output.

### RF VOLTMETER

A germanium diode in conjunction with a low-range milliammeter and a few resistors can be assembled to form an rf voltmeter suitable for connecting across the two conductors of a coaxial line, as shown in Fig. 1. It consists of a voltage divider, R1-R2, having a total resistance about 100 times the $Z_0$ of the line (so the power consumed will be negligible) with a diode rectifier and milliammeter connected across part of the divider to read relative rf voltage. The purpose of R3 is to make the meter readings directly proportional to the applied voltage, as nearly as possible, by "swamping" the resistance of D1, since the latter resistance will vary with the amplitude of the current through the diode.

The voltmeter may be constructed in a small metal box, indicated by the dashed line in the drawing, fitted with coax receptacles. R1 and R2 should be composition resistors. The power rating for R1 should be 1 watt for each 100 watts of carrier power in the matched line; separate 1- or 2-watt resistors should be used to make up the total power rating required, to a total resistance as given. Any type of resistor can be used for R3; the total resistance should be such that about 10 volts dc will be developed across it at full scale. For example, a 0-1 milliammeter would require 10 kΩ, a 0-500 microammeter would take 20 kΩ, and so on. For comparative measurements only, R3 may be a variable

Fig. 2 — A convenient method of mounting an rf ammeter for use in coaxial line. This is a metal-case instrument mounted on a thin bakelite panel, the diameter of the cutout in the metal being such as to clear the edge of the meter by about an eighth inch.

resistor so the sensitivity can be adjusted for various power levels.

In constructing such a voltmeter, care should be used to prevent inductive coupling between R1 and the loop formed by R2, D1 and C1, and between the same loop and the line conductors in the assembly. With the lower end of R1 disconnected from R2 and grounded to the enclosure, but without changing its position with respect to the loop, there should be no meter indication when full power is going through the line.

If more than one resistor is used for R1, the units should be arranged end to end with very short leads. R1 and R2 should be kept a half inch or more from metal surfaces parallel to the body of the resistor. If these precautions are observed the voltmeter will give consistent readings at frequencies up to 30 MHz. Stray capacitances and couplings limit the accuracy at higher frequencies but do not affect the utility of the instrument for comparative measurements.

### Calibration

The meter may be calibrated in rf voltage by comparison with a standard such as an rf ammeter. This requires that the line be well matched so the impedance at the point of measurement is equal to the actual $Z_0$ of the line. Since in that case $P = I^2Z_0$, the power can be calculated from the current. Then $E = \sqrt{PZ_0}$. By making current and voltage measurements at a number of different power levels, enough points may be secured to permit drawing a calibration curve for the voltmeter.

## RF AMMETERS

An rf ammeter can be mounted in any convenient location at the input end of the transmission line, the principal precaution in its mounting being that the capacitance to ground, chassis, and nearby conductors should be low. A bakelite-case instrument can be mounted on a metal panel without introducing enough shunt capacitance to ground to cause serious error up to 30 MHz. When installing a metal-case instrument on a metal panel it should be mounted on a separate sheet of insulating material in such a way that there is an eighth of an inch or more separation between the edge of the case and the metal.

A two-inch instrument can be mounted in a 2 × 4 × 4-inch metal box as shown in Fig. 2. This is a convenient arrangement for use with coaxial line.

Installed this way, a good-quality rf ammeter will measure current with an accuracy that is entirely adequate for calculating power in the line. As discussed above in connection with calibrating rf voltmeters, the line must be closely matched by its load so the actual impedance will be resistive and equal to $Z_0$. The scales of such instruments are cramped at the low end, however, which limits the range of power that can be measured by a single meter. The useful current range is about 3 to 1, corresponding to a power range of about 9 to 1.

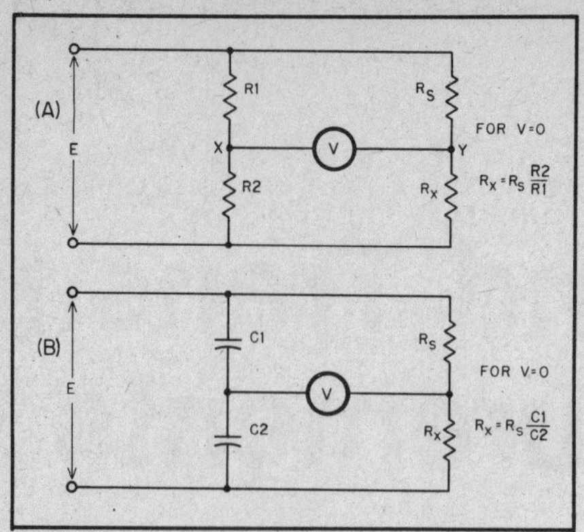


Fig. 3 — Bridge circuits suitable for SWR measurement. A — Wheatstone type using resistance arms. B — Capacitance-resistance bridge ("Micromatch"). Conditions for balance are independent of frequency in both types.

## SWR MEASUREMENTS

On parallel-conductor lines it is possible to measure the standing-wave ratio by moving a current (or voltage) indicator along the line, noting the maximum and minimum values of current (or voltage) and then computing the SWR from these measured values. This cannot be done with coaxial line since it is not possible to make measurements of this type inside the cable. The technique is, in fact, seldom used with open lines, because it is not only inconvenient but sometimes impossible to reach all parts of the line conductors. Also, the method is subject to considerable error from antenna currents flowing on the line.

Present-day SWR measurements made by amateurs practically always use some form of "directional coupler" or rf bridge circuit. The indicator circuits themselves are fundamentally simple, but considerable care is required in their construction if the measurements are to be accurate. The requirements for indicators used only for the adjustment of impedance-matching circuits, rather than actual SWR measurement, are not so stringent and an instrument for this purpose can be made easily.

### BRIDGE CIRCUITS

Two commonly used bridge circuits are shown in Fig. 3. The bridges consist essentially of two voltage dividers in parallel, with a voltmeter connected between the junctions of each pair of "arms," as the individual elements are called. When the equations shown to the right of each circuit are satisfied there is no potential difference between the two junctions, and the voltmeter indicates zero voltage. The bridge is then said to be in "balance."

Taking Fig. 3A as an illustration, if R1 = R2, half the applied voltage, E, will appear across each resistor. Then if Rs = Rx, 1/2 E will appear across each of these resistors and the voltmeter reading will be zero. Remember that a matched transmission line has a purely resistive input impedance, and suppose that the input terminals of such a line are substituted for Rx. Then if Rs is a resistor equal to the $Z_0$ of the line, the bridge will be balanced. If the line is not perfectly matched, its input impedance will not equal $Z_0$ and hence will not equal Rs, since the latter is chosen to equal $Z_0$. There will then be a difference in potential between points X and Y, and the voltmeter will show a reading. Such a bridge therefore can be used to show the presence of standing waves on the line, because the line input impedance will be equal to $Z_0$ only when there are no standing waves.

Considering the nature of the incident and reflected components of voltage that make up the actual voltage at the input terminals of the line, as discussed in Chapter 3, it should be clear that when $Rs = Z_0$, the bridge is always in balance for the incident component. Thus the voltmeter does not respond to the incident component at any time but reads only the reflected component (assuming that R2 is very small compared with the voltmeter impedance). The incident component can be measured across either R1 or R2, if they are equal resistances. The standing-wave ratio is then

$$\text{SWR} = \frac{E1 + E2}{E1 - E2}$$

where E1 is the incident voltage and E2 is the reflected voltage. It is often simpler to normalize the voltages by expressing E2 as

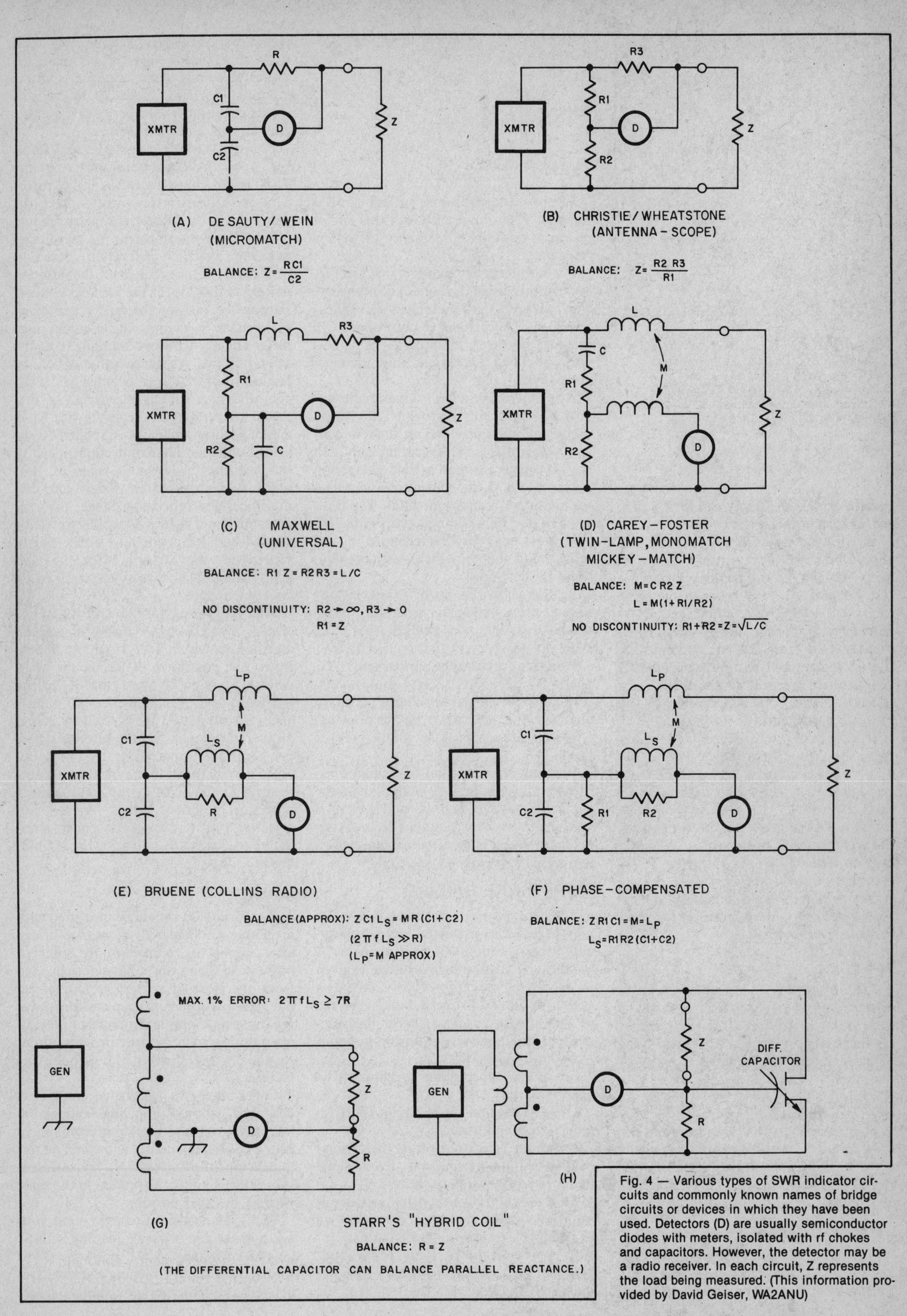


Fig. 4 — Various types of SWR indicator circuits and commonly known names of bridge circuits or devices in which they have been used. Detectors (D) are usually semiconductor diodes with meters, isolated with rf chokes and capacitors. However, the detector may be a radio receiver. In each circuit, Z represents the load being measured. (This information provided by David Geiser, WA2ANU)

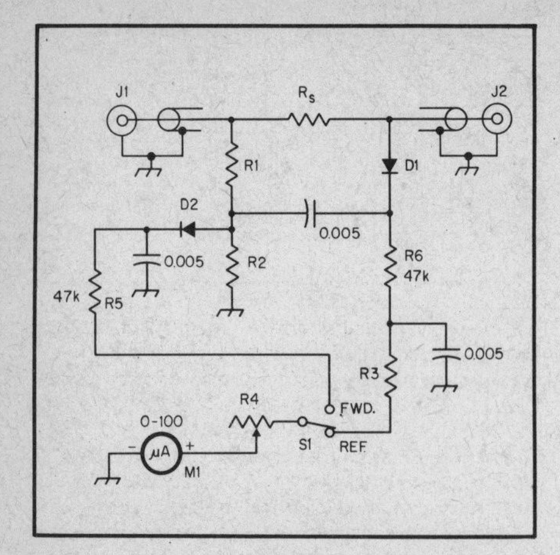


Fig. 5 — Resistance bridge for SWR measurement. Capacitors are disc ceramic. Resistors are 1/2-watt composition except as noted below.
D1, D2 — Germanium diode, high back resistance type (1N34A, 1N270, etc.).
J1, J2 — Coaxial connectors, chassis-mounting type.
M1 — 0-100 dc microammeter.
R1, R2 — 47 ohms, 1/2-watt composition (see text).
R3 — See text.
R4 — 50-kΩ volume control.
$R_s$ — Resistance equal to line $Z_0$ (1/2 or 1 watt composition).
S1 — Spdt toggle.

Fig. 6 — A 2 × 4 × 4-inch aluminum box is used to house this SWR bridge, which uses the circuit of Fig. 5. The variable resistor, R4, is mounted on the side. The bridge components are mounted on one side plate of the box and a miniature chassis formed from a piece of aluminum. The input connector is at the top in this view. $R_s$ is connected directly between the two center posts of the connectors. R2 is visible behind it and perpendicular to it. One terminal of D1 projects through a hole in the chassis so the lead can be connected to J2. R1 is mounted vertically to the left of the chassis in this view, with D2 connected between the junction of R1-R2 and a tie point.

a fraction of E1, in which case the formula becomes

$$SWR = \frac{1 + k}{1 - k}$$

where k = E2/E1.

The operation of the circuit in Fig. 3B is essentially the same, although this circuit has arms containing reactance as well as resistance.

It is not necessary that R1 = R2 in Fig. 3A; the bridge can be balanced, in theory, with any ratio of these two resistances provided Rs is changed accordingly. In practice, however, the accuracy is highest when the two are equal; this circuit is generally so used.

A number of types of bridge circuits appear in Fig. 4, many of which have been used in amateur products or amateur construction projects. All except that at G can have the generator and load at a common potential. At G, the generator and detector are at a common potential. The positions of the detector and transmitter (or generator) may be interchanged in the bridge, and this may be an advantage in some applications.

Bridges shown at D, E, F and H may have one terminal of the generator, detector and load common. Bridges at A, B, E, F, G and H have constant sensitivity over a wide frequency range. Bridges at B, C, D and H may be designed to show no discontinuity (impedance lump) with a matched line, as shown in the drawing. Discontinuities with A, E and F may be small.

Bridges are usually most sensitive when the detector bridges the midpoint of the generator voltage, as in G or H, or in B when each resistor equals the load impedance. Sensitivity also increases when the currents in each leg are equal.

## RESISTANCE BRIDGE

The basic bridge type shown in Fig. 3A may be home constructed and is reasonably accurate for SWR measurement. A practical circuit for such a bridge is given in Fig. 5 and a representative layout is shown in Fig. 6. Properly built, a bridge of this design can be used for measurement of standing-wave ratios up to about 15 to 1 with good accuracy.

Important constructional points to be observed are:

1) Keep leads in the rf circuit short, to reduce stray inductance.

2) Mount resistors two or three times their body diameter away from metal parts, to reduce stray capacitance.

3) Place the rf components so there is as little inductive and capacitive coupling as possible between the bridge arms.

In the layout shown in Fig. 6, the input and line connectors, J1 and J2, are mounted fairly close together so the standard resistor, Rs, can be supported with short leads directly between the center terminals of the connectors. R2 is mounted at right angles to Rs, and a shield partition is used between these two components and the others.

The two 47-kΩ resistors, R5 and R6 in Fig. 5, are voltmeter multipliers for the 0-100 microammeter used as an indicator. This is sufficient resistance to make the voltmeter linear (that is, the meter reading is directly proportional to the rf voltage) and no voltage calibration curve is needed. D1 is the rectifier for the reflected voltage and D2 is for the incident voltage. Because of resistor tolerances and small differences in diodes, the readings may differ slightly with two multipliers of the same nominal resistance value, so a correction resistor, R3, is included in the circuit. Its value should be selected so that the meter reading is the same with S1 in either position, when rf is applied to the bridge with the line connection open. In the instrument shown, a value of 1000 ohms was required in series with the multiplier for reflected voltage; in other cases different values probably would be needed and R3 might have to be put in series with the multiplier for the incident voltage. This can be determined by experiment.

The value used for R1 and R2 is not critical, but the two resistors should be matched to within 1 or 2 percent if possible. The resistance of Rs should be as close as possible to the actual $Z_0$ of the line to be used (generally 52 or 75 ohms). The resistor should be selected by actual measurement with an accurate resistance bridge, if one is available.

R4 is for adjusting the incident-voltage reading to full scale in the measurement procedure described below. Its use is not essential, but it offers a convenient alternative to exact adjustment of the rf input voltage.

### Testing

R1, R2 and Rs should be measured with a reliable ohmmeter or resistance bridge after wiring is completed, in order to make sure their values have not changed from the heat of soldering. Disconnect one side of the microammeter and leave the input and output terminals of the unit open during such measurements, in order to avoid stray shunt paths through the rectifiers.

Check the two voltmeter circuits as described above, applying enough rf (about 10 volts) to the input terminals to give a full-scale reading with the line terminals open. If necessary, try different values for R3 until the reading is the same with S1 in either position.

With J2 open, adjust the rf input voltage and R4 for full-scale reading with S1 in the reflected-voltage position. Then short-circuit J2 by touching a screwdriver between the center terminal and the frame of the connector to make a low-inductance short. Switch S1 to the incident-voltage position and readjust R4 for full scale, if necessary. Then throw S1 to the reflected-voltage position, keeping J2 shorted, and the reading should be full scale as before. If it is not, R1 and R2 are not the same value, or there is stray coupling between the arms of the bridge. It is necessary that the reflected voltage read full scale with J2 either open or shorted, when the incident voltage is set to full scale in each case, in order to make accurate SWR measurements.

The circuit should pass these tests at all frequencies at which it is to be used. It is sufficient to test at the lowest and highest frequencies, usually 1.8 or 3.5 and 28 or 50 MHz. If R1 and R2 are poorly matched but the bridge construction is otherwise good, discrepancies in the readings will be substantially the same at all frequencies. A difference in behavior at the low and high ends of the frequency range can be attributed to stray coupling between bridge arms, or stray inductance or capacitance in the arms.

To check the bridge for balance, apply rf and adjust R4 for full scale with J2 open. Then connect a resistor identical with Rs (the resistance should match within 1 or 2 percent) to the line terminals, using the shortest possible leads. It is convenient to mount the test resistor inside a cable connector (PL-259), a method of mounting that also minimizes lead inductance. When the test resistor is connected, the reflected-voltage reading should drop to zero. The incident voltage should be reset to full scale by means of R4, if necessary. The reflected reading should be zero at any frequency in the range to be used. If a good null is obtained at low frequencies but some residual current shows at the high end, the trouble may be the inductance of the test resistor leads, although it may also be caused by stray coupling between the arms of the bridge itself. If there is a constant low (but not zero) reading at all frequencies the cause is poor matching of the resistance values. Both effects can be present simultaneously. A good null must be obtained at all frequencies before the bridge is ready for use.

### Bridge Operation

The rf power input to a bridge of this type must be limited to a few watts at most, because of the power-dissipation ratings of the resistors. If the transmitter has no provision for reducing power output to a very low value — less than 5 watts — a simple "power absorber" circuit can be made up as shown in Fig. 7. The lamp DS1 tends to maintain constant current through the resistor over a fairly wide power range, so the voltage drop across the resistor also tends to be constant. This voltage is applied to the bridge, and with the constants given is in the right range for resistance-type bridges.

To make the measurement, connect the line to J2 and apply sufficient rf voltage to J1 to give a full-scale incident-voltage reading. R4 may be used to set to exactly full scale. Then throw S1 to the reflected-voltage position and note the meter reading. The SWR is then found by substituting the readings in the formula previously given.

For example, if the full-scale calibration of the dc instrument is 100 microamperes and the reading with S1 in the reflected-voltage position is 40 microamperes, the SWR is

$$SWR = \frac{100 + 40}{100 - 40} = \frac{140}{60} = 2.33 \text{ to } 1$$

Instead of determining the SWR value by calculations, the *voltage* curve of Fig. 8 may be used. In this example the ratio of reflected to forward voltage is 40/100 = 0.4, and from Fig. 8 the SWR value is seen to be about 2.3 to 1.

The meter scale may be calibrated in any arbitrary units so long as the scale has equal divisions, since it is the ratios of the voltages, and not the actual values, that determine the SWR.

## AVOIDING ERRORS IN SWR MEASUREMENTS

The principal causes of inaccuracies within the bridge are differences in the resistances of R1 and R2, stray inductance and capacitance in the bridge arms, and stray coupling between arms. If the checking procedure described above is followed through carefully, the bridge of Fig. 5 should be amply accurate for practical use. The accuracy is highest for low standing-wave ratios because of the nature of the SWR calculation; at high ratios the divisor in the equation above represents the difference between two nearly equal quantities, so a small error in voltage measurement may mean a considerable difference in the calculated SWR.

The standard resistor Rs must equal the actual $Z_0$ of the line. The actual $Z_0$ of a sample of line may differ by a few percent from the nominal figure because of manufacturing variations, but this has to be tolerated. In the 50- to 75-ohm range, the rf resistance of a composition resistor of 1/2 or 1 watt rating is essentially identical with its dc resistance.

### "Antenna" Currents

As explained in Chapter 5, there are two ways in which "parallel" or "antenna" currents can be caused to flow on the outside of a coaxial line — currents induced on the line because of its spatial relationship to the antenna, and currents that result from the direct connection between the coax outer conductor and (usually) one side of the antenna. The induced current usually will not be troublesome if the bridge and the transmitter (or other source of rf power for operating the bridge) are shielded so that any rf currents flowing on the *outside* of the line cannot find their way into the bridge. This point can be checked by "cutting in" an additional section of line (1/8 to 1/4 electrical wavelength, preferably) of the same $Z_0$. The SWR indicated by the bridge should not change except possibly for a slight decrease because of the additional line loss. If there is a marked change, better shielding may be required.

Parallel-type currents caused by the connection to the antenna will cause a change in SWR with line length even though the bridge and transmitter are well shielded and the shielding is maintained throughout the system by the use of

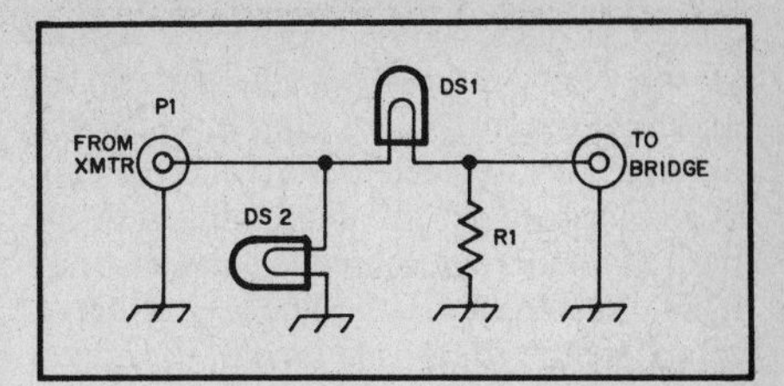


Fig. 7 — "Power absorber" circuit for use with resistance-type SWR bridges when the transmitter has no special provisions for power reduction. For rf powers up to 50 watts, DS1 is a 117-volt 40-watt incandescent lamp and DS2 is not used. For higher powers use sufficient additional lamp capacity at DS2 to load the transmitter to about normal output; for example, for 250 watts output DS2 may consist of two 100-watt lamps in parallel. R1 is made from three 1-watt 68-ohm resistors connected in parallel. P1 and P2 are cable-mounting coaxial connectors. Leads in the circuit formed by the lamps and R1 should be kept short, but convenient lengths of cable may be used between this assembly and the connectors.

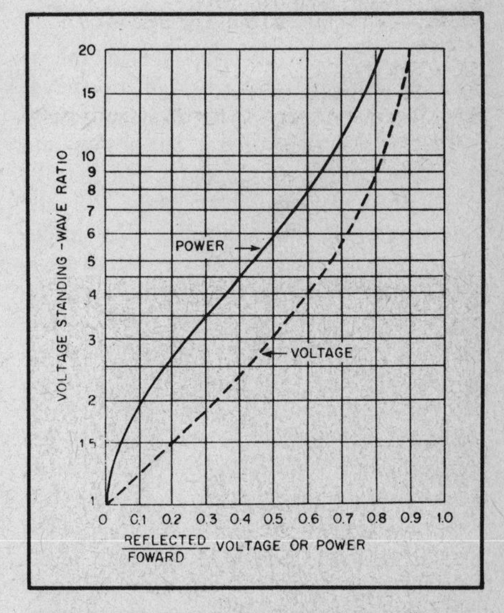


Fig. 8 — Chart for finding voltage standing-wave ratio when the ratio of reflected-to-forward voltage or reflected-to-forward power is known.

coaxial fittings. This is because the outside of the coax tends to become part of the antenna system, being connected to the antenna at the feed point, and so constitutes a load on the line, along with the desired load represented by the antenna itself. The SWR on the line then is determined by the composite load of the antenna and the outside of the coax, and since changing the line length changes one component of this composite load, the SWR changes too.

The remedy for such a situation is to use a good balun or to detune the outside of the line by proper choice of length. It is well to note that this is not a measurement error, since what the instrument reads is the actual SWR on the line. However, it is an undesirable condition since the line is operating at a higher SWR than it should — and would — if the parallel-type current on the outside of the coax were eliminated.

**Spurious Frequencies**

Off-frequency components in the rf voltage applied to the bridge may cause considerable error. The principal components of this type are harmonics and low-frequency subharmonics that may be fed through the final stage of the transmitter driving the bridge. The antenna is almost always a fairly selective circuit, and even though the system may be operating with a very low SWR at the desired frequency it is practically always mismatched at harmonic and subharmonic frequencies. If such spurious frequencies are applied to the bridge in appreciable amplitude, the SWR indication will be very much in error. In particular, it may not be possible to secure a null on the bridge with any set of adjustments of the matching circuit. The only remedy is to filter out the unwanted components by increasing the selectivity of the circuits between the transmitter final amplifier and the bridge.

**REFLECTOMETERS**

Low-cost reflectometers that do not have a guaranteed wattmeter calibration are not ordinarily reliable for accurate numerical measurement of standing-wave ratio. They are, however, very useful as aids in the adjustment of matching networks, since the objective in such adjustment is to reduce the reflected voltage or power to zero. Relatively inexpensive devices can be used for this, since only good bridge balance is required, not actual calibration. Bridges of this type are usually "frequency-sensitive" — that is, the meter response becomes greater with increasing frequency, for the same applied voltage. When matching and line monitoring, rather than SWR measurement, is the principal use of the device, this is not a serious handicap.

Various simple reflectometers, useful for matching and monitoring, have been described from time to time in *QST* and in *The Radio Amateur's Handbook*. Because most of these are frequency sensitive, it is difficult to calibrate them accurately for power measurement, but their low cost and suitability for use at moderate power levels, combined with the ability to show accurately when a matching circuit has been properly adjusted, make them a worthwhile addition to the amateur station.

## AN IN-LINE RF WATTMETER

Considerable attention was devoted to the resistance-type SWR bridge in the

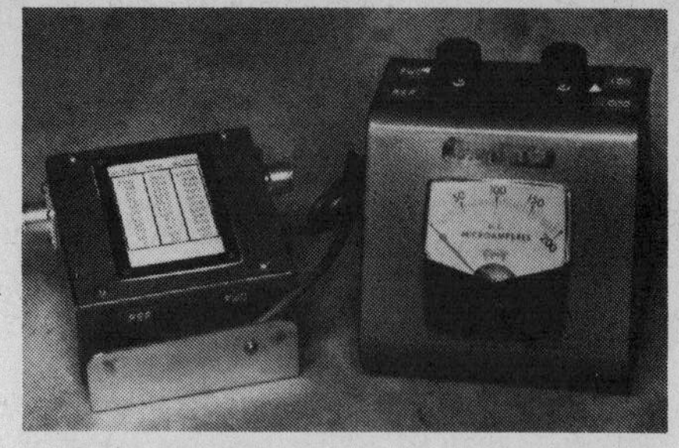

Fig. 9 — The rf wattmeter consists of two parts, the rf head (left), and the control-meter box (right). The paper scale affixed to the rf head contains the calibration information which appears in Fig. 10.

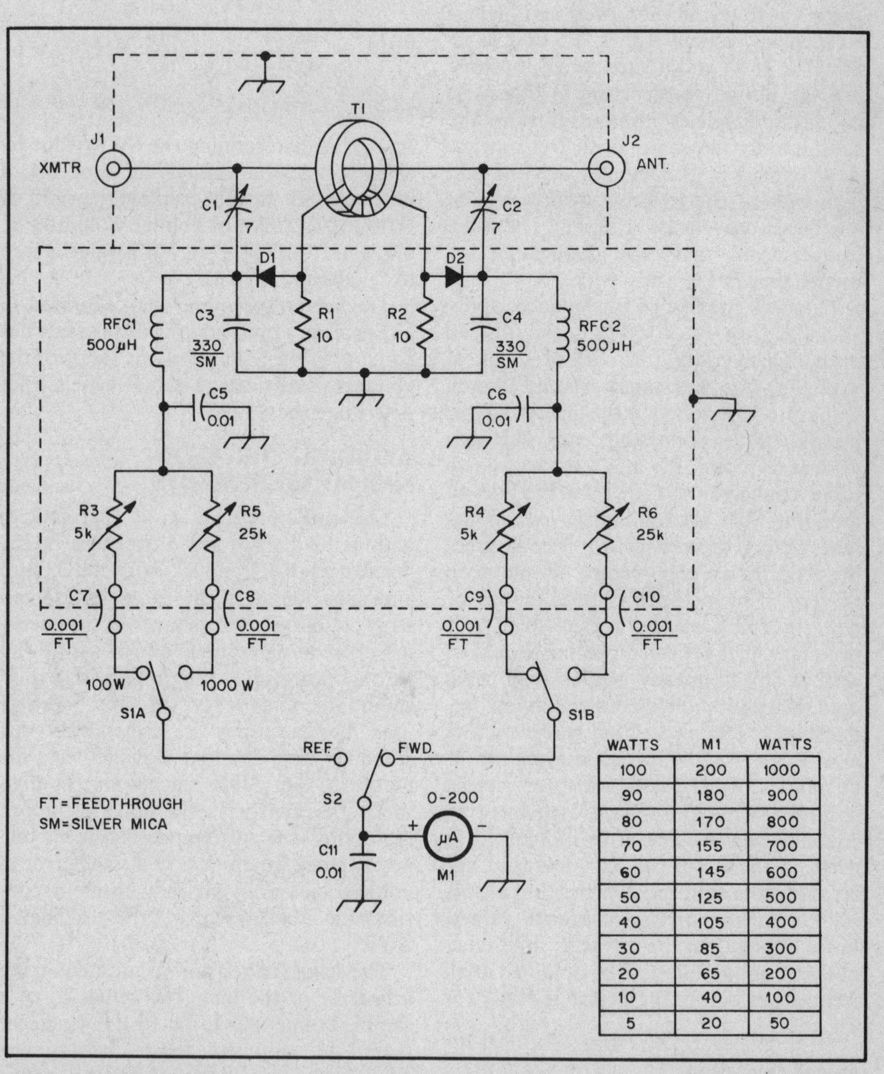


| WATTS | M1 | WATTS |
|---|---|---|
| 100 | 200 | 1000 |
| 90 | 180 | 900 |
| 80 | 170 | 800 |
| 70 | 155 | 700 |
| 60 | 145 | 600 |
| 50 | 125 | 500 |
| 40 | 105 | 400 |
| 30 | 85 | 300 |
| 20 | 65 | 200 |
| 10 | 40 | 100 |
| 5 | 20 | 50 |

Fig. 10 — Schematic diagram of the rf wattmeter. A calibration scale for M1 is shown also. Fixed-value resistors are 1/2-watt composition. Fixed-value capacitors are disc ceramic unless otherwise noted. Decimal-value capacitances are in microfarads. Others are picofarads. Resistance is in ohms; k = 1000.

C1, C2 — 1.3- to 6.7-pF miniature trimmer (E. F. Johnson 189-502-4, available from Newark Electronics, Chicago, Illinois)

C3-C11, incl. — Numbered for circuit-board identification.

D1, D2 — Matched small-signal germanium diodes, 1N34A, etc. (see text).

J1, J2 — Chassis-mount coax connector of builder's choice. Type SO-239 used here.

M1 — 0- to 200-µA meter (Triplett type 330-M used here).

R1, R2 — Matched 10-ohm resistors (see text).

R3, R4 — 5-kΩ printed-circuit carbon control (IRC R502-B).

R5, R6 — 25-kΩ printed-circuit carbon control (IRC R252-B).

RFC1, RFC2 — 500-µH rf choke (Millen 34300-500 or similar).

S1 — Dpdt single-section phenolic wafer switch (Mallory 3222J).

S2 — Spdt phenolic wafer switch (Centralab 1460).

T1 — Toroidal transformer; 35 turns of no. 26 enam. wire to cover entire core of Amidon T-68-2 toroid (Amidon Assoc., 12033 Otsego St., N. Hollywood, CA 91607, or Radiokit, Box 411, Greenville, NH 03048).

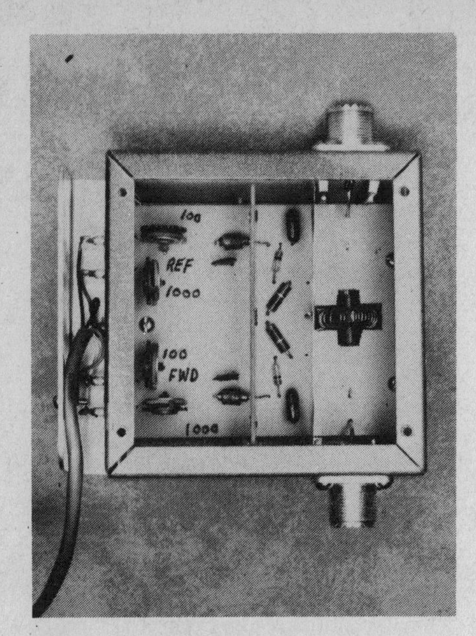

Fig. 11 — Top view of the rf head for the circuit of Fig. 10. A flashing-copper shield isolates the primary rf line and T1 from the rest of the circuit. The second shield (thicker) is not required and can be eliminated from the circuit. If a 2000-watt scale is desired, fixed-value resistors of approximately 22 kΩ can be connected in series with high-range printed-circuit controls. Instead, the 25-kΩ controls shown here can be replaced by 50-kΩ units.

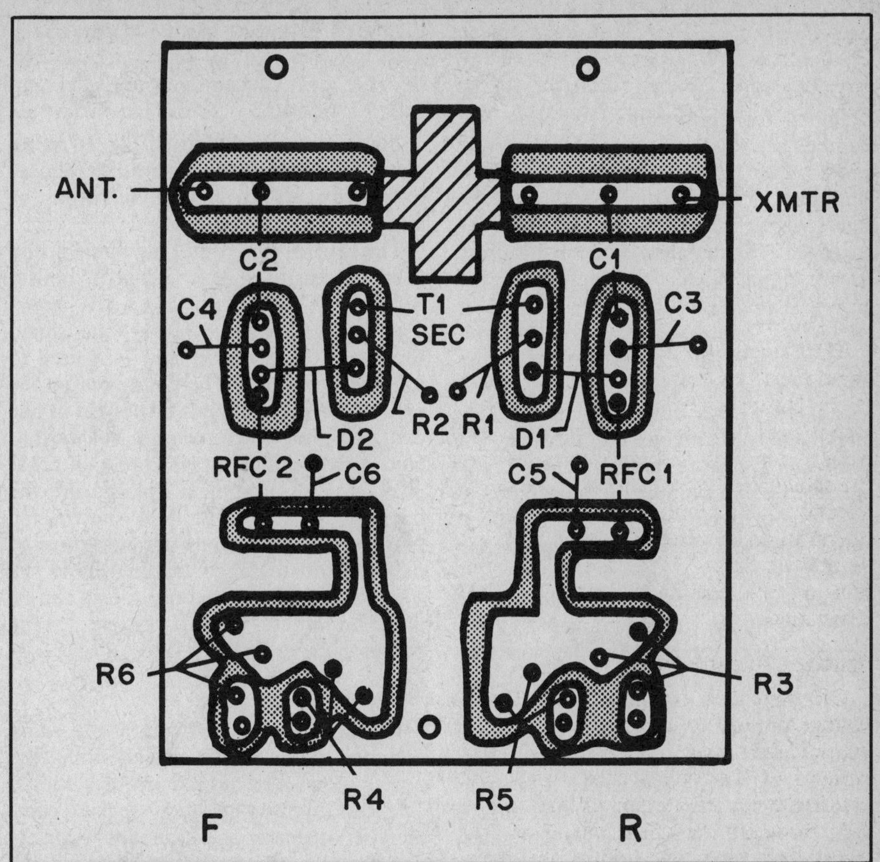


Fig. 12 — Etching pattern and parts layout for the rf wattmeter, as viewed from the foil side of the board. The etched-away portions of the foil are shown as darkened areas in this drawing. The area with diagonal lines is to be cut out for the mounting of T1.

preceding section because it is the simplest type that is capable of adequate accuracy in measuring voltage standing-wave ratio. Its disadvantage is that it must be operated at a very low power level, and thus is not suitable for continuous monitoring of the SWR in actual transmission. To do this the instrument must be capable of carrying the entire power output of the transmitter, and should do it with negligible loss. An rf wattmeter meets this requirement.

It is neither costly nor difficult to build an rf wattmeter. And, if the instrument is equipped with a few additional components, it can be switched to read reflected power as well as forward power. With this feature the instrument can be used as an SWR meter for antenna matching and Transmatch adjustments. The wattmeter shown in Figs. 9 through 12 meets these requirements. The instrument uses a directional type of coupler for sampling the energy on the transmission line. The indicator sensitivity of this instrument is not related to frequency, as is the case with some types of directional couplers. This unit may be calibrated for power levels as low as 1 watt, full scale, in any part of the hf spectrum. With suitable calibration, it has good accuracy over the 3-30 MHz range. It is built in two parts, an rf head for inserting in the coaxial line leaving the transmitter, and a control-meter box which can be placed in any location where it can be operated conveniently. Only direct current flows in the cable connecting the two pieces.

### Design Philosophy

See the circuit of Fig. 10. The transmission-line center conductor passes through the center of a toroid core and becomes the primary of T1. The multiturn winding on the core functions as the transformer secondary. Current flowing through the line-wire primary induces a voltage in the secondary which causes a current to flow through resistors R1 and R2. The voltage drops across these resistors are equal in amplitude, but 180 degrees out of phase with respect to common or ground. They are thus, for practical purposes, respectively in and out of phase with the line current. Capacitive voltage dividers, C1-C3 and C2-C4, are connected across the line to obtain equal-amplitude voltages *in phase* with the line voltage, the division ratio being adjusted so that these voltages match the voltage drops across R1 and R2 in amplitude. (As the current/voltage ratio in the line depends on the load, this can be done only for a particular value of load impedance. Load values chosen for this standardization are pure resistances that match the characteristic impedance of the transmission line with which the bridge is to be used, 50 or 75 ohms usually.) Under these conditions, the voltages rectified by D1 and D2 represent, in the one case, the vector *sum* of the voltages caused by the line current and voltage, and in the other, the vector *difference*. With respect to the resistance for which the circuit has been set up, the sum is proportional to the forward component of a traveling wave such as occurs on a transmission line, and the difference is proportional to the reflected component.

### Component Selection

R1 and R2 should be selected for the best null reading when adjusting the bridge into a resistive 50- or 75-ohm load. Normally, the value will be somewhere between 10 and 47 ohms. The 10-ohm value worked well with the instruments shown here. Half-watt composition resistors are suitable to 30 MHz. R1 and R2 should be as closely matched in resistance as possible. Their exact value is not critical, so an ohmmeter may be used to match them.

Ideally, C3 and C4 should be matched in value. Silver-mica capacitors are usually close enough in tolerance that special selection is not required, providing there is enough leeway in the ranges of C1 and

C2 to compensate for any difference in the values of C3 and C4.

Diodes D1 and D2 should also be matched for best results. An ohmmeter can be used to select a pair of diodes having forward dc resistances within a few ohms of being the same. Similarly, the back resistances of the diodes can be matched. The matched diodes will help to assure equal meter readings when the bridge is reversed. (The bridge should be perfectly bilateral in its performance characteristics.) Germanium diodes should be used to avoid misleading results when low values of reflected power are present during antenna adjustments. The SWR can appear to be zero when actually it isn't. The germanium diodes conduct at approximately 0.3 volt, making them more suitable for low-power readings than silicon diodes.

Any meter having a full-scale reading between 50 microamperes and 1 milliampere can be used at M1. The more sensitive the meter, the more difficult it will be to get an absolute reflected-power reading of zero. Some residual current will flow in the bridge circuit no matter how carefully the circuit is balanced, and a sensitive instrument will indicate this current flow. Also, the more sensitive the meter, the larger will have to be the calibrating resistances, R3 through R6, to provide high-power readings. A 0- to 200-microampere meter represents a good compromise for power ranges between 100 and 2000 watts.

**Construction**

It is important that the layout of any rf bridge be as symmetrical as possible if good balance is to be had. The circuit-board layout of Fig. 12 meets the requirement for this instrument. The input and output ports of the equipment should be isolated from the remainder of the circuit so that only the sampling circuits feed voltage to the bridge. A shield across the end of the box which contains the input and output jacks, and the interconnecting line between them is necessary. If stray rf gets into the bridge circuit it will be impossible to obtain a complete zero reflected-power reading on M1 even though a 1:1 SWR exists.

All of the rf-head components except J1, J2 and the feedthrough capacitors are assembled on the board. It is held in place by means of a homemade aluminum L bracket at the end nearest T1. The circuit-board end nearest the feedthrough capacitors is secured with a single no. 6 spade bolt. Its hex nut is outside the box, and is used to secure a solder lug which serves as a connection point for the ground braid in the cable which joins the control box to the rf head.

T1 fits into a cutout area of the circuit board. A 1-inch long piece of RG-8/U coax is stripped of its vinyl jacket and shield braid, and is snug-fit into the center hole of T1. The inner conductor is soldered to the circuit board to complete the line-wire connection between J1 and J2.

The upper dashed lines of Fig. 10 represent the shield partition mentioned above. It can be made from flashing copper or thin brass.

The control box, a sloping-panel utility cabinet measuring 4 × 5 inches, houses S1 and S2, and the meter, M1. Four-conductor shielded cable — the shield serving as the common lead — is used to join the two pieces. There is no reason the entire instrument cannot be housed in one container, but it is sometimes awkward to have coaxial cables attached to a unit that occupies a prominent place in the operating position. Built as shown, the two-piece instrument permits the rf pickup head to be concealed behind the transmitter, while the control head can be mounted where it is accessible to the operator.

**Adjustment**

Perhaps the most difficult task faced by the constructor is that of calibrating the power meter for desired wattage range. The least involved method is to use a commercial wattmeter as a standard. If one is not available, the power output of the test transmitter can be computed by means of an rf ammeter in series with a 50-ohm dummy load, using the standard formula, $P = I^2R$. The calibration chart of Fig. 10 is representative, but the actual calibration of a particular instrument will depend on the diodes used at D1 and D2. Frequently, individual scales are required for the two power ranges.

Connect a noninductive 50-ohm dummy load to J2. A Heath Cantenna or similar load will serve nicely for adjustment purposes. Place S2 in the FORWARD position, and set S1 for the 100-watt range. An rf ammeter or calibrated power meter should be connected between J2 and the dummy load during the tests, to provide power calibration points against which to plot the scale of M1. Apply transmitter output power to J1, gradually, until M1 begins to deflect upward. Increase transmitter power and adjust R4 so that a full-scale meter reading occurs when 100 watts is indicated on the rf ammeter or other standard in use. Next, switch S2 to REFLECTED and turn the transmitter off. Temporarily short across R3, turn the transmitter on, and gradually increase power until a meter reading is noted. With an insulated screwdriver adjust C2 for a null in the meter reading.

The next step is to reverse the coax connections to J1 and J2. Place S2 in the REFLECTED position and apply transmitter power until the meter reads full scale at 100 watts output. In this mode the REFLECTED position actually reads forward power because the bridge is reversed. Calibrating resistance R3 is set to obtain 100 watts full scale during this adjustment. Now, switch S2 to FORWARD and temporarily place a short across R4. Adjust C1 for a null reading on M1. Repeat the foregoing steps until no further improvement can be obtained. It will not be necessary to repeat the nulling adjustments on the 1000-watt range, but R5 and R6 will have to be adjusted to provide a full-scale meter reading at 1000 watts. If insufficient meter deflection is available for nulling adjustments on the 100-watt range, it may be necessary to adjust C1 and C2 at some power level higher than 100 watts. If the capacitors tune through a null, but the meter will not drop all the way to zero, chances are that some rf is leaking into the bridge circuit through stray coupling. If so, it may be necessary to experiment with the shielding of the through-line section of the rf head. If only a small residual reading is noted it will be of minor importance and can be ignored.

With the component values given in Fig. 10, the meter readings track for both power ranges. That is, the 10-watt level on the 100-watt range, and the 100-watt point on the 1000-watt range, fall at the same place on the meter scale, and so on. This no doubt results from the fact that the diodes are conducting in the most linear portion of their curve. Ordinarily, this desirable condition does not exist, making it necessary to plot separate scales for the different power ranges.

Tests indicate that the SWR caused by insertion of the power meter in the transmission line is negligible. It was checked at 28 MHz and no reflected power could be noted on a commercially built rf wattmeter. Similarly, the insertion loss was so low that it could not be measured with ordinary instruments.

**Operation**

It should be remembered that when the bridge is used in a mismatched feed line that has not been properly matched at the antenna, a reflected-power reading will result. The reflected power must be subtracted from the forward power to obtain the actual power output. If the instrument is calibrated for, say, a 50-ohm line, the calibration will not hold for other values of line $Z_0$.

If the instrument is to be used for determining SWR, the reflected-to-forward power ratio can easily be converted into the corresponding voltage ratio for use in the formula given earlier. Since power is proportional to voltage squared, the normalized formula becomes

$$VSWR = \frac{1 + \sqrt{k}}{1 - \sqrt{k}}$$

where k is the ratio of reflected power to forward power. The *power* curve of Fig. 8 is based on the above relationship, and may be used in place of the equation to determine the SWR.

## AN INEXPENSIVE VHF DIRECTIONAL COUPLER

Precision in-line metering devices capable of reading forward and reflected power over a wide range of frequencies are very useful in amateur vhf and uhf work, but their rather high cost puts them out of the reach of many vhf enthusiasts. The device shown in Figs. 13 through 16 is an inexpensive adaptation of their basic principles. It can be made for the cost of a meter, a few small parts, and bits of copper pipe and fittings that can be found in the plumbing stocks at many hardware stores.

### Construction

The sampler consists of a short section of handmade coaxial line, in this instance, of 50 ohms impedance, with a reversible probe coupled to it. A small pickup loop built into the probe is terminated with a resistor at one end and a diode at the other. The resistor matches the impedance of the loop, not the impedance of the line section. Energy picked up by the loop is rectified by the diode, and the resultant current is fed to a meter equipped with a calibration control.

The principal metal parts of the device are a brass plumbing T, a pipe cap, short pieces of 3/4-inch ID and 5/16-inch OD copper pipe, and two coaxial fittings. Other available tubing combinations for 50-ohm line may be usable. The ratio of outer-conductor ID to inner-conductor OD should be 2.4/1. For a sampler to be used with other impedances of transmission line, see Chapter 3 for suitable ratios of conductor sizes. The photographs and Fig. 15 show construction details.

Soldering of the large parts can be done with a 300-watt iron or a small torch. A neat job can be done if the inside of the T and the outside of the pipe are tinned before assembling. When the pieces are reheated and pushed together, a good mechanical and electrical bond will result. If a torch is used, go easy with the heat, as an over-heated and discolored fitting will not accept solder well.

Coaxial connectors with Teflon or other heat-resistant insulation are recommended. Type N, with split-ring retainers for the center conductors, are preferred. Pry the split-ring washers out with a knife point or small screwdriver. Don't lose them, as they'll be needed in the final assembly.

The inner conductor is prepared by making eight radial cuts in one end, using a coping saw with a fine-toothed blade, to a depth of 1/2 inch. The fingers so made are then bent together, forming a tapered end, as shown in Figs. 14 and 15. Solder the center pin of a coaxial fitting into this, again being careful not to overheat the work.

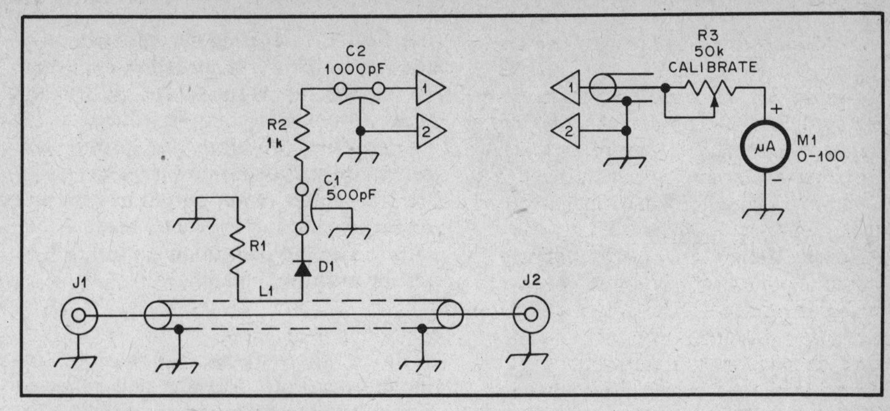


Fig. 13 — Circuit diagram for the line sampler.
C1 — 500-pF feedthrough capacitor, solder-in type.
C2 — 1000-pF feedthrough capacitor, threaded type.
D1 — Germanium diode 1N34, 1N60, 1N270, 1N295, or similar.
J1, J2 — Coaxial connector, type N (UG-58A/U).
L1 — Pickup loop, copper strap 1-inch long × 3/16-inch wide. Bend into "C" shape with flat portion 5/8-inch long.
M1 — 0-100 μA meter.
R1 — Composition resistor, 82 to 100 ohms. See text.
R3 — 50-kΩ composition control, linear taper.

In preparation for soldering the body of the coax connector to the copper pipe, it is convenient to use a similar fitting clamped into a vise as a holding fixture, with the T assembly resting on top, held in place by its own weight. Use the partially prepared center conductor to assure that the coax connector is concentric with the outer conductor. After being sure that the ends of the pipe are cut exactly perpendicular to the axis, apply heat to the coax fitting, using just enough so that a smooth fillet of solder can be formed where the flange and pipe meet.

Before completing the center conductor, check its length. It should clear the inner surface of the connector by the thickness of the split ring on the center pin. File to length; if necessary, slot as with the other end, and solder the center

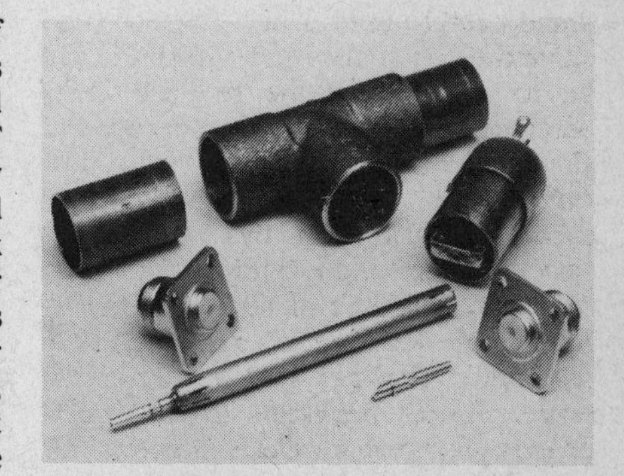

Fig. 14 — Major components of the line sampler. The brass T and two end sections are at the rear in this picture. A completed probe assembly is at the right. The N connectors have their center pins removed. The pins are shown with one inserted in the left end of the inner conductor and the other lying in the right foreground.

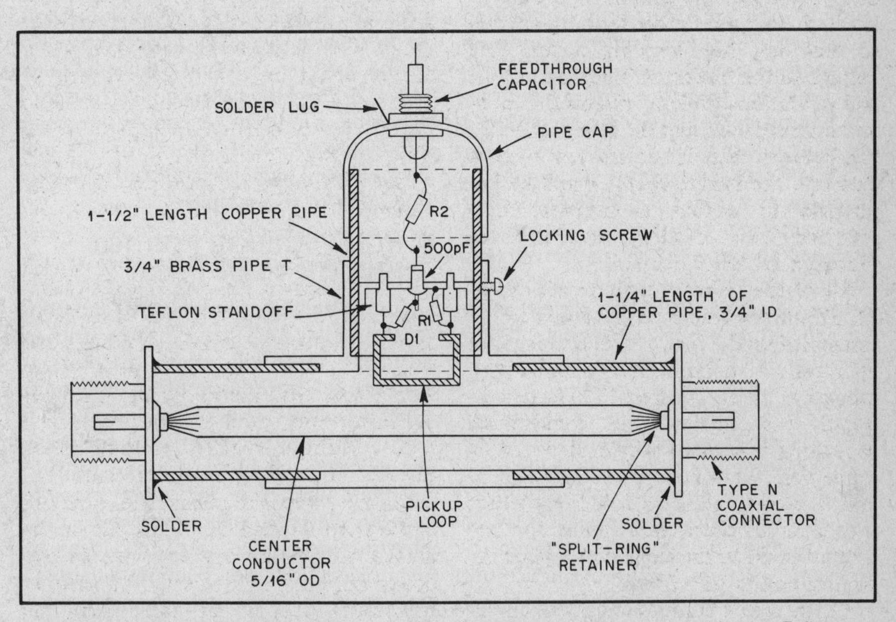


Fig. 15 — Cross-section view of the line sampler. The pickup loop is supported by two Teflon standoff insulators. The probe body is secured in place with one or more locking screws through holes in the brass T.

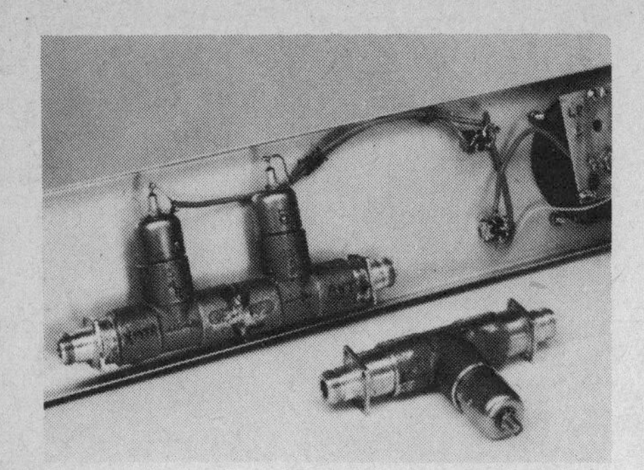

Fig. 16 — Two versions of the line sampler. The single unit described in detail here is in the foreground. Two sections in a single assembly provide for monitoring forward and reflected power without probe reversal.

pin in place. The fitting can now be soldered onto the pipe, to complete the 50-ohm line section.

The probe assembly is made from a 1-1/2-inch length of the copper pipe, with a pipe cap on the top to support the upper feedthrough capacitor, C2. The coupling loop is mounted by means of small Teflon standoffs on a copper disc, cut to fit inside the pipe. The disc has four small tabs around the edge for soldering inside the pipe. The diode, D1, is connected between one end of the loop and a 500-pF feedthrough capacitor, C1, soldered into the disc. The terminating resistor, R1, is connected between the other end of the loop and ground, as directly as possible.

When the disc assembly is completed, insert it into the pipe, apply heat to the outside, and solder the tabs in place by melting solder into the assembly at the tabs. The position of the loop with respect to the end of the pipe will determine the sensitivity of a given probe. For power levels up to 200 watts the loop should extend beyond the face of the pipe about 5/32 inch. For use at higher power levels the loop should protrude only 3/32 inch. For operation with very low power levels the best probe position can be determined by experiment.

The decoupling resistor, R2, and feedthrough capacitor, C2, can be connected, and the pipe cap put in place. The threaded portion of the capacitor extends through the cap. Put a solder lug over it before tightening its nut in place. Fasten the cap with two small screws that go into threaded holes in the pipe.

### Calibration

The sampler is very useful for many jobs, even if it is not accurately calibrated, although it is desirable to calibrate it against a wattmeter of known accuracy. A good 50-ohm dummy load is required.

The first step is to adjust the inductance of the loop or the value of the terminating resistor, for lowest reflected-power reading. The loop is the easier to change. Filing it to reduce its width will increase its impedance. Increasing the cross-section of the loop will lower it, and this can be done by coating it with solder. When the reflected-power reading is reduced as far as possible, reverse the probe and calibrate for forward power by increasing the transmitter power output in steps and making a graph of the meter readings obtained. Use the calibration control, R3, to set the maximum reading.

### Variations

Rather than to use one sampler for monitoring both forward and reflected power by repeatedly reversing the probe, it is better to make two assemblies by mounting two T fittings end-to-end, using one for forward and one for reflected power. The meter can be switched between the probes, or two meters can be used.

The sampler described was calibrated at 146 MHz, as it was intended for 2-meter repeater use. On higher bands the meter reading will be higher for a given power level, and it will be lower for lower frequency bands. Calibration for two or three adjacent bands can be achieved by making the probe depth adjustable, with stops or marks to aid in resetting for a given band. And, of course, more probes can be made, with each calibrated for a given band, as is done in some of the commercially available units.

Other sizes of pipe and fittings can be used by making use of information given in Chapter 3 to select conductor sizes required for the desired impedances. (Since it is occasionally possible to pick up good bargains in 72-ohm line, a sampler for this impedance might be desirable.)

Type N fittings were used because of their constant impedance, and their ease of assembly. Most have the split-ring retainer, which is simple to use in this application. Some have a crimping method, as do apparently all BNC connectors. If a fitting must be used that cannot be taken apart, drill a hole large enough to clear a soldering-iron tip in the copper-pipe outer conductor. A hole of up to 3/8-inch diameter will have very little effect on the operation of the sampler.

## A NOISE BRIDGE FOR 160 THROUGH 10 METERS

The noise bridge, sometimes referred to as an antenna (RX) noise bridge, is an instrument that will allow the user to measure the impedance of an antenna or other electrical circuits. The unit shown in Figs. 17 through 21, designed for use in the 160- through 10-meter range, provides adequate accuracy for most measurements. Battery operation and small physical size make this unit ideal for remote-location use. Tone modulation is applied to the wide-band noise generator as an aid for obtaining a null indication. A detector, such as the station receiver, is required for operation.

Fig. 17 — Interior and exterior views of the noise bridge. The unit is finished in red enamel. Press-on lettering is used for the calibration marks. Note that the potentiometer must be isolated from ground.

The noise bridge consists of two parts — the noise generator and the bridge circuitry. See Fig. 18. A 6.8-volt Zener diode serves as the noise source. U1 generates an approximate 50-percent duty cycle, 1000-Hz, square wave signal which is applied to the cathode of the Zener diode. The 1000-Hz modulation appears on the noise signal and provides a useful null detection enhancement effect. The broadband-noise signal is amplified by Q1, Q2 and associated components to a level that produces an approximate S-9 signal in the receiver. Slightly more noise is available at the lower end of the frequency range, as no frequency compensation is applied to the amplifier. Roughly 20 mA of current is drawn from the 9-volt battery, thus ensuring long battery life — providing the power is switched off after use!

The bridge portion of the circuit consists of T1, C1, C2 and R1. T1 is a trifilar-wound transformer with one of the windings used to couple noise energy into the bridge circuit. The remaining two windings are arranged so that each one is in an arm of the bridge. C1 and R1 complete one arm and the UNKNOWN circuit, along with C2, comprise the remainder of the bridge. The terminal labeled RCVR is for connection to the detector.

### Construction

The noise bridge is contained in a homemade aluminum enclosure that measures 5 × 2-3/8 × 3-3/4 inches. Many of the circuit components are mounted on a circuit board that is fastened to the rear wall of the cabinet. The circuit-board layout is such that the lead lengths to the board from the bridge and coaxial connectors are at a minimum. Etching pattern and parts-placement-guide information for the circuit board are shown in Figs. 20 and 21.

Care must be taken when mounting the potentiometer. For accurate readings the potentiometer must be well insulated from ground. In the unit shown this was accomplished by mounting the control on a piece of Plexiglas, which in turn was fastened to the chassis with a piece of aluminum angle stock. Additionally, a 1/4-inch control-shaft coupling and a length of phenolic rod were used to further isolate the control from ground where the shaft passes through the front panel. A high-quality potentiometer is required if good measurement results are to be obtained.

Mounting the variable capacitor is not a problem since the rotor is grounded. As with the potentiometer, a good grade of capacitor is important. If you must cut corners to save money, look elsewhere in the circuit. Two BNC-type female coaxial fittings are provided on the rear panel for connection to a detector (receiver) and to the UNKNOWN circuit. There is no reason why other types of connectors can't be used. One should avoid the use of plastic-insulated phono connectors, however, as these might influence the accuracy at the higher frequencies. A length of miniature coaxial cable (RG-174/U) is used between the RCVR connector and the appropriate circuit board foils. Also, C2 has one lead attached to the circuit board and the other connected directly to the UNKNOWN circuit connector.

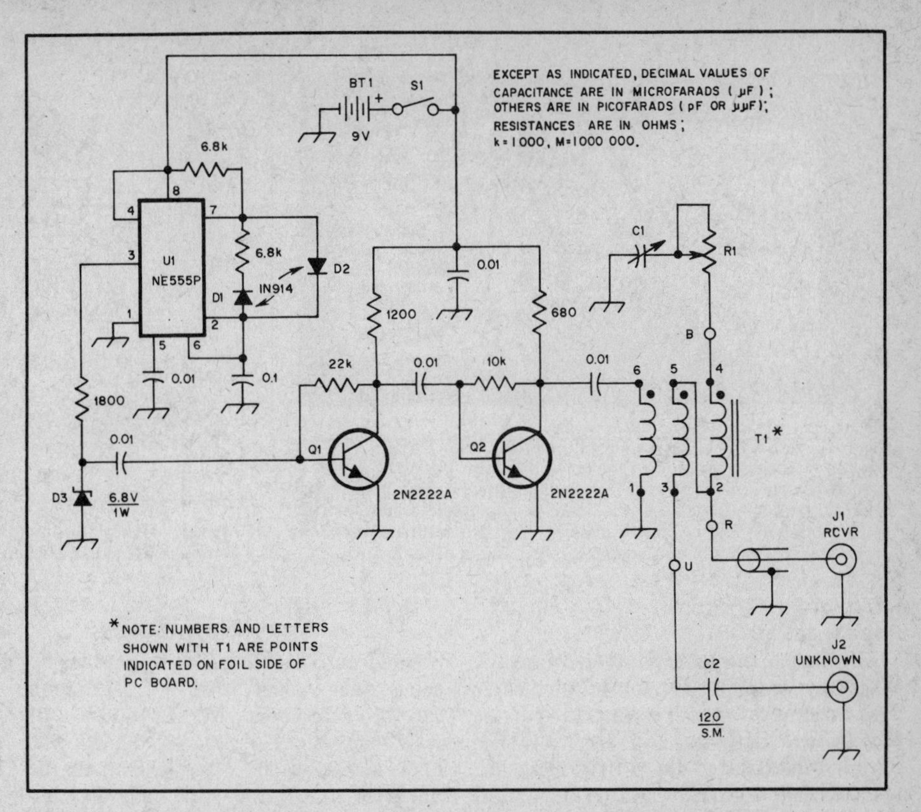


Fig. 18 — Schematic diagram of the noise bridge. Resistors are 1/4-watt composition types. Capacitors are miniature ceramic units unless indicated otherwise. Component designations indicated in the schematic but not called out in the parts list are for text and parts-placement reference only.

BT1 — 9-volt battery, NEDA 1604A or equiv.
C1 — Variable, 250 pF maximum. Use a good grade of capacitor.
C2 — Approximately 1/2 of C1 value. Selection may be necessary — see text.
J1, J2 — Coaxial connector, BNC type.
R1 — Linear, 250 ohm, AB type. Use a good grade of resistor.
S1 — Toggle, spst.
T1 — Broadband transformer, 8-trifilar turns of no. 26 enameled wire on an Amidon FT-37-43 toroid core.
U1 — Timer, NE555 or equiv.

## Calibration and Use

Calibration of the bridge is straightforward and requires no special instruments. A receiver tuned to any portion of the 15-meter band is connected to the RCVR terminal of the bridge. The power is switched on and a broadband noise with a 1000-Hz note should be heard in the receiver. Calibration of the resistance dial should be performed first. This is accomplished by inserting small composition resistors of appropriate values across the UNKNOWN connector of the bridge. The resistors should have the shortest lead lengths possible in order to mate with the connector. Start with 25 ohms of resistance (this may be made up of series or parallel connected units). Adjust the capacitance and resistance dials for a null of the signal as heard in the receiver. Place a calibration mark on the front panel at that location of the resistance dial. Remove the 25-ohm resistor and insert a 50-ohm resistor, 100-ohm unit and so on until the dial is completely calibrated.

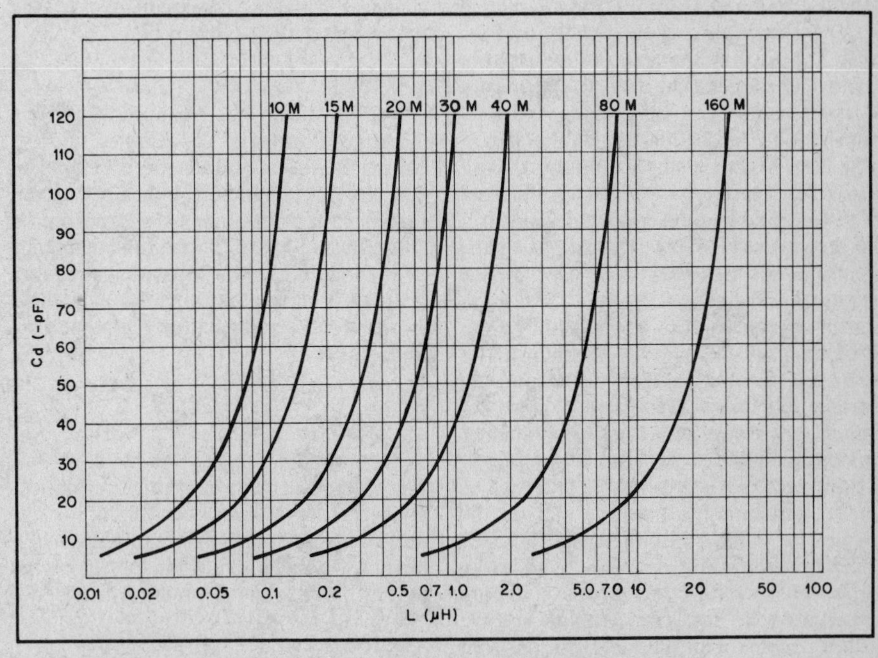


Fig. 19 — Graph for determining actual inductance from the calibration marks on the negative portion of the dial. These curves are accurate only for bridges having 120 pF at C2.

The capacitance dial is calibrated in a similar manner. Initially, this dial is set so that the plates of C1 are exactly half meshed. If a capacitor having no stops is used, orient the knob so as to unmesh the

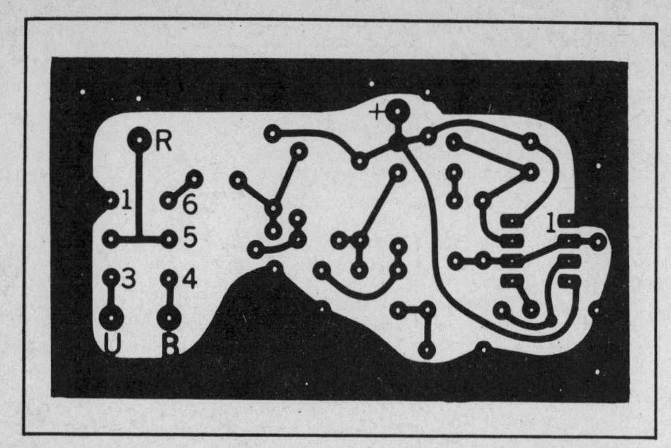


Fig. 20 — Etching pattern for the noise bridge pc board, at actual size. Black represents copper. This is the pattern for the bottom side of the board. The top side of the board is a complete ground plane with a small amount of copper removed from around the component holes. Mounting holes are located in two corners of the board.

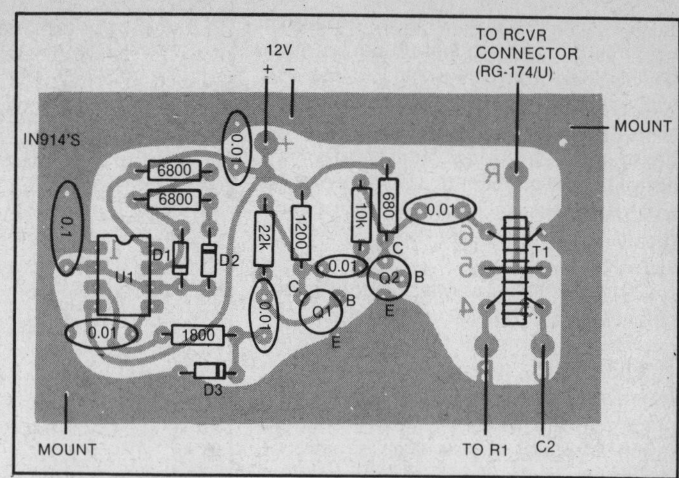


Fig. 21 — Parts-placement guide for the noise bridge as viewed from the component or top side of the board.

plates when the knob is rotated into the positive capacitance region of the dial. A 50-ohm resistor is connected to the UNKNOWN terminal and the resistance control is adjusted for a null. Next, the reactance dial is adjusted for a null and its position is noted. If this setting is significantly different than the half-meshed position, the value of C2 will need to be changed. Unit-to-unit value variations of 120-pF capacitors may be sufficient to provide a suitable unit. Other values can be connected in series or parallel and tried in place of the 120-pF capacitor. The idea is to have the capacitance dial null as close as possible to the half-meshed position of C1.

Once the final value of C2 has been determined and the appropriate component installed in the circuit, the bridge should be adjusted for a null. The 0 reactance point can be marked on the face of the unit. The next step is to place a 20-pF capacitor in series with the 50-ohm load resistor. Use a good grade of capacitor, such as a silver-mica type and keep the leads as short as possible. Null with the capacitance dial and make a calibration mark at that point. Remove the 20-pF capacitor and insert a 40-pF unit in series with the 50-ohm resistor. Again null the bridge and make a calibration mark for 40-pF. Continue on in a similar manner until that half of the dial is completely calibrated.

To calibrate the negative half of the scale, the same capacitors may be used. This time they must be placed temporarily in parallel with C2. Connect the 50-ohm resistor to the UNKNOWN terminal and the 20-pF capacitor in parallel with C2. Null the bridge and place a calibration mark on the panel. Remove the 20-pF unit and temporarily install the 40-pF capacitor. Again null the bridge and make a calibration mark at that point. Continue this procedure until the capacitance dial is completely calibrated. It should be pointed out that the exact resistance and capacitance values used for calibration can be determined by the builder. If resistance values of 20, 40, 60, 80, 100 ohms and so on are more in line with the builder's needs, the scale may be calibrated in those terms. The same is true for the capacitance dial. The accuracy of the bridge is determined by the components that are used in the calibration process.

Many amateurs use a noise bridge simply to find the resonant (nonreactive) impedance of an antenna system. For this service it is necessary only to calibrate the zero-reactance point of the capacitance dial. This simplification relaxes the stringent quality requirement for the bridge capacitors, C1 and C2.

## Operation

The resistance dial is calibrated directly in ohms, but the capacitance dial is calibrated in picofarads of capacitance. The +C half of the dial indicates that the load is capacitive and the −C portion is for inductive loads. To find the reactance of the load when the capacitance reading is positive, the dial setting must be applied to the standard capacitive reactance formula:

$$X = \frac{1}{2\pi fC}$$

The result will be a capacitive reactance.

Inductance values corresponding to negative capacitance dial readings may be taken from the graph of Fig. 19. The reactance is then found from the formula:

$$X = 2\pi fL$$

When using the bridge remember that the instrument measures the impedance of loads as connected at the UNKNOWN terminal. This means that the actual load to be measured must be directly at the connector rather than being attached to the bridge by a length of coaxial cable. Even a short length of cable will transform the load impedance to some other value. Unless the electrical length of line is known and taken into account, it is necessary to place the bridge at the load. An exception to this would be if the antenna were to be matched to the characteristic impedance of the cable. In this case the bridge controls may be preset for 50-ohms resistance and 0-pF capacitance. With the bridge placed at any point along the coaxial line, the load (antenna) may be adjusted until a null is obtained. If the electrical length of line is known to be an integral multiple of half-wavelengths at the frequency of interest, the readings obtained from the bridge will be accurate.

## Interpreting the Readings

A few words on how to interpret the measurements may be in order. For example, assume that the impedance of a 40-meter inverted-V antenna fed with a half-wavelength of cable was measured. The antenna had been cut for roughly the center of the band (7.150 MHz) and the bridge was nulled with the aid of a receiver tuned to that frequency. The results were 45-ohms resistive and 600 picofarads of capacitance. The 45-ohm resistance reading is close to 50 ohms as would be expected for this type of antenna. The capacitive reactance calculates to be 37 ohms from the equation:

$$X = \frac{1}{2\pi(7.15 \times 10^{6})(600 \times 10^{-12})}$$

$$= 37\ \Omega$$

When an antenna is adjusted for resonance, reactance will be zero. Since this antenna looks capacitive it is too short, and wire should be added to each side of the antenna. An approximation of how much wire to add can be made by tuning the receiver higher in frequency until a point is reached where the bridge nulls with the capacitance dial at zero. The percentage difference between this new frequency and the desired frequency indicates the approximate amount that the antenna should be lengthened. The same system will work if the antenna has been cut too long. In this case the capacitance dial would have nulled in the −C region, indicating an inductive reactance. This procedure will work for any directly fed single-element antenna.

## RF IMPEDANCE BRIDGE FOR COAX LINES

The bridge shown in Fig. 22, dubbed the "Macromatcher," may be used to measure unknown complex impedances. Measured values are of equivalent series form, R + jX. With suitable frequency coils, the Macromatcher can be used throughout the frequency range 1.8 to 30 MHz. The useful impedance range of the instrument is from about 5 to 400 ohms if the unknown load is purely resistive, or 10 to 150 ohms resistive component in the presence of appreciable reactance. The reactance range is from 0 to approximately 100 ohms for either inductive or capacitive loads. Although the Macromatcher cannot indicate impedances with the accuracy of a laboratory bridge, its readings are quite adequate for most amateur uses, including the taking of line lengths into account with a Smith chart. By inherent properties of the Macromatcher, its accuracy is best at the centers of the dial calibration ranges.

### The Basic Bridge Circuit

The basic circuit of the Macromatcher is that of Fig. 3B, earlier in this chapter. If C1 and C2 of that circuit are the sections of a differential capacitor, the bridge may be used over a wide impedance range, rather than for a single fixed impedance. A variable ratio in the C1-C2 arms is provided by two identical capacitor sections on the same frame, arranged so that when the shaft is rotated to increase the capacitance of one section, the capacitance of the other section decreases. With a fixed value for Rs, the settings of the capacitor may be calibrated in terms of resistance at Rx.

The circuit of Fig. 3B is modified slightly for use in the Macromatcher, as shown in Fig. 23. The differential capacitor is retained for C1 and C2 to measure resistance. L1 and C3 have been added in series in the "unknown" arm of the bridge, and it is these components which are used to measure the amount and type of reactance at the unknown load. (Both L1 and C3 are adjustable in the actual bridge circuit.) The Macromatcher is initially balanced at the frequency of measurement with a pure resistance in place of R2, Fig. 23, so that the reactances of L1 and of C3 at its midsetting are equal. Thus, these reactances cancel each other in this arm of the bridge, and no reactance is reflected into the remaining bridge arms. For measurement, an unknown complex-impedance load is then connected into the bridge in place of R2. The resistive component of the load is balanced by varying the C1-C2 ratio. The reactive component is balanced by varying C3 either to increase or decrease its capacitive reactance, as required, to cancel any reactance present in the load. If the load is inductive, more capacitive reactance (less capacitance) is required from C3 to obtain a balance. Less reactance (more capacitance) is needed from C3 if the load is capacitive. The end result, after C3 is properly adjusted for the particular unknown load, is that the overall R2 arm of the bridge again looks purely resistive, and a complete null is obtained on the null detector.

The settings of C3 are calibrated in terms of the value and type of reactance at the load terminals. Because of the relationship of capacitive reactance to frequency, the calibration for the dial of the reactance-measuring capacitor is valid at only one frequency. It is therefore convenient to calibrate this dial for equivalent reactances at 1 MHz. Frequency corrections may then be made simply by dividing the reactance dial reading by the measurement frequency in megahertz.

Fig. 24 is the complete schematic diagram of the Macromatcher. C1 is the resistance-measuring capacitor, and L1 and C2 the reactance-measuring components. R1 is the bridge "standard" resistor. Aside from the jacks, all other parts are associated with the null-detector metering section of the circuit. This portion of the circuit is adopted from that of the resistance bridge described earlier in this chapter, and the discussion and precautions which pertain to that instrument in general apply here, as well.

D1 rectifies rf energy present when the bridge is unbalanced, and this energy is filtered into direct current which is metered at M1. The 12-kΩ resistor provides a high-impedance input for the metering circuit, and the 4.7-kΩ resistor at J1 provides a return path for meter-current flow if the input source is capacitance coupled. J4 is for the connection of an external meter, in the event it is desired to observe readings remotely. D2,

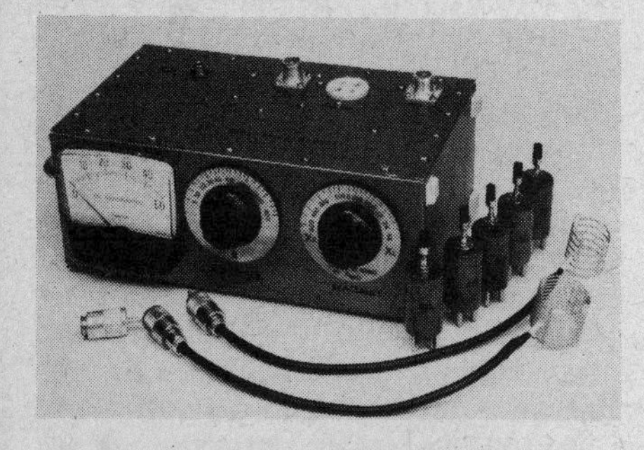

Fig. 22 — An RCL bridge for measuring unknown values of complex impedances. A plug-in coil is used for each frequency band. The bridge operates at an rf input level of about 5 volts; pickup-link assemblies for use with a dip oscillator are shown. Before measurements are made, the bridge must be balanced with a nonreactive load connected at its measurement terminals. This load consists of a resistor mounted inside a coaxial plug, shown in front of the instrument at the left. The aluminum box measures 4-1/4 × 10-3/4 × 6-1/8 inches and is fitted with a carrying handle on the left end and self-sticking rubber feet on the right end and bottom. Dials are Millen no. 10009 with skirts reversed and calibrations added.

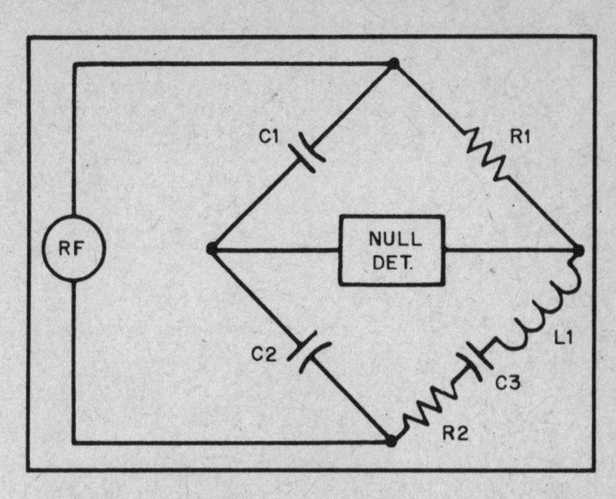


Fig. 23 — The basic circuit of the Macromatcher. In this circuit the bridge is balanced before measurements are made, by setting $X_{L1}$ and $X_{C3}$.

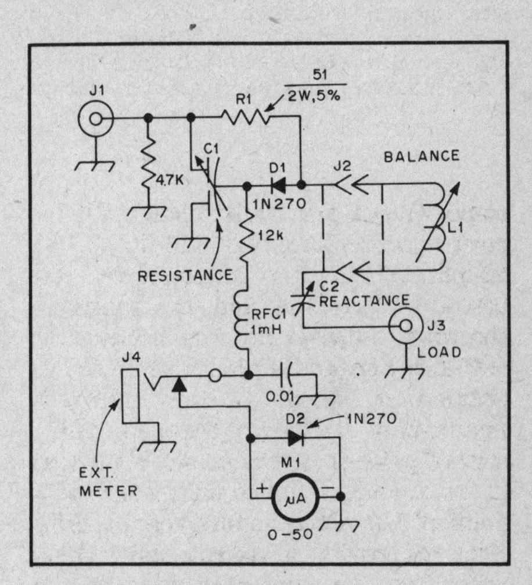


Fig. 24 — Schematic diagram of the Macromatcher. Capacitance is in microfarads; resistance is in ohms, k = 1000. Resistors are 1/2-W 10-percent tolerance unless otherwise indicated.
C1 — Differential capacitor, 11-161 pF per section (Millen 28801 or Jackson Bros. C702/5301 suitable.)
C2 — 17.5-327 pF with straight-line capacitance characteristic (Millen 19335 or Jackson Bros. C9/5070 suitable.)
D1, D2 — Germanium diode, high back resistance.
J1, J3 — Coaxial connector, chassis type.
J2 — To mate plug of L1, ceramic.
J4 — Phone jack, closed-circuit type.
L1 — See text and Table 1.
M1 — 0-50 μA dc (Simpson Model 1223 Bold-vue, Cat. no. 15560 used here).
R1 — For text reference.
RFC1 — Subminiature rf choke (Miller 70F103AI or equiv.).

Fig. 25 — All components except the meter are mounted inside the top of the box. C1 is visible inside the homemade shield at the left, with C2 at the right and J2 mounted between them. J1 is hidden beneath C1 in this view; a part of J3 may be seen in the lower right corner of the box. Components for the dc metering circuit are mounted on a tie-point strip which is affixed to the shield wall for C1; all other components are interconnected with very short leads. The 4.7-kΩ input resistor is connected across J1. This photograph was made before the diode was connected across the terminals of M1.

**Table 1**
**Coil Data for RF Impedance Bridge**

| *Band* | *Nominal Inductance Range, μH* | *Frequency Coverage, MHz* | *Coil Type or Data* |
|---|---|---|---|
| 160 | 27.5-58 | 1.6-2.3 | Miller 42A475CBI. |
| 80 | 6.5-13.8 | 3.2-4.8 | 28 turns no. 30 enam. wire close-wound on Miller form 42A000CBI. |
| 40 | 2.0-4.4 | 5.8-8.5 | Miller 42A336CBI or 16 turns no. 22 enam. wire close-wound on Miller form 42A000CBI. |
| 20 | 0.6-1.1 | 11.5-16.6 | 8 turns no. 18 enam. wire close-wound on Miller form 42A000CBI. |
| 15 | 0.3-0.48 | 18.5-23.5 | 4-1/2 turns no. 18 enam. wire close-wound on Miller form 42A000CBI. |
| 10 | 0.18-0.28 | 25.8-32.0 | 3 turns no. 16 or 18 enam. or tinned bus wire spaced over 1/4-inch winding length on Miller form 42A000CBI. |

placed directly across M1, protects the meter from over-current surges. Although it appears from the schematic diagram that this germanium diode will shunt out all meter current, such is not the case in actual operation because approximately 250 millivolts must be developed across the diode before it begins to conduct an appreciable amount of current. The internal resistance of a typical 50-μA meter is 1800 or 2000 ohms, and this means that more than 100 μA of current must be flowing through the meter before the diode shunting effect becomes appreciable. In operation, this diode prevents the meter needle from slamming against the peg if the load is disconnected while input rf excitation is still applied; the needle eventually reaches full scale, but travels more slowly with the diode in the circuit.

### Construction

In any rf-bridge type of instrument, the leads must be kept as short as possible to reduce stray reactances. Placement of component parts, while not critical, must be such that lead lengths greater than about 1/2 inch (except in the dc metering circuit) are avoided. Shorter leads are desirable, especially for R1, the standard resistance for the bridge. In the unit photographed, the body of this resistor just fits between the terminals of C1 and J2 where it is connected. C1 should be enclosed in a shield and connections made with leads passing through holes drilled through the shield wall. The frames of both variable capacitors, C1 and C2, must be insulated from the chassis, such as on ceramic pillars, with insulated couplings used on the shafts. As Fig. 25 indicates, all parts of the bridge except the meter and the calibrated dials are mounted inside the top panel of the box. The dials are front-panel mounted on shafts with panel bearings.

Band-switching arrangements for L1 complicate the construction and contribute to stray reactances in the bridge circuit. For these reasons plug-in coils are used at L1, one coil for each band over which the instrument is used. The coils must be adjustable, to permit initial balancing of the bridge with C2 set at the zero-reactance calibration point. Coil data are given in Table 1. Millen 45004 coil forms, with the coils supported inside, provide a convenient method of constructing these slug-tuned plug-in coils. A phenolic washer cut to the proper diameter is epoxied to the top or open end of each form, giving a rigid support for mounting of the coil by its bushing. Small knobs for 1/8-inch shafts, threaded with a no. 6-32 tap, are screwed onto the coil slug-tuning screws to permit ease of adjustment without a tuning tool. Knobs with setscrews should be used to prevent slipping. A ceramic socket to mate with the pins of the coil form is used for J2.

### A Nonreactive Termination

For calibrating the reactance dial and for initially balancing the Macromatcher each time it is used on a new frequency, a purely resistive load is required for connection at J3. A suitable load which is essentially nonreactive can be made by mounting a 51- or a 56-ohm 1-W composition (carbon) resistor inside a PL-259 plug.

The body of the resistor should be inserted as far as possible into the plug, with one resistor lead extending through the center-conductor pin. Solder this center-pin connection, and clip off any excess lead length. Make a 1/2-in.-dia copper or brass disc with a small hole at its center. Use a 1/16-in. or, preferably, a no. 60 drill to make this hole. (Initially the "disc" may be a square or rough-cut piece of metal. It may be rounded by filing or grinding after the assembly process is completed.) Place the shell of the plug over its body, and then slip the disc over the grounded-end lead of the resistor, so the resistor lead protrudes through the small hole. First solder the disc to the body of the plug and then clip off any excess lead length from the resistor. Next, solder the connection at the small hole. The disc, when assembled in this manner, completes the shielding, reduces lead inductance, and also prevents the shell of

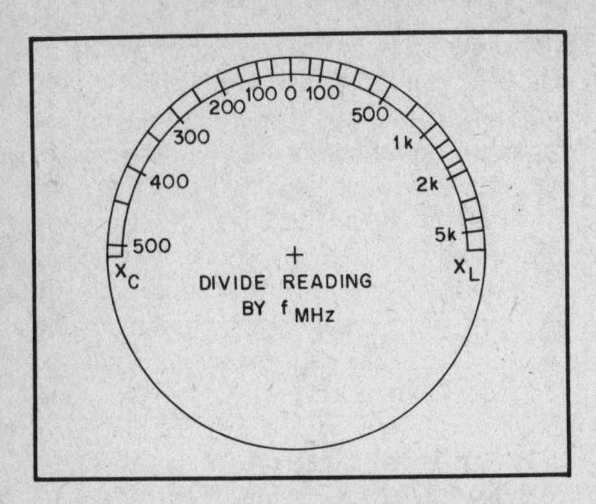


Fig. 26 — Calibration scale for the reactance dial used at C2 of the Macromatcher. See text.

the plug from being removed completely.

**Calibration**

The resistance dial of the bridge may be calibrated by using a number of 1/2- or 1-watt 5-percent-tolerance composition resistors of different values in the 5- to 400-ohm range as loads. For this calibration, the appropriate frequency coil must be inserted at J2 and its inductance adjusted for the best null reading on the meter *when C2 is set with its plates half meshed.* For each test resistor, C1 is then adjusted for a null reading. Alternate adjustment of L1 and C1 should be made for a complete null, without touching C2. The leads between the test resistor and J3 should be as short as possible, and the calibration preferably should be done in the 3.5-MHz band where stray inductance and capacitance will have the least effect.

If the constructional layout of the bridge closely follows that shown in the photographs, the calibration scale of Fig. 26 may be used for the reactance dial. This calibration was obtained by connecting various reactances, measured on a laboratory bridge, in series with a 47-ohm 1-W resistor connected at J3. The scale is applied so that maximum capacitive reactance is indicated with C2 fully meshed.

If it is desired to obtain an individual calibration for C2, known values of inductance and capacitance may be used in series with a fixed resistor of the same approximate value as R1. For this calibration it is *very important* to keep the leads to the test components as short as possible, and calibration should be performed in the 3.5-MHz range to minimize the effects of stray reactances. Begin the calibration by setting C2 at half mesh, marking this point as 0 ohms reactance. With a purely resistive load connected at J3, adjust L1 and C1 for the best null on M1. From this point on during calibration, do not adjust L1 except to rebalance the bridge for a new calibration frequency. The ohmic value of the known reactance for the frequency of calibration is multiplied by the frequency in megahertz to obtain the calibration value for the dial.

**Using the Impedance Bridge**

This instrument is a low-input-power device, and is *not* of the type to be excited from a transmitter or left in the antenna line during station operation. Sufficient sensitivity for all measurements results when a 5-V rms rf signal is applied at J1. This amount of voltage can be delivered by most dip oscillators. In no case should the power applied to J1 exceed 1 watt or damage to the instrument may result. The input impedance of the bridge at J1 is low, in the order of 50 to 100 ohms, so it is convenient to excite the bridge through a length of 52- or 75-ohm line such as RG-58/U or RG-59/U. If a dip oscillator is used, a link coupling arrangement to the oscillator coil may be used. Fig. 22 shows two pickup link assemblies. The larger coil, 10 turns of 1-1/4-inch-dia stock with turns spaced at 8 turns per inch (B&W 3018), is used for the 160-, 80-, 40- and 20-meter bands. The smaller coil, 5 turns of 1-inch-dia stock with turns spaced at 4 turns per inch (B&W 3013), is used for the 15- and 10-meter bands. Coupling to the oscillator should be as light as possible, while obtaining sufficient sensitivity, to prevent severe "pulling" of the oscillator frequency. Overcoupling may cause the oscillator to shift in frequency by a few hundred kilohertz, so for the most reliable measurements, a receiver should be used to check the oscillator frequency.

Before measurements are made, it is necessary to balance the bridge. Set the reactance dial at zero and adjust L1 and C1 for a null with a nonreactive load connected at J3. This null should be complete; if not, reduce the signal level being applied to the Macromatcher. The instrument must be rebalanced after any appreciable change is made in the measurement frequency, more than approximately 1 percent. After the bridge is balanced, connect the unknown load to J3 and alternately adjust C1 and C2 for the best null. Measured impedances are of equivalent series form, $R + jX/f$, where R and X are the Macromatcher dial readings, and f is the frequency in megahertz. When the reactive component, X, is divided by the frequency, the result is $R + jX$ in ohms.

As shown in Fig. 26, the calibration of the reactance dial is nonlinear, with a maximum indication for capacitive reactance of 500/f. The measurement range for capacitive loads may be extended by "zeroing" the reactance dial at some value other than 0. For example, if the bridge is initially balanced with the reactance dial set at 500 in the $X_L$ range, the 0 dial indication is now equivalent to an $X_C$ reading of 500/f, and the total range of measurement for $X_C$ has been extended to 1000/f.

When the Macromatcher is used at the antenna, excitation may be "piped" to the instrument through the coaxial line which normally feeds the antenna. Unless an assistant can check the oscillator frequency during each measurement, however, a dip oscillator is unsatisfactory for this type of work. A more stable frequency source, such as a signal generator or low-power transmitter capable of delivering approximately 100 to 200 milliwatts, is ideal, as it can be left running during the time measurements and adjustments are being made. (Alternatively, the "power absorber" circuit of Fig. 7 may be used with higher power transmitters.) Here is where the Macromatcher can really be of value for adjustment of matching networks such as the L, gamma and hairpin, because the resistive and reactive components of the load are indicated separately. In these networks one adjustable element affects primarily the resistive component (the rod length of the gamma or the physical length of the hairpin), while the other adjustment affects primarily the reactive component (gamma-capacitor setting or driven-element length with the hairpin match). Of course there is some amount of interaction in the two adjustments, but the effects of making just one adjustment can be seen immediately on the Macromatcher. Obtaining an acceptable match in a matter of a few minutes is simple — adjust one of the two variables for the proper resistance, adjust the other variable for zero reactance, perform a slight touchup on these adjustments, and the job is finished.

Of course it is not necessary to use the Macromatcher at the load to determine the impedance. Measurements may be performed through an electrical half wavelength of feed line. Disregarding attenuation (which should be negligible if the line is only a single half wave in length), the impedance will be the same at the input end of the line as it is at the load, no matter what the line impedance may be. Nor is it necessary to trim the coaxial line to an exact half wavelength (good for a single frequency only) in order to make "remote" measurements accurately. The line may be of any convenient physical length, but its electrical length must be known. Readings taken at the input end of the line can be converted into actual impedances at the termination point of the line by means of a Smith Chart, as described in Chapter 3. Line attenuation may also be taken into account.

## MEASURING SOIL CONDUCTIVITY

An important parameter for both vertical and horizontal antennas is soil conductivity. For horizontal antennas, the energy reflected from the earth beneath it affects the antenna impedance, thereby affecting the SWR and the current flowing in the antenna elements, which in turn affects the distant signal strength.

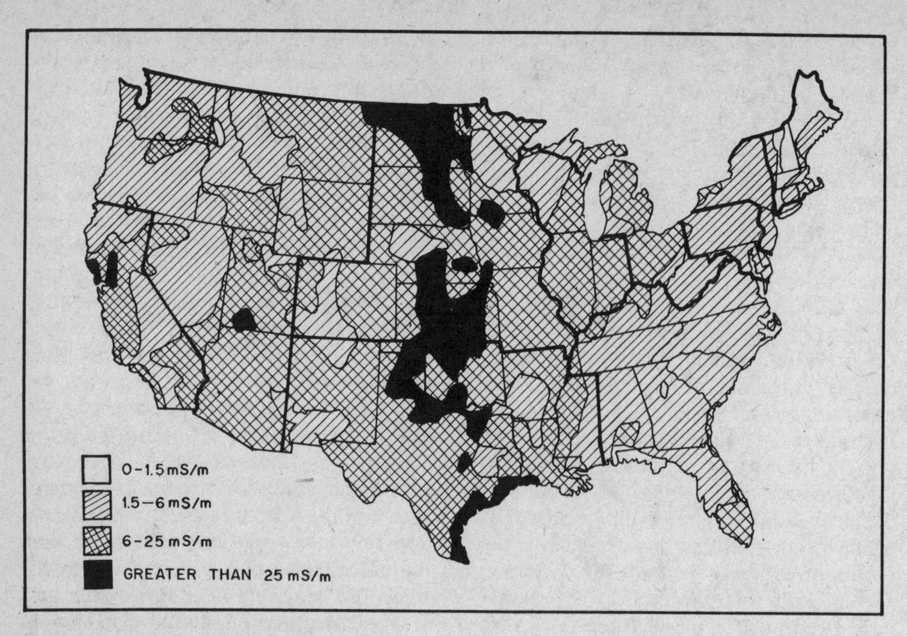


Fig. 27 — Approximate soil conductivity in the continental U.S. This information was adapted from U.S. government publications.

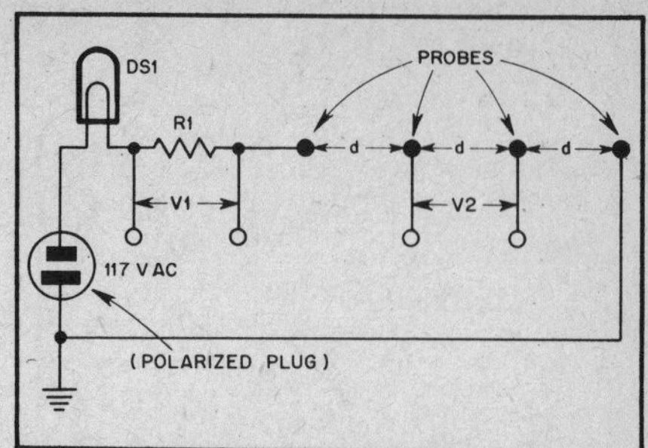


Fig. 29 — Schematic diagram, four-point probe method for measuring earth conductivity.
DS1 — 100-W electric light bulb.
R1 — 14.6 ohms, 5 W. A suitable resistance can be made by paralleling five 1-W resistors, three of 68 Ω and two of 82 Ω. (The dissipation rating of this combination will be 4.7 W.)
Probes — See text and Fig. 28.

**Table 2**
**General Classification of Conductivity**

| *Material* | *Conductivity (millisiemens per meter)* |
|---|---|
| Poor Soil | 1-5 |
| Average Soil | 10-15 |
| Very Good Soil | 100 |
| Salt Water | 5000 |
| Fresh Water | 10-15 |

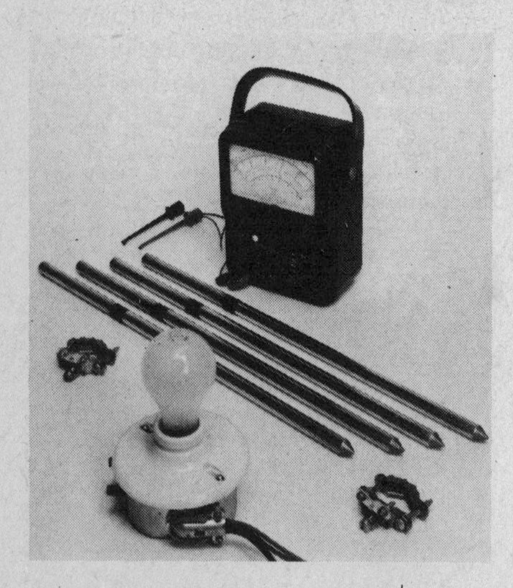

Fig. 28 — The complete soil conductivity measuring setup. The four probes are cut to 18 inches in length from an 8-foot copper-coated steel ground rod. (This length provides a measuring stick for spacing the probes when driving them into the soil.) The tip of each probe is ground to a point, and black electricians' tape indicates the depth to which it is to be driven for measurements. Two ground clamps provide for connections to the driven probes.

(This is discussed in more detail in Chapter 2.) The conductivity of the ground within several wavelengths of the antenna also affects the ground reflection factors discussed in Chapter 2.

The conductivity of the soil under and in the near vicinity of a vertical antenna is most important in determining the extent of the radial system required and the overall performance. Short verticals with very small radial systems can be surprisingly effective.

Most soils are nonconductors of electricity when completely dry. Conduction through the soil results from conduction through the water held in the soil. Thus, conduction is electrolytic. Dc techniques for measuring conductivity are impractical because they tend to deplete the carriers of electricity in the vicinity of the electrodes. The main factors contributing to the conductivity of soil are

1) Type of soil.
2) Type of salts contained in the water.
3) Concentration of salts dissolved in the contained water.
4) Moisture content.
5) Grain size and distribution of material.
6) Temperature.
7) Packing density and pressure.

Although the type of soil is an important factor in determining its conductivity, rather large variations can take place between locations because of the other factors involved. Generally, loams and garden soils have the highest conductivities. These are followed in order by clays, sand and gravel. Soils have been classified according to conductivity, as shown in Table 2. Although some differences are noted in the reporting of this mode of classification because of the many variables involved, the classification generally follows the values shown in the table. Approximate soil conductivities for the continental U.S. are shown in Fig. 27.

Fig. 30 — A standard 3-1/2-in. electrical outlet box and a porcelain ceiling fixture may be used to construct the soil conductivity test set. The resistors comprising R1 are mounted on a tie-point strip inside the box, and test-point jacks provide for measuring the voltage drop across the resistor combination. Leads exiting the box through the cable clamp are protected with several layers of electricians' tape. These leads run approximately 4 feet to the power plug and to small alligator clips for attachment to the ground clamps shown in Fig. 28. Large clips such as for connecting to automotive battery posts may be used instead of ground clamps.

### Making Conductivity Measurements

Since conduction through the soil is almost entirely electrolytic, ac measurement techniques are preferable. Many commercial instruments using ac techniques are available and described in the literature. But rather simple ac measurement techniques can be used that provide accuracies on the order of 25% and are quite adequate for the radio amateur.

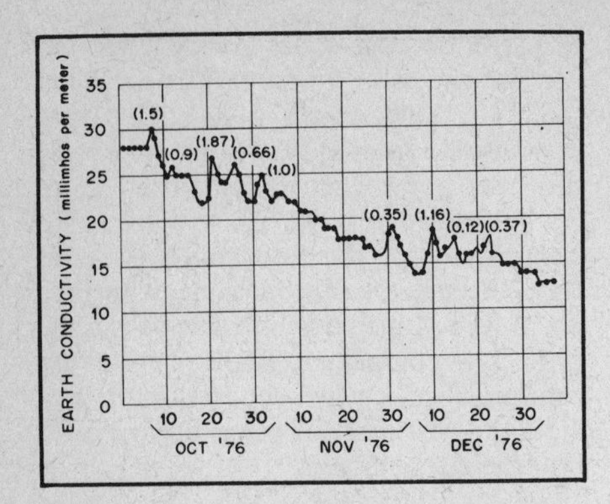


Fig. 31 — Earth conductivity at a central New Jersey location during a three-month period. Numbers in parentheses indicate inches of rainfall.

Such a setup was developed by Jerry Sevick, W2FMI, and M. C. Waltz, W2FNQ. It is shown in Figs. 28 through 30.

Four probes are used. Each is 9/16 inch in diameter, and may be made of either iron or copper. The probes are inserted in a straight line at a spacing of 18 inches (dimension d in Fig. 29). The penetration depth is 12 inches. *Caution*: Do not insert the probes with power applied! A shock hazard exists! After applying power, measure the voltage drops V1 and V2, as shown in the diagram. Depending on soil conditions, readings should fall in the range from 2 to 10 volts.

Earth conductivity, c, may be determined from

$$c = 21 \times \frac{V1}{V2} \text{ millisiemens per meter}$$

For example, assume the reading across the resistor (V1) is 4.9 V, and the reading between the two center probes (V2) is 7.2 V. The conductivity is calculated as $21 \times 4.9/7.2 = 14$ mS/m.

Soil conditions may not be uniform in different parts of your yard. A few quick measurements will reveal whether this is the case or not.

Fig. 31 shows the conductivity readings taken in one location over a period of three months. It is interesting to note the general drop in conductivity over the three months, as well as the short-term changes from periods of rain.

## A SWITCHABLE RF ATTENUATOR

A switchable rf attenuator is helpful in making antenna gain comparisons or plotting antenna radiation patterns; attenuation may be switched in or out of the line leading to the receiver to obtain an initial or reference reading on a signal-strength meter. Some form of attenuator is also required for locating hidden transmitters, where the real trick is pinpointing the signal source from within a few hundred feet. At such a close distance, strong signals may overload the front end of the receiver, making it impossible to obtain any indication of a bearing.

The attenuator of Figs. 32 and 33 is designed for low power levels, not exceeding 1/4 watt. If for some reason the attenuator will be connected to a transceiver, a means of bypassing the unit during transmit periods must be devised. An attenuator of this type is commonly called a step attenuator, because any amount of attenuation from 0 dB to the maximum available (81 dB for this particular instrument) may be obtained in steps of 1 dB. As each switch is successively thrown from the OUT to the IN position, the attenuation sections add in cascade to yield the total of the attenuator steps switched in. The maximum attenuation of any single section is limited to 20 dB because leak-through would probably degrade the accuracy of higher values.

Fig. 32 — A construction method for a step attenuator. Double-sided circuit-board material, unetched (except for panel identification), is cut to the desired size and soldered in place. Flashing copper may also be used, although it is not as sturdy. Shielding partitions between sections are necessary to reduce signal leakage. Brass nuts soldered at each of the four corners allow machine screws to secure the bottom cover. The practical limit for total attenuation is 80 or 90 dB, as signal leakage around the outside of the attenuator will defeat attempts to obtain much greater amounts. A kit of parts is available from Circuit Board Specialists, P.O. Box 969, Pueblo, CO 81002.

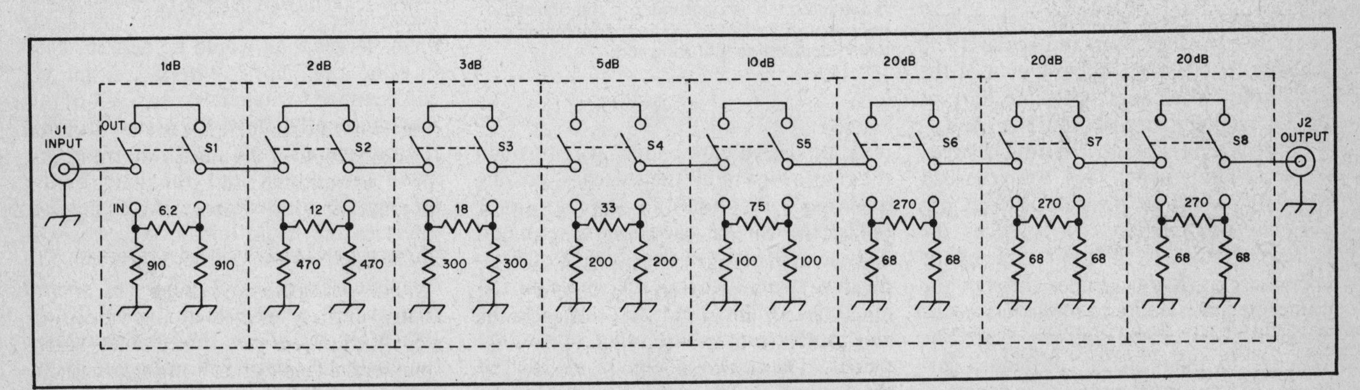


Fig. 33 — Schematic diagram of the step attenuator, designed for a nominal impedance of 52 ohms. Resistance values are in ohms. Resistors are 1/4-watt, carbon-composition types, 5% tolerance. Broken lines indicate walls of circuit-board material. A small hole is drilled through each partition wall to route bus wire. Keep all leads as short as possible. The attenuator is bilateral; i.e., the input and output ends may be reversed.

J1, J2 — Female BNC connectors, Radio Shack 278-105 or equiv.

S1-S8, incl. — Dpdt slide switches, standard size. (Avoid subminiature or toggle switches.) Stackpole S-5022CD03-0 switches are used here.

The tolerance of resistor values also becomes more significant regarding accuracy at higher attenuation values.

A good-quality commercially made attenuator will cost upwards from $150, but for less than $25 in parts and a few hours of work, an attenuator may be built at home. It will be suitable for frequencies up to 450 MHz. Double-sided pc board is used for the enclosure. The version of the attenuator shown in Fig. 32 has identification lettering etched into the top surface (or front panel) of the unit. This adds a nice touch and is a permanent means of labeling. Of course rub-on transfers or Dymo tape labels could be used as well.

Female BNC single-hole, chassis-mount connectors are used at each end of the enclosure. These connectors provide a means of easily connecting and disconnecting the attenuator.

**Construction**

After all the box parts are cut to size and the necessary holes made, scribe light lines to locate the inner partitions. Carefully tack-solder all partitions in position. A 25-watt pencil type of iron should provide sufficient heat. Dress any pc board parts that do not fit squarely. Once everything is in proper position, run a solder bead all the way around the joints. Caution! Do not use excessive amounts of solder, as the switches must later be fit flat inside the sections. The top, sides, ends and partitions can be completed. Dress the outside of the box to suit your taste. For instance, you might wish to bevel the box edges. Buff the copper with steel wool, add lettering, and finish off the work with a coat of clear lacquer or polyurethane varnish.

Using a little lacquer thinner or acetone (and a lot of caution), soak the switches to remove the grease that was added during their manufacture. When they dry, spray the inside of the switches lightly with a TV tuner cleaner/lubricant. Use a sharp drill bit (about 3/16 inch will do), and countersink the mounting holes on the actuator side of the switch mounting plate. This ensures that the switches will fit flush against the top plate. At one end of each switch, bend the two lugs over and solder them together. Cut off the upper halves of the remaining switch lugs. (A close look at Fig. 32 will help clarify these steps.)

Solder the series-arm resistors between the appropriate switch lugs. Keep the lead lengths as short as possible and do not overheat the resistors. Now solder the switches in place to the top section of the enclosure by flowing solder through the mounting holes and onto the circuit-board material. Be certain that you place the switches in their proper positions; correlate the resistor values with the degree of attenuation. Otherwise, you may wind up with the 1-dB step at the wrong end of the box — how embarrassing!

Once the switches are installed, thread a piece of no. 18 bare copper wire through the center lugs of all the switches, passing it through the holes in the partitions. Solder the wire at each switch terminal. Cut the wire between the poles of each individual switch, leaving the wire connecting one switch pole to that of the neighboring one on the other side of the partition, as shown in Fig. 32. At each of the two end switch terminals, leave a wire length of approximately 1/8-inch. Install the BNC connectors and solder the wire pieces to the connector center conductors.

Now install the shunt-arm resistors of each section. Use short lead lengths. Do not use excessive amounts of heat when soldering. Solder a no. 4-40 brass nut at each inside corner of the enclosure. Recess the nuts approximately 1/16-inch from the bottom edge of the box to allow sufficient room for the bottom panel to fit flush. Secure the bottom panel with four no. 4-40, 1/4-inch machine screws and the project is completed. Remember to use caution, always, when your test setup provides the possibility of transmitting power into the attenuator.

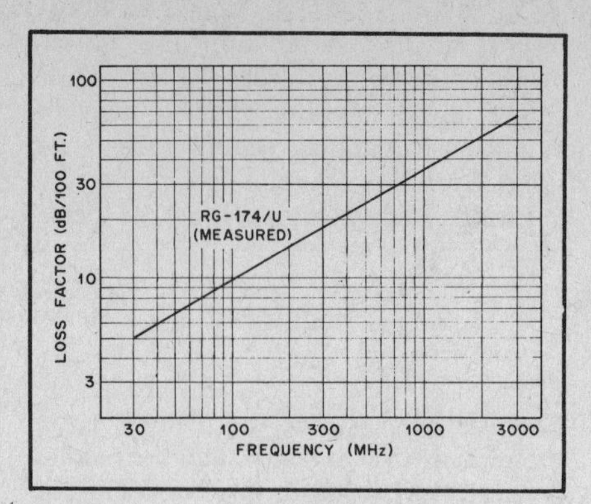


Fig. 34 — Loss factor of RG-174/U used in the calorimeter.

**Table 3**

**Calculated Input SWR for 68 Feet of Unterminated RG-174/U Cable**

| *Freq. (MHz)* | *SWR* |
|---|---|
| 50 | 2.08 |
| 144 | 1.35 |
| 220 | 1.20 |
| 432 | 1.06 |
| 1296 | 1.003 |
| 2304 | 1.0003 |

Fig. 35 — The calorimeter ready for use. The roll of cable is immersed in one quart of water in the left-hand compartment of the Styrofoam container. Also shown is the thermometer, which doubles as a stirring rod.

## A CALORIMETER FOR VHF AND UHF POWER MEASUREMENTS

A quart of water in a Styrofoam ice bucket, a roll of small coaxial cable and a thermometer are all the necessary ingredients for an accurate rf wattmeter, whose calibration is independent of frequency. The wattmeter works on the calorimeter principle: A given amount of rf energy is equivalent to an amount of heat, which can be determined by measuring the temperature rise of a known quantity of thermally insulated material. This principle is used in many of the more accurate high-power wattmeters. This procedure was developed by James Bowen, WA4ZRP.

The roll of coaxial cable serves as a dummy load to convert the rf power into heat. RG-174/U cable was chosen for use as the dummy load in this calorimeter because of its high loss factor, small size, and low cost. It is a standard 50-ohm cable approximately 0.11 inch in diameter. A prepackaged roll marked as 60 feet long, but measured to be 68 feet, was purchased at a local electronics store. A plot of measured RG-174/U loss factor as a function of frequency is shown in Fig. 34.

In use, the end of the cable not connected to the transmitter is left open circuited. Thus at 50 MHz, the reflected wave returning to the transmitter (after making a round trip of 136 feet through the cable) is 6.7 dB × 1.36 = 9.11 dB below the forward wave. A reflected wave 9.11 dB down represents an SWR to the transmitter of 2.08:1. While this value seems larger than would be desired, keep in mind that most 50-MHz transmitters can be tuned to match into an SWR of this magnitude efficiently. To assure accurate results, merely tune the transmitter for maximum power into the load before making the measurement. At higher frequencies the cable loss increases so the SWR goes down. Table 3 presents the calculated input SWR values at several frequencies for 68 feet of RG-174/U. At 1000 MHz and above, the SWR caused by the cable connector will undoubtedly exceed the very low cable SWR listed for these frequencies.

In operation, the cable is submerged in a quart of water and dissipated heat energy flows from the cable into the water, raising the water temperature. See Fig. 35. The calibration of the wattmeter is based on the physical fact that one calorie of heat energy will raise one gram of liquid water 1 degree Celsius. Since one quart of water contains 946.3 grams, the transmitter must deliver 946.3 calories of heat energy to the water to raise its temperature 1 degree Celsius. One calorie of energy is equivalent to 4.186 joules and a joule is equal to 1 watt for 1 second. Thus, the heat capacitance of 1 quart of water expressed in joules is 946.3 × 4.186 = 3961 joules/°C.

The heat capacitance of the cable is small with respect to that of the water, but nevertheless its effect should be included for best accuracy. The heat capacitance of the cable was determined in the manner described below.

The 68-foot roll of RG-174/U cable was raised to a uniform temperature of 100° C by immersing it in a pan of boiling water for several minutes. A quart of tap water was poured into the Styrofoam ice bucket and its temperature was measured at 28.7° C. The cable was then transferred quickly from the boiling water to the water in the ice bucket. After the water temperature in the ice bucket had ceased to rise, it measured 33.0° C. Since the total heat gained by the quart of water was equal to the total heat lost from the cable, we can write the following equation:

$$(\Delta T_{WATER})(C_{WATER}) = -(\Delta T_{CABLE})(C_{CABLE})$$

where

$\Delta T_{WATER}$ = the change in water temperature
$C_{WATER}$ = the water heat capacitance
$\Delta T_{CABLE}$ = the change in cable temperature
$C_{CABLE}$ = the cable heat capacitance

Substituting and solving:

$$(33.0 - 28.7)(3961) = -(33.0 - 100)(C_{CABLE})$$

$$\frac{(4.3)(3961)}{67} = C_{CABLE}$$

$$254 \text{ joules/° C} = C_{CABLE}$$

Thus the total heat capacitance of the water and cable in the calorimeter is 3961 + 254 = 4215 joules/° C. Since 1° F = 5/9° C, the total heat capacitance can also be expressed as 4215 × 5/9 = 2342 joules/° F.

**Materials and Construction**

The quart of water and cable must be thermally insulated to assure that no heat is gained from or lost to the surroundings. A Styrofoam container is ideal for this purpose since Styrofoam has a very low thermal conductivity and a very low thermal capacitance. A local variety store was the source of a small Styrofoam cold chest with compartments for carrying sandwiches and drink cans. The rectangular compartment for sandwiches was found to be just the right size for holding the quart of water and coax.

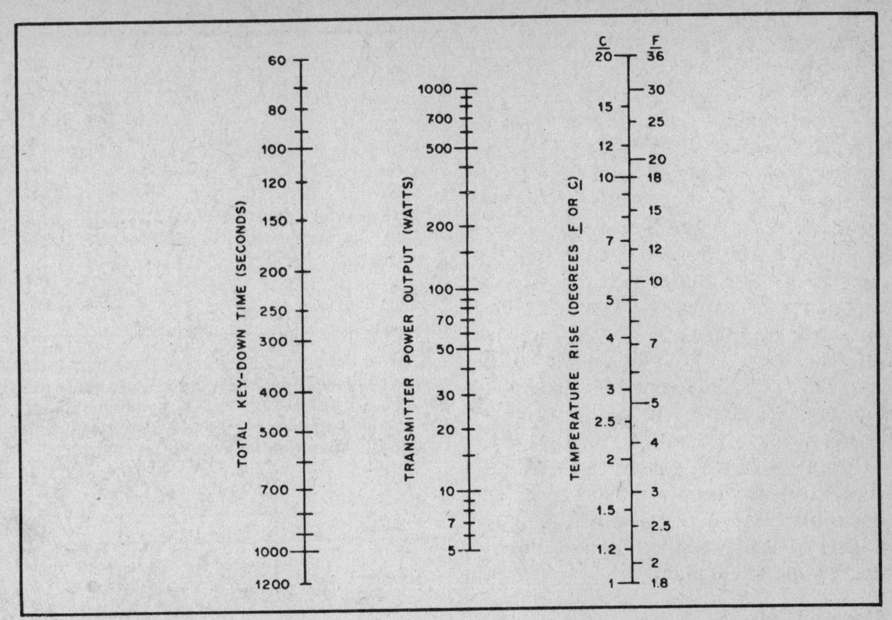


Fig. 36 — Nomogram for finding transmitter power output for the calorimeter.

The thermometer can be either a Celsius or Fahrenheit type, but try to choose one which has divisions for each degree spaced wide enough so that the temperature can be estimated readily to one-tenth degree. Photographic supply stores carry darkroom thermometers, which are ideal for this purpose. In general, glass bulb thermometers are more accurate than mechanical dial-pointer types.

The rf connector on the end of the cable should be a constant-impedance type. A BNC type connector especially designed for use on 0.11-inch diameter cable was located through surplus channels. If you cannot locate one of these, wrap plastic electrical tape around the cable near its end until the diameter of the tape wrap is the same as that of RG-58/U. Then connect a standard BNC connector for RG-58/U in the normal fashion.

Carefully seal the opposite open end of the cable with plastic tape or silicone caulking compound so that no water can leak into the cable at this point.

**Procedure for Use**

Pour 1 quart of water (4 measuring cups) into the Styrofoam container. As long as the water temperature is not very hot or very cold, it is unnecessary to cover the top of the Styrofoam container during measurements. Since the transmitter will eventually heat the water several degrees, water initially a few degrees cooler than air temperature is ideal because the average water temperature will very nearly equal the air temperature and heat transfer to the air will be minimized.

Connect the RG-174/U dummy load to the transmitter through the shortest possible length of lower-loss cable such as RG-8/U. Tape the connectors and adapter at the RG-8/U to RG-174/U joint carefully with plastic tape to prevent water from leaking into the connectors and cable at this point. Roll the RG-174/U into a loose coil and submerge it in the water. Do not bind the turns of the coil together in any way, as the water must be able to freely circulate among the coaxial cable turns. All the RG-174/U cable must be submerged in the water to ensure sufficient cooling. Also submerge part of the taped connector attached to the RG-174/U as an added precaution.

Upon completing the above steps, quickly tune up the transmitter for maximum power output into the load. Cease transmitting and stir the water slowly for a minute or so until its temperature has stabilized. Then measure the water temperature as precisely as possible. After the initial temperature has been determined, begin the test "transmission," measuring the total number of seconds of key-down time accurately. Stir the water slowly with the thermometer and continue transmitting until there is a significant rise in the water temperature, say 5 to 10 degrees. The test may be broken up into a series of short periods, as long as you keep track of the total key-down time. When the test is completed, continue to stir the water slowly and monitor its temperature. When the temperature ceases to rise, note the final indication as precisely as possible.

To compute the transmitter power output, multiply the calorimeter heat capacitance (4215 for C or 2342 for F) by the difference in initial and final water

Fig. 37 — The completed field-strength meter. A decibel calibration chart is affixed to the back of the instrument, as shown in Fig. 40.

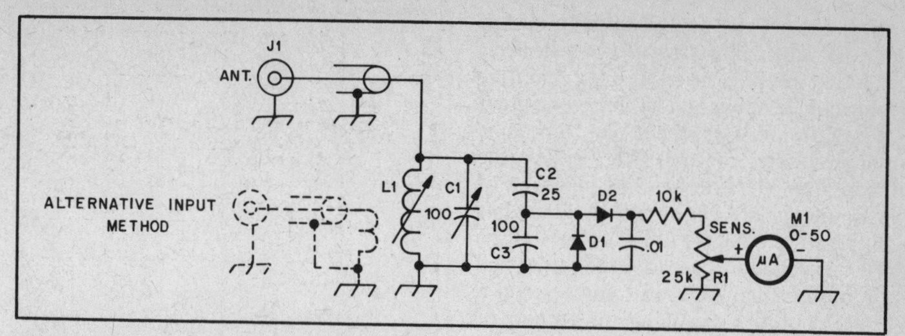


Fig. 38 — Schematic diagram of the simple field-strength meter. Fixed-value capacitors are disc ceramic unless otherwise noted. Fixed-value resistors are 1/2-W composition. C1 is a small 100-pF variable. D1 and D2 are 1N34A germanium diodes or equivalents. J1 is an antenna connector of the builder's choice. L1, C2 and C3 are selected values (see Table 4). M1 is a 50-μA meter, and R1 is a 25-kΩ, linear-taper composition control.

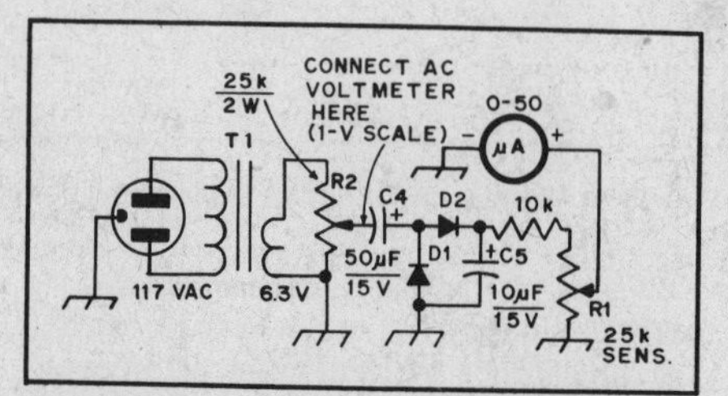


Fig. 39 — Schematic illustration of a suitable circuit for calibrating the field-strength meter in decibels (see text). T1 is a small 6.3-V filament transformer. C4 and C5 are electrolytic capacitors.

temperature. Then divide by the total number of seconds of key-down time. The resultant is the transmitter power in watts. A nomogram which can also be used to find transmitter power output is given in Fig. 36. With a straight line, connect the total number of key-down seconds in the time column to the number of degrees change (F or C) in the temperature rise column, and read off the transmitter power output at the point where the straight line crosses the power-output column.

### Power Limitation

The maximum power handling capability of the calorimeter is limited by the following. At very high powers the dielectric material in the coaxial line will melt because of excessive heating or the cable will arc over due to excessive voltage. As the transmitter frequency gets higher, the excessive-heating problem is accentuated, as more of the power is dissipated in the first several feet of cable. For instance, at 1296 MHz, approximately 10 percent of the transmitting power is dissipated in the first foot of cable. Overheating can be prevented when working with high power by using a low duty cycle to reduce the average dissipated power. Use a series of short transmissions, such as two seconds on, ten seconds off. Keep count of the total key-down time for power calculation purposes. If the cable arcs over, use a larger-diameter cable, such as RG-58/U, in place of the RG-174/U. The cable should be long enough to assure that the reflected wave will be down 10 dB or more at the input. It may be necessary to use more than one quart of water in order to submerge all the cable conveniently. If so, be sure to calculate the new value of heat capacitance for the larger quantity of water. Also you should measure the new coaxial cable heat capacitance using the method previously described.

## A RELATIVE FIELD-STRENGTH METER

A field-strength meter serves many useful purposes during antenna experiments and adjustments. Extreme meter sensitivity is not always a requisite, and for hf-band near-field checks the instrument of Figs. 37 through 40 should be quite suitable. The unit described here has ample sensitivity for most amateur work. The larger the pickup antenna, the greater the sensitivity. Far-field measurements can be made by using the alternative input circuit shown in Fig. 38. In that application a reference dipole, cut for the frequency of interest, is connected to the input link. Alternatively, a quarter-wavelength wire can be used as a far-field pickup antenna. The polarization of the two antennas involved in a test should be the same if meaningful results are to be obtained.

Most of the simple field-strength meters used by amateurs are capable of recording only *relative* signal levels, and such readings are useful in a number of tests. However, knowing the approximate decibel increase resulting from antenna adjustments can be helpful in evaluation work with matching networks, loading inductors, and the like. Reasonable accuracy can be had with the circuit of Fig. 38.

**Table 4**

**Component Values for Various Amateur Bands**

| *Freq. Band* | *1.8 MHz* | *3.5 MHz* | *7 MHz* | *10 MHz* | *14 MHz* | *18 MHz* | *21 MHz* | *25 MHz* | *28 MHz* |
|---|---|---|---|---|---|---|---|---|---|
| L1 (μH) | 100 (Nom.) | 25 (Nom.) | 10 (Nom.) | 4.4 (Nom.) | 2.2 (Nom.) | 1.6 (Nom.) | 1.3 (Nom.) | 0.8 (Nom.) | 0.5 (Nom.) |
| C2 (pF) | 25 | 25 | 15 | 15 | 15 | 10 | 10 | 10 | 10 |
| C3 (pF) | 100 | 100 | 68 | 68 | 68 | 47 | 47 | 47 | 4303 |
| Miller Coil | 4409 | 4407 | 4406 | 4405 | 4404 | 4404 | 4403 | 4403 | |

Miller coils can be ordered by mail from: J. W. Miller Co., 19070 Reyes Ave., P.O. Box 5825, Compton, CA 90224.

Fig. 40 — Rear view of the field-strength meter cabinet showing the calibration chart in decibels. Ranges 1, 2 and 3 are discussed in the text.

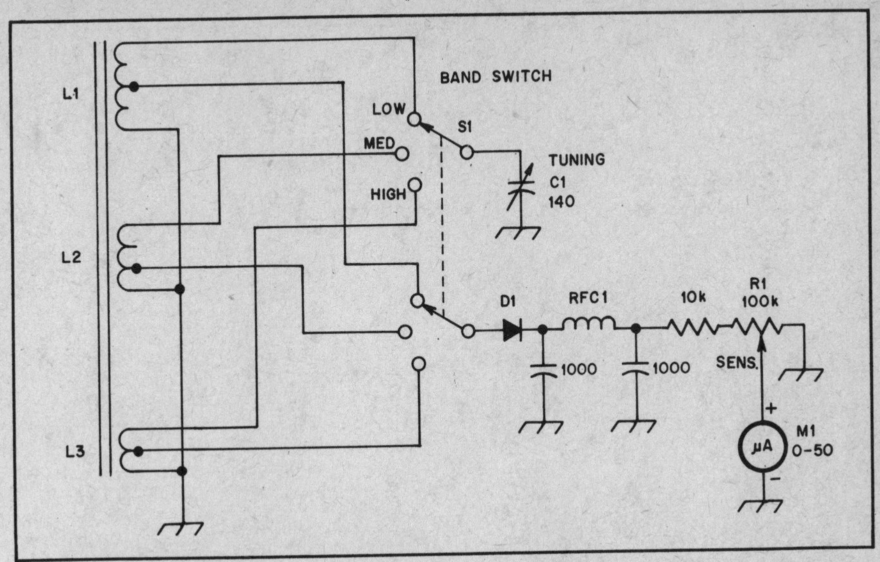


Fig. 42 — Schematic diagram of the rf current probe. Resistances are in ohms; k = 1000. Capacitances are in picofarads; fixed capacitors are silver mica. L1, L2 and L3 are each close-wound with no. 22 enam. wire on a single ferrite rod, 4 in. long and 1/2 in. dia, with µ = 125 (Amidon R61-50-400). Windings are spaced approximately 1/4 in. apart. The ferrite rod, the variable capacitor, and other components may be obtained from Radiokit, Box 411, Greenville, NH 03048.

C1 — Air variable, 6-140 pF; Hammarlund HF140 or equiv.
D1 — Germanium diode; 1N34A, 1N270 or equiv.
L1 — 1.6-5 MHz; 30 turns, tapped at 3 turns from grounded end.
L2 — 5-20 MHz; 8 turns, tapped at 2 turns from grounded end.
L3 — 17-39 MHz; 2 turns, tapped at 1 turn.
M1 — Any microammeter may be used. The one pictured is a Micronta meter, Radio Shack no. 270-1751.
R1 — Linear taper.
RFC1 — 1 mH; Miller no. 4642 or equiv. Value is not critical.
S1 — Ceramic rotary switch, 1 section, 2 poles, 2 to 6 positions; Centralab PA2002 or PA2003 or equiv.

In this model the tuned circuit, L1, C1, C2 and C3, was selected for the 160-meter operation. The constants can be changed for the amateur bands from 80 through 10 meters. Table 4 lists the inductance and capacitance values needed.

A capacitive divider (C2 and C3) is used to provide a low-impedance connection point for the voltage doubler, D1-D2. Since the rectifier diodes would otherwise load down the tuned circuit and degrade the Q — assuming they were connected at the high-impedance end of L1 — the capacitive divider is employed. Because D1 and D2 provide a square-law response as rectifiers, the meter deflection (respective to changes in signal-input level) will be nonlinear. Addition of the 10-kΩ resistor between D2 and R1 helps to linearize the meter response, but it also reduces the sensitivity of the instrument somewhat. This design trade-off is acceptable for most amateur work.

Fig. 41 — The rf current probe. The sensitivity control is mounted at the top of the instrument, with the tuning and band switches on the lower portion of the front panel. Frequency calibration of the tuning control was not considered necessary for the intended use of this particular instrument, but marks identifying the various amateur bands would be helpful. If the unit is provided with a calibrated dial, it can also be used as an absorption wavemeter.

It makes little difference what the size or shape of the enclosure is. It should be made of metal to prevent rf energy from entering the tuned circuit by any path other than that of the pickup antenna. A 4-1/2-inch meter was used in the model shown here. Physically smaller meters are quite acceptable, but the builder should use a 50-µA instrument if sensitivity is desired. Large-format meters are more suitable for viewing from a distance. For far-field observations one can use a telescope or field glasses to read the meter scale.

Component wiring inside the instrument box can be of the builder's choosing, provided all rf leads are kept as short and direct as practical. Although metal knobs are shown in the photographs, they aren't necessarily the most ideal kind to use. Touching them will affect the meter readings markedly if the meter case is not connected to an earth ground. Plastic knobs are, therefore, recommended.

## Calibration

A simple technique for obtaining calibration of M1 in decibels is shown in Fig. 39. D1 and D2 are used as rectifiers, but the tuned circuit of Fig. 38 is disconnected from them. T1, R2, and C4 are used temporarily to provide low-level 60-Hz voltage for calibrating the instrument. It is necessary to install C4 and C5 (electrolytic capacitors) for 60-Hz work. The capacitance values shown are suitable for low reactance at this frequency.

T1 and R2 are used in combination with a VOM or VTVM to supply and monitor 60-Hz energy between 0.1 and 1 volt. Midscale (25) on M1 represents 0 dB. All readings below 25 are *minus* dB, and the readings above 25 are *plus* dB. Thus, midscale, or 25, is the reference point for all measurements. Meter calibration is performed at three settings of the sensitivity control, R1. This is done because the meter readings versus field strength in decibels will vary somewhat as the ohmic value of R1 is changed. The greater the amount of resistance, the more linear the meter response.

During calibration, R2 (Fig. 39) is adjusted for midscale response of M1 at each of the settings of R1. Maximum sensitivity will occur when R1 is set at maximum resistance (position 1). Position 2 (moderate sensitivity) is established by using an ohmmeter to locate a position of the R1 arm that places 8 kΩ of resistance between the high end of R1 and M1. Position 3 (low sensitivity) is established when the arm of R1 is positioned to provide 16

Fig. 43 — The current probe just before final assembly. Note that all parts except the ferrite rod are mounted on a single half of the 3 × 4 × 5-in. Minibox (Bud CU-2105B or equiv.). Rubber grommets are fitted in holes at the ends of the slot to accept the rod during assembly of the enclosure. Leads in the rf section should be kept as short as possible, although those from the rod windings must necessarily be left somewhat long to facilitate final assembly.

kΩ of resistance between the high end of R1 and M1.

Decibel reference points are found by varying the voltage to D1 and D2 by means of R2. In each instance the value of ac voltage at *zero reference* is recorded. Then the setting of R2 is changed until the meter reading drops or increases to the next numbered point — 30, 35, 40 and so on, or 20, 15, 10 and downward. The change in decibels for each change in meter reading can be computed by:

$$dB = 20 \log \frac{E1}{E2}$$

where E2 = ac voltage at zero reference (25 on the meter scale), and E1 = the applied voltage at the arm of R2. Thus, if E1 was 0.9 volt, and E2 was 0.2 volt, the increase in decibels would equal 13. Example:

$$dB = 20 \times \log \frac{0.9}{0.2}$$

$$= 20 \times 0.6532 = 13$$

If E2 is larger than E1, our answer would be *minus* decibels.

A calibration chart is attached to the rear panel of the instrument, as shown in Fig. 40. If the circuit of Fig. 1 is used, the calibration points visible in the photograph will be valid for 50-μA meters. The scale may be protected with a coating of clear varnish or lacquer.

## AN RF CURRENT PROBE

The rf current probe of Figs. 41 through 43 operates on the magnetic component of the electromagnetic field, rather than the electric field. Since the two fields are precisely related, as discussed in Chapter 1, the relative field-strength measurements are completely equivalent. The use of the magnetic field offers certain advantages, however. The instrument may be made more compact for the same sensitivity, but its principal advantage is that it may be used near a conductor to measure the current flow without cutting the conductor.

In the average amateur location there may be substantial currents flowing in guy wires, masts and towers, coaxial-cable braids, gutters and leaders, water and gas pipes, and perhaps even drainage pipes. Current may be flowing in telephone and power lines as well. All of these rf currents may have an influence on antenna patterns or be of significance in the case of RFI.

The circuit diagram of the current probe appears in Fig. 42, and construction is shown in the photo, Fig. 43. The winding data given here apply only to a ferrite rod of the particular dimensions and material specified. Almost any microammeter can be used, but it is usually convenient to use a rather sensitive meter and provide a series resistor to "swamp out" nonlinearity arising from diode conduction characteristics. A control is also used to adjust instrument sensitivity as required during operation. The tuning capacitor may be almost anything that will cover the desired range.

As shown in the photos, the circuit is constructed in a metal box. This enclosure shields the detector circuit from the electric field of the radio wave. A slot must be cut with a hacksaw across the back of the box, and a thin file may be used to smooth the cut. This slot is necessary to prevent the box from acting as a shorted turn.

### Using the Probe

In measuring the current in a conductor, the ferrite rod should be kept at right angles to the conductor, and at a constant distance from it. In its upright or vertical position, this instrument is oriented for taking measurements in vertical conductors. It must be laid horizontal to measure current in horizontal conductors.

Numerous uses for the instrument are suggested in an earlier paragraph. In addition, the probe is an ideal instrument for checking the current distribution in antenna elements. It is also useful for measuring rf ground currents in radial systems. A buried radial may be located easily by sweeping the ground. Current division at junctions may be investigated. "Hot spots" usually indicate areas where additional radials would be effective.

Stray currents in conductors not intended to be part of the antenna system may often be eliminated by bonding or by changing the physical lengths involved. Guy wires and other unwanted "parasitic" elements will often give a tilt to the plane of polarization and make a marked difference in front-to-back ratios. When the ferrite rod is oriented parallel to the electric field lines, there will be a sharp null reading that may be used to locate the plane of polarization quite accurately. When using the meter, remember that the magnetic field is at right angles to the electric field.

The current probe may also be used as a relative signal strength meter. In making measurements on a vertical antenna, the meter should be located at least two wavelengths away, with the rod in a horizontal position. For horizontal antennas, the instrument should be at approximately the same height as the antenna, with the rod vertical.

## ANTENNA MEASUREMENTS

Of all the measurements made in Amateur Radio systems, perhaps the most difficult and least understood are various measurements of antennas. For example, it is relatively easy to measure the frequency and cw power output of a transmitter, the response of a filter, or the gain of an amplifier. These are all what might be called bench measurements because, when performed properly, all the factors that influence the accuracy and success of the measurement are under control. In making antenna measurements, however, the "bench" is now perhaps the backyard. In other words, the environment surrounding the antenna can affect the results of the measurement. Control of the environment is not at all as simple as it was for the bench measurement, because now the work area may be rather spacious. This section describes antenna measurement techniques which are closely allied to those used in an antenna measuring event or contest. With these procedures the measurements can be made successfully and with meaningful results. These techniques should provide a better understanding of the measurement problems, resulting in a more accurate and less difficult task. (Information provided by Dick Turrin, W2IMU.)

## SOME BASIC IDEAS

An antenna is simply a transducer or coupler between a suitable feed line and the environment surrounding it. In addition to efficient transfer of power from feed line to environment, an antenna at vhf or uhf is most frequently required to concentrate the radiated power into a particular region of the environment.

To be consistent in comparing different antennas, it is necessary that the environment surrounding the antenna be standardized. Ideally, measurements should be made with the measured antenna so far removed from any objects causing environmental effects that it is literally in outer space — a very impractical situation. The purpose of the measurement techniques is therefore to simulate, under practical conditions, a controlled environment. At vhf and uhf, and with practical-size antennas, the environment *can* be controlled so that successful and accurate measurements can be made in a reasonable amount of space.

The electrical characteristics of an antenna that are most desirable to obtain by direct measurement are: (1) gain (relative to an isotropic source, which by definition has a gain of unity); (2) space-radiation pattern; (3) feed-point impedance (mismatch) and (4) polarization.

### Polarization

In general the polarization can be assumed from the geometry of the radiating elements. That is to say, if the antenna is made up of a number of linear elements (straight lengths of rod or wire which are resonant and connected to the feed point) the polarization of the electric field will be linear and polarized parallel to the elements. If the elements are not consistently parallel with each other, then the polarization cannot easily be assumed. The following techniques are directed to antennas having polarization that is essentially linear (in one plane), although the method can be extended to include all forms of elliptic polarization.

### Feed-Point Mismatch

The feed-point mismatch, although affected to some degree by the immediate environment of the antenna, does *not* affect the gain or radiation characteristics of an antenna. If the immediate environment of the antenna does not affect the feed-point impedance, then any mismatch intrinsic to the antenna tuning reflects a portion of the incident power back to the source. In a receiving antenna this reflected power is reradiated back into the environment, "free space," and can be lost entirely. In a transmitting antenna, the reflected power goes back to the final amplifier of the transmitter if it is not matched. In general an amplifier by itself is *not* a matched source to the feed line, and, if the feed line has very low loss, the amplifier output controls are customarily altered during the normal tuning procedure to obtain maximum power transfer to the antenna. The power which has been reflected from the antenna combines with the source power to travel again to the antenna. This procedure is called conjugate matching, and the feed line is now part of a resonant system consisting of the mismatched antenna, feed line, and amplifier tuning circuits. It is therefore possible to use a mismatched antenna to its full gain potential, provided the mismatch is not so severe as to cause heating losses in the system, especially the feed line and matching devices. Similarly, a mismatched receiving antenna may be conjugately matched into the receiver front end for maximum power transfer. In any case it should be clearly kept in mind that the feed-point mismatch does *not* affect the radiation characteristics of an antenna. It can only affect the system efficiency wherein heating losses are concerned.

Why then do we include feed-point mismatch as part of the antenna characteristics? The reason is that for efficient system performance most antennas are resonant transducers and present a reasonable match over a relatively narrow frequency range. It is therefore desirable to design an antenna, whether it be a simple dipole or an array of Yagis, such that the final single feed-point impedance be essentially resistive and of magnitude consistent with the impedance of the feed line which is to be used. Furthermore, in order to make accurate, absolute gain measurements, it is vital that the antenna under test accept all the power from a matched-source generator, or that the reflected power caused by the mismatch be measured and a suitable error correction for heating losses be included in the gain calculations. Heating losses may be determined from information contained in Chapter 3.

While on the subject of feed-point impedance, mention should be made of the use of baluns in antennas. A balun is simply a device which permits a lossless transition between a balanced system — feed line or antenna — and an unbalanced feed line or system. If the feed point of an antenna is symmetric such as with a dipole and it is desired to feed this antenna with an unbalanced feed line such as coax, it is necessary to provide a balun between the line and the feed point. Without the balun, current will be allowed to flow on the outside of the coax. The current on the outside of the feed line will cause radiation and thus the feed line becomes part of the antenna radiation system. In the case of beam antennas where it is desired to concentrate the radiated energy in a specific direction, this extra radiation from the feed line will be detrimental, causing distortion of the expected antenna pattern.

## ANTENNA TEST SITE SET-UP AND EVALUATION

Since an antenna is a reciprocal device, measurements of gain and radiation patterns can be made with the test antenna used either as a transmitting or as a receiving antenna. In general and for practical reasons, the test antenna is used in the receiving mode, and the source or transmitting antenna is located at a specified fixed remote site and unattended. In other words the source antenna, energized by a suitable transmitter, is simply required to illuminate or flood the receiving site in a controlled and constant manner.

As mentioned earlier, antenna measurements ideally should be made under "free-space" conditions. A further restriction is that the illumination from the source antenna be a plane wave over the effective aperture (capture area) of the test antenna. A plane wave by definition is one in which the magnitude and phase of the fields are uniform, and in the test-antenna situation, *uniform over the effective area plane of the test antenna.* Since it is the nature of all radiation to expand in a spherical manner at great distance from the source, it would seem to be most desirable to locate the source antenna as far from the test site as possible. However, since for practical reasons the test site and source location will have to be near the earth and not in outer space, the environment must include the effects of the ground surface and other obstacles in the vicinity of both antennas. These effects almost always dictate that the test range (spacing between source and test antennas) be as short as possible consistent with maintaining a nearly error-free plane wave illuminating the test *aperture.*

A nearly error-free plane wave can be specified as one in which the phase and amplitude, from center to edge of the illuminating field over the test aperture, do not deviate by more than about 30 degrees and 1 decibel, respectively. These conditions will result in a gain-measurement error of no more than a few percent less than the true gain. Based on the 30-degree phase error alone, it can be shown that the minimum range distance is approximately

$$S_{min} = 2\frac{D^2}{\lambda}$$

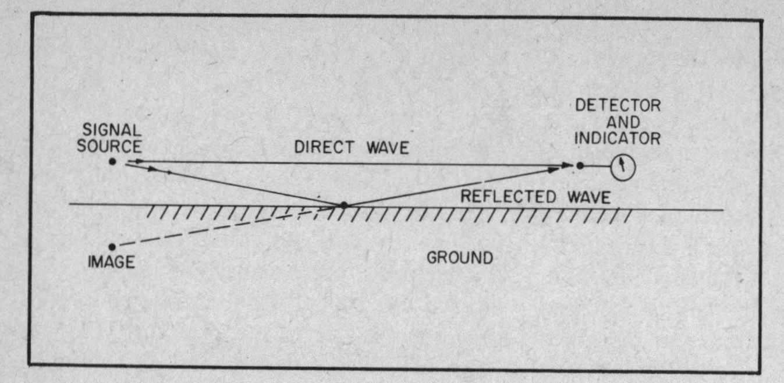


Fig. 44 — On an antenna test range, energy reaching the receiving equipment may arrive after being reflected from the surface of the ground, as well as by the direct path. The two waves may tend to cancel each other, or may reinforce one another, depending on their phase relationship at the receiving point.

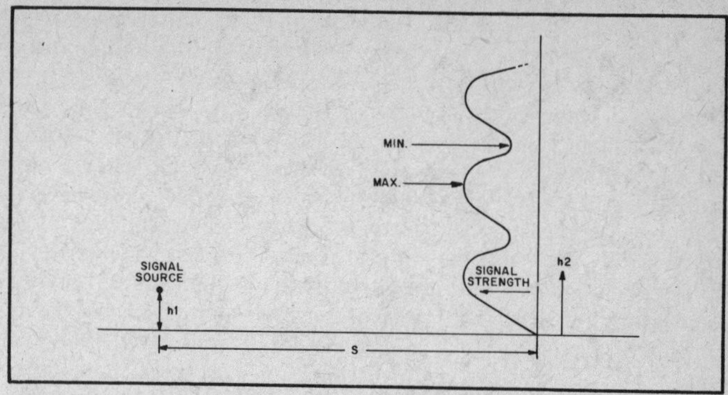


Fig. 45 — The vertical profile, or plot of signal strength versus test-antenna height, for a fixed height of the signal source above ground and at a fixed distance. See text for definitions of symbols.

where D is the largest aperture dimension and $\lambda$ is the free-space wavelength in the same units as D. The phase error over the aperture D for this condition is 1/16 wavelength.

Since aperture size and gain are related by

$$\text{Gain} = \frac{4\pi A_e}{\lambda^2}$$

where $A_e$ is the effective aperture area, the dimension D may be obtained for simple aperture configurations. For a square aperture

$$D^2 = G\frac{\lambda^2}{4\pi}$$

which results in a minimum range distance for a square aperture of

$$S_{min} = G\frac{\lambda}{2\pi}$$

and for a circular aperture of

$$S_{min} = G\frac{2\lambda}{\pi^2}$$

For apertures with a physical area that is not well defined or is much larger in one dimension than in other directions, such as a long thin array for maximum directivity in one plane, it is advisable to use the maximum estimate of D from either the expected gain or physical aperture dimensions.

Up to this point in the range development, only the conditions for minimum range length, $S_{min}$, have been established, as though the ground surface were not present. This minimum S is therefore a necessary condition even under "free-space" environment. The presence of the ground further complicates the range selection, not in the determination of S but in the exact location of the source and test antennas above the earth.

It is always advisable to select a range whose intervening terrain is essentially flat, clear of obstructions, and of uniform surface conditions, such as all grass or all pavement. The extent of the range is determined by the illumination of the source antenna, usually a beam, whose gain is no greater than the highest gain antenna to be measured. For gain measurements the range consists essentially of the region in the beam of the test antenna. For radiation-pattern measurements, the range is considerably larger and consists of all that area illuminated by the source antenna, especially around and behind the test site. Ideally a site should be chosen where the test-antenna location is near the center of a large open area and the source antenna located near the edge where most of the obstacles (trees, poles, fences, etc.) lie.

The primary effect of the range surface is that some of the energy from the source antenna will be reflected into the test antenna while other energy will arrive on a direct line-of-sight path. This is illustrated in Fig. 44. The use of a flat, uniform ground surface assures that there will be essentially a mirror reflection even though the reflected energy may be slightly weakened (absorbed) by the surface material (ground). In order to perform an analysis it is necessary to realize that horizontally polarized waves undergo a 180-degree phase reversal upon reflection from the earth. The resulting illumination amplitude at any point in the test aperture is the vector sum of the electric fields arriving from the two directions, the direct path and the reflected path. If a perfect mirror reflection is assumed from the ground (it is nearly that for practical ground conditions at vhf/uhf) and the source antenna is isotropic, radiating equally in all directions, then a simple geometric analysis of the two path lengths will show that at various points in the vertical plane at the test-antenna site the waves will combine in different phase relationships. At some points the arriving waves will be in phase, and at other points they will be 180 degrees out of phase. Since the field amplitudes are nearly equal, the resulting phase change due to path length difference will produce an amplitude variation in the vertical test site direction similar to a standing wave, as shown in Fig. 45.

The simplified formula relating the location of h2 for maximum and minimum values of the two-path summation in terms of h1 and S is

$$h2 = n\frac{\lambda}{4}\cdot\frac{S}{h1}$$

with n = 0, 2, 4 . . . for minimums and n = 1, 3, 5 . . . for maximums, and S is much larger than either h1 or h2.

The significance of this simple ground reflection formula is that it permits the approximate location of the source antenna to be determined to achieve a nearly plane-wave amplitude distribution *in the vertical direction* over a particular test *aperture size*. It should be clear from examination of the height formula that as h1 is decreased, the vertical distribution pattern of signal at the test site, h2, expands. Also note that the signal level for h2 equal to zero is always zero on the ground regardless of the height of h1.

The objective in using the height formula then is, given an effective antenna aperture to be illuminated from which a minimum S (range length) is determined and a suitable range site chosen, to find a value for h1 (source antenna height). The required value is such that the *first* maximum of vertical distribution at the test site, h2, is at a practical distance above the ground and at the same time the signal amplitude over the aperture in the vertical direction does not vary more than about 1 dB. This last condition is not sacred but is closely related to the particular antenna under test. In practice these formulas are only useful to initialize the range set-up. A final check of the vertical distribution at the test site must be made by direct measurement. This measurement should be conducted with a small low-gain but unidirectional probe antenna such as a corner reflector or 2-element Yagi that is moved along a vertical line over the intended aperture site. Care should be exercised to minimize the effects of local environment around the probe antenna and that the beam of the

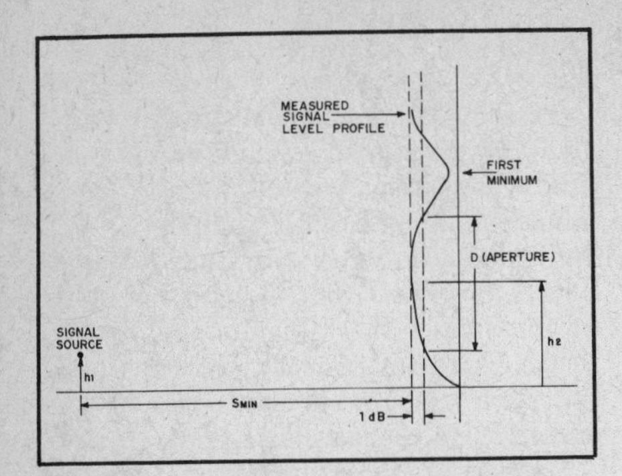


Fig. 46 — Sample plot of a measured vertical profile.

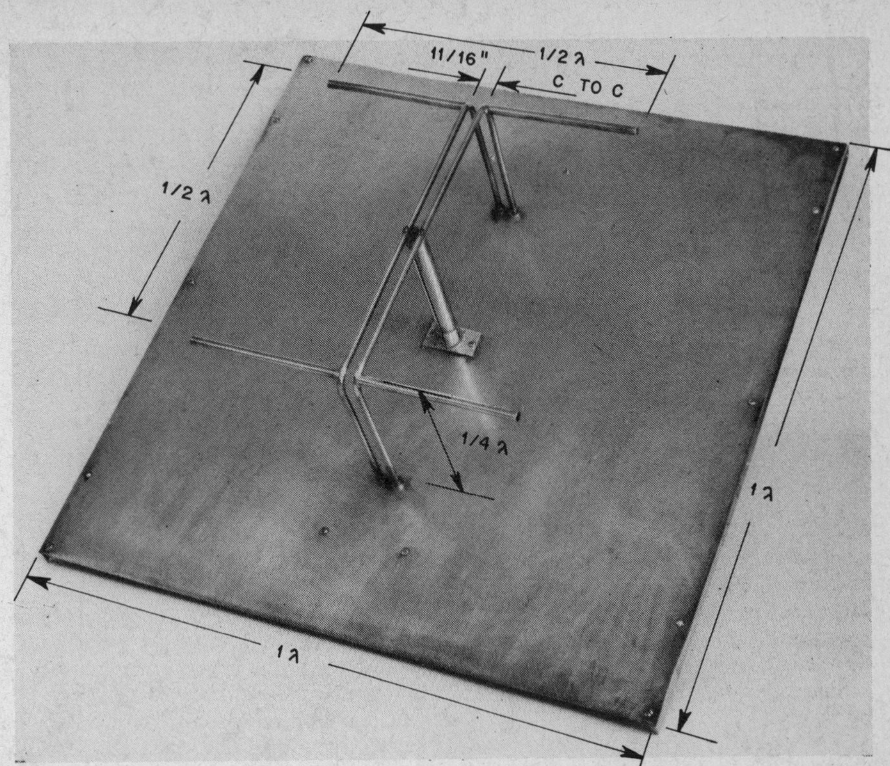


Fig. 47 — Standard-gain antenna. When accurately constructed for the desired frequency, this antenna will exhibit a gain of 7.7 dB over a dipole radiator, plus or minus 0.25 dB. In this model, constructed for 432 MHz, the elements are 3/8-inch dia tubing. The phasing and support lines are of 5/16-inch dia tubing or rod.

probe be directed at the source antenna at all times for maximum signal. A simple dipole is undesirable as a probe antenna because it is susceptible to local environmental effects.

The most practical way to instrument the vertical distribution measurement is to construct some kind of vertical track, preferably of wood, with a sliding carriage or platform which may be used to support and move the probe antenna. It is assumed of course that a stable source transmitter and calibrated receiver or detector are available so that variations of the order of 1/2 dB can be clearly distinguished.

Once these initial range measurements are completed successfully, the range is now ready to accommodate any aperture size less in vertical extent than the largest for which $S_{min}$ and the vertical field distribution were selected. The test antenna is placed with the center of its aperture at the height h2 where maximum signal was found. The test antenna should be tilted so that its main beam is pointed in the direction of the source antenna. The final tilt is found by observing the receiver output for maximum signal. This last process must be done empirically since the apparent location of the source is somewhere between the actual source and its image, below the ground.

An example will illustrate the procedure. Assume that we wish to measure a 7-foot-diameter parabolic reflector antenna at 1296 MHz ($\lambda = 0.75$ foot). The minimum range distance, $S_{min}$, can be readily computed from the formula for a circular aperture.

$$S_{min} = 2\frac{D^2}{\lambda} = 2 \times \frac{49}{0.75} = 130 \text{ ft}$$

Now a suitable site is selected based on the qualitative discussion given before.

Next determine the source height, h1. The procedure is to choose a height h1 such that the first minimum above ground (n = 2 in formula) is at least two or three times the aperture size, or about 20 feet.

$$h1 = n\frac{\lambda}{4}\frac{S}{h2} = 2 \times \frac{0.75}{4} \times \frac{130}{20}$$

$$= 2.4 \text{ feet}$$

Place the source antenna at this height and probe the vertical distribution over the 7-foot aperture location, which will be about 10 feet off the ground

$$h2 = n\frac{\lambda}{4}\frac{S}{h1} = 1 \times \frac{0.75}{4} \times \frac{130}{2.4}$$

$$= 10.2 \text{ feet}$$

The measured profile of vertical signal level versus height should be plotted. From this plot, empirically determine whether the 7-foot aperture can be fitted in this profile such that the 1-dB variation is not exceeded. If the variation exceeds 1 dB over the 7-foot aperture, the source antenna should be lowered and h2 raised. Small changes in h1 can quickly alter the distribution at the test site. Fig. 46 illustrates the points of the previous discussion.

The same set-up procedure applies for either horizontal or vertical linear polarization. However, it is advisable to check by direct measurement at the site for each polarization to be sure that the vertical distribution is satisfactory. Distribution probing in the horizontal plane is unnecessary as little or no variation in amplitude should be found, since the reflection geometry is constant. Because of this, antennas with apertures which are long and thin, such as a stacked colinear vertical, should be measured with the long dimension parallel to the ground.

A particularly difficult range problem occurs in measurements of antennas which have depth as well as cross-sectional aperture area. Long end-fire antennas such as long Yagis, rhombics, V-beams, or arrays of these antennas, radiate as volumetric arrays and it is therefore even more essential that the illuminating field from the source antenna be reasonably uniform in depth as well as plane wave in cross section. For measuring these types of antennas it is advisable to make several vertical profile measurements which cover the depth of the array. A qualitative check on the integrity of the illumination for long end-fire antennas can be made by moving the array or antenna axially (forward and backward) and noting the change in received signal level. If the signal level varies less than 1 or 2 dB for an axial movement of several wavelengths then the field can be considered satisfactory for most demands on accuracy. Large variations indicate that the illuminating field is badly distorted over the array depth and subsequent measurements are questionable. It is interesting to note in connection with gain measurements that any illuminating field distortion will always result in measurements that are lower than true values.

## ABSOLUTE GAIN MEASUREMENT

Having established a suitable range, the measurement of gain relative to an isotropic (point source) radiator is almost always accomplished by direct comparison with a calibrated standard-gain

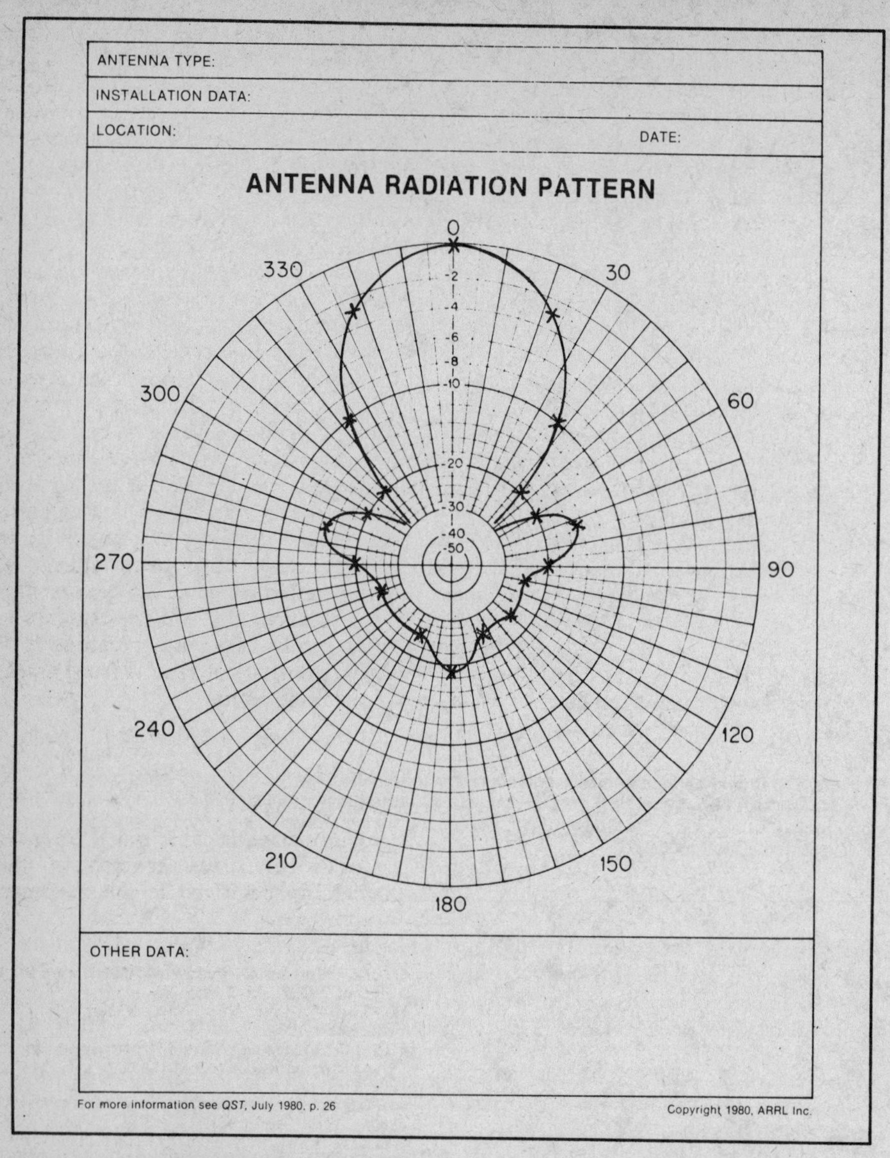


Fig. 48 — Sample plot of a measured radiation pattern, using techniques described in the text. The plot is on coordinate paper available from ARRL Hq. (See Fig. 11, Chapter 2.) The form provides space for recording significant data.

antenna. That is, the signal level with the test antenna in its optimum location is noted. Then the test antenna is removed and the standard-gain antenna is placed with its aperture at the center of location where the test antenna was located. The difference in signal level between the standard and the test antennas is measured and appropriately added to or subtracted from the gain of the standard-gain antenna to obtain the absolute gain of the test antenna, absolute here meaning with respect to a point source which has a gain of unity by definition. The reason for using this reference rather than a dipole, for instance, is that it is more useful and convenient for system engineering. It is assumed that both standard and test antennas have been carefully matched to the appropriate impedance and an accurately calibrated and matched detecting device is being used.

A standard-gain antenna may be any type of unidirectional, preferably planar-aperture, antenna, which has been calibrated either by direct measurement or in special cases by accurate construction according to computed dimensions.

A standard-gain antenna has been suggested. It consists of two in-phase dipoles one half wavelength apart and backed up with a ground plane one wavelength square. Such an antenna is shown in Fig. 47. When constructed accurately to scale for the frequency of interest, this type of standard will have an absolute gain of 7.7 dBd (dB gain over a dipole) with an accuracy of plus or minus. 0.25 dB.

## RADIATION-PATTERN MEASUREMENTS

Of all antenna measurements, the radiation pattern is the most demanding in measurement and most difficult to interpret. Any antenna radiates to some degree in all directions into the space surrounding it. Therefore, the radiation pattern of an antenna is a three-dimensional representation of the magnitude, phase and polarization. In general, and in practical cases for Amateur Radio communications, the polarization is well defined and only the magnitude of radiation is important. Furthermore, in many of these cases the radiation in one particular plane is of primary interest, usually the plane corresponding to that of the earth's surface, regardless of polarization.

Because of the nature of the range set-up, measurement of radiation pattern can only be successfully made in a plane nearly parallel to the earth's surface. With beam antennas it is advisable and usually sufficient to take two radiation pattern measurements, one in the polarization plane and one at right angles to the plane of polarization. These radiation patterns are referred to in antenna literature as the principal E-plane and H-plane patterns, respectively, E plane meaning parallel to the electric field which is the polarization plane and H plane meaning parallel to the magnetic field. The electric field and magnetic field are always perpendicular to each other in a plane wave as it propagates through space.

The technique in obtaining these patterns is simple in procedure but requires more equipment and patience than does making a gain measurement. First, a suitable mount is required which can be rotated in the azimuth plane (horizontal) with some degree of accuracy in terms of azimuth angle positioning. Second, a signal-level indicator calibrated over at least a 20-dB dynamic range with a readout resolution of at least 2 dB is required. A dynamic range of up to about 40 dB would be desirable but does not add greatly to the measurement significance.

With this much equipment, the procedure is to locate first the area of maximum radiation from the beam antenna by carefully adjusting the azimuth and elevation positioning. These settings are then arbitrarily assigned an azimuth angle of zero degrees and a signal level of zero decibels. Next, without changing the elevation setting (tilt of the rotating axis), the antenna is carefully rotated in azimuth in small steps which permit signal-level readout of 2 or 3 dB per step. These points of signal level corresponding with an azimuth angle are recorded and plotted on polar coordinate paper. A sample of the results is shown on ARRL coordinate paper in Fig. 48.

On the sample radiation pattern the measured points are marked with an X and a continuous line is drawn in, since the pattern is a continuous curve. Radiation patterns should preferably be plotted on a logarithmic radial scale, rather than a voltage or power scale. The reason is that the log scale approximates the response of the ear to signals in the audio range. Also many receivers have agc systems that are somewhat logarithmic in response; therefore the log scale is more representative of actual system operation.

Having completed a set of radiation-pattern measurements, one is prompted to ask, "Of what use are they?" The primary answer is as a diagnostic tool to determine if the antenna is functioning as it was intended to. A second answer is to know how the antenna will discriminate against interfering signals from various directions.

Consider now the diagnostic use of the radiation patterns. If the radiation beam is well defined, then there is an approximate formula relating the antenna gain to the measured half-power beamwidth of the E- and H-plane radiation patterns. The half-power beamwidth is indicated on the polar plot where the radiation level falls to 3 dB below the main beam 0-dB reference on either side. The formula is

$$\text{Gain} \approx \frac{25{,}000}{\theta_E \times \phi_H}$$

where $\theta_E$ and $\phi_H$ are the half-power beamwidths in degrees of the E- and H-plane patterns, respectively.

To illustrate the use of this formula, assume that we have a Yagi antenna with a boom length of two wavelengths. From known relations (described in Chapter 6) the expected gain of a Yagi with a boom length of two wavelengths is about 12 dB; its gain, G, equals 16. Using the formula, the product of $\theta_E \times \phi_H = 1600$ square degrees. Since a Yagi produces a nearly symmetric beam shape in cross section, $\theta_E \approx \phi_H = 40$ degrees. Now if the measured values of $\theta_E$ and $\phi_H$ are much larger than 40 degrees, then the gain will be much lower than the expected 12 dB.

As another example, suppose that the same antenna (a 2-wavelength-boom Yagi) gives a measured gain of 9 dB but the radiation pattern half power beamwidths are approximately 40 degrees. This situation indicates that although the radiation patterns seem to be correct, the low gain shows inefficiency somewhere in the antenna, such as lossy materials or poor connections.

Large broadside collinear antennas can be checked for excessive phasing-line losses by comparing the gain computed from the radiation patterns with the direct-measured gain. It seems paradoxical but it is indeed possible to build a large array with a very narrow beamwidth indicating high gain but actually having very low gain because of losses in the feed distribution system.

In general, and for most vhf/uhf Amateur Radio communications, gain is the primary attribute of an antenna. However, radiation in other directions than the main beam, called side-lobe radiation, should be examined by measurement of radiation patterns for effects such as non-symmetry on either side of the main beam or excessive magnitude of sidelobes (any sidelobe which is less than 10 dB below the main beam reference level of 0 dB should be considered excessive). These effects are usually attributable to incorrect phasing of the radiating elements or radiation from other parts of the antenna which was not intended, such as the support structure or feed line.

The interpretation of radiation patterns is intimately related to the particular type of antenna under measurement. Reference data should be consulted for the particular antenna type of interest, to verify that the measured results are in agreement with expected results.

To summarize the use of pattern measurements, if a beam antenna is first checked for gain (the easier measurement to make) and it is as expected, then pattern measurements may be academic. However, if the gain is lower than expected it is advisable to make the pattern measurements as an aid in determining the possible cause of low gain.

Regarding radiation-pattern measurements, it should be remembered that the results measured under proper range facilities will not necessarily be the same as observed for the same antenna at a home-station installation. The reasons may be obvious now in view of the preceding information on the range set-up, ground reflections, and the vertical-field distribution profiles. For long paths over rough terrain where many large obstacles may exist, these effects of ground reflection tend to become diffused, although they still can cause unexpected results. For these reasons it is usually unjust to compare vhf/uhf antennas over long paths.

## BIBLIOGRAPHY

Source material and more extended discussion of topics covered in this chapter can be found in the references given below.


Bruene, "An Inside Picture of Directional Wattmeters," *QST,* April 1959.

DeMaw, "In-Line RF Power Metering," *QST,* December 1969.

Hall and Kaufmann, "The Macromatcher, an RF Impedance Bridge for Coax Lines," *QST,* January 1972.

McMullen, "The Line Sampler, an RF Power Monitor for VHF and UHF," *QST,* April 1972.

Yang, R. F. H., "A Proposed Gain Standard for VHF Antennas," *IEEE Transactions on Antennas and Propagation,* November 1966.

# Chapter 16

# Antenna Orientation

Anyone laying out a fixed directive array does so in order to put his signal into certain parts of the world; in such cases, it is essential to be able to determine the bearings of the desired points. Too, amateurs with a rotatable directive array like to know where to aim if they are trying to pick up certain countries. And even amateurs with the single wire are interested in the directive pattern of the lobes when the wire is operated harmonically at the higher frequencies, and often are able to vary the direction of the wire to take advantage of the lobe pattern.

### Which Direction?

It is probably no news to most people nowadays that true direction from one place to another is not what it appears to be on the old Mercator school map. On such a map, if one starts "west" from Pratt in central Kansas, he winds up in the neighborhood of Peking, Peoples Republic of China. Actually, as a minute's experiment with a strip of paper on a small globe will show, a signal starting due west from Pratt never hits China at all but rather passes over Perth, in western Australia.

"The shortest distance between two points is a straight line" is true only on a flat surface. The determination of the shortest path between two points on the surface of a sphere is a bit more complicated. Imagine a plane that intersects two points on the surface and the center of the sphere. The intersection of the plane and the sphere describe a circle on the surface of the sphere that is defined as a great circle. The shortest distance between the points follows the path of the great circle. The direction or bearing from your location to another point on the earth is the direction of a great circle as it passes through your location on its way to the other point.

If, therefore, we want to determine the direction of some distant point from our own location, the ordinary Mercator projection alone is utterly useless. True bearing, however, may be found in several ways: by using a special type of world map that does show true direction from a specific location to other parts of the world, by working directly from a globe or by using mathematics.

## DETERMINING TRUE NORTH

Determining the direction of distant points is of little use to amateurs erecting a directive array unless they can put up the array itself in the desired direction. This, in turn, demands a knowledge of the direction of *true* north (as against magnetic north), since all our directions from a globe or map are worked in terms of true north.

A number of ways may be available to amateurs for determining true north from their location. Frequently, the streets of a city or town are laid out, quite accurately, in north-south and east-west directions. A visit to the office of your city or county engineer will enable you to determine whether or not this is the case for the street in front of or parallel to your own lot. Or from such a visit it is often possible to locate some landmark, such as a factory chimney or church spire, that lies true north with respect to your house.

If you cannot get true north by such means, three other methods are available: compass, pole star and sun.

### By Compass

Get as large a compass as you can; it is difficult though not impossible, to get satisfactory results with the "pocket" type. In any event, the compass *must* have not more than 2 degrees per division.

It must be remembered that the compass points to *magnetic* north, not true north. The amount by which magnetic north differs from true north in a particular location is known as *variation.* Your city engineer's office or the flight office at a nearby airport can tell you the magnetic variation for your locality. The information is also available from U.S. Geological Survey topographic maps for you locality; these may be on file in your local library. When correcting your "compass north," do so *opposite* to the direction of the variation. For instance if the variation for your locality is 12 degrees west (meaning that the compass points 12 degrees west of north) then true north is found by counting 12 degrees *east* of north as shown on the compass.

When taking the bearing, make sure that the compass is located well away from ironwork, fencing, pipes, etc. Place the instrument on a wooden tripod or support of some sort, at a convenient height as near eye level as possible. Make yourself a sighting stick from a flat stick about 2 feet long with a nail driven upright in each end (for use as "sights") and then, after the needle of the compass has settled down, carefully lay this stick across the face of the compass — with the necessary allowance for variation — to line it up on true north. *Be sure you apply the variation correctly.*

This same sighting-stick and compass rig can also be used in laying out directions for supporting poles for antennas in other directions — provided, of course, that the compass dial is graduated in degrees.

### By the Pole Star

Many amateurs use the pole star, Polaris, in determining the direction of true north. An advantage is that the pole star is never more than 0.8° from true north, so that in practice no corrections are necessary. Disadvantages are that some people have difficulty identifying the pole star, and that because of its comparatively high angle above the horizon at high northerly latitudes, it is not always easy to "sight" on it accurately. Polaris is not visible in the southern hemisphere. In

**Table 1**

**Time Correction for Various Dates of the Year**

Apply to clock time as indicated by the sign, to get time of true noon

| *Date* | | *Min.* | *Date* | | *Min.* |
|---|---|---|---|---|---|
| Jan. | 1 | + 4 | July | 10 | + 5 |
| | 10 | + 8 | | 20 | + 6 |
| | 20 | +11 | | 30 | + 6 |
| | 30 | +13 | | | |
| Feb. | 10 | +14 | Aug. | 10 | + 5 |
| | 20 | +14 | | 20 | + 3 |
| | 28 | +13 | | 30 | + 1 |
| Mar. | 10 | +10 | Sept. | 10 | − 3 |
| | 20 | + 8 | | 20 | − 7 |
| | 30 | + 4 | | 30 | −10 |
| Apr. | 10 | + 1 | Oct. | 10 | −13 |
| | 20 | − 1 | | 20 | −15 |
| | 30 | − 3 | | 30 | −16 |
| May | 10 | − 4 | Nov. | 10 | −16 |
| | 20 | − 4 | | 20 | −14 |
| | 30 | − 3 | | 30 | −11 |
| June | 10 | − 1 | Dec. | 10 | − 7 |
| | 20 | + 1 | | 20 | − 2 |
| | 30 | + 4 | | 30 | + 3 |

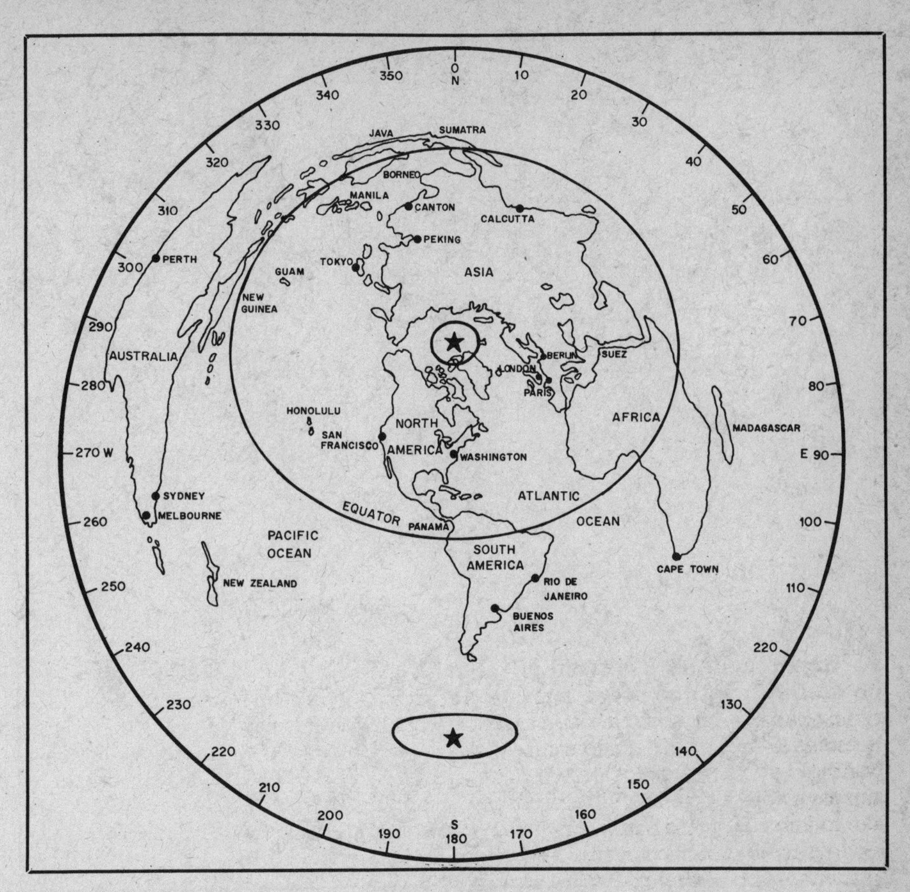


Fig. 1 — Azimuthal map centered on Washington, DC.

any event, if visible, it is a handy check on the direction secured by other means.

### By the Sun

With some slight preparation, the sun can be used easily for determination of true north. One of the most satisfactory methods is described below. The method is based on the fact that exactly at noon, local time, the sun bears due south, so that at that time the shadow of a vertical stick or rod will bear north. The resulting shadow direction, incidentally, is *true* north.

Two corrections to your Standard Time must be made to determine the exact moment of local true noon.

The first is a longitude correction. Standard Time is time at some particular meridian of longitude: EST is based on the 75th meridian; CST on the 90th meridian; MST on the 105th meridian; and PST on the 120th meridian. From an atlas or perhaps Table 3 or 4, determine the difference between your longitude and the longitude of your time meridian. Getting this to the nearest 1/4 degree of longitude is close enough. Example: Newington, Connecticut, which runs on 75th meridian time (EST) is at 72.75° longitude, or a difference of 2.25°. Now for each 1/4° of longitude, figure one minute of time; thus 2.25° is equivalent to nine minutes of time (there are 60 "angle" minutes to a degree, so that each degree of longitude equals four minutes of time). *Subtract* this correction from noon if you are *east* of your time meridian; *add* if you are *west*.

To the resulting time, apply a further correction for the date from Table 1. The resulting time is the time, by Standard Time, when it will be true noon at your location. Put up your vertical stick (use a plumb bob to make sure it is actually vertical), check your watch with Standard Time, and, at the time indicated from your calculations, mark the position of the shadow. That is true north.

In the case of Newington, if we wanted correct time for true noon on October 20: First, subtracting the longitude correction — because we are east of the time meridian — we get 11:51 A.M.; then, applying the further correction of − 15 minutes, we get 11:36 A.M. EST (12:36 P.M. EDST) as the time of true noon at Newington on October 20.

## AZIMUTHAL MAPS

While the Mercator projection does not show true directions, it is possible to make up a map that will show true bearings for all parts of the world from any single point. Three such maps are reproduced in this chapter. Fig. 1 shows directions from Washington, DC, Fig. 2 gives directions from San Francisco and Fig. 3 (a simplified version of the ARRL Amateur Radio map of the world) gives directions from the approximate center of the United States — Wichita, Kansas.

For anyone living in the immediate vicinity (within 150 miles) of any of these three reference points, the directions as taken from the maps will have a high degree of accuracy. However, one or the other of the three maps will suffice for any location in the United States for all except the most accurate work; simply choose the map whose reference point is nearest you. Greatest errors will arise when your location is to one side or the other of a line between the reference point and the destination point; if your location is near or on the resulting line, there will be little or no error.

By tracing the directional pattern of the antenna system on a sheet of tissue paper, then placing the paper over the azimuthal map with the origin of the pattern at one's location, the "coverage" of the antenna will be readily evident. This is a particularly useful stunt when a multi-lobed antenna, such as any of the long single-wire systems is to be laid out so that the main lobes cover as many desirable directions as possible. Often a set of such patterns will be of considerable assistance in determining what length antenna to put up, as well as the direction in which it should run.

The current edition of the ARRL Amateur Radio Map of the World, entirely different in concept and design from any other radio amateurs' map, contains a wealth of information especially useful to amateurs. A special type of azimuthal projection made by Rand-McNally to ARRL specifications, it gives great-circle bearings from the geographical center of the United States, as well as the great-circle distance measurement in miles and kilometers, within an accuracy of two percent. The map shows principal cities of the world; local time zones; WAC divisions; more than 265 countries, indexed; and amateur prefixes throughout the

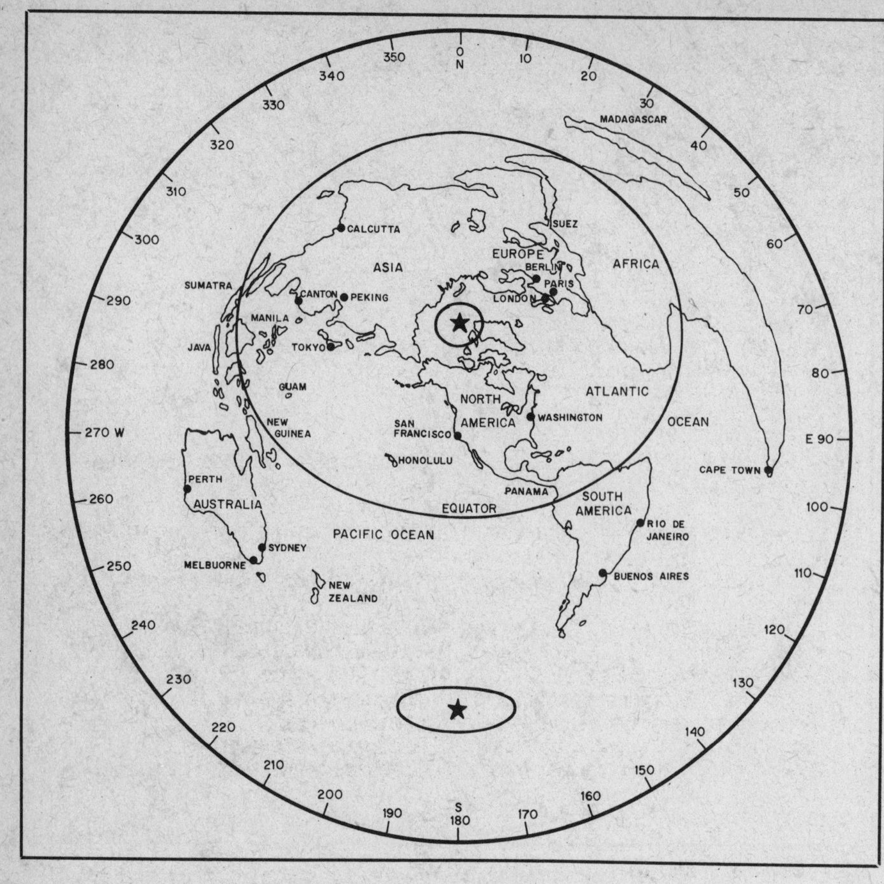


Fig. 2 — Azimuthal map centered on San Francisco, California.

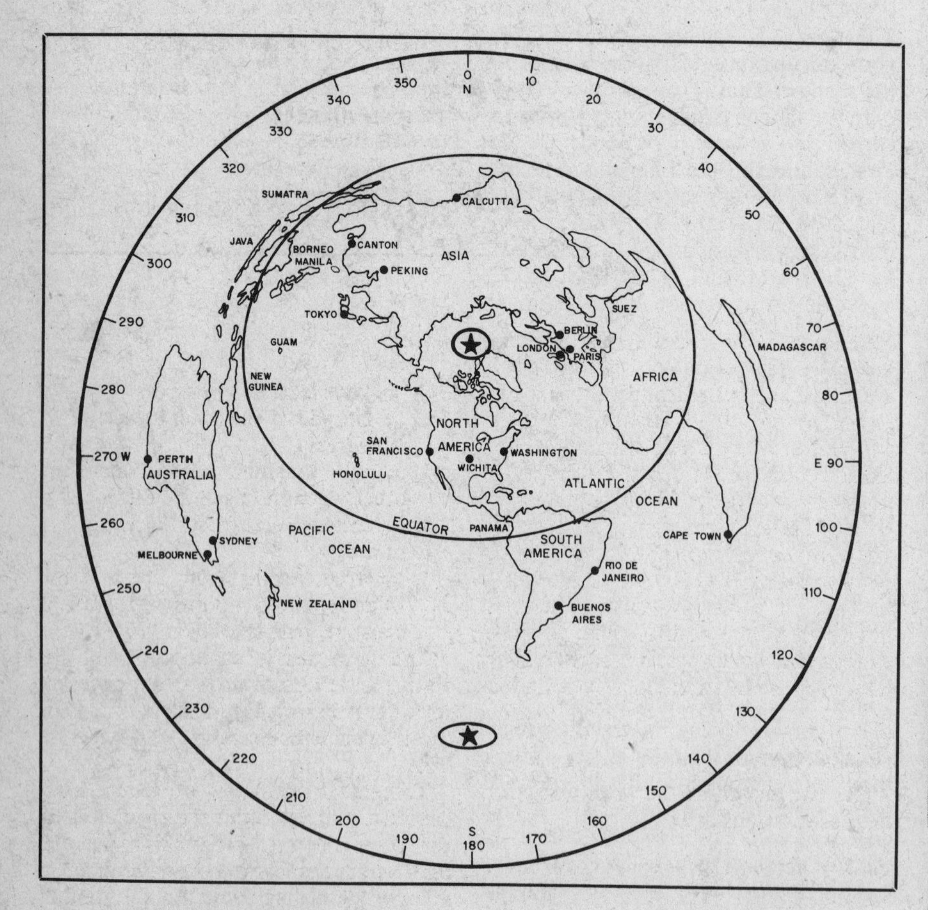


Fig. 3 — Azimuthal map centered on Wichita, Kansas. Copyright by Rand McNally & Co., Chicago. Reproduction License No. 394.

world. The map is large enough to be easily readable from the operating position, 31 × 41 inches: and is printed in six colors on heavy paper. The map is available from ARRL Headquarters, 225 Main St., Newington, CT 06111.

The *Radio Amateur's Callbook* includes great-circle maps and tables, and another *Callbook* publication, *The Radio Amateur's World Atlas,* features a polar-projection world map, maps of the continents and world amateur prefixes. The maps are in color.

Bill Johnston, N5KR, offers computer-calculated and -drawn great-circle maps. An 11 × 14 inch map is custom made for your location. (See Fig. 13, Chapter 1.) Write to 1808 Pomona Dr., Las Cruces, NM 88001.

## WORKING FROM A GLOBE

Bearings for beam-heading purposes may be determined easily from an ordinary globe with nothing more complicated than a small school protractor of the type available in any school-supply or stationery store. For best results, however, the globe should be at least 8 inches in diameter. A thin strip of paper may be used for a straightedge to determine the great-circle path between your location and any other location on the earth's surface. The bearing from your location may be determined with the aid of the protractor. For convenience, a paper-scale circle calibrated in degrees of bearing may be made and affixed over the point representing your location on the globe. The 0° mark of this scale should point toward the north pole.

## A SIMPLIFIED DIRECTION FINDER

A simplified direction finder may be made by removing a globe from its brazen

Fig. 4 — A simple direction finder made by modifying a globe. Bearing and distance to other locations from yours may be determined quickly after modification, no calculations being required.

meridian (semi-circular support) and remounting it in the manner shown in Fig. 4. Drill a hole that will accept the support at your location on the globe, and another hole directly opposite the first. This second hole will have the same latitude as yours but will be on the other side of the equator (north latitude vs. south latitude). Its longitude will be opposite in direction from yours from the Greenwich or 0° meridian, east vs. west, and will be equal to 180° minus your longitude. For example, if your location is 42° N. lat., 72° W. long., the point opposite yours on the globe is 42° S. lat., 180 − 72 or 108° E. long.

Once the holes are drilled, remount the globe with your location in the position formerly occupied by the north pole. By rotating the globe until the distant point of interest lies beneath the brazen meridian, this support may be used to indicate the great-circle path. A new "equator" calibrated in a manner to indicate the bearing may be added with India ink, as shown in Fig. 4, or a small protractor-like scale may be added at the top of the globe, over your location. A distance scale can be affixed to the brazen meridian so that both the bearing and distance to other locations may be readily determined (12,500 miles or 40,000 km to the semicircle).

**Table 2**
**BASIC Language Program for Determining Bearings and Distances**

```
10 REM * * *  BEARINGS/BAS  * * *
15 REM A=YOUR LAT.
20 REM B=OTHER STATION LAT.
25 REM C=BEARING ANGLE
30 REM D=DEGREES OF ARC
35 REM E=INTERMEDIATE VALUE
40 REM K=CONVERSION CONST., ARC TO KILOMETERS
45 REM L=DIFF  IN LONGITUDES
50 REM L1=YOUR LONG.
55 REM L2=OTHER STATION LONG.
60 REM M=CONVERSION CONSTANT, DEGREES TO RADIANS
65 REM N=CONVERSION CONST., ARC TO NAUT. MI.
70 REM S=CONVERSION CONST., ARC TO STATUTE MI.
75 REM V=STRING VARIABLE
100 CLS
110 PRINT"PROGRAM TO CALCULATE GREAT CIRCLE DISTANCES AND BEARINGS"
120 PRINT
130 PRINT"BY J. HALL AND C. HUTCHINSON, ARRL HQ., JULY 1981"
140 PRINT"THIS PROGRAM IS NOT COPYRIGHTED AND MAY BE REPRODUCED FREELY"
150 DEFDBLA,B,C,D,E,L,M:DEFSTRV
160 D=1:K=111.11:M=57.29577951308238:N=60:S=69.041:V="######.#"
170 PRINT
180 PRINT"ENTER NEGATIVE VALUES FOR SOUTHERLY LATITUDES"
190 PRINT"ENTER NEGATIVE VALUES FOR EASTERLY LONGITUDES"
200 PRINT:IFD<>1THEN230
210 INPUT"YOUR LATITUDE (DEGREES AND DECIMAL)";A:A=A/M
220 INPUT"YOUR LONGITUDE (DEGREES AND DECIMAL)";L1
230 INPUT"OTHER LATITUDE (DEGREES AND DECIMAL)";B:B=B/M
240 INPUT"OTHER LONGITUDE (DEGREES AND DECIMAL)";L2
250 L=(L1-L2)/M
260 E=SIN(A)*SIN(B)+COS(A)*COS(B)*COS(L)
270 D=-ATN(E/SQR(1-E*E))+1.57079
280 C=(SIN(B)-SIN(A)*E)/(COS(A)*SIN(D))
290 IFC>=1THENC=0:GOTO310ELSEIFC<=-1THENC=180/M:GOTO310
300 C=-ATN(C/SQR(1-C*C))+1.57079
310 C=C*M
320 IFSIN(L)<0THENC=360-C
330 PRINT"THE BEARING IS ";:PRINTTAB(29)USINGV;C;:PRINT" DEGREES"
340 PRINT"THE GREAT CIRCLE DISTANCE IS ";
350 PRINTUSINGV;K*D*M;:PRINT" KILOMETERS"
360 PRINTTAB(29)USINGV;N*D*M;:PRINT" NAUTICAL MILES"
370 PRINTTAB(29)USINGV;S*D*M;:PRINT" STATUTE MILES"
380 PRINT
390 D=0:PRINT"TO CONTINUE PRESS ENTER"
400 INPUT"TO CALCULATE BEARING FROM A DIFFERENT LOCATION ENTER 1";D
410 GOTO170
```

## DIRECTION AND DISTANCE BY TRIGONOMETRY

The methods to be described will give the bearing and distance as accurately as one cares to compute them. All that is required is a table of latitude and longitude information, such as is found in tables at the end of this chapter, and an electronic calculator or computer with trigonometric functions. The latitude and longitude for any other location can be taken from a map of the area in question.

Fig. 5 will help you to visualize the nature of the situation. That sketch represents the path between points situated relatively such as Pratt, Kansas, USA (at point A), and Perth, Western Australia, (at point B). In using these equations, northerly latitudes are taken as positive, and southerly latitudes are taken as negative. Also, westerly longitudes are taken as positive, and easterly longitudes are taken as negative. *In all calculations,* the appropriate signs are to be retained. *All additions and subtractions throughout the procedure are to be made algebraically.* Thus, if a negative-value number is subtracted from a positive-value number, the resultant will be positive, and it will be the *sum* of the two absolute values, and so on.

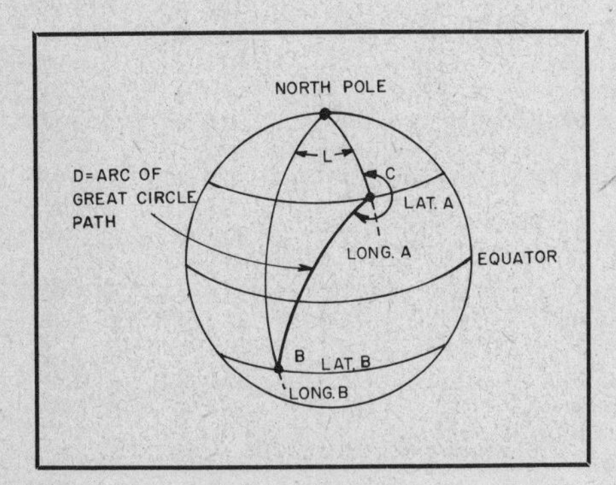


Fig. 5 — The various terms used in the equations for determining bearing and distance. North latitudes and west longitudes are taken as positive, while south latitudes and east longitudes are taken as negative.

### The Calculations

The two equations we'll be using for these calculations are:

$$\cos D = \sin A \sin B + \cos A \cos B \cos L \quad \text{(Eq. 1)}$$

$$\cos C = \frac{\sin B - \sin A \cos D}{\cos A \sin D} \quad \text{(Eq. 2)}$$

where

A = *your* latitude in degrees
B = latitude of the other location in degrees
L = *your* longitude minus that of the other location (algebraic difference).
D = distance along path in degrees of arc
C = true bearing from north if the value for sin L is positive. If sin L is negative, true bearing is 360 − C.

The term cos is an abbreviation for cosine, and the term sin is an abbreviation for sine. A knowledge of the meanings of these terms isn't necessary for their use here.

The actual calculating procedure uses, first, Eq. 1 to determine the angular value for D, in degrees. From this value the path-length distance may be determined in miles or kilometers. Next, Eq. 2 is used to determine the bearing angle.

Using the Pratt-to-Perth example mentioned earlier, refer to Fig. 5 to see how the equations are used. From tables at the end of this chapter it can be seen that the location of Pratt is 37.7° N. lat., 98.7° W. Long. Similarly, Perth is located at 32° S. lat., 115.9° E. long. Your location is in Pratt. Values for use in the equations are as follows:

$A$ = lat. A = +37.7°
$B$ = lat. B = −32°
$L$ = long. A − long. B
= 98.7° − (−115.9°) = 214.6°

Solving Eq. 1, $\cos D = \sin 37.7° \sin(-32°) + \cos 37.7° \cos(-32°) \cos 214.6°$. $D = 151.21°$. Each degree along the path equals 60 nautical miles. Therefore, 151.2° of arc is equivalent to $60 \times 151.2 = 9073$ nautical miles. To convert to statute miles, multiply degrees by 69.041. If the distance is desired in kilometers, multiply degrees by 111.111. This means that the Pratt-to-Perth distance is 10,440 miles or 16,801 kilometers.

Solving Eq. 2,

$$\cos C = \frac{\sin(-32°) - \sin 37.7° \cos 151.2°}{\cos 37.7° \sin 151.2°}$$

$C = 89.1°$. Because the sin of L (214.6°) is negative, however, the correct value for C is $360° - 89.1° = 270.9°$. Thus, the true bearing from Pratt to Perth is 270.9° and the distance is 10,440 statute miles. If the bearing from Perth were desired it would be necessary only to work through Eq. 2, interchanging latitude values for A and B. Because of the way L is defined, sin L will be positive in this case, and it will not be necessary to subtract from 360° to get the true bearing at Perth, which is 68.9°

These equations give information for the great-circle bearing and distance for the shortest path. For long-path work, the bearing will be 180° away from the answers obtained.

The equations described above may be used for any two points on the earth's surface — both locations in the northern hemisphere, both locations in the southern hemisphere, either or both on the equator, and so on. The equations themselves are exact, not being based on any approximations. However, there are some cases where practical limitations exist in the accuracy of the results obtained from Eq. 2, in relation to the number of significant figures used during calculations. (Round-off errors in calculators and computers during computations will effectively reduce the number of significant figures in the resulting answers.) These cases are where both locations are near or at exact opposite points on the earth (antipodes), where the locations are close together, or where *your* location is at or near one of the poles. (At the poles, all directions are either south or north, anyway.) More specifically, these situations exist when lat. A is near ±90°, or where D is near 0° or 180°.

Table 2 is a BASIC language program for calculating bearings and distances by computer. The program is written for a Radio Shack TRS-80 Level II computer. There are numerous versions of BASIC, and it may be necessary for you to modify portions of the program for use with your computer system. Statements 15 through 75 indicate by way of remarks what variables are used in the program and the information they represent.

Tables 3 and 4 show latitude and longitude for various U.S. and Canadian cities and other areas of the world. The data are arranged alphanumerically by call prefixes; Table 4 closely follows the ARRL DXCC List.


### Acknowledgments

Data for Table 3 was compiled from various sources, including *The World Almanac and Book of Facts 1981*, Newspaper Enterprise Associates, Inc., New York. Special thanks for latitude and longitude data for Table 4 go to Daryl Kiebler, WB8EUN, to Stan Horzepa, WA1LOU, and to Michael Kaczynski, W1OD.


## BIBLIOGRAPHY

Source material and more extended discussions of topics covered in this chapter can be found in the references given below.


Davis, "A Simplified Direction Finder," Hints and Kinks, *QST*, May 1972.

Hall, "Bearing and Distance Calculations by Sleight of Hand," *QST*, August 1973.

Klopf, "A Bearing and Distance Calculator," *QST*, March 1971.

Norton, ***Norton's Star Atlas and Reference Handbook***, Gail and Inglis, London, England; also published in the U.S. by Sky Publishing Corp., Cambridge, Massachusetts.


**Table 3**
**Latitudes and Longitudes of Various U.S. and Canadian Cities**

| *Call Area* | *State or Province and City* | *Lat.* | *Long.* |
|---|---|---|---|
| VE1 | New Brunswick, Saint John | 45.3 N | 66.1 W |
| | Nova Scotia, Halifax | 44.6 N | 63.6 W |
| | Prince Edward Island, Charlottetown | 46.2 N | 63.1 W |
| VE2 | Quebec | | |
| | Montreal | 45.5 N | 73.6 W |
| | Quebec City | 46.8 N | 71.2 W |
| VE3 | Ontario | | |
| | London | 43.0 N | 81.3 W |
| | Ottawa | 45.4 N | 75.7 W |
| | Sudbury | 46.5 N | 81.0 W |
| | Thunder Bay | 48.4 N | 89.2 W |
| | Toronto | 43.7 N | 79.4 W |
| VE4 | Manitoba, Winnipeg | 49.9 N | 97.1 W |
| VE5 | Saskatchewan | | |
| | Regina | 50.5 N | 104.6 W |
| | Saskatoon | 52.1 N | 106.7 W |
| VE6 | Alberta | | |
| | Calgary | 51.0 N | 114.1 W |
| | Edmonton | 53.5 N | 113.5 W |
| VE7 | British Columbia | | |
| | Prince George | 53.9 N | 122.8 W |
| | Prince Rupert | 54.3 N | 130.3 W |
| | Vancouver | 49.3 N | 123.1 W |
| VE8 | Northwest Territories, Yellowknife | 62.5 | 114.4 W |
| | Yukon, Whitehorse | 60.7 N | 135.1 W |
| VO1 | Newfoundland, St. John's | 47.6 N | 52.7 W |
| VO2 | Labrador, Goose Bay | 51.3 N | 60.4 W |
| W1 | Connecticut, Hartford | 41.8 N | 72.7 W |
| | Maine | | |
| | Bangor | 44.8 N | 68.8 W |
| | Portland | 43.7 N | 70.3 W |
| | Massachusetts, Boston | 42.4 N | 71.1 W |
| | New Hampshire, Concord | 43.2 N | 71.5 W |
| | Rhode Island, Providence | 41.8 N | 71.4 W |
| | Vermont, Montpelier | 44.3 N | 72.6 W |
| W2 | New Jersey, Atlantic City | 39.4 N | 74.4 W |
| | New York | | |
| | Albany | 42.7 N | 73.8 W |
| | Buffalo | 42.9 N | 78.9 W |
| | New York City | 40.8 N | 74.0 W |
| | Syracuse | 43.1 N | 76.2 W |
| W3 | Delaware, Wilmington | 39.7 N | 75.5 W |
| | District of Columbia, Washington | 38.9 N | 77.0 W |
| | Maryland, Baltimore | 39.3 N | 76.6 W |
| | Pennsylvania | | |
| | Harrisburg | 40.3 N | 76.9 W |
| | Philadelphia | 39.9 N | 75.2 W |
| | Pittsburgh | 40.4 N | 80.0 W |
| | Scranton | 41.4 N | 75.7 W |
| W4 | Alabama, Montgomery | 32.4 N | 86.3 W |
| | Florida | | |
| | Jacksonville | 30.3 N | 81.7 W |
| | Miami | 25.8 N | 80.2 W |
| | Pensacola | 30.4 N | 87.2 W |
| | Georgia | | |
| | Atlanta | 33.8 N | 84.4 W |
| | Savannah | 32.1 N | 81.1 W |
| | Kentucky | | |
| | Lexington | 38.0 N | 84.5 W |
| | Louisville | 38.2 N | 85.8 W |
| | North Carolina | | |
| | Charlotte | 35.2 N | 80.8 W |
| | Raleigh | 35.8 N | 78.6 W |
| | Wilmington | 34.2 | 77.9 W |
| | South Carolina, Columbia | 34.0 N | 81.0 W |
| | Tennessee | | |
| | Knoxville | 36.0 N | 83.9 W |
| | Memphis | 35.1 N | 90.1 W |

| Call Area | State or Province and City | Lat. | Long. |
|---|---|---|---|
| | Nashville | 36.2 N | 86.8 W |
| | Virginia | | |
| | Norfolk | 36.9 N | 76.3 W |
| | Richmond | 37.5 N | 77.4 W |
| W5 | Arkansas, Little Rock | 34.7 N | 92.3 W |
| | Louisiana | | |
| | New Orleans | 29.9 N | 90.1 W |
| | Shreveport | 32.5 N | 93.7 W |
| | Mississippi, Jackson | 32.3 N | 90.2 W |
| | New Mexico, Albuquerque | 35.1 N | 106.7 W |
| | Oklahoma, Oklahoma City | 35.5 N | 97.5 W |
| | Texas | | |
| | Abilene | 32.5 N | 99.7 W |
| | Amarillo | 35.2 N | 101.8 W |
| | Dallas | 32.8 N | 96.8 W |
| | El Paso | 31.8 N | 106.5 W |
| | San Antonio | 29.4 N | 98.5 W |
| W6 | California | | |
| | Los Angeles | 34.1 N | 118.2 W |
| | San Francisco | 37.8 N | 122.4 W |
| W7 | Arizona | | |
| | Flagstaff | 35.2 N | 111.7 W |
| | Phoenix | 33.5 N | 112.1 W |
| | Idaho | | |
| | Boise | 43.6 N | 116.2 W |
| | Pocatello | 42.9 N | 112.5 W |
| | Montana | | |
| | Billings | 45.8 N | 108.5 W |
| | Butte | 46.0 N | 112.5 W |
| | Great Falls | 47.5 N | 111.3 W |
| | Nevada | | |
| | Las Vegas | 36.2 N | 115.1 W |
| | Reno | 39.5 N | 119.8 W |
| | Oregon, Portland | 45.5 N | 122.7 W |
| | Utah, Salt Lake City | 40.8 N | 111.9 W |
| | Washington | | |
| | Seattle | 47.6 N | 122.3 W |
| | Spokane | 47.7 N | 117.4 W |
| | Wyoming | | |
| | Cheyenne | 41.1 N | 104.8 W |
| | Sheridan | 44.8 N | 107.0 W |
| W8 | Michigan | | |
| | Detroit | 42.3 N | 83.0 W |
| | Grand Rapids | 43.0 N | 85.7 W |
| | Sault Ste. Marie | 46.5 N | 84.4 W |
| | Traverse City | 45.8 N | 85.6 W |
| | Ohio | | |
| | Cincinnati | 39.1 N | 84.5 W |
| | Cleveland | 41.5 N | 81.7 W |
| | Columbus | 40.0 N | 83.0 W |
| | West Virginia, Charleston | 38.4 N | 81.6 W |
| W9 | Illinois, Chicago | 41.9 N | 87.6 W |
| | Indiana, Indianapolis | 39.8 N | 86.2 W |
| | Wisconsin | | |
| | Green Bay | 44.5 N | 88.0 W |
| | Milwaukee | 43.0 N | 87.9 W |
| WØ | Colorado | | |
| | Denver | 39.7 N | 105.0 W |
| | Grand Junction | 39.1 N | 108.6 W |
| | Iowa, Des Moines | 41.6 N | 93.6 W |
| | Kansas | | |
| | Pratt | 37.7 N | 98.7 W |
| | Wichita | 37.7 N | 97.3 W |
| | Minnesota | | |
| | Duluth | 46.8 N | 92.1 W |
| | Minneapolis | 45.0 N | 93.3 W |
| | Missouri | | |
| | Columbia | 39.0 N | 92.3 W |
| | Kansas City | 39.1 N | 94.6 W |
| | St. Louis | 38.6 N | 90.2 W |
| | Nebraska | | |
| | North Platte | 41.1 N | 100.8 W |
| | Omaha | 41.3 N | 95.9 W |
| | North Dakota, Fargo | 46.9 N | 96.8 W |
| | South Dakota | | |
| | Rapid City | 44.1 N | 103.2 W |

**Table 4**
**Latitudes and Longitudes of Various Areas of the World (corrected in 2nd printing)**

| Prefix | Country (and City or Area) | Lat. | Long. |
|---|---|---|---|
| A2 | Botswana | 22.0 S | 25.0 E |
| A3 | Tonga, Rep of | 21.0 S | 175.0 W |
| A4X | Oman, Masqat | 23.5 N | 59.0 E |
| A5 | Bhutan | 28.0 N | 90.0 E |
| A6X | United Arab Emirates | 25.3 N | 55.3 E |
| A7X | Qatar | 25.3 N | 51.5 E |
| A9X | Bahrein | 26.0 N | 50.5 E |
| AP | Pakistan, Karachi | 25.0 N | 67.0 E |
| BV | Taiwan, Taipei | 25.0 N | 122.0 E |
| BY | China | | |
| | Chunking | 29.8 N | 106.5 E |
| | Peking | 40.0 N | 116.4 E |
| | Shanghai | 31.2 N | 121.5 E |
| C2 | Nauru, Rep. of | 0.5 S | 166.9 E |
| C3 | Andorra | 42.5 N | 1.5 E |
| C5 | The Gambia | 13.0 N | 17.0 W |
| C6 | Bahamas, Nassau | 25.0 N | 77.5 W |
| C9 | Mozambique | | |
| | Lourenco Marques | 26.0 S | 33.0 E |
| | Mozambique | 15.0 S | 41.0 E |
| CE | Chile, Santiago | 33.5 S | 70.8 W |
| CEØA | Easter Island | 28.0 S | 109.0 W |
| CEØX | San Felix | 26.0 S | 80.0 W |
| CEØZ | Juan Fernandez | 33.6 S | 78.8 W |
| CM, CO | Cuba, Havana | 23.1 N | 82.4 W |
| CN | Morocco, Casablanca | 33.0 N | 8.0 W |
| CP | Boliva, La Paz | 16.5 S | 68.4 W |
| CR9 | Macao | 22.0 N | 114.0 E |
| CT | Portugal, Lisbon | 38.7 N | 9.2 W |
| CT2 | Azores Is. | 38.0 N | 25.0 W |
| CT3 | Madeira Is. | 33.0 N | 17.0 W |
| CX | Uruguay, Montevideo | 34.9 S | 56.2 W |
| D2 | Angola, Luanda | 8.8 S | 13.2 E |
| D4 | Cape Verde, Rep. Of | 16.0 N | 24.0 W |
| D6 | Comoros, Moroni | 11.8 S | 43.7 E |
| DA-DL | Germany, Fed. Rep. Of (W), Bonn | 51.0 N | 6.0 E |
| DU | Philippines, Manila | 14.6 N | 121.0 E |
| EA | Spain, Madrid | 40.4 N | 3.7 W |
| EA6 | Balearic Is. | 39.5 N | 3.0 E |
| EA8 | Canary Is. | 28.0 N | 15.0 W |
| EA9 | Ceuta & Melilla | | |
| | Ceuta | 36.7 N | 5.4 W |
| | Melilla | 35.3 N | 3.0 W |
| EI | Ireland, Dublin | 53.3 N | 6.3 W |
| EL | Liberia, Monrovia | 7.0 N | 11.0 W |
| EP | Iran, Tehran | 35.8 N | 51.8 E |
| ET | Ethiopia | | |
| | Addis Ababa | 9.0 N | 39.0 E |
| | Asmara | 15.0 N | 38.0 E |
| F | France, Paris | 48.8 N | 2.3 E |
| FB8W | Crozet | 46.5 S | 52.0 E |
| FB8X | Kerguelen Is. | 50.0 S | 70.0 E |
| FB8Y | Antarctica, Dumont D'Urville | 67.0 S | 140.0 E |
| FB8Z | Amsterdam & St. Paul Is. | 37.0 S | 77.6 E |
| FC | Corsica | 42.0 N | 9.0 E |
| FG | Guadeloupe | 17.0 N | 62.0 W |
| FH | Mayotte | 13.0 S | 45.3 E |
| FK | New Caledonia, Noumia | 22.0 S | 167.0 E |
| FM | Martinique | 15.0 N | 61.0 W |
| FO | Clipperton | 11.0 N | 110.0 W |
| FO | Fr. Polynesia, Tahiti | 18.0 S | 150.0 W |
| FP | St. Pierre & Miquelon | 47.0 N | 56.5 W |
| FR | Glorioso Is. | 10.6 S | 47.3 E |
| FR | Juan De Nova Is. | 21.5 S | 40.0 E |
| FR | Reunion Is | 21.0 S | 55.5 E |
| FR | Tromelin Is. | 15.5 S | 54.5 E |
| FS | St. Martin Is. | 18.0 N | 63.0 W |
| FW | Wallis & Futuna Is. | 14.0 S | 176.0 W |
| FY | Fr. Guiana, Cayenne | 5.0 N | 52.0 W |
| G | England, London | 51.5 N | 0.1 W |
| GD | Isle of Man | 54.0 N | 4.5 W |
| G1 | Ireland, Northern, Belfast | 54.6 N | 5.9 W |
| GJ | Jersey | 49.3 N | 2.2 W |
| GM | Scotland, Glasgow | 55.8 N | 4.3 W |
| GU | Guernsey | 49.5 N | 2.7 W |
| GW | Wales | 52.5 N | 3.5 W |
| H4 | Solomon Is. | 9.4 S | 160.0 E |
| HA | Hungary, Budapest | 47.5 N | 19.1 E |
| HB | Switzerland, Bern | 47.0 N | 7.0 E |
| HBØ | Liechtenstein | 47.0 N | 9.5 E |
| HC | Ecuador, Quito | 0.2 S | 78.0 W |
| HC8 | Galapagos Is. | 2.0 S | 92.0 W |
| HH | Haiti, Port-Au-Prince | 19.0 N | 72.0 W |
| HI | Dominican Rep., Santo Domingo | 18.5 N | 70.0 W |

| Prefix | Country (anb City or Area) | Lat. | Long. |
|---|---|---|---|
| HK | Colombia, Bogota | 5.0 N | 74.0 W |
| HKØ | Bajo Nuevo | 15.9 N | 78.7 W |
| HKØ | Malpelo Is. | 4.0 N | 81.1 W |
| HKØ | San Andres Is. | 12.5 N | 81.7 W |
| HKØ | Serrana Bank | 14.3 N | 81.3 W |
| HM, HL | Korea, Seoul | 37.5 N | 127.0 E |
| HP | Panama | 9.0 N | 79.5 W |
| HR | Honduras, Tegucigalpa | 14.0 N | 87.0 W |
| HS | Thailand, Bangkok | 13.8 N | 100.5 E |
| HV | Vatican City | 41.9 N | 12.5 E |
| HZ, 7Z | Saudi Arabia | | |
| | Dharan | 26.3 N | 50.0 E |
| | Mecca | 22.0 N | 40.0 E |
| I | Italy | | |
| | Rome | 41.9 N | 12.5 E |
| | Trieste | 45.5 N | 14.0 E |
| IS | Sardinia | 40.0 N | 9.0 E |
| IT | Italy, Sicily | 37.5 N | 14.0 E |
| J2 | Djibouti | 12.0 N | 43.0 E |
| J3 | Grenada | 12.0 N | 61.8 W |
| J5 | Guinea-Bissau | 12.0 N | 15.0 W |
| J6 | St. Lucia | 14.0 N | 61.0 W |
| J7 | Dominica | 15.4 N | 61.3 W |
| JA-JN | Japan, Tokyo | 35.7 N | 139.8 E |
| JD | Minami Torishima | 25.0 N | 154.0 E |
| JD | Ogasawara, Kazan Is. | 27.5 N | 141.0 E |
| JT | Mongolia, Ulan Bator | 48.0 N | 107.0 E |
| JW | Svalbard Is. | 78.0 N | 16.0 E |
| JX | Jan Mayen | 71.0 N | 8.0 W |
| JY | Jordan, Amman | 32.0 N | 36.0 E |
| KC4 | Antarctica | | |
| | Byrd Station | 80.0 S | 120.0 W |
| | McMurdo Sound | 77.5 S | 167.0 E |
| | Palmer Station | 65.0 S | 64.0 W |
| KC6 | Caroline Is., E., Ponape | 8.0 N | 158.0 E |
| KC6 | Caroline Is., W., Yap | 9.5 N | 138.2 E |
| KG4 | Guantanamo Bay | 20.0 N | 75.0 W |
| KH1 | Baker, Howland, Am. Phoenix | 0.5 N | 176.0 W |
| KH2 | Guam | 14.0 N | 145.0 E |
| KH3 | Johnston Is. | 17.0 N | 170.0 W |
| KH4 | Midway Islands | 28.0 N | 177.5 W |
| KH5 | Palmyra Is. | 6.0 N | 162.0 W |
| KH5K | Kingman Reef | 7.5 N | 162.8 W |
| KH6 | Hawaii | | |
| | Hilo | 19.7 N | 155.1 W |
| | Honolulu | 21.3 N | 157.9 W |
| KH7 | Kure Is. | 28.5 N | 178.5 W |
| KH8 | American Samoa | 14.0 S | 171.0 W |
| KH9 | Wake Is. | 19.0 N | 167.0 E |
| KHØ | Mariana Is., Saipan | 16.0 N | 146.0 E |
| KL7 | Alaska | | |
| | Adak | 51.8 N | 176.6 W |
| | Anchorage | 61.2 N | 150.0 W |
| | Fairbanks | 64.8 N | 147.9 W |
| | Juneau | 58.3 N | 134.4 W |
| | Nome | 64.5 N | 165.4 W |
| KP1 | Navassa Is. | 18.0 N | 75.0 W |
| KP2 | Virgin Is. | 18.0 N | 64.5 W |
| KP4 | Desecheo Is. | 18.3 N | 67.5 W |
| KP4 | Puerto Rico, San Juan | 18.5 N | 66.2 W |
| KX | Marshall Is. | 9.2 N | 167.0 E |
| LA-LJ | Norway, Oslo | 60.0 N | 10.7 E |
| LU | Argentina, Buenos Aires | 34.6 S | 58.4 W |
| LX | Luxembourg | 49.0 N | 6.5 E |
| LZ | Bulgaria, Sofia | 42.7 N | 23.3 E |
| OA | Peru, Lima | 12.0 S | 77.0 W |
| OD | Lebanon, Beirut | 33.9 N | 35.5 E |
| OE | Austria, Vienna | 48.2 N | 16.3 E |
| OH | Finland, Helsinki | 60.2 N | 25.0 E |
| OHØ | Aland Is. | 60.2 N | 20.0 E |
| OJØ | Market | 60.3 N | 19.0 E |
| OK | Czechoslovakia, Prague | 50.1 N | 14.4 E |
| ON | Belgium, Brussels | 50.9 N | 4.4 E |
| OX,XP | Greenland, Godthaab | 63.0 N | 52.0 W |
| OY | Faeroe | 62.0 N | 7.0 W |
| OZ | Denmark, Copenhagen | 55.7 N | 12.6 E |
| P2 | Papua New Guinea | | |
| | Madang | 5.2 S | 145.6 E |
| | Port Moresby | 9.4 S | 147.1 E |
| PA-PI | Netherlands, Amsterdam | 52.4 N | 4.9 E |
| PJ | Neth. Antilles | 12.0 N | 69.0 W |
| PJ | St. Maarten, Saba | 18.0 N | 63.0 W |
| PT2 | Brazil, Brasilia | 16.0 S | 48.0 W |
| PY6 | Brazil | | |
| | Rio De Janeiro | 23.0 S | 43.2 W |
| | Natal | 6.0 S | 35.2 W |
| PYØ | Fernando De Noronha | 3.0 S | 32.0 W |

| Prefix | Country (and City or Area) | Lat. | Long. |
|---|---|---|---|
| PYØ | St. Peter & Paul's Rocks | 1.5 N | 30.0 W |
| PYØ | Trindade & Martin Vaz Is. | 19.0 N | 29.0 W |
| PZ | Surinam | 6.0 S | 55.0 W |
| S2 | Bangladesh, Dacca | 24.0 N | 90.5 E |
| S7 | Seychelles Is. | 4.0 S | 56.0 E |
| S9 | Sao Tome | 0.3 N | 6.0 E |
| SK-SM | Sweden, Stockholm | 59.3 N | 18.1 E |
| SP | Poland | | |
| | Cracow | 50.0 N | 20.0 E |
| | Warsaw | 52.2 N | 21.0 E |
| ST | Sudan, Khartoum | 15.5 N | 32.5 E |
| STØ | Southern Sudan, Juba | 5.0 N | 31.6 E |
| SU | Egypt, Cairo | 30.0 N | 31.4 E |
| SV | Crete | 35.4 N | 25.2 E |
| SV | Dodecanese | 36.5 N | 27.5 E |
| SV | Greece, Athens | 38.0 N | 23.7 E |
| SV | Mount Athos | 40.2 N | 24.3 E |
| T2 | Tuvalu, Funafuti | 8.7 S | 178.6 E |
| T3 | Kiribati, Cent. | 4.0 S | 171.0 W |
| T3 | Kiribati, West | 5.0 S | 175.0 E |
| T3 | Kiribati, East | 10.0 S | 150.0 W |
| TA | Turkey, Ankara | 39.9 N | 32.9 E |
| TF | Iceland, Reykjavik | 64.1 N | 22.0 W |
| TG | Guatemala, Guatemala City | 14.6 N | 90.5 W |
| TI | Costa Rica | 10.0 N | 84.0 W |
| TI9 | Cocos Is. | 5.0 N | 87.0 W |
| TJ | Cameroon, Yaounde | 4.0 N | 12.0 E |
| TL | Central Africa Rep., Bangui | 4.5 N | 18.0 E |
| TN | Republic of Congo, Brazzaville | 4.0 S | 15.0 E |
| TR | Gabon, Libreville | 0.5 N | 9.0 E |
| TT | Chad, Ft. Lamy | 12.0 N | 15.0 E |
| TU | Ivory Coast, Abidjan | 5.5 N | 4.0 W |
| TY | Benin | 6.3 N | 2.3 E |
| TZ | Mali Republic, Bamako | 13.0 N | 8.0 W |
| UA6 | Russia, European | | |
| | Leningrad (UA1) | 60.0 N | 30.0 E |
| | Moscow (UA3) | 56.0 N | 37.5 E |
| | Kuibyshev (UA4) | 53.5 N | 50.5 E |
| | Rostov (UA6) | 47.0 N | 40.0 E |
| UA1 | Franz Josef Land | 80.0 N | 50.0 E |
| UA2 | Kaliningrad | 55.0 N | 20.5 E |
| UA9,Ø | Russia, Asiatic | | |
| | Novosibirsk (UA9) | 55.0 N | 83.0 E |
| | Perm (UA9) | 57.5 N | 56.0 E |
| | Khabarovsk (UAØ) | 48.0 N | 135.0 E |
| | Krasnoyarsk (UAØ) | 56.0 N | 93.0 E |
| | Yakutsk (UAØ) | 62.0 N | 130.0 E |
| UB5 | Ukraine, Kiev | 50.5 N | 31.0 E |
| UC2 | White Russia, Minsk | 54.0 N | 27.5 E |
| UD6 | Azerbaijan, Baku | 40.5 N | 50.0 E |
| UF6 | Georgia, Tbilisi | 41.5 N | 45.0 E |
| UG6 | Armenia, Erivan | 40.5 N | 44.5 E |
| UH8 | Turkoman, Ashkhabad | 38.0 N | 58.0 E |
| UI8 | Uzbek, Bukhara | 40.0 N | 64.0 E |
| UJ8 | Tadzhik, Samarkand | 39.5 N | 67.0 E |
| UL7 | Kazakh, Alma-ata | 53.0 N | 76.0 E |
| UM8 | Kirghiz | 42.0 N | 75.0 E |
| UO5 | Moldavia, Kishinev | 47.0 N | 29.0 E |
| UP2 | Lithuania, Vilna | 54.5 N | 25.5 E |
| UQ2 | Latvia, Riga | 57.0 N | 24.0 E |
| UR2 | Estonia, Tallinn | 59.0 N | 25.0 E |
| VE1 | Sable Is. | 43.8 N | 60.0 W |
| VE1 | St. Paul Is. | 47.2 N | 60.1 W |
| VK | Australia | | |
| | Canberra (VK1) | 35.5 S | 149.0 E |
| | Sydney (VK2) | 34.0 S | 151.0 E |
| | Melbourne (VK3) | 38.0 S | 145.0 E |
| | Brisbane (VK4) | 27.5 S | 153.0 E |
| | Adelaide (VK5) | 34.9 S | 138.6 E |
| | Perth (VK6) | 32.0 S | 115.9 E |
| | Hobart, Tasmania (VK7) | 42.9 S | 147.3 E |
| | Darwin (VK8) | 12.5 S | 130.9 E |
| VK2 | Lord Howe Is. | 31.6 S | 159.1 E |
| VK9 | Christmas Is. | 10.5 S | 105.7 E |
| VK9 | Cocos (Keeling) Is. | 12.2 S | 96.8 E |
| VK9 | Mellish Reef | 17.6 S | 155.8 E |
| VK9 | Norfolk Is. | 29.0 S | 168.0 E |
| VK9 | Willis Is. | 16.0 S | 149.5 E |
| VKØ | Heard Is. | 53.0 S | 73.4 E |
| VKØ | Macquarie Is. | 54.7 S | 158.8 E |
| VP1 | Belize | 17.5 N | 88.3 W |
| VP2A | Antigua | 17.0 N | 59.8 W |
| VP2E | Anguilla | 18.3 N | 63.0 W |
| VP2K | St. Kitts & Nevis | 17.3 N | 62.6 W |
| VP2M | Montserrat | 16.8 N | 62.2 W |
| VP2S | St. Vincent | 13.3 N | 61.3 W |
| VP2V | British Virgin Is., Tortola | 18.4 N | 64.6 W |

| Prefix | Country (and City or Area) | Lat. | Long. |
|---|---|---|---|
| VP5 | Turks & Caicos Is. | 22.0 N | 72.0 W |
| VP8 | Falkland Is. | 52.0 S | 60.0 W |
| VP8 | So. Georgia Is. | 54.0 S | 37.0 W |
| VP8 | So. Orkney Is. | 60.0 S | 46.0 W |
| VP8 | So. Sandwich Is. | 57.0 S | 28.0 W |
| VP8 | So. Shetland Is., King George Is. | 62.0 S | 58.5 W |
| VP9 | Bermuda | 32.3 N | 64.7 W |
| VQ9 | Chagos, Diego Garcia | 7.3 S | 72.4 E |
| VR6 | Pitcairn Is. | 25.1 S | 130.1 W |
| VS5 | Brunei | 5.0 N | 115.0 E |
| VS6 | Hong Kong | 22.5 N | 114.0 E |
| VS9K | Kamaran Is. | 15.3 N | 42.8 E |
| VU | India | | |
| | Bombay | 19.0 N | 72.8 E |
| | Calcutta | 22.6 N | 88.4 E |
| | New Delhi | 28.0 N | 77.0 E |
| VU7 | Andaman Is. | 10.0 N | 94.0 E |
| VU7 | Laccadive Is. | 10.0 N | 70.0 E |
| XE | Mexico | | |
| | Mexico City (XE1) | 19.4 N | 99.1 W |
| | Chihuahua (XE2) | 28.7 N | 106.0 W |
| | Merida (XE3) | 21.0 N | 89.7 W |
| XF4 | Revilla Gigedo Is. | 18.0 N | 112.0 W |
| XT | Upper Volta | 12.3 N | 1.7 W |
| XU | Khmer Rep., Phnom Penh | 11.5 N | 105.0 E |
| XV | Vietnam, Saigon | 10.8 N | 106.7 E |
| XW | Lao Peoples Dem. Rep. | 18.0 N | 102.5 E |
| XZ | Burma, Rangoon | 16.8 N | 96.0 E |
| Y2-9 | German Dem. Rep. (E), Berlin | 52.5 N | 13.4 E |
| YA | Afghanistan, Kandahar | 32.0 N | 65.0 E |
| YB, YC | Indonesia | 6.0 S | 107.0 E |
| YI | Iraq, Baghdad | 33.0 N | 44.5 E |
| YJ | New Hebrides, Villa | 18.0 S | 168.0 E |
| YK | Syria | 34.0 N | 36.5 E |
| YN, HT | Nicaragua, Managua | 12.0 N | 86.0 W |
| YO | Romania, Bucharest | 44.4 N | 26.1 E |
| YS | El Salvador, San Salvador | 14.0 N | 89.0 W |
| YU | Yugoslavia, Belgrade | 44.9 N | 20.5 E |
| YV | Venezuela, Caracas | 10.5 N | 67.0 W |
| YVØ | Aves Is. | 12.0 N | 67.5 W |
| ZA | Albania, Tirana | 41.5 N | 19.5 E |
| ZB | Gibraltar | 36.0 N | 5.5 W |
| ZD7 | Saint Helena | 16.0 S | 6.0 W |
| ZD8 | Ascension Is. | 8.0 S | 14.0 W |
| ZD9 | Tristan Da Cunha | 37.1 S | 12.3 W |
| ZE | Zimbabwe, Salisbury | 18.0 S | 31.0 E |
| ZF | Cayman Is. | 19.5 N | 81.2 W |
| ZK1 | Cook Is. No., Manihiki | 10.5 S | 161.0 W |
| ZK1 | Cook Is. So., Raratonga | 21.0 S | 159.5 W |
| ZK2 | Niue Is. | 19.0 S | 170.0 W |
| ZL | New Zealand | | |
| | Auckland (ZL1) | 36.9 S | 174.8 E |
| | Wellington (ZL2) | 41.3 S | 174.8 E |
| | Christchurch (ZL3) | 43.0 S | 172.5 E |
| | Dunedin (ZL4) | 46.0 S | 170.0 E |
| ZL5 | Antarctica, Scott Base | 78.0 S | 165.0 E |
| ZL | Auckland Is. & Campbell Is. | 52.5 S | 169.0 E |
| ZL | Chatham Is. | 44.0 S | 176.0 W |
| ZL | Kermadec Is. | 30.0 S | 178.0 W |
| ZM7 | Tokelaus, Atafu | 8.4 S | 172.7 W |
| ZP | Paraguay, Asuncion | 25.3 S | 57.7 W |
| ZS | So. Africa | | |
| | Cape Town (ZS1) | 33.9 S | 18.4 E |

| Prefix | Country (and City or Area) | Lat. | Long. |
|---|---|---|---|
| | Durban (ZS5) | 29.9 S | 30.9 E |
| | Johannesburg (ZS6) | 26.2 S | 28.1 E |
| ZS2 | Prince Edward & Marion Is. | 46.8 S | 37.8 E |
| ZS3 | **Namibia** | 22.6 S | 17.1 E |
| 1A | S.M.O.M. | 41.9 N | 12.4 E |
| 1S | Spratly Is. | 8.8 N | 111.9 E |
| 3A | Monaco | 44.0 N | 7.5 E |
| 3B6,7 | Agalega & St. Brandon | 10.4 S | 56.6 E |
| 3B8 | Mauritius | 20.3 S | 57.5 E |
| 3B9 | Rodriguez Is. | 19.7 S | 63.4 E |
| 3C | Equatorial Guinea | 1.8 N | 10.0 E |
| 3CØ | Annobon | 1.5 N | 5.6 E |
| 3D2 | Fiji Is. | 17.0 S | 178.0 E |
| 3D6 | Swaziland | 27.0 S | 31.5 E |
| 3V | Tunisia, Tunis | 36.8 N | 10.2 E |
| 3X | Guinea, Rep. of, Conakry | 10.0 N | 13.0 W |
| 3Y | Bouvet Is. | 54.5 S | 3.0 E |
| 4S | Sri Lanka, Colombo | 7.0 N | 79.9 E |
| 4U | I.T.U. Geneva | 46.2 N | 6.2 E |
| 4U | United Nations Hq. | 40.8 N | 74.0 W |
| 4W | Yemen | 15.0 N | 44.0 E |
| 4X, 4Z | Israel | 32.0 N | 35.0 E |
| 5A | Libya, Tripoli | 32.5 N | 12.5 E |
| 5B, ZC | Cyprus | 35.0 N | 33.0 E |
| 5H | Tanzania, Dar Es Salaam | 7.0 S | 39.5 E |
| 5N | Nigeria, Lagos | 6.5 N | 3.0 E |
| 5R | Malagasy Rep., Tananarive | 18.5 S | 47.0 E |
| 5T | Mauritania, Nouakchott | 18.0 N | 16.0 W |
| 5U | Niger, Niamey | 13.5 N | 2.0 E |
| 5V | Togo | 6.0 N | 1.5 E |
| 5W | Western Samoa | 13.0 S | 172.0 W |
| 5X | Uganda | 1.0 N | 32.5 E |
| 5Z | Kenya, Nairobi | 1.5 S | 37.5 E |
| 6O | Somali, Mogadisho | 2.0 N | 46.0 E |
| 6W | **Senegal, Dakar** | 15.0 N | 18.0 W |
| 6Y | Jamaica | 18.0 N | 76.0 W |
| 70 | Yemen People's Dem., Aden | 13.0 N | 45.0 E |
| 7P | Lesotho | 29.5 S | 28.0 E |
| 7Q | Malawi, Zomba | 15.0 S | 35.0 E |
| 7X | Algeria, Algiers | 36.7 N | 3.0 E |
| 8P | Barbados | 11.5 N | 59.5 W |
| 8Q | Maldive Is. | 4.4 N | 73.4 E |
| 8R | Guyana, Georgetown | 6.8 N | 58.2 W |
| 9A, M1 | San Marino | 44.0 N | 13.0 E |
| 9G | Ghana, Accra | 5.5 N | 0.2 W |
| 9H | Malta | 36.0 N | 14.4 E |
| 9J | Zambia, Lusaka | 15.0 S | 28.0 E |
| 9K | Kuwait | 29.0 N | 48.0 E |
| 9L | Sierra Leone, Freetown | 8.5 N | 13.2 W |
| 9M2 | Malaysia, West | 3.0 N | 102.0 E |
| 9M6, 8 | Malaysia, East | | |
| | Sabah (9M6) | 5.0 N | 117.0 E |
| | Sarawak (9M8) | 2.0 N | 113.0 E |
| 9N | Nepal, Katmandu | 27.5 N | 85.0 E |
| 9Q | Zaire, Rep. of | | |
| | Kinshasa | 4.3 S | 15.3 E |
| | Kisangani | 1.0 N | 25.0 E |
| | Lubumbashi | 12.0 S | 27.5 E |
| 9U | Burundi | 3.0 S | 29.0 E |
| 9V | Singapore | 1.3 N | 103.8 E |
| 9X | Rwanda | 1.5 S | 30.0 E |
| 9Y | Trinidad & Tobago Is. | 11.0 N | 62.5 W |
| | Abu Ail | 14.1 N | 42.8 E |

# Appendix

### Table 1 — Glossary of Antenna Terms

*Actual Ground* — The point within the earth's surface where effective ground conductivity exists. The depth for this point varies with frequency, the condition of the soil, and the geographical region.

*Antenna* — An electrical conductor or array of conductors that radiates signal energy (transmitting) or collects signal energy (receiving).

*Aperture, Effective* — An area enclosing an antenna, on which it is convenient to make calculations of field strength and antenna gain. Sometimes referred to as the "capture area."

*Apex* — The feed-point region of a V type of antenna.

*Apex Angle* — The enclosed angle in degrees between the wires of a V, an inverted V, and similar antennas.

*Balanced Line* — A symmetrical two-conductor feed line that can have uniform voltage and current distribution along its length.

*Balun* — A device for feeding a balanced load with an unbalanced line, or vice versa. May be a form of choke, or a transformer that provides a specific impedance transformation (including 1:1). Often used in antenna systems to interface a coaxial transmission line to the feed point of a balanced antenna, such as a dipole.

*Base Loading* — A coil of specific reactance value that is inserted at the base (ground end) of a vertical antenna to cancel capacative reactance and resonate the antenna.

*Bazooka* — A transmission-line balancer. It is a quarter-wave conductive sleeve (tubing or flexible shielding) placed at the feed point of a center-fed dipole and grounded to the shield braid of the coaxial feed line at the end of the sleeve farthest from the feed point. It permits the use of unbalanced feed line with balanced-feed antennas.

*Beamwidth* — Related to directive antennas. The width, in degrees, of the major lobe between the two directions at which the relative radiated power is equal to one half its value (half-power or −3-dB points) at the peak of the lobe.

*Beta Match* — Sometimes called a "hairpin match." It is a U-shaped conductor that is connected to the two inner ends of a split dipole for the purpose of creating an impedance match to a balanced feeder.

*Bridge* — A circuit with two or more ports that is used in measurements of impedance, resistance or standing waves in an antenna system. When the bridge is adjusted for a balanced condition, the unknown factor can be determined by reading its value on a calibrated scale or meter.

*Capacitance Hat* — A conductor of large surface area that is connected to an antenna to lower its resonant frequency. It is sometimes mounted directly above a loading coil to reduce the required inductance for establishing resonance. It usually takes the form of a series of wheel spokes or a solid circular disc. Sometimes referred to as a "top hat."

*Capture Area* — See aperture.

*Center Fed* — Transmission-line connection at the electrical center of an antenna radiator.

*Center Loading* — A scheme for inserting inductive reactance (coil) at or near the center of an antenna element for the purpose of resonating it. Used with elements that are less than 1/4 wavelength.

*Coax Cable* — Any of the coaxial transmission lines that have the outer shield (solid or braided) on the same axis as the inner or center conductor. The insulating material can be air, helium or solid-dielectric compounds.

*Collinear Array* — A linear array of radiating elements (usually dipoles) with their axes arranged in a straight line. Popular at vhf and higher.

*Conductor* — A metal body such as tubing, rod or wire that permits current to travel continuously along its length.

*Counterpoise* — A wire or group of wires mounted close to ground, but insulated from ground, to form a low-impedance, high-capacitance path to ground. Used at mf and hf to provide an effective ground for an antenna.

*Current Loop* — The point of maximum current (antinode) on an antenna. The minimum point is called a "node."

*Delta Loop* — A full-wave loop shaped like a triangle or delta.

*Delta Match* — Matching technique used with half-wave radiators that are not split at the center. The feed line is fanned near the radiator center and connected to the radiator symmetrically. The fanned area is delta-shaped.

*Dielectrics* — Various insulating materials used in antenna systems, such as found in insulators and transmission lines.

*Dipole* — An antenna that is split at the exact center for connection to a feed line. Usually a half wavelength in dimension. Also called a "doublet."

*Directivity* — The property of an antenna that concentrates the radiated energy to form one or more major lobes.

*Director* — A conductor placed in front of a driven element to cause directivity. Frequently used singly or in multiples with Yagi or cubical-quad beam antennas.

*Direct Ray* — Transmitted signal energy that arrives at the receiving antenna directly rather than being reflected from the ionosphere, ground or a man-made passive reflector.

*Doublet* — See dipole.

*Driven Array* — An array of antenna elements which are all driven or excited by means of a transmission line.

*Driven Element* — The radiator element of an antenna system. The element to which the transmission line is connected.

*E Layer* — The ionospheric layer nearest earth from which radio signals can be reflected to a distant point, generally a maximum of 2000 km (1250 mi.).

*E Plane* — The plane containing the axis of the antenna and the electric field vector of an antenna.

*Efficiency* — The ratio of useful output power to input power, determined in antenna systems by losses in the system, including in nearby objects.

*Elements* — The conductive parts of an antenna system that determine the antenna characteristics. For example, the reflector, driven element and directors of a Yagi antenna.

*End Effect* — A condition caused by capacitance at the ends of an antenna element. Insulators and related support wires contribute to this capacitance and effectively lower the resonant frequency of the antenna. The effect increases with diameter and must be considered when cutting an antenna element to length.

*End Fed* — An end-fed antenna is one to which power is applied at one end rather than at some point between the ends.

*F Layer* — The ionospheric layer that lies above the *E* layer. Radio waves can be reflected from it to provide communications ranges of several thousand miles by means of single- or double-hop skip.

*Feeders* — Transmission lines of assorted types that are used to route rf power from a transmitter to an antenna, or from an antenna to a receiver.

*Feed Line* — See feeders.

*Field Strength* — The intensity of a radio wave as measured at a point some distance from the antenna. This measurement is usually made in terms of microvolts per meter.

*Front to Back* — The ratio of the radiated power off the front and back of a directive antenna. A dipole would have a ratio of 1, for example.

*Front to Side* — The ratio of radiated power between the major lobe and the null side of a directive antenna.

*Gain* — Increase in effective radiated power in the desired direction of the major lobe.

*Gamma Match* — A matching system used with driven antenna elements to effect a match between the transmission line and the feed point of the antenna. It consists of an adjustable arm that is mounted close to the driven element and in parallel with it near the feed point.

*Ground Plane* — A man-made system of conductors placed below an antenna to serve as an earth ground.

*Ground Screen* — A wire mesh gound plane.

*Ground Wave* — Radio waves that travel along the earth's surface to the receiving point.

*H Plane* — Related to a linearly polarized antenna. The plane that is perpendicular to the axis of the elements and contains the magnetic field vector.

*Harmonic Antenna* — An antenna that will operate on its fundamental frequency and the harmonics of the fundamental frequency for which it is designed. An end-fed half-wave antenna is one example.

*Helical* — A helical or helically wound antenna is one that consists of a spiral conductor. If it has a very large winding length to diameter ratio it provides broadside radiation. If the length/diameter ratio is small, it will operate in the axial mode and radiate off the end opposite the feed point. The polarization will be circular for the axial mode, with left or right circularity, depending on whether the helix is wound clockwise or counter-clockwise.

*Image Antenna* — The imaginary counterpart of an actual antenna. It is assumed for mathematical purposes to be located below the earth's surface beneath the antenna, and is considered symmetrical with the antenna above ground.

*Impedance* — The ohmic value of an antenna feed point, matching section or transmission line. An impedance may contain a reactance as well as a resistance component.

*Inverted V* — A half-wavelength dipole erected in the form of an upside-down V, with the feed point at the apex. It is essentially omnidirectional, and is sometimes called a "drooping doublet."

*Isotropic* — An imaginary or hypothetical antenna in free space that radiates equally in all directions. It is used as a reference for the directive characteristics of actual antennas.

*Lambda* — Greek symbol ($\lambda$) used to represent a wavelength with reference to electrical dimensions in antenna work.

*Line Loss* — The power lost in a transmission line, usually expressed in decibels.

*Line of Sight* — Transmission path of a wave that travels directly from the transmitting antenna to the receiving antenna.

*Load* — The electrical entity to which power is delivered. The antenna is a load for the transmitter. A dummy load is a nonradiating substitute for an antenna.

*Loading* — The process of a transferring power from its source to a load. The effect a load has on a power source.

*Lobe* — A defined field of energy that radiates from a directive antenna.

*Log Periodic Antenna* — A broadband directive antenna that has a structural format which causes its impedance and radiation characteristics to repeat periodically as the logarithm of frequency.

*Long Wire* — A wire antenna that is one wavelength or greater in electrical length. When two or more wavelengths long it provides gain and a multilobe radiation pattern. When terminated at one end it becomes essentially unidirectional off that end.

*Marconi Antenna* — Any type of vertical monopole operated against ground or a radial system.

*Matching* — The process of effecting an impedance match between two electrical circuits of unlike impedance. One example is matching a transmission line to the feed point of an antenna. Maximum power transfer to the load (antenna system) will occur when a matched condition exists.

*Null* — A condition during which an electrical property is at a minimum. The null in an antenna radiation pattern is that point in the 360-degree pattern where minimum field intensity is observed. An impedance bridge is said to be "nulled" when it has been brought into balance.

*Open-Wire Line* — A type of transmission line that resembles a ladder, sometimes called "ladder line." Consists of parallel, symmetrical wires with insulating spacers every few inches to maintain the line spacing. The dielectric is principally air, making it a low-pass type of line.

*Parabolic Reflector* — An antenna reflector that is a portion of a parabolic revolution or curve. Used mainly at uhf and higher to obtain high gain and a relatively narrow beamwidth when excited by one of a variety of driven elements placed in the plane of and perpendicular to the axis of the parabola.

*Parasitic Array* — A directive antenna that has a driven element and independent directors, a reflector, or both. The directors and reflector are not connected to the feed line. A Yagi antenna is one example. See Driven Array.

*Phasing Lines* — Sections of transmission line that are used ensure correct phase relationship between the bays of an array of antennas. Also used to effect impedance transformations while maintaining the desired array phase.

*Polarization* — The polarization of the wave radiated by an antenna. This can be horizontal, vertical, elliptical or circular (left- or right-hand circularity), depending on the design and application.

*Q Section* — Term used in reference to transmission-line matching transformers and phasing lines.

*Quad* — Rectangular or diamond-shaped full-wave wire-loop antenna. Most often used with a parasitic loop director and a parasitic loop reflector to provide approximately 8 dB of gain and good directivity. Often called the "cubical quad." Another version uses delta-shaped elements, and is called a "delta loop" beam.

*Random Wire* — A random length of wire used as an antenna and fed at one end by means of a Transmatch. Seldom operates as a resonant antenna unless the length happens to be correct.

*Radiation Pattern* — The radiation characteristics of an antenna as a function of space coordinates. Normally, the pattern is measured in the far-field region and is represented graphically.

*Radiation Resistance* — The ratio of the power radiated by an antenna to the square of the rms antenna current, referred to a specific point and assuming no losses. The effective resistance at the antenna feed point.

*Radiator* — A discrete conductor in an antenna system that radiates rf energy. The element to which the feed line is attached.

*Reflected Ray* — A radio wave that is reflected from earth, the ionosphere or a man-made medium, such as a passive reflector.

*Reflector* — A parasitic antenna element or a metal assembly that is located behind the driven element to enhance forward directivity. Hillsides and large manmade structures such as buildings and towers may reflect radio signals.

*Refraction* — Process by which a radio wave is bent and returned to earth from an ionospheric layer or other medium after striking the medium.

*Rhombic* — A rhomboid or diamond-shaped antenna consisting of sides (legs) that are each one wavelength or greater in electrical length. The conductors are made from wire, and the antenna is usually erected parallel to the ground. A rhombic antenna is bidirectional unless terminated by a resistance, at which time it is predominently unidirectional. The greater the leg length, the greater the gain.

*Shunt Feed* — A method of feeding an antenna driven element with a parallel conductor mounted adjacent to a low-impedance point on the radiator. Frequently used with grounded quarter-wave vertical antennas (Marconis) to provide an impedance match to the feeder. Series feed is used when the base of the vertical is insulated from ground.

*Source* — The point of origination (transmitter or generator) for rf power supplied to an antenna system.

*Stacking* — The process of placing similar directive antennas atop or beside one another, forming a "stacked array."

*Stub* — A section of transmission line used to tune an antenna element to resonance or to aid in obtaining an impedance match.

*SWR* — Standing-wave ratio on a transmission line in an antenna system. More correctly, "VSWR," or *voltage standing wave ratio.* The ratio of the forward to reflected voltage on the line, and not a power ratio. A VSWR of 1:1 occurs when all parts of the antenna system are matched correctly to one another.

*Tilt Angle* — Half the angle included between the wires at the sides of a rhombic antenna.

*T Match* — Method for matching a transmission-line to an unbroken driven element. Attached at the electrical center of the driven element in a T-shaped manner. In effect it is a double gamma match.

*Top Hat* — Capacitance hat used at the high-impedance end of a quarter-wave driven element to effectively increase the electrical length. See Capacitance Hat.

*Top Loading* — Addition of inductive reactance (coil) and/or a capacitance hat at the end of a driven element opposite the feed point to increase the electrical length of the radiator.

*Traps* — Parallel L-C networks inserted in an antenna element to provide multiband operation with a single conductor.

*Velocity Factor* — That which affects the speed of radio waves in accordance with the dielectric medium they are in. A factor of 1 is applied to the speed of light and radio waves in free space, but the velocity is reduced in various dielectric mediums, such as transmission lines. When cutting a transmission line to a specific electrical length, the velocity factor of the particular line must be taken into account.

*VSWR* — Voltage standing-wave ratio. See SWR.

*Wave* — A disturbance that is a function of time or space, or both. A radio wave, for example.

*Wave Angle* — The angle above the horizon of a radio wave as it is launched from an antenna.

*Wave Front* — A continuous surface that is a locus of points having the same phase at a specified instant.

*Yagi* — A directive, gain type of antenna that utilizes a number of parasitic directors and a reflector. Named after one of the inventors (Yagi and Uda).

*Zepp Antenna* — A half-wave wire antenna that operates on its fundamental and harmonics. Fed at one end (end-fed Zepp) or at the center (center-fed Zepp) by means of open-wire feeders. The name evolved from its popularity as an antenna on zeppelins.

## The Decibel

The decibel and its use are discussed in Chapter 2. Table 2 below shows the number of decibels corresponding to various power and voltage ratios.

The decibel value is read from the body of the table for the desired ratio, including decimal increment. For example, a *power* ratio of 2.6 is equivalent to 4.15 dB. A *voltage* ratio of 4.3 (voltages measured across like impedances) is equivalent to 12.67 dB. Values from the table may be extended, as indicated at the lower left in each section. For example, a *power* ratio of 17, which is the same as 10 × 1.7, is equivalent to 10 + 2.30 = 12.30 dB. Similarly, a power ratio of 170 (100 × 1.7) = 20 + 2.30 = 22.30 dB.

### Table 2
### Power Ratio to Decibel Conversion

| | Decimal Increments | | | | | | | | | |
|---|---|---|---|---|---|---|---|---|---|---|
| *Ratio* | *0.0* | *0.1* | *0.2* | *0.3* | *0.4* | *0.5* | *0.6* | *0.7* | *0.8* | *0.9* |
| *1* | 0.00 | 0.41 | 0.79 | 1.14 | 1.46 | 1.76 | 2.04 | 2.30 | 2.55 | 2.79 |
| *2* | 3.01 | 3.22 | 3.42 | 3.62 | 3.80 | 3.98 | 4.15 | 4.31 | 4.47 | 4.62 |
| *3* | 4.77 | 4.91 | 5.05 | 5.19 | 5.32 | 5.44 | 4.56 | 5.68 | 5.80 | 5.91 |
| *4* | 6.02 | 6.13 | 6.23 | 6.34 | 6.44 | 6.53 | 6.63 | 6.72 | 6.81 | 6.90 |
| *5* | *6.99* | *7.08* | *7.16* | *7.24* | 7.32 | 7.40 | 7.48 | 7.56 | 7.63 | 7.71 |
| *6* | 7.78 | 7.85 | 7.92 | 7.99 | 8.06 | 8.13 | 8.20 | 8.26 | 8.33 | 8.39 |
| *7* | 8.45 | 8.51 | 8.57 | 8.63 | 8.69 | 8.75 | 8.81 | 8.86 | 8.92 | 8.98 |
| *8* | 9.03 | 9.08 | 9.14 | 9.19 | 9.24 | 9.29 | 9.34 | 9.40 | 9.44 | 9.49 |
| *9* | 9.54 | 9.59 | 9.64 | 9.68 | 9.73 | 9.78 | 9.82 | 9.87 | 9.91 | 9.96 |
| *10* | 10.00 | 10.04 | 10.09 | 10.13 | 10.17 | 10.21 | 10.25 | 10.29 | 10.33 | 10.37 |
| *× 10* | + 10 | | | | | | | | | |
| *× 100* | + 20 | | | | | | | | | |
| *× 1000* | + 30 | | | | | | | | | |
| *× 10,000* | + 40 | | | | | | | | | |
| *× 100,000* | + 50 | | | | | | | | | |

### Voltage Ratio to Decibel Conversion

| | Decimal Increments | | | | | | | | | |
|---|---|---|---|---|---|---|---|---|---|---|
| *Ratio* | *0.0* | *0.1* | *0.2* | *0.3* | *0.4* | *0.5* | *0.6* | *0.7* | *0.8* | *0.9* |
| *1* | 0.00 | 0.83 | 1.58 | 2.28 | 2.92 | 3.52 | 4.08 | 4.61 | 5.11 | 5.58 |
| *2* | 6.02 | 6.44 | 6.85 | 7.23 | 7.60 | 7.96 | 8.30 | 8.63 | 8.94 | 9.25 |
| *3* | 9.54 | 9.83 | 10.10 | 10.37 | 10.63 | 10.88 | 11.13 | 11.36 | 11.60 | 11.82 |
| *4* | 12.04 | 12.26 | 12.46 | 12.67 | 12.87 | 13.06 | 13.26 | 13.44 | 13.62 | 13.80 |
| *5* | 13.98 | 14.15 | 14.32 | 14.49 | 14.65 | 14.81 | 14.96 | 15.12 | 15.27 | 15.42 |
| *6* | 15.56 | 15.71 | 15.85 | 15.99 | 16.12 | 16.26 | 16.39 | 16.52 | 16.65 | 16.78 |
| *7* | 16.90 | 17.03 | 17.15 | 17.27 | 17.38 | 17.50 | 17.62 | 17.73 | 17.84 | 17.95 |
| *8* | 18.06 | 18.17 | 18.28 | 18.38 | 18.49 | 18.59 | 18.69 | 18.79 | 18.89 | 18.99 |
| *9* | 19.08 | 19.18 | 19.28 | 19.37 | 19.46 | 19.55 | 19.65 | 19.74 | 19.82 | 19.91 |
| *10* | 20.00 | 20.09 | 20.17 | 20.26 | 20.34 | 20.42 | 20.51 | 20.59 | 20.67 | 20.75 |
| *× 10* | + 20 | | | | | | | | | |
| *× 100* | + 40 | | | | | | | | | |
| *× 1000* | + 60 | | | | | | | | | |
| *× 10,000* | + 80 | | | | | | | | | |
| *× 100,000* | + 100 | | | | | | | | | |

## Length Conversions

Throughout this book, equations may be found for determining the design length and spacing of antenna elements. For convenience, the equations are written to yield a result in feet. (The answer may be converted to meters simply by multiplying the result by 0.3048.) If the result in feet is not an integral number, however, it is necessary to make a conversion from a decimal fraction of a foot to inches and fractions before the physical distance can be determined with a conventional tape measure. Table 3 may be used for this conversion, showing inches and fractions for increments of 0.01 foot. For example, if a calculation yields a result of 11.63 feet, Table 3 indicates the equivalent distance is 11 feet 7-9/16 inches.

Similarly, Table 4 may be used to make the conversion from inches and fractions to decimal fractions of a foot. This table is convenient for using measured distances in equations.

### Table 3
### Conversion, Decimal Feet to Inches (Nearest 16th)

| | Decimal Increments | | | | | | | | | |
|---|---|---|---|---|---|---|---|---|---|---|
| | *0.00* | *0.01* | *0.02* | *0.03* | *0.04* | *0.05* | *0.06* | *0.07* | *0.08* | *0.09* |
| *0.0* | 0-0 | 0-1/8 | 0-1/4 | 0-3/8 | 0-1/2 | 0-5/8 | 0-3/4 | 0-13/16 | 0-15/16 | 1-1/16 |
| *0.1* | 1-3/16 | 1-5/16 | 1-7/16 | 1-9/16 | 1-11/16 | 1-13/16 | 1-15/16 | 2-1/16 | 2-3/16 | 2-1/4 |
| *0.2* | 2-3/8 | 2-1/2 | 2-5/8 | 2-3/4 | 2-7/8 | 3-0 | 3-1/8 | 3-1/4 | 3-3/8 | 3-1/2 |
| *0.3* | 3-5/8 | 3-3/4 | 3-13/16 | 3-15/16 | 4-1/16 | 4-3/16 | 4-5/16 | 4-7/16 | 4-9/16 | 4-11/16 |
| *0.4* | 4-13/16 | 4-15/16 | 5-1/16 | 5-3/16 | 5-1/4 | 5-3/8 | 5-1/2 | 5-5/8 | 5-3/4 | 5-7/8 |
| *0.5* | 6-0 | 6-1/8 | 6-1/4 | 6-3/8 | 6-1/2 | 6-5/8 | 6-3/4 | 6-13/16 | 6-15/16 | 7-1/16 |
| *0.6* | 7-3/16 | 7-5/16 | 7-7/16 | 7-9/16 | 7-11/16 | 7-13/16 | 7-15/16 | 8-1/16 | 8-3/16 | 8-1/4 |
| *0.7* | 8-3/8 | 8-1/2 | 8-5/8 | 8-3/4 | 8-7/8 | 9-0 | 9-1/8 | 9-1/4 | 9-3/8 | 9-1/2 |
| *0.8* | 9-5/8 | 9-3/4 | 9-13/16 | 9-15/16 | 10-1/16 | 10-3/16 | 10-5/16 | 10-7/16 | 10-9/16 | 10-11/16 |
| *0.9* | 10-13/16 | 10-15/16 | 11-1/16 | 11-3/16 | 11-1/4 | 11-3/8 | 11-1/2 | 11-5/8 | 11-3/4 | 11-7/8 |

### Table 4
### Conversion, Inches and Fractions to Decimal Feet

| | Fractional Increments | | | | | | | |
|---|---|---|---|---|---|---|---|---|
| | *0* | *1/8* | *1/4* | *3/8* | *1/2* | *5/8* | *3/4* | *7/8* |
| *0-* | 0.000 | 0.010 | 0.021 | 0.031 | 0.042 | 0.052 | 0.063 | 0.073 |
| *1-* | 0.083 | 0.094 | 0.104 | 0.115 | 0.125 | 0.135 | 0.146 | 0.156 |
| *2-* | 0.167 | 0.177 | 0.188 | 0.198 | 0.208 | 0.219 | 0.229 | 0.240 |
| *3-* | 0.250 | 0.260 | 0.271 | 0.281 | 0.292 | 0.302 | 0.313 | 0.323 |
| *4-* | 0.333 | 0.344 | 0.354 | 0.365 | 0.375 | 0.385 | 0.396 | 0.406 |
| *5-* | 0.417 | 0.427 | 0.438 | 0.448 | 0.458 | 0.469 | 0.479 | 0.490 |
| *6-* | 0.500 | 0.510 | 0.521 | 0.531 | 0.542 | 0.552 | 0.563 | 0.573 |
| *7-* | 0.583 | 0.594 | 0.604 | 0.615 | 0.625 | 0.635 | 0.646 | 0.656 |
| *8-* | 0.667 | 0.677 | 0.688 | 0.698 | 0.708 | 0.719 | 0.729 | 0.740 |
| *9-* | 0.750 | 0.760 | 0.771 | 0.781 | 0.792 | 0.802 | 0.813 | 0.823 |
| *10-* | 0.833 | 0.844 | 0.854 | 0.865 | 0.875 | 0.885 | 0.896 | 0.906 |
| *11-* | 0.917 | 0.927 | 0.938 | 0.948 | 0.958 | 0.969 | 0.979 | 0.990 |

# INDEX

### T



### V



### W



### Y



### Z

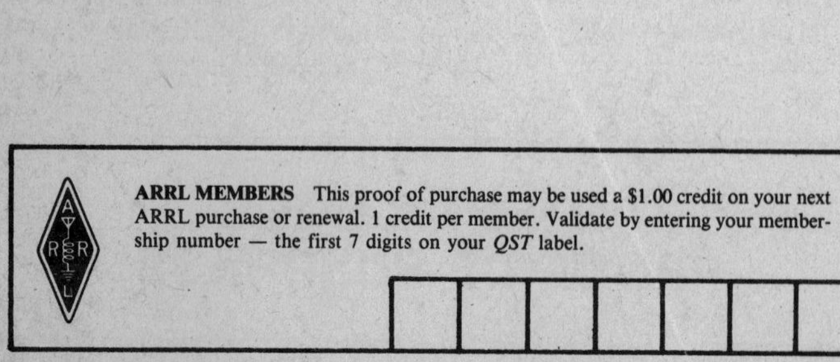
**ARRL MEMBERS** This proof of purchase may be used a $1.00 credit on your next ARRL purchase or renewal. 1 credit per member. Validate by entering your membership number — the first 7 digits on your *QST* label.

Mail to ARRL Headquarters